THE COMPLETE BOOK OF

ENZYME

THERAPY

THE COMPLETE BOOK OF
ENZYME
THERAPY

DR. ANTHONY J. CICHOKE

AVERY PUBLISHING GROUP
Garden City Park • New York

The therapeutic procedures in this book are based on the training, personal experiences, and research of the author. Because each person and situation are unique, the author and publisher urge the reader to check with a qualified health professional before using any procedure when there is any question regarding the presence or treatment of any abnormal health condition.

The publisher does not advocate the use of any particular diet or health program, but believes the information presented in this book should be available to the public.

Because there is always some risk involved, the author and publisher are not responsible for any adverse effects or consequences resulting from the use of any of the suggestions, preparations, or procedures described in this book. This book is as timely and accurate as its publisher and author can make it; nevertheless, they disclaim all liability and cannot be held responsible for any problems that may arise from its use. Please do not use the book if you are unwilling to assume the risk. Feel free to consult with a physician or other qualified health professional. It is a sign of wisdom, not cowardice, to seek a second or third opinion.

Cover designer: Eric Macaluso
In-house editor: Dara Stewart
Typesetters: Elaine V. McCaw and Helen Contoudis

Avery Publishing Group
120 Old Broadway
Garden City Park, NY 11040
1-800-548-5757

EASE is a registered trademark of Dr. Anthony J. Cichoke.

Library of Congress Cataloging-in-Publication Data

Cichoke, Anthony J.
    The complete book of enzyme therapy : a complete and up-to-date reference to effective remedies using enzymes, vitamins, and minerals / Anthony J. Cichoke.
      p.   cm.
    Includes bibliographical references and index.
    ISBN 0-89529-817-1
    1. Enzymes—Therapeutic use. 2. Vitamin therapy. 3. Trace elements—Therapeutic use. 4. minerals—therapeutic use.
    I. Title.
    RM666.E55c447 1998
    615' .35—dc21

98-21195
CIP

Printed in the United States of America

10   9   8   7   6   5   4   3   2   1

CONTENTS

Acknowledgments, ix

Foreword, xi

Preface, xiii

How to Use This Book, xvii

Part One   Enzymes

1. A Look at Enzymes and Their Functions, 1
2. Enzymes and Digestion, 10
3. Enzyme Depleters in Our Lives, 17
4. Maximizing Enzymes in Our Diets, 26
5. Enzyme Supplements, 37
6. Enzyme Helpers, 55

Part Two   The Conditions

Introduction, 80
Trouble-Shooting for Symptoms
  and Possible Causes, 83
Abrasions, 89
Abscess, 89
Acidosis, 91
Acne, 93
Acquired Immune Deficiency Syndrome, 96
Adenoiditis, 96
Adnexitis, 98
Age Spots, 100
Aging, 102
AIDS, 104
Alcoholism, 106
Alkalosis, 109
Allergies, 110
Alzheimer's Disease, 113
Anemia, 116
Angina Pectoris, 118
Ankylosing Spondylitis, 120
Arteriosclerosis, 122
Arthritis, 125
Asthma, 125
Atherosclerosis, 127
Athlete's Foot, 127
Backache, 127
Bacterial Infections, 131
Bamboo Spine, 133
Bed Sores, 133
Bee Stings, 133
Bell's Palsy, 134
Benign Prostate Hyperplasia, 135

Bladder Infection, 135
Blood Clots, 137
Boils, 137
Bone Fractures, 140
Brain Tumors, 142
Breast Cancer, 144
Bronchitis, 146
Bruises and Hematomas, 148
Burns, 150
Bursitis and Synovitis, 153
Cancer, 156
Candidiasis, 161
Canker Sores, 161
Carbohydrate Intolerance, 163
Carbuncles, 165
Cardiovascular Disorders, 165
Carpal Tunnel Syndrome, 169
Cataracts, 172
Celiac Disease, 174
Chickenpox, 176
Cholesterol, Elevated, 179
Chronic Fatigue Syndrome, 181
Cirrhosis, 183
Cold Sores, 186
Colds, 186
Colitis, Ulcerative, 188
Colorectal Cancer, 188
Conjunctivitis, 190
Constipation, 192
Cramps, Menstrual, 195
Cramps, Muscle, 195
Crohn's Disease, 195

Cuts, 195
Cystic Fibrosis, 195
Cystic Mastitis, 197
Cystitis, 197
Dermatitis, 197
Dermatomyositis, 200
Diabetes, 201
Diarrhea, 206
Digestive Disorders, 208
Diverticular Disease, 208
Diverticulitis, 210
Diverticulosis, 210
Dysmenorrhea, 210
Ear Infection, 213
Eczema, 215
Embolism, 216
Empyemas, 218
Epididymitis, 220
Eye Problems, 221
Fever Blisters, 221
Fibrocystic Breast Disease, 221
Fibroids, 223
Fibromyalgia, 225
Flatulence, 225
Flu, 225
Fungal Skin Infections, 225
Furuncles, 228
Gas, Abdominal, 228
Gastritis, 230
Gastroesophageal Reflux Disease, 232
Gingivitis, 232
Glaucoma, 234
Glomerulonephritis, 237
Gluten Enteropathy/Intolerance, 239
Gout, 239
Guillain-Barré Syndrome, 241
Hangover, 242
Hay Fever, 244
Headache, 246
Heart Disease, 249
Heartburn, 249
Hematomas, 251
Hemorrhoids, 251
Hepatitis, 252
Herniated Disc, 255
Herpes Simplex Virus, 255
Herpes Zoster, 257
High Blood Pressure, 259
HIV, 261
Hives, 261
Hypercholesterolemia, 264
Hypertension, 264
Hypochlorhydria, 264

Hypoglycemia, 266
Indigestion, 268
Inflammatory Bowel Disease, 271
Influenza, 273
Insect Bites and Stings, 276
Intermittent Claudication, 278
Irritable Bowel Syndrome, 280
Jock Itch, 282
Kidney Disorders, 282
Lactase Deficiency, 284
Laryngitis, 286
Leaky Gut Syndrome, 288
Leiomyomas, 290
Lung Cancer, 290
Lung Infection, 293
Lupus, 293
Lymphedema, 293
Macular Degeneration, 295
Marie-Strümpell Disease, 296
Measles, 297
Menstrual Cramps, 299
Migraines, 299
Mononucleosis, 299
Multiple Sclerosis, 301
Mumps, 304
Muscle Cramping, 306
Myofascial Pain Syndrome, 308
Nausea, 310
Nephrotic Syndrome, 311
Neuritis, 311
Obesity, 314
Osteoarthritis, 317
Osteoporosis, 319
Pancreatic Cancer, 321
Pancreatic Insufficiency, 324
Pancreatitis, 325
Peptic Ulcers, 328
Pharyngitis, 330
Phlebitis, 330
Pinkeye, 330
Pleurisy, 330
Pneumonia, 332
Polymyositis, 333
Post-Childbirth Complications, 335
Post-Thrombotic Syndrome, 337
Premenstrual Syndrome, 337
Prostate Cancer, 339
Prostate Disorders, 341
Prostatatitis, 343
Psoriasis, 343
Radiation Sickness, 345
Raynaud's Disease and Phenomenon, 347
Reactive Arthritis, 348

Reiter's Syndrome, 349
Retinopathy, 351
Rheumatic Fever, 353
Rheumatoid Arthritis, 354
Ringworm, 357
Rosacea, 357
Rubella, 359
Ruptured Disc, 359
Scars, 359
Sciatica, 361
Scoliosis, 362
Shingles, 364
Sinusitis, 364
Sjögren's Syndrome, 367
Skin Cancer, 368
Skin Problems, 371
Skin Rash, 371
Skin Ulcers, 373
Slipped Disc, 375
Sore Throat, 377
Sports Injuries, 379
Sprains and Strains, 384
Sprue, Nontropical, 386
Staphylococcal Infections, 386
Steatorrhea, 388
Sties, 390
Stress, 390
Stroke, 393

Subluxation, 395
Surgical-Related Problems, 397
Synovitis, 400
Systemic Lupus Erythematosus, 400
Temporomandibular Joint Dysfunction, 404
Tendonitis, 406
Testicular Inflammation, 408
Thrombophlebitis, 408
Thrombosis, 408
Thrush, 411
Tinnitus, 411
Tonsillitis, 411
Tooth Decay, 413
Torticollis, 415
Truck Driver's Syndrome, 417
Ulcers, 417
Underweight, 417
Varicella, 419
Varicose Veins, 419
Verrucae, 421
Vertigo, 421
Viral Infections, 422
Warts, 425
Whiplash, 426
Wounds, 429
Wryneck, 429
X-Ray Hangover, 429
Yeast Infections, 429

Part Three   Complementary Therapies

Introduction, 432
Baths, 432
Detoxification Methods, 432
  Coffee Retention Enema, 433
  Enzyme Retention Enema, 433
  Enzyme Kidney Flush, 433
  Enzyme Toxin Flush, 434
  The Vitamin C/Enzyme Flush, 434
  The Two-Day Enzyme Juice Fast, 434
Diet, 435
Exercise, 437
The Five-Step Jump-Start Enzyme
   Program, 438

Gargles, 438
  Echinacea/Enzyme Gargle, 438
  Enzymatic Gargle, 438
  Garlic/Enzyme Gargle, 439
Light Thereapy, 439
Meditation, 439
Positive Mental Attitude, 439
Relaxation, 440
Skin Care, 440
  Enzyme Skin Exfoliants, 441
  Enzymatic Skin Salves, 441
  General Recommendations, 442
Water, 442

Appendix

Glossary, 446
Digestive Enzyme Products, 448
Enzyme Companies, 456

Treatment Centers, 465
Resource Groups, 467
Bibliography, 477

Index, 480

ACKNOWLEDGMENTS

The author wishes to thank the following researchers and physicians for their cooperation, advice, and assistance in writing this book: Raul Ahumada, Edward Alstat, Motoyuki Amano, Vladimir Badmaev, Wayne Battenfield, W. Bartsch, D.J.A. Cole, Tony Collier, Michael Culbert, F.W. Dittmar, James Duke, Charles Fox, Naritada Fujiki, G. Gallacchi, G. Gebert, Wilhelm Glenk, Yoshihide Hagiwara, Clare M. Hasler, John Heinerman, A. Hoffer, Rudolph Inderst, Hans Jager, Peter Karnezos, Leslie Kenton, Shigeki Kimura, Franz Klaschka, Gert Klein, Michael W. Kleine, Stephen E. Langer, Benjamin Lau, Robert I. Lin, D.A. Lopez, T. Pearse Lyons, Muhammed Majeed, E.J. Menzel, Mark Messina, John Mills, Earl Mindell, Daniel B. Mowrey, Michael Murray, Tracey Mynott, Sven Neu, Christine Neuhoffer, Richard A. Passwater, Otto Pecher, Barbara Pfannenschmidt, Joseph Pizzorno, Joan Priestley, H.D. Rahn, Karl Ransberger, Corey Resnick, Robert Rosen, W. Scheef, Ed Schuler, Art Sears, J. Seifert, Lendon Smith, G. Stauder, C. Steffen, Peter Streichhan, Koichi Suzuki, Mitsuru Takiura, Steven Taussig, G.P. Tilz, Ron Tominga, Klaus Uffelmann, Wolf Vogler, W. Von Schaik, Morton Walker, R. Michael Williams, Heinrich Wrba, Janet Zand, and finally Ms. Katie Cichoke.

Special thanks to my secretary, Mrs. Karen Hood, for all of her assistance in the preparation of this book, including doing research, typing, and editing; and to Ms. Dara Stewart of Avery Publishing for her special work as an editor. Further, I would like to express my thanks to Mr. Rudy Shur of Avery Publishing for his constant encouragement and never-ending faith in my work.

Finally, my eternal thanks to my wife, Margie, for her support and for enduring the mountains of paper and manuscripts in our house for these past three years.

FOREWORD

Life is a wonderful, incomprehensible complex of thousands of interrelated chemical reactions using about fifty basic nutrients as the raw material. A plant needs only minerals and a place to grow; and using air, water, and sunlight, it manufactures everything it needs. All the organic nutrients needed by animal life are made by plants, and the minerals are available from the oceans and soil. It is impossible to conceive the multiplicity of reactions in the body and how they are self-regulated and controlled, even in one cell. Yet these reactions continue and are controlled, accelerated, or slowed when this is essential. None of this would have been possible if nature had not invented enzymes. Why, then, have we in English-speaking medical therapeutic literature neglected these remarkable substances for so long? Well, this is now coming to an end, for with this remarkable book on enzymes, Dr. Cichoke brings us up to date with respect to the medical uses of enzymes to prevent and treat a large series of medical diseases.

Enzymes are organic catalysts. They accelerate reactions in the body that are essential for life. And in so doing, they are not themselves destroyed, being used over and over with remarkable rapidity. Enzyme-catalyzed reactions can be accelerated by changing the chemical conditions, or inhibited by enzyme inhibitors. A reaction consists of molecules, which are transformed by the reaction into a different set of molecules. In the process, energy is required or released and substances are constructed. The original molecules are called substrates. The final molecules may, in turn, be converted into something else that is used in construction, or into waste and must be eliminated. These large enzyme molecules contain smaller molecules, which are essential for their activity, such as vitamins and minerals.

The older views of nutrition concluded that the body can make all the enzymes it needs. Therefore, giving enzymes by mouth would be of no value, since these large molecules would be destroyed in the stomach and in the gut, and if they were not destroyed could not be transferred into the blood because of their large size. Since enzymes are organic, they would create difficulty if injected. The same idea existed that vitamin supplements were not needed. The old paradigm depended on the view that vitamins were needed only in very small amounts—the amounts that would be found in the average diet. In fact, this was the article of faith transmitted by nutritional societies, by nutritionists, by nutritional bio-

chemists, by government, and by the medical profession. Vitamins were needed only to prevent deficiency diseases, such as scurvy or beriberi, and, therefore, they had no use for anything else, and certainly must not be used in doses larger than those recommended by the Recommended Daily Allowances. The old view was that foods were adequate, and that if we did not remain well on these diets, the fault was our own for not following the recommended food rules.

Linus Pauling, following the discovery that vitamin B3 could be used in megadoses for treating schizophrenia and for lowering cholesterol levels and that vitamin C could be used in huge doses for a variety of conditions with impunity, developed the concept of orthomolecular medicine, i.e., the use of natural molecules with which the body was familiar in optimum amounts. The use of megadoses initiated the modern paradigm, which is that vitamins must be used in optimum amounts, which may be large or small, even for conditions that are not vitamin deficiency diseases. The resistance to these views was amazing, and only after about twenty-five years has the medical profession accepted that this is the correct paradigm and is now grudgingly using vitamin E, vitamin C, and folic acid, as well as other vitamins, for conditions other than deficiency diseases. This was intolerable many years ago. Thus the orthomolecular concepts brought in the proper use of vitamins and minerals in prevention and treatment.

Enzymes are not as readily available. They are not synthesized but must be extracted from plants and from animal organs. This means they are more difficult to make. They came into general use in Europe but not North America where we, the medical profession, remained ignorant of their roles and value. This author will do for enzyme medicine what the pioneers in orthomolecular medicine did for the vitamins.

According to Dr. Cichoke, enzyme therapy is very valuable for a large number of disparate conditions. This may trouble many physicians who have been brought up with the belief that each disease must have its one medicine for treatment, such as penicillin for pneumonia, insulin for diabetes mellitus, and so on. Why are enzymes so helpful for these conditions when so many of them also respond to vitamins and minerals? If one accepts the view that the nutrients enable the enzymes to do their work then the difficulty disappears.

Many years ago, Dr. D. Rudin proposed that there are several forms of pellagra. He maintained that pellagra is due to a deficiency of prostaglandins. These are made from essential fatty acids. Therefore a deficiency may occur: (1) if there is a deficiency of these essential fatty acids, and (2) if there is a deficiency of some of the vitamins and minerals needed for the enzymes that make those transformations. He termed the second type substrate pellagra, when the substrate, the essential fatty acids, were lacking. Or it occurs when vitamin B_3 or vitamin B_6 is missing, as these are needed for the activity of these enzymes. If he is correct, as I think he is, it follows that many of the conditions described by Dr. Cichoke are caused by a deficiency of some substances, and that the production of these compounds can be increased by the use of enzymes that catalyze the reactions or by the deficiency of those cofactors needed by the enzymes. Ideally, the body should have optimum amounts of the substrates and the cofactors—vitamins and minerals—to ensure that the essential reactions are maintained. This is why in this book, Dr. Cichoke recommends that, in addition to the enzymes, one should also pay attention to the food and to the use of vitamins, minerals, and essential fatty acids. It is a broad complex regimen for orthomolecular health, meaning good health.

—A. Hoffer, M.D., Ph.D., FRCP (C)

PREFACE

My father once told me, "Tony, life is a struggle and the victory is in that struggle." What he meant was that the experiences gained from a struggle, even a tragedy, can give focus and help crystallize one's direction in life.

So it was with the tragedy and ultimate triumph of my son David. David was diagnosed as having acute cerebellar ataxia at three years of age. With no cure in sight and no hope or help from orthodox medicine, my family and I began a struggle with David that became a survival voyage and a discovery of the advantages better nutrition and alternative health care can offer.

As I write these words, my mind's eye goes back to that early Sunday morning in Pittsford, New York when my wife, Margie, and I were abruptly jolted from our sleep by the crashing, thunderous sounds of something careening violently down the stairs. Instantly, we jumped out of bed and ran to investigate.

There at the bottom of the stairs lay our three-year-old son, David. His eyes were wide and filled with fear. His arms flailed helplessly as he repeatedly attempted, and failed, to get up. Vomit spewed from his mouth. Each time he tried to stand up, his arms and legs gave way, and his head crashed into the stairs with a sickening thud.

We rushed David to Strong Memorial Hospital, where he received a brain scan and spinal tap. His screams of pain echoed through the hospital halls, terrified us, and remain fused in my memory even to this very day.

Days passed and David's movements remained limited as he lay helplessly in bed. His vomiting persisted. A constant bedside vigil was necessary, so either my wife (pregnant with our fourth child, Katie) or I remained at his bedside night and day. Time and time again his frail back muscles would tighten and his little body would arch like a bow as he retched.

David could not function on his own. He had to be fed and cleaned and be carried to the bathroom. David could not walk normally. His arms and legs moved in spasms, like a child with cerebral palsy. His head flopped uncontrollably from side to side and his eyes would roll back. Ultimately, David was diagnosed as suffering from *acute cerebellar ataxia*, the result of the head injuries he suffered in the fall.

Finally, he was released from the hospital, not because he was cured but because there were no rehabilitation programs for him. The doctors were uncertain about David's future. He could develop like a child with cerebral palsy. Possibly, he would have to wear braces and use crutches the rest of his life. Possibly, he could become a vegetable! Possibly, possibly . . . and more possibly.

The days passed. Capable of minimal movement and hopelessly bedridden, David had a very bleak-looking future. Though we thanked God for allowing David to remain with us, we knew we had to do something. We had to take charge of David's and our lives. Since organized medicine offered no help, I knew it would be up to me to acquire any information possible on nutrition and physical rehabilitation or whatever it took to make David better.

At the time of David's accident I was teaching and doing research in the Department of Orthodontics at Eastman Dental Center in Rochester, NY, working amongst the staff at Strong Memorial Hospital (University of Rochester Medical School), and working as a therapist at the Al Sigl Rehabilitation Center in Rochester.

Since my background was in anatomy and physiology (specifically of the head and neck), I knew an enormous amount of work would be necessary to rehabilitate David. Drawing on my experience as a researcher, I developed a huge study with David as the subject. Since organized medicine had no answers, I looked to alternative health care and nutrition. David's accident drove me back into the labyrinths and hallowed halls of research libraries and institutions.

FIRST PROTECT THE BODY—CAUSE NO HARM

Because of his poor coordination, David kept falling and hitting his head. When falling backward, a normal child hunches his back, but not David. His head would swing helplessly and smack the wall or floor. David had no protective mechanism. He did not put his hands out when he fell forward, or "hunch" when falling backward. To protect him, we devised a "Magic Room," covering the floor with six-inch thick sponge rubber and attaching carpet remnants to all four walls. The sponge floor became David's "Magic Carpet."

We devised a new scheme to reprogram David's protective mechanisms. When reprimanding the children, we had them do pushups. We hoped this would teach David to put his hands out in front of him when he fell. At about the same time, I read in the newspaper of a revolutionary new process called "cross crawl" (or cross patterning) developed by

Temple Fay, M.D., (Department of Neurology, Temple University) to reprogram neuropathways of brain-injured children. "Cross patterning" requires five assistants (one working each extremity and the head). Not having that help, I had to do it myself. By strapping my arms to his arms and my legs to his legs, he was forced to move as I moved. We started with the basic dolphin-type movement, which is similar to the butterfly stroke in swimming (moving the legs and hips in an undulating movement, while keeping the arms still). As we progressed, I would lie on my back with David lying on his back on my chest (his legs and arms still strapped to mine) and practiced moving the right arm and leg simultaneously, then the left arm and leg. We then progressed to a crawling position, and proceeded ultimately to an upright, walking position.

COURAGE KNOWS NO AGE

Repeatedly, our little David would fall down and get up, then fall down and get up again. Every time it happened it made me sick to my stomach. But do you know what was so impressive? The fact that David kept trying. He would not give up!

Each time he fell, I helped him up. "David! David are you OK?" I would ask.

"I'm OK, Dad. Don't worry," David would say as he struggled to his feet. Then he would squint his eyes, set his little jaw, and look me straight in the eye. "Daddy," he would say. "I can do it. I know I can!"

And somewhere down inside, I knew he could.

Using the walls for support, David first conquered walking inside the house. Then the stairs became a challenge to conquer. Finally, the outdoors with its uneven grass surface became the next challenge. The rough ground was more difficult because he had no walls or furniture to grab for support. But he conquered each challenge that came his way.

Even at three years of age, David would not quit! And in those fragile moments of quiet desperation, I knew what it meant to be a real, true champion—to face whatever came and to never, never give up! I learned so much from David.

My family and I rededicated ourselves to fight for David, and to help him reach whatever potential he was capable of attaining—to help him reach for the sky and grab a handful of stars! The road was not easy. It was filled with many potholes, plus unexpected curves and setbacks. But we never gave up. We knew David needed much more. In addition to cross patterning, we started an extensive nutrition and exercise program.

BETTER NUTRITION WAS CRITICAL FOR DAVID'S IMPROVEMENT

I contacted the giants in nutrition from around the world—giants such as Drs. Roger Williams, Abram Hoffer, Linus Pauling, and Emanuel Cheraskin. With their help and encouragement, we devised a nutritional program for David. These men are great for a reason. I followed their guidance and advice and can never thank them enough for their unending reinforcement.

So that David could function at his optimal level, we eliminated all refined and processed foods (including food additives, preservatives, and artificial colors and flavors) from his diet. We instead emphasized whole grains and whole, uncooked foods (high in enzymatic activity), such as fresh fruits and vegetables, in his diet. We started David on a meganutritional program of vitamins and minerals, with special emphasis on such nutrients as vitamins B, C, and E; magnesium; calcium; bioflavonoids; protein; and wheat germ oil.

DAVID BEGINS TO IMPROVE

As David began to regain his mobility, we went for daily walks, then we began to do quick, short wind sprints down the driveway, then jogs. Wanting to learn still more about David's problem, I entered graduate school at Case Western Reserve University and worked toward my doctoral degree with an emphasis on anatomy and physiology of the head and neck.

In order to improve David's neuromuscular coordination, we enrolled him in a developmental gymnastics program. We chose a Montessori School to encourage conceptual development plus hand-eye coordination.

In my unceasing quest for further insights, I accepted a teaching position at the National College of Chiropractic, became director of the Sacroiliac Research Project, and ultimately obtained my chiropractic degree.

ENZYMES TO THE RESCUE!

We left no stone unturned in our quest for David's rehabilitation. Fortunately, I had heard of Drs. Max Wolf and Karl Ransberger and their outstanding enzyme research. Wolf and Ransberger are considered pioneers in systemic enzyme therapy. Therefore, I contacted Dr. Ransberger and learned more about enzymes, such as how they are required for every activity of our body, are necessary for proper functioning of minerals and vitamins, and help fight chronic disorders. Immediately, these findings were put into action. David was given enzyme supplements daily. With the added enzymes, I saw a decided improvement in David's health status and knew I had really found something terrific.

Eventually, David not only crawled, walked, and ran again, but became a nationally ranked swimmer and runner, an Oregon Scholar, and a national honor society student. He was a member of two David Douglas High School Oregon state championship swim teams and helped lead Jesuit High School to third place in the Oregon state cross country championships. He also helped lead Santa Clara University to a fourth place national ranking in football. Not bad for a child who doctors thought might never walk again! At Santa Clara University, David was already being recruited by professional football teams such as the Green Bay Packers and the Indianapolis Colts.

Once we realized the positive value of improved nutrition (especially enzyme-rich foods and enzyme supplements) in David's rehabilitation and ultimate recovery, we knew we had something special. We had seen firsthand the power of good nutrition (especially enzymes) in action. If David could be rehabilitated, so could others. If David could become an outstanding athlete with the help of enzymes, so could others. I have repeatedly witnessed this in my own practice and research.

Since that time, I have attained a diplomate's status in nutrition (through the American Chiropractic Association) and my work with enzymes has taken me around the globe, riding the crest of the enzyme wave. Through my roles as chairman of the Amateur Athletic Union (AAU) Committee on Sports Medicine (Oregon), a member of the National AAU Sports Medicine Steering Committee, a member of the three-man Research Committee of the American Swim Coaches' Association, and chiropractic team physician at Portland State University, I was able to help athletes and nonathletes alike profit incredibly from the use of enzymes. From world-class athletes and weekend warriors to couch potatoes, the infirm and the healthy have all benefited from enzymes.

WHY THIS BOOK AND WHY NOW?

Fortunately, almost daily more and more information is becoming available on enzymes. There is an enzyme information explosion. We're learning more every minute of every day. Many scientific and lay journals are publishing articles on enzymes and their wondrous benefits. But with that increased awareness comes, unfortunately, increased *misinformation* about enzymes as well.

Therefore, the purpose of this book is to give you clear, easy-to-understand, up-to-date information about enzymes —to clarify the enzyme mystery and to help you take charge of your life. Based upon scientific research, this book will show you how enzymes can help or cure a multitude of conditions.

Just as nutrition and enzymes have helped David, my family, my patients, and many others to get well, stay well, recover at incredible rates, and reach unprecedented levels of performance (naturally); so, too, can enzymes help you and yours live longer, healthier, more disease-free lives. This book will show you the way. So, let's take charge of our lives.

HOW TO USE THIS BOOK

Over the last ten years, science has made great strides in understanding the importance of enzymes in our bodies. Researchers are also learning how supplemental enzymes can improve our health and help us overcome various disorders.

This book has been designed to provide you with a clear understanding of what enzymes are, how they work, and how to use them on an everyday basis to help you achieve optimum health. The book is divided into three parts. Part One provides you with a clear picture of the nature of enzymes, the jobs they perform in the body, how they are depleted in our foods, how to maximize the performance of enzymes in your diet, the types of enzyme supplements available (as well as when, why, and how to take them), and other nutrients that can assist your enzymes. Part Two provides you with specific enzyme treatment plans for over 150 conditions. Each entry begins with a description of the condition, pro-vides you with a list of the necessary enzymes for the treatment of the condition, and also includes vitamins, minerals, and other nutrients that may help make the enzymes more available to your body. Other helpful suggestions and considerations are also included. Part Three provides further information on diet, therapies, and cures to be used in conjunction with the treatment plans in Part Two. Following Part Three, you will find a list of enzyme companies and their products, a glossary, and a suggested reading list.

This book was designed to be a reference. It is not meant to replace the services of your health-care provider. When appropriate, I suggest you take this book to your health-care provider and discuss the information in the book that is pertinent to your situation. In this way, you and your physician can work together to improve your health by considering an overall treatment program that can include the use of enzymes.

PART ONE

ENZYMES

1. A Look at Enzymes and Their Functions

No plant, animal, or human could exist without enzymes. During every moment of our lives, enzymes keep us going. At this very instant, millions of tiny enzymes are working throughout your body causing reactions to take place. You couldn't breathe, hold or turn the pages of this book, read its words, eat a meal, taste the food, or hear a telephone ring without enzymes. You'd be dead, pushing up daisies and fertilizing some living plant, if not for enzymes.

So far, researchers have identified more than 3,000 kinds of enzymes in the human body. There are millions of these energizers that renew, maintain, and protect us. Every second of our lives, these enzymes are constantly changing and renewing, sometimes at an unbelievable rate.

ENZYMES—THE SPARK OF LIFE

You may be trying to eat right; get plenty of exercise; and take vitamin, mineral, and herbal supplements. But what if you still feel constantly rundown and wrung out? Is each step you take a drag? Is indigestion, heartburn, or gas a frequent and unwanted companion? Do you have skin or weight problems? Allergies? Maybe your body is saying "enough is enough!" Foods that lack enzymes and unhealthy lifestyles could be the problems, and the answer is in this book—*enzymes!*

Historically, the best sources of enzymes were fresh fruits, vegetables, and grains. And by now, most of us know the importance of eating these foods every day. Even the USDA's (United States Department of Agriculture's) food pyramid says we should eat three to five servings of vegetables and two to three servings of fruit daily. These foods are rich in vitamins, minerals, fiber, and *enzymes.*

But, unfortunately, few of us actually follow the government's guidelines—even though it's a proven fact that those whose diets are rich in vegetables, fruits, and grains have a significantly lower risk of cancer and heart disease. Sadly, fewer than 10 percent of Americans eat two servings of fruit, or three servings of vegetables per day. That's right, fewer than 10 percent! Fifty percent eat no vegetables at all, 70 percent eat no vegetables or fruits rich in vitamin C, and 80 percent eat no vegetables or fruits rich in carotenoids per day.

So, what are we living on? French fries, sugar-loaded soft drinks, coffee, and fast-food hamburgers. Further, our foods are fried, baked, canned, frozen, dried, or irradiated—all processes that kill enzymes. No wonder we're sick. Our bodies have no fuel—they're trying to run on fumes. Our bodies' engines are too pooped to pop!

Enzymes are essential for everything that goes on in the body, including digestion, breathing, circulation . . . everything. Your body uses enzymes to fight disease and inflammation and to slow down the aging process. Plus, enzymes can help you look good and live longer.

As we age, the numbers of our enzymes and their activity levels decrease. Our bodies just can't produce as many enzymes, and those that remain have lost their power-laden punch. This is when enzymes from fresh fruits and vegetables, food concentrates, and supplements from animal, microbial, and plant sources can help.

But even if we *do* try to eat only fresh foods, did you know that the enzymatic level of food is reduced by long-term storage, as well as by pesticides and other toxins in our water, soil, and air? Augmenting your diet with enzyme supplements may be your only answer.

Supplemental enzymes can aid digestion, dissolve blood clots, fight back pain, decrease swelling, speed up healing, fight wrinkles, clean surfaces of dirty wounds, help in delicate surgery, ease hindered breathing, stimulate the immune system, and help fight cancer and HIV/AIDS and other viruses. In other words, enzymes can do an awful lot.

A whole new world is opening up for enzymes and their applications. Enzymes are also used in tests that help doctors determine what may be ailing us. Today, enzymes are being added to toothpaste for more cleaning power and to skin and nail products to make you look better, and they are widely used in the baking and brewing industries for a variety of applications.

What Are Enzymes?

Enzymes are protein-based substances found in every cell of every living plant and animal, including the human body. Without enzymes, the grass or trees would not grow, seeds would not sprout nor would flowers bloom, and beer and wine could not ferment. Even the autumn leaves would not burst forth in glorious colors without the help of enzymes. Mother Nature has blessed us with her wondrous magic in providing enzymes to ripen bananas from green to yellow-gold, or tomatoes from green to robust, juicy red.

Enzymes are the powerhouses of each and every cell. They either start chemical reactions, or they make them run faster. Enzymes remain unchanged even after a reaction is completed. Enzymes appear throughout nature, even in the food we eat. When the thinly sliced cabbage that your grandmother placed in a crock pot fermented into sauerkraut, or your grandfather's elderberries slowly turned into wine, enzymes were at work.

The use of enzymes in food preparation probably began long before anyone knew what these little dynamos were or what they could do. Like a travelogue of enlightened civilizations, the history of enzymes takes us from ancient Egypt to Greece, Germany, Denmark, and Japan . . . around the world. All of these civilizations, knowingly or unknowingly, used enzymes in their food preparations. From the first bag of wine, round of cheese, keg of beer, vat of vinegar, or loaf of bread—in fact, any food that required fermentation (the word enzyme comes from the Greek word *enzymos*, which means leavened or fermented)—the action of enzymes helped our ancestors prepare and preserve the food they needed.

The ancients felt that the secret of life (and the difference between living and nonliving structures) was a certain "vitality." This is why the changes that occurred in foods as they fermented were considered to be almost magic—they possessed that special vitality. The ancients knew that milk could be magically changed into cheese, and grapes into wine. Couldn't this elusive vitality, this "magic," be harnessed and used to transform iron into gold? Such was the attempt of the early alchemists.

Though they probably didn't understand the process, our earliest ancestors knew that fermentation would change their food. Even cave dwellers discovered that aged meat had a more pleasing flavor and was more tender than the freshly killed variety. Your grandmother may have used *rennin*, an enzyme from calf's stomach, to clot milk—the first step in making cheese. Grapes left to ferment turned into wine, while grains, with a little coaxing (and the addition of malt, which is rich in an enzyme called *amylase*), produced beer.

Today, we know that fermentation is not magic but a chemical change caused by bacteria, microscopic yeasts, and molds. In some instances, fermentation is used to alter or change a material in a way that would be very costly or difficult if other methods were used. For example, modern medicine uses fermentation under controlled conditions to produce a number of antibiotics.

Although fermentation has been used for centuries in the processing of food, our knowledge of the science behind the way it works is relatively new. We now know that it is the action of enzymes in yeast (and not the yeast itself) that causes alcoholic fermentation. Eduard Buchner received the Nobel Prize in 1907 for this discovery. Enzymes act as natural catalysts that cause a chemical change without themselves being affected.

Now that scientists know how enzymes can affect our foods and our food-processing techniques, they have drawn on this knowledge to concoct better, faster, and cheaper ways to manufacture what we eat. And we see the results practically every day in almost every food or drink that we consume. For example, many beer drinkers, trying to cut down on calories, are opting for light beer, a product impossible to make without enzymes. Enzymes are also used in the production of cheese, wine, baked goods, soy sauce, fruit and vegetable juices, and food ingredients such as fructose, aspartame, modified oils and fats, and emulsifiers.

The Enzymes in Your Life

All things living have enzymatic activity, whether it be the grass in front of your house, your dog Fido, or your food (if the enzymes have not been killed yet). Whatever is alive has enzymes in it. That includes *you*. In fact, enzymes make your body work.

All life processes, such as digestion, breathing, even thinking, consist at least in part of a complex series of chemical reactions called *metabolism*. But that explanation is a little too simple. Actually metabolism is composed of two parts: *anabolism* and *catabolism*. Anabolism is any process in which simpler substances are combined to form more complex substances. It is the process of building up (such as new tissue growth). Catabolism is the flip side. It is any process in which living cells break down substances into simpler substances (such as that which occurs in digestion). The sum of these two processes is metabolism. Enzymes are the catalysts (the jump-starters) that make these chemical reactions possible. In fact, many of the body's chemical reactions would never take place without enzyme catalysts.

In addition to their roles in metabolism, enzymes are also food potentiators. All foods have potential nutrients. Enzymes have the ability to turn these potential nutrients into available nutrients. For example, carrots contain beta-carotene. But we can't get the beta-carotene out if we can't unlock it from the carrot's cells. Therefore, we won't benefit from the carrot's nutrients. It's like Fort Knox—you've got to unlock the door in order to get the treasure out.

As long as our bodies can make enzymes, we live. But our bodies' production of enzymes can be decreased by illness, injury, stress, or aging. If the body can't produce them fast enough or with enough activity level, then enzymes must be acquired from an outside source. This is similar to a factory that uses a certain part faster than it can be made. Production can come to a standstill until the missing part is provided. Further, as the years go by, the machinery making the part begins to wear out and produces fewer parts. Therefore, more and more parts must be obtained from outside the factory in order to maintain the same production level. Your body works much the same way. When enzyme production falters or ceases, you're in trouble and enzyme supplements may be necessary. How long can your bodily functions remain active when breathing, circulation, or other systems are at a standstill?

Our bodies are magnificent machines containing over 3,000 kinds of enzymes, with each enzyme performing a different job. And without them, there would be no breathing, no digestion, no growth, no blood coagulation, and no reproduction. There are millions and millions of enzymes found in the lungs, liver, digestive system, and brain—in fact, in every system of our bodies. Some of these enzymes are secreted in an active form, while others (for instance, some enzymes involved in digestion) are secreted in an inactive form. They are activated when needed, sometimes by other enzymes, and are then ready to do their jobs.

From the top of your head to the tips of your toes, these enzymes are everywhere in your body. They help keep us alive and functioning. They keep us physically and mentally healthy, and they slow down that inevitable process of aging. Enzymes are so important that when their quantities drop and activity levels fall, illness is just around the corner.

The Best Kept Secret

As we surf through shelves of books on health care, we see volumes on proteins, fats, carbohydrates, vitamins, minerals, herbs, fruits, vegetables, and the like. But how much is written about *enzymes*? Little or nothing! They are the best kept secret in health and in fighting disease. Yet, nothing works without enzymes—no plants, animals, or humans. Where can you go for enzyme information?

This book is the answer to your thoughts and questions. It's all about enzymes—those energizing proteins that buck up your body and jump start your life.

AMINO ACIDS

Enzymes are composed of amino acids. Amino acids are the structural units of all protein. Think of an enzyme as a long chain of sausage links with each sausage representing an amino acid. There are approximately twenty different amino acids, which occur in each enzyme in different numbers and combinations. Enzymes differ in the order and number of amino acids. The body can make many of these amino acids by itself. But nine of them, called the *essential amino acids*, cannot be made by the body and must be obtained through our diet. (See Table 1.1 for a list of the twenty most common amino acids.)

HOW ENZYMES ARE NAMED

Enzymes don't have parents to give them names like Margie, Bill, Kate, Tony, David, Karen, or Rudy. When enzymes were first discovered and isolated, scientists named them.

Table 1.1. The Twenty Most Common Amino Acids

AMINO ACID	ESSENTIAL	NON-ESSENTIAL
Alanine		✔
Arginine		✔
Asparagine		✔
Aspartic Acid		✔
Cysteine		✔
Glutamic Acid		✔
Glutamine		✔
Glycine		✔
Histidine	✔	
Isoleucine	✔	
Leucine	✔	
Lysine	✔	
Methionine	✔	
Phenylalanine	✔	
Proline		✔
Serine		✔
Threonine	✔	
Tryptophan	✔	
Tyrosine		✔
Valine	✔	

Although they added the "-in" suffix to most (such as trypsin and pepsin), there was no uniformly accepted method of naming enzymes. The confusion increased as the numbers of enzyme discoveries grew. Because of this, the International Commission on Enzymes was established in 1956 to develop a nomenclature system. Today, enzymes are named for the substance (or substrate) that they work on and break down. The suffix "-ase" is then affixed to the end of the word. For example, an enzyme that breaks down protein is called *protease*, lipid- or fat-digesting enzymes are called *lipases*, and so on.

SIX CLASSES OF ENZYMES

Because enzymes have so many applications, scientists have found it helpful to classify them based on what they do, what substances they act upon (substrates), and the reaction they start or accelerate. There are six main groups of enzymes, each having fundamentally different activities (see Table 1.2). *Hydrolases* break down proteins, carbohydrates, and fats such as during the process of digestion. They do this by adding a water molecule, thus the name *hydrolases*. *Isomerases* catalyze the rearrangement of chemical groups within the same molecule. The *ligases* catalyze the formation of a bond between two substrate molecules through the use of an energy source. *Lyases* catalyze the formation of double bonds between atoms by adding or subtracting chemical groups. *Oxidoreductases* make oxidation-reduction (the process by which an atom loses an electron to another atom) possible. *Transferases* transfer chemical groups from one molecule to another.

3

Table 1.2. The Six Classes of Enzymes

CLASS	FUNCTION
Hydrolases	
proteases	Break down the peptide bonds in proteins.
amylases	Break down carbohydrates.
lipases	Break down fats (lipids).
Isomerases	Break down the rearrangement of chemical groups within the same molecule.
Ligases	Catalyze the formation of a bond between two substrate molecules through the use of an energy source.
Lyases	Split the double bonds between atoms withthe accumulation or dissociation of chemical groups.
Oxidoreductases	Make oxidation and reduction possible.
Transferases	Transfer chemical groups from one molecule to another.

Your body contains many enzymes from each group. In this book, we will primarily discuss two groups: hydrolases and oxidoreductases. Why? Because hydrolases play a number of important roles in our bodies, including aiding digestion and fighting inflammation, while some oxidoreductases, such as the antioxidant enzymes superoxide dismutase, catalase, and glutathione peroxidase, help fight free radicals. Although enzymes from all six groups are important, these two groups are of critical importance in maintaining your health and, unlike many enzymes, are also widely available in supplements.

HOW ENZYMES WORK

Most enzymes work by helping take something apart. For instance, your digestive enzymes are the forces that help break down that hamburger you ate last night into its smallest components, that is amino acids, mono- and disaccharides, esters, etc. Your teeth only do part of the work. The enzymes in your digestive tract actually snip apart the bonds that hold the various components of that hamburger together. Most enzymes work by taking bonds apart. Only about 3 to 5 percent of enzymes synthesize instead of breaking apart. These are the anabolic enzymes, not the cleaving catabolic enzymes.

In order to digest that hamburger, certain enzymes in your gastrointestinal tract break apart (lyse) the protein in the meat, while others work on the bread in the bun, and still others attack the onion, lettuce, pickles, ketchup, and mustard. Why so many enzymes? Because, with very few exceptions, each enzyme works on only one kind of substrate and in a specific way. Enzymes are very specialized. They are "substrate specific."

How does this work? There are at least two theories: the *lock and key theory* and the *induced fit theory*. The first theory—lock and key—likens the substrate to a key that must fit into a specific shape on the enzyme in order to activate that enzyme (much the same way a key fits into a lock to unlock a door). The induced fit theory states that the shape of the enzyme actually changes to allow the substrate to bind, similar to a glove conforming to the shape of a hand. The shape of the hand makes the glove's shape alter a little. Whatever the mechanism, the enzyme and the substrate come together and the enzyme is able to begin its work. The site of the connection between the enzyme and the substrate is called the *active site*. In order for the enzyme to do its work, the substrate (the molecule that is to be altered) must come into contact with the enzyme's active site. This recognition process ensures that only a specific molecule is recognized by an enzyme as being the proper substrate.

How Quickly Do Enzymes Work?

Each enzyme works under unique conditions and at its own speed—and they're fast. We can get some idea of the speed of enzymes by considering the slowest known enzyme, lysozyme. Lysozyme helps destroy bacteria and can process about thirty substrate molecules per minute. That is one substrate every two seconds! And as fast as that seems, it's nothing compared with the enzyme carboanhydrase, which processes an astonishing 36 million substrate molecules in one minute!

The speed of an enzyme is influenced by its work environment. It's like the work conditions in a factory. If a worker has pleasant conditions, he or she will be happier and work harder. If it's too cold or too hot, the workers will suffer and so will production. So it is with enzymes. They must have optimal working conditions.

Every enzyme also works best when in a specific range of pH—a measure of acidity and alkalinity—Some enzymes work better in an acid environment (like that of your stomach), while others need a more alkaline environment to do their jobs efficiently. There is a more complete discussion of pH in Chapter 5 of Part One.

Coenzymes and Cofactors

Although enzymes stimulate a variety of chemical reactions, they can do so only in association with small molecules called coenzymes and cofactors. *Cofactors* are substances that must be present for an enzyme to function. Minerals, such as zinc, magnesium, copper, and calcium, are some cofactors. *Coenzymes* are organic substances that combine with an inactive enzyme (an apoenzyme) to form an active enzyme (a holoenzyme). A coenzyme may be a cofactor. Some coenzymes include the B vitamins and vitamin C. For more information on cofactors and coenzymes, see Chapter 6 of Part One.

Enzyme Inhibitors

Though some substances help enzymes work better, there are others that actually inhibit the activity of an enzyme. Sometimes these enzyme inhibitors are *competitive*. That is, they actually compete with the substrate, preventing it from getting to the active site where it bonds to the enzyme. Other inhibitors are *noncompetitive* and work by retarding the conversion of the substrate by an enzyme. Either way, enzyme inhibitors can terminate or retard enzyme activity.

We encounter a number of substances every day that can inhibit our bodies' enzymes, this includes most medicines (even aspirin). An example of other inhibitors would be organic solvents, the most frequently produced chemicals in the United States. Primarily obtained from petroleum or natural gas, these chemicals are used extensively in the chemical industry and in many manufacturing processes and are known to inhibit a wide variety of enzymes. Examples of organic solvents include methanol, ethanol, propenol, formic acid, ethylene glycol, hexane, benzene, and butanol. These and other organic solvents are used in the manufacture of, or are found in, a variety of products you encounter every day, including paints and numerous household cleaners.

How Long Do Enzymes Last?

Just like any other protein, enzymes do not live forever. They age and die. When an enzyme begins to show signs of wear and tear, another enzyme comes along and makes short work of it. The worn-out enzyme is broken down, dissolved, and transported away.

Some enzymes have a life of only about twenty minutes. After this, the enzyme is replaced by a newly produced enzyme of the same type. Other enzymes remain active for several weeks before they are replaced.

One of the most fascinating properties of all enzymes is their ability to work with each other to form cooperatives when necessary and to continually exchange information with other enzyme cooperatives. The equilibrium of all systems that they maintain and the mutual effort toward a common goal are all positive properties of enzymes.

ARE YOU ENZYME-DEFICIENT?

Our bodies' ability to function, repair when injured, and ward off disease is directly related to the strength and numbers of our enzymes. That's why an enzyme deficiency can be so devastating.

Disease, diets consisting of foods with dead enzymes, chemotherapy, stress, physical injuries, illness, aging, or digestive problems can all affect our enzyme levels. But sometimes eating right and living a healthy lifestyle are not enough. Certain individuals have genetic or inborn problems affecting their enzyme production or activity (see Table 1.3).

You've probably seen the label on diet soda warning that the soda contains phenylalanine. Most diet sodas use aspartame as a sweetener. But aspartame is made from two amino acids: phenylalanine and aspartic acid. Those individuals who are *phenylketonurics* lack the enzyme phenylalanine hydroxylase, which is necessary to break down phenylalanine. And if it can't be broken down, it will build up in the bloodstream leading to neurological symptoms and mental retardation. The only recognized treatment for this condition is to avoid taking phenylalanine—hence, the warning labels.

Those who suffer from *lactose intolerance* don't produce enough of the enzyme lactase, which digests lactose, so drinking milk, which contains lactose, can lead to diarrhea and pain. *Gaucher's disease* is a rare familial disorder of fat metabolism. This disease, which usually begins in childhood, is due to lack of the enzyme glucocerebrosidase.

Some of these enzyme deficiency diseases can be corrected with supplemental enzymes, some cannot. But an enzyme deficiency might not have blatant symptoms or be life-threatening. Many of us are suffering from suboptimal health *only* because we're enzyme-deficient.

Signs of Enzyme Deficiency

The first sign that you're not getting enough enzymes will probably be disturbed digestion. You know—indigestion, stomach upset . . . gas. Many people notice a bloated feeling after eating particular foods, such as beans or cauliflower. This could be a sign that they don't have the enzymes necessary to adequately digest that food. Many foods, including beans, contain complex sugars. If these sugars cannot be broken down, they will sit in the large intestine and putrefy, leading to a bloated feeling and gas. This is easily corrected by taking such enzyme supplements as BeSure; Beano; Beans, Beans . . . and More Beans Rx; or other digestive enzyme supplements.

Table 1.3. Enzyme-Deficiency Diseases

DISEASE	DEFICIENT ENZYME
Acatalasemia	Catalase
Fabry's disease	Alpha-galactosidase A
Gaucher's disease	Glucocerebrosidase
Glycogen storage disease	Various, including glucose-6-phosphatase, glucose-6-phosphatase translocase, lysosomal glucosidase, liver phosphorylase, and phosphofructokinase
Niemann-Pick disease	Sphingomyelinase
Phenylketonuria	Phenylalanine hydroxylase
Tay-Sachs disease	Hexosaminidase A
Wolman's disease	Acid cholesteryl ester hydrolase

Another sign of an enzyme shortage that's not so easy to see is free-radical formation. In a telephone interview, Hans Kugler, Ph.D., international authority on aging and former president of the National Health Federation, stated that we should be living to the age of 120 or better. But the average American lives to be about 72 years old. Why does this happen? Pollutants and free radicals in our environment—as well as free radicals produced within the body—cause our bodies to "rust" and age. Wrinkling is an external sign of free-radical damage. Certain enzymes are antioxidants, that is free-radical scavengers (some of the best known are superoxide dismutase, catalase, and glutathione peroxidase). These enzymes fight the free radicals that destroy our bodies. We'll discuss antioxidant enzymes in more detail later in the book.

Illness is probably the most obvious sign that you're not getting enough enzymes or that your body enzyme levels are depleted. As mentioned previously, enzymes make your body work. Any illness or disease process, such as cardiovascular disease, degenerative diseases, cancer, or even a slow recovery rate after an injury are all indications that your body's enzymes are not working optimally.

Where to Find Enzymes to Jump-Start Your Life

In the past, our food was our primary source of enzymes. In theory, this works (take food = get enzymes); however, in practice, many of today's foods are loaded with additives, preservatives, and artificial colors and flavors, plus they are radiated, heated, canned, dried, and stored for months or years. This kills the enzymes in food. Even the heat of cooking can kill enzymes. Is it any wonder that the actual enzyme activity level of our foods is depleted?

So what's the answer to the problem? The answer is to eat live, fresh, organically grown, enzyme-rich foods. Fortunately, uncooked foods such as raw fruits and vegetables are usually high in enzyme activity and taste good, too. Some foods, such as pineapple and papaya, are particularly rich sources of food enzymes. But even if you're getting the best diet possible, you might need enzyme supplements if you have a problem digesting or absorbing the nutrients in your foods. Remember, you're not what you eat, you are what you absorb.

But what if you're eating correctly, digesting properly, and also absorbing the nutrients? That's great, *unless* your immune system is depleted or you have a cholesterol build-up (which leads to heart problems), or HIV/AIDS, herpes, cancer, etc. Any of these conditions and others can severely impact your health status and, therefore, your enzyme systems. Supplemental enzymes might be required. You can buy supplemental enzymes in health-food stores, drug stores, grocery stores, and through mail-order and multi-level marketing companies.

This book is designed to explain which supplemental enzymes are available, what they do, and how you can use them.

Which Conditions Can Be Treated With Enzymes?

Many people are familiar with enzymes as digestive aids. But enzymes can also be used to treat a wide variety of conditions through *systemic enzyme therapy*. In systemic enzyme therapy the enzymes are distributed throughout the body to help restore the body to health. Conditions that can be treated with systemic enzyme therapy include premature aging, arthritis and other inflammatory conditions, back pain, circulatory problems, myofascial pain syndrome, gynecological problems, herpes, injuries, multiple sclerosis, skin problems, systemic lupus erythematosus and other autoimmune diseases, viruses, and weight problems. And that's only scratching the surface. Volumes of scientific research from Germany, Japan, Italy, and the United States now exist on the use of enzyme therapy. Plus, new applications are being discovered every day.

In addition, enzymes can be taken in formulations made with vitamins, minerals, herbs, phytochemicals (plant nutrients), and other nutrients. I call such combinations Enzyme Absorption System Enhancers (EASE). These combinations are beneficial because they improve the absorption and bioavailability of other nutrients, maximize enzyme activity when combined with these nutrients, and reduce the drain of the body's own digestive enzymes.

IS ENZYME CARE NEW?

The use of enzymes in health care is nothing new. They're even mentioned in the Bible. For instance, when Hezekiah was dying, Isaiah told him to "take a lump of figs. And they took and laid it on the boil, and he recovered." (2 Kings 20:7). It was the enzyme *ficin*, derived from the fig, that caused the tumor to be cured. Further, the Indians of Middle and South America have used the leaves and fruit of papaya trees and the fruit of pineapples since time immemorial to treat inflammatory problems.

During the Middle Ages in Europe, the juice from plants of the spurge family (*Euphorbiaceae*) was used topically for the treatment of boils, warts, and decubitus ulcers. This form of topical enzyme therapy is still effectively employed today in treating leg ulcers.

However, modern enzyme therapy really began at the turn of the twentieth century when Scottish physician and embryologist John Beard was looking for a new, more effective way to treat cancer patients. Dr. Beard injected purified enzyme juices (freshly extracted from the pancreatic tissues of young calves and lambs) into the veins or even directly into the malignant tumors of his cancer patients. Dr. Beard found that some of the tumors stopped spreading while others actually regressed. Beard published the results of his findings in his book *Enzyme Treatment of Cancer* (London: Chatto and Windus, 1911). Other researchers, trying to duplicate his experiments, were not always successful. This is because Dr. Beard used freshly extracted pancreatic juice from young animals, which is high in enzyme activity, while his colleagues used pancreatic extracts that were a number of hours (or even days) old, unaware that the important enzyme activity already had been lost. Therefore, their experiments failed. At the time, no one understood why. As a result, research using systemic enzyme therapy was halted, and Dr. Beard's enzymatic therapy fell into disfavor.

At about the same time John Beard was working with pancreatic enzymes and cancer, Japanese researcher Jokiche Takamine (1854–1922) was experimenting with fermentation concepts and developing microbial enzymes for treatment of digestive disorders. He received a U.S. patent for his diastase, called Taka-Diastase, in 1894. Takamine was also instrumental in arranging a gift of 3,000 cherry trees from the mayor of Tokyo to Washington, D.C., something thousands of springtime visitors delight in seeing every year. Takamine also obtained the patent rights to adrenaline (first crystallized by a chemist in Takamine's labs).

During the mid-twentieth century, Max Wolf, M.D., (1885–1976) rediscovered the therapeutic value of systemic enzyme therapy and developed a method for treating various conditions with enzymes. For this achievement, he is considered to be the father of systemic enzyme therapy.

Dr. Wolf was born in Vienna, Austria in 1885. After obtaining his medical degree in 1919 at Fordham University, he was appointed professor of medicine there and taught at Fordham for a number of years. Through his research in gene-manipulated bacteria and plants for human protein nutrition, Wolf became increasingly aware of enzymes' key roles in the body. Publications by his friend Dr. Ernst Freund, M.D., further aroused his interest in enzyme therapy. Dr. Freund, a Viennese researcher, had observed that the blood serum of healthy individuals was able to destroy tumor cells (a characteristic not present in the serum of cancer patients). Freund assumed that the blood of healthy individuals contained a substance that could recognize and destroy cancer cells. This substance is lacking, or only present in extremely small amounts, in the blood of cancer patients. Freund called this "normal substance"; Wolf later identified it as enzymes.

With his friend and patient John Foster Dulles (American Secretary of State from 1953 to 1959), Professor Wolf founded the Biological Research Institute and engaged the well-known cell biologist Helen Benitez as his most important collaborator. Before working with Wolf, Benitez worked at Columbia University as the head of its neurosurgical department's laboratory for biocyto-culture technology.

As a result of his research with Benitez, it was soon clear to Wolf that adding enzymes to the blood of cancer patients helped fight cancer. This effect intensified when he combined enzymes from a variety of vegetable and animal sources.

Wolf also believed that enzyme deficiency could lead to premature aging. In 1960, he began treating a large number of elderly patients with enzyme combinations. He also thought it important to normalize their weight and regulate bowel movements. He achieved this through dietary changes, such as reducing animal fat intake while increasing the consumption of fish, vegetables, and raw fruit. He eliminated enzyme depleters including smoking and excessive quantities of coffee or tea. In addition, his patients took a balanced supply of vitamin and mineral supplements known to be essential for enzyme activity and, depending on age and state of health, he instructed his patients to be physically active.

Wolf was astounded by the results and observed that his treatment had a positive effect on vascular diseases, lymphedema, and zoster illnesses, as well as in healing of injuries and various inflammatory conditions. As a result of his investigations, he and German biologist Karl Ransberger, Ph.D., wrote the book *Enzymtherapie* (Vienna: Maudrich-Verlag, 1970).

Dr. Ransberger was born in Rosenheim, Germany in 1931. For many years, he cooperated with Dr. Wolf in cell-culture investigations, animal experiments, and clinical studies using enzymes. In 1967, Wolf and Ransberger formed the Medical Enzyme Research Institute in Munich, Germany.

Over the course of his life, Wolf treated a number of celebrities and dignitaries, including Pablo Picasso, Marilyn Monroe, Charlie Chaplin, Marlene Dietrich, members of the Kennedy family, the Roosevelts, J. Edgar Hoover, and Aldous Huxley. After Wolf's death (at 90 years of age) Ransberger assumed this scientific heritage and has continued a vast array of research studies in hospitals, clinics, and universities throughout the world.

Another pioneer in enzymes was Dr. Edward Howell (1898-1988). Many believe that Dr. Howell is single-handedly responsible for popularizing digestive enzymes and keeping them in view of the American scientific community and the general population. His landmark book *Enzyme Nutrition* (Garden City Park, NY: Avery Publishing Group, Inc., 1985) explains the concept and theories of food enzymes and human health with an emphasis on digestion from a historical perspective. It has been used as a bible for those advocating the use of microbial (plant-derived) enzymes.

Dr. Howell began his study of human health and food enzymes more than sixty years ago. Trained as an osteopath, physical therapist, and chiropractor, he spent six years on the professional staff of the well-known "nature cure" hospital Lindlahr Sanitarium. Utilizing nutritional and physical therapies, Dr. Howell established his own facility for the treatment of chronic ailments in 1930. He devoted his time to food and soil enzyme research and to his private practice until his retirement in 1970.

Other Uses of Enzymes

In addition to their important roles in the body and in health care, enzymes are valuable components used in medical testing, a number of industries, the production of many foods, coal processing, leather tanning, and the animal feed industry.

Medical Testing

Glucose test strips are used by many diabetics to measure glucose in the urine. These strips use a combination of enzymes (glucose oxidase and peroxidase) and change colors depending on the amount of glucose in the patient's urine. Of the estimated 12 million diabetics in the U.S., over 1 million use these test strips. Although strips for testing levels of cholesterol and theophylline have now been introduced, their markets are tiny in comparison.

Pulp and Paper Industry

To turn your typing paper and toilet paper white, the pulp and paper industry uses bleaching chemicals (especially chlorine and chlorine-based chemicals). Enzymes are now being used to decrease our reliance on deadly chlorine. Enzymes are also used as de-inking agents to remove the ink from recycled paper.

If you recycle your office paper, printouts, or bad copies from your laser printer or copy machine, you know the importance of recycling. But did you know that much of what you think you're recycling is actually ending up with the rest of the garbage in the local landfill? This is because the toners used in most printers and copiers use a thermal nylon polymer that fuses with the paper fibers.

In exploring this problem, Thomas W. Jeffries (a microbiologist at the University of Wisconsin at Madison) and his colleagues have found that they could remove the toner by treating the office paper pulp with simple enzymes. In their tests, the researchers found that nearly all enzyme trials were slightly more efficient than the chemical treatment for ink removal. In laboratory tests, the enzymes removed up to 96 percent of the toner, while in industrial tests, 94 percent was removed. In addition, researchers noted that enzymes were cheap, safe, and biodegradable.

Enzyme-Rich Detergents

Washing clothes has come a long way since women stood near the river bashing their clothes on rocks. Now, we add detergent to our dirty clothes in the washing machine, the washer goes swish-swash, tossing the clothes back and forth, and in a few minutes, your favorite blouse, shirt, or once grimy socks are all clean again . . . courtesy of those energetic enzymes.

Many modern detergents contain lipases, proteases, amylases, and cellulases to get the dirt out. Even many super-concentrated machine dishwashing detergents contain enzymes to remove both dried-on food and tough stains.

Why put enzymes in detergents? Let's use lipase as an example. Remember, lipase is an enzyme that takes fat molecules apart, breaking them into their smallest components. Water can then more easily lift the fat particles off of your favorite shirt.

Food Production

Enzymes are used in the production of baked goods; beer; wine; cheese; whey; food ingredients, such as fructose, aspartame, modified oils and fats, and emulsifiers; soy sauce . . . the list is endless. They also act as tantalizing taste enhancers and help to clarify fruit and vegetable juices.

You may already be using an enzyme in your cooking and not know it. Adolph's Meat Tenderizer contains the enzyme papain (from papaya), an enzyme that helps break down protein.

Enzymes in Coal Processing

Enzymes are even effective as biocatalysts in coal processing. Researchers have come up with a technique to convert solid coal into two types of liquid fuel: one of excellent quality and the other combustible.

Leather Tanning

Enzymes are now used in many stages of leather tanning, including soaking, dehairing, dewooling, and bating (a process that used to be particularly disagreeable because it used animal feces). The process of bating is necessary to make leather soft and supple.

Enzymes in Animal Feed

Animal feed contains mainly vegetable and plant materials. But sometimes the animal can't digest all the calories or nutrients in a food. This is especially the case with nonstarch polysaccharides. Enzymes can help break down these polysaccharides. Supplemental enzymes can also be used to assist the animals' own enzymes and are particularly useful for young animals who don't yet have mature digestive systems.

Supplementing enzymes in animal feed can also reduce the load of pollutants on the local environment (and water supplies) from the nitrogen and phosphorus excreted in animal manure. Enzymes reduce the quantity of manure, can decrease the odor, and can make the manure more biodegradable by decreasing the phosphorus content.

Dr. Howell believed that the body's ability to produce enzymes promotes rapid growth and prevents illness, but with age, our internal enzymes become depleted, leading to obesity, acute health problems, and chronic illnesses. He noted that enzymes make digestion possible and that fresh fruits and vegetables, raw sprouted grains, and enzyme supplements were necessary for optimal health.

Research by these and other visionaries has promoted interest in enzymes. As a result, many medical schools and hospitals currently have departments of enzymology whose purpose is to study and use enzymes to fight diseases.

The potential applications of enzymes are limitless. Recently, researchers have discovered new areas for which enzymes can be used successfully. This includes the treatment of many autoimmune diseases, such as rheumatoid arthritis, systemic lupus erythematosus, and multiple sclerosis. It seems that enzymes are on the edge of a breakthrough in the control of these diseases.

As you can see, the importance of enzymes, while known by many people in the scientific community, is still a relative secret to the average individual. The more you understand about enzymes and how they work, the more likely you will be to use enzymes for improving your own health.

2. Enzymes and Digestion

A healthy digestive system is the gateway to better health, and enzymes are the keys that open the door. Unfortunately, most people don't give digestion much thought—that is, until something goes wrong. Then it can really ruin your day—and your life.

The purposes of the gastrointestinal system are simple:

• To extract nutrients from foods.
• To digest nutrients into units small enough to be absorbed.
• To eliminate waste products.

Poor digestion can do more than give you a stomachache or gas because digestion is the mechanism that makes your body work. It converts the fuel provided by your food into a usable form. Faulty plumbing in the body can interfere with your digestive system's ability to turn the food you eat into nutrients to fuel your engine. A clog in the works may be only the tip of the iceberg. Chronic fatigue, premature aging, arthritis, poor skin and hair quality, toxicity, allergies, cancer, and many other diseases can all result from faulty digestion because poor digestion interferes with nutrient breakdown, absorption, and metabolism; allows toxins to remain in the body and accumulate; and overstresses the body.

Why is it that some people eat nourishing foods yet are always tired, age prematurely, and are more vulnerable to illness? Could their bodies be inefficient at digesting and absorbing food nutrients? No matter how good the gas is that you put in your car or how much you put in, if the fuel line is blocked, the engine will perform poorly or will stop. So it is with the body. You can eat a highly nutritious meal, but if the food nutrients can't reach the body's engine, your health will diminish. It's called *dis-ease* of function. Ease of function is *health*, while dis-ease is lack of normal function. Dis-ease leads to *disease*—that is, declining health, illness, and possibly death.

According to the National Digestive Diseases Information Clearinghouse, digestive disorders are a major problem in America. In fact, some 60 to 70 million of us suffer from some type of digestive disease. In the United States, we spent $107 billion in 1992 fighting digestive diseases. But worse than the financial cost is the human suffering that resulted. In 1987, some 10 million people were hospitalized due to digestive diseases—that's 13 percent of all hospitalizations—and 1.4 million people were disabled due to digestive illnesses.

DO YOU SUFFER FROM POOR DIGESTION?

What did you have for breakfast? What about for dinner last night? Did you avoid something on your plate because you know you can't eat it? Does the food you love not love you back? If so, is it because of an allergy or (more likely) faulty digestion? Do certain foods give you stomach upset? indigestion? gas?

Gas occurs when we can't digest what we eat. If foods don't break down, they reach our intestines in large sizes where they become a banquet for bacteria. They begin to putrefy, giving off carbon dioxide and methane . . . and you have gas!

Unfortunately, Americans live in a cooked-food, fast-food society. Cooked foods are "enzyme-dead"—that is the enzymes are killed during the cooking process—and pass through the digestive tract more slowly than do raw foods. As time passes, cooked foods collect on the walls of the large intestine. Here, they can putrefy and feed harmful bacteria, producing toxic by-products that could result in cancer and other serious diseases. At the very least, it will cause stomach discomfort and gas.

Gas is a symptom of dis-ease. It tells us that our bodies aren't functioning properly. According to the National Cancer Institute, we can improve our health by following the food pyramid and eating fresh fruits, vegetables, and whole grains—foods rich in enzymes. But the very foods the National Cancer Institute recommends to fight cancer, such as broccoli, Brussels sprouts, cabbage, cauliflower, rutabagas, turnips, and high-fiber cereals and breads, are great gas-producers, since they contain indigestible components that may ferment in the intestines. Even pretzels, bagels, and pastas can cause gas! Also, adapting to a fresh-food diet may be hard if you're not used to it. In fact, any sudden change in diet can affect the digestive tract. You know what happens when you start eating cabbage or broccoli . . .

And what about indigestion? It, too, occurs when we can't digest what we eat or drink. Some people take antacids, but over-the-counter, traditional antacids don't solve the problem, they only cover it up, which in some cases may be the worst thing you can do. Why not treat the *problem* instead of the *symptom*? Perhaps when we have gas or indigestion we should be thinking, "I can't produce enough enzymes to break down my food."

Carbohydrates

The human diet is composed of a wide range of carbohydrates, which provide energy to feed our brains, nervous tissues, lungs, and hemoglobin. In fact, carbohydrates are the body's primary energy source. Most carbohydrates come from plant sources (except for lactose and small amounts of glycogen).

There are three general types of carbohydrates—monosaccharides, disaccharides, and polysaccharides. Monosaccharides are composed of only one sugar molecule and include glucose (a sugar found widely in most plant and animal tissue) and fructose (a sugar found in fruit). Disaccharides are composed of two sugar molecules and include sucrose (table sugar, composed of one glucose and one fructose molecule) and lactose (milk sugar, composed of one glucose molecule and one galactose molecule). Mono- and disaccharides are considered sugars or simple carbohydrates.

Polysaccharides are carbohydrates that are composed of three or more sugar molecules. These would include starch, cellulose, and glycogen. Polysaccharides are termed complex carbohydrates. Fiber is an indigestible type of carbohydrate. Fiber provides no energy to the body, but it aids the absorption of sugars into the bloodstream and helps the intestines to function efficiently.

Most Americans' diets consist of too many simple carbohydrates and not enough complex carbohydrates. In fact, complex carbohydrates, particularly fiber, should compose most of one's diet. Most Americans consume approximately 12 grams of fiber per day, but should be consuming from 40 to 60 grams of fiber per day.

As explained in Chapter 1, enzymes are the great extractors and potentiators. Lack of these digestive and energizing catalysts can mean that essential fats, proteins, carbohydrates, minerals, and vitamins will fail to be extracted from your foods and put to work in your body. Enzymes are very specific. That is, each enzyme acts in a specific way and on a specific food component. Imagine a person at a commuter train stop waiting for a *specific* train—but it never comes. Other trains come and go, but our passenger can't get home without taking one specific train. The food we eat is like this passenger. The food will never be digested if the right kind of enzyme never comes along. A deficiency or absence of one enzyme could lead to improper digestion, fatigue, and a wide range of diseases.

The proper chemical balance in enzymes makes the digestive system work. Enzymes are present in every phase of digestion, and without them we just can't process what we eat. Enzymes break foods down into smaller sizes, making them easier to digest. In this way, enzymes release the nutrients in our foods. Without digestive enzymes, food would just go in your mouth and out the back end, undigested (see Table 2.1).

In order to change our foods into materials we can use, our bodies require more than twenty enzymes from the following groups:

• Proteases, which break down proteins.
• Amylases, which work on carbohydrates.
• Lipases, which lyse fats.

These three major types of enzymes are naturally manufactured by your stomach, pancreas, or small intestine, and they are also available in the fruits, vegetables, and grains you eat, if the food is fresh and the enzymes are not yet killed by cooking, frying, or microwaving.

Our bodies have factories that produce and dispense enzymes for proper digestion. This includes our salivary glands, stomach walls, pancreas, liver, and intestines. Even after the food particles have been broken down in the small intestine and delivered to the blood and lymphatic vessels, our body provides other enzymes to keep the delivery pipes to the cells clean and in working order. Plus, each cell is like a factory unto itself with enzymes to further metabolize and transform foodstuffs into energy.

LOOKING AT THE PROCESS

The digestive system is like a long tube or a conveyor belt. From the time food hits your tongue to the time it reaches various body cells or passes out of the body as waste, your food is going through different changes.

The Mouth

The first time your food encounters enzymes is in the mouth (see Figure 2.1). Saliva contains the enzyme *ptyalin* (salivary amylase), which begins the digestive process. Amylase attacks the carbohydrates in your food and starts to break them up into smaller particles.

Have you ever noticed how hard it is to swallow or talk when your mouth is dry? Keeping your mouth moist is one of saliva's main jobs. It also lubricates the food we eat, making it easier to swallow. As you might have guessed, saliva is mostly water (97 to 99.5 percent). Since digestive enzymes are hydrolytic, they need water molecules in order to function. Saliva is also a little bit acidic (pH 6.75 to 7.0), but its pH can vary.

Other substances in saliva include the enzyme *lysozyme*, which may help to prevent tooth decay because it guards against bacteria; *kallikrein*, an enzyme that may regulate blood flow in the salivary glands; *lactoferrin*, an iron-binding protein that works as an anitimicrobial agent; *peroxidase*, an antioxidant enzyme; and *growth factor*, which is present in the saliva of many animals and promotes clot formation (necessary for the repair of wounds.) Have you ever seen a dog or cat lick its wounds? It may be growth factor that helps wounds to heal. Growth factor was recently identified in human saliva.

The longer you chew your foods, the more surface area will be exposed to enzymatic activity. This is important because enzymes can act only on the surface of food particles. Think of the mouth as a cement mixer, rolling the food around, chopping it up, and breaking it down. The result—better digestion. Some experts say that you should chew each piece of food fifty times before swallowing. In any case, the more you chew, the better your food will be broken down.

Do you and chew on only one side of your mouth? You could unknowingly be affecting your enzyme levels and activity. A recent study measured saliva secretion and enzyme activity in patients who suffered from hemiplegia (paralysis on one side of the body) after a stroke. They discovered that greater secretion and higher activity of enzymes were found on the chewing side of the mouth. So chew on both sides of your mouth for best oral enzymatic activity.

Enzymes in Your Stomach—The Voyage Continues

Food travels from the mouth, down the esophagus, to the stomach. Think of your stomach as an expandable sack, shaped like the letter "J." It is located in the upper left por-tion of your abdomen. The lining of the stomach is loaded with glands supplying hormones, hydrochloric acid, and enzymes—all essential for digestion.

In the stomach, food is broken down into increasingly smaller particles. The stomach acts as a food reservoir; mixes food with gastric juices, including pepsinogen (the inactive form of pepsin) and hydrochloric acid, which activates pepsin; and slowly empties the food, now called *chyme*, into the small intestine. (See Figure 2.2.)

Digestion in the Small Intestine

The small intestine is a 23-foot-long coiled tube, about an inch in diameter, that runs from your stomach to the large intestine. The greatest amount of digestion and absorption take place in the small intestine. It consists of three parts: the *duodenum*, the *jejunum*, and the *ileum*. Each part plays an important role in digestion and each section is responsible for the absorption of different nutrients.

Chyme travels from the stomach into the duodenum, the first part of the small intestine. Bile, supplied by the liver, and pancreatic enzymes are added to the chyme here to aid digestion and absorption. Calcium, vitamin A, and the B vit-amins riboflavin and thiamin are also absorbed in the duo-denum and beta-carotene is converted into vitamin A by enzymes and bile salts. Most fats are absorbed from the jejunum, and vitamin B_{12} is absorbed in the last section of the small intestine, the ileum.

By the time they reach your small intestine, carbohydrates and protein have been partially broken down by the stom-ach. But only a small portion of fat is digested in the stomach; most fat digestion occurs in the small intestine. At various times during the digestive process, hormones send signals to the pancreas, gall bladder, and liver requesting assistance.

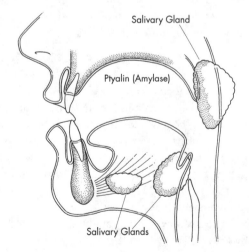

Figure 2.1. Digestion in the oral cavity. The enzyme ptyalin (salivary amy-lase) begins to break down carbohydrates into the sugars maltose and dextrin.

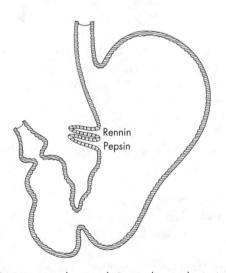

Figure 2.2. Digestion in the stomach. Protein digestion begins in the stom-ach with the action of pepsin, rennin, hydrochloric acid, and small quantities of amylase and lipase. Polypeptides and casein begin to be formed.

Table 2.1 Enzymes in Digestion

ENZYME	LOCATION OF ENZYME ACTIVITY	SOURCE OF ENZYMES	OPTIMAL PH RANGE	ACTION
Salivary amylase (ptyalin)	Mouth	Salivary glands	6.0–7.0	Breaks down carbohydrates.
Gastric lipase	Stomach	Stomach glands	approx. 6.0	Breaks down fats into fatty acids and glycerol, initiates triglyceride digestion.
Pepsin	Stomach	Stomach glands	1.0–2.0	Breaks down proteins. Secreted in the form of pepsinogen, but hydrochloric acid converts it to the active enzyme pepsin.
Rennin	Stomach	Stomach glands	4.0–5.35	Breaks down milk casein into milk curd (coagulates milk) Releases minerals (calcium, iron, phosphorus, and potassium) from milk.
Enterokinase	Small intestine	Pancreas	5.2–6.0	Transforms trypsinogen into trypsin in the duodenum.
Proteolytic enzymes, including trypsin, chymotrypsin, and carboxypeptidase	Small intestine	Pancreas	7.9–9.7	Breaks protein and polypeptides into dipeptides and some amino acids.
Amylolytic enzymes, including amylase	Small intestine	Pancreas	6.7–7.2	Attacks starches and other carbohydrates.
Lipolytic enzymes, including lipase phospholipase A1 and A2, and esterase	Small intestine	Pancreas	8.0	Splits fats into glycerol and fatty acids.
Amylolytic enzymes including sucrase, maltase, and lactase	Small intestine	Intestinal lining	5.0–7.0	Break down carbohydrate fragments (sucrose, maltose, lactose).

The pancreas is located just behind the stomach. Its number one job is to produce enzymes that pass through the pancreatic duct into the duodenum. When I think of this system, I think of the subterranean canals and pipes lodged below the streets of Paris, France—an intricate, yet essential, maze. The pancreas also produces hormones, including insulin and glucagon, which play important roles in the metabolism of carbohydrates.

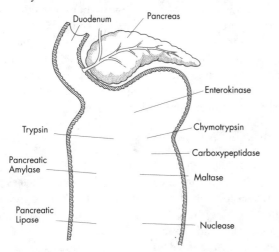

Figure 2.3. Digestion in the small intestine by pancreatic juices. Proteins, fats, and carbohydrates are further broken down: proteins by trypsin, chymotrypsin, and carboxypeptidase; fats by pancreatic lipase; and carbohydrates by maltase and nuclease. Nutrients are also absorbed into the bloodstream.

Pancreatic juice consists of enzymes, water, and electrolytes (primarily bicarbonate—necessary for neutralizing the acidic chyme). Our bodies produce approximately 1,200 to 1,500 ml of pancreatic juice daily. (See Figure 2.4)

Some of the enzymes, such as trypsinogen, chymotrypsinogen, and procarboxypeptidase, are secreted from the pancreas in their inactive form (they are called *proenzymes*). When the pancreatic juice reaches the small intestine, the inactive forms of the enzymes are converted to their active forms (trypsin, chymotrypsin, and carboxypeptidase, respectively). If activated while still in the pancreas, they would digest the organ. To assure that this does not happen, the pancreas secretes a substance called *trypsin inhibitor*. This substance keeps trypsinogen from converting into the active enzyme trypsin too early. Since trypsin can activate the conversion of other proenzymes into their active forms, inhibiting trypsin effectively inhibits the other proenzymes too. However, self-digestion can occur if the pancreatic duct becomes blocked or the pancreas is damaged. In these instances, the proenzymes will overwhelm the inhibitor, causing large amounts of trypsin, chymotrypsin, and carboxypeptidase to become active in the pancreas. This condition, called *acute pancreatitis*, can result in a lifetime of *pancreatic insufficiency* (a condition in which the pancreas does not produce an adequate amount of enzymes), and can even

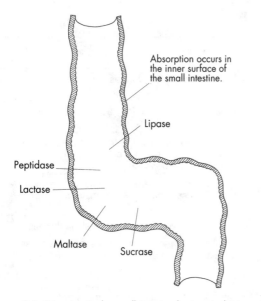

Figure 2.4. Digestion in the small intestine by intestinal juices. Proteins, fats, and carbohydrates are further broken down by intestinal juices.

be dangerous or life-threatening due to its accompanying shock. This self-digestion does not occur in the small intestine because it has a protective mucous lining to prevent self-digestion from occurring.

Enzymes from the pancreas, the mucous membranes of the small intestine, and the liver get together to break down foods for absorption of nutrients into the bloodstream and for expulsion of waste products into the large intestine. Some pancreatic enzymes, such as lipases, amylases, and nucleases, are secreted in active form, but for their optimal activity, bile must be present in the intestinal opening. Bile is a bitter brownish-green or yellowish-brown liquid secreted by the liver. It is essential for emulsifying lipids (fats). It dissolves the products of fat digestion (fatty acids and monoglycerides) and serves to alkanize (make less acidic) the intestinal contents. Bile is stored in the gall bladder until it is needed in the small intestine.

The inner surface of the small intestine breaks down and absorbs nutrients into the body. It's more than just a protective membrane for your intestines. The lining of the small intestine contains many folds, bunched up like a rug. These folds increase the surface area that is exposed to the food products. The lining of the intestinal folds is further increased by many microscopic fingerlike projections called *villi*. These villi are large in the upper part of the small intestine (the duodenum) and gradually become shorter and narrower in the jejunum and ileum. Also present are even smaller projections (*microvilli*) that give the surface a "fuzzy" appearance. This area is sometimes called the "brush border." In addition to increasing the absorptive surface, the membranes of the microvilli contain enzymes. The peptidases in the brush border break proteins down into their com-

ponent amino acids, while the disaccharidases change complex sugars into simple ones.

The intestinal lining also contains a diverse array of transport systems that move nutrients such as sugars, small peptides, and amino acids into the cells' cytoplasm and subsequently into the circulatory system. These transport systems are the mechanisms that make sure the nutrients from your food eventually get to the various cells in your body that need those nutrients.

Enzyme Absorption

Though research continues to attempt to find out just how the intestinal lining does its job, we do know that the lining is an active tissue that breaks down nutrients, selects those to be absorbed by the blood, and controls the speed with which this action takes place.

Previously, it was thought that enzymes (whether in food or in supplement form) could not be absorbed intact and instead were completely broken down into their smallest parts (such as amino acids) before being absorbed from the small intestine into the circulatory system. Today, we know this is untrue. Growing amounts of research by a number of highly respected scientists has shown that a significant amount of enzymes are, in fact, absorbed intact. The primary method of absorption is *pinocytosis*, a method in which whole molecules are engulfed by cells and absorbed. After connection to a receptor in the intestinal wall, the enzymes are guided through the intestinal cells in vesicles (small sacs), and released into the blood. The absorption rate of orally ingested enzymes is about 20 percent within six hours. In pinocytosis, the enzyme is like a passenger stepping into an elevator on the first floor. The elevator closes (similar to the way the intestinal lining encloses an enzyme) and travels to the second floor where the door opens, releasing the passenger.

A second method of absorption is transportation to the bloodstream by way of the lymphatic system. The lymphocytes help the enzymes into the bloodstream by binding to them. A third method is transcellular absorption (persorption). In this process, large nutrient particles are pushed between the intestinal cells by the contractions of the intestines (peristalsis). They are then forced into the bloodstream.

Into the Large Intestine

By the time the chyme makes it to the large intestine, very few nutrients remain. The large intestine is about 5 feet long and two-and-one-half inches in diameter. Divided into several subsections (the ascending, transverse, descending, and sigmoid colon and the rectum) the large intestine's job is to help you get rid of food wastes, those portions of your food that your body couldn't break down and absorb.

Enzymes that act on the remaining food residues, fiber,

Lipids

Lipids (fats and oils) are composed of carbon, hydrogen, and oxygen. Fats are solid at room temperature and oils are liquid at room temperature. Lipids are important in our diets because they contain the fat-soluble vitamins A, D, E, and K, and they are sources of the essential fatty acids, which are very important in maintaining our cell membranes.

Once in our bodies, lipids serve many important functions. They provide a cushion under the skin against trauma. They build healthy cell membranes. They aid in the absorption of the fat-soluble vitamins. They provide insulation. They protect vital internal organs. They provide energy. And they act as building blocks for other molecules.

The intake of lipids has become quite controversial over the years. Lipids have gained the reputation of being a bad component of one's diet and as something to be avoided. As you can see, however, some lipids are absolutely essential to health.

---

cells, and mucus discarded from the upper intestinal tract are produced by a large number of bacteria found in the colon. These bacteria live in the colon and ferment some of the undigested carbohydrates. It is estimated that more than 400 bacterial species are located in the intestinal flora of any given person. Most of these bacteria are anaerobic, which means that they don't need oxygen in order to live and grow. In the large intestine, the remaining unusable bulk is processed, water is absorbed, and waste excreted through the rectum and out the anal canal.

THE EFFECTS OF ALTERED ENZYME ACTIVITY

So what happens if Mother Nature has forgotten to give you enough or any of these enzymes? You'll have poor digestion. Poor digestion can also be a result of a lifestyle that does not allow for proper nutrition. In this fast-paced world, twenty-four hours each day doesn't seem to be enough to get everything done. And because we don't have enough time, we make sacrifices in our quality of life—this includes our diet. We eat on the run. We eat fast foods. We eat too much fat and too little fiber. This all has a devastating impact on our diges-

tion and, therefore, our health. Improper diet can even lead to cancer. In fact, according to the National Cancer Institute, at least 35 percent of all cancer cases are directly related to diet.

Aging is another problem. As we age, our bodies' enzymes decrease in quantity and activity level. The speed of aging is greatly influenced by our lifestyles and diets. An enzyme-poor diet can overtax an already deficient system, causing us to age faster.

Stress can also affect our digestion because it reduces the secretion of digestive juices and diverts blood from digestive muscles to other muscles in the body. Any form of nervous anxiety, agitation, worry, or fatigue can lead to disturbances in the gastrointestinal tract, interfering with digestion and, therefore, nutrient absorption.

Sometimes, something goes wrong and our bodies either don't make the enzymes we need to digest our foods or don't make quite enough of a particular type of enzyme. Lactase deficiency is a good example. If you suffer from diarrhea, gas, or intestinal bloating one to four hours after drinking one or two cups of milk, you might be deficient in the enzyme lactase. Milk contains a sugar called lactose, which

---

Proteins

The proteins in our foods are varying combinations of amino acids, linked together in chains ranging from several to thousands in length. Each plant and animal species (and therefore, each food) has its own characteristic proteins, distinguished by their amino acid sequence.

Because the content and balance of amino acids can vary in different foods, some are considered *incomplete* and others *complete* proteins. Complete proteins contain all the essential amino acids, while incomplete proteins lack one or more. You can compensate for this deficiency by combining foods based on their amino acid content.

For instance, eat beans with rice or corn tortillas. When consumed together or within a short time (though scientists haven't agreed on the exact length of time), such proteins can meet requirements for essential amino acids. In this way, strict vegetarians can stay healthy without eating food derived from animals.

Protein is necessary in the body for a variety of reasons. It is essential for the proper growth and repair of the cells in our bodies. It is also necessary for the production of enzymes, hormones, and DNA in the body. The individual amino acids also serve important roles in the body.

you can't digest or absorb if you don't have enough (or any) lactase. So it sits in the colon and ferments, and you experience gas, bloating, diarrhea, and cramps. Luckily, this problem can now be helped with supplemental lactase enzymes. (Supplemental enzymes will be discussed in Chapter 5.)

There are a number of conditions due to enzyme deficiency that can cause, or that result from, poor digestion. These conditions can result in detrimental effects both locally and systemically. *Hypochlorhydria* is a condition in which the sufferer does not produce enough hydrochloric acid in his or her stomach. Hydrochloric acid triggers the activation of pepsinogen, converting it to the active enzyme pepsin. It also kills some of the microorganisms we ingest. So a shortage of this gastric secretion can have a dramatic effect on your digestion. *Achlorhydria* is even worse than hypochlorhydria because the sufferer's stomach produces no hydrochloric acid at all.

Pancreatitis is inflammation of the pancreas, often caused when pancreatic enzymes are activated before they reach the small intestine. Pancreatitis can result in *pancreatic insufficiency*, in which the pancreas does not produce enough pancreatic juice. This will result in improper digestion and can cause very serious illness, or even death. Pancreatic insufficiency can also be caused by illness (such as cancer or cystic fibrosis) or injury.

Intestinal toxemia occurs when foods sit undigested in the colon, fermenting and producing toxins that can be absorbed by the bloodstream. This can happen as a result of an enzyme deficiency. *Malabsorption*, impaired absorption of nutrients, can also occur due to deficient enzymes.

Symptoms such as gas, indigestion, constipation, diarrhea, or poor skin (to name just a few) may be indications of faulty digestion. Digestion should be the key to your health—not the root of all your problems. Improper digestion or *dis-ease* of function can affect body metabolism and lead to devastating chronic diseases, such as cancer and cardiovascular disorders. Enzymes are critical if your digestive system is to extract and release the nutrients in your food that your body needs. Unfortunately, our environment and diets can negatively affect our body's enzymes. We'll cover these topics in the next chapter.

3. ENZYME DEPLETERS IN OUR LIVES

Now you know a little more about those jump-starting enzymes and their importance in our bodies. Unfortunately, a number of factors in our day-to-day lives can inhibit and destroy important enzymes in our food and in our bodies. In many cases, our enzymes are subjected to toxins that either annihilate them or hamper their function. We are surrounded by enzyme depleters. They are in our air and in our food. This chapter will take a look at some of the enzyme depleting and inhibiting processes and additives in our foods, enzyme inhibitors that naturally occur in some of the foods you eat, and some enzyme inhibitors in our environment.

THE LINK BETWEEN
ENZYME-DEAD DIETS AND DISEASE

Do you want cancer? No? Neither did my grandfather. Cancer of the colon killed him. Before surgery, his abdomen became quite distended and painful. Following surgery, Grandpa was fitted with a colostomy sack and lived three miserable years after the ordeal.

Before his illness, Grandpa had walked over five miles to and from work seven days a week, which is good exercise for anyone. But exercise isn't everything. Unfortunately, Grandpa had a history of poor bowel movement and eating practices. Like many of us, Grandpa was a victim of habit. His diet was loaded with fatty foods, and his food was usually cooked. Breakfast primarily consisted of fried eggs, American fried potatoes, and greasy bacon or sausage, with ample servings of cream-laden coffee. Lunch and dinner were just as fatty and unhealthy. No matter how much Grandpa exercised, his body's organs were overburdened with fatty oil slicks. He seldom ate anything that wasn't baked, boiled, fried, canned, or processed in some way—very few fresh fruits or fresh vegetables and, therefore, woefully few enzymes.

I don't know about you, but I don't want to develop my grandfather's digestive problems or die from painful colon cancer. Could a better diet have given Grandpa a longer, happier, healthier life? Research says yes. The inclusion of live, enzyme-rich foods in your diet can have a direct inhibitory effect on cancer, arthritis, atherosclerosis, multiple sclerosis, and a long list of other conditions. Let's look at the oldest enzyme depleter—cooking.

COOKING

For thousands of years, humans ate their food raw. But with the advent of fire, we learned how to cook meat, boil vegetables, and bake bread. Now, instead of eating predominantly uncooked food, our diet is primarily (if not exclusively) composed of cooked food, with only an occasional raw fruit or vegetable thrown in. Have you eaten anything fresh today? If you have, you're one of the few.

It is generally agreed that temperatures over 140° F (some say as low as 107° F) will kill the enzymes in food. Yet every cooking method ever devised is hot enough to destroy enzymes. Water boils at 212° F. In order for many foods to become soft enough to eat, they must boil for several minutes. For instance, cooking potatoes requires that their internal starch molecules reach the "gelatinization stage," which is at about 150° F. This is only accomplished after several minutes in boiling water. So any food that is cooked by boiling has been sufficiently heated to kill all enzymes. Frying and deep-fat frying require even higher temperatures. For example, to properly fry an egg, the temperature needs to be somewhere between 255° F and 280° F. Even baking will destroy the enzymes in food. Most foods are baked at 350° F, and some require even higher temperatures. Broiling is typically conducted at temperatures between 450° F and 500° F. Even though internal temperatures do not reach oven temperature, food temperatures do get high enough to kill the enzymes in it when baked. (See "Cooking Temperatures" on page 18.)

Cooking certain foods is sometimes necessary. Heat produces a chemical change, altering and sometimes intensifying flavor and aroma. Cooking also breaks down the tough fibers in many foods, making vitamins and minerals more accessible, softening food, and making some (such as grains) easier to chew. Eating wheat or rice would be impossible if they weren't softened by cooking first. The heat of cooking also makes some foods (particularly meat) safer by killing harmful microorganisms, including *Staphylococcus aureus*, *Salmonella*, *Escherichia coli* (*E. coli*), *Clostridium perfringens*, and a host of other bacteria, as well as parasites and other contaminants. This is why it is extremely important to thoroughly cook any meat to be consumed. Unfortunately, this ensures that virtually all of the enzymes naturally occurring in the meat will also be killed. To eat meat in its enzyme-rich state is just not safe—so I do not advocate it.

Cooking Temperatures

Temperatures as low as 107° F may be hot enough to kill the enzymes in your food. Below are the temperatures reached with different methods of cooking.

TYPE OF COOKING	TEMPERATURE USED
Boiling	212° F
Frying	250° F and up
Deep-fat frying	400° F and up
Baking	350° F and up
Broiling	450°–500° F

MILLING AND REFINING

Heating food isn't the only way that their enzymes can be destroyed. Some processing practices literally strip the active enzymes from the foods. The milling and refining processes that several grains (including wheat and rice) undergo after harvesting are examples of such practices.

Until about 1870, white bread was considered a luxury and reserved for the wealthy since refining flour was an expensive, labor-intensive practice. The peasants actually ate healthier bread than the wealthy did because whole wheat bread still contains the germ and the bran—storehouses of vitamins, minerals, protein, fat, fiber, and enzymes naturally found in the wheat berry. White bread and white flour lose these nutrients during the milling and refining processes.

Wheat starts out as a highly nutritious grain, loaded with enzymes, including alpha- and beta-amylases, proteases, peptidases, hemi-cellulases, and oxidases. However, in order to make "modern" white flour, the wheat is milled, a process that removes the nutrient- and enzyme-rich bran. The remainder is then ground into smaller particles, chemically bleached, and then aged (often with the assistance of chlorine dioxide or potassium bromate, harmful chemicals that can cause damage to the respiratory tract and the kidney, respectively). The few enzymes that remain will be destroyed when the loaf of bread is baked.

Are there any easily eaten grains when uncooked that still have active enzymes? Oats, a common component of granola, would seem to be an excellent choice. Unfortunately, most commercially prepared oats not only have the bran and germ removed, but are usually steam treated as well. Oats contain more fat than wheat (up to five times more), as well as large amounts of the enzyme lipase, which acts upon fat. Unless the enzyme is inactivated by steam treatment, the oats continue their maturation process and eventually become rancid. Steam treating ensures that the oats will stay fresh longer on the store shelf (and in your cupboard). Unfortunately, it also ensures that the oats will be enzyme-dead. So we have a choice: longer shelf-life or foods that are still alive enzymatically. An alternative to steam treating would be to refrigerate the oats immediately after rolling or cutting. This would slow the enzymes' activity without destroying them.

The modern milling and refining methods used on wheat, rice, corn, oats, and other grains not only deplete the enzymes in those foods, but also many of the vitamins and minerals that your body's enzymes need in order to function. For example, removing the bran and germ from wheat also reduces the amount of B vitamins (required as coenzymes by many of your body's enzymes). In fact, milling of white flour from wheat reduces the thiamin content by about 90 percent and the iron, riboflavin, and niacin by about 70 to 80 percent. In fact, the overall mineral content is reduced by as much as 70 percent. In addition, 10 percent of the protein is lost. According to *The Wellness Encyclopedia of Food and Nutrition* (Sheldon Margen, ed., New York: Rebus, 1992) of "the 22 nutrients decreased during the milling of white flour, only four—niacin, thiamin, riboflavin, and iron are replaced when flour is enriched." In fact, "enriched" white bread contains about 75 percent less fiber than wheat bread and is also lacking in zinc and copper. This is why bread is "fortified"—to put back some of the nutrients that were removed during processing. But even fortifying doesn't put back the enzymes.

FOOD PRESERVATION METHODS

For centuries, humans have faced the challenge of ensuring a steady, reliable, and sufficient food supply. Over time, we've learned the necessity of preserving food and the tech-

Heat and Enzymes in Action

To visualize how heat inactivates enzymes, try the following experiment:

Prepare an instant gelatin mix according to the directions on the package, add some canned pineapple to the mixture, and then refrigerate. What happens? The gelatin will become firm after a period of time. Now repeat the same experiment with freshly cut pineapple. What happens? Will the gelatin firm with freshly cut pineapple? No! The active proteolytic enzymes (especially bromelain) in fresh pineapple keep lysing the gelatin, preventing it from becoming firm. The canned pineapple won't interfere with gelatin's action because the enzymes were killed by the high temperature used in the canning process.

Now you've conducted a home experiment showing the effect of high temperature and canning on foods. You're a research scientist.

niques of doing so, thereby providing calories and nutrients during months when nothing fresh could possibly grow.

Drying

Drying was probably the first food preservation method, possibly discovered by accident when early humans found dried food did not deteriorate as rapidly as fresh food and could, therefore, be put away for a rainy (or sub-zero) day. Drying is a very efficient way of preserving food that works by removing water from food. This makes it harder for microorganisms and other living organisms, which require water for their metabolism, to survive. Drying effectively inhibits the microorganisms that might cause food to spoil. Dried vegetables retain only about 5 percent of their original moisture, while dried fruits retain 15 to 20 percent. This is why dried fruits and vegetables are so much lighter in weight than when in their original forms.

Most food is no longer sun-dried. Instead, commercial processors use enormous food dryers where fruits are dried at about 135° F and vegetables at about 125° F. Cooler temperatures would make the food dry too slowly and may lead to spoilage, while higher temperatures may actually cook and harden the outside of the food, leaving the interior too moist.

In addition to reducing the water content, drying also inhibits enzyme actvity—but it doesn't stop it entirely. Fresh fruits and vegetables are loaded with enzymes. These catalysts occur naturally and are necessary for that fruit or vegetable to mature and ripen. But food processors need to halt that maturation and ripening if they are to ensure a product that will last a long time on the store shelves or in your pantry. So before the dehydration process, commercially dried food is first pretreated to kill any enzymes in the food. The most common methods used to inactive the enzymes are blanching the food with hot water or steaming it followed by rapid cooling. These steps can also set the color, kill microorganisms, and sometimes shorten the drying time. Another way to inactive enzymes is to dip the food in, or spray the food with, sulfur compounds. This step effectively kills the enzymes that are responsible for turning the fruit or vegetable brown. Whichever method is used, the end product is guaranteed to be enzyme-dead. To make matters worse, salt or sugar is often added to the product to assist in the drying process. Salt and sugar are both enzyme depleters.

Canning

Canning is a more recently invented food preservation method. Nicolas Appert, a French brewer, confectioner, and pickler, discovered canning in about 1804 and won a prize from Napoleon for his discovery. Appert thought his process preserved food because the high temperatures drove air from the glass containers. He assumed that oxygen was the reason for food spoilage. Louis Pasteur challenged that claim, noting that some microorganisms can survive without air. In fact, many organisms, including *Clostridium botulinum*—the bacterium that causes botulism, are anaerobic (able to live without oxygen) and thrive in airless cans, producing deadly toxins. Scientists now know that the high temperatures used in canning also kill any bacteria present in food. Unfortunately, this high heat also kills the food's enzymes.

Once the food is canned, it must be heated in its sealed container (usually glass or metal containers). The seal keeps air (and any microorganisms) out of the container, while the high heat kills any microorganisms in the food. Some foods, such as tomatoes and fruits, are acidic. The acid is an extra aid in inhibiting microbial growth. If you do any home canning, you know that these foods can be sufficiently canned when boiled (in their sealed jars) for about twenty to thirty minutes. Other foods, including most vegetables, are not acidic and require a longer processing time at higher temperatures to ensure elimination of harmful organisms. In fact, many vegetables must be pressure cooked for as long as ninety minutes at temperatures of about 240° F.

Both canning methods require lengthy processing at temperatures of at least 212° F or above. Both processes also kill all of the enzymes and deplete most of the nutrients in the food.

Freezing

What about frozen food? Isn't that the closest thing to fresh food? The theory behind freezing is that microbes cannot grow at temperatures below their freezing point. However, freezing can leave some enzymes more or less unaltered. As the temperature drops, the activity of enzymes decreases. The temperature necessary to reach inactivity varies depending upon the enzymes in question. However, some enzymatic activity can occur at temperatures as low as 100° F below zero (most home freezers are set at about 32° F).

To inactivate these enzymes and ensure longer storage, food processors usually treat vegetables by blanching them in steam or boiling water for a few minutes before freezing. This same process is applied to a number of fruits, including apples, peaches, and plums. Unfortunately, blanching can kill enzymes and results in a 10 to 44 percent loss of vitamin C in vegetables, and losses of beta-carotene, folic acid, and pantothenic acid as well.

Blanching can make some fruits (especially berries and melons) too soft because it weakens the cell walls. These foods simply can't stand up to the blanching process and would disintegrate into a pulpy mass. So instead of blanching, these foods often have sugar or sugar syrup added to them before freezing to retard enzymatic activity.

Irradiation

A recently devised aid to preserving food is *irradiation*, a process in which food is exposed to as much as 300,000 rads (about the same amount of radiation as 30 million chest X-rays). According to the United States Department of Agriculture (USDA), "food is irradiated to make it safer and more resistant to spoilage. Irradiation destroys insects, fungi, or bacteria that cause human disease or cause food to spoil. Irradiation makes it possible to keep food longer and in better condition." In fact, most of the *E. coli* bacteria and *Salmonella* that may be present in food can be killed by food irradiation. However, proper cooking can also kill the pathogens without the health and environmental risks associated with irradiation.

In this process, food is passed through an irradiator, an enclosed chamber where it is exposed to an ionizing energy source. The gamma rays penetrate the food and its packaging. There is definite evidence that irradiated foods lose vitamin content, especially some of the vitamin B complex, plus vitamins A, C, and E. Your body's enzymes require many of these nutrients in order to function. The USDA says nutritional losses from irradiating food are about the same as, or less than, those that occur when food is cooked or frozen.

Currently, irradiation is used in the United States primarily on dried spices for processed foods. However, the Food and Drug Administration (FDA) has approved irradiation for all fresh vegetables, fruits, poultry, pork, teas, spices, and nuts. Therefore, irradiation of these foods could begin at any time. Irradiation of beef also appears imminent as the USDA is currently seeking approval. (See "No Irradiation" on page 21.)

The FDA requires that all irradiated whole foods be labeled with a written warning that the food has been radiated, along with the radura symbol, consisting of green petals (representing the food) in a broken circle (representing the rays from the energy source). It's important to note that no label of any kind is required if those same foods are used as ingredients in a finished product. So an irradiated tomato must be labeled if it's sold as a tomato, but if used in a can of tomato soup, the warning label is not required.

Irradiation is also used to destroy bacteria in cosmetics, to purify wool, to make nonstick cookware coating, to perform security checks on hand luggage at airports, to make tires more durable, and to sterilize medical products, including surgical gloves.

The jury is still out on whether or not eating irradiated food is dangerous. But we do know that the irradiation process causes new chemicals, called radiolytic products, to be created in food. A number of these radiolytic products are known mutagens or carcinogens.

FOOD ADDITIVES

As you can see, all of our food preservation methods inactivate the natural enzymes in our foods. But these preservation methods are not the only ways that our food's enzymes are killed before we buy it. The addition of chemical preservatives, additives, and artificial flavorings and colors also deplete our food of enzymes.

Food additives are generally considered to be anything not naturally present in food. Although we tend to think of additives as recent inventions, they have been used in food preparation for centuries. Technically, even salt is an additive and is effective in preserving food because it helps remove water and, if used in high enough concentrations, can denature (break the bonds maintaining structure) proteins. Remember, enzymes are proteins, so salt effectively destroys them.

Though some additives get into our food accidentally (such as with animal hairs or bits of plastic, glass, or other packaging material), many are put there intentionally. In fact, according to the American Academy of Allergy Asthma & Immunology, more than 2,000 additives are commonly used in our food today. These additives are used to prolong shelf-life (preservatives), improve flavor (flavorings or flavor enhancers), aid in processing (texturizers, stabilizers, emulsifiers), or simply make the product look more appealing. Some chemicals do double or even triple duty. For instance, sodium nitrite and sodium nitrate can preserve color, inhibit bacterial growth, and enhance flavor.

In the United States, the use of food additives is regulated by the FDA. All food additives not listed as *generally recognized as safe* (GRAS) must receive approval by the FDA before use. However, lack of adequate testing is still a problem, that is, few additives have received neurotoxicity testing, including many on the GRAS list.

Preservatives

Preservatives are added to foods to extend storage life. They do so by damaging the cell membranes of microorganisms that could spoil the food. Because microorganisms are alive, they, too, have enzymes in their cells. By damaging the microorganism's cells, preservatives damage their enzymes and, therefore, the microorganism. Unfortunately, preservatives also damage the enzymes in your foods.

Commonly used preservatives include BHA (butylated hydroxyanisole), BHT (butylated hydroxytoluene), sodium nitrate, sodium nitrite, and sulfites (which also act as bleaching agents and dough-conditioning agents). Some of these additives have serious health effects. For instance, BHA has been associated with cancer, and sodium nitrite and sodium nitrate can be converted in the body to nitrosamines, which have also been found to cause cancer in animals. In addition, sodium nitrate can reduce the activity of digestive enzymes in the small intestine. The ingestion of sulfites can cause tightness in the chest, lowered blood pressure, abdominal cramps, hives, an elevated pulse rate, lightheadedness, and weakness. Sulfites can cause toxicity because they adversely react with cellular components. In sulfite-sensitive asthmatics, sulfites may trigger asthma attacks, and they can lead to

No Irradiation

According to Food and Water (a nonprofit public interest organization that works to educate government officials, the food industry, and consumers about the health and environmental risks of food irradiation), more than eighty major food industry corporations, as well as thousands of retailers and food wholesalers have issued statements that they will not use irradiation in their products. The list of participating companies includes Kellogg's, General Foods, Gerber, Kraft, Borden, International Beef Processors, Coca-Cola, Pepsi, Campbell's, Quaker Oats, A & P Supermarkets, Perdue, and McDonald's. For more information, contact Food and Water, Inc., RR1 Box 68D, Walden, VT 05873 (802) 563-3300.

seizures and death. This is why the FDA banned the use of sulfites on fruits and vegetables intended to be sold or served raw (such as in salad bars) in 1986.

Coloring Agents

The term "FD & C" (food dye and coloring) was first used in The Food, Drug and Cosmetic Act of 1938. A variety of dyes were approved in this act for use in foods and beverages. These dyes are identified on labels by color and number—for example, FD & C Red No. 3 (erythrosine) or FD & C Yellow No. 5 (tartrazine).

Synthetic dyes make up the great bulk of artificial colorings used in food (as opposed to those that are derived from natural ingredients). Used widely in the production of everything from ice cream to breakfast cereals, many of these dyes adversely affect our enzymes and have been suspected of being carcinogenic or toxic for decades. Therefore, a number of them have been banned, while others are under review. For example, Red No. 3 has been banned for some purposes because it caused thyroid tumors in male rats. Yellow No. 5 (America's second most popular food color after Red No. 40) is associated with allergic reactions leading to stuffy or runny noses, hives, or severe breathing difficulties in sensitive individuals (as many as 94,000 Americans may be sensitive to Yellow No. 5).

Flavorings and Flavor Enhancers

Throughout history, humans have added herbs and spices to their foods to alter or improve taste. Most flavorings are added to increase or decrease sweetness, bitterness, saltiness, astringency, or pungency. Though some flavorings used in modern food processing are derived from plant substances, many are synthetic and made to resemble a natural flavor.

Synthetic flavors are cheaper and more widely available and withstand processing better than their natural counterparts. Long thought to be relatively safe, scientists are now discovering that many artificial flavorings can cause allergic reactions in some and may also inactivate many of your body's enzymes. For instance, the flavoring agent benzaldehyde, made synthetically through the oxidation of toluene or from benzal chloride and lime, effectively inactivates glutathione peroxidase, an antioxidant enzyme. This enzyme is responsible for removing hydrogen peroxide from the brain. Inactivating the enzyme may interfere with nerve transmission.

Salt is probably the most commonly used flavoring. But in sufficient concentrations, salt is an enzyme inhibitor because it denatures proteins. MSG (monosodium glutamate) is a flavor enhancer that was approved for use as a food additive by the FDA in 1963. However, it has been implicated as a culprit in "Chinese Restaurant Syndrome," a condition marked by gastrointestinal pain, headache, palpitations, and/or diarrhea that occurs in some individuals after eating Chinese food or any food prepared with MSG. MSG also appears to affect the normal activity of the enzymes in the liver responsible for metabolizing drugs, which may lead to toxicity when taking certain drugs. When fetuses are exposed to MSG, the additive may block the secretion of growth hormone, necessary for proper growth and development.

MSG occurs naturally in soybeans, sugar beets, and seaweed. It can also be commercially produced by fermenting starch, sugar, or sugar beets. More than 150 million pounds of MSG are now consumed per year.

PESTICIDES IN OUR FOODS

Pesticides are all around us in our soil and air, but they are also in our food. (See "Why Farmers Use Pesticides" on page 22.) If you eat pesticide-sprayed, factory-produced food, you are consuming enzyme-deficient (or enzyme-dead) substances. Pesticides can interfere with the plant's absorption of minerals from the soil. Many of your body's enzymes require these minerals (as well as some vitamins) as cofactors and, in fact, cannot function without them. In addition, the pesticide residues remaining in some foods may inhibit your body's own enzyme systems.

What can you as a consumer do to eliminate or control your intake of pesticides? First, buy certified organically grown food. If your grocer does not offer these for sale, encourage him or her to buy them and to indicate which foods are grown organically and which are grown with pesticides, so that you can make intelligent food choices. If you are unsuccessful, most large chain supermarkets now offer some organically grown produce.

Buy locally grown fruits and vegetables. This will decrease the transportation and storage times of the produce, and thus should reduce the use of pesticides and preservatives in your produce. These chemicals are repeatedly sprayed on many foods that are stored for long periods. This is also why you should buy all fruits and vegetables in season.

It is also helpful to wash fruits and vegetables thoroughly. Washing can remove 90 percent of the residues of the fungicide thiabendazole from potatoes. Washing can remove 97 percent of carbaryl (an insecticide) from tomatoes, 87 percent of the residue from spinach, and 77 percent of carbaryl from broccoli. In one study on tomatoes, washing reduced the amount of the fungicide benomyl by 80 percent. Canning and blanching can also reduce pesticide residues—but they also kill any enzymes present in your food.

ENZYME-INHIBITING FOODS AND BEVERAGES

We've discussed how additives in your food can affect your enzyme stores, but certain foods and beverages are inherently detrimental to our enzymes and thus should be avoided or, at least, limited.

Alcohol

A glass or two of red wine daily can be beneficial in lowering cholesterol and fighting cardiovascular problems; however, in excess, alcohol consumption can have detrimental effects. Unfortunately, for many people alcohol constitutes a major part of their total daily caloric intake. High alcohol intake can impair the digestion and absorption of nutrients, especially the B vitamins (needed as coenzymes); vitamins A, C, D, E, and K; iron; magnesium; calcium; phosphorus; and zinc (required for the activity of hundreds of enzymes).

In addition, alcohol produces undesirable fermentation in the bowel and stimulates improper metabolism of zinc, chromium, and insulin.

Alcohol is an irritating chemical that can damage the mucous linings, causing gastritis or esophagitis. Excessive alcohol intake can also weaken the capillaries, leading to fragility, which can cause gastric bleeding. Prolonged alcohol abuse can increase the risk of cancer (especially of the pharynx, mouth, esophagus, and larynx), can enlarge the liver, and can lead to cirrhosis or hepatitis, as well as nutritional deficiencies, pancreatitis, personality changes, sensory impairments, and sleep disturbances. All of these effects are due to the by-products of alcohol metabolism, which are detrimental to our enzymes and the rest of our bodies.

Readily absorbed from the gastrointestinal tract, alcohol is metabolized in the liver with the help of several enzymes, including alcohol dehydrogenase and acetaldehyde dehydrogenase. These enzymes first metabolize alcohol into acetaldehyde, then to acetate, and finally to carbon dioxide and water. Some individuals (especially alcoholics) rapidly convert alcohol to acetaldehyde, but only slowly break acetaldehyde down further into acetate. This is particularly dangerous because acetaldehyde is very toxic to the brain and the heart. Acetaldehyde causes B vitamin deficiencies, interferes with energy production and normal brain function, and deactivates an enzyme involved in prostaglandin production.

The enzymes that work on alcohol exist in several forms and metabolize alcohol and acetaldehyde at different rates in different people. These variations in enzyme activity may be the reason some individuals of different genetic backgrounds may have lower tolerances for alcohol consumption.

A number of enzymes, especially trypsin and chymotrypsin, are inhibited by alcohol and acetaldehyde. And the higher the alcohol concentration, the greater the enzyme inhibition. As mentioned in Chapter 2, trypsin and chymotrypsin play important roles in digestion, helping break proteins down into their component parts so that their amino acids may be absorbed into the bloodstream.

Why Farmers Use Pesticides

Though pesticides are harmful to our bodies, farmers still use them on their produce. Though there may be more healthful alternatives, pesticides do serve some type of purpose. Pesticides can:

• Reduce crop losses due to insects.

• Control fungal toxins and diseases caused by them, such as ergot poisoning, or St. Anthony's fire, an epi-

demic in the Middle Ages that began because bread was often made from rye containing the fungus ergot.

• Repel rodents or other animals that would damage or consume stored food.

• Eliminate weeds that compete for nutrients and water.

• Produce a better looking product that is not marked and disfigured by insects.

Coffee

Just like any other plant product, green coffee beans contain enzymes, including lipase, protease, amylase, catalase, polyphenol oxidase, and peroxidase. In fact, there is some indication that the quality of the coffee can be related to the activity of polyphenol oxidase in the green beans.

But to turn green coffee beans into the robust beans many of us know and love, roasting is required. In fact, roasting (at about 500° F) helps develop coffee aroma and taste. However, these temperatures inactivate all enzymes in the beans.

There is some indication that daily coffee consumption can lower levels of some liver enzymes. In laboratory tests, a decrease in alkaline phosphatase, one liver enzyme, is indicative of pernicious anemia, hypoparathyroidism, and low phosphorus levels in the blood, while a decrease in another liver enzyme alanine aminotransferase is linked to pyridoxine (vitamin B_6) deficiency.

In addition, coffee contains caffeine, which can overstimulate your nervous system. This stimulation ultimately leads to overactivity and then to fatigue, not only of the nervous system, but also of vital organs and glands that are important when you are fighting disease. Caffeine can be a major factor leading to hypoglycemia and adrenal gland fatigue, and it can overstimulate the kidneys. It tends to cause restlessness, and thereby prohibits you from obtaining the proper amount of rest and sleep. Coffee is definitely to be avoided as much as possible. However, it can be employed in the form of an enema to facilitate the liver's detoxification processes (see COFFEE RETENTION ENEMA in Part Three).

While you may think that you can avoid these problems caused by the caffeine in coffee by drinking decaffeinated coffe, consumption of decaffeinated coffee may be even worse than drinking regular coffee. To remove the caffeine, the unroasted green beans are softened (usually with water or steam), treated with a solvent to draw the caffeine from the beans, heated, steam treated, and dried. This process is sometimes repeated up to twenty-four times until the beans are deemed sufficiently "decaffeinated." Solvents (such as methylene chloride or ethyl acetate) are enzyme inhibitors. There is some concern that sufficient solvent residues remaining in the beans can cause health problems. FDA regulations stipulate that methylene chloride residues cannot exceed ten parts per million; however, some residue, no matter how little, may remain. The FDA estimates that your risk of developing cancer after a lifetime of drinking decaffeinated coffee is less than one in 250 million—that's great, unless you're that "one."

Fruits and Vegetables

Though it is recommended that you eat lots of raw fruits and vegetables for optimal enzyme activity, there are some that may actually inhibit your enzyme activity. All plants have natural defense mechanisms to protect them from attack by bacteria, fungi, and sometimes animals. Often the protective mechanism is external, such as thorns or spines that keep you from gaining access to the fruit or vegetable. In order to guarantee the perpetuation of a particular plant species, sometimes the food itself contains compounds that keep you from being able to digest it—enzyme inhibitors, that is, compounds that protect the food by interfering with the ability of your body's enzymes to digest that food.

Protease inhibitors, those inhibitors that interfere with your ability to digest protein, are probably the best-known enzyme inhibitors in plants. Beans, peas, peanuts, soybeans, and other legumes contain protease inhibitors, as do seeds, grains, and some vegetables. Soybeans are well known for their ability to inhibit the action of the protease enzyme trypsin.

Scientists have known for years that feeding animals untreated soy meal reduces their growth, decreases fat absorption, and increases the size of the pancreas while stimulating increased secretion of pancreatic enzymes. This is because the protease inhibitors naturally occurring in the unheated soy interfere with trypsin activity and, therefore, with the digestion and absorption of the soy's beneficial nutrients. Heating the soybean product destroys most of the trypsin inhibitor.

One study on humans found that a single dose of the Bowman Birk inhibitor (one type of protease inhibitor in soybeans) increased the pancreas' secretion of the digestive enzymes trypsin and chymotrypsin. By forcing the pancreas to produce an increased number of enzymes, according to enzyme expert Dr. Edward Howell, these protease inhibitors can overstimulate the pancreas, which causes the pancreas to enlarge and ultimately limit its enzyme potential.

Because protease inhibitors interfere with your body's digestion of protein, raw foods of this type should be avoided. These inhibitors can be inactivated by the heat of cooking, thereby improving their nutritional value. However, heat also kills enzymes. Fortunately, it's possible to sprout seeds and beans, a process that not only inactivates the inhibitors, but actually increases the amount of enzymes in the food. We'll cover sprouting in Chapter 4.

OTHER ENZYME DEPLETERS

In addition to our diets, our lifestyle choices can drain us of energy and destroy our health. Smoking, excessive sunshine, and drugs (both legal and illicit) generate free radical activity and destroy our bodies' enzymes.

Free Radicals

Free radicals are incomplete, highly unstable, reactive compounds or molecules. Their electron arrangements are out of "spin" balance, like your washing machine when it's weighted on one side. Free radicals lead to faulty metabolism of proteins, including DNA and enzymes, by oxidizing cells so that they practically rust. Unlike a stable molecule, in which every atom is ringed by pairs of electrons, free radicals carry an unmated electron that desperately wants to pair up with another. By snaring an electron from a neighboring molecule, it can set off a chain reaction that wreaks havoc on cells. Free radicals can inactivate enzymes in the cell membrane. This damages the membrane, interfering with the cells' ability to take in nutrients and expel wastes. They can cause lipid peroxidation in cell membranes, in which the protective lipid layer of the cell is oxidized, which damages it. This causes body fat compounds to become rancid and release even more free radicals. Free radicals can also cause cross-linking to occur, a process in which proteins, including DNA, fuse together. This process damages genetic material. If their destructive "rage" is not stopped, free radicals can weaken the whole body, causing illness and premature aging.

Antioxidants protect the body from free-radical damage by helping the body repair cellular damage caused by free radicals or by intercepting the free radicals before they can do any harm. We have many antioxidants at our disposal to fight free radicals, including alpha-tocopherol (vitamin E); ascorbic acid (vitamin C); coenzyme Q_{10}; beta-carotene; trace elements, such as zinc and selenium; food concentrates, such as aged garlic extract, green barley extract, chlorella, and wheat grass extract; and enzymes, including glutathione peroxidase, superoxide dismutase (SOD), and catalase.

Tobacco Smoke

The biggest indoor air polluter is probably cigarette and other tobacco smoke. Is there anyone around who does not yet realize the harmful effects of smoking? Yet we're still puffing away and, worse yet, harming innocent bystanders. There is abundant evidence linking cigarette smoking with increased risk of cancer, atherosclerosis, cardiovascular disease, chronic bronchitis, and emphysema. A recent nationwide survey found that 43 percent of the children in this country aged 2 months to 11 years live with at least one smoker and 37 percent of adult nontobacco users either live with a smoker or are exposed to smoke at work. This is particularly worrisome because tobacco smoke contains a wide range of toxic particles and vapor toxins that can injure not only the smoker, but also those around him or her due to secondhand smoke (also called passive smoking). Numerous studies show that exposure to secondhand smoke is linked to cardiovascular damage and lung cancer. During pregnancy, passive smoking is a health hazard to mother and fetus alike. During childhood, secondhand smoke exposure can predispose a person to lung and circulatory problems that may not appear until adulthood.

Smokers have an increased susceptibility to free-radical damage. Children of smoking parents seem to be under similar oxidant stress, which may result in lower enzyme levels (particularly of antioxidant enzymes, including glutathione peroxidase and superoxide dismutase).

Smoking can affect the status of a number of vitamins and minerals in your body, many of which are required for proper enzyme activity. For instance, the antioxidant enzyme glutathione peroxidase requires selenium in order to function. But smoking can lower selenium levels in the blood, plasma, and red blood cells, directly affecting glutathione peroxidase whose job is to remove hydrogen peroxide (and other hydroperoxides) formed during cellular metabolism. This enzyme prevents lipid peroxides and free radicals from accumulating and damaging cell membranes. Insufficient selenium intake means this enzyme cannot do its job. People over the age of thirty who smoke have significantly reduced glutathione peroxidase activity levels.

Smoking can also have a detrimental effect on the coenzyme vitamin B_6. In fact, smokers have significantly lower levels of some of the compounds in vitamin B_6. This reduction can mean that enzymes dependent upon vitamin B_6 for proper function cannot do their jobs. Cigarette smoke can also deplete the body's ascorbic acid (vitamin C) reserves.

Among other things, cigarette smoke contains the carcinogen benzene. Benzene is known to inhibit those enzymes involved in DNA repair and replication. Benzene also decreases the activity of superoxide dismutase and glutathione peroxidase, antioxidant enzymes that help your body fight free radicals.

The Sun's Rays

Oh, the lazy days of summer—lying in the sun with a good paperback in one hand, suntan oil in the other. While you're trying to get a tan, your body is busy trying to provide natural protection against the sun. It does this by thickening the epidermis (the outermost layer of skin) and producing melanin at an increased rate. You may think tanning makes you look healthier, but you wouldn't believe what it's doing to your body and your enzyme levels.

The body needs a certain amount of sunshine to form vitamin D. But by now it should be no surprise that skin is aged more rapidly by chronic exposure to sunlight. Those penetrating rays can also encourage free-radical formation in the skin, causing your skin to oxidize like rust on a railroad track, which can lead to runaway wrinkles, pigment alteration, and tissue breakdown. The scariest result can be skin cancer.

Antioxidant enzymes (like superoxide dismutase, glutathione peroxidase, and catalase) fight free radicals and the ravages of aging. But antioxidants can be weakened or destroyed by excessive exposure to the sun.

A recent study measured the effect of ultraviolet B (UVB) radiation on the body's enzymes. (Ultraviolet A and ultraviolet B radiation make up the sun's rays.) The study analyzed the activities of the antioxidant enzymes superoxide dismutase (SOD) and catalase in the body one-half hour after UVB radiation and found enzyme activity decreased to almost 60 percent of the control values. Catalase and SOD act by trapping hydrogen peroxide and superoxide, respectively. In this way, the enzymes protect cells from peroxidation. The destruction of SOD and catalase is important because it can lead to lipid peroxidation after repeated UVB irradiation and may open the door for skin cancer.

Drugs

Many think that they are not on drugs when, in fact, they might have a very serious problem. Are you on drugs? You are if you take laxatives, appetite suppressants, antibiotics, antacids, contraceptives, anti-inflammatories (such as aspirin and ibuprofen), alcohol, anticonvulsants, diuretics, or antihypertensive drugs. Remember—it's not just the "hard" drugs that can cause problems or kill you.

Drugs (legal and otherwise) can affect your enzymes and your nutritional status in many ways because they can increase or decrease your appetite, alter food digestion, and/or interfere with the absorption of nutrients. They can affect your metabolism and excretion. In addition, drugs can directly affect specific enzymes and enzyme systems.

What does modern medicine design drugs to do? According to Professor Thomas Devlin, Ph.D., in his book *Textbook of Biochemistry* (New York: Wiley-Liss, 1993), enzyme inhibition is the goal of most, if not all, modern drug therapy. Many drugs, including antibacterial, antiviral, and antitumor drugs, are designed to inhibit specific enzymes and, therefore, interfere with certain metabolic processes. But these drugs also cause toxicity in the body because most bacteria, viruses, and tumors have the same metabolic pathways as the systems that keep your body alive. Therefore, drugs designed to kill an organism often also kill the host cells as well. Since viruses or bacteria have relatively short life spans, the hope is that the drug will kill the "bad guys" before it kills you. However, you may win the battle and lose the war. Take a drug and wait—Will it kill me or "cure" me?

Even antacids can destroy your body's enzymes in a number of ways. The very nature of an antacid is to neutralize your stomach's acid. But the hydrochloric acid in your stomach is there for a reason; it is necessary to activate the enzyme pepsin. By neutralizing that acid, you are effectively inhibiting your stomach's enzyme activities and, therefore, your digestion.

Illicit drugs are even more dangerous to your body's enzymes. One popular illicit drug, cocaine, was considered a wonder drug at the turn of the century. But growing awareness of its toxicity led to its loss in favor. Cocaine increases blood pressure and heart rate and can create feelings of increased alertness and euphoria. Unfortunately, it also decreases the activity of certain liver enzymes that are important for helping oxygen to be catalyzed. These enzymes help the liver break down and excrete foreign toxins and drugs. However, cocaine can change cellular enzyme activity resulting in liver damage.

Cocaine use is particularly devastating during pregnancy because it interferes with the baby's normal development. It does this by inhibiting the enzyme ornithine decarboxylase (ODC). ODC is critical for growth regulation, so its loss during the early developmental stages of the fetus can lead to growth suppression and, ultimately, illness and death.

Unlike our caveman ancestors, today's humans no longer eat food that is truly fresh and enzyme-rich—and we have the modern diseases to show for it. Most of our diet is composed of food that has been cooked, dried, canned, frozen, processed, or adulterated in some way. Unlike raw foods, which have within them enzymes necessary for their digestion, processed foods are enzyme-dead because the heat used in their preparation destroys all the active enzymes. In fact, any heating over 140° F (some say as low as 107° F) will destroy all the active enzymes. So if your food has been fried, boiled, baked, or even lightly steamed, the enzymes are gone. This puts added stress on your digestive system, which has to produce even more enzymes to digest "dead" food. Our lifestyle choices can also impact our bodies' enzymes, depleting the very enzymes whose job it is to protect us.

What can we do to maximize the enzymes in our diets? That's the subject of Chapter 4.

4. MAXIMIZING ENZYMES IN YOUR DIET

The whole world is a maze of living enzymes . . . they are all around us. Every living plant and animal must have active enzymes pulsating through it. But there are numerous factors in the environment and in our diets that either deplete or completely destroy the enzymes in our foods and in our bodies. However, to jump-start your day and your life, you need sufficient enzymes. How can we get these enzymes? Why not try the Enzyme Two-Step Program? (No, it's not a dance.) Step one involves eliminating all enzyme-dead foods from your diet. Step two includes ways to maximize your enzyme intake.

STEP ONE: ELIMINATE ENZYME-DEAD FOODS

The first step in maximizing the enzymes in your diet is to eliminate all foods in which the enzymes have been killed. This includes all cooked foods, as well as any canned, dried, frozen, irradiated, or processed foods. In order to ensure a long, stable shelf- (or freezer-) life, food processors treat food with chemicals, heat, or additives. This guarantees the death of all enzymes in that food, thereby guaranteeing that the enzymes won't keep the food alive and changing. Thus, the "food-like products" are enzyme-dead and the cardboard surrounding the food is probably as nutritious and enzyme-rich as the food itself.

These enzyme-dead foods overwork your body and tax your enzyme supply. They force your pancreas and other organs to produce more enzymes in a desperate effort to digest what you eat. Any energy your body expends digesting food is energy that can't be used in maintaining your immune system, your circulation, or any other process.

Is it possible to remove all the enzyme-dead foods from your life? Yes. Is it difficult? Certainly! Nothing worth doing is easy. Take a stroll through your pantry and gaze into your refrigerator and freezer with an eye toward eliminating anything and everything that is enzyme-dead. What would you throw out?

- All canned goods.
- All commercially prepared frozen foods and any foods blanched before they were frozen.
- All commercially dried foods and packaged foods (because the producers blanched them first).
- All baked goods.
- Any foods that have been treated with preservatives, additives, or other chemicals, or those that have been irradiated.
- All foods grown with the use of pesticides.

What's left? If you eat a typical American diet, probably not much. Time to move to step two.

STEP TWO: MAXIMIZE YOUR ENZYME INTAKE

Now that you've eliminated all the foods that are enzyme-dead or that tax your body's own enzymes, the second step is to increase your intake of enzyme-rich foods, including fresh, raw fruits and vegetables (and their juices); and fermented foods and enzyme "enhancers"—foods usually eaten in small quantities but that pack a big enzyme punch (such as sprouts, raw honey, and wheat germ).

Fresh Fruits and Vegetables

Besides providing us with necessary calories, carbohydrates, proteins, fats, minerals, vitamins, and phytochemicals (the beneficial chemicals naturally found in plants), raw fruits and vegetables also contribute life-giving, energizing enzymes. In fact, they are among the best sources of enzymes. This is because any plant needs enzymes to live, just like any human. But as explained in Chapter 3, these enzymes are lost when your food is cooked. Is there a "good" way to cook these foods and minimize the enzyme destruction? Maybe. Because enzymes are so highly individualized, with each having different properties, there are few absolutes. Some enzymes are destroyed at lower or higher temperatures than others. Lightly steaming or stir-frying your foods *might* help minimize enzyme loss, but remember, most enzymes are denatured (the bonds that hold them together broken) at about 140º F (and some at even lower temperatures). The degree to which you destroy enzymes when you cook food will vary depending on how thickly you slice the food (because thin slices cook faster than thick ones) and how long and at what temperature you cook it. To minimize enzyme loss, stir-fry vegetables rapidly and for only a short time, if you must cook them at all.

When increasing your intake of raw fruits and vegetables, try to spread your consumption out over the course of the day. You should eat four to five times per day. I like to call this method of eating the "food drip method." By spreading your food out over many feedings, your digestive system is not overtaxed and is able to digest and absorb the nutrients more easily.

Enzyme Activity in Plants

Plants live for as long as they have enzymes. And just like any other living thing, plants are constantly changing, renewing, metabolizing, and maturing. Your body is not quite the same today as it was yesterday—your cells have been multiplying and dividing, and some cells are being destroyed while others are being replaced. This is the process of life. Plants, too, use enzymes throughout their growth processes to carry on metabolism. Enzymes are responsible for every single activity that occurs in the germination, growth, and ripening of every fruit and vegetable (in fact, every plant).

To regulate metabolic activities, plants contain a number of enzymes, produced during different times of the plant's growth cycle. In green plants, chloroplasts (the specialized cell structures in the cytoplasm that carry out photosynthesis) contain a vast array of enzymes, as well as chlorophyll. These enzymes are critical for the many reactions involved in photosynthesis.

Plants also produce enzymes as defense mechanisms. For instance, plants naturally produce the enzyme chitinase, which digests chitin (a component of crustacean shells, insect exoskeletons, and the cell walls of certain fungi) and effectively wards off potentially devastating fungal diseases. In fact, researchers have created plants that manufacture greater amounts of chitinase, thus improving the plant's natural resistance and decreasing the need for pesticides.

Germinating seeds probably contain the richest source of enzymes in plant organs (that's why sprouts are recommended as "enzyme-enhancers" later in this chapter). It's important to note that without moisture, some plant enzymes in seeds can sustain their activity (potency) for extended periods of time. For example, Dr. H. Miehe in 1923 found that even after 112 years, rye seeds still contained amylase, which was active enough to hydrolyze starch.

Enzymes are widely distributed in the tissues and organs of plants, and enzymes or, at least their proenzymes (inactive precursors of enzymes, as mentioned in Chapter 2) occur in every living cell. Some organs of a plant may contain a particular set of enzymes, while other organs of that same plant, may contain a different set. For example, the enzymes sucrase, amylase, and maltase are found in the leaves of garden beets, while the stems contain amylase, sucrase, emulsin, and inulase, and the roots contain amylase, emulsin, and inulase. Each enzyme plays a different role.

The most widely studied hydrolytic enzymes in plants are lipases, amylases, and proteases. Lipase is particularly abundant in germinating seeds that contain relatively large quantities of fats (such as the soybean, castor bean, flaxseed, coconut, rapeseed, hemp seed, and corn) but is also present in an inactive form in the resting seeds of many species. In addition, this enzyme is also found in some species of bacteria and in humans and other animals.

The amylases, which digest carbohydrates such as starch and sugar, are very important to plants as well as to humans. Although amylase is found in nongreen tissues and organs, it is almost always found in the green parts of plants. This enzyme is also produced by many animals, as well as fungi and bacteria. Malt (made by germinating barley seeds) is rich in the enzyme amylase. A microbial form of this enzyme, Takadiastase, is prepared commercially from colonies of a mold called *Aspergillus oryzae*, grown on a substrate of steamed wheat bran or rice. Fermentation of microorganisms will be discussed more extensively in Chapter 5.

Other amylase enzymes abundant in plant tissues are sucrase, maltase, cellulase, and hemicellulases. Not as widely available is inulase (which breaks down inulin to fructose). Though inulin is only found in a few plants, including the Jerusalem artichoke (*Helianthus tuberosus*), it is also produced in some animals (especially of the invertebrate group) and by certain fungi (*Penicillium glaucum* and *Aspergillus niger*).

Of all the enzymatic reactions catalyzed by plant enzymes, those catalyzed by proteases generally proceed at the slowest rate. In fact, protein hydrolysis by plant enzymes may take several days, while a plant amylase will hydrolyze a fairly concentrated starch in only a few minutes. This difference in enzyme activity makes it more difficult to study protease action and sometimes even the presence of proteases in plant tissues.

If each plant contains within it the enzymes necessary for its digestion, then why do some fruits contain such high amounts of proteolytic activity when they don't contain large amounts of protein? This is the case with pineapples, figs, papayas, and cantaloupes and other melons. One theory is that the enzymes work as a defense mechanism, protecting the plant from fungi, bacteria, or insects by breaking down their cell walls. Whatever the reason, we're lucky to have these fruits as they are effective aids in digesting protein foods.

In order to work, an enzyme must attach to its substrate. Most of the time, especially in plants, the enzyme and its substrate are located within the same cell but stored in separate compartments (vacuoles). They have to be stored separately, otherwise the enzymes would hydrolyze the contents of the cell. Plant tissues rich in a certain substrate will also contain high quantities of the corresponding enzyme (or at least its zymogen) and often vice versa (although, as previously mentioned, some plants have high amounts of protease, even though they do not contain high amounts of protein). This is a key point, because it means that any food you eat (if in its raw, uncooked, unprocessed state) already contains the enzymes within it necessary for its digestion.

Sometimes, however, the enzyme and its substrate are not in the same cell. This is true in bitter almond seeds where the substrate (amygdaline) and the enzyme (emulsin) both are present, but only in adjoining cells. The substrate and its enzyme come in contact with each other only when the bitter almond seeds are crushed.

A good example of how enzymes act in plants can be seen in the case of the tomato, which, just like any other fruit or vegetable, continues to breathe and metabolize (and is, therefore, alive) long after it is picked. The tomato's cellular machinery whirls away with purpose, even as the tomato sits on the kitchen counter, hard as an apple and green as grass. Enzymes in the tomato are produced in a specific sequence. The outside world and the tomato exchange gases. Tomatoes breathe, in fact, this breathing becomes heavier as the tomato ripens. When a certain gene is turned on, ethylene gas is produced, and ripening begins in earnest. Ethylene (produced by many vegetables and fruits) actually triggers a cascade of events. Scientists suspect dozens of enzymes are manufactured under the direction of ethylene. This action then turns on the action of other genes. Commercial growers have learned to capitalize on ethylene's action and can speed the ripening of tomatoes and other produce by placing them in ethylene gas chambers.

The enzymes in a tomato perform at least three crucial jobs. They soften the tomato by weakening the rigid cell walls and the pectin that glues the cells together. At the same time, with the help of enzymes, the green chlorophyll in the tomato is replaced with other pigments that give the tomato its orange and subsequent red color, and the green tomato turns red. But Mother Nature waits until the last minute of ripening before performing possibly her greatest act. The main ingredient that gives a tomato its distinctive aroma takes action and is produced in quantity only when the tantalizing tomato is sliced.

Chewing or cutting a tomato disrupts cells, releasing enzymes that break down a special fatty acid in the tomato. The result is that the fatty acid is converted into an aromatic compound (Z-9-hexenal). Almost instantly (within seconds of cell breakdown), the air is filled with the aromatic chemical. Within about three minutes of slicing the tomato, the aroma peaks. Some believe the bouquet of scents we recognize as "tomato" is possibly dozens of other fruit perfumes mingling with the fatty acid hexenal.

But all this enzymatic activity comes to a screeching halt once you heat the tomato because heat kills enzymes. You can see why enzymes must be killed if food is to be stored for extended periods of time. Though stored food can still provide calories, along with some vitamins and minerals, unfortunately, it does not provide enzymes.

The activation of aroma by cell disruption is also what happens when you crush a clove of garlic. In fact, a garlic bulb really doesn't smell until you cut or crush it, thus enabling an enzyme called *allinase* to come in contact with the *alliin* in the garlic. When this happens, allinase catalyzes the conversion of alliin into allicin, resulting in the characteristic pungent garlic odor and burning sensation it creates in the mouth (effective chemical weapons against fungi and insects). But the end product, allicin, is a strong oxidant and in humans may cause undesirable effects such as skin irritation, and if taken internally, can cause stomach disorders and hemolytic anemia. Allicin is a highly unstable and transient compound. Luckily, allicin decomposes rapidly and is almost completely decomposed within one day (according to Dr. Robert I-San Lin, chairman of the First World Congress on Garlic, Washington, D.C., 1990).

Many fruits and vegetables contain polyphenol oxidase. This enzyme is responsible for aerobic darkening. When you cut an apple, it is the polyphenol oxidases that oxidize phenolic compounds (those substances essential for plant growth and reproduction), turning that apple brown. Polyphenol oxidases also contribute to the ruby hues and astringent flavors of red wines. And there is some indication that this group of enzymes has a role in protecting plants from insect attack. In addition to polyphenol oxidases, many foods contain pectic enzymes. The pectic enzymes are responsible for breaking down pectin, and so are involved in such natural processes as ripening of fruit, spoilage of fruits and vegetables, and the invasion of plant tissues by pathogens.

The Benefits of Eating Fruits and Vegetables

As stated earlier, fresh, raw fruits and vegetables contain within them the enzymes necessary for their digestion. That's why eating these foods can directly contribute to our enzyme supply and help the stomach and small intestine digest what is eaten. Fruits and vegetables also induce our body to produce enzymes, actually assisting the body in "kick-starting" enzyme production. For example, eating a "living food" (uncooked fruit-and-vegetable-based) diet has been shown to increase concentrations of the antioxidant enzymes superoxide dismutase (SOD) and glutathione peroxidase.

One study measured blood concentrations of antioxidants, including beta-carotene and vitamins C and E, and the activities of selenium-dependent glutathione peroxidase and zinc and copper-dependent SOD in people who eat no animal products of any kind (vegans). Researchers compared these levels with those of individuals who eat both plant and animal foods (omnivores). Although the vegans had a significantly lower selenium intake, they had significantly higher levels of superoxide dismutase activity than the omnivores did. The vegans also had higher intakes of beta-carotene, vitamins C and E, and copper than did the ominvorous control subjects (see Table 4.1). Researchers concluded that significantly more dietary antioxidants are obtained from a living-food diet than from an omnivorous, cooked-food diet.

Many fruits and vegetables contain components that can actually stimulate the enzymes involved in breaking down and eliminating foreign chemicals from the body. In fact, many dietary foods can change the metabolism of carcinogens because of their effect on Phase II enzymes, the enzymes that detoxify carcinogens. This is one well-established way to reduce the risk of cancer. For instance, broccoli contains a powerful ingredient called *sulforaphane* that catalyzes critical cell enzyme activity and, therefore, helps guard against tumors. These protective enzymes are stimulated by many synthetic and natural compounds. Cruciferous vegetables, such as broccoli, also inhibit chemically caused cancer. One way they do this is by stimulating certain cancer-fighting enzymes such as glutathione S-transferases (GST), epoxide-hydrolases (EH), or cytochrome P450-dependent monooxygenases (what a mouthful!). These enzymes change carcinogens to forms that are easier for the body to excrete.

We know that fruits and vegetables are loaded with vitamins, minerals, phytochemicals, and enzymes. All of these nutrients can affect your health. You probably know the importance of vitamins and minerals, but did you know that without them, many enzymes and enzyme systems in your body cannot function? In many ways, vitamins and minerals are nothing more than fuel for enzymes.

Eating more fresh fruits and vegetables will also mean that you're eating more fiber. In the past two decades, dietary fiber has become a leading factor in the treatment and prevention of many chronic diseases and conditions. Dietary fiber is the indigestible cell-wall component of plant materials. Our bodies can't digest this fiber because we don't have the necessary enzymes.

Generally, fiber is classified as either insoluble or soluble. Insoluble fibers pass through the digestive tract relatively unchanged except when broken up during the chewing process. Since insoluble fibers absorb large amounts of water (sometimes up to fifteen times their weight), they create a bulky, soft stool. The most insoluble fiber is cellulose, which is found in whole grain products, wheat bran, and fruit and vegetable skins. The major role of insoluble fiber seems to be to prevent constipation by helping move the stool through the colon. Insoluble fiber can help protect against digestive disorders and colon cancer but apparently has little or no effect on blood cholesterol.

Soluble fibers (such as those in barley, oat bran, and dried beans and other legumes) are sticky and combine with water to form gels. Guar, mucilages, and pectin are all examples of soluble fibers. Because of their ability to lower blood cholesterol, the soluble fibers have gained increased attention in recent years.

There is mounting evidence that sufficient fiber intake can improve bowel function, help control weight, and aid in the prevention of diverticular disease, irritable bowel syndrome, colon cancer, heart disease, and diabetes. Dietary fiber also helps protect against chemical toxicity.

Eating a diet high in fruits and vegetables will also mean a low intake of fat. With the exception of such fruits as coconuts and avocados, most fruits and vegetables have very little natural fat. This is a plus because a high fat intake has been linked to heart disease, obesity, and cancer (especially colon and breast cancer).

Many experiments on rodents have examined the effects of dietary fat on the development of cancer (including skin, mammary gland, and colon cancer). Repeatedly, studies have shown that rats fed diets in which 40 percent of the calories come from fat develop a greater number of mammary tumors sooner than rats fed diets in which 10 percent of the calories come from fat.

In addition to all of the health problems associated with high-fat diets, eating too much fat can directly affect your body's enzyme levels. In one study, rats fed high-fat diets showed depressed activity of the antioxidant enzyme glutathione peroxidase.

Researchers have known for a long time that people with high intakes of fruits and vegetables have a decreased risk for a number of diseases and conditions, including coronary heart disease, diabetes, hypertension, obesity, and several forms of cancer. In fact, after analyzing over 200 studies, researchers have found that fruits and vegetables can offer significant protection against cancers of the colon, lung, cervix, breast, oral cavity, esophagus, bladder, stomach, ovary, and pancreas. What they discovered was that people who had low intakes of fruits and vegetables had about *twice the risk* of cancer compared with those with high intake. In particular, fruits significantly protect against cancers of the oral cavity, larynx, and esophagus.

It is generally agreed that many forms of cancer are initiated because of genetic mutation (mutagenesis). Researchers recently tested the antimutagenic activity of extracts from about twenty vegetables and fruits. In the presence of Benzo[a]pyrene (a carcinogen and mutagen), about 80 percent of the vegetables and fruits showed antimutagenic activity. Those fruits and vegetables showing more than 50-percent inhibition of genetic mutation included juices from such fruits and vegetables as raw celeriac, broccoli, red cabbage, carrots, green peppers, lettuce, asparagus, apricots,

Table 4.1. Intake of Antioxidants by Vegans

Vitamin E	313%
Vitamin C	305%
Copper	120%
Vitamin A	247%
Zinc	92%
Selenium	49%

(based on percentages of the USRDA)

red currants, gooseberries, raspberries, and pineapple. Less effective were leek, kohlrabi, cucumber, zucchini, French beans, fennel leaves, rhubarb, and sweet cherries. Onions, Chinese cabbage, radish, and white cabbage showed no antimutagenic activity.

One interesting note: The researchers discovered that *cooking* considerably reduced the antimutagenic activity of several foods (including celeriac, leek, broccoli, French beans, carrots, asparagus, cherries, and pineapple). We know that cooking kills the enzymes in foods. Could these enzymes have been the "magic" forces in the fruits and vegetables responsible for the antimutagenic activity?

There are many theories about the cancer-protective factors in fruits and vegetables. It is unlikely that any one factor is totally responsible for cancer-protective properties, but rather that many factors work together. Some of the agents possibly involved include:

- Polyphenols that scavenge carcinogens and mutagens.
- Carotenoids and ascorbates that quench free radicals.
- Sulfur-containing compounds that can stimulate DNA repair.
- Phytochemicals that can alter hormone metabolism.
- Detoxifying enzymes.

Juicing

An easy way to ingest more fruits and vegetables is to drink their juices. If you juice fresh fruits and vegetables at home, you know you're getting nutritious juices whose enzymes have not been destroyed by commercial canning. These enzyme-rich juices can easily be digested by the body.

Juices, like their whole fruit and vegetable counterparts, have all-around protective action. Juices have been used to fight rheumatic, toxic, and degenerative conditions, including cancer, kidney problems, cardiovascular diseases, hypertension, obesity, and high cholesterol. Further, juices are valuable in treating conditions of acidosis because of their potentially high buffering capacities.

Juices from raw fruits and vegetables are a must. Remember, heat destroys enzymes. Therefore, it's important to drink fresh, untreated, and uncooked juices from raw foods, or to obtain the vegetables and fruits as a concentrated tablet or powder (see Chapter 6). Something I like to do is to combine the benefits of fresh juices with vegetable and/or fruit concentrates (available as powders or in tablet form). It's the best of both worlds.

For additional information on juicing, see the Bibliography for books on the subject.

Fermented Foods and Enzyme Enhancers

For centuries, humans have fermented food as a way to preserve the harvest. Fermented foods, which should be eaten in small quantities daily, include yogurt, kefir, sauerkraut, soy sauce, miso, tempeh, natto, kimchi, and cheese. These foods contain beneficial enzymes and bacteria.

For at least 4,000 years, lactic acid bacteria (LAB) have been used to ferment a number of the above foods, as well as sausage, ham, wine, cider, beer, olives, pickles, butter, buttermilk, and koumiss. Widespread in nature, lactic acid bacteria are found mainly in soil but also occur in milk, vegetables, meat, and the human body.

Lactic acid bacteria contain proteolytic enzymes in the cell wall, cell membrane, and cytoplasm. In addition, the bacteria also contain lactose-digesting enzymes. So fermented dairy products (such as yogurt) contain the enzymes necessary to digest lactose. This is why individuals who lack the enzyme lactase can still eat fermented dairy products. The food itself is supplying the enzymes necessary for its digestion.

Adding enzyme enhancers to your food is another way to maximize your enzyme intake. Enzyme enhancers, small foods with a big enzyme punch, include sprouts, raw honey, and wheat germ.

Yogurt

Yogurt is a wonderful enzyme-rich food. This fermented milk,

Top Ten Fruits and Top Ten Vegetables

As you have read, eating fresh fruits and vegetables and drinking their juices are critical for health. They bolster your enzymes and can even help your body fight a number of degenerative diseases. The following list indicates the ten most often eaten fruits and vegetables in America according to 1995 figures from the Produce Marketing Association.

NUMBER	FRUIT	VEGETABLE
1	Banana	Potato
2	Apple	Iceberg lettuce
3	Watermelon	Tomato
4	Orange	Onion
5	Cantaloupe	Carrot
6	Grape	Celery
7	Grapefruit	Corn
8	Strawberry	Broccoli
9	Peach	Green cabbage
10	Pear	Cucumber

which has been an essential food for people of the Balkan countries for hundreds of years, can reportedly extend life, protect against tooth decay, fight premature aging, and prevent hair loss. Yogurt is an excellent source of predigested protein, calcium for stronger teeth and bones, plus important vitamin B complex and vitamin K. Because milk proteins are largely broken down during the culturing process, yogurt is particularly well-tolerated by elderly and ill individuals.

Familiar to most of us, yogurt is made from whole, low-fat, or skim milk that has been cultured with friendly bacteria (*Lactobacillus bulgaricus* and *Streptococcus thermophilus*). Eating yogurt or taking supplements containing live cultures of these beneficial bacteria can help fight a number of conditions. This is primarily because of their effect on the digestive tract.

Our bodies contain billions of live bacteria. If we could weigh all the bacteria in our digestive tract they would weigh approximately three-and-a-half pounds, so states Dr. Leon Chaitlow and Natasha Trenev (*Probiotics*, New York: Harper Collins, 1990). These bacteria are divided into several hundred species, some are bad, others are good. When the "good" bacteria are outnumbered by the "bad," illness develops. Balance is the key to life and to health. Therefore, our systems must be in balance with sufficient good bacteria to keep the bad guys in check. How can balance be achieved? By introducing "good" bacteria into the system. Yogurt and other fermented dairy products could be the answer.

In Chapter 2 it was explained how bacteria in the colon produce enzymes that help to digest any remaining food residues. These bacteria (and there are more than 400 different kinds) colonize the large intestine and ferment some of the undigested carbohydrates (such as cellulose). The bacteria can metabolize components of your diet, as well as substances made by your body (including digestive enzymes, bile acids, and urea). The microflora in your gastrointestinal tract must stay balanced and stable if you are to remain healthy. After all, the intestinal flora not only protect you from disease, they can also affect the state of your nutrition. But this balance is easily upset. Antibiotics kill bacteria indiscriminately—they can't tell the good guys from the bad guys, so they kill all of them. Stress can affect the flora as can changes in your diet. But changing your diet by adding yogurt can also help *correct* a gastrointestinal problem by encouraging the growth of helpful bacteria.

Many of us are only familiar with the sweetened, flavored yogurt widely sold in this country. This type of yogurt, though it may contain beneficial bacteria, may also contain sugar, artificial flavors, stabilizers, and other additives. A review of a few labels showed that an 8-ounce container of commercial yogurt can vary from 100 calories to a whopping 280, primarily because of the sugar content. To know what you're getting, either buy plain yogurt with live cultures, or make your own. (See "Making Yogurt at Home" below, and "Jazzing Up Your Yogurt" on page 32.)

Kefir

Kefir (sometimes called kifir or kephir) is another fermented dairy food that is considered to be one of the richest sources of enzymes. Delicious kefir will jump-start digestive enzyme function in a number of ways. Kefir stimulates salivary enzyme flow and the waves of muscular contraction (peristalsis) in the intestinal tract. It also increases the flow of digestive enzyme juices. In restoring enzymatic power to your intestinal tract, kefir can help fight constipation.

Kefir originated in the Balkan states and has been used for centuries. It is fizzy and refreshing. Many believe that kefir is the most powerful enzyme source known today. This is because the method of preparing kefir does not destroy its large enzyme stores.

Making Yogurt at Home

Most commercially prepared yogurts contain additives, including sugar, that you just don't need. Here is a recipe for making wholesome and nutritious yogurt, without all the additives, at home.

Ingredients:
2 cups of skim milk powder
5 cups of water
3 tablespoons plain yogurt (with live cultures) for the starter

1. Add the powder to the water in a glass or ceramic bowl and stir until thoroughly mixed. Mix in the starter.

2. Cover with a towel, and wrap in a blanket to keep it warm.

3. Let the mixture rest overnight in a warm place (about 85-90 degrees).

4. The next morning, your yogurt will be ready.

5. Refrigerate.

6. Enjoy—but save a few tablespoons to use as "starter" for your next batch of yogurt.

The starter can be kept refrigerated for about three weeks.

Since kefir has a very low curd tension (the amount of time it takes for curd to break up into small particles), it is superior to yogurt in enzymate activity. The enzymes are instantly released to perform their life-giving duties. On the other hand, curds of yogurt seem to break into lumps or hold together. Therefore, enzymatic release in yogurt is not as great and is slower than in kefir.

Kefir is credited with anti-aging qualities, including increased physical and mental capabilities. People living in the Caucasus region, where kefir has been used for centuries, are known to live one hundred years or more. In addition to kefir, there are a number of similar fermented dairy products traditionally used around the world, including kishk (Egypt), Laban zeer (Egypt), m'bannick (Senegal), zabaday (Egypt), zincica (Czechoslovakia), and smetanka (Russia).

See "Making Kefir" on page 33 for directions for making your own kefir at home. Hoegger's (160 Providence Road, Fayetteville, Georgia 30215 800–221–4628) offers a catalog that furnishes kefir grains, as well as starters for yogurt, and cheese-making supplies. Call them for a catalog.

In the past, I have recommended raw, certified milk for making fermented dairy products because it still contains active, live enzymes. However, now that several cases of *E. coli* infections have been traced to raw, unpasteurized milk, I find that I can no longer recommend it. Use pasteurized milk in the recipes in this book using milk. Fortunately, the lactic acid and other bacteria produced during the fermentation process will provide many helpful enzymes.

Sauerkraut

Sauerkraut is literally "sour cabbage." Although most of us think of sauerkraut as being German in origin, according to one report, laborers building the Great Wall of China ate it regularly (the Great Wall was built over 2,000 years ago).

To make sauerkraut, cabbage is shredded and packed in a container with a small amount of salt (approximately 6 tablespoons of salt for every 10 pounds of shredded cabbage), which drains the juices from the cabbage. Because the cabbage must be kept submerged in the resulting liquid to properly ferment and not spoil, a plate weighted down with a heavy object is often placed on top of the cabbage. The mixture is allowed to ferment for three to four weeks, ideally at about 70 to 75 degrees fahrenheit.

In addition to its active enzymes, sauerkraut is a great source of many B vitamins and vitamin C. But some commercially produced sauerkraut contains an abundance of salt and may be canned (which will kill all active enzymes).

Sauerkraut or sauerkraut juice is reportedly helpful in digestion and is an effective laxative. Sauerkraut may also be helpful in promoting a clear complexion.

Soy Sauce

Soy sauce is a popular Asian condiment made from fermented soybeans. Although enzyme-rich soy sauce tastes salty, it may be lower in sodium than table salt depending on the type of soy sauce you purchase. Several kinds of soy sauce are available, including tamari (made from soybeans only), shoyu (a blend of soybeans and wheat), and teriyaki (a sweeter product often made from soy sauce, sugar, and sake and mirin—two types of rice wine).

Shoyu is made by cooking soybeans and mixing them with wheat. The mixture is then inoculated with a mold. Within three to four days (under specific conditions of moisture and temperature) the mold's enzymes transform this soybean/wheat mix into shoyu-koji (a mold fermented product). After a second lengthier fermentation, the final product, shoyu (a clean, brownish, salty liquid) is produced. It is used throughout Asia as a flavoring and takes the place of vegetable and meat extracts or table salt in many countries.

Much of the soy sauce sold in the United States is synthetic, made with caramel, corn syrup, and hydrolyzed vegetable protein. This type of soy sauce takes only three to four days to ferment, unlike the four to six month minimum for natural soy sauce.

Does soy sauce have any health benefits? A 1992 article in *Cancer Research* (A. Nagahara, H. Benjamin, J. Storkson, J. Krewson, K. Sheng, W. Lui, and M.W. Pariza, 52(7):1754-1756, April 1, 1992) indicates that soy sauce can inhibit the

Jazzing Up Your Yogurt

Plain yogurt can be quite tasty, but some people find it a little boring. Here are some ways to "jazz up" your yogurt:

• Add raisins, nuts, wheat germ, honey, or fresh fruit to yogurt.
• Use plain yogurt instead of sour cream on potatoes or fruit.
• Add herbs, spices, onions, and garlic for dips or salad dressings. For instance, combine 1 cup yogurt; 2 cups mayonnaise; 1 cup crushed, drained tomatoes; 2 teaspoons horseradish; and vinegar to taste for an excellent Russian dressing.
• Make frozen yogurt by adding equal amounts of whipped cream and yogurt, and fold in puréed fruit and some honey for sweetening. Freeze (ice cube trays work well), blend, and enjoy.

development of cancer (forestomach neoplasia) in mice when those animals were exposed to the potent carcinogen benzo[a]pyrene. So, when you flavor your favorite foods with soy sauce, you could be the recipient of its cancer-fighting effects.

Miso

The essence of Japanese cooking is characterized by miso, a rich, salty condiment. Miso has an honored and long history of use in Japan. In fact, miso has been used in Japan for at least 1,000 years to flavor a variety of foods and to make soup. Miso is a smooth paste, made from a grain such as barley or brown rice, fermented soybeans, an enzyme mold culture called *koji*, which contains *Aspergillus oryzae*, and salt. The mixture is aged for one to three years in cedar vats. Miso is used to flavor sauces, marinades, dressings, and patés. It is used as a base for soup and is also a seasoning for grain, bean, and vegetable dishes. Miso is very popular in macrobiotic cooking.

According to Michio Kushi, in his book *The Book of Macrobiotics* (Tokyo: Japan Publications, Inc., 1992), the enzymes in miso (along with those in tamari soy sauce) represent the most basic form of life. Reported health benefits of miso include cancer prevention, antioxidant activity, and cholesterol-lowering ability. Miso is available in Japanese markets, as well as many grocery stores.

Tempeh

Tempeh (pronounced tem-pay) is a tender, chunky cake of fermented soybeans. Originally developed in Indonesia, delectable and chewy tempeh can be sliced, marinated, roasted, grilled, dried, fried in oil, or added to soups or other foods. Michio Kushi notes that tempeh is attractive to those trying to cut down on animal foods because it resembles pork and chicken in texture and taste.

Although a variety of delicious fermented foods exists today, tempeh is one of the most highly researched and accepted. Tempeh is made by soaking and cooking soybeans, then inoculating the beans with a mold. The mold synthesizes enzymes that help to hydrolyze the soybean constituents and to develop a desirable flavor, aroma, and texture. The enzymes also help decrease or even eliminate antinutritional factors, such as the enzyme inhibitors naturally found in soybeans. After fermentation, the soybeans are bound in a compact "cake" that is held together by a dense threadlike mycelium (a web of fibers produced by the mold).

Natto

Natto, made of fermented, cooked whole soybeans, looks a lot like baked beans connected by long, sticky strands. Natto has a cheesy texture and strong odor. Traditionally, natto is served in miso soups, as a topping for rice, or with vegetables.

To make natto, soybeans are soaked in water until they are soft. The length of soaking time varies depending on season, water temperature, and the kind of soybean used. The soybeans are then boiled in a pressure cooker until soft enough to be easily crushed between one's fingers. A special microbe, *Bacillus natto*, is then inoculated and the beans are placed in a warm environment to ferment (for about twenty-four hours).

Natto contains several powerful enzymes that aid in food digestion. In fact, natto reportedly is easier to digest than whole soybeans because the fermentation process breaks down the complex proteins in the soybeans. Natto also contains a great deal of vitamin B, and appears to stimulate the production of a large amount of vitamin K in the intestinal tract. In addition, natto is particularly rich in genistein (229.1

Making Kefir

How can I make kefir and jump-start my life, you ask? Happily, you need no special conditions or equipment to prepare kefir:

1. Place 1 teaspoon of kefir grains (colonies of milk-fermenting enzymes and yeasts) in a 6 to 8 ounce glass of milk. Mix thoroughly.
2. Allow the mixture to sour (usually overnight at room temperature).
3. Check with a spoon to ensure that the kefir is velvety smooth and thick.
4. Eat and enjoy.
5. Refrigerate any remaining kefir, which will remain edible for about three weeks.

For a booster loaded with enzymes, try the nutritiously delicious combination of kefir and brewer's yeast flakes in milk:

Ingredients:
6 ounces of milk
1 to 2 tablespoons brewer's yeast flakes
1 to 2 tablespoons kefir grains

1. Mix all ingredients well in an 8-ounce glass.
2. Cover the glass.
3. Allow to stand overnight at room temperature.
4. The next morning, enjoy the booster before breakfast or anytime throughout the day. This will jump-start your entire digestive system.
5. Refrigerate any remaining booster and keep for about three weeks.

micrograms per gram) and genistin (492.8 micrograms per gram). These two phytochemicals found in soy products may play a preventive role against cancer in humans. Natto has been found to be effective against hangovers, osteoporosis, and high blood pressure.

Researchers have been able to purify a strong fibrinolytic agent, *Nattokinase* (NK), from natto. The purified nattokinase can digest several synthetic substrates, as well as fibrin. Although fibrin forms the essential portion of a blood clot (so is necessary for normal clotting of blood), there are some conditions in which a clot can be life threatening. This is particularly true in the case of an embolism or thrombus, a sudden blocking of an artery by a clot . To normalize blood flow, it is critical that the clot (which is composed of fibrin) be dissolved. The oral administration of natto or nattokinase produces a mild improvement of plasma fibrinolytic activity, that is, it can assist in the breakdown of a blood clot. In addition, when NK capsules were given orally to dogs with experimentally induced thrombosis (the development of a blood clot), there was a breakdown of the thrombi. The results suggest that NK could be used to treat and also prevent blood clots in the bloodstream. Further, NK can be mass produced and is proven to be safe.

Kimchi (Kimchee)

A favorite dish of Koreans, kimchi (pickled vegetables) is made hundreds of ways using various vegetables, including cabbage, radish, cucumber, garlic, ginseng, hot peppers, and Indian mustard leaves or sesame leaves. The special variations in taste of kimchi depend on the individual vegetables used.

This enzyme-rich, fermented food is a stamina builder. It is high in vitamins (especially the B vitamins, particularly B_1 and B_2, and vitamins C and A, including carotene), minerals (calcium and iron), protein, and carbohydrates. Further, the number of essential amino acids, which act as our bodies' building blocks appear to be increased because kimchi encourages the breakdown of protein into its components.

Kimchi can help fight nutrient depletion and be a primary source of nutrients during Jack Frost's winter visit, when fresh vegetables are not readily available. See "How to Prepare for Winter" below. Kimchi often contains garlic (with all its healing values). Garlic is a wonderful resource for helping to improve digestion, increase stamina, and reduce fatigue, plus it has cardiovascular and cancer-fighting benefits.

Kimchi is often served as an appetizer or a side dish. In Korea, no table is complete without it. It is used with fried rice, stew, soup, and much more. It fits in quite nicely with Japanese, Chinese, or American and other Western meals.

How to Prepare for Winter

During winter, truly fresh vegetables are not as readily available as they are during summer. Those sold in grocery stores have been stored for months or have been shipped (sometimes with the addition of preservatives) from more amenable climates. But we need fresh, enzyme-rich vegetables. What's the answer? Fermented foods! Kimchi, for example, is an excellent source of enzymes.

In the United States, kimchi is available in the vegetable section of most grocery stores, or you can make it at home. The following is just one of several kimchi recipes:

Ingredients:

1 head cabbage, cut in 2-inch pieces
I pound white daikon radish, sliced
2 tablespoons salt
5 ½ cups water
2 tablespoons minced fresh ginger
2 cloves of garlic, minced
2 chopped green onions
1 teaspoon cayenne
1 teaspoon honey

1. Combine the cabbage, radish, salt, and water in a large bowl and set aside overnight.
2. Remove the cabbage and radish and place in a 3-quart jar or crock.
3. Combine the remaining liquid with the ginger, garlic, green onions, cayenne, and honey. Mix well and pour over the vegetables (to within an inch of the top).
4. Cover with a clean cloth and let ferment for three to seven days.

Homemade kimchi can be stored in cool or refrigerated areas (such as the refrigerator or root cellar). Store the kimchi in a number of smaller crocks instead of one large crock (to minimize the risk of spoilage). The temperature should remain constant at about 40 to 50 degrees F during the entire winter season. Remember that at lower temperatures the enzymatic activity is decreased. Therefore, the kimchi will remain fresh longer.

In order to minimize possible mold formation or continued fermentation, care must be taken not to disturb the kimchi crocks unnecessarily. The principles applied to kimchi storage can also be used for many other fermented products.

In Korea, kimchi's taste can vary according to the region of the country. In the warmer climate of the South, kimchi's flavor is much richer and less juicy (more pickled fish and salt is added). On the other hand, the kimchi produced in the cooler North is less salty and has a stronger taste of fresh vegetables.

Cheese

Raw milk contains about fifty enzymes, but as mentioned previously, these enzymes are destroyed when the milk is pasteurized. Fortunately, the cheese-making process, as well as the bacteria used in some cheese production, can reintroduce some enzymes into the final product.

The first step in making cheese is milk coagulation using the enzyme rennet from either animals or microbial sources. This enzyme helps to "clot" milk. Other enzymes (such as the lipases) play a part in flavor formation (this is especially true with Roquefort, Gorgonzola, blue, and Stilton cheeses). The strongest smelling cheese has the greatest enzymatic activity.

According to Ted Whitehead, Ph.D., quality assurance/quality control manager of the Tillamook County Creamery Association, "In terms of active enzymes, cheese is loaded. These can be enzymes naturally present in milk (i.e., plasmin, plasminogen, lipases), they may be added to milk for cheese making (i.e., chymosin, pepsin, microbial proteases), or they may be produced by starter cultures and other adventitious microflora (i.e., proteases, peptidases, lipases, various other metabolic enzymes)."

Sprouts

Sprouted seeds, beans, grains, and nuts are high in nutritive value. In addition, they are easy to store. For these reasons, they should be a welcomed addition to your everyday diet. There are a number of advantages in preparing sprouted foods. Many dried foods cook faster if they are first sprouted. For example, kidney beans cook in about 30 minutes if sprouted first, but can take hours otherwise.

But sprouts are also enzyme-rich. For instance, when the seeds of a cereal grain, such as wheat, germinate (begin to sprout), the proteins reserved in the endosperm are broken down into their constituent amino acids and peptides by increasing proteolytic enzyme levels. In fact, enzyme activity is increased many times through sprouting. In the case of wheat, protease activity increases fifteen times when its grains are germinated. Aminopeptidases, carboxypeptidases, endopeptidases, and other peptidases are some of the proteolytic enzymes present in wheat seeds.

When a grain has been moistened, allowed to germinate, and then dried, it is called malt (barley is probably the most commonly malted grain). Similar to the original grain in appearance, malt is rich in enzymatic activity. Malt is also stable, so it can be stored for long periods of time and shipped over long distances. Barley malt is widely used in brewing, partly because barley is an excellent source of enzymes (including phytase, hemicellulases, beta-glucanases, proteases, and amylases), which are formed during germination.

To sprout seeds at home, remember that dry seeds require a great deal of moisture to soften the outer covering and swell the seed. They then require an even, constant supply of moisture to grow. This moisture can be provided by spraying a fine mist of water on top of the sprouts or by placing the seeds on or between moist cloths. Sprouts can be added to salads, sandwiches, and soups. An energizing and enzyme-rich drink called *Rejuvelac* can also be made with sprouts. See "Rejuvelac" on page 36.

Raw Honey

Honey bees produce honey from plant nectar and, to a lesser extent, from honeydew. Unlike sugar, which is made from sugar cane or sugar beets, honey can be stored and eaten just as it is produced without further processing. It is a high-density food, it remains stable in storage, and it was the primary dietary sweetener until beet and cane sugar became widely available about 100 years ago.

To make honey, the honeybee secretes enzymes that convert the sucrose in the nectar to fructose and glucose. The

Some Fermented Foods

PRODUCT	MADE FROM
Cheese	Milk curd with the addition of *Streptococcus spp.* or *Leuconostoc spp.*
Kefir	Milk with the addition of *Streptococcus lactis, Lactobacillus bulgaricus, Lactobacillus lactis, Lactobacillus kefir.*
Kimchi	Cabbage and other vegetables with the addition of lactic acid bacteria.
Kumiss	Raw mare's milk with the addition of *L. bulgaricus, Lactobacillus leichmanni, Candida spp.*
Miso	Brown rice, soybeans, or barley with the addition of *Aspergillus oryzae, Saccharomyces rouxii.*
Olives	Green olives with the addition of *Leuconostoc mesenteroides, Lactobacillus plantarum.*
Pickles	Vegetables, usually cucumber, with lactic acid bacteria.
Sauerkraut	Cabbage with the addition of *L. mesenteroides, L. plantarum.*
Soy sauce	Soybeans with the addition of *A. oryzae* or *Aspergillus soyae; S. rouxii, Lactobacillus delbrueckii.*
Tempeh	Soybeans with the addition of *Rhizopus ofigosporus; Rhizopus oryzae.*
Yogurt	Milk, milk solids, with the addition of *Streptococcus thermophilus, L. bulgaricus.*

Rejuvelac

Rejuvelac was developed by Ann Wigmore, author of *The Wheatgrass Book* (Garden City Park, NY: Avery Publishing Group, Inc., 1985) in the 1960s. It is a fermented wheatberry drink that is packed with enzymes, B complex vitamins, and vitamins C and E. It is also relatively easy to make.

1. Put two cups of wheatberries in a half-gallon jar and cover the mouth of the jar with nylon mesh or cheesecloth secured with a rubber band.

2. Fill the remainder of the jar with spring or filtered water, and allow the wheatberries to soak for eight to ten hours.

3. Drain the berries, rinse them, and then drain them again. Place the jar at an angle, so that the berries will continue to drain. The wheatberries will begin to sprout.

4. After two days, rinse the sprouted wheatberries thoroughly, then fill the jar with spring or filtered water and allow the berries to soak for forty-eight hours. This liquid is your first batch of Rejuvelac. Pour it into another container, and drink or refrigerate it.

5. Refill the jar and soak another day (only twenty-four hours this time) for another batch of Rejuvelac.

6. You may soak the berries a third time for twenty-four hours for another batch of Rejuvelac. Then throw the berries in your compost pile.

Rejuvelac will be tart, resembling lemonade in flavor. The longer it is allowed to ferment the more sour the flavor becomes. The Rejuvelac will also be slightly carbonated with a layer of foam on top. This is natural and quite harmless. Rejuvelac can be kept refrigerated for up to a week—as long as the taste is still agreeable.

---

honey is then stored in honeycombs where it ripens and decreases in water content. Enzymes play a major role during the ripening period, turning the honey into a complex sugar. Enzymes also dictate the flavor and color of honey and are partly responsible for the antiseptic and healing properties of honey. The antiseptic property of honey is due to the enzyme glucose oxidase, since it catalyzes the formation of hydrogen peroxide. This helps stabilize the honey against microorganisms.

To destroy yeasts, commercial honey is often heated or pasteurized. It is then filtered, bottled, and ultimately sold. However, this heat also destroys the honey's natural enzymes, including amylase, catalase, and invertase. Buy raw honey instead. But do not feed raw honey to infants under one year of age because of the risk of infant botulism (a problem that does not occur in older children or adults).

Wheat Germ

As mentioned in Chapter 3, wheat (before it is processed) is a storehouse of enzymes, including alpha-amylase (up to twenty-two forms), beta-amylase, protease, hemicellulase, and oxidase. But to make white flour, these enzymes are removed. Wheat germ, however, still has the active enzymes, along with a high level of lipase and other enzymatic activity (which decreases during storage).

Unfortunately, most of the wheat germ on the market is steamed to kill those enzymes and, therefore, prolong shelf-life. I was able to locate one source of raw, natural wheat germ that has not been steam treated: Bob's Red Mill Natural Foods, 5209 S.E. International Way, Milwaukie, Oregon 97222. You can order wheat germ through their mail-order department at (800) 553-2258.

Enzyme depletion is one of the greatest tragedies in today's society. As mentioned in Chapter 3, if your food has been heated or processed, it is enzyme-dead. We can maximize our enzyme intake by increasing consumption of fresh fruits and vegetables and by eating fermented products and enzyme enhancers. But most Americans rarely eat fresh fruits and vegetables. In fact, fewer than 10 percent of Americans eat two servings of fruit or three servings of vegetables per day. And just how much sauerkraut, yogurt, wheat germ, or sprouts can a person eat? Chapter 5 will provide you with more information on supplementing your diet with enzymes.

5. ENZYME SUPPLEMENTS

Enzymes are critical for life and living. But life today is complex. Because of today's fast-paced lifestyle, poor eating habits, environmental pollution, and manufacturing and cooking processes, our food is enzyme-dead. Fortunately, we can replenish those enzyme losses with additional enzymes in two ways: in fresh, raw foods, and with enzyme supplements. In order to survive in today's society, enzyme supplements are critical. This chapter will explain the basics of enzyme supplements, their forms, types, characteristics, and uses. The second part of the chapter will list the most frequently used therapeutic enzymes for wellness and fighting diseases and their sources; the foods that contain the highest levels of these enzymes; the actions, uses, and benefits of each enzyme; and the way in which the enzyme can be taken.

ENZYME SOURCES

Supplemental enzymes can be obtained from a variety of sources, including animals (usually hog or calf pancreas), plants (such as pineapple and papaya), and microbial fermentations (from bacteria or fungi—these are often called "plant-derived"), in combinations of the above, and in concentrated food or plant extracts, called "functional" foods. (See Chapter 6 for further discussion of functional foods.)

Animal Enzymes

The human body produces and secretes a number of enzymes, many of which are also made by other animals. For instance, the enzyme trypsin, found in pancreatic juice, is also found in the pancreas of many vertebrates, as well as in insects, crayfish, and microorganisms such as *Streptomyces griseus*. Each form of trypsin (regardless of the source) is a hydrolase and catalyzes the same reaction. Enzymes taken from animal sources are commonly extracted from the pancreas, liver, or stomach of beef, oxen, and pigs. These enzymes include proteases, amylases, and lipases. Some of the best known are trypsin, chymotrypsin, pepsin, rennin, and pancreatin (which is not a single enzyme, but rather a combination of primarily amylase, lipase, and protease enzymes).

Protomorphogens (or glandulars) are organ- and glandular-based food supplements that contain a mixture of enzymes that naturally occur in the particular organ or gland from which the extract was obtained. They are taken from animal organs and glands, such as the pancreas, thy-roid, ovaries, testicles, brains, and so on. The enzyme composition of protomorphogens varies depending on the animal and organ or gland used as a source, and the treatment the organ or gland received during processing into an extract. For this reason and because the individual enzymes contained in the mix cannot easily be ascertained, protomorphogens will be discussed in Chapter 6—Enzyme Helpers.

Plant Enzymes

In addition to enzymes taken from animal sources, many enzyme supplements are derived from plants. Although every plant has enzymatic activity, some plants are particularly rich enzyme sources. The enzymes used in supplements are primarily those from pineapple (bromelain), papaya (papain), fig (ficin), and barley (malt diastase). Although available in purified form, most enzymes from plants are also found in plant or food extracts. In this book, concentrated food extracts are discussed and classified as "functional foods" (which will be covered in Chapter 6). These food concentrates often contain the amylase, lipase, and protease enzymes that naturally occur in the fruit or vegetable used in the product, as well as phytochemicals (beneficial plant chemicals), vitamins, and minerals. For example, pineapple extract contains the enzyme bromelain as well as vitamins A, B, C, D, E, and K; calcium; iron; phosphorus; potassium; sodium; and magnesium. Pineapple is one of the most nutrient-rich, enzymatically active fruits in the world and is considered an excellent source of natural antioxidants. In addition to pineapple, many other fruits, especially papaya, kiwi, figs, guava, and ginger root are known to have a high protease content.

Microbial Enzymes

Microbial (sometimes called "plant-derived") enzymes are of bacterial or fungal origin and are produced through fermentation using these microorganisms. In recent years, microorganisms have increasingly been used as a source of enzymes for supplements because they are relatively inexpensive and provide an abundant supply. In fact, microbial enzymes now represent about 90 percent of all enzymes produced commercially for any purpose, according to Tony Godfrey and Stuart West in their book *Industrial Enzymology*, 2nd Edition (New York: Macmillan Press, Ltd., 1996).

But before you back away in horror, be aware that fermentation is a source of many modern medicines. For instance, *Bacillus subtilis* is a common bacterium found in soil and water and is used to produce the antibiotic bacitracin. Even penicillin is derived from *Penicillium* and other soil-inhabiting fungi. Industrial microbiology is also used to produce vitamins (such as B_{12}, thiamin, and riboflavin), amino acids (such as lysine, arginine, and glutamic acid), and antibiotics (such as streptomycin, tetracycline, and cloramphenicol).

The best known sources of the many microbial enzymes used in supplements are *Aspergillus oryzae* (a fungus), *Aspergillus niger* (a fungus), *Rhizopus niveus* (a fungus), *Bacillus licheniformis* (a bacterium), *Bacillus subtilis* (a bacterium), and several *Saccharomyces* species (yeast).

Enzyme Combinations

Sometimes, manufacturers combine different types of enzymes from a number of animal, plant, and/or microbial sources. Because of their variety of origins and substrates, wide ranges of optimal temperatures and pH levels, synergism, increased percentage of absorption, and increased level of effectiveness, enzyme mixtures have a wider range of therapeutic advantages than do individual enzymes, according to Dr. Peter Streichhan, a world-renowned enzyme researcher from Geretsried, Germany.

Plant and microbial enzymes are still highly active at higher temperatures and in acidic environments, whereas most animal enzymes function best at body temperature and at a neutral to alkaline pH range. Very few animal enzymes demonstrate maximum activity at higher temperatures and in acidic environments (such as during fever or inflammation). This is why combining animal enzymes with plant or microbial enzymes is of value. The enzyme mixtures support one another in their range of activities. By combining enzymes from many sources, you can get wider pH, temperature, and substrate ranges. Because of this, you will have a broader range of activity and application. I like to think of these enzyme combinations as "cocktails."

There are a number of enzyme combinations on the market. In fact, most products sold in health-food stores are enzyme combinations, rather than individual enzymes. The following are a number of enzyme combinations available to you. I have categorized them according to their enzyme content and given each formulation a number. These formulations will be referred to throughout Part Two. For example, when Formula One is recommended, you may choose any of the products listed under Formula One.

Formula One:
Contains: bromelain (25–45 mg), chymotrypsin (.5–1 mg), pancreatin (100 mg), papain (25–60 mg), and trypsin (24 mg).
Brand names: Lyso-Lyph from Nutri-West, Medi-zyme from Gero Vita, Mega-Zyme from Enzymatic Therapy, Poly-Zyme 021 from General Research Laboratories, Somzyme from Life Plus, Wobenzym N from Naturally Vitamin Supplements, Intenzyme Forte from Biotics Research Corporation, Product #01161 Enzymes from Michael's Natuopathic Programs, Multi-Enzym and Multi-Enzym Forte from Energetica Natura Benelux B.V., Ultra-Zyme from Nature's Plus, and Bio-Zyme from PhytoPharmica.

Formula Two:
Contains: A mixture of pancreatin, microbial lipase, and cellulase.
Brand names: Mult-E-Zyme from Enzyme Process, Zyme-Aid from Country Life.

Formula Three:
Contains: superoxide dismutase, catalase, and glutathione peroxidase.
Brand names: Antioxzyme from Enzyme Process, S.O.D. Lozenge from Nutri-West, Opti-Gaurd from Optimal Nutrients, S.O.D. Lozenges (with S.O.D. and catalase only), SUPEROXYM (S.O.D.) from general Research Laboratories.

As mentioned, functional foods contain a combination of enzymes, the specific types vary depending on the foods used as components in the product. Unfortunately, only a minimum of research has been conducted to identify the specific enzymes within these food products.

ENZYME USES

Enzymes are wondrous aids used to improve health, maintain wellness, and fight disease and injuries. Scientific investigations of enzymes taken from plants, animals, and microbial sources have been ongoing for over forty years. Acting as catalysts, enzymes are best known for their ability to aid digestion and ease digestive problems. Treatment of some conditions requires that enzymes be absorbed into the bloodstream. This treatment is referred to as *systemic enzyme therapy*.

Digestive Aids

Supplemental enzymes are frequently used as digestive aids. In fact, papaya has been used as a digestive aid for centuries because of its enzyme papain. The primary goal of digestive enzymes is to break food down into smaller particles so that the body can more easily absorb the nutrients in the food. The majority of this work is completed in the stomach and subsequently the small intestine.

The enzymes used most frequently to treat digestive problems include proteases, amylases, and lipases. As mentioned, each digestive enzyme is very specific and works on a different substrate. All the carbohydrate-digesting enzymes in the world will not begin to affect any protein or fat that you ingest. If your diet is high in fat, taking a lipase (a fat-digesting) enzyme may help you better digest that fat. People often find that protease enzymes can improve the digestion of protein (such as beef, chicken, poultry, or fish). Amylases break down carbohydrates, such as bread, pasta, potatoes, fruits, vegetables, and sugars, so are particularly beneficial in aiding the digestion of these types of foods. A specific amylase, cellulase, breaks down cellulose, the indigestible fiber found in many fruits and vegetables. For this reason, cellulase is often used in digestive enzyme formulations.

But digestive enzymes can do more than aid digestion. Because they are so efficient at breaking down protein, carbohydrates, and fats, they are able to stimulate the good bacteria in the gut, detoxify and cleanse the colon, and therefore improve a number of digestion-related conditions, including food and other allergies. They help to mobilize and expel harmful products from the body. This process occurs at the cellular level, in organs, and in various systems of the body. For further information on colon cleansing and detoxification, see Part Three of this book.

Systemic Enzyme Therapy

Enzymes are critical for all life processes. When we are lacking enzymes, illness results. Systemic enzyme therapy is the treatment of disorders with the use of enzymes. Hundreds of disorders, including cancer, cardiovascular diseases, and infections, can be treated with systemic enzyme therapy. It is often most effective when used in conjunction with standard medical therapies; although for the treatment of minor conditions, enzyme therapy may preclude the need for medical intervention.

Unlike digestive enzymes, enzymes used for systemic purposes *must* be absorbed into the bloodstream in order to be effective. When made for systemic use, most enzymes from animal and plant sources are enterically coated or protected in some way with an acid-resistant coating so that they can pass through the acidic stomach intact. Microbial enzymes can be designed and produced to be acid-resistant and, therefore, can resist the low pH of the stomach and pass onto the small intestine to do their work.

FACTORS AFFECTING ENZYME EFFECTIVENESS

An enzyme's environment and the conditions in which it is used can greatly influence its effectiveness. Such factors include pH level in the body, body temperature, substrate concentration, rate of absorption, and the amount of coenzymes and enzyme-inhibitors present.

pH Levels

The acidity or alkalinity of any substance can be measured on a 15-step scale (0 to 14) known as pH (potential hydrogen). A pH of 7 is neutral (neither acidic nor alkaline). Anything above 7.0 is alkaline, and anything below 7.0 is acidic. For example, a pH of 1.0 is much more acidic than a pH of 3.0. A pH of 14.0 is more alkaline than a pH of 10.0

```
0 ————————————— 7 ————————————14
Acidic            Neutral          Alkaline
```

Usually, the pH of the mouth is neutral, while the stomach is acidic and the small intestine is alkaline. See Table 5.1 for pH levels of common foods and household products to get a better understanding of acidity and alkalinity.

Characteristically, each enzyme has an optimal pH range, which may be broad or narrow. At this optimum pH level, the enzymatic catalytic reaction occurs most rapidly.

The pH of a particular enzyme may vary depending on the processing method or, in the case of microbial enzymes, the fungus or bacteria used. For instance, a microbial lipase from *Aspergillus oryzae* may have an optimum pH of 6.0 to 6.5, while one from *Candida cylindracea* may have an optimum pH closer to 7.0 to 7.5, one from *Rhizopus arrhizus* or *Aspergillus niger* may be 5.6, and one from *Geotrichum candidum* may be closer to 6.0.

Absorption

Enzymes used to improve digestion do not need to be absorbed into the bloodstream; they do their work primarily in the stomach and small intestine. Enzymes used in systemic therapy must be absorbed as whole enzymes and not broken down into their component parts in the digestive tract.

There are at least three methods by which enzymes are absorbed from the small intestine. The first is called pinocytosis. In pinocytosis, the membranes of intestinal cells develop pouchlike sacs and engulf enzymes and other large molecules, and guide them into the bloodstream. Transcellular

Table 5.1. The pH of Various Foods and Household Products

PRODUCT	PH LEVEL
Lye	13
Household ammonia	12
Milk of magnesia	11
Baking soda	9
Egg white	8
Human blood, tears	7+
Saliva, milk	6.5
Normal rainwater	6.2
Black coffee	5
Tomato juice	4
Orange juice	3.5
Cola, apple juice	3
Vinegar	2.6
Lemon juice	2
Hydrochloric acid	1
Battery acid	.5

absorption (persorption) is the second menthod of enzyme absorption. In this process, large nutrient particles are pushed between the intestinal cells by the contractions of the intestines (peristalsis). They are then forced into the blood-stream. Enzymes can also be absorbed into the blood-stream from the small intestine by being transported by the lymphatic system. The epithelium of the small intestine con-tains M cells, which are the cells along the microfolds of the intestine. These M cells transport enzymes to the lymphatic system, where lymphocytes stick to them, transporting them through the lymphatic vessels and into the blood-stream.

These mechanisms of absorption function for all large mol-ecules, not just enzymes. Unfortunately, these mechanisms have a limited capacity and transport molecules on a first-come first-served basis, which means that other large mole-cules may be absorbed, leaving enzymes behind. For this rea-son, as much enzyme as possible must be present at the site of absorption in order to ensure maximum enzyme absorption. This is why high doses of oral enzymes should be taken. This limited capacity is also why it is important to take enzymes used for systemic purposes between meals. If they are taken during meals their action may be directed toward the digestive process, and they will not be available for absorption.

Immunity can also affect enzyme absorption. In tests, increased enzyme absorption has been noted in immuno-suppressed animals, but decreased enzyme absorption resulted when animals were immunized. Therefore, if you have been immunized against certain diseases, you may not absorb as many enzyme molecules as someone who has not been immunized.

But before an enzyme can even be absorbed from the small intestine, it must first complete an extremely difficult obstacle course through the acidic (low pH) stomach, then the alkaline (high pH) small intestine, and finally be absorbed across the brush border of the small intestine into the circulatory system. Many an enzyme has been weakened or destroyed in the acid environment of the stomach.

Since enzymes for systemic purposes must first pass through the acidic stomach, their ability to survive in an acid environment is critical. Enzyme manufacturers have ad-dressed this challenge through the enteric coating, microen-capsulation, or special formulation of supplements. These techniques protect the enzyme from acidic environments. In this way, the enzymes are safely protected until reaching the section of the intestine where absorption is possible. Only in the small intestine is the tablet's covering dissolved and the enzyme absorbed into the bloodstream.

According to Dr. J. Seifert and his associates at Christian-Albrechts-Universität, Kiel, Germany, the rate of enzyme absorption can range from 44 percent for amylase to 7 per-cent for papain (see Table 5.2).

Table 5.2. Absorption Rates of Certain Enzymes

	AMOUNT PER COATED TABLET	AMOUNT ABSORBED	ABSORPTION RATE
Amylase	10 mg	4.49 mg	44 percent
Bromelain	45 mg	17.86 mg	39 percent
Chymotrypsin	1 mg	0.16 mg	16 percent
Pancreatin	100 mg	19.03 mg	19 percent
Papain	24 mg	6.90 mg	7 percent
Trypsin	24 mg	6.90 mg	28 percent

Source: J. Seifert, et al., "Quantitative Untersuchungen zur Resorption von Trypsin, Chymotrypsin, Amylase, Papain und Pankreatin aus dem magen-Darm-Trakt nach oraler Application," *Allgemeinarzt*, 19:132, 1990.

Temperature

In addition to pH and coating concerns, the viability of an enzyme supplement (whether coated or uncoated) can be affected by temperature. For instance, the enzymes used to treat humans work best at about body temperature and lose activity quickly if exposed to extremes. Therefore, don't take enzyme supplements with hot beverages and don't add them to hot foods. In addition, if you're suffering from a fever, your enzymes will work harder until they poop out and are destroyed. So you may need additional enzymes in such a situation. Slight elevations in body temperature increase the activity of the body's enzymes critical in com-bating the health crisis. However, if the temperature increas-es too much (exceeding the optimal temperature by even a small amount), enzymes are destroyed and an enzyme imbalance can result, leading to serious consequences. This loss in enzyme activity can lead to illness and ultimately death. Luckily, the body heats up only in cases of emergency.

On the other hand, as temperatures decrease, enzyme activity also decreases. This is why cheese, butter, and other foods stay fresh in the refrigerator. The cool temperature reduces enzymatic activity, helping the butter and cheese to last longer. But don't store your enzyme supplements in the refrigerator to make them last longer, because they may draw moisture from the refrigerator and lose activity.

Other Factors

Many enzymes also need helpers in order to function opti-mally. These helpers are called cofactors and coenzymes and include many vitamins and minerals (such as the B vitamins, zinc, copper, and selenium). Our bodies are incapable of pro-ducing coenzymes from the components available in the body, so we must consume them in our diets. This will be discussed in more detail in Chapter 6.

Enzyme inhibitors are another factor that can affect your enzymes' effectiveness. As mentioned in Chapter 3, we encounter enzyme inhibitors every day, not only in our air and water, but also in our foods. If your diet contains enzyme inhibitors, such as alcohol, salt, sugar, additives, and preservatives, as well as seemingly healthy foods that contain enzyme inhibitors such as beans, peas, peanuts, soybeans, and other legumes, you may need to increase your enzyme intake.

ENZYME THERAPY

Research has shown that a number of conditions, including acne, aging, allergies, autoimmune conditions, cancer, circulatory diseases, health conditions requiring detoxification, infections, inflammation, injuries, rheumatic diseases, skin problems, sports and other injuries, and viruses can all be treated with enzymes, according to Drs. D.A. Lopez, R.M. Williams, and M. Miehlke, authors of *Enzymes: The Foundation of Life* (Charleston, SC: Neville Press, Inc., 1994).

Although there is variation among individuals, our bodies' enzymatic activity generally begins to decline somewhere between the ages of sixteen and twenty, as our growth peaks. At this time, the rate of tissue reproduction tends to decrease, and what we refer to as aging begins. Since every activity in the body requires enzymes, replenishing the enzyme supply lost with aging can restore your body's natural enzyme balance. This, in turn, can help fight disease and improve your health status. This is why systemic enzyme therapy is effective against such a wide variety of conditions.

The type of enzyme you take, the dosage level, individual variables (such as height, weight, general health, lifestyle, and dietary patterns), and the condition being treated can all influence how rapidly you respond to enzyme therapy. For instance, if you are taking enzymes to improve digestion, you should notice an improvement within a couple of hours. If not, perhaps you need to increase the dose. If improvement still is not seen, analyze your diet and make sure you're taking the proper type of enzyme to digest what you eat, or that you don't have some other illness (for instance, a heart condition) that is mimicking a digestive problem.

Generally speaking, when enzymes are taken to treat an inflammatory condition or one involving pain, some improvement will be noted within three to seven days. Chronic conditions such as rheumatoid arthritis may require one to three months (or more) before you notice a change in your symptoms. In addition, some conditions have periods of waxing and waning (such as rheumatoid arthritis or multiple sclerosis)—times when the symptoms may be markedly worse or, perhaps, a little better. When the condition is worse (or when you notice an exacerbation of symptoms), it's important to increase the dose.

Choosing an Enzyme Supplement

Enzymes are specialized, and because the human body contains over 3,000 types of these catalysts (many of which act synergistically) it is impossible to say which particular enzyme is most important. However, a shortage of just one enzyme can mean poor health.

The enzyme supplement you choose will depend on your goal. If taking enzymes as digestive aids, you should choose enzymes based on your diet. For instance, those who have trouble digesting the protein in beef should take a proteolytic enzyme, such as pancreatin, trypsin, chymotrypsin, papain, or a microbial protease. Lipases assist in digesting fats, and a number of enzymes work on carbohydrates and sugars (including lactase, amylase, cellulase, alpha-galactosidase, and similar products).

If taking enzymes to fight free radicals, take an antioxidant enzyme (such as superoxide dismutase, glutathione peroxidase, and catalase). These enzymes are often enterically coated to improve absorption. Enterically coated enzymes are generally used for systemic purposes, although some (such as pancreatin) are also effective at improving digestion. Each enzyme has certain advantages, depending on what you want to achieve. (For guidelines on which enzymes to use for treatment of a particular condition, see Part Two of this book.)

Enzyme supplements are available in many different forms—as tablets (uncoated, enterically coated, or microencapsulated), coated granules, powder, capsules, liquid, chewing gum, injections, creams and ointments, and retention enemas. They may be formulated individually or in combinations of enzymes. You can choose your supplement based on your likes and needs. (See "Choosing a Supplement That's Right for You" on page 42.)

Although many supplemental enzymes are taken orally, some can be taken sublingually (under the tongue), or rectally with an enema. Enzymes can also be applied topically in ointments, creams, salves, and lotions.

There are also some enzymes administered primarily by injection, which is done in a doctor's office or hospital. When administered in this way, absorption is not of critical importance, since they are injected directly into the bloodstream or into tissue. Those enzymes administered by injection include brinase, chymopapain, collagenase, hyaluronidase, lysozyme, plasmin, streptokinase, streptodornase, and urokinase. However, collagenase and hyaluronidase can also be administered topically. Other enzymes that can be taken by injection include carbohydrase, chymotrypsin, superoxide dismutase (SOD), and trypsin. It should be noted, however, that injectable enzymes should be used with caution. They can cause very serious side effects, such as anaphylactic shock, and are only used by physicians in severe life-threatening situations.

Enzymes are also of value in face and body creams. Papain and bromelain are probably the best known topical enzymes. Zia Wesley-Hosford, author of *Face Value* (San Francisco, Zia Cosmetics, Inc., 1990) believes that green papaya enzymes assist the healing of uneven pigmentation,

---

Choosing a Supplement That's Right for You

Choosing an enzyme supplement is not always easy. There are many factors to consider in choosing the supplement that is best for you. Often, a quick look at the label will help you decide whether or not to purchase a supplement.

What to look for on the label:

• Directions for use.

• Formulation (coated or uncoated).

• The enzymes included in the formulation and their sources. For example, a vegetarian would want to avoid enzymes taken from animal sources, and those with allergies would want to ensure that there are no potential allergens in the formulation.

• Weight of the ingredients in milligrams (mg).

• Expiration date.

• Price.

---

fine lines, and brown spots by fighting free-radical damage and boosting cell production and are considered the "natural alternative to Retin-A." In fact, enzymes, especially papain, are being used in a number of facial products. Many of the larger cosmetic companies, such as Estée Lauder, are now beginning to add enzymes to their facial products.

Sometimes skin creams, salves, or exfoliants made with enzymes are purchased in powdered form and must be mixed with oil or other fluid to activate before application (follow the instructions on the bottle or box).

As previously mentioned, food extracts contain many of the enzymes naturally occurring in the plants used in each particular formula. In addition, formulas containing enzymes combined with herbs, vitamins, minerals, and other nutrients are effective at improving the absorption and bioavailability of the nutrients, maximizing enzyme activity, and reducing the drain on the body's own digestive enzymes. I call these formulations Enzyme Absorption System Enhancers (EASE).

The price of enzyme supplements can vary widely depending on the activity level, the amount of enzyme in the tablet, and bottle size. The price of a bottle of enzymes can range from two or three dollars for a bottle of papain to $80 or more for certain enzyme combinations. Although the price is not always a guarantee of quality or potency, it can be one indication. Just like any other commodity, the bulk price for enzymes is fairly uniform. For this reason, you would expect to pay more for a bottle of one hundred 100 mg bromelain tablets than you would for a bottle of one hundred 50 mg bromelain tablets, regardless of the manufacturer, assuming that the enzymes in question had the same activity level. Comparing price gets a little trickier if the activity levels are different. Generally speaking, I compare brands and choose the tablets or pills with the highest activity level regardless of price.

Keep in mind that when choosing an enzyme supplement, almost any choice is a good choice because the enzymes will help your body to function in a more effective fashion. Further, they will help you overcome the plight of enzyme-dead food, environmental problems (such as today's pollution and toxins), lifestyle changes, stress, premature aging, and the development of chronic disorders.

Dosage Considerations

The enzyme dose needed to treat any condition will vary depending on the quantity (amount in milligrams) and quality (activity level) of the enzymes in each tablet. It will also vary depending upon the condition being treated and whether or not the enzyme is designed to be systemic in nature. In most cases, you can follow the directions on the label. Dosages for the treatment of individual conditions will be discussed in Part Two.

In systemic enzyme therapy, initially, the enzyme dosage should be relatively high and then gadually reduced until a maintenance dose has been reached. It may be necessary to ingest a large number of pills. This is because enzyme molecules are very large, so it is difficult to pack a sufficient amount of them in one tablet. In addition, the acid-resistant coating often requires a lot of space. Do not worry that you may be overdosing when taking so many pills. Enzymes do not have toxicity levels like many vitamins, minerals, and other supplements and medication. They are rapidly eliminated from the body and are undetectable in the blood after twenty-four to forty-eight hours.

Keep in mind that individual differences can affect the dosage level. What works best for you might not do the job for the next person. Such factors as weight, age, sex, body metabolism, health status, lifestyle, and even the composition of your meals can all influence what is best for you. Keep in tune with your body in order to develop the best enzyme therapy program.

Taking Your Enzyme Supplement

Enterically coated tablets should be swallowed whole and not chewed or ground up. Chewing the tablet would destroy the coating, exposing the enzymes to the stomach's acid. However, other tablets that are not enterically coated can be chewed (check the instructions on the bottle first). It's also important to take enzymes with sufficient water (at least one 8-ounce glass). This is because water is required to activate the "water-craving" hydrolytic enzymes. The moisture in your mouth may not be sufficient.

If you are taking enzymes to improve digestion, they should be taken just before, during, or just after meals. Therefore, if you eat three meals a day, take digestive enzymes three times per day. Some people also take digestive enzymes before retiring, so their bodies can complete the digestive process while they sleep. When in doubt, follow the instructions on the label.

When being used for systemic therapy, enzymes should not be taken just before, with, or just after a meal. If this were done, the enzymes would become mixed with the nutrient "mash." It's perfectly safe to take enzymes on an empty stomach. In fact, this is the best time to take enzymes when using them for systemic therapy. Taken between meals, enzymes are more readily absorbed and assist various systems in fighting inflammation, as well as acute and chronic diseases. If taken with a meal, the enzymes would support digestion, but only a reduced portion (if any) would be absorbed into the bloodstream for work in your bodily systems.

In order to obtain the best level of enzyme concentration within the body, it is better to spread the dose out over the day. For instance, rather than taking fifteen tablets all at once, it is better to take five tablets three times a day. Enzymes for systemic enzyme therapy should be taken either one hour before or at least two hours after a meal. Topical enzymes should be applied according to the directions on the label, and injectable enzymes are used only at a physician's discretion.

Enzymes have a synergistic effect with vitamins, minerals, herbs, food extracts, and other nutrients. That is, enzymes can help the digestion and absorption of these nutrients, as well as improve their systemic function. In addition, the absorption and effectiveness of a number of medications, including antibiotics, anticancer agents, and cortisone can actually be improved when enzymes are taken at the same time you take them.

Measuring Enzyme Activity

Enzymes can be measured by weight and activity level. The label should indicate the weight in milligrams (mg), but this gives the consumer no indication of the enzyme's activity. As the quality of an enzyme can vary, the activity level of that enzyme can also vary a great deal. Therefore, the therapeutic value is directly dependent upon its activity. For this reason, it is very important that the activity of the specific enzyme (as well as the full activity of any enzyme mixture), be stated on

the label. This is the only way the consumer can make an intelligent decision regarding enzyme purchase and usage.

Activity is usually indicated in "units," however there is no one standard unit of measurement for enzyme activity level. Manufacturers may follow any of a number of recognized guidelines for measuring enzyme activity, such as those listed in *Food Chemicals Codex* (FCC), *U.S. Pharmacopoeia* (USP), *Federation Internationale du Pharmaceutiques* (FIP), *British Pharmacopoeia* (BP), and *Japanese Pharmacopoeia* (JP). Although the measurement methods indicated in FCC, USP, and FIP give the consumer an excellent way to compare activity between various enzyme products, they are not interchangeable. Dr. G. J. Peschke, in discussing active components of enzyme preparations (*Pancreatic Enzymes in Health and Disease*, Paul G. Lankisch, Ed., Berlin: Springer-Verlag, 1991) states that, although European Pharmacopoeia units of protease, amylase, and lipase activity are equivalent to BP and FIP units, they are not interchangeable with USP units. Table 5.3 (below) can be used as a guide for roughly converting USP units into EP, FIP, or BP units.

Unfortunately, many labels on enzyme products sold in this country do not indicate the actual activity levels of the enzymes. Adding to the confusion, various enzyme manufacturers use many different "in-house" methods to analyze enzymes. If I wanted to produce enzyme products, I could devise "Cichoke Protease Units" or "Cichoke Activity Units" and set my own method of measuring proteolytic (or any other enzyme) activity. Unfortunately, this doesn't give the consumer the information he or she needs to make an intelligent decision when comparing and purchasing enzymes. The confusion the manufacturers create by speaking different "languages" about enzymes is similar to what would happen if the United Nations tried to operate without interpreters. Fortunately, the supplement industry is currently formulating guidelines regarding the measurement of enzyme activity. And more and more progressive companies are now listing source, enzyme content, substrates, pH range, and activity per capsule or tablet on their products or in their literature.

Storing Enzyme Supplements

As mentioned, enzymes are alive. Since enzymes easily become denatured and lose catalytic activity, careful handling and storage are essential. If not treated carefully,

Table 5.3. Conversion Table of Enzyme Activity Measurements

ENZYME	USP UNITS	EP UNITS	FIP UNITS	BP UNITS
Protease	62.50	1	1	1
Amylase	4.15	1	1	1
Lipase	1	1	1	1

enzymes can be destroyed. They should be stored in tight containers with moisture-proof liners, preferably in a dry, cool place. In general, pH extremes specific to each enzyme, temperatures above 140º F (maybe even above 107° F), and moisture should be avoided at all times.

If protected from heat, moisture, and light, enzyme supplements will retain potency for months, and sometimes years, before they begin to lose activity level. You should purchase only those enzyme supplements sold in opaque or dark containers, which will protect them from light. Most bottles have an expiration date printed on the label. Unless the label says otherwise, as a general rule, if unopened, keep enzymes on your shelf no longer than twelve months. Once opened, enzyme supplements are more easily exposed to moisture, air, and heat, all of which can decrease activity. If kept in a cool, dry, dark place, an open bottle of enzyme supplements will probably last for about two to three months.

Side Effects

There appear to be no side effects of long-term duration when taking oral enzymes. Temporary side effects (that will disappear when therapy is discontinued or reduced) include changes in the color, consistency, and odor of the stool. Gastrointestinal disturbances, such as flatulence, a feeling of fullness, diarrhea, or nausea, may occur in individual cases, and minor allergic reactions (reddening of the skin) are rare and occur only with high doses. Long-term usage of supplemental enzymes has not been found to be detrimental, but varying the enzyme dosage level periodically as well as the time of day the enzymes are taken, seem to improve results.

Studies indicate that enzymes taken for their systemic action cannot be detected in blood analysis after twenty-four to forty-eight hours. This means there should be no side effects of long-term duration. But it also means you should continue to take enzymes daily to maintain their level in the body and to achieve your wellness goals.

Drs. Max Wolf and Karl Ransberger state in their book *Enzyme Therapy* (Los Angeles: Regent House, 1972) that high doses of an enzyme mixture containing pancreatin, trypsin, chymotrypsin, amylase, lipase, and rutin have been taken after injuries without long-term side effects. In one case, a patient took seventy supplemental enzyme tablets (17.5 grams of enzymes) immediately after a fall. The patient had landed full force on the left side of the face. Initially, the injury resulted in marked facial hemorrhage and edema. However, except for a small cut at the eyebrow, the only visible sign of the injury the day following the enzyme therapy was a slight discoloration around the eye. There were no side effects other than loose bowels and flatulence, which disappeared as the dosage was decreased.

In a controlled, double-blind study I conducted on sports injuries at Portland State University, along with Professor Leo Marty (PSU'sdirector of sports medicine), we gave thirty-two football players ten supplemental enzyme combination tablets per day as a preventative maintenance dose during PSU's spring football practice season. In a control group, thirty-two players were given a placebo. We wanted to see if giving players enzymes before they sustained an injury would help them recover more rapidly if an injury occurred.

When athletes taking maintenance doses of enzymes were injured during football practice, they recovered twice as quickly as those who had not taken the enzymes. And only one player complained of any side effects—a feeling of "fullness" felt during practice. This side effect went away immediately follwing the cessation of the supplementation. For the PSU study, a ten-enzyme-tablet regimen was used because that same number had been used effectively with athletes in Germany. Eight tablets may have worked just as well, but twelve might have been better. There are individual differences, and each person must discover what works best for his or her own body.

Periodically, I receive letters or phone calls from individuals concerned that proteolytic enzymes will digest the protein parts of their bodies. There seems to be no danger under normal circumstances because the body has protective mechanisms against self-digestion. However, when these mechanisms break down, irritation might develop. For instance, in stomach ulcers, there is a possibility that proteolytic enzymes will irritate the ulcer, so proteolytic enzyme therapy should be discontinued until the ulcer is healed. Mucus normally protects the stomach lining from acid secretion. An ulcer occurs when that protective lining has been damaged. Treatment usually involves neutralizing or decreasing gastric acidity, thus allowing the lining to repair. When in doubt, see a well-trained metabolic physician.

However, there are some individuals who should not use enzyme therapy, especially with proteolytic enzymes. These include those with hereditary clotting disorders such as hemophilia, those suffering from coagulation disturbances because the enzymes could thin the blood too much, those undergoing dialysis, those who are just about to undergo, or who just underwent, surgery (because proteolytic enzymes have a fibrinolytic effect, which might cause excessive bleeding), those on anticoagulant therapy, anyone suffering from protein allergies (those allergic to pork should avoid enzymes derived from pork sources), and pregnant women or those breast-feeding (we just don't know enough about the effects of supplemental enzymes on an unborn or nursing child).

Are enzymes safe for pre-school children? There has been no research done on the safety of enzymes for children. Although enzymes would appear to be safe for children, it is particularly important to first emphasize a healthy diet before putting any child on supplements of any kind. A wholesome diet can go a long way in correcting a digestive, or any other, problem. If the child has a serious health problem, a well-trained metabolic physician should be consulted. However, as children age and become young adults, they suffer from a number of conditions that are not life-threatening—just annoying—such as acne, indigestion, allergies, and sports injuries, which can be helped by using enzyme supplements.

Working With Your Health-Care Practitioner

When seeking a health-care practitioner who will incorporate enzyme therapy into your wellness program, be sure to choose someone who has a clear understanding of the use of enzyme therapy and who has experience with its use with other patients. You may consider seeking the services of a physician, chiropractor, nutritionist, homeopath, naturopath, osteopath, or other qualified healers. Once you choose a doctor with whom you feel comfortable, make sure to ask questions specific to your situation.

In a number of the disorders covered in Part Two, you will find that some recommended enzymes are best used in the form of injectables or retention enemas. In such cases, only a medical doctor should be used to provide such treatment. Should you find that your doctor is not willing to work with you in the use of enzyme therapies, consider finding a qualified doctor who will.

Whenever you question the appropriateness of any treatment—those contained herein or otherwise—always consult with your health-care provider.

NEW AND FUTURE INNOVATIONS IN ENZYME USE

Every day, researchers throughout the world are discovering new enzymes and new therapeutic applications for them. For instance, researchers from Genentech, Inc. recently developed a recombinant human deoxyribonuclease called Pulmozyme that has shown great promise in treating the respiratory problems so often suffered by cystic fibrosis patients. Administered by inhalation of an aerosol mist, Pulmozyme hydrolyzes the DNA in sputum and reduces sputum viscoelasticity, thereby reducing the incidence of respiratory tract infections in cystic fibrosis patients.

Several enzyme products are offered as ointments, such as Travase, Accuzyme, Elase, Santyl, and Panafil. These products break down proteins and are used to debride the dead tissue resulting from burns, decubitus ulcers, and wounds.

Researchers are even studying enzyme replacement therapy and devising new treatment methods for diseases that they have historically been unable to treat. This includes many diseases that result from inborn errors of metabolism, such as the inability of the body to make a particular enzyme. People suffering from Gaucher's disease (a deficiency of the enzyme glucocerebrosidase) have new hope since scientists discovered and began industrial production of the enzyme alglucerase, which can work as a substitute for glucocerebrosidase. Other innovative research is helping to grow new tissue, new skin, and new hair through the use of enzymes.

COMMON ENZYME SUPPLEMENTS

The following enzymes are those most frequently found in health-food stores and drug stores, through mail order, or from multi-level marketing companies. Included are the sources from which they are taken, their actions and benefits, the best way to take them, and any additional comments.

ALPHA-GALACTOSIDASE (or Melibiase) (Commonly sold as Beano.)

The following are the actions and uses of alpha-galactosidase:
- Breaks down carbohydrates, such as raffinose, stachyose, and verbascose.
- Used as a digestive aid.
- Prevents the gas and other gastrointestinal symptoms that occur after eating a high-fiber diet of beans, grains, and other vegetables.

This enzyme is best taken orally.

SUPPLEMENT SOURCES	
Microbial	Aspergillus niger, Aspergillus oryzae.

FOOD SOURCES	
	Cucumbers and legumes, such as soy beans and cowpeas (black-eyed peas).

ALPHA-GLUCOSIDASE. See Maltase.

AMYLASE (or Carbohydrase, Glycogenase)

The following are the actions and uses of amylase:
- Breaks down carbohydrates, such as starch, glycogen, and related polysaccharides and oligosaccharides.
- Used to aid digestion (usually in combination with other enzymes).
- Used in pancreatic enzyme replacement therapy.

This enzyme is best taken orally.

SUPPLEMENT SOURCES	
Animal	Bovine and porcine pancreas.
Plant	Barley (Hordeum vulgare).
Microbial	Aspergillus aureus, Aspergillus niger, Aspergillus oryzae, Bacillus licheniformis, Bacillus stearothermophilus, Bacillus subtilis, Rhizopus niveus, and Rhizopus oryzae.

FOOD SOURCES	
	Amylase is the most frequently found enzyme in plants and occurs most abundantly in raw sweet potato; corn, barley, wheat, oats, rice, and other grains; red lingzhi (or reishi) mushrooms; beet roots, leaves, and stems; banana; cabbage; egg; kidney bean; maple sap; milk; mushrooms; raw honey; and sugar cane.

BETA-FRUCTOFURANOSIDASE. See Invertase.

BETA-GALACTOSIDASE. See Lactase.

BETA-GLUCOSIDASE (or Emulsin)

The following are the actions and uses of beta-glucosidase:
- Breaks down carbohydrates, such as sugars and starches.
- Used as a digestive aid (often in conjunction with amylase, as it facilitates its activity).

This enzyme is best taken orally.

SUPPLEMENT SOURCES	
Plant	Sweet almonds.
Microbial	*E. coli, Aspergillus niger*, and *Trichoderma longibrachiatum*.

FOOD SOURCES	
	Almonds and green plants.

BRINASE

The following are the actions and uses of brinase:
- Breaks down proteins.
- Lyses intravascular fibrin and thus is used as a thrombolytic agent in the treatment of arterial thrombosis, thromboangiitis obliterans, and similar conditions; is used to treat advanced limb ischemia (a deficiency of blood in a part of a limb as a result of blood vessel constriction or obstruction); fights arteriosclerosis; and fights chronic peripheral arterial disease.
- *Injectable enzymes should be used with caution. May cause severe, life-threatening side effects.*

This enzyme is best taken by injection.

SUPPLEMENT SOURCES	
Microbial	*Aspergillus oryzae*

BROMELAIN

The following are the actions and uses of bromelain:
- Breaks down protein.
- Aids in overall digestion and absorption of nutrients, particularly of protein.
- Used in pancreatic enzyme replacement therapy.
- Because of its wide pH range, can be used as a substitute for pepsin and trypsin in cases of digestive deficiency.
- Fights inflammation; reduces swelling; inhibits fibrin synthesis; degrades fibrin and fibrinogen; used to treat conditions such as cellulitis, epididymitis, hypostatic and diabetic ulcers, numerous inflammatory conditions, furniculosis (boils), and epinephrine-caused pulmonary swelling.

- Speeds recovery from such injuries resulting from trauma, childbirth, sports, or surgery as sprains, strains, contusions, abrasions, hematomas, ecchymoses (small hemorrhagic spots), lacerated and/or perforated wounds, and fractures; also reduces the swelling and hematomas that often follow surgery.
- Improves respiratory conditions including throat infections, pharyngitis, sinusitis, bronchitis, and pneumania.
- Fights arthritis and other degenerative bone and joint disease.
- Used to topically treat skin conditions, including infections and burns (accelerates the elimination of burn debris and promotes healing), and used in many cosmetics and personal care products (such as facial cleansers, bath preparations, and exfoliants).
- Fights cardiovascular problems, such as blood platelet aggregation, phlebitis (inflammation of a vein), varicose ulcers, peripheral venous disease, thromboses (in a variety of sites, including central retinal vein), and heart attacks.
- Bolsters the immune system, uncovers the membranes of antigens (such as viruses and bacteria) and assists your body in better identifying and attacking these antigens, helps break up antigen-antibody complexes (immune complexes), improves antibiotic absorption.
- Helps fight cancer and activates tumor necrosis factor (a tumor-fighting substance produced by the body).
- Helps prevent dysmenorrhea, allergies, and oral infections.
- Can be used for thyroid therapy.
- Can inhibit appetite.
- Prevents intestinal bacterial infections, which often cause diarrhea.
- Can help extend life.

This enzyme is best taken orally and topically.

SUPPLEMENT SOURCES	
Plant	Stem or fruit of the pineapple (*Ananas comosus* or *Ananas sativus*).

FOOD SOURCES	
	Fresh, raw pineapple.

CARBOHYDRASE. See Amylase.

CARBOXYPEPTIDASE

The following are the actions and uses of alpha-galactosidase
- Breaks down proteins.
- Used as a digestive aid.
- Used in pancreatic replacement therapy.
- One of four proteolytic enzymes found in pancreatin along with trypsin, chymotrypsin, and elastase.

This enzyme is best taken orally.

SUPPLEMENT SOURCES	
Animal	Bovine and porcine pancreas.
Plant	Wheat.
Microbial	*Pseudomonas* sp., *Penicillium janthinellum*, and yeast.

CATALASE

The following are the actions and uses of catalase:
• Breaks hydrogen peroxide down into water and oxygen.
• Reportedly lowers serum cholesterol.
• One of the most potent antioxidant enzymes.

This enzyme is best taken orally.

SUPPLEMENT SOURCES	
Animal	Bovine liver.
Microbial	*Aspergillus niger* and *Micrococcus lysodeikticus.*
FOOD SOURCES	
	Peas, maize, soybeans, grape, mango, milk, mushroom, raw honey, and sugar cane. Found in almost all plant tissues.

CELLULASE

The following are the actions and uses of cellulase:
• Breaks down cellulose (an indigestible fiber found in many fruits and vegetables).
• Used as a digestive aid.
• Used in pancreatic enzyme replacement.
• Used in treating gastric bezoars (hard masses composed of hair and/or fruit and vegetable fibers that can form in the alimentary canal).

This enzyme is best taken orally.

SUPPLEMENT SOURCES	
Microbial	*Aspergillus oryzae, Aspergillus niger, Rhizopus* sp., *Trichoderma longibrachia tum* (formerly *reesei*).
FOOD SOURCES	
	Avocado, peas, oat sprouts, and red lingzhi (reishi) mushrooms.

CHYMOPAPAIN

The following are the actions and uses of chymopapain:
• Breaks down proteins.
• Used to treat herniated lumbar intervertebral discs.
• Used to treat degenerative intervertebral disc disorders (low back pain, sciatica, etc.).

• *Injectable enzymes should be used with caution. May cause severe, life-threatening side effects.*

This enzyme is best taken by injection.

SUPPLEMENT SOURCES	
Plant	Latex from the papaya [*Carica papaya*]

CHYMOSIN. See Rennin.

CHYMOTRYPSIN

The following are the actions and uses of chymotrypsin:
• Breaks down proteins.
• Used as a digestive aid.
• Fights inflammation and reduces swelling.
• Fights arthritis (osteo- and traumatic).
• Treats soft tissue injuries, acute traumatic injuries, sprains, contusions, hematomas, ecchymosis, infection, edema (eyelids, genitalia), charley horse, and sports injuries.
• Aids in surgical recovery.
• Can be used in debridement, treatment of ulcerations and abscesses, and the liquefaction of mucous secretions.
• Can be used against enterozoic worms.
• Used to treat cancer.
• One of at least four proteolytic enzymes found in pancreatin.

This enzyme is best taken orally, tobically, or by injection.

SUPPLEMENT SOURCES	
Animal	Bovine or porcine pancreas.

COLLAGENASE

The following are the actions and uses of collagenase:
• Breaks down proteins, specifically collagen.
• Used in debridement of dermal ulcers, necrotic tissue, and burns.
• Used in the treatment of herniated discs.
• Used experimentally to transplant pancreatic islet cells, thereby alleviating diabetic symptoms.
• *Injectable enzymes should be used with caution. May cause severe, life-threatening side effects.*

This enzyme is best taken by injection or topically.

SUPPLEMENT SOURCES	
Microbial	*Clostridium histolyticum.*

DIASTASE (or malt diastase)

The following are the actions and uses of diastase:

• Breaks down carbohydrates.
• Used as a digestive aid.

This enzyme is best taken orally.

SUPPLEMENT SOURCES	
Plant	Barley malt.

FOOD SOURCES	
	Barley.

ELASTASE

The following are the actions and uses of elastase:
• Breaks down proteins, including elastin, fibrin, hemoglobin, albumin, soybean proteins, and casein.
• Used as a digestive aid.
• Often used in conjunction with other enzymes, such as trypsin, chymotrypsin, and collagenase.
• One of at least four proteolytic enzymes found in pancreatin.

This enzyme is best taken orally.

SUPPLEMENT SOURCES	
Animal	Bovine and porcine pancreas.

EMULSIN. See Beta-glucosidase.

ENDOGLUCONASE. See Cellulase.

ENTEROKINASE

The following are the actions and uses of enterokinase:
• Breaks down proteins.
• Used as a digestive aid.
• Used in pancreatic enzyme replacement therapy.

This enzyme is best taken orally.

SUPPLEMENT SOURCES	
Animal	Porcine or bovine intestine (duodenum).

ESTERASE

The following are the actions and uses of esterase:
• Breaks down ester bonds.
• Used as a digestive aid.

This enzyme is best taken orally.

SUPPLEMENT SOURCES	
Animal	Bovine and porcine.
Microbial	*Aspergillus niger.*

FOOD SOURCES	
	Found in many plant foods.

EXOGLUCONASE. See Cellulase.

FIBRINOLYSIN. See Plasmin.

FICIN (or Ficain)

The following are the actions and uses of ficin:
• Action similar to papain.
• Breaks down proteins.
• Used as a digestive aid.
• Fights inflammatory activity.
• Reduces swelling (edema).
• Fights intestinal worms in veterinary care.

This enzyme is best taken orally.

SUPPLEMENT SOURCES	
Plant	Latex of the fig tree *[Ficus sp.].*

FOOD SOURCES	
	Figs.

GLUCOAMYLASE

The following are the actions and uses of glucoamylase:
• Breaks down carbohydrates, specifically polysaccharides.
• Used as a digestive aid.

This enzyme is best taken orally.

SUPPLEMENT SOURCES	
Microbial	*Aspergillus niger, Aspergillus oryzae, Rhizopus oryzae,* and *Rhizopus niveus.*

GLUTATHIONE PEROXIDASE

The following are the actions and uses of glutathione peroxidase:
• Potent antioxidant enzyme.
• Requires selenium.

This enzyme is best taken orally.

SUPPLEMENT SOURCES	
Animal	Bovine liver.

HEMI-CELLULASE

The following are the actions and uses of hemi-cellulase:
• Breaks down carbohydrates, especially polysaccharides,

such as hemi-celluloses, which are found in plant foods.
• Used as a digestive aid.
• Used in pancreatic enzyme replacement therapy.

This enzyme is best taken orally.

SUPPLEMENT SOURCES	
Microbial	Aspergillus niger and Trichoderma longibrachiatum.
FOOD SOURCES	
	Green plants and plant seeds.

HYALURONIDASE

The following are the actions and uses of hyaluronidase:
• Breaks down fibrinous scar tissue formation in connective tissue and skin.
• Used to treat acute myocardial infarction (to decrease the severity of myocardial necrosis).
• Used in the prevention of hematomas and in specialized orthopedic, ophthalmologic, and dental problems.
• *Injectable enzymes should be used with caution. May cause severe, life-threatening side effects.*

This enzyme is best taken by injection or topically.

SUPPLEMENT SOURCES	
Animal	Bovine testes.

INVERTASE (or Beta-fructofuranosidase, Saccharase)

The following are the actions and uses of invertase:
• Breaks down carbohydrates, especially sucrose.
• Used as a digestive aid.

This enzyme is best taken orally.

SUPPLEMENT SOURCES	
Microbial	Aspergillus oryzae and Saccharomyces sp. (Kluyveromyces).
FOOD SOURCES	
	Cucumbers, green plants, potato, and sugar cane.

KALLIKREIN (or Kininogenin)

The following are the actions and uses of kallikrein:
• Action is similar to trypsin.
• Breaks down protein.
• Used as a digestive aid.
• Used as a vasodilator to lower blood pressure.
• Improves capillary permeability.
• Improves blood flow in all cases of coronary artery and peripheral vascular diseases, migraine headaches, fractures, and delayed wound healing.

• Used in treatment of idiopathic infertility in men.

This enzyme is best taken orally.

SUPPLEMENT SOURCES	
Animal	Porcine pancreas.

LACTASE (or Beta-galactosidase) (Contained in Lactaid.)

The following are the actions and uses of lactase:
• Breaks down lactose (milk sugar).
• Used as a digestive aid.
• Used to treat lactase insufficiency.

This enzyme is best taken orally.

SUPPLEMENT SOURCES	
Animal	Bovine liver.
Microbial	Aspergillus niger, Aspergillus oryzae, Saccharomyces lactis, Candida pseudotropicalis, Kluyveromyces lactis, and E. coli.
FOOD SOURCES	
	Tomatoes, persimmons, apples, peaches, almonds, and milk.

LIPASE

The following are the actions and uses of lipase:
• Breaks down lipids and improves fat utilization in the body, and thus improves poor fat digestion in cases of lipid malabsorption due to liver or gall bladder insufficiency or surgical intervention.
• Used as a digestive aid.
• Used in pancreatic enzyme replacement therapy.
• Decreases fat level in stools.
• Synergistically intensifies activity lipase in blood.

This enzyme is best taken orally.

SUPPLEMENT SOURCES	
Animal	Bovine and porcine pancreas and calf or lamb forestomachs.
Microbial	Rhizopus arrhizus, Rhizopus japonicus, Aspergillus oryzae, Aspergillus niger, Candida rugosa (formerly cylindracea), Staphylococcus aureus, Welchia perfringens, Candida paralipolytica, Mycotorula lipolytica, Geotrichum candidum, Pseudomonas, Rhizomucor (mucor) miehei, and Chromobacter viscosum.
FOOD SOURCES	
	Avocado, wheat germ, rice, maize, green plants, soybeans, coconuts, flaxseeds, rape seeds, corn, and other germinating plants containing relatively large amounts of fats.

LYSOZYME (or Muramidase)

The following are the actions and uses of lysozyme:
- Breaks down carbohydrates.
- Has antibiotic properties.
- Used to stimulate immune activity and potentiate antibiotic therapy.
- Fights viral infections.
- Used with cancer patients for its analgesic effect.
- Used in cold medicines often in combination with herbs, vitamins, and minerals.
- Relieves dry throats due to colds, flu, and strep throat.
- Used to break down mucus.
- Injectable enzymes should be used with caution. May cause severe, life-threatening side effects.

This enzyme is best taken orally and by injection.

SUPPLEMENT SOURCES	
Animal	Hen egg white.
FOOD SOURCES	
	Fresh juice of carica papaya, ficus, bromeliads, and red lingzhi (reishi) mushrooms.

MALTASE (or Alpha-glucosidase)

The following are the actions and uses of maltase:
- Breaks down the carbohydrates maltose and starch.
- Used as a digestive aid.

This enzyme is best taken orally.

SUPPLEMENT SOURCES	
Plant	Barley (Hordeum vulgare).
Microbial	Aspergillus niger, Saccharomyces sp., and various yeast species.
FOOD SOURCES	
	Usually found in plant tissues that contain amylase, such as brewer's yeast, rice, barley and other grains, beet leaves, green plants, sugar cane, banana, and mushrooms.

NUCLEASE

The following are the actions and uses of nuclease:
- Breaks down nucleic acids.
- Used as a digestive aid.
- Requires zinc to function.

This enzyme is best taken orally.

SUPPLEMENT SOURCES	
Microbial	Aspergillus oryzae.
FOOD SOURCES	
	Mung beans.

PANCREATIN

The following are the actions and uses of pancreatin:
- Breaks down proteins, carbohydrates, and fats because it contains many enzymes, principally trypsin, chymotrypsin, amylase, and lipase. There are at least four proteolytic enzymes (with different substrates) found in pancreatin.
- Used as a digestive aid.
- Used to treat pancreatic insufficiency or after pancreas removal.
- Used to treat steatorrhea (excessive fat in stools due to malabsorption).
- Used after gastrectomy.
- Used in the treatment of cystic fibrosis.
- Calcium can increase activity.

This enzyme is best taken orally.

SUPPLEMENT SOURCES	
Animal	Bovine or porcine pancreas.

PANCRELIPASE

The following are the actions and uses of pancrelipase:
- Contains principally protease, amylase, and lipase enzymes, so breaks down proteins, carbohydrates, and fats.
- Similar to pancreatin, but with a higher ratio of lipase.
- Used as a digestive aid.
- Used to treat pancreatic insufficiency.
- Used in treatment of chronic pancreatitis, pancreatectomy, cystic fibrosis, and steatorrhea.

This enzyme is best taken orally.

SUPPLEMENT SOURCES	
Animal	Porcine pancreas.

PAPAIN

The following are the actions and uses of papain:
- Similar to chymotrypsin in actions and uses.
- Breaks down proteins (although other components degrade fats and carbohydrates).
- Used as a digestive aid (especially to digest protein-rich foods).
- Treats chronic diarrhea and celiac disease.
- Used in pancreatic enzyme replacement therapy.
- Treats gastrointestinal discomfort due to intestinal parasites (nematodes).
- Used as a sedative.
- Used as a diuretic.
- Fights allergies, infections, and inflammation.
- Treats soft tissue injuries, including strains, sprains,

hematomas, contusions and abrasions, acute athletic injuries, charley horse, and pulled muscles.
• Used in skin-care products, such as face creams, cleansers, moisturizers, exfoliants, and face-lift formulations; treats psoriasis, warts, corns, skin cancer, and various skin ailments.
• Used in many types of surgery to decrease inflammation, pain, and swelling.
• Treats ureteral obstruction, peritoneal adhesions, and children's enteritis.
• Treats infected wounds, sores, ulcers, tumors, hay fever, and catarrh.
• Used to treat intoxication caused by stings of insects and jellyfish.
• Used in dentrifices.
• Used in ophthalmology to prevent corneal scar malformation.
• Used to accelerate wound healing.

This enzyme is best taken orally, topically, and as a retention enema.

SUPPLEMENT SOURCES	
Plant	Latex of the unripe papaya (Carica papaya).
FOOD SOURCES	
	Papaya.

PECTIN DEPOLYMERASE. See Pectinase.

PECTINASE (or Polygalacturonase, Pectin depolymerase)

The following are the actions and uses of pectinase:
• Breaks down carbohydrates, such as pectin (found in many fruits, such as apples).
• Used as a digestive aid.

This enzyme is best taken orally.

SUPPLEMENT SOURCES	
Microbial	Aspergillus niger and Rhizopus oryzae.
FOOD SOURCES	
	Citrus fruit, cucumbers and green plants.

PEPSIN

The following are the actions and uses of pepsin:
• Breaks down protein.
• Used as a digestive aid, especially when pepsin secretion is deficient (an acid solution may be used in pepsin's administration to increase gastric juice's effectiveness).
• Used in pancreatic enzyme replacement.

This enzyme is best taken orally.

SUPPLEMENT SOURCES	
Animal	Porcine stomach.

PEPTIDASE. See Proteases.

PEROXIDASE (See also Glutathione Peroxidase.)

The following are the actions and uses of peroxidase:
• One of the most effective antioxidant enzymes.
• Capable of causing low-density lipoprotein oxidation.
• May help fight idiopathic Parkinsonism.

This enzyme is best taken orally.

SUPPLEMENT SOURCES	
Plant	Horseradish roots and soybeans.
FOOD SOURCES	
	Horseradish, peas, oats, apple, egg, grape, mango, milk, and sugar cane. Widely found in fruits and vegetables.

PHYTASE

The following are the actions and uses of phytase:
• Breaks down carbohydrates, specifically phytates (phytic acid), present in the leaves of plants.
• Used as a digestive aid because it improves protein digestion and digestive enzyme activities.
• Can increase mineral absorption and the bioavailability of iron, zinc, calcium, and magnesium.
• Can contain other enzymes, such as cellulase, pectinase, and xylanase.

This enzyme is best taken orally.

SUPPLEMENT SOURCES	
Microbial	Aspergillus oryzae, Aspergillus niger, and Aspergillus ficuum.
FOOD SOURCES	
	Wheat.

PLASMIN (or Fibrinolysin)

The following are the actions and uses of plasmin:
• Breaks down proteins.
• Converts fibrin into soluble products.
• Used to treat acute peripheral arterial embolism, thrombosis (including deep vein), myocardial infarction, and pulmonary embolism.
• Used to break up blood clots.
• Injectable enzymes should be used with caution. May cause severe, life-threatening side effects.

This enzyme is best taken by injection.

SUPPLEMENT SOURCES	
Animal	Bovine or porcine sources.

POLYGALACTURONASE. See Pectinase.

PROTEASES (See also Bromelain, Chymopapain, Chymotrypsin, Papain, Pancreatin, and Trypsin.)

The following are the actions and uses of proteases:
• Breaks down proteins.
• Used as a digestive aid.
• Used in pancreatic enzyme replacement therapy.
• Fights inflammation.
• Fights acute conditions, such as sports injuries, surgery, wounds.
• Fights chronic conditions, such as cancer and arthritis.

This enzyme is best taken orally.

SUPPLEMENT SOURCES	
Animal	Bovine and porcine pancreas.
Plant	Various, including pineapple, papaya.
Microbial	*Aspergillus oryzae, Aspergillus niger, Aspergillus melleus, Aspergillus saitoi, Rhizopus niger, Bacillus subtilis, Bacillus licheniformis, Streptomyces caespitosus, Bacillus polymyxa, Streptomyces griseus, Staphylococcus aureau, Aspergillus sojae,* and *Serratia species.*
FOOD SOURCES	
	Pineapple, papaya, figs, guava, kiwi, ginger root, green plants, mushrooms, soybean, wheat, kidney bean.

RENNIN (or Chymosin, Rennase)

The following are the actions and uses of rennin:
• Breaks down proteins.
• Used as a digestive aid.
• Coagulates (curdles) milk and converts casein (the principal protein in milk) into insoluble curds that can be further digested by pepsin.
• Releases the valuable mineral elements (potassium, phosphorus, calcium, and iron) in milk. By doing this, the body can use rennin to strengthen bones and teeth, to stabilize water balance, to build nutrient-rich red blood cells in the circulatory system, and to aid in thinking more clearly by strengthening your nervous system.

This enzyme is best taken orally.

SUPPLEMENT SOURCES	
Animal	Calf stomach.

RIBONUCLEASE

The following are the actions and uses of ribonuclease:
• Used as a digestive aid (liberates nucleic acid).

This enzyme is best taken orally.

SUPPLEMENT SOURCES	
Animal	Bovine pancreas and calf thymus.
Microbial	*Aspergillus oryzae* and *Bacillus subtilis.*

SACCHARASE. See Invertase.

SERRATIOPEPTIDASE (or Serratia Protease)

The following are the actions and uses of serratiopeptidase:
• Breaks down proteins.
• Has fibrinolytic and anti-edema activity.
• Fights inflammation.
• Stimulates immune activity.
• Reduces sputum viscosity.
• Used in treatment of arthritis, fibrocystic breast disease, carpal tunnel syndrome, atherosclerosis, sinusitis, bronchitis, tuberculosis, bronchial asthma, cystitis, epididymitis, allergies, psoriasis, uveitis, ulcerative colitis, multiple sclerosis, some forms of cancer, bronchopulmonary secretions, traumatic injury (i.e., sprains and torn ligaments), and vaginal hysterectomy.
• Facilitates effects of antibiotics in treatment of infections.

This enzyme is best taken orally.

SUPPLEMENT SOURCES	
Microbial	*Serratia* sp.

SFERICASE

The following are the actions and uses of sfericase:
• Breaks down protein.
• Used to treat acute and chronic inflammation.
• Used to treat bronchitis, pneumonia, and sinusitis and other bronchopulmonary diseases, since it reduces viscosity of sputum.
• Requires calcium for activity.

This enzyme is best taken orally.

SUPPLEMENT SOURCES	
Microbial	*Bacillus sphaericus* (bacteria).

STREPTODORNASE

The following are the actions and uses of streptodornase:

- Breaks down fibrin.
- Fights inflammation.
- Used to treat cystic bronchiectasis (chronic dilatation of the bronchi).
- Used in thrombolytic therapy.
- Speeds local wound healing.
- *Injectable enzymes should by used with caution. May cause severe, life-threatening side effects.*

This enzyme is best taken by injection.

SUPPLEMENT SOURCES	
Microbial	*Streptococcus* sp.

STREPTOKINASE

The following are the actions and uses of streptokinase:
- Breaks down protein.
- Fights inflammation, swelling (edema), hematoma, and pain.
- Fights rheumatoid arthritis.
- Fights soft tissue injuries, such as surgical wound infections, leg ulcers, dermal ulcers, acne vulgaris, abscesses, and cellulitis.
- Debrides second- and third-degree burns.
- Treats respiratory conditions, such as bronchiectasis and chronic bronchitis.
- Fights circulatory problems including thromboembolism and inflammation, retinal vein thrombosis due to diabetes, thrombophlebitis (acute and chronic), and hematoma.
- Exhibits antithrombin and fibrinolytic activity.
- *Injectable enzymes should be used with caution. May cause severe, life-threatening side effects.*

This enzyme is best taken by injection.

SUPPLEMENT SOURCES	
Microbial	*Streptococcus* sp.

SUCRASE (or Sucrose alpha-glucosidase, Sucrase isomaltase)

The following are the actions and uses of sucrase:
- Breaks down carbohydrates, specifically sucrose and maltose.
- Used as a digestive aid.
- A liquid form of sucrase has shown effectiveness in enzyme replacement therapy for those suffering from congenital sucrase-isomaltase deficiency.

This enzyme is best taken orally.

SUPPLEMENT SOURCES	
Microbial	*Saccharomyces* sp.

FOOD SOURCES
Green plants, beet leaves and stems, and banana.

SUCROSE ALPHA-GLUCOSIDASE. See Sucrase.

SUPEROXIDE DISMUTASE (SOD)

The following are the actions and uses of superoxide dismutase:
- Breaks down superoxide.
- Requires copper and zinc to function.
- Fights inflammation.
- Fights cataracts.
- Used to treat temporomandibular joint dysfunction (by injection).
- One of the most potent antioxidant enzymes.
- *Injectable enzymes should by used with caution. May cause severe, life-threatening side effects.*

This enzyme is best taken orally and by injection.

SUPPLEMENT SOURCES	
Animal	Bovine liver and kidney.
Microbial	*Bacillus stearothermophilus* and *E. coli*.

FOOD SOURCES
Horseradish root, green pea, wheat germ, avocado, peas, regular and sweet potatoes, spinach, tomatoes, and numerous other fruits, vegetables, and grains.

TRYPSIN

The following are the actions and uses of trypsin:
- Breaks down proteins.
- Used as a digestive aid.
- Aids gastric retention due to malfunctioning stomach, pancreatic insufficiency, and intestinal obstruction.
- Used in debridement of ulcerations, empyemas, fistulas, necrotizing wounds, abscesses, hematomas, and decubitus ulcers.
- Used as an auxiliary agent in meningitis therapy.
- Fights inflammation as occurs in intercostal neuritis, urticaria (hives), postoperative parotitis, decubitus ulcers, pleural effusion, infected wounds, old scars, pancreatitis, trench mouth, and eczematoid dermatitis.
- Treats circulatory problems, such as thromboembolic diseases, peripheral arteriosclerosis, pulmonary infarcts, peripheral vascular diseases, milk leg (phlegmasia alba dolens), and ischemic purulent leg ulcers.
- Assists in surgical repair, reduces postoperative swelling from many surgeries.
- Speeds healing (inflammation, swelling) from injuries,

including strains, sprains, contusions, fractures, black eyes, bruises, hematomas, tendonitis, and bursitis.
- Treats arthritis, such as acute rheumatoid arthritis and acute gouty arthritis.
- Treats skin disorders, including atopic dermatitis, pustular eczema, and acute eczematoid dermatitis.
- Helps respiratory and throat conditions, including influenza, bronchitis, tuberculous adenitis, lung abscess, infected bronchopleural fistula, bronchial asthma, sinusitis, peritonsillar abscess, pulmonary diseases, pulmonary emphysema, unresolved pneumonia-atelectasis, pulmonary tuberculosis, traumatic hyperemia, and viral pneumonia.
- Treats urogenital conditions, including lymphogranuloma venereum, acute gonorrheal urethritis, and proteus vulgaris infection of urinary tract.
- Used in eye care, such as in glaucoma, acute iridocyclitis, and thrombosis of central retinal vein.
- Used in diabetes management to prevent such problems as cellulitis, infected leg ulcers, and carbuncle.
- Helps fight cancer and such associated problems as ascites due to cancer or cirrhosis and lymphosarcoma with infection.
- One of at least four proteolytic enzymes found in pancreatin.
- *Injectable enzymes should by used with caution. May cause severe, life-threatening side effects.*

Best Way to Take: Orally, topically, by injection, or by retention enema.

SUPPLEMENT SOURCES	
Animal	Bovine or porcine pancreas.

UROKINASE (Plasminogen activator)

The following are the actions and uses of urokinase:
- Breaks down proteins.
- Used in thrombolytic therapy.

- Treats pulmonary embolism.
- Treats acute myocardial infarct.
- Treats occlusion of retinal vessels.
- Injectable enzymes should be used with caution. May cause severe, life-threatening side effects.

This enzyme is best taken by injection.

SUPPLEMENT SOURCES	
Animal	Human urine.

You are dealing with an extremely powerful genie in a bottle when using enzyme supplements. The healing benefits of enzymes have been documented for centuries. The Greeks, Romans, Egyptians, Chinese, Germans, and Japanese have all used enzymes to maintain health and fight disease.

Because of our enzyme-dead, processed foods, polluted air and water, and high-stress lifestyles, supplements may be the only way we can ensure a sufficient daily enzyme intake. So, who needs enzyme supplements? Everyone. But especially anyone who:

- Eats cooked food.
- Eats fast food.
- Has poor eating habits.
- Is overweight.
- Is underweight.
- Wants to stay energetic.
- Suffers from stomach upset.
- Wants to stay "young" and live longer.
- Wants to fight diseases, both acute and chronic.
- Cares about nutrition and their health.
- Wants to recover faster from injury or surgery.

Read on for more information on supplements that will increase your enzymes' effectiveness in Chapter 6—Enzyme Helpers.

6. ENZYME HELPERS

Enzymes are essential for every plant and animal to function. However, vitamins and minerals are required if enzymes are to do their jobs effectively. Other helpers found in functional foods, phytochemicals, glandulars, herbs, metabolites, and amino acids, can also benefit your body's enzymes. This chapter will discuss these nutrients, how they help enzymes (and how enzymes help them), their functions, and their sources (both in foods and in supplements).

You can stimulate enzyme activity by making the required vitamins and minerals available to the body by eating enzyme-, vitamin-, and mineral-rich fresh foods, or by taking commercial supplements. In this way, enzymes and other nutrients are under your control. Although this is especially important for individuals with increased enzyme needs (such as the elderly, those under stress, and those with chronic disorders or acute injuries), we all need enzyme helpers. Unfortunately, food preservation processes strip our foods of these helpers. Further, because of poor soil quality and environmental toxins, even "fresh" vegetables and fruits may be deficient in the vitamins or minerals your body's enzymes need (not to mention that the enzymes in these fruits or vegetables may be dead). Therefore, increased supplementation is necessary.

The Recommended Dietary Allowances (RDAs) are the levels of essential nutrients that the Food and Nutrition Board believes are adequate to meet the nutrient needs of almost all healthy people. The RDAs are established for calories, carbohydrates and fiber, lipids, and most vitamins and minerals. Unfortunately, these recommendations are not intended to address the special nutritional needs of people with individual differences in metabolism or health situations (which may require restricted intakes or increased supplemental intake), nor have any been established for enzymes, herbs, or amino acids.

Remember, the Recommended Dietary Allowances were designed to be met through dietary intake from natural foods (not supplements). However, as I have mentioned previously, the typical American diet isn't as healthy as it should be. Because of our fast-, processed-food society, our diets can be lacking in essential and helpful nutrients. Supplements can help. Although earlier generations survived without supplements, our forefathers ate three square meals a day with healthy portions of fresh fruits and vegetables. Today, we grab our food on the run, eat one meal a day, or stuff ourselves in the evening, thus decreasing efficient digestion.

To maintain good health and proper body function, some amounts of supplemental vitamins, minerals, amino acids, and enzymes are essential. Ideally, a well-balanced diet would be sufficient. However, the fact is that most of us do not eat well-balanced diets. Further, deficiencies, illness, aging, injuries, and stress all take their toll and demand an increased supply of enzymes, amino acids, vitamins, and minerals to meet our bodies' requirements. In addition, as we age, our bodies lose their abilities to absorb and use nutrients as efficiently as in younger years. Once again, supplements seem critical.

We need supplementation for a number of reasons—to ensure an adequate nutrient intake to maintain wellness, to aid in repair (from injuries, illnesses, and stress), and to allow for individual differences and deficiencies due to problems in digestion or absorption.

VITAMINS (COENZYMES)

The human body works at its most optimal level when every system is working at peak efficiency. But nothing in the body works alone. Every system in the body requires enzymes and enzyme "helpers" to do its job. Many vitamins (or *coenzymes*) serve as enzyme helpers.

Our bodies cannot manufacture coenzymes. They must be obtained from our food or through supplements. Coenzymes in humans are usually the B vitamins. These vitamins serve as helpers to enzymes by releasing energy from carbohydrates, fat, and protein. Enzymes help vitamins to function, and at the same time, vitamins help enzymes in their functions. This is an example of synergistic action.

Coenzymes and cofactors (minerals) are consumed by the body when doing their jobs, so without constant replacement you can suffer a deficiency in a vitamin or mineral. This will mean that certain enzymes in your body cannot and will not function. The result? Disease!

Vitamins are organic food substances that are required for a wide variety of bodily activities. These include:

• Blood coagulation.
• Antibody formation.
• Growth and reproduction.
• Bone and tooth formation.
• Resistance to infection.
• Formation and maintenance of skin and nerves.

Vitamins are grouped according to their solubility in water or fat. The water-soluble vitamins (the B vitamins and vitamin C) are absorbed from the intestine and excreted in urine. This is why they must be replenished daily.

The fat-soluble vitamins (vitamins A, D, E, and K) are absorbed, transported, and metabolized along with fat. So a low fat intake may interfere with the absorption and utilization of these vitamins. Unlike water-soluble vitamins, fat-soluble vitamins are not excreted (to any great extent) in urine. Instead, they are either eliminated in the feces, or are absorbed and stored in fat and liver tissue. And since excretion of these vitamins is minimal, caution should be exercised when taking fat-soluble vitamin supplements, as toxicity can result. Follow directions on the label for dosage information.

Although the fat-soluble vitamins are not usually considered to be coenzymes, they are nevertheless involved in many bodily functions. Because they are essential for life and act synergistically in the body with other nutrients, I will consider them *indirect coenzymes*.

Vitamin A and Beta-Carotene

Vitamin A is a fat-soluble vitamin that is an indirect enzyme helper, required by digestive glands to produce enzymes. Beta-carotene and other carotenoids are precursors of vitamin A—the body converts them into vitamin A. When vitamin A is taken in the form of beta-carotene and other carotenoids, it is nontoxic. Vitamin A and its precursors promote healthy vision, mucous linings, and skin and cell membrane structure; a strong immune system; growth; hair, bone, teeth, and nail development; and normal taste, appetite, and hearing. It is a potent antioxidant and cancer fighter. It is needed by digestive glands to produce enzymes. In one study, a vitamin A deficiency reduced the activity of disaccharidase, an enzyme in the small intestine that plays a role in carbohydrate digestion.

Food Sources

Active vitamin A is found only in animal sources. Fish liver oils from cod, halibut, salmon, and shark are particularly high in vitamin A. It can also be found in dairy products (such as milk, cheese, and butter), egg yolks, liver, and fish (such as herring, sardines, and tuna). Beta-carotene (provitamin A) and other carotenoids can be found in dark green and yellow-orange fruits and vegetables, such as broccoli, carrots, sweet potatoes, yellow squash, tomatoes, cantaloupe, apricots, peaches, pumpkins, spinach, kale, and collard greens.

Supplements

Preformed vitamin A is taken from fish liver oil or lemon grass oil, or is made from synthetic forms. It is available in liquid or capsules. Emulsified vitamin A (also called retinol palmitate) is an excellent form because chance of toxicity is reduced, even at high levels of intake (available in capsule or cream). This form is water-soluble and is preferable for people who have difficulty absorbing fats. It should be taken with lipase. Beta-carotene is usually taken from palm oil or from the algae Dunaliella, or it is made synthetically.

Vitamin B$_1$ (Thiamin)

Thiamin is a water-soluble vitamin. As part of TPP (thiamin pyrophosphate) it serves as a coenzyme by assisting many enzymes in energy metabolism. It is essential for carbohydrate metabolism, proper energy production in the brain, nerve function, appetite, and heart function. It supports muscles, hair, eyes, and hearing.

Food Sources

Organ meats are particularly high in thiamin. Other good sources include carbohydrate-rich foods (such as legumes, whole grains, enriched breads and cereals, seeds, nuts, and the germ of cereal grains), brewer's yeast, and pork.

Supplements

Thiamin supplements can be found as thiamin hydrochloride, thiamin mononitrate, and thiamin pyrophosphate. They can be found in both capsule and tablet form.

Vitamin B$_2$ (Riboflavin)

Riboflavin is a water-soluble vitamin that the body requires to help enzymes break down starches and sugars to receive energy. The coenzyme forms of riboflavin are FMN (flavin mononucleotide) and FAD (flavin adenine dinucleotide).The enzyme glutathione reductase requires FAD as a coenzyme. Riboflavin supports vision, mucous membrane integrity, and healthy skin, nails, and hair. It is involved in oxidation-reduction reactions—the complex respiratory processes occurring in the mitochondria of cells.

Food Sources

Food sources of riboflavin include meat (especially organ meats); milk; yogurt; cheese; eggs; whole grains; leafy, green vegetables; and beans.

Supplements

Riboflavin can be found in supplement form as riboflavin or riboflavin 5-phosphate. It is sold in tablet and capsule form.

Vitamin B$_3$ (Niacin and Niacinamide)

Vitamin B$_3$ is a water-soluble vitamin that is a component of the coenzymes NAD (nicotinamide adenine dinucleotide) and NADP (nicotinamide adenine dinucleotide phosphate). These coenzymes take part in numerous metabolic activities, including energy production, production of some hormones, and blood sugar regulation. Niacin is involved in the metabolism of fats and, therefore, can be used to reduce cholesterol and triglyceride levels in the the blood. (Niacinamide, however, is not effective in this capacity.) Vitamin B_3 also stimulates gastric juices and hydrochloric acid; maintains the brain and nervous system; and supports healthy skin, liver, and gastrointestinal tract. It is also important in biological oxidation reduction.

Food Sources

Vitamin B_3 is widely distributed in animal and plant foods, including meats, milk, vegetables, such as legumes and green, leafy vegetables, whole grains (except corn), brewer's yeast, and even tea and coffee.

Supplements

Vitamin B_3 can be found in supplement form as niacin (nicotinate or nicotinic acid) or niacinamide (nicotinamide) in capsules or tablets.

Vitamin B_5 (Pantothenic Acid)

Pantothenic acid (from the Greek word meaning "universal") is a water-soluble vitamin that is converted in the body into coenzyme A, which is important in energy metabolism and the metabolism of protein, fats, and carbohydrates. It is involved in the synthesis of hormones, hemoglobin, steroids, and neurotransmitters. It is needed in the digestive system, skin, adrenal glands, and brain.

Food Sources

Food sources of pantothenic acid include meat, organ meats, fish, poultry, eggs, milk, cereal, whole grains, potatoes, legumes, and brewer's yeast.

Supplements

Pantothenic acid is available as calcium pantothenate and as pantethine. Supplements can be found in capsule and tablet form.

Vitamin B_6 (Pyridoxine)

Pyridoxine is a water-soluble vitamin. Its coenzyme forms are pyridoxal phosphate (PLP—used by over sixty enzymes) and pyridoxamine phosphate (PMP). Vitamin B_6 occurs in foods in one of three forms: pyridoxine (pyridoxol), pyridoxal, and pyridoxamine. All three can be converted to the coenzyme PLP (needed for amino acid metabolism, the conversion of tryptophan into niacin, the formation of antibod-

ies and hemoglobin, and the metabolism of protein and urea). Vitamin B_6 also assists the enzymes that metabolize amino acids and fats, is necessary for proper immune function, is involved in the synthesis of several neurotransmitters and neurohormones (including epinephrine, serotonin, dopamine, and melatonin), assists in the transport of potassium into cells, and is needed for healthy nerves, muscle, blood, and skin.

Food Sources

Food sources of vitamin B_6 include meat, organ meats, poultry, fish, fruits (especially bananas), whole grains, legumes, nuts, seeds, and brewer's yeast.

Supplements

Vitamin B_6 is available as pyridoxine hydrochloride and pyridoxal-5-phosphate in tablets or capsules.

Vitamin B_{12} (Cyanocobalamin)

Vitamin B_{12} is a water-soluble vitamin that is part of the coenzymes methyl-cobalamin and deoxyadenosylcobalamin, which play a part in the synthesis of new cells and DNA, assist in the breakdown of some amino acids and fatty acids; function in the gastrointestinal tract; and help maintain nerve cell function. Vitamin B_{12} is needed for the activation of folic acid (and vice versa), and it is essential for the synthesis of choline, methionine, and nucleic acids and in brain, nerve, and red blood cell function.

Food Sources

Food sources of vitamin B_{12} include meat, organ meats, poultry, fish, milk, eggs, and shellfish.

Supplements

Supplements of vitamin B_{12} can be found as cobalamine concentrate, cyanocobalamin, and hydroxycobalamin It can be found in capsule, tablet, sublingual, nasal drop, and injection form.

Folic Acid

Folic acid, or folate, is a B-complex, water-soluble vitamin. It is the basis of the coenzymes THF (tetrahydrofolate) and DHF (dihydrofolate), which are involved in the formation of new red blood cells and the synthesis of DNA, methionine and other amino acids, and choline. Folic acid is required for proper cell division, normal neural tube development in the fetus, and synthesis of nucleotides. It maintains good appetite and improves lactation. It is important for red blood cell formation, body growth, protein metabolism, reproduction, cell division, and hydrochloric acid production.

Food Sources

Food sources of folic acid include meat, liver and other organ meats, eggs, legumes, seeds, whole grains, wheat germ, brewer's yeast, mushrooms, nuts, asparagus, broccoli, lettuce, deep green, leafy vegetables, and lima beans.

Supplements

Folic acid supplements are available as folinic acid (5-methyl-tetra-hydrofolate) and folic acid (folate) in capsules and tablets.

Biotin

Biotin is a B-complex, water-soluble vitamin. It is a coenzyme for several enzymes, including acetyl-CoA carboxylase and many enzymes that play important roles in metabolism, glycogen and fatty acid synthesis, and amino acid breakdown. It is essential for Krebs cycle and energy production, and it fights baldness, arteriosclerosis, alcoholism, high cholesterol, ringing in the ears, constipation, eczema, dizziness, headaches, insomnia, hypoglycemia, and high blood pressure.

Food Sources

Food sources of biotin include egg yolk, milk, organ meats, beef, cauliflower, mushrooms, nuts, legumes, whole wheat, brewer's yeast, fruit, unpolished rice (rice before the hulls are removed), and soybeans.

Supplements

Biotin is available as d-biotin or as biocytin (from brewer's yeast) in tablets and capsules.

Para-Aminobenzoic Acid (PABA)

PABA is a B-complex, water-soluble vitamin. PABA plays an important role in protein breakdown and utilization and has a function in blood cell formation, intestinal health, and skin health. It decreases pain from burns and aids in restoring color to graying hair. PABA can be manufactured by intestinal bacteria and can stimulate the bacteria to produce folic acid (so it often occurs in combination with folic acid).

Food Sources

Food sources of PABA include brewer's yeast, liver, wheat germ, molasses, kidney, whole grains, rice bran, peas, green vegetables, egg yolks, peanuts, and beans.

Supplements

PABA can be found as para-aminobenzoic acid in tablets and capsules.

Inositol

Inositol is a B-complex, water-soluble vitamin that serves as a coenzyme in metabolism, working closely with pantothenic acid, PABA, folic acid, and vitamin B_6 in their various activities. It plays a role in promoting cell membrane integrity, it is involved in nervous system function, it lowers cholesterol, it has a calming effect, and it fights eczema. Inositol occurs in animal tissues as a component of phospholipids and in plant cells as phytic acid.

Food Sources

Food sources of inositol include animal sources, whole grains, brewer's yeast, molasses, seeds, nuts, legumes, citrus fruits, raisins, cantaloupe, peanuts, wheat germ, and milk.

Supplements

Inositol supplements are available as inositol monophosphate and inositol nicotinate in tablets, capsules, and powder.

Choline

Choline is a B-complex, water-soluble vitamin that serves as a coenzyme in metabolism and is involved in cholesterol and fat utilization. Choline is a basic component of lecithin (an emulsifying agent). It is essential for the health of the kidneys, liver, and the myelin sheath (the covering) of nerve fibers. It plays an important role in nerve impulse transmission, as it is essential for the synthesis of the neurotransmitter acetylcholine.

Food Sources

Grains, egg yolks, liver, legumes, wheat germ, brewer's yeast, and green leafy vegetables are food sources of choline.

Supplements

Choline is available as a soluble salt (as either choline dihydrogen citrate, choline chloride, or choline bitartrate) or as phosphatidylcholine (lecithin), each in tablets and capsules.

Vitamin C

Vitamin C is a water-soluble vitamin that serves as a coenzyme for a number of enzymes, including lysyl oxidase and proline hydroxylase—two enzymes that convert the amino acids lysine and proline into hydroxylysine and hydroxyproline, respectively, which help maintain the structure of collagen. Vitamin C helps the body to fight infections, diseases, allergies, the common cold, and cancer. It is essential to vascular function, wound healing, bone and joint formation, and red blood cell formation. It stimulates the adrenal glands to produce hormones (which also act as coenzymes), aids in detoxification, is a potent antioxidant, improves the function of many enzymes, is important for converting tryptophan into serotonin, enhances the absorption of iron, assists in converting cholesterol into bile salts, is required for converting folic acid into its active form, and is essential for collagen production.

Food Sources

Broccoli, Brussels sprouts, turnip greens, red and green peppers, black currants, and guava are particularly high in vitamin C. Other sources of vitamin C include citrus fruits (such as oranges, tangerines, grapefruits, lemons, and limes), rose hips (used in tea), apples, cantaloupe, strawberries, papayas, mangoes, persimmons, dark green vegetables, tomatoes, potatoes, and cauliflower.

Supplements

Vitamin C supplements are available as ascorbic acid, calcium ascorbate, and polyascorbate in tablets, capsules, and powder.

Vitamin D

Vitamin D is a fat-soluble indirect enzyme helper, as it is involved in regulating the transport and absorption of calcium and phosphorus. It plays a part in bone formation and resorption (loss of tissue), is involved in the processes of insulin secretion and the production of red blood cells, and is produced in the skin after exposure to sunlight.

Food Sources

Food sources of vitamin D include oily fish (such as sardines, herring, and salmon) and their oils, liver, butter, vegetable oils, egg yolks, and vitamin-D fortified milk.

Supplements

Vitamin D comes in the form of vitamin D_2 (ergocalciferol) and vitamin D_3 (cholecalciferol) and is sold in tablet, capsule, or liquid form.

Vitamin E

Vitamin E is a fat-soluble vitamin and is an indirect enzyme helper. It consists of eight substances, including alpha-tocopherol, beta-tocopherol, delta-tocopherol, and gamma-tocopherol. Of these, alpha-tocopherol is the most active. Vitamin E acts as an intercellular antioxidant, improves circulation, helps resist diseases at the cellular level, and aids in stimulating heart function. It fights arteriosclerosis, blood clots, varicose veins, phlebitis, thrombosis, diabetes, migraine headaches, allergies, sinusitis, baldness, menstrual and

menopausal disorders, myopia, stress, wounds, scars, wrinkles, and warts. Vitamin E works closely with the mineral selenium.

Food Sources

Food sources include vegetable oils, margarine, wheat germ, nuts, seeds, legumes, whole grains, green, leafy vegetables (such as spinach), brewer's yeast, Brussels sprouts, soybeans, and eggs.

Supplements

Vitamin E supplements can be found as d-alpha tocopherol, d-alpha tocopheryl acetate, d-alpha tocopheryl succinate, (dl-alpha tocopherol, dl-alpha tocopheryl acetate, dl-alpha tocopheryl succinate, and mixed tocopheryl in capsules, tablets, liquids, and emulsions. Emulsified vitamin E is an excellent form because the chance of toxicity is reduced, even at high levels of intake. Emulsified vitamin E is available in capsules or cream.

Vitamin K

Vitamin K is a fat-soluble vitamin that is important for normal blood clotting. It has a function in the formation of proteins containing gamma carboxyglutamic acid, it activates proteins in the bones and kidneys. It is a cosubstrate for certain liver enzymes that convert vitamin K-dependent protein precursors into the active forms; it activates energy-producing tissue; and it fights bleeding gums, varicose veins, ulcers, dizziness caused by high blood pressure, edema, miscarriages, rheumatism, hemorrhaging, bruising ulcers, celiac disease, menstrual problems, and postsurgical bleeding.

Food Sources

Green, leafy vegetables (such as spinach, turnip greens, and kale), kelp, cauliflower, broccoli, Brussels sprouts, tomatoes, and liver are the richest sources of vitamin K. Soybean, safflower, vegetable, and fish liver oils; pork; and egg yolks are also good sources.

Supplements

Vitamin K can be found as K_1 (phylloquinone), which is found in foods; K_2 (menaquinone), which is made by our intestinal bacteria; and K_3 (menadione), a synthetic form that is available only by prescription. Vitamin K supplements can be found in tablet, capsule, and liquid form.

MINERALS (COFACTORS)

Minerals perform a number of roles in the body. Some function as cofactors for enzymes, others regulate fluid and electrolyte balance or provide rigidity to the skeleton. Still others regulate the function of muscles and nerves. Minerals also work together with vitamins, hormones, peptides, and other substances to regulate the body's metabolism.

Just like vitamins, we must get our minerals from outside sources (through food or supplements) because our bodies cannot manufacture them. Minerals are found in a variety of foods, but only in limited amounts. Unfortunately, most of us don't get all the minerals we need. For example, although 300 enzymes require zinc as a coenzyme for optimal function, studies show that 68 percent of the American population consumes less than the RDA of zinc. Further, 74 percent of Americans take in too little magnesium, two out of three Americans consume less than the RDA of calcium, 81 percent of Americans consume less than the RDA of copper, 58 percent consume too little iron, and 50 percent consume less manganese than the RDA requires. In addition, oxalates (substances found in many fruits and vegetables that may bind to certain minerals) and phytates (substances found in the outer layer of cereal grains that may bind to certain minerals) may interfere with the absorption of many minerals. Some medications may interfere with mineral absorption and metabolism, leading to deficiencies, as can gastrointestinal disease, renal disease, or alcoholism. Thus, supplements may be necessary if we are to meet daily requirements because a deficiency in just one mineral can reduce or impair enzyme function, just as the body experiences disruption if you are deficient in just one enzyme.

Most minerals found in our foods are transported into the bloodstream and made bioavailable by a method known as *chelation*. Chelation is a process by which a mineral becomes surrounded by and bonded to amino acids. Chelation is one natural means for the body to transport minerals across the intestinal wall into the bloodstream, thus improving absorption.

In order for the body to absorb some mineral supplements, however, they must first undergo complex digestive and transport processes. This can decrease the rate of absorption and the resulting bioavailability of the mineral. One form of mineral supplement, called amino acid chelates or chelated mineral supplements, helps sidestep part of the digestive process. The body is very efficient at absorbing amino acids (of those broken down during digestion, approximately 95 percent are absorbed). So chelating minerals to amino acids allows them to be "smuggled" across the intestinal wall into the bloodstream.

Research shows that amino acid chelates are among the most highly bioavailable forms of minerals. This means that a greater amount of mineral is available to the body's cells when needed.

Calcium

Calcium is essential for the activity of certain enzymes, such as glycogen phosphorylase kinase (involved in energy pro-

duction), and adenosine triphosphatase (ATPase, involved in muscle contraction), and several enzymes involved in the healing process. It builds strong bones and teeth, maintains bone mass as we age, and plays important roles in muscle and nerve cell function. You need sufficient enzymes to generate calcium utilization, even if you have enough calcium.

Food Sources

Milk and other dairy products (such as yogurt, cheese, and cottage cheese) are excellent sources of calcium. Other calcium sources include meat, fish, eggs, beans, and fruits and vegetables (especially green, leafy vegetables).

Supplements

Calcium supplements may be found as calcium amino acid chelate, calcium alpha keto glutarate, calcium ascorbate, calcium aspartate, calcium carbonate, calcium gluconate, calcium lactate, calcium hydroxyapatite, calcium lysinate, calcium malate dihydrate, calcium succinate, calcium glycinate, calcium citrate, calcium fumarate, calcium yeast, or dolomite (a combination of calcium and magnesium) in capsules or tablets.

Chromium

Chromium is an indirect enzyme helper because it stimulates the activity of enzymes involved in cholesterol and fatty acid synthesis and of those involved in glucose metabolism. It is essential to human nutrition and health, and is required for normal carbohydrate, protein, and fat metabolism; muscle development; energy production; and blood sugar regulation. Chromium works with the hormone insulin and plays a vital role as a glucose tolerance factor.

Food Sources

Liver, oysters and other seafood, red meats, chicken, black pepper, cheeses, nuts, whole grains, brewer's yeast, and beer are sources of chromium.

Supplements

Chromium supplements can be found in the form of chromium polynicotinate, (also called niacin-bound chromium) chromium dinicotinate, chromium picolinate, GTF (glucose tolerance factor) chromium, inorganic chromium salts (chromium oxide, chromium acetate, and chromium chloride), chromium amino acid chelate, chromium ascorbate, chromium aspartate, and chromium yeast in capsules or tablets.

Cobalt

Though cobalt is a mineral, it is also a component of vitamin B_{12}, and it activates several enzymes. It can substitute for zinc or manganese as a cofactor for many enzymes.

Food Sources

Vegetables, fruits, and buckwheat are good sources of cobalt.

Supplements

Cobalt supplements can be found in the form of cobalt yeast and are available in tablets.

Copper

Copper is a constituent of many enzymes involving reactions that consume oxygen or oxygen radicals, including superoxide dismutase (SOD). Copper plays a role in protein metabolism, phospholipid synthesis, and wound healing. It assists in the conversion of iron into hemoglobin and works with manganese in effecting an immune response and with zinc to keep the arteries flexible.

Food Sources

Meats and organ meats, shellfish (including oysters), nuts, legumes, and whole grains are good sources of copper.

Supplements

Copper supplements can be found as copper aspartate, copper amino acid chelate, copper salicylate, copper yeast, copper citrate, copper gluconate, and copper sulfate. They can be found in capsule or tablet form.

Iodine

Iodine is an indirect enzyme helper. It helps form the thyroid hormones triiodothyronine (T3) and thyroxine (T4) and is important in regulating energy control mechanisms in the body and the cellular metabolic rate.

Food Sources

Iodized salt, seafood, shellfish, cod liver oil, dairy products, and vegetables grown in iodine-rich soil are good sources of iodine.

Supplements

Iodine supplements may be purchased in tablets as sea kelp, or iodine yeast is available in capsule, tablet, or liquid form.

Iron

Iron is a part of several enzymes required for the production of energy in cells throughout the body, including cytochrome oxidase, catalase, and peroxidase, and it is involved in hemoglobin biosynthesis.

Food Sources

Meat and organ meats, shellfish, dried fruit, nuts, whole-grain or iron-enriched breads and cereals, egg yolks, blackstrap molasses, and legumes are good sources of iron.

Supplements

Iron supplements are available as iron sulfate, iron glycinate, iron ascorbate, iron aspartate, iron citrate, iron peptonate, iron yeast, ferrous fumarate (coated tablets or powder), ferrous gluconate, and ferrous succinate. These supplements can be found in capsule or tablet form.

Magnesium

Magnesium is involved in more than 300 enzyme systems. It plays important roles in protein building, muscle contraction, nerve function, energy production, calcium assimilation, and bone and tooth formation. It is necessary for glucose metabolism; the synthesis of fats, proteins, and nucleic acid; and blood clotting.

Food Sources

Milk and other dairy products; nuts; legumes; leafy, green vegetables; whole grain cereals; and seafood are good sources of magnesium.

Supplements

Magnesium supplements are available as magnesium amino acid chelate, magnesium alpha-ketoglutarate, magnesium ascorbate, magnesium aspartate, magnesium citrate, magnesium carbonate, magnesium malate, magnesium fumarate, magnesium gluconate, magnesium glycinate, magnesium lysinate, magnesium nicotinate, magnesium oxide, magnesium salicylate, magnesium stearate, magnesium succinate, magnesium yeast, mono magnesium di-L-aspartate. Supplements are available in capsules, tablets, and powder.

Manganese

Manganese serves both as part and activator of several enzymes, including arginase, ribonucleotide reductase pyruvate carboxylase, superoxide dismutase (SOD), glycosyltransferases, and enolase. It is necessary for normal neural function; for protein and carbohydrate breakdown; and for fatty acid, cholesterol, and hemoglobin synthesis. It assists in urea synthesis; prevents lipid peroxidation; and activates enzymes needed for vitamin C, biotin, thiamin, and choline use.

Food Sources

Whole grains; nuts; dried fruits; green, leafy vegetables; and tea are good sources of manganese.

Supplements

Manganese supplements can be found as manganese ascorbate, manganese amino acid chelate, manganese aspartate, manganese citrate, manganese gluconate, manganese sulfate, manganese picolinate, and manganese yeast in capsules and tablets.

Molybdenum

Molybdenum is part of a complex called molybdenum cofactor, which is required for the three mammalian enzymes sulfite oxidase (involved in sulfur-containing amino acid metabolism), aldehyde oxidase (catalyzes the conversion of aldehydes into acids), and xanthine oxidase (participates in purine—uric acid compound—metabolism). Molybdenum is a cofactor for several metalloenzymes and plays a role in iron metabolism

Food Sources

Milk and other dairy products, dried legumes, organ meats, and whole grains are sources of molybdenum.

Supplements

Molybdenum supplements can be found as molybdenum amino acid chelate, molybdenum ascorbate, molybdenum aspartate, molybdenum citrate, and molybdenum yeast in capsules and tablets.

Phosphorus

Phosphorus activates many enzymes and is required for carbohydrate, fat, and protein oxidation. It is necessary for bone and tooth formation, the maintenance of acid-base balance, energy production, cell permeability, and nerve and muscle activity. It is a component of proteins, phospholipids, nucleic acids, ATP (adenosine triphosphate), and phosphates. Many enzymes and the B vitamins become active only when a phosphate group is attached.

Food Sources

Meat, fish, poultry, dairy products, eggs, legumes, carbonated drinks, and whole grains are good sources of phosphorus.

Supplements

Phosphorus supplements can be found in the form of phosphorus proteinate and phosphorus yeast in tablets and capsules.

Potassium

Potassium is part of the enzyme pyruvate kinase. It is essential for normal growth. It maintains proper water balance between body fluids and cells and is important in nerve transmission, muscle activity, and protein synthesis. Potassium is necessary for cellular enzymes to work properly, and it works together with sodium to normalize heartbeat.

Food Sources

Sources of potassium include milk, meat, poultry, fish, whole grains, fruits (including oranges, grapefruit, bananas, prunes, cantaloupe, raisins) and vegetables (such as asparagus, potatoes, watercress, and peppers).

Supplements

Potassium supplements can be found in the form of potassium amino acid chelate, potassium ascorbate, potassium aspartate, potassium citrate, potassium fumarate, potassium gluconate, potassium magnesium aspartate, potassium malate, potassium salicylate, potassium succinate, and potassium yeast. Supplements can be found in the form of tablets and capsules.

Selenium

Selenium is best known for its role as part of the enzyme glutathione peroxidase, which works as an antioxidant. It is essential for some metabolic processes. It sometimes acts as an enzyme activator and sometimes as an enzyme inhibitor. It also works with vitamin E in its role as an antioxidant.

Food Sources

Organ meats (such as liver and kidney) and seafood are the richest sources of selenium, followed by cereal and grains, dairy products, and fruits and vegetables.

Supplements

Selenium supplements can be found as selenium amino acid chelate, selenium ascorbate, selenium aspartate, selenium citrate, selenium lysinate, and selenium yeast in tablet form.

Silica

Silica is required for the activity of several enzymes involved in bone, cartilage, and collagen synthesis. It is essential for proper growth and development and fights osteoporosis and atherosclerosis. It occurs in nature as silicon dioxide.

Food Sources

Sources of silica include whole grain cereal products, unrefined fiber, root vegetables, alfalfa, brown rice, mother's milk, horsetail grass, soybeans, leafy green vegetables, parsley, chasteberry, ginger root, and bell peppers.

Supplements

Silica supplements can be found as silica amino acid chelate and silicon yeast in tablets and capsules.

Sodium

Sodium is an indirect enzyme helper that works with potassium to normalize the heartbeat, help maintain cellular fluid balance, and aid in maintaining proper blood pH levels. However, when taken in excess, sodium can inhibit enzyme activity. It is involved in muscle contraction and nerve transmission.

Food Sources

Cured meats, pork, sardines, cheese, sauerkraut, green olives, and snack foods are rich sources of sodium.

Supplements

Sodium is available in buffered tablet form as sodium chloride, commonly termed "salt pills." However, most recieve more than enough sodium from their diets.

Zinc

Zinc is needed by over 300 enzymes. Many zinc enzymes work to metabolize carbohydrates, alcohol, and essential fatty acids; to synthesize proteins; to dispose of free radicals; and to manufacture heme (a constituent of hemoglobin). Zinc is needed for white blood cell immune function, wound healing, growth, development, and thyroid hormone function. It is a component of insulin, interacts with platelets in blood clotting, and affects behavior and learning. Zinc is needed to produce retinal (a pigment in the retina formed by the oxidation of vitamin A alcohols); is essential to normal taste, sperm production, and fetal development; and helps protect the body from heavy metal poisoning.

Food Sources

Food sources of zinc include red meats, seafood, wheat germ, legumes, and nuts. Plant sources are less bioavailable than animal sources because of the presence of phytic acid in plant sources, which can impair mineral absorption.

Supplements

Zinc supplements can be found as zinc amino acid chelate, zinc ascorbate, zinc aspartate, zinc citrate, zinc gluconate, zinc glycinate, zinc lysinate, zinc picolinate, zinc succinate, and zinc yeast in tablets and capsules.

Mineral Yeasts

Minerals are essential for the body to function properly. But according to Dr. T. Pearse Lyons, Ph.D., in *Biotechnology in the Feed Industry* (Nicholasville, Kentucky: Alltech Technical Publications, 1993), the key to a mineral's effectiveness is not only its biological availability, but also its biological activity. That is, in order to benefit the body, a mineral must not only be present in the bloodstream, it must also retain its activity and ability to function. Mineral yeasts are natural products, easily digested and rapidly absorbed into the body, that are not only biologically available, but biologically active as well. Mineral yeasts are produced by adding individual minerals directly to a live yeast fermentation. This process provides a reliable and excellent source of organically bound mineral yeast.

A few mineral yeasts include calcium yeast, chromium yeast, cobalt yeast, copper yeast, iodine yeast, magnesium yeast, manganese yeast, molybdenum yeast, potassium yeast, selenium yeast, and zinc yeast.

FLAVONOIDS

Another group of enzyme helpers are flavonoids, a family of naturally occurring compounds that can help the function of your enzymes. You may have heard them referred to as bioflavonoids, but the terms are virtually interchangeable. Flavonoids are sometimes referred to as "vitamin P," but strictly speaking, they are not true vitamins. However, since our body cannot produce flavonoids, they must be provided by our diets or in supplement form.

There are more than 2,000 individual members of the flavonoid group, widely distributed in the plant kingdom (especially in vegetables and fruits). They are found in the rinds of citrus fruits including oranges, grapefruit, and lemons; in cherries, prunes, apricots, and berries; in the seeds and skins of red grapes; and in red wine and green tea. They are usually found in flowering plants, clover blossoms, buckwheat, soybeans, rose hips, hawthorn berries, black currants, elderberries, peppers, shepherd's purse, horsetail, seeds of milk thistle, and leaves of the eucalyptus and pagoda trees.

Flavonoids are probably best known for their ability to improve capillary strength. Capillaries are tiny vessels that fan out throughout our bodies, supplying nutrients and removing wastes from every cell. When they weaken, nutrients can't get in and waste products can't get out—a traffic jam occurs. At this point, disease and other disasters are just around the corner.

Flavonoids are also effective antioxidants, and they help vitamin C by increasing its absorption and preventing it from being oxidized, thus preserving its action. Flavonoids have many therapeutic applications, among them are the treatment of heart disorders and circulatory problems, ulcers, diabetes related cataracts, injuries (including strains, sprains, and bruises), inflammation, allergies, viruses, cancer, high cholesterol, and liver disease.

Flavonoid supplements come in tablet form as individual supplements or in a number of combinations. These include flavonoids and enzymes (such as pancreatin, trypsin, chymotrypsin, bromelain, papain, and the flavonoid rutin), flavonoid mixtures (including rutin, quercetin, naringin, and hesperidin), flavonoids and vitamins and/or minerals, and flavonoids in combinations with other antioxidants. See Table 6.1 for information on some individual flavonoids and their therapeutic applications.

COENZYME Q_{10}

Another nutrient sometimes referred to as a vitamin is coenzyme Q_{10} (also known as CoQ_{10} or ubiquinone). CoQ_{10} is similar in structure to vitamin E but is not classified as a vitamin because it can be made in the body from the amino acids phenylalanine and tyrosine and vitamins B_1, B_6, E, and folic acid.

There are ten types of coenzyme Q, but CoQ_{10} is the only one found in humans. It occurs in especially high concentrations in the heart (ten times higher than in any other region of the body), the liver, the immune system, and elsewhere to a lesser degree. Unfortunately, as we age, our bodies make less and less of this important nutrient. In a personal interview, Dr. Stephen T. Sinatra, M.D., (a world-renowned expert on CoQ_{10}) stated, "I believe that coenzyme Q_{10} is perhaps one of the greatest medicinal discoveries of the 20th century. It has tremendous application in patients with congestive heart failure, systolic and diastolic dysfunction, cardiomyopathy, and high blood pressure. Coenzyme Q_{10} is also extremely important in acute ischemic syndromes, that is, unstable angina and myocardial infarction. I personally take coenzyme Q_{10} as a way of helping to prevent lipid peroxidation."

In addition, CoQ_{10} is a potent antioxidant, plays an important role in immune system function, and may be effective in treating AIDS, preventing heart disease, and generating energy. Various studies indicate that CoQ_{10} can improve physical performance in patients suffering from muscular dystrophies, can decrease blood pressure and total serum cholesterol, can limit the severity of ophthalmoplegia (eye muscle paralysis), and, when applied topically, can improve adult periodontitis. CoQ_{10} may help people live longer and have shinier hair and brighter eyes without the normal signs of advanced aging.

CoQ_{10} can be found in whole grains, organ meats, oily fish (such as sardines, mackerel, and salmon), nuts, and seeds. Humans and other vertebrates can also produce CoQ_{10} in their cells. It is available in capsules and softgels. Q-Gel Softsules is a special soft-gel form that provides high coenzyme Q_{10} serum levels.

FUNCTIONAL FOODS

For centuries, humans have used foods such as fruits, vegetables, herbs, grains, and animal products to keep well, fight disease, and extend life. Our forefathers believed that these foods contained mysterious ingredients capable of keeping us healthy or curing us when sick. Today, extensive research is identifying the components responsible for the natural, health-giving properties of these foods. Scientists can now concentrate and standardize extracts from these fresh foods. Called by a number of different terms, including functional foods, pharmafoods, nutraceuticals, and designer foods, these extracts are excellent sources of enzymes, vitamins, minerals, chlorophyll, protein, phytochemicals, hormones, and other nutrients. These functional food products are available as tablets, capsules, powders, granules, liquids, elixirs, lozenges, and extracts.

The universe has a natural order. When we deviate from, or interfere with, this order (as our modern lifestyles cause us to do), illness results. Foods contain components that will keep our bodies in balance and help rectify these disorders whenever they occur. Therefore, functional foods, which contain these health-giving components, are effective in fighting diseases such as cardiovascular disease, diabetes, hypertension, cancer, and arthritis, improving wellness, and extending life.

Functional foods are helpful to our enzymes in many ways. They often serve as a direct source of enzymes. They provide coenzymes and cofactors (vitamins and minerals) needed for the proper function of your body's enzymes. Functional foods contain nutrients known to "kick-start" your body's enzyme production or function. For instance, among other nutrients, plant functional foods contain phytochemicals, and animal-derived functional foods contain hormones necessary for your body's enzyme function.

Plant Functional Foods

Because plants are excellent natural sources of live enzymes, functional foods from plants are usually high in enzymatic activity and also contain enzyme helpers. Therefore, concentrates from these enzyme-rich foods, grown in pesticide-free, healthy environments, can be used to ensure wellness and fight disease.

To retain their enzyme activity, extracts from plants should be concentrated at temperatures that do not exceed normal body temperature. Osmotic pressure and rapid freeze-drying (processes not requiring excessive heat) are most effective in retaining the natural enzyme activity of fruits, vegetables, grains, and herbs.

The enzymes found in food concentrates will vary depending on whether extracted from the root, leaf, stem, or fruit of the plant. The garden beet is a good example. The enzymes amylase, maltase, and sucrase are found in the leaves, while amylase, emulsin (or beta-glycosidase), and inulase are found in the roots. Check your supplement label for the plant source. The following are a few enzymes found in concentrated plant extracts (information on the individual enzymes can be found in Chapter 5).

- Amylase (also called diastase).
- Bromelain.
- Catalase.
- Cellulase.

Table 6.1. Flavonoids and Their Actions

FLAVONOID	SOURCE	APPLICATIONS
Anthocyanidins	Flowering plants.	Potent antioxidants; fight capillary permeability.
Catechins	Red wine, green tea.	Potent antioxidants; reduce capillary permeability.
Epicatechins	Oolong and green tea.	Fight infection; lower blood pressure.
Genistein	Soybeans.	Improves night vision; potent cancer fighter and antioxidant.
Glycyrrhizins	Licorice.	Inhibit tumors; fight viruses, inflammation, and allergies; have interferon activity; inhibit cancer of the prostate; good for skin care.
Hesperidin	Citrus fruits, milk thistle seeds.	Protects capillaries; antioxidant; helps fight liver diseases; strengthens cell membranes; protects against ultraviolet rays.
Isoflavonoids	Soybeans.	Improve night vision; potent cancer fighters and antioxidants.
Naringin	Citrus fruits, milk thistle seeds.	Potent antioxidant; stabilizes cells; fights liver diseases.
Proanthocyanidins	Red grape skin extract, elderberry extract, hawthorn, berries, bilberry red cabbage.	Antioxidants; protect against cancer; fight high cholesterol and influenza; strengthen blood vessel cell walls.
Pycnogenol	Seeds and skins of grapes, inside portion of bark of Pinus Maritimus trees, which grow near Bordeaux, France.	Potent antioxidant; slows aging; fights arthritis, cancer, and heart disease.
Quercetin	Onions, broccoli, squash, apples, seeds and skins of grapes, inside portion of bark of pine trees, clover blossoms, red wine, green tea. Quercetin is one of the most abundant flavonoids.	Potent antioxidant; fights cancer and arthritis; stabilizes membranes; protects against heart disease, cataracts, and allergies; helps fight diabetes and its major complications; normalizes blood pressure; decreases production of harmful prostaglandins; helps lower cholesterol; slows aging; fights capillary fragility.
Rutin	White pulp of all citrus fruits, grapes, red wine, apricots, cherries, blackberries, and the leaves	Strengthens blood vessel walls; potent antioxidant; decreases bruising; fights colds, flu, asthma, inflammation, and vision; protects against
Rutin (continued)	of the pagoda and eucalyptus trees.	radiation; speeds healing of bleeding gums, varicose veins, hemorrhoids, edema, high blood pressure, and rheumatism.
Tannins	The seeds and skins of grapes, red wine.	Potent antioxidants; show antiviral activity.

- Beta-glycosidase or emulsin.
- Ficin.
- Glutathione peroxidase.
- Lipase.
- Maltase.
- Papain.
- Peroxidase.
- Protease.
- Superoxide dismutase.

As mentioned earlier in this chapter, functional foods serve as excellent sources of vitamins and minerals—the coenzymes and cofactors your enzymes require in order to function. Functional foods are sort of an "Irish stew" containing enzymes and their helpers all mixed into a healthy blend from Mother Nature.

For example, green barley extract (a popular functional food) contains vitamins B_1, B_2, B_3, B_6, C, E, biotin, folic acid, pantothenic acid, and choline, plus potassium, calcium, magnesium, phosphorus, manganese, zinc, and several amino acids. Another functional food, wheat grass extract, contains vitamin C, beta-carotene, several B vitamins, vitamin E, calcium, iron, sodium, potassium, magnesium, selenium, and amino acids (including alanine, arginine, aspartic acid, glutamic acid, glycine, proline, serine, and tyrosine). See inset on page 69 for different ways to use your powdered plant functional foods.

Phytochemicals in Plant Functional Foods

Phytochemicals are simply the beneficial naturally occurring chemicals found in plants ("phyto" means plant). These plant nutrients can be isolated and concentrated from plant foods. Phytochemicals are an exploding area of nutrition with new discoveries coming almost daily. All plants worthy of their names have energetic enzymes and feisty phytochemicals within their cells.

Some phytochemicals are effective antioxidants, others can help fight cancer, stimulate immune function, decrease cholesterol levels, reduce arterial plaque, stimulate production of enzymes inside your body, or block DNA damage at the cellular level. See Table 6.2 for some of the more commonly occurring phytochemicals found in food extracts and their sources and actions.

65

Table 6.2. Functional Food Phytochemicals

PHYTOCHEMICAL	PLANT SOURCE	DISEASE-FIGHTING PROPERTIES
Alkaloids	Goldenseal, cocoa beans, legumes, tea, coffee, alfalfa, and cranberry.	Prevent yeast overgrowth and maintain healthy bacteria levels in the gastrointestinal and urinary tracts; promote proper ecology in the colon; support immune system function.
Canthaxanthines	Paprika.	Potent antioxidants; fight cancer.
Carotenoids (including alpha- and beta-carotene, xanthophylls, cryptoxanthin, lutein, and zeaxanthin)	Dark green, yellow, and orange fruits and vegetables, such as carrots, winter squash, apricots, sweet potatoes, spinach, yams, kale, turnip greens, parsley, cantaloupe, citrus fruits, spirulina, marine algae, and marigold flowers.	Potent antioxidants; protect against and fight cancer; reduce arterial plaque; block DNA damage.
Chlorophyll	All green plants.	Fights bacteria, burns, and wounds; excellent source of naturally occurring vitamin K; fights cancer.
Coumarins	Parsley, carrots, citrus fruits.	May inhibit cancer growth; help cells dispose of cancer cells; prevent blood clotting.
Curcumin	Turmeric.	Has partial flavonoid structures. Used to regulate menstrual cycle, relieve menstrual cramps, and decrease fevers; also used externally for wound healing.
Diterpene	Rosemary extract.	Potent antioxidant, anticarcinogenic, antimutagenic, and antihepatotoxic agent; inhibits lipid peroxidation and scavenges peroxyl radicals.
Eleutherosides	Siberian ginseng.	Help stamina, appetite, metabolism, digestion, and physical and mental vigor; body toners; strengthen glandular and immune systems; stimulate central nervous system; improve irregular periods, hot flashes, and other problems arising from menopause.

PHYTOCHEMICAL	PLANT SOURCE	DISEASE-FIGHTING PROPERTIES
Ellagic acid	Berries, walnuts, grapes, apples, pomegranates, tea.	Stimulates production of glutathione S-transferase, which detoxifies certain cancer compounds; fights skin and lung cancer.
Essential fatty acids (EFA)	Flaxseed oil, chasteberry, saw palmetto, spirulina, evening primrose, black currant, canola, and borage oils.	Improve myelin sheath health (fatty material that protects nerve fibers); stimulate prostaglandin production; lower cholesterol levels; lower blood pressure; inhibit platelet aggregation; help fight diabetes, breast cancer, menstrual cycle disorders, multiple sclerosis, inflammation, alcoholism, and hyperactivity; strengthen immune system; help strengthen cell membranes.
Flavonglycosides/terpene lactones	Ginkgo biloba, black tea.	Potent antioxidants; dilate blood vessels; improve mental perception; helpful in cerebrovascular function and kidney and heart disorders; improve clarity and mental perception; helpful with senile dementia, vision, tinnitus (ringing in the ears), hearing, memory loss, and depression.
Fructooligo-saccharides (FOS)	Chicory root, Jerusalem artichokes, garlic, asparagus, honey, brown sugar, bananas, onion, barley, rye, tomato, and wheat.	Increase peristaltic action of intestines; improve liver function; used in elimination diets; promote growth of beneficial bacteria in intestines; may help control triglyceride levels.
Gallic acid	Green and oolong tea, red wine, roasted malt extract.	Potent antioxidant; inhibits infections; fights atherosclerosis; lowers blood pressure; strengthens capillaries; destroys viruses and bacteria.
Gamma-glutamyl allylic cysteines	Aged garlic extract.	Lower blood pressure; strengthen immune system.
Gingerols	Ginger.	Potent antioxidants; improve digestion of proteins and fat; soothe the stomach; fight liver toxicity and inflammation.
Ginkolic acid	Ginkgo biloba.	Potent antioxidant; protects against cancer; improves circulation and mental

Table 6.2. Functional Food Phytochemicals (continued)

PHYTOCHEMICAL	PLANT SOURCE	DISEASE-FIGHTING PROPERTIES
Ginkolic acid (continued)		clarity; relieves cerebral insufficiency; treats depression; keeps blood platelets from aggregating; improves capillary blood flow.
Glucosinolates	Cruciferous vegetables.	Activate liver detoxification enzymes; regulate white blood cells.
Hypericin (or hypericine)	St. John's wort.	May help support healthy emotional and mental function (possibly by regulating brain neurotransmitter levels, including norepinephrine and serotonin); improves mood.
Indoles	Brussels sprouts, cabbage, kale, cauliflower, mustard greens, and collard greens.	Induce Phase II enzymes, which protect against cancer.
Isothiocyanates	Broccoli, cabbage, cauliflower, mustard, horseradish, radishes.	Induce production of protective enzymes; reduce breast cancer risk; inhibit DNA cell damage.
Lactones	Hawthorn berry extract, kava kava root extract.	Protect the body against cancer; detoxify; eliminate carcinogens.
Lignans	Flaxseed, walnuts.	Block prostaglandins and inhibit estrogen.
Limonoids	Citrus fruits and peels.	Induce production of Phase I and Phase II enzymes, which protect lung tissue and eliminate carcinogens.
Lipoic acid	In many plant foods.	Potent antioxidant; detoxifies heavy metals; normalizes blood sugar levels; key compound in energy production; slows aging; protects against cancer and heart disease; key coenzyme for several enzymes.
Lycopene	Red grapefruit, tomatoes, watermelon, apricots.	Potent antioxidant; helps fight cancer.
Monoterpenes	Basil, broccoli, cabbage, carrots, citrus fruits, cucumbers, eggplant, mint, parsley, peppers, squash, tomatoes, yams.	Fight cancer; potent antioxidants; aid the activity of protective enzymes; inhibit the production of cholesterol.

PHYTOCHEMICAL	PLANT SOURCE	DISEASE-FIGHTING PROPERTIES
Organosulfur compounds (including allium and allylic sulfides)	Aged garlic extract (AGE), garlic, onions, scallions, shallots, chives, leeks.	Potent free radical fighters; immune system stimulators; help lower cholesterol; fight circulatory disorders; help the body eliminate carcinogens and inhibit tumor cell growth.
Pectin (polygalacturonic acid)	Grapefruit rinds, apples, and numerous fruits.	Lowers plasma cholesterol levels; reduces arterial blockage caused by oxidized low-density lipoproteins.
Phenolic acids	Berries, broccoli, peppers, tomatoes, cabbage, carrots, citrus fruits, eggplant, parsley, whole grains, all flowering plants.	Inhibit nitrosamine formation; antioxidants; affect enzyme activities.
Phthalides	Parsley, carrots, celery.	Stimulate beneficial enzyme production; detoxify carcinogens.
Phytic acid	Wheat, lima beans, rice, sesame seeds, soybeans, peanuts, rye.	Potent antioxidant; prevents colon cancer; improves natural killer cell activity.
Phytosterols	Broccoli, cabbage, cucumbers, pepper eggplant, squash, soy products, whole grains, tomatoes, yams.	Prevent colon cancer; block absorption of cholesterol; inhibit breast cancer occurrence due to estrogen promotion.
Polyacetylenes	Carrots, parsley, celery.	Regulate prostaglandin production, protect against carcinogens.
Rosemarinic acid	Rosemary extract.	Fights nausea, intestinal gas and indigestion; treats headaches.
Salin (or salicin)	White willow bark extract.	Fights inflammation; relieves pain; lowers fever; relieves symptoms of arthritis, bursitis, headaches, chills, influenza, and eczema.
Saponins	Ginseng root (from Panax species), licorice, black cohosh, sea cucumber, horse chestnut seeds, yucca.	Fight inflammation, ulcers, atherosclerosis, and cancer formation; enhance wound healing; reduce blood cholesterol; fight bacteria and fungi; improve gastrointestinal tract activity.
Silymarin	Milk thistle extract.	Potent antioxidant; protects the liver; supports new liver cell growth.

67

Table 6.2. Functional Food Phytochemicals
(continued)

PHYTOCHEMICAL	PLANT SOURCE	DISEASE-FIGHTING PROPERTIES
Sulforaphanes	Brussels sprouts, broccoli, cauliflower, cabbage, horseradish, mustard greens, kale, radish.	Detoxify and eliminate carcinogens by encouraging Phase II enzyme production.
Terpenes—one of the largest phytonutrient classes (includes carotenoids)	Ginkgo biloba, citrus fruits, caraway seed oil, green foods, soy products, grains.	Inhibit cancer; potent antioxidants.
Thioallyl compounds	Aged garlic extract.	Immune system stimulators; protect against cancer; cardiovascular system; protect potent antioxidants.
Thiocyanates	Cabbage, broccoli, cauliflower.	Reduce risk of breast cancer; inhibit DNA damage.
Triterpenoids	Citrus fruits, reishi mushrooms, licorice root extract, gotu kola, soy products.	Fight ulcers; prevent dental decay; inhibit cancer; fight liver toxicity.
2"-0-Glycosylisovitexin (GIV)	Green barley extract.	Potent antioxidant.

What's on the Market?

There are many functional foods from plants for sale in health-food stores. They are available in tablet, powder, or liquid form. The following list is a sample of the many products found on the market:

- Kyo-Green from Wakunaga of America Co., Ltd., is a powdered drink mix made from a natural organic combination of young barley grass and wheat grasses (grown in the pristine Nasu Highlands in Japan), blended with Bulgarian chlorella, brown rice, and kelp from the Northern Pacific. Kyo-Green is high in minerals, vitamins, carotene, amino acids, chlorophyll, and enzymes.

- Green Magma from Green Foods Corporation is a natural powder made from the juice of young barley leaves, organically grown in the United States and Japan. It contains numerous enzymes important in metabolism, such as cytochrome oxidase (an oxidation-reduction enzyme required for cell respiration), catalase (a potent antioxidant), peroxidase (decomposes hydrogen peroxide), transhydrogenase (important for the muscular tissue of the heart), superoxide dismutase, nitrogen oxyreductase, transferase, and proteinase, among others. Green Magma contains vitamin B_1, nicotinic and linoleic acids, plus the potent antioxidant 2-0-GIV (2"-0-Glycosylisovitexin). Also try Dr. Hagiwara's Magma Plus and Dr. Hagiwara's Veggie Magma.

- Green-Zymes (from Nikken) is a mixture of extracts from vegetables (including cabbage, lettuce, red beets, spinach, celery, and carrots), fruits (including apple, banana, pineapple, and papaya), grains, wheat germ, and wheat grass, along with papain, lipase, and protease. This green pill can be chewed or crushed and mixed with food.

- Naturally's Phyto Vita Boost (in tablet form) contains many enzymes, vitamins, and minerals, and beta-carotene, coenzyme Q_{10}, ginkgo biloba, green tea extract, unsaturated fatty acids (borage oil), phosphate-containing lipids from soy, spirulina algae, alfalfa concentrate, parsley powder, and a citrus bioflavonoid complex.

- Green Barley Essence from Green Foods Corporation is a powder containing a combination of barley juice, alfalfa, aloe, broccoli, brown rice, carrots, celery, coix (a South Asian cereal grass), garlic, green pepper, kelp, shiitake mushrooms, lecithin, spinach, and wheat germ. Many of these ingredients are rich in enzymes, chlorophyll, vitamins, and minerals.

- Carrot • Barley • Spirulina from Futurebiotics is a powder made from freeze-dried pineapple juice concentrate, carrot juice concentrate, barley juice concentrate, spirulina, sprouted barley malt, schizandra berries, and enzymes (including protease, amylase, superoxide dismutase, and many others).

- Juice Plus offers two food concentrates in tablet form: Orchard Blend (AM) and Garden Blend (PM). Orchard Blend is made primarily from fruits and contains several enzymes, including bromelain, papain, lipase, amylase, protease, and cellulase. Garden Blend (PM) is made from grains and vegetables. This product also contains such enzymes as lipase, amylase, protease, and cellulase.

- Body Oxygen is a liquid functional food. It is a concentrated, enhanced dietary supplement containing numerous enzymes. The enzyme-rich yeast cells in Body Oxygen are cultivated in fruit juice concentrate along with B complex. Body Oxygen contains a number of enzymes, including hydrolases, transferases, oxidases, demolases, and phosphorylases. This liquid is easily mixed with beverages or consumed as is.

- Manda Enzyme is a commercial fermented product made from oranges, pineapples, bananas, apples, soybeans, wild vines, akebi (a sweet Japanese fruit), silvervine, edible seaweeds, carrots, garlic, and persimmons, along with other raw fruits, vegetables, and herbs. Manda is rich in a number of enzymes (such as protease, amylase, lipase, superoxide dismutase, and glutathione peroxidase), minerals (including potassium, calcium, and phosphorus), and amino acids.

- Earth Source Greens & More (from Solgar) is a powder made from four organic grasses (kamut, wheat, barley, alfalfa); reishi, shiitake, and maitake mushrooms; sea algae; Hawaiian blue green spirulina; Chinese chlorella; several vegetables (including carrots, broccoli, and red beets); and

Ways to Use Functional Foods From Plants

Most concentrated food supplements are available in tablet or powder form. Tablets can be chewed or swallowed whole and are not enterically coated. Tablets can also be ground to a powder by placing between two tablespoons and applying a little grinding pressure.

Powdered plant extracts can be used in many ways:

- Stir powder in water or juices (some people like it mixed in milk).
- Sprinkle on salads, cereals, meat, or spaghetti.
- Sprinkle over cookies.

- Sprinkle over ice cream, cottage cheese, or yogurt (or swirled into yogurt).
- Sprinkle over gelatin or pudding after refrigeration. Remember, high temperatures will destroy all active enzymes. Therefore, enzymes will be denatured if powdered extracts are placed in boiling water.
- Substitute for parsley as a garnish.
- Use in salad dressings (combine with natural salad oil).
- Mix into any sandwich spread.
- Use as a spice substitute or taste-enhancer.

---

fruit powders (strawberry, orange juice, apple, and acerola). This mixture provides proteins, lipids, vitamins (including B vitamins and vitamin E), minerals (such as zinc and magnesium), chlorophyll, essential fatty acids, and fiber.

Animal Functional Foods
(Gland and Organ Concentrates)

Just like humans, all animals require enzymes. Therefore, concentrates from the glands and organs of young animals (whose enzymes are at their peak of activity), raised in pesticide-free, healthy environments, can be used to improve wellness and fight disease. These concentrates (called glandulars or protomorphogens) are usually obtained from calves or young pigs and are rich in vitamins, minerals, hormones, and live enzymes. Organ- and glandular-based food supplements could be considered "whole food concentrates" because they are so nutrient-rich.

Although gland and organ concentrates have been used for centuries to treat illness, they remain a controversial issue. Some say glandular therapy (also called organotherapy or live-cell therapy) is useless, while others feel "like cures like"—that concentrates can be very beneficial in health care, improving glandular and organ function. According to Dr. Hans Kugler (former president of the National Health Federation), animal glands can assist similar cells in the human body because they are genetically similar. "An individual suffering from adrenal failure might be given raw adrenal extract because its inherent qualities help support human adrenal functions," says Dr. Kugler.

Many major international enzyme companies use animal organs as sources for enzymes, including bovine liver (for catalase, glutathione peroxidase, lactase, and superoxide dismutase), bovine kidney (superoxide dismutase), and bovine and porcine pancreas (for amylase, carboxypeptidase, chymotrypsin, elastase, lipase, pancreatin, proteases, and trypsin). Porcine stomach produces pepsin, calf stomach produces ren-

nin, while porcine and bovine intestines are sources of enterokinase and peptidase. Other animal organs, such as the heart and thymus glands produce numerous enzymes, too. Table 6.3 on page 70 lists sources of glandular and organ extracts and their applications.

"The quality of any glandular formula depends upon the process of preparation as well as the quality of the glandular extracts used," according to Dr. Michael Murray in his book *Encyclopedia of Nutritional Supplements* (Rocklin, CA: Prima Publishing, 1996). He further states that the biologically active material in a glandular (such as enzymes, natural lipid factors, soluble proteins, hormone precursors, minerals, and vitamins) will be eliminated or destroyed if not properly prepared. In fact, processing temperatures over 37º C (body temperature) will destroy many of the active ingredients. Freeze-drying conserves the active principles of the organ or gland.

Gland and organ concentrates can benefit a number of conditions, including allergies, arteriosclerosis, circulatory disorders, chronic fatigue syndrome, Epstein-Barr syndrome, diabetes, herpes, hypertension, hypoglycemia, immune weakness, impotence, kidney and chronic liver problems, obesity, eye problems, Parkinson's disease, and prostatic hypertrophy. These concentrates may also slow down the aging process. As dietary supplements, they are among the best choices for rejuvenation, according to Dr. Kugler. He uses a glandular that contains concentrates from a number of glands and organs as an essential part of his daily supplementation program.

What's on the Market?

Glandular and organ concentrates are plentiful on the shelves of health-food stores in tablets, capsules, soft gel capsules, and protein powders. Here are just a few of those available.

- American Laboratories offers extracts from the brain, duodenum, eye, heart, hypothalamus, kidney, liver, lung, lymph, mammary glands, ovaries, pancreas, parathyroid, parotid, pituitary (anterior or whole), placenta, prostate, spleen, stomach, testes, and adrenal and thymus glands. Combinations are also available.

- Highland Labs has a desiccated liver tablet processed at a low temperature.
- Enzymatic Therapy offers Gland-H, ThymuPlex, Thymus, Raw Adrenal, Ovary-Uterus Complex, Kidney-Liver Complex, Lung Complex, Adrenal-Cortex Complex, and Lymph-Spleen Complex.
- Bioactive Cell Complex from Gero Vita is a combination of glandular extracts.
- Enzyme Process Laboratories provides a number of bovine gland and organ products, including adrenal, bone, connective tissue, heart, intestine, kidney, liver, lung, lymph, pancreas, parathyroid, prostate, spleen, thymus, and thyroid.
- Futurebiotics carries several products, including Adrenal Plus, Prostabs Plus (with raw prostate concentrate), and Male Power (containing over nine difference glandulars).

AMINO ACIDS

As mentioned in Chapter 1, enzymes are proteins. Amino acids are the structural units of all proteins. Enzymes vary depending on the amino acids in their composition, their number, and their order. Think of each enzyme as a series of box cars on a train. The amino acids are like individual box cars. The characteristics of the train (enzyme) will differ according to the sequence of each car (amino acid) and the contents therein.

Although the body can make many amino acids by itself, nine of them are essential and must be obtained from the diet or supplements, since they cannot be produced by the body. These amino acids include histidine, isoleucine, leucine, lysine, methionine, phenylalanine, threonine, tryptophan, and valine. Further, stress (including disease or injury) may cause increased demand for certain amino acids.

Amino acids, those "building blocks" of our entire body, play a number of important roles in the body. They are used to form enzymes, peptides, and hormones; to help the body maintain proper fluid balance and acid/base balance; and to form antibodies. They are the foundation of the body's structure and the source of thousands of enzymes required for a multitude of bodily functions. Vitamins and minerals can't do their jobs effectively without amino acids. In fact, a deficiency in any of the amino acids can lead to health problems and a shortage of the enzymes that require that particular amino acid. Amino acids are critical for the composition and function of every enzyme. For example, researchers have located some 201 total amino acids in the proteolytic enzyme trypsin alone.

Many athletes use amino acid supplements in an effort to replace the muscle tissue broken down during exercise, stimu-

Table 6.3. Gland and Organ Concentrates and Their Uses

EXTRACT	APPLICATION
Adrenal	Helps fight stress and adrenal exhaustion; fights inflammation, chronic fatigue, allergies, hypoglycemia, and diabetes; restores normal carbohydrate balance; good for health and sexual maturation.
Aorta	Helps improve blood vessel function; improves peripheral and cerebral circulation; helps protect against atherosclerosis and other problems of aging, such as varicose veins; aids in wound healing and fighting cancer.
Brain	Helps improve brain function.
Heart	Improves heart function.
Kidney	Helps fight kidney damage due to infections, kidney stones, cadmium toxicity, or high blood pressure.
Liver	Treats chronic liver disease; improves liver function; promotes liver cell regeneration; used in cancer therapy; supplies folic acid, iron, and B_{12}; improves immune defense and impaired neurological function. Enzymes found in the liver include amylase, arginase, catalase, esterase, lipase, superoxide dismutase, and oxidase.
Lung	Improves lung function and respiratory distress syndrome.
Ovary	Improves ovarian function.
Pancreas	Decreases allergies to foods; aids digestion; fights steatorrhea and viral problems; assists the pancreatic enzymes trypsin and chymotrypsin. Contains amylase, lipase, and protease.
Pituitary	Helps form new blood vessels; fights tumor growth; aids wound healing.
Spleen	Helps normalize capillary permeability and bleeding time; inhibits inflammation; supports the immune system; has antitumor and antibacterial activities.
Thymus	Assists thymus gland (important in immune resistance); helps improve immune responses to allergies, hay fever, rheumatoid arthritis, viruses, and chronic or frequent infections; fights migraine headaches and cancer.

late the production of new muscle tissue, and provide energy. Of all the amino acids, the branched-chained amino acids (BCAAs—isoleucine, leucine, and valine) have probably received the most attention. This is probably because a significant amount of muscle tissue is composed of these amino acids. In fact, one-third of the amino acids in muscle protein are BCAAs. A great deal of research on BCAAs has been conducted with athletes who are concerned with their muscle mass.

The human body does not actually require protein per se, but rather, the component amino acids and enzymes contained in protein. A food known to be high in protein will contain many kinds of amino acids. For example, peanuts are considered to be high in protein, because they contain several amino acids, including phenylalanine, isoleucine,

leucine, valine, methionine, cystine, tryptophan, lysine, threonine, tyrosine, arginine, and histidine. Animal foods are usually high in several amino acids, and so are referred to as "complete" proteins. See the discussion on complete and incomplete proteins in Chapter 2.

Amino acids are available individually or in combinations, in protein mixtures, functional food extracts, or in combination with numerous multivitamin formulas. The majority of amino acid supplements are taken from animal, egg, or yeast protein sources. Usually grains and grain extracts are used to obtain crystalline free-form amino acids (the purest amino acids). Although cold-pressed yeast and milk proteins are also used, brown rice bran is a primary source of crystalline-free amino acids. Amino acid supplements can be purchased as tablets, powders, or capsules. See Table 6.4 on page 72 for a list of amino acids and their therapeutic roles in the body.

Caution: Many amino acids should not be taken if you are pregnant or are taking MAO drugs (monoamineoxidases). If either of these conditions applies to you, check with your health-care provider before taking amino acid supplements. Further, some amino acids may present a health threat when taken over prolonged periods of time at high doses. When in doubt, seek the advice of a well-trained health-care practitioner.

ANTIOXIDANTS

Antioxidants are critical enzyme helpers. Antioxidants are a number of naturally occurring compounds (including enzymes, vitamins, minerals, amino acids, and phytochemicals) found in plant and animal foods, herbs, functional food concentrates, and supplements (individually or in combinations) that either help the body protect itself against free radicals or are themselves protective. Antioxidants can help inhibit or delay oxidation reactions caused by free radicals.

Oxidation reactions can occur anywhere in our body (in any organ or cell). In fact, they are essential for life. But problems occur when oxidation reactions get out of control, and slip past the constraints usually put in place by the body's protective mechanisms. This results in free radicals.

These reactive molecules that damage tissue can be controlled (not stopped) through antioxidant supplementation. To stop all cell oxidation would mean death. As with enzymes, antioxidants can be taken as individual supplements or in combinations. Taking a combination of antioxidants will affect several tissues in the body, while individual antioxidants would target specific areas. It is my opinion that both approaches are necessary.

Your body is composed of many kinds of tissues, which are attacked by more than one kind of free radical. There are different ways that antioxidants fight the free radical attack on our bodies and their various tissues:

- They help prevent the formation of free radicals.
- They stop the domino effect of free radicals.
- They repair or clean up after the damage has been done.

Antioxidants are found in tablets, capsules, powders, granules, liquids, extracts, and gelatins. Antioxidants can be compounded individually or in combinations, along with vitamins, minerals, and enzymes, or as functional foods containing fruits, vegetables, and herbs. See Table 6.5 for a listing of antioxidants and their target areas.

OTHER ENZYME HELPERS

There are many other nutrients that our bodies need to help our enzymes, in addition to the vitamins, minerals, and phytochemicals already discussed. There are herbs, plants, and other supplements that we can take to give our enzymes a boost. Herbs and food extracts are considered foods, and foods contain vitamins, minerals, and enzymes. The vitamins and minerals help activate the enzymes in the foods, and the activated enzymes in foods work synergistically with enzymes in the body. This makes nearly all herbs and supplements enzyme helpers.

Alfalfa

Since alfalfa roots burrow ten to twenty feet deep, they can draw many minerals from the soil. In addition, alfalfa is a good source of enzymes, vitamins A, B_6, E, and K (it contains small amounts of vitamin C), fats, protein, fiber, and phytochemicals (such as chlorophyll and saponins).

Enzyme-rich alfalfa is of benefit in fighting allergies; improving digestion and assimilation; increasing vitality and strength; and fighting weight gain, prostate inflammation, cystitis, insomnia, acidosis, lower back pains, and rheumatic aches and pains. It can increase mother's milk, reduce fevers, and help regulate the bowels. There is even some indication that alfalfa may be of benefit for cancer patients.

In Chapter 4, I mentioned that sprouting grains increases their enzyme activity dramatically. Alfalfa seeds are an excellent choice for sprouting. Further, Dr. T. Pearse Lyons, Ph.D. (president of Alltech, Incorporated—one of the world's largest researchers of animal enzymes) suggests that supplementing alfalfa with enzymes improves the activity of both.

Algae

Among the most chlorophyll-rich organisms, algae are rich in enzymes, vitamins, minerals, and eight essential amino acids. Cyanobacteria (blue-green algae) are nutrient-dense foods containing significant amounts of chlorophyll, carotenoids, vitamins, minerals, lipids, enzymes, and proteins. In fact, according to one distributor, blue-green algae's

Table 6.4. The Role of Amino Acids

AMINO ACID	ROLES IN THE BODY/ THERAPEUTIC USES	SIGNS OF A DEFICIENCY
Alanine	Essential in the metabolism of tryptophan and pyridoxine; can reduce cholesterol (when combined with arginine and glycine); may stabilize blood glucose; enhances fat metabolism.	Nervousness, loss of energy.
Arginine	Fights hypertension; accelerates wound healing; enhances thymus activity; increases fat metabolism; aids insulin production and glucose tolerance.	Overweight, infertility (especially in men), premature aging, free radical damage.
Asparagine	Fights chronic fatigue, cirrhosis, and drug addiction.	Nervousness, chronic fatigue, cirrhosis.
Aspartic acid	Plays important role in metabolism; increases endurance; treats drug addiction, cirrhosis, fatigue.	Chronic fatigue, cirrhosis.
Cysteine	Promotes healing and improves disease resistance; detoxifies body.	Delayed healing; copper toxicity.
Glutamic acid	Influences brain health; detoxifies body.	Excessive grouchiness and other personality disorders.
Glutamine	Treats alcoholism; helps maintain gastrointestinal tract health; prevents malabsorption.	Fatigue, poor attention span, depression, alcoholism.
Glycine	Detoxifies liver; part of glucose tolerance factor.	Acid stomach, premature aging.
Histidine	Needed for tissue growth and repair and blood cell production; treats digestive disorders, arthritis, and allergies.	Allergies, poor hearing, psychological disorders, rheumatoid arthritis, anemia.
Isoleucine	Aids healing from burns.	Fatigue, blood sugar problems.
Leucine	Aids healing from burns; treats Duchenne muscular dystrophy.	Slow healing; elevated blood sugar.
Lysine	Fights herpes; aids calcium absorption, tissue repair, and collagen formation; essential for growth in infants; maintains nitrogen equilibrium	Anemia, hair loss, irritability, energy loss, visual problems.

AMINO ACID	ROLES IN THE BODY/ THERAPEUTIC USES	SIGNS OF A DEFICIENCY
Lysine (cont.)	in adults; plays a role in enzyme, hormone, and antibody production; treats cold sores.	
Methionine	Helps prevent liver and artery fat build-up; maintains blood flow to kidneys, heart, brain; helps protect cells from free radicals; detoxifies harmful agents; aids digestion; helps prevent brittle hair; aids muscle weakness; fights osteoporosis and allergies.	Hair loss; poor skin tone; liver problems; atherosclersis.
Phenyl-alanine	Prevents and treats depression; treats hyperactivity and attention deficit disorder; produces neurotransmitters; improves memory; aids weight loss; elevates mood; powerful pain reliever.	Emotional upset, poor circulation, increased appetite, some eye problems. Phenylketo-nurics should not ingest phenylalanine.
Proline	Needed for formation and maintenance of collagen and healthy skin.	Poor skin and soft tissue texture.
Serine	Involved in biosynthesis of pyrimidine, purine (uric acid compounds), creatine (a nitrogen compound found mainly in muscle tissue), and porphyrin (organic compounds), forms cystine (an amino acid) with homocysteine.	Reduced immune function.
Threonine	Aids digestion; improves absorption and assimilation of nutrients; important to formation of collagen and elastin; helps control epileptic seizures.	Gastrointestinal disorders, malabsorption disorders, malnutrition.
Tryptophan	Important in energy production, glycolysis (the energy-producing process in which sugar is broken down into lactic acid), tissue respiration, and fat synthesis; relieves pain; aids sleep; precursor of niacin and serotonin.	Indigestion, insomnia, brittle nails, premature aging.
Tyrosine	Helps form antibodies and nourish blood; treats Parkinson's disease, narcolepsy, and hypertension; a melanin precursor; functions in synthesis of hormones (thyrosine and epinephrine), and neurotransmitters (norepinephrine, dopamine).	Depression, low blood pressure, low body temperature, "restless" legs, fatigue, irritability.
Valine	Speeds healing from burns; aids normal metabolism; important to muscular coordination, mental energy, and nervous system function.	Insomnia; slow tissue healing; nail biting; poor mental health.

amino acid profile is virtually identical to that which is required by the human body.

Algae have been valued for thousands of years by many cultures because of their ability to regenerate health. There are reports that algae can help with many conditions, including hypoglycemia, chronic fatigue syndrome, depression, diabetes, ulcers, anemia, and hepatitis.

Aloe Vera

Aloe vera is probably best known for its ability to relieve the pain of a burn or sunburn when applied topically. It is high in enzymatic action, including such enzymes as amylase, catalase, and cellulose, among others. Aloe also contains a number of phytochemicals including sterols, and saponins.

In addition to aloe's pain-relieving effect, it is an excellent colon cleanser and effective laxative, and is used in the treatment of hemorrhoids. It purportedly works with your immune system to keep you strong, vibrant, and healthy. *Warning:* Do not take aloe internally if you are pregnant.

Astragalus

Astragalus is an Asian herb that contains immune-modulating polysaccharides and flavonoids that enhance blood circulation. It strengthens the immune system, promotes healing, and is used to treat high blood pressure, diabetes, and heart disease and to improve digestion. There is some indication that astragalus may prevent the spread of malignant cells in cancer patients.

Astragalus is often combined with other herbs that are classified as tonics (such as ginseng and licorice). These combinations have been found useful in treating both immune deficiency and autoimmune disorders.

Bee Pollen

Bee pollen is considered a miracle food rich in enzymes, minerals, protein, and vitamins (including B_1, B_2, B_6, niacin, pantothenic acid, C, and beta-carotene). Some people may be allergic to bee pollen, so begin with small amounts, and discontinue if discomfort occurs.

Bee pollen contains a natural antibiotic, so it is effective against infections. It is also used to treat allergies and fatigue, to decrease protein cravings, as a sexual rejuvenate, and for prostate gland disorders.

Bilberry

The bilberry is a botanical cousin to the blueberry. Bilberries contain antioxidants called anthocyanosides, which aid night vision, improve visual clarity and image detail, protect and strengthen the blood vessels that feed the eyes, reduce visual fatigue, and support arteries and capillaries in the eyes of diabetics. Bilberries are also rich in enzymes. In addition, the bilberry is helpful in the treatment of stomach problems, liver conditions, and diarrhea. Bilberry has also shown to lower blood sugar levels in diabetics.

Brewer's Yeast

Brewer's yeast is an excellent food supplement and one of the best natural sources of the entire B vitamin complex—coenzymes to many of the body's enzymes. It is a superb source of concentrated protein, including all the essential amino acids, and is rich in nucleic acid. Brewer's yeast is also rich in minerals, such as chromium, calcium, copper, iron, magnesium, manganese, phosphorus, potassium, selenium, sodium, and zinc.

Brewer's yeast is a powerful health source, effective at reducing serum cholesterol and elevating high-density lipoprotens (HDLs—the good cholesterol). Because it is rich in glucose tolerance factor it improves glucose tolerance.

Chlorella

Chlorella is a whole food that is high in nutritive value and rich in chlorophyll, protein, enzymes, and minerals. It contains beta-carotene; vitamins C, E, K, B_1, B_2, B_6, B_{12}, niacin, pantothenic acid, folic acid, biotin, choline, and inositol; and essential fatty acids.

Chlorella is used to regulate cholesterol, improve bowel function, and stimulate the growth of healthful bacteria. It has potent antitumor and antiviral activity, enhances immune function, and can repair DNA. Its chlorella growth factor (CGF) is purported to be a unique group of substances that has the potential to help repair damaged tissues and organs. In addition, chlorella's cell wall materials may have the capacity to absorb intestinal poisons and to promote normal peristalsis.

Echinacea

The herb echinacea is native to the United States and Canada and was first used by Native Americans (especially the Plains tribes) as a remedy for snakebites and other skin wounds. Of echinacea's nine species, the most commonly used is the purple coneflower (*Echinacea purpurea*).

According to Dr. Edward Alstat, R.Ph., N.D., of the Eclectic Institute, echinacea is one of the most promising immune enhancers known to humans, mostly due to its high molecular weight polysaccharides. Further, echinacea has had more positive clinical and scientific studies done than any other herb.

Over the past forty years, studies have shown that echinacea has strong wound-healing action. It fights inflammation, stimulates the immune system, and is effective against certain viral and bacterial infections. It shows potential in

Table 6.5. Antioxidants

ANTIOXIDANT	TARGET AREAS
ENZYMES	
Catalase	Liver, cells throughout the body.
Glutathione peroxidase	Liver, cells throughout the body.
Superoxide dismutase	All cells.
VITAMINS AND COQ$_{10}$	
Beta-carotene (vitamin A precursor)	Mammary tissue, liver, eyes, cells throughout the body.
Coenzyme Q$_{10}$	Heart, blood vessels, cells throughout the body.
Vitamin A	Mammary tissue, liver, eyes, cells throughout the body.
Vitamin C (ascorbic acid)	Plasma, cellular fluids, capillaries, connective tissue, immune system.
Vitamin E (tocopherol)	Brain, heart, blood vessels, lungs, cell membranes, any tissue coming in contact with oxygen.
MINERALS	
Copper	Liver, cells throughout the body.
Manganese	All cells.
Selenium	Cell membranes and cells throughout the body.
Zinc	Reproductive organs, cells throughout the body.
AMINO ACIDS	
Cysteine	Skin, collagen, liver, cells throughout the body.
Glutathione	Cell membranes, liver, spleen, eyes, kidneys, stomach lining, cells throughout the body.
Methionine	All cells.
PHYTOCHEMICALS	
Allium compounds	Skin, capillaries, cells throughout the body.
Anthocyanidins	Capillaries, blood vessels, connective tissue, eyes, muscles, nerves, heart, cells throughout the body.
Carotenoids	Cardiovascular and immune systems, cells throughout the body.
Chlorophyll	Skin, gastrointestinal tract, cells throughout the body.
Curcumin	Circulatory system, cells throughout the body.
Diterpenes	Liver, cells throughout the body.
Ellagic acid	Endothelium, skin, lung.

ANTIOXIDANT	TARGET AREAS
Epicatechin	Capillaries, blood vessels.
Essential fatty acids	Nerves, organs, urinary tract.
Flavonglycosides	Blood vessels, cells throughout the body.
Flavonoids	Connective tissue, cardiovascular system, capillaries, immune system, muscles, cells throughout the body.
Gallic acid	Blood vessels.
Genistein	Brain, prostate, breasts, cells throughout the body.
Gingerols	Liver, digestive system.
Glycyrrhizin	Skin, immune, digestive, and respiratory systems.
Hesperidin	Capillaries, circulatory system, skin.
Limonene	Stomach, lung, skin, mammary glands.
Lipoic acid	All cells.
Lycopene	Blood vessels, cells throughout the body.
Monoterpenes	Blood vessel walls.
Naringin (bioflavonoid)	Capillaries.
Organosulfur compounds	Colon, lung, liver, stomach, gastrointestinal and immune systems.
Phenolic acids	Liver, cells throughout the body.
Phytic acid	Digestive tract.
Polygalacturonic acid	Blood vessels.
Polyphenols	Liver, skin, lungs, digestive tract.
Polysaccharides and saponins	Immune system, skin, cells throughout the body.
Proanthocyanidins	Cell wall, circulatory system.
Quercetin (a flavonoid)	Adrenals, circulatory system.
Rosemarinic acid	Digestive tract, circulatory system.
Rutin (a flavonoid)	Blood vessels, circulatory system.
S-allyl cysteine	All systems, especially liver.
Silymarin	Liver, skin, digestive tract.
Sulforaphane	All cells.
Tannins	Blood vessel walls, cells throughout the body.
Terpenes	Blood vessel walls, cells throughout the body.
Triterpenes	Immune system, urinary tract, gastrointestinal tract.
2-0-GIV	All cells.

treating a number of diseases from asthma to cancer. In Europe, echinacea has been used to treat candidiasis, prostatitis, impetigo, upper respiratory tract infections, tonsillitis, and other ailments and diseases.

Garlic

Garlic has been around for over 5,000 years and is credited with curing everything from cancer to the common cold (and driving away Dracula). Garlic (*Allium sativum*) contains several enzymes (including allinase, peroxidase, and myrosinase), as well as many vitamins and minerals.

Researchers around the world have been subjecting garlic in its many forms to extensive scientific research. Current garlic research has centered around a number of areas, including cardiovascular care (including cholesterol and blood pressure-lowering effects), cancer prevention and care, antioxidant capabilities, memory improvement, and antibacterial and antiviral effects.

Though most of us are familiar with garlic in its raw form and may use it as an aromatic and tasty addition to any recipe, garlic is also available as a powder, a tablet, an encapsulated garlic oil supplement, and as aged garlic extract. Many of these products can give you the health benefits of garlic without the bad breath.

In aged garlic extract (AGE), the pungent aftertaste and odor have been removed by the aging process without affecting its potent therapeutic effects. Aged garlic extract is the most highly researched form of garlic. Research shows AGE (with S-allyl cysteine) is a potent antioxidant, is effective in preventing and treating many types of cancer and in fighting cardiovascular disorders, and has cholesterol-lowering effects and immune system stimulating and modulating properties. Studies have shown AGE prolongs lifespan, retards aging, and restores memory deficit. Further, AGE prevents reduction in physical strength caused by chemical and physical stress.

Ginger

Ginger root is widely used in Asian cooking because of its pungent and refreshing flavor. It is often sliced and served in Japanese meals to aid digestion. Ginger contains high concentrations of the protease zingibain, whose beneficial effect might even exceed that of papain. In fact, ginger could be used for applications similar to those of papain or other plant proteases. Ginger also dramatically increases the concentration of the digestive enzyme amylase in saliva and stimulates the flow of saliva.

This gnarly root that is available at most grocery stores has been used for centuries to calm upset stomachs, treat nausea (it is especially effective in treating morning sickness), relieve symptoms of a cold, and soothe minor burns. Ginger is now also available in capsules and as an extract.

Ginkgo Biloba

Ginkgo (from the ginkgo biloba tree) contains a number of active components, including flavonoids and terpenoids (another class of phytochemicals), as well as some organic acids. It has been shown to retard aging, reduce inflammation, improve visual acuity, improve brain function, improve memory, reduce depression, reduce risk of heart disease, inhibit blood clot formation, inhibit free radicals, reduce inner ear problems (including vertigo, tinnitus, and hearing loss), enhance energy, reduce risk of phlebitis, improve glucose utilization, promote faster healing, reduce wrinkling of the skin, fight Alzheimer's disease, subdue allergic reactions (including asthma), reduce fatigue, and improve circulation. Ginkgo is effective against these conditions because it restores capillary integrity and improves circulation. This makes it easier for each cell to get necessary nutrients.

Ginseng

Since the dawn of history, ginseng (also called the root of immortality) has ranked among Asia's most revered and prestigious herbs. Although there are many kinds of ginseng, probably the most widely used are American ginseng (*Panax quinquefolium*), Asian ginseng (*Panax ginseng*), and Siberian ginseng (*Eleutherococcus senticosus*).

Ginseng contains numerous essential oils, vitamins, minerals, enzymes, saccharides (mono-, di- and tri-), and saponins, which include a number of the phytochemicals ginsenosides, which have antineoplastic, antiaging, and immunologic enhancing actions, among others. Many people take ginseng to fight fatigue. This has particular attraction for athletes who frequently push their endurance capacity to the limit. The presence or absence of ginsenosides seems to be a key to ginseng's applications. It is effective against fatigue and daily stresses. This may be due to ginseng's effect on the adrenal system.

American ginseng (*Panax quinquefolius*) reduces cholesterol, enhances physical and mental performances, helps the body to adapt to stress, normalizes body functions, acts as an aphrodisiac and mild stimulant, and may inhibit the growth of cancerous tumors.

Asian ginseng (*Panax ginseng*) reduces cholesterol, normalizes body functions, helps the body adjust to stressful situations, increases endurance (physical and mental), increases energy, may enhance sexual desire, may reduce discomfort caused by menopause, and may inhibit the growth of cancerous tumors.

Siberian ginseng (*Eleutherococcus senticosus*) helps prevent heart disease by reducing cholesterol and blood pressure, helps the body withstand stresses, helps cure colds and infections, improves mental alertness, and improves overall health. It contains no ginsenosides, but does contain other saponins.

Glucosamine Sulfate

Glucosamine sulfate is a highly concentrated amino sugar naturally formed in joint structures, where it stimulates cartilage repair; provides one of the basic building blocks for healthy tendons, ligaments, and cartilage; provides cushioning support and lubrication to help sustain flexibility; and helps maintain synovial fluid levels.

Available in Italy for decades, glucosamine sulfate is beginning to attract some interest here in the United States for the treatment of arthritis. Indications are that it is as effective as, yet safer, than, corticosteroid products.

Goldenseal

The herb goldenseal functions as a natural antibiotic and is good for respiratory conditions (including colds and flu) and infections. It is also effective against nausea, indigestion, and constipation. *Caution*: Do not use during pregnancy.

Gotu Kola

Used for centuries in China, India, and the South Pacific as both an herb and nutritious food, gotu kola is related to the carrot. It helps maintain the integrity of collagen, skin, and blood vessels; helps heal minor wounds and skin irritations; and promotes circulation throughout the body, including the brain.

Green Barley

One of the most highly researched green foods is green barley. World-famous researcher and lecturer Dr. Yoshihide Hagiwara believes that green barley is the ideal "fast food." Juice of young barley plants contains sodium, potassium, calcium, magnesium, iron, copper, phosphorus, manganese, zinc, enzymes, and chlorophyll. This is a quick and easy way to get many of the nutrients your body needs.

Dr. Hagiwara discovered the health benefits of young barley leaf extract because it helped restore his own health. He has reported detailed case histories of patient relief for a variety of conditions, including arthritis, asthma, skin problems, obesity, anemia, constipation, impotence, high blood pressure, diabetes, heart disease, cancer, and kidney problems.

According to Hagiwara, green barley extract contains all necessary enzymes for biological activities. He believes there may be over 3,000 enzymes in green barley extract. Some enzymes confirmed in experiments to be in green barley extract include superoxide dismutase, cytochrome oxidase, catalase, peroxidase, fatty acid oxidase, and transhydrogenase among others.

Professor Takayuki Shibamoto, Ph.D., chairman of Environmental Toxicology at the University of California, Davis, is studying an antioxidant compound found in young green barley leaves called 2-0-GIV (2"-0-Glycosylisovitexin). Professor Shibamoto says that 2-0-GIV is as potent as any other antioxidant, including beta-carotene, vitamin C, or vitamin E.

Hawthorn

Hawthorn contains the phytochemical vitexin-2"-rhamnoside and a variety of biologically active flavonoids, which help stabilize collagen. Hawthorn is a potent antioxidant, promotes overall cardiovascular health, supports the heart's metabolic processes by increasing blood supply and oxygen utilization, helps decrease arterial plaques, and helps maintain healthy cholesterol and blood pressure levels.

Kelp

Kelp (laminaria or seaweed) is high in minerals (calcium, magnesium, potassium, and iodine) and enzymes. Kelp is used to treat gastrointestinal, respiratory, and genitourinary disorders; reduce cholesterol and blood pressure; and fight cancer.

A few years ago, an epidemiological study reviewed the increased incidence of breast cancer in Japan during recent decades. The investigators compared different lifestyles and concluded that the Japanese women of today who are not eating seaweed had an increased incidence of breast cancer. That is, women in rural areas whose diets still include seaweed had much lower incidences of breast cancer than those eating a modern diet.

Licorice

Licorice has been used for centuries as a flavoring and sweetening ingredient in candies and various medicines. Recent studies indicate that licorice may have cancer-preventing properties, as well as anti-HIV, anti-inflammatory, and immune-stimulating activity.

Maitake Mushrooms

In Japan, Maitake (pronounced *my-tah-keh*) mushrooms (Grifola frondosa) are highly prized both as food and medicine. People with serious degenerative illnesses travel long distances in search of the enzyme-rich Maitake mushroom. However, Western medicine has not historically viewed mushrooms as a treatment option for illness or disease. But the Maitake mushroom has recently received increased clinical and research attention.

Maitake mushrooms (which means dancing mushroom in Japanese) not only taste good, but seem to be effective against a variety of conditions, including cancer, HIV/AIDS, high blood pressure, diabetes, and obesity. Maitake can improve immune system function, and reduce uterine fibroids. The "King of Mushrooms," maitake brings its healing powers from Japan to you.

Milk Thistle

Silymarin (extract of milk thistle weed) helps to repair the liver. Milk thistle seed extract (Silymarin or Silybum Marianum) is used to treat chronic liver necroses, cirrhosis, and hepatitis (A and B); lowers fat and lipid deposits; and protects and regenerates the liver even after exposure to a virulent liver toxin (such as the death cap mushroom).

Mistletoe

Extracts of European mistletoe are commonly used with enzymes in Germany to treat malignant tumors; however, its use during chemotherapy may cause some damage to white blood cells, so use mistletoe only under the supervision of a physician well-trained in its use.

Onion

Related to garlic, onions are also enzyme-rich and are used to relieve gas, reduce blood cholesterol and blood pressure, prevent blood clots, fight intestinal infections, prevent stomach cancer, and aid digestion. Onions are also an effective antiseptic.

Pau D'Arco

Pau d'arco (also called Lapacho, taheebo, trumpet bush, and Ipe Roxo) is taken from the inner bark of a tree that grows abundantly along the Andes Mountains of South America and deep in the Amazon jungle. For over a thousand years, tribes of the Inca Empire used this marvelous tea for its healing powers.

Pau d'arco is especially rich in phytochemicals called quinones—primarily lapachol, considered one of the most important antitumor and antibacterial agents.

Enzyme-rich pau d'arco has been used in South America to treat a myriad of illnesses, including cancer, flu, colds, fevers, infections, gonorrhea, syphilis, and malaria. In addition, various reports indicate that tea made from the tree's inner bark has been successful in treating diabetes, leukemia, external wounds, boils, chlorosis (believed to be associated with iron-deficiency anemia), allergies, candidiasis, ulcers, and rheumatism.

Probiotics

According to probiotics expert Natasha Trenev, "Probiotics is a category of dietary supplements consisting of beneficial microorganisms. Beneficial microorganisms limit the proliferation of disease-causing microorganisms by competitive exclusion in the gastrointestinal tract of man and animals."

Billions of bacteria live in our gastrointestinal tracts (some bad, some good). For our bodies to function normally, beneficial microorganisms are essential. However, all forms of stress to our body (such as, disease, infection, injury, toxic pollution, etc.) can cause a reduction in these needed bacteria.

Supplementation with probiotics may be necessary to help our bodies stay in balance and, when ill, return to health. Probiotic supplements are a safe, natural, and effec-

tive way to help maintain this balance by reintroducing helpful bacteria into the gastrointestinal tract.

Probiotics are responsible for several activities in the gut, including the manufacture of some B vitamins, including biotin, niacin, folic acid, and pyridoxine; the production of lactase, an enzyme that helps digest the lactose in milk products; the recycling of the female hormone estrogen; and the production of antibacterial substances that kill disease-causing bacteria (they also make the environment more acidic, depriving the "bad" guys of the nutrients they need). These activities result in the improvement of digestive tract function.

Probably the best known probiotic, *Lactobacillus acidophilus* is the primary inhabitant of the small intestine. It manufactures the enzyme lactase, which digests the sugars in milk and produces lactic acid. This increased acid environment kills candida yeasts and inhibits other unwanted bacteria. *L.acidophilus* is especially sensitive to antibiotics, as well as poor diet and stress. Other probiotics include several other lactobacillus bacteria, as well as bifidobacteria and streptococcal bacteria.

Reishi Mushrooms

For four thousand years, enzyme-rich reishi mushrooms (*Ganoderma Lucidum*) have been used as part of Japanese and Chinese medicine, especially in the treatment of chronic hepatitis, nephritis, hepatopathy, neurasthenia, arthritis, bronchitis, asthma, gastric ulcer, and insomnia. According to Christopher Hobbs (*Medicinal Mushrooms*. Santa Cruz, CA: Botanica Press, 1995), whole reishi extracts have demonstrated a number of pharmacological effects (either *in vivo* or in *vitro*), including analgesic, anti-allergy, bronchitis preventative, anti-inflammatory, antibacterial, antioxidant, antitumor, antiviral, blood-pressure lowering, bone marrow enhancing, cardiotonic, central nervous system depressant and peripheral anticholinergic, expectorant and antitussive immune-stimulating, anti-HIV and adrenocortical effects.

Royal Jelly

Royal jelly is the milky-white substance produced in the glands of worker honey bees to feed the queen bee. The queen bee grows 40- to 60-percent larger than the worker bees and lives five or more years longer than her genetically identical sisters.

Royal jelly contains all of the B complex vitamins, including a high concentration of pantothenic acid (B_5) and pyridoxine (B_6), and is the only natural source of pure acetylcholine. Royal jelly also contains minerals, vitamins A, C, D, and E, enzymes, hormones, eighteen amino acids, and antibiotic components.

Royal jelly has antibacterial properties; has antibiotic activity against a variety of microorganisms; aids in treating bronchial asthma, liver disease, pancreatitis, insomnia, stomach ulcers, kidney disease, bone fractures, and skin disorders; decreases total serum lipids in some people; and is a potentiator of the immune system. It spoils easily, so keep it refrigerated and tightly sealed.

In Japan, royal jelly has been taken orally. Topically, it has been used to fight certain skin diseases. It is also used in cosmetics. *Warning:* Although rare, royal jelly may cause allergic reactions in some people.

St. John's Wort (Hypericum)

St. John's wort owes its name to the fact that it flowers on the feast of St. John the Baptist on June 24th. Recent studies of this herb have focused on the antiretroviral effects of hypericin and pseudohypericin, two key components of the plant. This enzyme-rich herb is used to treat gastrointestinal disorders, HIV/AIDS, and depression, and when applied topically can relieve minor burns and skin irritations.

Saw Palmetto

Widely used in Europe, enzyme-rich saw palmetto berries have gained a lot of attention in the United States lately as a cure for prostate problems, especially benign prostatic hyperplasia (enlargement of the prostate). But it is also effective in treating respiratory conditions (such as coughs, bronchitis, asthma) and urinary tract infections.

Shiitake Mushrooms

Shiitake mushrooms (*Lentinula Edodes*) contain lentinan (a polysaccharide), believed to have antitumor activity. Shiitake also has antiviral activity, is immune enhancing, lowers cholesterol and blood pressure, helps produce antibodies to hepatitis B, protects the liver from immunological damage, improves liver function, builds vitality in the elderly, protects physically active people from exhaustion and overwork, reduces bronchial inflammation, regulates urinary incontinence, and inhibits growth of HIV. It is also rich in enzymes.

Spirulina

Spirulina is a green food that grows wild in lakes in certain warm weather climates. Spirulina is a whole food rich in minerals, vitamins, enzymes, proteins, essential amino acids, and essential fatty acids, especially gamma-linolenic acid. Spirulina is a good source of beta-carotene. In addition, the iron in spirulina is easily absorbed. Recent research studies have indicated that spirulina is effective in certain therapeutic applications, including improving immune function, decreasing serum cholesterol levels, improving liver function, and increasing beneficial bacteria in the gastrointestinal tract (when eaten regularly).

Wheat Grass

Wheat grass contains enzymes (including the antioxidant enzymes superoxide dismutase, catalase, and glutathione peroxidase), chlorophyll, amino-acid chains (polypeptides), and bioflavonoids. Wheat grass is an excellent natural source of minerals. It is especially high in magnesium, calcium, and potassium and antioxidant vitamins A (as beta-carotene), C, and E. Wheat grass is purported to stimulate blood circulation, promote liver function, and guard against degenerative disease. Fresh wheat grass juice may be taken orally or rectally.

Enzymes cannot function without helpers, such as vitamins and minerals. These nutrients, along with those found in herbs, functional foods, glandulars, and amino acids can also benefit your body's enzymes. Don't shortchange yourself. Take enzymes and their helpers daily. This dynamic combination can also help fight a multitude of disorders that we will discuss in Part Two of this book.

PART TWO

THE CONDITIONS

INTRODUCTION

Part One examined our body's enzymes, enzyme applications, and enzyme helpers. Because of age, stress, injury, environmental pollution, our fast-paced lifestyles, poor eating habits, and nutrient-poor foods, our bodies require enzyme supplementation for energy, to function, and to survive.

Although each organ has a specific task, no organ is an isolated entity. Each organ is part of a mutually interacting network. This is why a local disturbance can affect the entire body. For example, a chronic tooth abscess can trigger the mechanism for circulatory disturbances, joint inflammation, and stomach ulcers.

As mentioned in Chapter 5 of Part One, supplemental enzymes are used as digestive aids, as systemic enzyme therapy, or as a means to increase the absorption and bioavailability of other nutrients. Enzymes are probably best known for their ability to improve digestion by breaking down proteins, fats, and carbohydrates in our foods. In this way, digestive enzymes improve health by allowing the nutrients in our foods to be extracted, absorbed, and carried through the bloodstream to the organs and cells of the body. This is why improving digestion can benefit so many conditions that, at first glance, don't appear to have anything to do with digestion (such as allergies, acne, aging, headaches, and gout, to name a few). By augmenting the body's enzymes, supplemental digestive enzymes free pancreatic enzymes to perform other health functions in the body, such as boosting immune function, decreasing inflammation, and improving circulation. Digestive enzymes aid normal digestion by helping to replace enzymes that are lost when food is processed or cooked. Further, digestive enzymes are essential for proper detoxification and for maintaining healthy flora in the colon. Improving digestion can enhance health and help your body fight just about every illness.

Systemic enzyme therapy takes the use of enzymes one step further by allowing the enzyme to enter the bloodstream intact and be carried to every cell in the body. In this way, systemic enzyme therapy can fight inflammation and stimulate the body's own enzymatic processes, decrease swelling and pain, combat free radicals, improve circulation, and bolster immunity. Because nearly every disease or disorder involves one or more of these processes, systemic enzyme therapy is effective against a wide range of diseases. Systemic enzyme therapy is an integrative, holistic approach to health and healing.

In addition to digestive and systemic uses, enzymes can also be used in formulas containing herbs, vitamins, minerals, and food extracts. I call these combinations Enzyme Absorption System Enhancers (EASE). Combining enzymes in these formulations is beneficial because they provide a number of unique opportunities:

- They improve the absorption of herbs, plant food constituents, vitamins, minerals, and other nutrients.
- They improve the bioavailability of these same nutrients.
- They maximize enzyme activity.
- They reduce the drain on the body's own digestive enzymes.

Food extracts contain their own enzymes, as well as phytochemicals. These enzyme-rich "designer" foods should be used daily to maintain health and to fight injury and disease whenever they occur.

All four of these supplemental enzyme uses—supplemental digestive enzymes, systemic therapy enzymes, Enzyme Absorption System Enhancers, and food extracts—are essential for health and work synergistically.

Part Two discusses the disorders in the body that can be helped or healed with enzyme therapy. The discussion of each condition contains a description including its symptoms, as well as a treatment plan including essential enzymes and supportive enzymes.

In the "Essential Enzymes" charts in the discussion of each condition, those enzymes from which you must select one are listed in order of effectiveness. All of the enzyme choices are generally similar in activity and effectiveness, so, if you are unable to obtain or take the first enzyme listed, feel free to take the second, third, or any subsequent enzyme.

As mentioned in Chapter 6 of Part One, vitamins and minerals and other enzyme helpers are required if enzymes are to do their jobs effectively. For this reason, the discussion of each condition also includes information on helpful vitamins, minerals, EASE formulas, food extracts, and other helpers, and other relevant comments. A certain amount of helpers should be taken every day, regardless of the condition or the enzymes used. The ideal multivitamin, multimineral, and multiglandular enzyme-helper complexes are listed below. Please note and incorporate these helpers into the treatment programs. The treatment plan for each condition will only list those necessary supplements not already mentioned in the multivitamin, multimineral, and multi-glandular enzyme-helper complexes listed below.

MULTIVITAMIN ENZYME HELPER COMPLEX

VITAMIN	DOSE
Vitamin A	10,000 IU
Beta carotene	17,500 IU
Vitamin B_1 (thiamin)	50–100 mg
Vitamin B_2 (riboflavin)	50–100 mg
Vitamin B_3 or	100 mg
Niacin or niacin-bound chromium	2–4 mg
Vitamin B_5 (pantothenic acid)	50–100 mg
Vitamin B_6 (pyridoxine)	50–100 mg
Vitamin B_{12} (cobalamin)	100 mcg
Biotin	100–300 mcg
Folic acid	400 mcg
Para-aminobenzoic acid	50–100 mg
Choline	25–100 mg
Inositol	25–50 mg
Vitamin C	1,000–2,000 mg
Vitamin D	400 IU
Vitamin E	800 IU
Vitamin K	50 mcg
Bioflavonoids	100–300 mg
Essential fatty acid complex	100 mg

MULTIMINERAL ENZYME HELPER COMPLEX

MINERAL	DOSE
Boron	3 mg
Calcium	1,000 mg
Chromium	100–200 mcg
Copper	2–3 mg
Iodine (kelp)	150 mcg
Magnesium	500 mg
Manganese	5–10 mg
Molybdenum	50–100 mcg
Phosphorus	200 mg
Potassium	99 mg
Selenium	100–200 mcg
Silica	100 mg
Vanadium	10 mcg
Zinc	50 mg

MULTIGLANDULAR ENZYME HELPER COMPLEX

GLANDULAR	DOSE
Raw adrenal	100 mg
Raw liver	25 mg
Raw pancreas	200 mg
Raw pituitary	120 mg
Raw spleen	10 mg
Raw thymus	10 mg
Raw thyroid substance	16 mg

Be sure to take your vitamins, minerals, herbs, and food extracts with enzymes. This increases the absorption and bioavailability of the nutrients. If the enzymes are not already included in the supplement, buy proteolytic, amy-lolytic, and lipolytic enzymes.

Food extracts can still contain enzymes after processing (if not overheated). Some companies combine food extracts with enzymes, vitamins, minerals, and herbs. These combinations are helpful.

Choosing the right products for your nutritional program can be overwhelming. Use these four easy steps to help you address your nutritional needs:

1. Choose enzymes, vitamins, minerals, and herbs that are most appropriate for you. Most people vary in their needs; for instance, it is not advisable for pregnant and lactating women to take many supplements, including enzymes and herbs, without their doctors' approval.
2. Identify the areas of your health status that have specific needs and require extra support.
3. Identify symptoms and disorders that relate specifically to you.
4. Choose the appropriate treatment program (or programs) from this book. Do not forget the importance of phytonu-trients, herbs, food extracts, and their associated enzymes to enhance well-being.

During the initial stages of enzyme therapy, the symptoms of a chronically ill person may occasionally increase in severity. This is because toxins are being drawn from the cells and entering the bloodstream for elimination. This temporary increase in symptoms can be a positive sign that a therapeutic reaction is taking place. The enzymes should not be discontinued. However, a temporary dose reduction may be considered.

If you see no change after taking the enzymes for two to three days, especially with digestive disorders, acute trauma and minor soft tissue injuries, or inflammatory conditions, it may be necessary to increase the dosage. On the other hand, rheumatoid arthritis and many other chronic conditions may not improve for four to six weeks. This is because it takes time to remove the antigen-antibody complexes from the affected tissues. In addition, because of individual differences, some people may need more enzymes than others.

If you have any questions about the appropriateness of any of these recommended therapies, seek the advice of an alternative health-care physician or a nutrition professional well trained in the use of enzymes. This book and the programs contained herein are not intended to replace the services of your health-care professional.

Troubleshooting for Symptoms and Possible Causes

Part Two will provide you with treatment programs for a number of conditions. Sometimes, however, you may experience certain symptoms, but do not know what disorder may be causing the condition. The table below lists some symptoms and provides you with some possible causes of those symptoms. Do not, however, try to diagnose yourself and treat whatever condition you think you may have. Use this table as a guide to help you and your physician determine the disorder that may be affecting you. Once the disorder is determine, you and your physician can incorporate the appropriate enzyme therapy into your overall treatment program.

SYMPTOMS	POSSIBLE CAUSES
Abdominal cramps or pain	Acidosis (especially in diabetes mellitus); Alcoholism; Allergies; Asthma; Bladder Infection; Carbohydrate Intolerance; Cardiovascular Disorders; Celiac Disease; Colorectal Cancer; Constipation; Diarrhea; Diverticular Disease; Fibroids or polyps of the uterus; Gas, Abdominal; Gastritis; Hangover; Hay Fever; Hepatitis; Indigestion; Inflammatory Bowel Disease; Lactase Deficiency; Leaky Gut Syndrome; Pancreatic Cancer; Pancreatitis; Peptic Ulcer; Pleurisy; Pneumonia; Premenstrual Syndrome; Rheumatic Fever; Steatorrhea; Stress; Thrombosis.
Abscesses, cysts, pustules	Acne; Allergies; Boils; Chickenpox; Eczema; Gout; Herpes Zoster; Psoriasis; Rheumatic Fever; Rheumatoid Arthritis; Skin Cancer; Staphylococcal Infection; Warts.
Aches, general	Cancer; Chronic Fatigue Syndrome; Infection; Influenza; Mumps; Myofascial Pain Syndrome; Osteoarthritis; Pneumonia; Rheumatic Fever; Rheumatoid Arthritis; Sprains and Strains; Subluxations; Systemic Lupus Erythematosus; Whiplash.
Anxiety	Alcoholism; Alzheimer's Disease; Premenstrual Syndrome.
Appetite loss	Abscess; Alcoholism; Allergies; Anemia; Ankylosing Spondylitis; Cancer; Chickenpox; Cirrhosis; Constipation; Gastritis; Glomerulonephritis; Hay Fever; Inflammatory Bowel Disease; Influenza; Kidney Disorders; Mumps; Radiation Sickness; Stress.
Appetite, increased	Diabetes; Gastritis; Hypoglycemia; Peptic Ulcers; Premenstrual Syndrome; Stress.
Back Pain	Ankylosing Spondylitis; Bone Fractures (of the vertebrae) Bronchitis; Cancer (of the spinal cord and surrounding areas, vertebrae, prostate, and other organs); Cardiovascular Disorders; HIV/AIDS; Infections; Kidney Disorders; Muscle Cramps; Myofascial Pain Syndrome; Obesity; Osteoarthritis; Osteoporosis; Peptic Ulcer; Pneumonia; Premenstrual Syndrome; Rheumatoid Arthritis; Scoliosis; Slipped Disc; Sprains and Strains; Subluxation; Whiplash.
Bad breath	Abscess; Adenoiditis; Alcoholism; Cancer (of the mouth, tonsils, or throat); Constipation; Diabetes; Gastritis; Indigestion; Measles; Sinusitis; Tonsillitis; Tooth Decay.
Bed wetting (enuresis)	Adenoiditis; Cancer (of the prostate, spinal cord, or internal organs); Multiple Sclerosis; Prostate Disorders.
Belching and burping, frequent	Acidosis; Alcoholism; Carbohydrate Intolerance; Gastritis; Hangover; Heartburn; Indigestion; Nausea; Pancreatic Insufficiency; Stress.
Blinking, repeated	Aging; Allergies; Alzheimer's Disease; Cataracts; Conjunctivitis; Glaucoma; Infection; Retinopathy; Sjögren's Syndrome; Stroke.
Bloating, abdominal	Alcoholism; Allergies; Cancer (of the stomach, abdomen, kidneys, or colon); Carbohydrate Intolerance; Cardiovascular Disease; Celiac Disease; Cirrhosis; Constipation; Diverticular Disease; Gas, Abdominal; Gastritis; Hangover; Indigestion; Inflammatory Bowel Disease; Kidney Disorders; Lactase Deficiency; Leaky Gut Syndrome; Obesity; Pancreatic Cancer; Pancreatic Insufficiency; Pancreatitis; Peptic Ulcer; Premenstrual Syndrome; Steatorrhea.
Blood in the sputum	Abscess (in a lung); Bronchitis; Cancer (of the oral cavity, lungs, or throat).
Blood in the stool	Abscess; Allergies; Cirrhosis; Colorectal Cancer; Constipation; Diarrhea; Diverticular Disease; Hemorrhoids; HIV/AIDS; Infections; Inflammatory Bowel Disease; Peptic Ulcers; Warts (genital). The darker the blood, the deeper the source of the blood is in the body. With bright red blood and streaked stools, the bleeding is closer to the anus.

83

SYMPTOMS	POSSIBLE CAUSES
Blood in the urine	Anemia (pernicious); Cancer (of the bladder, kidneys, uterus, urethra, vagina, prostate or penis, or leukemia); Glomerulonephritis; Infections; Kidney Disorders; Prostate Disorders.
Body odor	Diabetes; Indigestion; Infection.
Breast lumps	Abscess; Boils; Cancer; Fibrocystic Breast Disease; Premenstrual Syndrome.
Breast tenderness	Abscess; Breast Cancer; Fibrocystic Breast Disease; Boils; Premenstrual Syndrome.
Breathing difficulties	Allergies; Anemia; Asthma; Bronchitis; Cancer; Cardiovascular Disorders; Cystic Fibrosis; Diabetes; Empyema; Hay Fever; Pleurisy; Pneumonia; Sinusitis.
Bruising	Aging; Alcoholism; Allergies; Anemia; Cancer; HIV/AIDS; High Blood Pressure; Infection; Measles; Sprains and Strains; Surgery; Varicose Veins.
Chest pain	Abscess; Allergies; Anemia; Angina Pectoris; Asthma; Bone Fracture (of the sternum or ribs); Bronchitis; Bruises (of the sternum or ribs); Cancer (especially lung); Cardiovascular Disorders; Embolism (pulmonary); Gas, Abdominal; Gastritis; Heartburn; Herpes Zoster; Indigestion; Influenza; Osteoarthritis; Pleurisy; Pneumonia; Scoliosis; Steatorrhea; Strains; Stress.
Chills	Anemia; Bronchitis; Cancer (of the stomach); Ear Infection; Glomerulonephritis; Herpes Zoster; Infection; Influenza; Kidney Disorders; Mumps; Pneumonia; Stress.
Cold hands and feet	Arteriosclerosis; Cardiovascular Disorders; Diabetes; Embolism; Raynaud's Disease; Stress; Thrombosis.
Cold sweats	AIDS; Angina Pectoris; Asthma; Cancer; Cardiovascular Disorders; Diabetes; High Blood Pressure; Influenza; Mononucleosis; Stress.
Constipation	Aging; Alcoholism; Carbohydrate Intolerance; Celiac Disease; Colorectal Cancer; Diverticular Disease; Embolism; Indigestion; Pancreatic Insufficiency; Premenstrual Syndrome;Thrombosis.
Coughing, persistent	Adenoiditis; Allergies; Asthma; Bronchitis; Cardiovascular Disorders; Cystic Fibrosis; Hay Fever; Influenza; Laryngitis; Lung Cancer; Measles; Pharyngitis; Pleurisy; Pneumonia.
Decreased resistance to infection	Anemia; HIV/AIDS.
Delirium	Aging; Alcoholism; Alzheimer's Disease; Cardiovascular Disorders; Neuritis (alcoholic); Pneumonia; Stroke.
Depression	Aging; Alcoholism; Alzheimer's Disease; Anemia; Hay Fever; Headache; HIV/AIDS; Hypoglycemia; Premenstrual Syndrome; Stress.
Diarrhea	Alcoholism; Allergies; Asthma; Carbohydrate Intolerance; Celiac Disease; Colorectal Cancer; Diabetes; Diverticular Disease; Gastritis; Hay Fever; Hepatitis; Indigestion; Inflammatory Bowel Disease; Influenza; Pancreatic Insufficiency; Radiation Sickness; Steatorrhea.
Disorientation	Aging; Alcoholism; Alzheimer's Disease; Anemia; Angina Pectoris; Cancer (of the brain); Cardiovascular Disorders; Celiac Disease; Embolism; Hypoglycemia; Stroke.
Dizziness	Allergies; Anemia; Angina Pectoris; Asthma; Brain Tumor; Cardiovascular Disorders; Diabetes; High Blood Pressure; Infections; Hypoglycemia; Osteoarthritis; Premenstrual Syndrome; Stress; Stroke; Temporomandibular Joint Dysfunction (TMJ); Vertigo; Whiplash.
Drooling	Brain Tumor and other Cancer; Infection; Stroke.
Drowsiness	Alcoholism; Anemia (due to hemorrhage); Arteriosclerosis; Brain Tumors or other Cancer; Diabetes; Obesity; Stroke; Thrombosis.
Drug sensitivity, increased	Alcoholism; Allergies; Cirrhosis.
Dryness of the mouth	Aging; Alkalosis; Brain Tumor; Diabetes; Gastritis; Glomerulonephritis; Indigestion; Mumps; Sjögren's Syndrome.
Ear discharge	Dermatitis (chronic of the external ear canal); Infection.

SYMPTOMS	POSSIBLE CAUSES
Ear infections, recurrent	Abscess; Adenoiditis; Allergies; Cancer; Herpes Zoster.
Eczema	Allergies; Varicose Veins.
Eye bulging	Glaucoma; High Blood Pressure; Infection; Sinusitis; Thrombosis.
Eyelid drooping	Aging; Bell's Palsy; Brain Tumors; Diabetes; Headache (migraine); Hematoma; Thrombosis.
Fainting	Alcoholism; Stroke.
Fatigue	Abscess; Aging; Anemia; Angina Pectoris; Ankylosing Spondylitis; Arteriosclerosis; Cardiovascular Disorders; Celiac Disease; Chronic Fatigue Syndrome; Diabetes; Embolism; Hepatitis; High Blood Pressure; Hypoglycemia; Inflammatory Bowel Disease; Influenza; Insect Bites and Stings; Leaky Gut Syndrome; Multiple Sclerosis; Pleurisy; Pneumonia; Polymyositis; Premenstrual Syndrome; Raynaud's Disease; Rheumatic Fever; Scoliosis; Sjögren's Syndrome; Stress; Stroke; Thrombosis.
Fever	Abscess; Adenoiditis; Anemia; Angina Pectoris; Ankylosing Spondylitis; Asthma; Bladder Infection; Bone Fractures; Bronchitis; Cancer (particularly kidney cancer, leukemia, and lymphoma); Cardiovascular Disorders; Chickenpox; Cirrhosis; Colds; Diabetes; Ear Infection; Gout; Guillain-Barré Syndrome; Headache (migraine); Hepatitis; Herpes Zoster; HIV/AIDS; Infections; Inflammatory Bowel Disease; Influenza; Kidney Disorders; Measles; Mononucleosis; Mumps; Pleurisy; Pneumonia; Polymyositis; Rheumatic Fever; Systemic Lupus Erythematosus.
Gas, flatulence	Alcoholism; Allergies; Asthma; Carbohydrate Intolerance; Celiac Disease; Colorectal Cancer; Gastritis; Hangover; Indigestion; Inflammatory Bowel Disease; Leaky Gut Syndrome; Pancreatic Insufficiency; Pancreatitis; Steatorrhea; Stress.
Headache	Abscess (of scalp); Alcoholism; Allergies; Anemia; Arteriosclerosis; Asthma; Bone Fracture (skull); Brain Tumor; Celiac Disease; Chickenpox; Ear Infection; Glaucoma; Hangover; Hay Fever; High Blood Pressure; Hypoglycemia; Infection; Influenza; Leaky Gut Syndrome; Mumps; Premenstrual Syndrome; Radiation Sickness; Retinopathy; Sinusitis; Stress; Stroke; Subluxation; Temporomandibular Joint Dysfunction (TMJ); Whiplash.
Hearing loss	Aging; Brain Tumors; Cancer; Ear Infection; Eczema (of the external ear); Glomerulonephritis; Hay Fever; Measles; Mumps; Neuritis; Stroke.
Heartbeat, irregular	Aging; Alcoholism; Anemia; Angina Pectoris; Arteriosclerosis; Asthma; Cancer; Cardiovascular Disorders; Glomerulonephritis; High Blood Pressure; Obesity; Radiation Sickness; Stress.
Heartburn (acid stomach)	Alcoholism; Cancer (of the stomach); Gas, Abdominal; Gastritis; Indigestion; Pancreatic Insufficiency.
Hives (urticaria)	Allergies; Asthma; Hay Fever; Hepatitis; Insect Bites and Stings.
Hoarseness/harshness	Alcoholism; Laryngitis; Pharyngitis.
Hyperactivity	Hypoglycemia.
Incontinence	Aging; Alzheimer's Disease; Diabetes; Multiple Sclerosis; Prostate Cancer; Prostate Disorders; Stroke.
Indigestion (dyspepsia)	Acidosis; Alcoholism; Alkalosis; Allergies; Anemia (pernicious), Angina Pectoris; Arteriosclerosis; Cancer (of the stomach, pancreas, or liver); Cardiovascular Disorders; Cirrhosis; Constipation; Cystic Fibrosis; Gas, Abdominal; Gastritis; Hepatitis; Hypochlorhydria; Infections; Pancreatic Insufficiency; Pancreatitis; Peptic Ulcer; Rheumatoid Arthritis; Stress.
Inflammation	Bursitis and Synovitis; Conjunctivitis; Dermatitis; Dermatomyositis; Diverticular Disease; Ear Infections; Epididymitis; Gingivitis; Gout; Inflammatory Bowel Disease; Laryngitis; Myofascial Pain Syndrome; Neuritis; Osteoarthritis; Pancreatitis; Polymyositis; Reiter's Syndrome; Rheumatoid Arthritis; Sinusitis; Sprains and Strains; Systemic Lupus Erythematosus; Tendonitis; Tonsillitis.
Insomnia	Adenoiditis; Aging; Alcoholism; Alzheimer's Disease; Anemia; Arteriosclerosis; Asthma; Bronchitis; Cardiovascular Disorders; Cirrhosis; Constipation; Diabetes; Gastritis; Hay Fever; High Blood Pressure; Indigestion; Pneumonia; Stress.
Intercourse, painful	HIV/AIDS; Infection; Warts (genital).

SYMPTOMS	POSSIBLE CAUSES
Irritability and agitation	Aging; Alcoholism; Allergies (food); Alzheimer's Disease; Anemia; Brain Tumor; Diabetes; Hypoglycemia; Premenstrual Syndrome; Stress; Stroke.
Itching	Allergies; Hemorrhoids; Insect Bites and Stings; Warts.
Joint aches and pains	Aging; Ankylosing Spondylitis; Bone Fractures; Bursitis and Synovitis; Cancer; Carpal Tunnel Syndrome; Cirrhosis; Diabetes; Gout; Hepatitis; Infections; Inflammatory Bowel Disease; Kidney Disorders; Neuritis; Osteoarthritis; Osteoporosis; Polymyositis; Premenstrual Syndrome; Reiter's Syndrome; Rheumatic Fever; Rheumatoid Arthritis; Scoliosis; Slipped Disc; Sprains and Strains; Stress; Subluxation; Systemic Lupus Erythematosus (SLE); Temporomandibular Joint Dysfunction (TMJ); Tendonitis; Whiplash.
Loss of libido	Aging; Alzheimer's Disease; Cirrhosis; Prostate Cancer; Prostate Disorders; Stress.
Memory loss (dementia)	Aging; Alcoholism; Alzheimer's Disease; Celiac Disease; Embolism; Stroke.
Mood swings	Alcoholism; Alzheimer's Disease; Cancer; Cardiovascular Disorders; Hypoglycemia; Premenstrual Syndrome.
Morning stiffness	Ankylosing Spondylitis; Myofascial Pain Syndrome; Osteoarthritis; Rheumatoid Arthritis; Slipped Disc; Sprains and Strains.
Mouth breathing	Adenoiditis; Allergies; Asthma; Hay Fever; Sinusitis.
Mouth sores	Anemia (pernicious); Cancer; Canker Sores; Chickenpox; Measles.
Muscle control, loss of	Alcoholism; Bell's Palsy; Brain Tumor; Diabetes; Embolism (brain); Multiple Sclerosis; Neuritis; Sprains and Strains.
Muscle pain	Anemia; Bone Fracture; Chronic Fatigue Syndrome; Dermatomyositis; Diabetes; Glomerulonephritis; Infection; Influenza; Myofascial Pain Syndrome; Neuritis; Obesity; Polymyositis; Sprains and Strains; Stress; Subluxation; Tendonitis; Whiplash.
Muscle weakness	Bell's Palsy; Bone Fractures; Cancer (Hodgkin's disease); Cardiovascular Disorders; Embolism; HIV/AIDS; Influenza; Multiple Sclerosis; Myofascial Pain Syndrome; Neuritis; Osteoarthritis (of the spine); Polymyositis; Rheumatoid Arthritis; Sprains and Strains; Stroke; Whiplash.
Nasal congestion	Allergies; Asthma; Bronchitis; Cancer; Empyema; Hay Fever; Measles; Pharyngitis; Pneumonia; Premenstrual Syndrome; Sinusitis.
Nausea	Alcoholism; Allergies; Asthma; Cancer (brain, colorectal); Carbohydrate Intolerance; Cardiovascular Disorders; Celiac Disease; Chickenpox; Cirrhosis; Diabetes; Gas, Abdominal; Gastritis; Hay Fever; Headache (migraine); Hepatitis; HIV/AIDS; Indigestion; Inflammatory Bowel Disease; Influenza; Kidney Disorders; Mononucleosis; Pancreatic Insufficiency; Pancreatitis; Peptic Ulcers; Radiation Sickness; Sinusitis; Stress; Vertigo; Whiplash.
Night sweats	Bronchitis; Cancer; Cardiovascular Disorders; Hepatitis; HIV/AIDS; Influenza; Pneumonia; Rheumatic Fever; Stress.
Numbness	Anemia (pernicious); Arteriosclerosis; Brain Tumor; Carpal Tunnel Syndrome; Diabetes; Embolism; Gout; Multiple Sclerosis; Neuritis; Osteoarthritis (of the spine); Raynaud's Disease; Rheumatoid Arthritis; Sciatica; Stroke; Thrombosis; Whiplash.
Pain, general	Bone Fracture; Diverticular Disease; Gastritis; Heartburn; Indigestion; Inflammatory Bowel Disease; Kidney Disorders; Muscle Cramps; Pancreatitis; Peptic Ulcers; Sprains and Strains; Whiplash.
Pain, arm	Angina Pectoris; Bone Fracture (humerus, radius, or ulna); Herpes Zoster; Infection; Osteoarthritis; Slipped Disc; Whiplash.
Pain, facial	Bell's Palsy; Headache; Herpes Zoster; Infection; Neuritis; Sinusitis (maxillary).
Pain, leg	Abscess; Arteriosclerosis; Bone Fractures; Cancer; Diabetes; Guillain-Barré Syndrome; Infections; Intermittent Claudication; Multiple Sclerosis; Muscle Cramping; Neuritis; Osteoarthritis; Osteoporosis; Sciatica; Slipped Disc; Sprains and Strains; Tendonitis; Thrombosis; Varicose Veins.
Pain, nasal	Abscess (of the nasal septum); Boils; Bone Fracture; Hay Fever; Sinusitis.

SYMPTOMS	POSSIBLE CAUSES
Pale or sallow skin, lips, fingernails, palms, or linings of the eyelids	Anemia; Arteriosclerosis; Cancer; Cardiovascular Disorders; Cirrhosis; Peptic Ulcers; Raynaud's Disease; Rheumatoid Arthritis.
Post-nasal drip	Adenoiditis; Allergies; Asthma; Bronchitis; Colds; Hay Fever; Sinusitis.
Rash, or flushing skin	Acne; Alcoholism; Allergies; Arteriosclerosis; Asthma; Burns; Chickenpox; Cirrhosis; Constipation; Dermatitis; Gout; Hay Fever; Herpes Simplex; Herpes Zoster; High Blood Pressure; Insect Bites and Stings; Measles; Polymyositis; Psoriasis; Radiation Sickness; Raynaud's Disease; Rheumatic Fever; Rosacea; Stress.
Seizures	Aging; Alcoholism; Alzheimer's Disease; Brain Tumor; Stroke.
Ringing in the ear (tinnitus)	Allergies; Anemia; Arteriosclerosis; Cancer; Cardiovascular Disorders; Ear Infections; Glomerulonephritis; High Blood Pressure; Temporomandibular Joint Dysfunction (TMJ); Whiplash.
Runny nose (nasal discharge)	Adenoiditis; Allergies; Asthma; Bone Fracture (in the nose); Cold or other Viral Infection; Empyema; Influenza; Hay Fever; Measles; Sinusitis.
Shortness of breath	Abscess (on the lung); Allergies; Anemia; Angina Pectoris; Asthma; Bone Fracture (ribs); Bronchitis; Cardiovascular Disorders; Cholesterol, Elevated; Empyema; Lung Cancer; Pleurisy; Pneumonia.
Skin hemorrhages	Allergies; Cancer (leukemia); Measles.
Skin redness	Allergies; Bruises and Hematomas; Bursitis; Dermatitis; Insect Bites and Stings; Psoriasis; Sprains and Strains; Tendonitis.
Skin ulcers	Thrombosis; Varicose Veins.
Sneezing	Allergies; Asthma; Cold; Hay Fever; Influenza; Measles; Sinusitis; Stress.
Snoring	Adenoiditis.
Sore throat	Abscess; Adenoiditis; Angina; Bronchitis; Cold or other Viral Infection; Influenza; Laryngitis; Mononucleosis; Pharyngitis; Pneumonia; Tonsillitis.
Speech difficulties	Abscess; Aging; Alcoholism; Alzheimer's Disease; Brain Tumor; Embolism; Headache (migraine); Hypoglycemia; Multiple Sclerosis; Stroke; Thrombosis.
Stomachache (or stomach discomfort)	Alcoholism; Allergies; Anemia; Angina Pectoris; Arteriosclerosis; Cancer (liver, stomach); Carbohydrate Intolerance; Cardiovascular Disorders; Constipation; Gas, Abdominal; Gastritis; Hepatitis; Indigestion; Infections; Pancreatic Cancer; Pancreatic Insufficiency; Pancreatitis; Peptic Ulcer; Rheumatoid Arthritis.
Swallowing difficulties	Adenoiditis; Allergies; Cancer (of the throat); Heartburn; Indigestion; Laryngitis; Pharyngitis; Sjögren's Syndrome; Stress; Tonsillitis.
Sweating, excessive	Alcoholism; Allergies (food); Cancer (Hodgkin's disease, lymphoma); Cardiovascular Disorders; Cystic Fibrosis; Diabetes; HIV/AIDS; Infection; Kidney Disorders; Pneumonia; Stress.
Swelling (edema)	Allergies; Arthritis; Bursitis; Cardiovascular Disorders; Cirrhosis; Diabetes; Infection; Lymphedema; Premenstrual Syndrome; Sports Injuries; Sprains and Strains; Steatorrhea; Systemic Lupus Erythematosus (SLE); Tendonitis; Thrombosis; Varicose Veins.
Swollen joints	Bursitis; Dermatomyositis; Gout; Osteoarthritis; Rheumatoid Arthritis; Subluxations.
Swollen legs, ankles, feet, and hands	Allergies (food); Cardiovascular Disorders; Kidney Disorders; Sprains and Strains; Thrombosis.
Swollen lips	Abscess; Allergies; Asthma; Cancer; Dermatitis; Hay Fever; Herpes Simplex; HIV/AIDS; Insect Bites and Stings.
Swollen lymph nodes	Anemia (sickle-cell); Cancer; Herpes Simplex; Herpes Zoster; HIV/AIDS; Infection; Lymphedema; Measles; Mononucleosis; Mumps.
Teeth grinding	Stress; Temporomandibular Joint Dysfunction.
Tenderness	Bruises; Cardiovascular Disorders; Gout; Osteoarthritis; Rheumatoid Arthritis; Sprains and Strains; Tendonitis.

SYMPTOMS	POSSIBLE CAUSES
Thirst, excessive	Allergies; Burns (severe); Diabetes; Diarrhea (excessive); Gastritis; Glomerulonephritis; Gout; Hangover; Infection; Pharyngitis.
Tremors	Alcoholism; Multiple Sclerosis; Stress; Stroke.
Urination, frequent	Aging; Alzheimer's Disease; Angina Pectoris; Bladder Infection; Cancer (bladder); Glomerulonephritis; Kidney Disorders; Obesity; Prostate Cancer; Prostate Disorders.
Urination, painful	Bladder Infection; Cancer (bladder); Kidney Disorders; Prostate Cancer; Prostate Disorders.
Urine, dark or cloudy with unpleasant or strong odor	Bladder Infection; Kidney Disorders.
Vaginal discharge and itching	Cancer; Herpes Zoster; Herpes Simplex; Fibroids.
Varicose veins in the legs	Aging; Obesity; Thrombosis.
Vision problems	Aging; Alcoholism; Alzheimer's Disease; Bone Fracture (around eye); Brain Tumors; Cataracts; Conjunctivitis; Diabetes; Embolism; Glaucoma; Macular Degeneration; Multiple Sclerosis; Retinopathy; Rheumatoid Arthritis; Sjögren's Syndrome; Stroke; Whiplash.
Vomiting	Alcoholism; Allergies; Asthma; Brain Tumor; Cancer (throat, stomach); Celiac Disease; Cirrhosis; Colorectal Cancer; Gastritis; Hangover; Hay Fever; Headache (migraine); Hepatitis; Indigestion; Influenza; Kidney Disorders; Pancreatitis; Radiation Sickness; Whiplash.
Watery eyes	Allergies; Asthma; Conjunctivitis; Hay Fever.
Weakness, general	Anemia; Cardiovascular Disorders; Celiac Disease; Gastritis; Insect Bites and Stings; Stroke.
Weight gain (sudden or unexplained)	Aging; Cardiovascular Disorders; Diabetes; Kidney Disorders; Premenstrual Syndrome.
Weight loss, unexplained	Abscess; Aging; Alcoholism; Alzheimer's Disease; Ankylosing Spondylitis; Cancer; Cirrhosis; Dermatomyositis; Diabetes; Hepatitis; HIV/AIDS; Infection; Inflammatory Bowel Disease; Mononucleosis; Polymyositis.
Wheezing	Adenoiditis; Allergies; Asthma; Bronchitis; Cardiovascular Disorders; Hay Fever; Lung Cancer; Pneumonia.
Wrist pain, numbness, swelling	Bone Fracture; Cancer (sarcoma in the wrist); Carpal Tunnel Syndrome; Neuritis; Osteoporosis; Rheumatic Fever; Rheumatoid Arthritis; Sprains and Strains; Subluxation (of wrist bones).

Abrasions

See SCARS; SKIN ULCERS.

Abscess

An abscess is a localized accumulation of pus, caused by infection, that can occur in a tissue, organ, or confined space in the body. It often begins as cellulitis (inflamed cells). Abscesses can occur internally or externally due to injury or the spread of infection from another area of the body. An abscess can form on almost any part of the body, including the skin, teeth, gums, liver, kidneys, breasts, and lungs.

Physicians often prescribe antibiotics for the treatment of abscesses; however, there are at least two reasons why that may not be the best treatment. First, antibiotics kill indiscriminately. As mentioned in Part One, your gastrointestinal tract relies on thousands of beneficial bacteria in order to keep you healthy. Antibiotics can't tell the good guys from the bad guys, so they kill all of them. This may lead to a yeast infection. Second, bacteria are getting smarter and stronger. In fact, in the last few years, many bacteria have actually developed a resistance to antibiotics.

Symptoms

Abscesses, whether on the surface or just below the skin, are marked by inflammation, swelling, heat, redness over the affected site, tenderness, and possibly fever. The symptoms of an internal abscess include tenderness and local pain, systemic symptoms such as fever, and nonspecific complaints such as fatigue, anorexia, and weight loss.

Enzyme Therapy

It is important to remove the abscess if possible. For example, a dentist may treat a tooth abscess by performing a root canal or extracting the tooth. Some abscesses require drainage or surgery. Systemic enzyme therapy can be used as a type of complementary care postoperatively to decrease inflammation, control the infection, dry up and eliminate pus, and encourage rapid healing from surgery. Enzyme therapy can also improve circulation, stimulate the immune system, help speed tissue repair by bringing nutrients to the damaged area and removing waste products, enhance wellness, and prevent future abscess formation.

Digestive enzyme therapy is used to improve digestion of food, reduce stress on the gastrointestinal mucosa, and aid in flushing waste products out of the colon. Enzymes help maintain normal pH levels, detoxify the body, and promote the growth of healthy intestinal flora. In this way, the body's immune system is strengthened and better able to fight infection. Digestive enzymes also serve as replacements for the body's pancreatic enzymes, thereby leaving the pancreatic enzymes free to perform other functions in the body, such as boosting immunity and decreasing inflammation.

Enzyme Absorption System Enhancers (EASE) maximize enzyme activity. They can also improve the absorption and bioavailability of various nutrients and decrease the drain on the body's own digestive enzymes, thus prolonging their lives.

In addition to the standard multivitamin, multimineral, and multiglandular complexes recommended in the introduction to Part Two, the following supplements are recommended for the treatment of abscesses. The following recommended dosages are for adults. Children between the ages of two and five should take one-quarter the recommended dosage, those between the ages of six and twelve should take one-half the recommended dosage, and those between the ages of thirteen and seventeen should take three-quarters the recommended adult dosage.

ESSENTIAL ENZYMES

Enzyme	Suggested Dosage	Actions
Superoxide dismutase	Take 150 mcg per day.	Fights inflammation; potent antioxidant.

AND SELECT ONE OF THE FOLLOWING:

Enzyme	Suggested Dosage	Actions
Formula One—See page 38 of Part One for a list of products.	Take 5 tablets twice per day between meals until infection resolves.	Fights inflammation, pain, and swelling; speeds healing; stimulates the immune system; strengthens capillary walls.
Bromelain	Take 750 mg two times per day between meals until infection resolves.	Fights inflammation, pain, and swelling; speeds healing; stimulates the immune system; strengthens capillary walls.
Serratiopeptidase	Take 10 mg two times per day between meals until infection resolves.	Fights inflammation; reduces pain and swelling.
Trypsin with chymotrypsin	Take 250 mg of trypsin and 10 mg of chymotrypsin three times per day between meals until infection resolves.	Fights inflammation, pain, and swelling; increases rate of healing; stimulates the immune system.
Pancreatin	Take 500 mg six times per day on an empty stomach until infection resolves.	Fights inflammation, pain, and swelling; increases rate of healing; stimulates the immune system; strengthens capillary walls.
Papain	Take 300 mg two times per day between meals until infection resolves.	Fights inflammation, pain, and swelling; increases rate of healing; stimulates the immune system; strengthens capillary walls.
A microbial protease	Take 300 mg two times per day between meals until infection resolves.	Fights inflammation; speeds healing.

SUPPORTIVE ENZYMES

Enzyme	Suggested Dosage	Actions
Catalase and glutathione peroxidase	Take as indicated on the labels.	Potent antioxidant enzymes.
A digestive enzyme product containing protease, amylase, and lipase. See the list of digestive enzyme products in Appendix.	Take as indicated on the label with meals.	Improves the breakdown and absorption of food nutrients; fights toxic build-up by helping to flush the gastrointestinal tract of toxins, thereby strengthening the body's immune system.

ENZYME ABSORPTION SYSTEM ENHANCERS (EASE)

Combination	Suggested Dosage	Actions
Advanced Nutritional System from Rainbow Light *or*	Take as indicated on the label.	Potent free-radical fighter and immune system stimulator.
Protease Concentrate from Tyler *and*	Take as indicated on the label.	Breaks down fibrin; fights inflammation.
Bromelain Complex from PhytoPharmica *and*	Take as indicated on the label.	Decreases swelling; improves circulation; helps break up circulating immune complexes.
Cardio-Chelex from Life Plus *or*	Take as indicated on the label.	Supports healthy circulation; improves energy.
NESS Formula #416 *and*	Take as indicated on the label.	Breaks down fibrin; fights inflammation.
NESS Formula #301	Take 3 capsules three times per day between meals.	Helps support lymphatic system.

ENZYME HELPERS

Nutrient	Suggested Dosage	Actions
Acidophilus (and other bacteria, such as bifidobacteria)	Take as indicated on the label.	Improves gastrointestinal function, thereby strengthening the immune system; needed for production of biotin, niacin, folic acid, and pyridoxine, which are needed by many of the body's enzymes; helps produce antibacterial substances; stimulates enzyme activity.
Adrenal glandular	Take 300 mg per day.	Fights inflammation and adrenal exhaustion.
Coenzyme Q_{10}	Take 30–60 mg one to two per day with meals; or follow the advice of your health-care professional.	Increases circulation; potent antioxidant.
Thymus glandular	Take 90 mg per day.	Assists the thymus gland in immune resistance; fights infections.

Vitamin A	Take 5,000 IU per day.	Needed for strong immune system and healthy mucous membranes; potent antioxidant.
Vitamin C	Take 8,000–13,000 mg per day.	Fights infection; improves healing; stimulates the adrenal adrenal glands; promotes immune system; powerful antioxidant.
Vitamin E	Take 400 IU per day.	Potent antioxidant; supplies oxygen to cells; improves circulation and energy level.

Comments

❑ Eat plenty of enzyme-rich fruits and vegetables with an emphasis on foods rich in proteolytic enzymes (such as pineapples, papayas, and figs). See Chapter 4 of Part One and Part Three for dietary guidelines.

❑ Drink at least six to eight 8-ounce glasses of water or juice per day.

❑ Limit your intake of sweets and sugars. Sugar reduces your white blood cells' ability to destroy viruses and bacteria.

❑ Prevent dental abscesses by brushing and flossing regularly. Visit your dentist every six months for a check-up.

❑ Chewable papain enzyme tablets are of benefit in reducing the inflammation that accompanies dental abscesses. Chew at least one tablet three times per day between meals in addition to the papain in the Essential Enzymes chart.

❑ Skin abscesses can be prevented by scrupulously keeping all wounds, even minor ones, clean. Watch for signs of infection, and treat immediately.

❑ Topical applications of enzyme salves, tea tree oil, or honey can encourage drainage. Apply an enzymatic skin salve at least three times a day.

❑ Break up vitamin A and E capsules and rub the contents on the abscess.

❑ Apply ice or heat to help alleviate the pain.

❑ Detoxification is especially important in cases of infection. Follow directions in Part Three of this book for information on cleansing, fasting, and juicing.

❑ Perform the Enzyme Kidney Flush (see Part Three for directions) every day for one week. During the following week, perform the Enzyme Toxin Flush (see directions in Part Three). Perform these procedures no more than once per month.

❑ Get plenty of rest.

❑ Keep a positive mental attitude. For more information, see Part Three.

Acidosis

The maintenance of an acid-base balance in the blood is critical for life. A slight variation in pH level can be devastating, possibly causing shock, coma, and even death. When there is too much acid in the blood as compared with the level of base, the condition is called *acidosis*. When there is too much base in the blood compared with the level of acid, the condition is *alkalosis*. (*See* ALKALOSIS in Part Two.)

The body does all it can to maintain a constant blood pH range of 7.35 to 7.45 (See the discussion of pH in Chapter 5 of Part One.) Its first line of defense against changes in pH levels is the kidneys. The kidneys excrete excess acid from the bloodstream if the blood's pH level falls too low (acidic). This process, however, takes several days to rectify the problem. There are also pH buffers in the blood, such as bicarbonate and carbon dioxide. Buffers are substances that neutralize any excess acid or alkali in the blood to maintain a constant pH level. Bicarbonate is a basic substance and carbon dioxide is acidic. These two compounds exist in an equilibrium in the blood. However, if excess acid enters the bloodstream or excess base is lost from the bloodstream, more bicarbonate is produced. If more base enters the bloodstream or acid is lost, more carbon dioxide is produced. The third mechanism the body uses to control the blood pH level is the excretion of carbon dioxide (which is acidic) through the lungs. Excess levels of carbon dioxide in the blood cause the body to breathe deeper and faster. When respiration increases, the level of carbon dioxide in the blood decreases. Likewise, when respiration decreases, the level of carbon dioxide in the blood increases.

When illness causes any of these mechanisms to malfunction, acidosis or alkalosis will result. There are two types of acidosis—metabolic and respiratory.

Metabolic acidosis occurs when there is not enough bicarbonate in the blood to counteract increases in acid. It can be caused by ingesting a poisonous acid substance or a substance that is metabolized into acid, such as wood alcohol, antifreeze, ammonia, or an overdose of aspirin; certain illnesses that cause the body to produce excess acid, such as diabetes or lactic acidosis; or kidney malfunction that affects the kidneys' ability to excrete acid, a condition called renal tubular acidosis. Metabolic acidosis can also occur due to loss of bicarbonate from the body, diarrhea, an ileostomy, or a colostomy.

Mild cases of metabolic acidosis may produce nausea and vomiting and fatigue, with deeper and faster breathing; although some may experience no symptoms. As it worsens, it may cause extreme weakness, fatigue, and nausea, and possibly disorientation. Severe cases of acidosis will lead to shock, coma, and ultimately death.

Respiratory acidosis results from slow breathing or respiratory disorders, resulting in a build-up of carbon dioxide in the blood. It can be caused by such disorders as asthma, chronic bronchitis, emphysema, or pneumonia. It can also occur if one's respiration is slowed due to taking strong tranquilizers or sedatives. Respiratory acidosis causes headache and drowsiness initially. As it gets worse, the person may fall into a coma.

Acidosis is not a condition, but rather a symptom of an underlying problem. So, treatment involves removing the cause or treating the causative disease. However, acidosis is a very serious symptom, so medical care is necessary. Asthma, bronchitis, diabetes, diarrhea, kidney disorders, and pneumonia are all conditions that can result in acidosis. See these conditions in Part Two for their treatment programs.

Enzyme Therapy

Digestive enzyme therapy is used to strengthen the body as a whole and build general resistance. Enzymes improve the digestion and absorption of food nutrients, reduce stress on the gastrointestinal mucosa, help maintain normal pH levels, detoxify the body, and promote the growth of healthy intestinal flora.

Systemic enzyme therapy is used to reduce inflammation, stimulate the immune system, and improve circulation. Enzymes help transport nutrients through the body, remove waste products, and improve health.

Enzyme Absorption System Enhancers (EASE) maximize enzyme activity. They can also improve the absorption and bioavailability of various nutrients and decrease the drain on the body's own digestive enzymes, thus prolonging their lives.

In addition to the standard multivitamin, multimineral, and multiglandular complexes recommended in the introduction to Part Two, the following supplements are recommended for the treatment of acidosis.

ESSENTIAL ENZYMES

Enzyme	Suggested Dosage	Actions
A digestive enzyme product containing protease, amylase, and lipase. For a list of digestive enzyme products, see Appendix. *or*	Take as indicated on the label with meals.	Can normalize digestion by improving the breakdown and absorption of food nutrients; fights toxic build-up.
NESS Formula #18 *or*	Take 2 to 4 capsules with each meal.	Helps the body utilize food nutrients; replaces enzyme activity lost from food during processing and cooking.
Fiber Enzyme Formula from Prevail	Take as indicated on the label.	Contains high-potency alkalizing enzymes.

91

SUPPORTIVE ENZYMES

Enzyme	Suggested Dosage	Actions
Formula Three—See page 38 of Part One for a list of products.	Take as indicated on the label.	Potent antioxidant enzymes.

ENZYME ABSORPTION SYSTEM ENHANCERS (EASE)

Combination	Suggested Dosage	Actions
Acid-Ease from Prevail or	Take as indicated on the label.	Helps relieve acidosis; helps break down and absorb foods in sensitive digestive systems.
NESS Formula #2	Take as indicated on the label; or take two capsules with each meal.	Helps the body utilize food nutrients; replaces enzyme activity lost from food during processing and cooking.

ENZYME HELPERS

Nutrient	Suggested Dosage	Actions
Bioflavonoids	Take 500 mg per day.	Potent antioxidants; help balance the body's pH level.
Calcium	Take as indicated on the label.	An alkaline-forming element.
Green-food products, such as Kyo-Green, Green Magma, Crystal Star Energy Green Drink, or Salute Aloe Vera Juice with ginger.	Take as indicated on the label.	Contain alkaline-forming foods; help balance the body's pH level.
Kelp	Take 500 to 750 mg twice per day.	Decreases acidity in the body; maintains proper mineral balance.
Kyolic Formula 102 (with enzymes)	Take as indicated on the label.	Helps relieve acidosis.
Magnesium	Take as indicated on the label.	An alkaline-forming element.
Pantothenic acid	Take 1,000 mg per day.	Needed for proper digestion; important coenzyme in balancing pH level.
Potassium	Take 150 mg per day.	An alkaline-forming element.
Probiotics, such as Dr. Dophilus Powder from Professional Nutrition, Life Start from Natren, or Acido-philase from Wakunaga.	Take as indicated on the label.	Helps maintain healthy intestinal flora.
Vitamin A	Take 25,000 IU twice per day for 4 to 6 weeks; then decrease dosage to 25,000 IU once per day. If you are pregnant, do not take more than 10,000 IU per day.	Essential for healthy digestion.
Vitamin B complex	Take 250 mg twice per day.	Essential for healthy digestion; the B vitamins are needed as coenzymes.

Comments

❑ Decrease your intake of acid-forming foods. They should constitute only 20 to 25 percent of your diet. See the inset on page 110 for a list of acid-forming foods. See the inset on page 93 for a list of alkali-forming foods to decrease the levels of acid in your blood.

❑ Citrus fruits actually become alkaline in the body; however, they will become acidic if sugar is added to them.

❑ Avoid meats, coffee and other caffeine products, dairy products (excluding kefir and yogurt), cheese, fish and other seafoods, legumes, eggs, peanuts, refined foods, and processed foods.

❑ Deep breathing can help increase the excretion of carbon dioxide.

❑ To rid the body of excess acid waste products, try the Twenty-Four Hour Liquid Diet. Drink eight 8-ounce glasses of pure filtered water, enzyme-rich fresh juice, or herbal tea per day. In addition, for breakfast, drink a mixture of the juice of two freshly squeezed lemons, one tablespoon of honey, and eight ounces of pure, filtered water. At midmorning, drink two heaping tablespoons of a green food product (see the list of green food products in the Enzyme Helpers chart) mixed in an eight-ounce glass of pure filtered water. For lunch, drink a juice made from three to four stalks of celery, four carrots, a bunch of parsley, and a bunch of spinach. At midafternoon, drink eight ounces of juice made from any of the fruits or vegetables listed in the inset "Alkali-Forming Foods" on page 93. For dinner, drink eight ounces of fresh juice made from pineapples or papayas blended with one apple. Late in the evening, mix two heaping tablespoons of a green food product in an eight-ounce glass of pure filtered water and drink. The next three days, eat only alkaline-forming fresh fruits and vegetables or drink their juices. After this, your diet should consist of 70 to 80 percent fresh alkalizing foods, such as sprouts, green drinks, fruits and their juices, and vegetables and their juices.

❑ Instead of the Twenty-Four Hour Liquid Diet, you may try drinking the Enzyme-Alkalizing Punch for two days. Each day, combine the juices of six grapefruits, six lemons, and twelve oranges in a gallon container. Fill the rest of the container with distilled water. Drink only one full gallon of the mixture throughout each day for two days. Repeat this process every four weeks. If weakened by the cleanse, discontinue for eight weeks. During the cleanse, certain uncomfortable sensations may be noted, such as cramps, dizziness, headaches, and nausea, as a result of the shock of the sudden removal of toxins from the body. These symptoms should last

Alkali-Forming Foods

Most fruits (with the exception of cranberries, plums, and prunes), vegetables, milk, and nuts create alkaline reactions in the body. In order to maintain a favorable alkaline balance in the body, the foods listed below should always be consumed in adequate amounts. It should be noted that the list of alkaline foods includes citrus fruits. Because these fruits have a distinct acidic taste and high acidic content, they are usually avoided by individuals having acidosis. However, the acids in these fruits (chiefly malic and citric acids), combine quickly to form salts (with such alkaline substances as potassium and sodium), and are quickly oxidized by the body to form carbonic acid. Carbonic acid, in turn, is exhaled and leaves an alkaline ash.

Fruits
- Apples
- Apricots
- Avocados
- Berries (all types)
- Cantaloupes
- Coconuts
- Currants
- Dates
- Figs
- Grapefruit
- Grapes
- Lemons
- Olives (ripe)
- Oranges
- Peaches
- Pears
- Persimmons
- Pineapples
- Raisins
- Rhubarb
- Tomatoes

Vegetables
- Artichokes
- Asparagus
- Beans
- Beets and beet greens
- Broccoli
- Cabbage
- Carrots
- Cauliflower
- Celery
- Chickory
- Corn
- Cucumbers
- Dandelions
- Eggplant
- Endive
- Garlic
- Horseradish
- Kale
- Kohlrabi
- Leek
- Lettuce

- Mushrooms
- Okra
- Onions
- Parsley
- Parsnips
- Peas
- Peppers
- Radishes
- Rutabagas
- Savory
- Sorrel
- Soybeans
- Spinach
- Sprouts
- Squash, summer
- Swiss chard
- Turnips
- Watercress

only a short time, disappearing as the toxins are eliminated. Note: The two-day Enzyme Alkalizing Punch should be used under the supervision of a well-trained health-care practitioner and should not be used by anyone suffering from diabetes or any other potentially life-threatening condition.

❏ Take two teaspoons daily of Vital K Plus from Future Biotics.

❏ Knowing whether your blood pH level is too acidic is a very important measure of health. Test strips for measuring urine pH levels are widely available at pharmacies, and are a quick and easy way to measure pH levels. Follow instructions on the container. Your blood pH level should be between 7.35 and 7.45. You can also check your body's pH levels using red and blue litmus paper. Blue litmus paper turns red in an acid environment, and red litmus paper turns blue in an alkaline environment.

❏ One or two glasses of wine (especially red wine) before dinner can help to decrease the body's acidity.

❏ Exercise every day for twenty to thirty minutes. Try walking, bicycling, or swimming at a moderate pace.

❏ If you are in a state of acidosis because of severe diarrhea, see a physician for immediate treatment of the diarrhea.

Acne

Acne is a skin condition caused by inflammation of the sebaceous glands. The sebaceous glands are located just beneath the skin and secrete sebum (an oil) to lubricate the skin. When oil becomes trapped in these glands, bacteria multiply, and inflammation ensues. Of the many types of acne, acne vulgaris is the common form that affects so many teenagers. Acne appears particularly in younger people because glandular activity increases as adolescence begins, due to high hormonal activity. In fact, as many as 80 percent of

Americans between the ages of twelve and twenty-four are affected by acne of one type or another.

Acne can be either superficial or deep. Although deep acne occurs most often on the face, it can also occur on the back, chest, and shoulders. If unattended, acne can become infected. This can result in the formation of scar tissue and pockmarks (permanent depressions in the facial tissue).

The exact cause of the trapping of oil in the glands, which causes acne, is unknown, but many factors can contribute to it. In addition to increased glandular activity, other factors include allergies, stress, oral contraceptives, heredity, and consumption of foods high in sugar or fats. Topical applications of skin creams and make-up can often block the sebaceous glands, leading to acne.

Symptoms

Symptoms of superficial acne include pimples, blackheads, whiteheads, inflamed nodules, superficial cysts, and pustules. Symptoms of deep acne include deep, inflamed nodules and pus-filled cysts, which may become abscesses, and scarring.

Enzyme Therapy

Enzymes are excellent for treating acne vulgaris. Topical applications of enzymes should be used to exfoliate dead skin, to reduce inflammation, and to remove toxins. Oral enzymes work from the inside out and are effective in treating acne, pimples, blackheads, rashes, and itching.

Digestive enzyme therapy is used to improve the digestion of food, reduce stress on the gastrointestinal mucosa, help maintain normal pH levels, detoxify the body, and promote the growth of healthy intestinal flora. In this way, the body's immune system is strengthened and better able to fight infection and inflammation. Digestive enzymes also serve as replacements for the body's pancreatic enzymes, leaving the pancreatic enzymes free to perform other functions in the body, such as decreasing inflammation.

Systemic enzyme therapy is used to reduce inflammation, stimulate the immune system, improve circulation, help speed tissue repair, bring nutrients to the affected area, remove waste products, and enhance wellness.

Our bodies are constantly eliminating the waste products of digestion, metabolism, and other processes. These waste products are excreted through the lungs, intestines, kidneys, and skin. Even our blood lends a hand in cleaning out cells by transporting and eliminating debris. This elimination is critical if the body is to function properly.

One of the best ways to maintain this excretion mechanism is through a properly functioning intestine. Constipation and other intestinal conditions can allow toxins to linger in the colon, be reabsorbed back into the overburdened bloodstream, and be carried to the skin for elimination. This can result in acne. Therefore, the first step in fighting acne is to improve digestion and thereby improve the breakdown, elimination, and excretion of toxins and wastes from the body. At the same time, it is important to ensure that necessary nutrients are absorbed and delivered to the various tissues. Enzymes can help accomplish both objectives by improving digestion and, therefore, elimination, and by improving circulation and promoting blood flow to and from the tissues. This will help the body to better eliminate toxins and improve the delivery of beneficial nutrients.

Enzyme Absorption System Enhancers (EASE) maximize enzyme activity. They can also improve the absorption and bioavailability of various nutrients and decrease the drain on the body's own digestive enzymes, thus prolonging their lives.

In addition to the standard multivitamin, multimineral, and multiglandular complexes recommended in the introduction to Part Two, the following supplements are recommended for the treatment of acne vulgaris. Though there are different types of acne, this enzyme therapy program is recommended only for acne vulgaris. The following recommended dosages are for adults. Children between the ages of two and five should take one-quarter the recommended dosage, those between the ages of six and twelve should take one-half the recommended dosage, and those between the ages of thirteen and seventeen should take three-quarters the recommended adult dosage.

ESSENTIAL ENZYMES

Enzyme	Suggested Dosage	Actions
Superoxide dismutase	Take 50 mcg per day.	Fights inflammation; potent antioxidant.

AND SELECT ONE OF THE FOLLOWING:

A digestive enzyme product containing protease, amylase, and lipase enzymes. See the list of digestive enzyme products in Appendix.	Take as indicated on the label with meals.	Improves the breakdown and absorption of food nutrients; fights toxic build-up.
Formula One— See page 38 of Part One for a list of products.	Take 5 tablets twice per day.	Fights inflammation, pain, and swelling; speeds healing; stimulates the immune system; strengthens capillary walls.
Bromelain	Take 130 mg two times per day between meals.	Fights inflammation, pain, and swelling; speeds healing; stimulates the immune system; strengthens capillary walls.
Trypsin with chymotrypsin	Take 75–100 mg of trypsin and 3 mg of chymotrypsin three times per day between meals.	Fights inflammation, pain, and swelling; increases rate of healing; stimulates the immune system.
Pancreatin	Take 100 mg six times per day on an empty stomach.	Fights inflammation, pain, and swelling; increases rate of healing; stimulates the immune system; strengthens capillary walls.

Papain	Take 180 mg two times per day between meals.	Fights inflammation; speeds healing.
Microbial protease	Take 300 mg two times per day between meals.	Fights inflammation; speeds healing.

SUPPORTIVE ENZYMES

Enzyme	Suggested Dosage	Actions
Catalase and glutathione peroxidase	Take as indicated on the labels.	Potent antioxidant enzymes.

ENZYME ABSORPTION SYSTEM ENHANCERS (EASE)

Combination	Suggested Dosage	Actions
Akne-Zyme from Enzymatic Therapy	Take as indicated on the label.	Stimulates immune function; fights free radicals.
and		
Detox-Zyme from Rainbow Light	Take as indicated on the label.	Detoxifies.
or		
Inner Act from Life Plus	Take as indicated on the label.	Detoxifies.
and		
ZymeDophilus from Enzymatic Therapy	Take as indicated on the label.	Promotes the growth of friendly gastrointestinal bacteria; and stimulates enzymatic activity to better break down proteins, fats, and carbohydrates, thus improving the overall health of the body.
or		
Inflamzyme from PhytoPharmica	Take as indicated on the label.	Immune system activator.
or		
NESS Formula #416	Take 3–4 tablets three times per day between meals.	Helps reduce inflammation.
and		
NESS Formula #301	Take 3 capsules three times per day between meals.	Helps support lymphatic system.

ENZYME HELPERS

Nutrient	Suggested Dosage	Actions
Adrenal glandular	Take 300 mg per day.	Fights inflammation and adrenal exhaustion.
Aged garlic extract	Take 2 capsules three times per day.	Fights inflammation; stimulates the immune system; potent antioxidant.
Amino-acid complex	Take as indicated on the label on an empty stomach.	Required for enzyme, peptide, and hormone production; helps the body maintain proper fluid and acid/base balance.
Coenzyme Q_{10}	Take 30–60 mg one to two times per day with meals; or follow the advice of your health-care professional.	Potent antioxidant.
Green barley extract	Take as indicated on the label.	Potent free-radical fighter; rich in vitamins, minerals, and enzyme activity.

Lecithin	Take as indicated on the label.	Improves digestion and absorption of essential fatty acids, known to improve many skin conditions.
Pantothenic acid	Take 100 mg per day.	Important for protein, fat, and carbohydrate metabolism; important for synthesis of hormones, hemoglobin, steroids, and neurotransmitters; needed for a healthy digestive system and skin; fights stress.
Thymus glandular	Take 90 mg per day.	Assists the thymus gland in immune resistance; improves immune response; fights infections.
Vitamin A	Take 20,000–35,000 IU per day. Take no more than 10,000 IU per day from all sources if pregnant.	Needed for strong immune system and healthy mucous membranes; potent antioxidant.
Vitamin C	Take 8,000–13,000 mg per day.	Fights infection; improves healing; stimulates the adrenal glands; promotes immune system; powerful antioxidant.
Vitamin E	Take 400 IU per day.	Potent antioxidant; supplies oxygen to cells; improves circulation and energy level.
Zinc	Take 50 mg per day. Do not take more than 100 mg per day from all sources.	Needed by over 300 enzymes; potent antioxidant.

Comments

❑ Several herbs can help fight the inflammation that occurs with acne, including alfalfa concentrate, echinacea, ginger root, ginkgo biloba, licorice, pau d'arco, and white willow bark. Gotu kola supports collagen and skin and promotes circulation.

❑ For optimum effectiveness, use whey powder and acidophilus with the digestive enzymes.

❑ Eat a well-balanced diet with plenty of fresh fruits and vegetables. Eat fiber-rich complex carbohydrates and eliminate refined carbohydrates, including alcohol, refined white flour, and sugar to improve the overall health of your body.

❑ Increase your intake of yogurt (a fermented dairy product) and other lactobacilli-containing foods, in order to increase the beneficial bacteria in your intestines. Avoid nonfermented dairy products (and avoid dairy products altogether if you are allergic to or intolerant them).

❑ Reduce your fat intake.

❑ Drink plenty of water, but avoid chlorinated water. Distilled, bottled spring, or even well water is recommended.

❑ Gargle with a combination of four ounces of fresh juiced pineapple and four ounces of lemon or grapefruit juice in an 8-ounce glass.

❏ Perform the Enzyme Kidney Flush and the Enzyme Toxin Flush as directed in Part Three.

❏ Apply an enzymatic skin conditioner three times per day. If you have purchased an enzyme skin product, follow directions on the label. There are a number of fine enzyme products on the market. *See* SKIN CARE in Part Three.

❏ Break open and spread the contents of two to four vitamin A and vitamin E capsules on affected areas.

❏ Greasy skin-care products should be avoided. Some cosmetics (especially those with comedogenic ingredients, such as isopropyl myristate) can actually aggravate acne. Scrupulously keep skin clean.

❏ Exercise regularly to help skin cells stay healthy. Toxins are released through the sweat glands during exercise, and exercise promotes blood circulation.

❏ Zinc is just as effective as tetracycline in treating acne, without its harmful side effects, according to a recent Swedish study.

❏ To maintain healthy skin, try supplements that can combine a group of different digestive enzymes, such as Zymase from Tyler (containing protease, amylase, lipase, cellulase, lactase, sucrase, and maltase, plus probiotics). A product from NESS combines protease, amylase, and lipase enzymes with calcium lactate.

❏ You may consider consulting a skin expert. Dermatologists, metabolic physicians, nutritional consultants, skin-care specialists, and cosmetologists can be of great help.

Acquired Immune Deficiency Syndrome

See AIDS.

Adenoiditis

Adenoiditis is inflammation, pain, and swelling of the adenoids—two doughnut-shaped objects located on the back part of the throat (called the posterior pharyngeal wall). It occurs primarily in children and may be secondary to an allergy or infection. The condition can result from a chronically infected middle ear (it can also cause a middle ear infection). Left untreated it can cause mastoiditis (inflammation of the bone behind the ear) or serous otitis media (an ear infection in which the ear secretes fluid) and permanent hearing loss. Adenoiditis can also be caused by nose or throat infections, breathing through the mouth, chronic sinusitis, or an obstruction of the eustachian tubes (the two tubes that run from the back of the throat to the middle ear).

The adenoids and tonsils protect the throat from foreign particles or viral and bacterial attacks; therefore, your adenoids and tonsils are essential as important parts of your lymphatic immune system.

Symptoms

The symptoms of adenoiditis include pain, redness, swelling, and difficulty swallowing. There may be a fever present as well.

Enzyme Therapy

Systemic enzyme therapy is used to decrease pain and swelling and reduce inflammation and infection of the adenoids. Enzymes stimulate the immune system, improve circulation, help speed tissue repair, bring nutrients to the damaged area, remove waste products, enhance wellness, and build general resistance.

Digestive enzyme therapy is used to improve the digestion of food, reduce stress on the gastrointestinal mucosa, help maintain normal pH levels, detoxify the body, promote the growth of healthy intestinal flora, and strengthen the body as a whole. In this way, the body's immune system is strengthened and better able to fight adenoiditis. Digestive enzymes also serve as replacements for the body's pancreatic enzymes, leaving the pancreatic enzymes free to perform other functions in the body, such as boosting immunity and decreasing inflammation.

Enzyme Absorption System Enhancers (EASE) maximize enzyme activity. They can also improve the absorption and bioavailability of various nutrients, and decrease the drain on the body's own digestive enzymes, thus prolonging their lives.

In addition to the standard multivitamin, multimineral, and multiglandular complexes recommended in the introduction to Part Two, the following supplements are recommended for the treatment of adenoiditis. The following recommended dosages are for adults. Children between the ages of two and five should take one-quarter the recommended dosage, those between the ages of six and twelve should take one-half the recommended dosage, and those between the ages of thirteen and seventeen should take three-quarters the recommended adult dosage.

ESSENTIAL ENZYMES

Enzyme	Suggested Dosage	Actions
Superoxide dismutase	Take 150 mcg per day.	Potent antioxidant; fights inflammation.
Papain lozenges	Take as indicated on the label.	Fights inflammation; stimulates the immune system.

AND SELECT ONE OF THE FOLLOWING:

Formula One—See page 38 of Part One for a list of products.	Take 5 tablets two times per day.	Stimulates the immune system; decreases inflammation; increases rate of recovery; improves circulation.
Bromelain	Take 750 mg two times per day between meals.	Reduces inflammation, pain, and swelling; speeds recovery after injury or surgery; stimulates the immune system; strengthens capillary walls.
Lysozyme	Take as indicated on the label.	Fights viral and bacterial infections; stimulates immune activity; breaks down mucus.
Serratiopeptidase	Take 10 mg two times per day between meals.	Fights inflammation; reduces pain and swelling.
Trypsin with chymotrypsin	Take 250 mg of trypsin and 10 mg of chymotrypsin three times per day between meals.	Decreases inflammation, pain, and swelling; increases rate of healing; stimulates the immune system.
Pancreatin	Take 500 mg six times per day on an empty stomach.	Fights inflammation, pain, and swelling; increases rate of healing; stimulates the immune system; strengthens capillary walls.
A microbial protease	Take 300 mg two times per day between meals.	Fights inflammation; speeds recovery from injuries and surgery.
Papain	Take 300 mg two times per day between meals.	Breaks up antigen-antibody complexes; fights inflammation; speeds recovery from injuries and surgery; stimulates the immune system.

SUPPORTIVE ENZYMES

Enzyme	Suggested Dosage	Actions
Catalase and glutathione peroxidase	Take as indicated on the labels.	Potent antioxidant enzymes.
A digestive enzyme product containing protease, amylase, and lipase enzymes. See the list of digestive enzyme products in Appendix.	Take as indicated on the label with meals.	Improves the breakdown and absorption of food nutrients; Fights toxic build-up.

ENZYME ABSORPTION SYSTEM ENHANCERS (EASE)

Combination	Suggested Dose	Actions
Combat from Life Plus *or*	Take as indicated on the label.	Helps support respiratory system; stimulates immune function; helps control infections.
Inflamzyme from PhytoPharmica *or*	Take as indicated on the label.	Gives support for tissues; activates immune system.
Connect-All from Nature's Plus *or*	Take as indicated on the label.	Supports tissue repair.

Protease Concentrate from Tyler *and*	Take as indicated on the label.	Fights inflammation.
Bromelain Complex from PhytoPharmica *or*	Take as indicated on the label.	Decreases swelling; improves circulation; helps break up ciruculating immune complexes.
NESS Formula #301 *and*	Take 3 capsules three times per day between meals.	Helps support lymphatic system.
Traumagesic Complex from Tyler	Take 3 tablets three times per day between meals.	Helps decrease pain and inflammation.

ENZYME HELPERS

Nutrient	Suggested Dosage	Actions
Acidophilus and other probiotics	Take as indicated on the label.	Improves gastrointestinal function, thus improving the overall health of the body; needed for production of biotin, niacin, folic acid, and pyridoxine, which are needed by many of the body's enzymes; helps produce antibacterial substances; stimulates enzyme activity.
Adrenal glandular	Take 100–400 mg per day.	Fights inflammation and adrenal exhaustion.
Amino-acid complex	Take as indicated on the label on an empty stomach.	Required for enzyme, peptide, and hormone production; helps the body maintain proper fluid and acid/base balance.
Bioflavonoids, such as rutin or quercetin	Take 500 mg three times per day between meals.	Strengthen capillaries; potent antioxidants.
Coenzyme Q_{10}	Take 30–60 mg one to two times per day with meals; or follow the advice of your health-care professional.	Potent antioxidant.
Kyolic Aged Garlic Liquid Extract with Vitamin B_1 and B_{12}	Take as indicated on the label; or take 3 teaspoons per day.	Fights inflammation and free radicals; stimulates immune function.
Pantothenic acid	Take 50 mg three times per day.	Plays a role in the formation of antibodies; improves immune function.
Spirulina	Take as indicated on the label.	Rich in minerals, vitamins, enzymes, proteins, essential amino acids, and essential fatty acids—especially gamma-linolenic acid; good source of beta-carotene; improves immune function.
Thymus glandular	Take 100 mg per day.	Assists the thymus gland in immune resistance; improves immune response; fights infections.
Vitamin A	Take 20,000–40,000 IU per day. If pregnant, do	Needed for strong immune system and healthy mucous

	not take more than 10,000 IU per day from all sources.	membranes; potent antioxidant.
Vitamin B complex	Take 50 mg three times per day, or as indicated on the label.	B vitamins act as coenzymes and support neuromusculo-skeletal function.
Vitamin C	Take 10,000–15,000 mg per day.	Fights infections, diseases, and allergies; improves wound healing; stimulates the adrenal glands; promotes immune system; powerful antioxidant.
Zinc lozenges	Take as indicated on the label. Take no more than 100 mg of zinc per day from all sources.	Enhances immune function; works with enzymes involved in immune function.

Comments

❏ Herbs, including alfalfa, echinacea, ginger, ginkgo biloba, licorice, and white willow bark, can help fight inflammation.

❏ Using an eye dropper, squirt about ten drops of liquid echinacea onto the adenoids (they are located on either side of the back of the throat). Do this every other hour (or at least twice per day) until symptoms resolve. Gargle with the Enzymatic Gargle two to three times per day. See Part Three for instructions.

❏ Juicing is essential to flush your body of toxins. Drink at least eight 8-ounce glasses of freshly juiced fruits and vegetables per day. Combine fruits and vegetables, such as pineapple, apples, parsley, garlic, grapes, celery, and cranberries—all high in enzymatic activity. Dandelion tea is very high in vitamins A, B_1, B_2, B_6, B_{12}, C, and E.

❏ Use cough and cold formulas such as Cold Zzap from Naturally Vitamin Supplements. This product contains lysozyme, zinc, vitamin A, and beta-1-3-glucan, which supports healthy immunity.

❏ Perform the Enzyme Kidney Flush every day for one week. See Part Three for instructions.

Adnexitis

Adnexitis is an inflammation of the ovaries and fallopian tubes, usually caused by an infection. This can occur when bacteria enter the body through the vagina and uterus (frequently after surgery or as a result of the use of a diaphragm or other contraceptive device). Though less frequent, adnexitis can also occur by invasion of bacteria through the bloodstream or lymphatic system.

If an acute case of adnexitis is not properly taken care of, chronic adnexitis will develop. Chronic adnexitis may also develop as a result of an autoimmune disease.

Symptoms

Adnexitis is characterized by pain, swelling, and inflammation of the ovaries and fallopian tubes. The ligaments of the uterus may also be involved. If untreated, adnexitis could cause sterility. The condition can even lead to tumor formation and cancer, so prompt care is essential.

Enzyme Therapy

Systemic enzyme therapy is used to decrease pain and swelling, reduce inflammation in the ovaries and fallopian tubes, and stimulate the immune system. Enzymes improve circulation, help speed tissue repair, bring nutrients to the damaged area, remove waste products, enhance wellness, and build general resistance.

Digestive enzyme therapy is used to improve the digestion of food, reduce stress on the gastrointestinal mucosa, help maintain normal pH levels, detoxify the body, promote the growth of healthy intestinal flora, and strengthen the body as a whole. In this way, the body's immune system is strengthened and better able to fight adnexitis. Digestive enzymes also serve as replacements for the body's pancreatic enzymes, leaving the pancreatic enzymes free to perform other functions in the body, such as boosting immunity and decreasing inflammation.

Enzyme Absorption System Enhancers (EASE) maximize enzyme activity. They can also improve the absorption and bioavailability of various nutrients and decrease the drain on the body's own digestive enzymes, thus prolonging their lives.

Enzymes reduce inflammation and inhibit the development of adhesions on the ovaries and fallopian tubes. Adhesions in the uterus are frequently the cause of sterility. Nonsteroidal anti-inflammatories (NSAIDs) are sometimes used as treatment, but they have side effects including immune system disturbances and disturbances in cell reproduction and organ regeneration. In addition to enzymes, antibiotics should always be used, under a doctor's care and with caution, to treat acute adnexitis. Since enzymes do not inhibit immune system function but rather support it, enzymes can be taken during all stages of the illness.

The average duration of antibiotic therapy can be reduced substantially through enzyme therapy. Complete relief of complaints due to acute adnexitis usually occurs after about fourteen days of combined antibiotic and enzyme treatment. Further, a relapse of the condition occurs much less frequently when enzyme therapy is used.

In addition to the standard multivitamin, multimineral, and multiglandular complexes recommended in the introduction to Part Two, the following supplements are recommended for the treatment of adnexitis. The following recommended dosages are for adults. Children between the ages of two and five should take one-quarter the recommended dosage, those between the ages of six and twelve should take one-half the recommended dosage, and those between the ages of thirteen and seventeen should take three-quarters the recommended adult dosage.

ESSENTIAL ENZYMES

Enzyme	Suggested Dosage	Actions
Superoxide dismutase	Take 150 mcg per day.	Fights inflammation; potent antioxidant.

AND SELECT ONE OF THE FOLLOWING:

Formula One— See page 38 of Part One for a list of products.	Take 5 tablets twice a day for 14 days or until symptoms are relieved.	Fights pain, swelling, and inflammation; stimulates immune function.
Bromelain	Take 750 mg two times per day between meals for 14 days or until symptoms are relieved.	Fights inflammation, pain, and swelling; speeds recovery after injury or surgery; stimulates the immune system.
Trypsin with chymotrypsin	Take 250 mg of trypsin and 10 mg of chymotrypsin three times per day between meals for 14 days or until symptoms are relieved.	Fights inflammation, pain, and swelling; increases rate of healing; stimulates the immune system.
Pancreatin	Take 500 mg six times per day on an empty stomach for 14 days or until symptoms are relieved.	Fights inflammation, pain, and swelling; increases rate of healing; stimulates the immune system.
Papain	Take 300 mg two times per day between meals for 14 days or until symptoms are relieved.	Fights inflammation; speeds recovery from injuries and surgery.
A microbial protease	Take 300 mg two times per day between meals for 14 days or until symptoms are relieved.	Fights inflammation; speeds recovery from injuries and surgery.

SUPPORTIVE ENZYMES

Enzyme	Suggested Dosage	Actions
Catalase and glutathione peroxidase	Take as indicated on the labels.	Potent antioxidant enzymes.
A digestive enzyme product containing protease, amylase, and lipase enzymes. See the list of digestive enzyme products in Appendix.	Take as indicated on the label with meals.	Improves the breakdown and absorption of food nutrients; fights toxic build-up.

ENZYME ABSORPTION SYSTEM ENHANCERS (EASE)

Combination	Suggested Dosage	Actions
Vita-C-1000 from Life Plus *or*	Take as indicated on the label.	Boosts immune function; fights infection.
Inflamzyme from PhytoPharmica *or*	Take as indicated on the label.	Immune system activator; reduces inflammation.
Hepazyme from Enzymatic Therapy *or*	Take as indicated on the label.	Stimulates immune function.
Ecology Pak from Life Plus	Take 2–3 tablets once or twice a day. The first dose should be taken in the morning no later than noon.	Assists with natural detoxification processes; supports the immune system.

ENZYME HELPERS

Nutrient	Suggested Dosage	Actions
Acidophilus and other probiotics	Take as indicated on the labels.	Improve gastrointestinal function; needed for production of biotin, niacin, folic acid, and pyridoxine, which are needed by many enzymes; help produce antibacterial substances; stimulate enzyme activity.
Adrenal glandular	Take 100–400 mg per day.	Fights inflammation and adrenal exhaustion.
Aged garlic extract	Take 2 capsules three times per day.	Fights inflammation; stimulates the immune system; potent antioxidant.
Amino-acid complex	Take as indicated on the label on an empty stomach.	Required for enzyme, peptide, and hormone production; helps the body maintain proper fluid and acid/base balance.
Bioflavonoids, such as rutin or quercetin	Take 500 mg three times per day between meals.	Strengthen capillaries; potent antioxidants.
Coenzyme Q_{10}	Take 30–60 mg one to two times per day with meals; or follow the advice of your health-care professional.	Potent antioxidant.
Raw ovary glandular	Take 50 mg per day.	Improves ovarian function.
Raw uterus glandular	Take 25 mg per day.	Improves uterine function.
Thymus glandular	Take 100 mg per day.	Assists the thymus gland in immune resistance; improves immune response; fights infections.
Vitamin A	Take 5,000 IU per day.	Needed for strong immune system and healthy mucous membranes; potent antioxidant.
Vitamin B complex	Take 50 mg three times per day; or take as indicated on the label.	B vitamins act as coenzymes and support neuromuscu-loskeletal function.
Vitamin C	Take 10,000–15,000 mg per day.	Fights infections, diseases, and allergies; improves wound healing; stimulates the adrenal glands; promotes immune system; powerful antioxidant.
Zinc	Take 15–50 mg per day. Do not take over 100 mg per day.	Needed by over 300 enzymes; necessary for white blood cell immune function; improves wound healing.

Comments

❏ Perform the Enzyme Kidney Flush every day for one week. During the following week, perform the Enzyme Toxin Flush. Perform these flushes only once per month. See Part Three for instructions.

❏ Follow the dietary recommendations as outlined in Part Three.

Age Spots

Age spots (or liver spots or chloasma) are areas of the skin in which pigment is overproduced. Prolonged exposure to sunlight is usually the cause of this pigment overproduction. The skin has already been irreversibly damaged by the time the brown spots appear.

Usually, these brown spots are harmless in and of themselves, but they can be a sign of a more serious underlying problem. As we age, our bodies' enzyme production decreases and the activity level of the enzymes produced decreases as well. Free radicals are normally destroyed by our bodies' defense mechanisms. But when our enzymes and other protectors can no longer keep pace with free-radical formation, the end result is a build-up of free radical wastes and by-products. Age spots may be full of these free radical wastes resulting in toxic build-up in the skin (a prime area for cancer).

Factors that may lead to the formation of age spots include toxic build-up, poor bowel habits, poor diet, inadequate or depressed liver function, excessive intake of fatty foods, intake of rancid oils, little or no exercise, smoking, excessive drinking, stress, and excessive exposure to sunlight.

Symptoms

Age spots are painless, flat, brown spots on the skin. They usually occur on the hands, face, and neck, but can appear anywhere on the body.

Enzyme Therapy

Enzyme therapy is an excellent tool in combating age spots. Such therapy would include enzyme peels, systemic enzyme therapy, and digestive enzyme therapy. Enzyme treatment is primarily geared toward stimulating enzyme and cellular activity. Topical enzyme creams are used to dissolve dead and damaged skin cells and exfoliate the skin, while leaving healthy skin cells intact.

Enzyme peels contain active proteolytic enzymes, such as papain. According to world-famous skin authority Zia Wesley-Hosford, "This is the most effective gentle type of at-home peel. The live enzyme 'eats' the dead skin cells, without harming the new, younger cells." She says that, used every day for nine to twelve months, an enzyme peel will gently decrease or eliminate sun damage, age spots, and signs of aging.

Systemic enzyme therapy is used to fight free radicals, stimulate the immune system, improve circulation, help speed tissue repair, bring nutrients to the damaged area, remove waste products, enhance wellness, strengthen the body as a whole and build general resistance.

Digestive enzyme therapy is used to improve the digestion of food, reduce stress on the gastrointestinal mucosa, help maintain normal pH levels, detoxify, and promote the growth of healthy intestinal flora. To be most effective, skin conditions must be treated from the inside out and from the outside in.

Enzyme Absorption System Enhancers (EASE) maximize enzyme activity. They can also improve the absorption and bioavailability of various nutrients and decrease the drain on the body's own digestive enzymes, thus prolonging their lives.

In addition to the standard multivitamin, multimineral, and multiglandular complexes recommended in the introduction to Part Two, the following supplements are recommended for the treatment of age spots.

ESSENTIAL ENZYMES

Enzyme	Suggested Dosage	Actions
Formula Three—See page 38 of Part One for a list of products.	Take as indicated on the label.	Potent antioxidant enzymes.

AND SELECT ONE OF THE FOLLOWING:

Formula One—See page 38 of Part One for a list of products.	Take 10 tablets 3 times per day between meals for six months. Then decrease dosage to 5 tablets three times per day between meals.	Bolsters the immune system; breaks up antigen-antibody complexes; fights free-radical formation.
Papain	Take 500 mg three times per day between meals for six months. Then decrease dosage to 300 mg three times per day between meals.	Stimulates immune function; breaks up circulating immune complexes; fights free radicals.
Bromelain	Take 400 mg three times per day between meals for six months. Then decrease dosage to 250 mg three times per day between meals.	Bolsters the immune system; breaks up antigen-antibody complexes; fights free-radical formation.
Trypsin with chymotrypsin	Take 200 mg of trypsin and 10 mg of chymotrypsin three times per day between meals for six months. Then decrease dosage to 125 mg of trypsin and 5 mg of chymotrypsin three times per day between meals.	Stimulates immune function; breaks up circulating immune complexes; fights free radicals.

A proteolytic enzyme	Take 10 tablets 3 times per day for six months. Then decrease dosage to 5 tablets three times per day.	Bolsters the immune system; breaks up antigen-antibody complexes; fights free-radical formation.

SUPPORTIVE ENZYMES

Enzyme	Suggested Dosage	Actions
A digestive enzyme product containing protease, amylase, and lipase enzymes. See the list of digestive enzyme products in Appendix.	Take as indicated on the label with meals.	Improves the breakdown and absorption of food nutrients; fights toxic build-up.

ENZYME ABSORPTION SYSTEM ENHANCERS (EASE)

Combination	Suggested Dosage	Actions
Detox Enzyme Formula from Prevail *or*	Take as indicated on the label.	Helps detoxify the gastrointestinal tract and the body.
Detox-Zyme from Rainbow Light *or*	Take as indicated on the label.	Helps detoxify the gastrointestinal tract and the body.
Vita-C-1000 from Life Plus *or*	Take as indicated on the label.	Fights free radicals; helps detoxify the liver.
Zyme Dophilus from PhytoPharmica *or*	Take as indicated on the label.	Promotes growth of friendly bacteria; helps stimulate enzymatic activity.
Metabolic Liver Formula from Prevail	Take as indicated on the label.	Improves liver function.

ENZYME HELPERS

Nutrient	Suggested Dosage	Actions
Acidophilus and other probiotics	Take as indicated on the label.	Stimulates enzyme activity; improves gastrointestinal function.
Adrenal glandular	Take 100–400 mg per day.	Helps fight stress and adrenal exhaustion; fights inflammation.
Aged garlic extract	Take 2 capsules three times per day.	Stimulates the immune system; fights free radicals.
Coenzyme Q$_{10}$	Take 30–60 mg one to two times per day with meals.	Important to immune system function; has antioxidant activity.
DHEA	Take 25 mg per day.	Supports immune and cardiovascular function, thus improving elimination of free radicals and delivery of nutrients to the skin.
Essential fatty acid complex	Take 100–200 mg two to three times per day with meals; or follow the advice of your health-care professional.	Strengthens immunity; helps strengthen cell membranes.

Kidney glandular	Take as indicated on the label.	Improves kidney function, thus improving elimination of toxins and free radicals from the body.
Thymus glandular	Take 100 mg per day, or take as indicated on the label.	Assists the thymus gland in immune resistance; improves immune response; fights infections.
Vitamin A	Take up to 50,000 IU for no more than one month (for higher doses, see a physician). If pregnant, do not take more than 10,000 IU.	Needed for strong immune system and healthy mucous membranes; potent antioxidant.
Vitamin B complex	Take 100–150 mg three times per day.	B vitamins act as coenzymes.
Vitamin C	Take 5,000 mg per day.	Fights infections, diseases, and allergies; improves wound healing; stimulates the adrenal glands; promotes immune system; powerful antioxidant.
Vitamin E	Take 1,200 IU per day from all sources (including the multivitamin complex in the introduction to Part Two).	Acts as an intercellular antioxidant; improves circulation; helps resist diseases at the cellular level.
Zinc	Take 50 mg per day. Do not take more than 100 mg per day from all sources.	Needed by over 300 enzymes; necessary for white blood cell immune function; improves healing.

Comments

❑ Eat plenty of fresh enzyme-rich fruits and vegetables (in season, if possible). If you cannot tolerate raw fruits and vegetables, increase your intake of digestive enzymes, or steam or sauté your produce. Increase your intake of foods with proteolytic enzymes, such as fresh pineapple and papaya.

❑ Avoid red meats, foods high in saturated fats, alcohol, refined sugar, and white flour. These foods produce toxic by-products that the body must eliminate through the kidneys, liver, and skin. An excessive amount of toxins overloads the system and can smother the skin, upset its chemical balance, and lead to skin problems, such as age spots.

❑ Avoid chlorinated water. Use bottled, distilled, or well water instead.

❑ Apply lemon juice to the area. Over time, this will lighten the age spots.

❑ Apply an enzyme exfoliant, such as Super Skin Zyme from Louise Bianco Skin Care, Inc., or Fresh Papaya Enzyme Peel from Zia Cosmetics, to the affected area three times per day.

❑ It is important to cleanse the system of toxins. Use the Enzyme Kidney Flush and the Enzyme Toxin Flush as outlined in Part Three.

❑ If the spots change appearance in any way, see a physician.

❑ *See* SKIN CARE in Part Three.

101

Aging

Everyone wants to live longer, look younger, be healthier, and avoid aging. Aging is a progressive decrease in the body's ability to function. Though aging is not a disease, the body does become more susceptible to disease due to it. Aging is directly related to the body's decreased enzyme production. According to Drs. Edward Howell and Max Wolf, not only does the body produce fewer enzymes with age, but those enzymes that are produced can't do their jobs as well. Enzyme potential declines with aging. Therefore, as each year passes, the body requires increased supplemental enzyme support to fight degenerative conditions associated with aging, such as arthritis; autoimmune disorders, such as rheumatism; glomerulonephritis, Alzheimer's disease; cancer; cardiovascular diseases; and stroke.

Unfortunately, many older people have at least one of these chronic conditions (some have several). According to the American Association of Retired Persons (AARP), in 1986, 23 percent of people 65 years of age or older in America (over 6 million people) had health-related problems that interfered with their daily living.

Many of the conditions suffered by older people are caused by a decrease in digestive ability, increased sclerosis (scar tissue formation) and fibrosis (abnormal increase in fibrous connective tissue formation), and an increase in free radical production. These factors all increase the rate of aging and are directly related to decreased enzyme levels and activity.

Numerous studies indicate that pancreatic enzyme output gradually decreases with aging. Pepsin output may be reduced by as much as 40 percent in the elderly. Gastric acid secretion also diminishes. In one study involving individuals aged 80 to 91, 80 percent were diagnosed as having hypochlorhydria (insufficient hydrochloric acid production). Hydrochloric acid helps destroy harmful bacteria that you ingest and triggers the conversion of pepsinogen into the active enzyme pepsin. Insufficient hydrochloric acid or digestive enzyme levels can severely reduce the body's ability to digest, extract, absorb, and utilize nutrients in the foods you eat.

An increase in fibrous or scar tissue development can interfere with normal function of organs and tissues. In treating old age, it is very important to inhibit or stop fibrosis and sclerosis, since so many conditions relate to these processes. In addition to cardiovascular disorders such as atherosclerosis, fibrosis can lead to lymphedema and other conditions that interfere with normal function.

Free radicals are very important factors in the aging process. They are incomplete, highly unstable, reactive compounds or molecules. Free radicals can lead to faulty metab-

olism of proteins, can damage DNA, can cause lipid peroxidation in the cell membranes, and can interfere with the cell's ability to take in nutrients and expel wastes. They can inactivate enzymes in the cell membranes, oxidizing cells so that they practically rust. They play a vital role in the formation of cross linkages—the undesired links between protein chains in connective tissues including the skin. Cross-linkages reduce connective tissue elasticity. The most apparent damage of cross linking is the formation of wrinkles.

If their destructive "rage" is not stopped, free radicals can weaken the whole body causing illness and premature aging. In fact, free radicals can contribute to the development of a number of diseases, including premature aging, cancer, atherosclerosis, asthma, inflammatory joint disease, degenerative eye disease, senile dementia, and diabetes.

Symptoms

Signs of aging include wrinkling and sagging skin, aching joints, whitening and/or loss of hair, increased fatigue, increased frequency of urination, constipation, varicose veins, memory loss, increased frequency of chronic disorders, and diminishing eyesight.

Enzyme Therapy

Enzymes can improve digestion, inhibit fibrosis and sclerosis, and fight free radicals. All of these actions help combat premature aging. Thirty years ago, Dr. Max Wolf asked twenty-six men (mostly doctors) of varying ages to take twenty enzyme tablets (containing bromelain, papain, trypsin, chymotrypsin, and the bioflavonoid rutin) every day for the rest of their lives. Every year they have annual physical examinations, including blood chemistry and immune system evaluations. The purpose of this still ongoing study is to determine whether or not enzymes are valuable in avoiding old age diseases, they have any adverse effects over such a long period, long-term oral enzyme intake can cause the body's own enzyme production to diminish, or long-term oral enzyme consumption can cause antibody reactions or allergies.

Of the original twenty-six men, three are dead (two from traumatic accidents and one from circulatory disease at the age of 91). The other twenty-three are still alive and healthy. To date, there is no evidence of any long-term side effects, no autoimmune diseases, no immune disorders, no atherosclerosis, and no other severe chronic disease. This despite the fact that some of the men are heavy smokers, are overweight, and consume a great deal of alcohol.

Enzyme Absorption System Enhancers (EASE) maximize enzyme activity. They can also improve the absorption and bioavailability of various nutrients and decrease the drain on the body's own digestive enzymes, thus prolonging their lives.

In addition to the standard multivitamin, multimineral,

and multiglandular complexes recommended in the introduction to Part Two, the following preventative enzyme treatment program can slow the aging process and prevent illnesses associated with aging. If illness or injury should occur, refer to that specific condition in this book for its recommended enzyme therapy program. Because enzyme production declines with advancing age, older individuals will probably require higher doses than their younger counterparts. Using the following treatment plan as a guideline, work with a health-care practitioner well trained in the use of enzyme therapy to develop an enzyme therapy program appropriate for you.

ESSENTIAL ENZYMES

Enzyme	Suggested Dosage	Actions
Formula Three—See page 38 of Part One for a list of products.	Take as indicated on the label.	Potent antioxidant enzymes.
A digestive enzyme product containing protease, amylase, and lipase enzymes. See the list of digestive enzyme products in Appendix.	Take as indicated on the label with meals.	Improves breakdown and absorption of food nutrients; fights toxic build-up.

AND SELECT ONE OF THE FOLLOWING:

Formula One—See page 38 of Part One for a list of products.	Take 5 tablets twice per day between meals.	Stimulates immune function; breaks up antigen-antibody complexes; fights free-radical formation and inflammation.
WobeMugos E from Mucos	Take 4 tablets twice per day between meals.	Stimulates immune function; breaks up antigen-antibody complexes; fights free-radical formation and inflammation.
Mulsal from Mucos	Take 5 tablets twice per day between meals.	Stimulates immune function; breaks up antigen-antibody complexes; fights free-radical formation and inflammation.
Phlogenzym from Mucos	Take 5 tablets twice per day between meals.	Stimulates immune function; breaks up antigen-antibody complexes; fights free-radical formation and inflammation.
Bromelain	Take four to six 230–250 mg capsules daily.	Stimulates immune function; breaks up antigen-antibody complexes; fights free-radical formation and inflammation.
A proteolytic enzyme	Take as indicated on the label.	Stimulates immune function; breaks up antigen-antibody complexes; fights free-radical formation and inflammation.

ENZYME ABSORPTION SYSTEM ENHANCERS (EASE)

Combination	Suggested Dosage	Actions
Somniset from	Take 1–5 tablets 30–	Improves quality of sleep, nec-
Life Plus or	40 minutes before bedtime.	essary for rejuvenation; improves immune and endocrine function; helps stabilize electrical activity in the central nervous system.
Food 4 Life from Rainbow Light or	Take as indicated on the label.	Increases energy.
Endocryn DHEA from Life Plus or	Take one 30 mg tablet once or twice a day as a dietary supplement; or take as indicated on the label.	Assists prolongation of life span; DHEA is a metabolic precursor of many hormones.
Food for Thought from Life Plus	Take as indicated on the label.	Improves mental health; contains anti-aging nutrients.

ENZYME HELPERS

Nutrient	Suggested Dosage	Actions
Acidophilus and other probiotics	Take as indicated on the label.	Stimulates enzyme activity; improves gastrointestinal function.
Adrenal glandular	Take 100–400 mg per day.	Helps fight stress and adrenal exhaustion; fights inflammation.
Coenzyme Q_{10}	Take 30–60 mg one to two times per day with meals.	Important to immune system function; has antioxidant activity.
Essential fatty acid complex	Take 100–200 mg two to three times per day with meals; or follow the advice of your health-care professional.	Strengthens immunity; helps strengthen cell membranes.
A green food product, such as Green Magma, Kyo-Green, or Green Kamut	Take as indicated on the label.	Improves digestion; fights the aging process; excellent source of chlorophyll.
Kidney glandular	Take as indicated on the label.	Helps fight kidney damage due to infections, kidney stones, cadmium toxicity, and high blood pressure.
Kyolic Aged Garlic Liquid Extract with Vitamin B_1 and B_{12}	Take 10 drops in a beverage three times per day.	Stimulates immune system function.
Thymus glandular	Take 100 mg per day, or as indicated on the label.	Assists the thymus gland in immune resistance; improves immune response; fights infections.

Comments

❑ Scientists believe increased life span is accomplished by getting the body to make more of its own life-extending enzymes. Fortunately, there is a way to do just that—by eating more fresh, enzyme-rich vegetables and fruits.

❑ Increase yogurt, cabbage, and fish liver oil intake.

❏ Add more gel-forming fiber, such as oat bran and flax-seed to your diet.

❏ Eat less, live longer. Dr. Roy Walford, a research immunologist at UCLA, has studied the effect of diet on aging in laboratory mice. He has discovered that restricting their caloric intake can lengthen their life spans two to three times. The National Toxicology Laboratory in Little Rock, Arkansas found that laboratory rats and mice whose caloric intake was cut by 40 percent lived twice as long as normal. However, if you are not careful, reducing your caloric intake may result in a reduction of nutrient intake. Make every calorie count by eating nutrient-dense foods.

❏ Scientists studying fruit flies have found that increasing their natural production of superoxide dismutase and catalase (two potent antioxidant enzymes) can extend life span by more than 30 percent.

❏ Minimize your intake of alcohol, salt, red meat, and coffee. If you smoke, quit.

❏ Enzymes are effective at treating a number of conditions that often plague senior citizens. If any of the conditions associated with aging affect you, see the enzyme therapy program for such conditions in this book.

❏ Daily mental and physical activity can keep you young. Try to exercise thirty to forty minutes every day. A positive mental attitude and acceptance of yourself can help you achieve a healthy and long life.

❏ Drink three 8-ounce glasses of fresh fruit and vegetable juice every day. See Part Three for information on juicing.

❏ Avoid anything that encourages the production of free radicals, such as excessive sunlight, radiation, nicotine, saturated fats, and alcohol.

❏ Since free radicals and decreased enzyme production and activity accelerate aging, detoxification is essential. See Part Three for instructions on detoxification, the Enzyme Kidney Flush, and the Enzyme Toxin Flush.

❏ Try taking 25 mg per day of DHEA, a hormone precursor that can improve immune, cardiovascular, and brain function. Levels of DHEA drop as one ages. DHEA supplmentation appears to slow the aging process.

❏ See the list of Resource Groups in the Appendix for a list of organizations that can provide you with more information about aging.

AIDS

AIDS (acquired immune deficiency syndrome) is an immune system disorder that inhibits the body's ability to defend itself against viruses, bacteria, and other infectious organisms.

It develops as a result of infection with the human immunodeficiency virus (HIV), a transmissible retrovirus that infects the cells of the immune system—the T cells. The viral cells are replicated during normal cell division, and ultimately cause a breakdown in the immune system, leaving the body vulnerable to opportunistic infections.

Someone is infected with HIV every thirteen minutes in the United States. The U.S. Centers for Disease Control estimated that as of 1996, between 650,000 and 900,000 Americans were living with HIV. AIDS has become the leading cause of death among Americans aged 25 to 44 (there is one AIDS-related death every eleven minutes in America). As of June 1996, more than 340,000 Americans died of AIDS complications. UNAIDS (the United Nations program on HIV/AIDS) estimates that 30.6 million people are living with HIV worldwide. Nearly 16,000 people become infected every day. Since the HIV pandemic began, more than 4.5 million people have died of AIDS complications worldwide.

Transmission of HIV requires contact with body fluids infected with HIV. This means that HIV can be transmitted sexually, through contaminated intravenous needles, through contaminated blood transfusions, and through the blood or breast milk of an infected mother to her fetus or newborn baby. HIV cannot be transmitted through casual contact.

A positive test for HIV does not mean that one has AIDS. In fact, a person can be infected with the human immunodeficiency virus for years and not experience symptoms of AIDS. However, once one tests positive for HIV, it is believed that he or she will ultimately develop AIDS.

Although researchers may never know how the AIDS epidemic began, they have found evidence of AIDS infection as early as the mid-1950s (blood specimens collected then, stored frozen, and tested recently, have shown evidence of the infection). In June of 1981, the Centers for Disease Control reported the first cases of what would eventually be called AIDS. Since then, the number of AIDS cases in the United States has increased rapidly. It is important to remember that the disease has a long incubation period (time between first becoming infected with the virus and developing clinical symptoms). Many people developing AIDS today were infected with HIV more than ten years ago. People infected since then may still be without symptoms.

Although we are far from conquering AIDS, there is reason to be optimistic, as scientists are learning more and more about AIDS every day. Many researchers believe that discovering a vaccine for the virus may be necessary before we can completely halt the spread of AIDS, and they are working toward that end.

Symptoms

Initial infection with HIV, called the acute stage, causes symptoms similar to the flu, as well as depression, rash, hives, diarrhea, and fever. After a few days, the body's immune system forms antibodies to attack the foreign invaders. This is called the asymptomatic stage. At this stage, the body is able to fend off the virus, and thus no symptoms are felt by the patient. In fact, patients can remain symptom-free for years, even though the virus survives in the immune system cells.

However, the viral cell population continues to increase to a point where the body can no longer fight it off. Patients may begin to develop symptoms known as the AIDS-related complex (ARC), which include fatigue, weight loss, swollen lymph nodes, fever, and diarrhea. One is determined to have full-blown AIDS, the final stage of the disease, when he or she develops one or more of the opportunistic infections or cancers known to be associated with HIV infection, such as *pneumocystitis carinii* pneumonia, aseptic meningitis, Kaposi's sarcoma, or non-Hodgkin's lymphoma.

Some other diseases and conditions to which HIV patients may be particularly susceptible include various types of respiratory infections, encephalitis, bacterial enteric infections, candidiasis, histoplasmosis, cytomegalovirus disease, herpes simplex, herpes zoster, and human papillomavirus infection.

Anyone who may have been exposed to HIV should be tested for the virus and then tested again six months later. Early treatment is critical for long-term survival. If you have any of these symptoms or have reason to believe you may be infected by the HIV virus, see a doctor. However, just because you experience one or some of the previously mentioned symptoms does not mean that you are infected with HIV. A simple blood test performed by a competent medical technician can determine whether or not you are infected with HIV.

Enzyme Therapy

Systemic enzyme therapy is used to stimulate and strengthen the immune system, to fight antigen-antibody (immune) complexes, and help normalize CD4 + lymphocyte levels (levels of cells of the immune system that diminish after HIV infection). Enzymes reduce inflammation, improve circulation, help speed tissue repair, transport nutrients throughout the body, remove waste products, and improve health. Enzyme therapy can increase the time between the infection and the outbreak of the disease.

Digestive enzyme therapy is helpful in combating HIV/AIDS because it can improve the digestion of food, reduce stress on the gastrointestinal mucosa, help maintain normal pH levels, detoxify the body, and promote the growth of healthy intestinal flora. Supplementation with digestive enzymes can decrease stress on the body's own enzyme systems, allowing the body's enzymes to fight HIV/AIDS.

Enzyme Absorption System Enhancers (EASE) maximize enzyme activity. They can also improve the absorption and bioavailability of various nutrients and decrease the drain on the body's own digestive enzymes, thus prolonging their lives.

In addition to the standard multivitamin, multimineral, and multiglandular complexes recommended in the introduction to Part Two, the following supplements are recommended for the treatment of AIDS. The following recommended dosages are for adults. Children between the ages of two and five should take one-quarter the recommended dosage, those between the ages of six and twelve should take one-half the recommended dosage, and those between the ages of thirteen and seventeen should take three-quarters the recommended adult dosage.

ESSENTIAL ENZYMES

Enzyme	Suggested Dosage	Actions
WobeMugos E from Mucos	Take 15 tablets per day.	Bolsters the immune system; breaks up antigen-antibody complexes; fights free radicals; fights inflammation.
or		
Formula One— See page 38 of Part One for a list of products.	Take 10 tablets three times per day.	Stimulates immune function; breaks up circulating immune complexes; stimulates immune function at the cellular level; fights inflammation.

SUPPORTIVE ENZYMES

Enzyme	Suggested Dosage	Actions
Formula Three—See page 38 of Part One for a list of products.	Take as indicated on the label.	Potent antioxidant enzymes.
A digestive enzyme product containing protease, amylase, and lipase. See the list of digestive enzyme products in Appendix.	Take as indicated on the label with meals.	Improves breakdown and absorption of food nutrients; fights toxic build-up.

ENZYME ABSORPTION SYSTEM ENHANCERS (EASE)

Combination	Suggested Dosage	Actions
Vita-C-1000 from Life Plus	Take as indicated on the label.	Boosts immune function; fights infectious diseases, such as AIDS.
or		
Inflamzyme from PhytoPharmica	Take as indicated on the label.	Immune system activator; reduces inflammation; fights free radicals.
or		
Hepazyme from Enzymatic Therapy	Take as indicated on the label.	Stimulates immune function.
or		
Ecology Pak from Life Plus	Take 2–3 tablets once or twice per day. Take the first dose in the morning, no later than noon.	Assists with natural detoxification; supports the immune system; fights unfriendly bacteria and other organisms.
or		
Metabolic Liver Formula from Prevail	Take as indicated on the label.	Improves liver function.

ENZYME HELPERS

Nutrient	Suggested Dosage	Actions
Acidophilus and other probiotics	Take as indicated on the label.	Stimulate enzyme activity; improve gastrointestinal function.
Adrenal glandular	Take 100–400 mg per day.	Helps fight stress and adrenal exhaustion; fights inflammation.
Coenzyme Q$_{10}$	Take 30–60 mg one to two times per day with meals.	Important for immune system function; has antioxidant activity.
Kyolic Aged Garlic Extract with Vitamins B$_1$ and B$_{12}$	Take 10 drops in water or juice three times per day.	Stimulates immune system function.
Thymus glandular	Take 100 mg per day; or take as indicated on the label.	Assists the thymus gland in immune resistance; improves immune response; fights infections.
Vitamin A	Take up to 50,000 IU for no more than one month (for higher doses, see a physician). If pregnant, do not take more than 10,000 IU.	Needed for strong immune system and healthy mucous membranes; potent antioxidant.
Vitamin B complex	Take 100–150 mg three times per day.	B vitamins act as coenzymes.
Vitamin C	Take 10,000 mg per day. Higher dosages should be taken only under the supervision of a physician.	Fights infections, diseases, and allergies; improves wound healing; stimulates the adrenal glands; promotes immune system; powerful antioxidant.
Vitamin E	Take 1,200 IU per day from all sources (including the multivitamin complex in the introduction to Part Two).	Acts as an intercellular antioxidant; improves circulation; helps resist diseases at the cellular level; aids in stimulating heart function.
Zinc	Take 50 mg per day. Do not take more than 100 mg per day from all sources.	Needed by over 300 enzymes; necessary for white blood cell immune function; improves healing.

Comments

❑ Enzyme-rich, fresh fruits and vegetables (particularly green vegetables) should comprise 60 percent of your diet. If your body cannot tolerate raw fruits and vegetables, increase your intake of digestive enzymes, or sauté or steam your produce. For additional dietary recommendations, see Part Three.

❑ To improve overall health, avoid alcohol, white flour, and refined sugar completely. Also avoid drugs (except those prescribed by your physician), cigarettes, food additives and preservatives, caffeine, and chlorinated water (use bottled, distilled, or well-water, instead).

❑ According to Joan Priestley, M.D. (a world-renowned AIDS specialist), maitake mushrooms can improve Kaposi's sarcoma and other complications of AIDS.

❑ Perform the Enzyme Kidney Flush every day for one week. During the following week, perform the Enzyme Toxin Flush. Perform these flushes only once per month. It is essential to restore the intestinal flora and re-establish intestinal peristalsis. Cleansing methods, including juicing, fasting, detoxification diets, exercise, and water therapy are helpful (see Part Three for more information).

❑ It has been said that HIV/AIDS is a disease of choice; you can choose not to get it by avoiding unprotected sex and by *not* sharing needles. HIV is infectious. Individuals in every stage of HIV disease are able to infect others. A person cannot tell whether it is safe to have sex with, or to share needles with, another person by looking for signs of illness or by asking the individual if she or he is healthy. Most infected persons have no outward sign of illness and most are unaware that they are infected.

❑ Enzyme injections will provide a higher concentration of enzymes directly into the bloodstream. Enzyme injections are administered in doctors' offices and hospitals only. Physicians may also prescribe enzyme retention enemas to enhance absorption of enzymes. For further information, see your physician.

❑ Because AIDS patients can develop so many other diseases and opportunistic infections, refer to such conditions in this book for treatment recommendations of their symptoms.

❑ See the list of Resource Groups in the Appendix for a list of organizations that can provide you with more information about AIDS.

Alcoholism

Alcoholism is a chronic illness characterized by an overwhelming physical and psychological dependence on alcoholic beverages. Alcoholism is also marked by the development of a pattern of deviant behaviors characteristic of excessive drinking. Included in this pattern of behaviors is frequent intoxication that interferes with the patient's social, work, and even family life.

Alcohol directly affects the brain cells and even at low doses can alter our coordination. It alters our perception of odors and taste and even our sensitivity to pain. It can act as a stimulant at lower doses and as a depressant at higher doses. What's worse, we appear to build up a tolerance to alcohol's effects, needing more and more alcohol to produce the same effects.

The physical dependence associated with alcoholism causes the patient to experience withdrawal symptoms when he or she attempts to stop drinking. Such symptoms range from mild—tremors, sweating, weakness, and stomach upset—to moderate—persistent hallucinations—to extreme—delirium tremens, cirrhosis, and, perhaps, death.

Alcohol is nutrient-poor and calorie-dense. A person who drinks 25 ounces of a 100-proof alcohol per day takes in about 2,000 calories just from alcohol—but without the proteins, trace minerals, and vitamins we all need. This is part of the reason that malnutrition is common in alcoholics. To make matters worse, heavy alcohol consumption can actually reduce the ability of the intestines to absorb ingested nutrients (especially vitamins B_1 and B_{12} and some amino acids).

In addition, alcohol can increase the workload of the heart, causing an irregular heartbeat and high blood pressure. It can irritate the linings of the throat and mouth, irritate the digestive system, poison the liver, interfere with the ability of the kidneys to maintain a proper acid/base balance in the body, dilate blood vessels, cause muscle weakness, and reduce the production of blood cells.

Alcoholism causes severe damage to the liver, pancreas, duodenum, immune system, and nervous system. Alcohol is the primary cause of liver cirrhosis. In 1985, cirrhosis was the ninth leading cause of death in the U.S. In 1983, the costs of alcohol problems in America were estimated to exceed $70 billion per year, and costs are rising every year. Nearly half of all homicides and suicides, as well as half of all deaths from motor vehicle accidents, are due to alcohol. In approximately one-third of all boating deaths and drownings, the victims are intoxicated. The leading preventable cause of birth defects is alcohol use during pregnancy.

Recent research shows that there are certain biological and genetic factors that make some people more likely to become alcoholics than others. Children of alcoholic parents are more likely to develop the disease than are children of nonalcoholic parents and adopted children of alcoholic parents. This makes some type of genetic defect a more likely cause than learned behavior. In addition, certain groups, including some Asian groups and Native Americans, may have lower tolerances to alcohol's effects, causing them to drink more. Researchers believe that this may be due to differences in the rates of alcohol metabolism caused by variations in the enzymes that metabolize alcohol.

Researchers say there is also a difference between the ways men and women metabolize alcohol in the stomach. Men make far greater amounts and better use of an enzyme called alcohol dehydrogenase, that breaks down alcohol before allowing it to reach the bloodstream. The result is that when drinking the same amount of alcohol, men don't get as intoxicated as women. Further, when one has a full stomach, alcohol dehydrogenase works best.

Symptoms

The most obvious signs of alcoholism are frequent intoxication and prolonged consumption of large amounts of alcohol. Other signs include work absenteeism and problems with relationships. After long-term drinking, the patient may suffer delirium tremens and cirrhosis.

Enzyme Therapy

The treatment of alcoholism includes decreasing alcohol intake, increasing the intake of those nutrients deficient in alcoholics (particularly the B vitamins), strengthening the body as a whole, and building general immune resistance. Digestive enzyme therapy is used to improve the digestion and assimilation of food nutrients, reduce stress on the gastrointestinal mucosa, help maintain normal pH levels, detoxify the body, and promote the growth of healthy intestinal flora. By improving the absorption of necessary nutrients, alcohol cravings will also decrease.

Systemic enzyme therapy is used to reduce inflammation, stimulate the immune system, improve circulation, help repair any damaged tissue, bring nutrients to any damaged area, remove waste products, and enhance wellness.

Enzyme Absorption System Enhancers (EASE) maximize enzyme activity. They can also improve the absorption and bioavailability of various nutrients and decrease the drain on the body's own digestive enzymes, thus prolonging their lives.

In addition to the standard multivitamin, multimineral, and multiglandular complexes recommended in the introduction to Part Two, the following supplements are recommended for the treatment of alcoholism.

ESSENTIAL ENZYMES

Enzyme	Suggested Dosage	Actions
A digestive enzyme formula containing protease, amylase, and lipase enzymes. For a list of digestive enzyme products, see Appendix. *or*	Take as indicated on the label.	Aids digestion by breaking down proteins, carbohydrates, and fats.
Pancreatin	Take as indicated on the label.	Contains protease, amylase, and lipases, which aid digestion.

SUPPORTIVE ENZYMES

Enzyme	Suggested Dosage	Actions
Formula Three—See page 38 of Part One for a list of products.	Take as indicated on the label.	Potent antioxidant enzymes.

ENZYME ABSORPTION SYSTEM ENHANCERS (EASE)

Combination	Suggested Dosage	Actions
Vita-C-1000 from Life Plus *or*	Take as indicated on the label.	Boosts immune function; fights free radicals; helps detoxify the body.

Zyme Dophilus from PhytoPharmica *or*	Take as indicated on the label.	Promotes the growth of beneficial bacteria in the intestines; helps stimulate enzymatic activity.
Detox-Zyme from Rainbow Light *or*	Take as indicated on the label.	Helps detoxify the gastroin-testinal tract and the body.
Detox Enzyme Formula from Prevail	Take as indicated on the label.	Helps detoxify the gastroin-testinal tract and the body.

ENZYME HELPERS

Nutrient	Suggested Dosage	Actions
Acidophilus or other probiotics, such as bifidobacterium; or Acidophilasé from Wakunaga, which contains *L. acidophilus, B. bifidum,* and protease, lipase, and amylase enzymes.	Take as indicated on the label.	Improves gastrointestinal function; needed for production of biotin, niacin, folic acid, and pyridoxine, which are needed by many enzymes; helps produce antibacterial substances; stimulates enzyme activity.
Adrenal glandular	Take 100–400 mg per day.	Fights adrenal exhaustion; restores normal carbohydrate balance.
Aged garlic extract	Take 2 capsules three times per day.	Potent antioxidant; improves digestion; stimulates healthy intestinal bacteria.
Amino-acid complex	Take as indicated on the label.	Amino acids serve as building blocks for the body's enzymes.
Betaine hydrochloric acid	Take 150 mg three times per day with meals.	Increases gastric acidity. If gastric upset occurs, discontinue use.
Fiber	Take as indicated on the label.	Provides bulk; needed for proper elimination.
Fructooligo-saccharides	Take as indicated on the label.	Increases peristaltic action of intestines; improves liver function.
A green food product, such as Green Kamut, Kyo-Green, or Green Magma	Take as indicated on the label.	Green foods are rich in enzymes, vitamins, minerals, chlorophyll, and amino acids.
Lecithin	Take 1,000 mg three times per day with meals.	Emulsifies fat.
Liver glandular	Take 50 mg per day.	Improves liver function; promotes liver cell regeneration; improves folic acid, iron, and vitamin B_{12} function.
Ox bile extract	Take 120 mg per day.	Aids digestion.
Vitamin A	Take 5,000 IU per day.	Needed by digestive glands to produce enzymes; promotes normal taste and appetite; required for healthy mucous linings, skin and cell membrane structure, and immune system; potent antioxidant.
Vitamin B_1 (thiamin)	Take 150 mg three times per day.	Essential for carbohydrate metabolism, proper energy production in the brain, nerve function, appetite, and heart function; supports muscles, hair, eyes, and hearing.
Vitamin B_3 (niacin) *or* Niacin-bound chromium	Take 50 mg three times per day. Take 10 mg three times per day.	Stimulates gastric juices and hydrochloric acid; important in energy metabolism; maintains nervous system and brain; supports healthy skin, liver, and gastrointestinal tract; important in biological oxidation reduction.
Vitamin B_6 (pyridoxine)	Take 50 mg three times per day.	Assists enzymes that metabolize amino acids and fats; necessary for proper immune function; involved in neurotransmitter synthesis; assists in transport of potassium into cells; needed for healthy nerves, muscle, blood, and skin.
Vitamin C	Take 5,000 mg per day.	Stimulates the adrenal glands to produce hormones (which also act as coenzymes); aids in detoxification; is a potent antioxidant; improves the function of many enzymes; enhances absorption of iron; required for converting folic acid into its active form and for collagen production.
Vitamin E	Take 1,200 IU per day from all sources (including the multivitamin complex in the introduction to Part Two).	Acts as an intercellular antioxidant; improves circulation; helps resist diseases at the cellular level.
Whey	Take 1 heaping tablespoon three times per day with meals.	Improves digestion.
Zinc	Take 10–50 mg per day. Take no more than 100 mg per day from all sources.	Needed by over 300 enzymes, many of which work to metabolize carbohydrates, alcohol, and essential fatty acids; synthesize proteins; dispose of free radicals; and manufacture heme (a constituent of hemoglobin).

Comments

❑ Avoid alcohol completely, as well as white flour and refined sugar.

❑ Eat plenty of enzyme-rich fresh fruits and vegetables (particularly green vegetables). They should comprise 60 percent of your diet. If your body cannot tolerate raw fruits and vegetables, increase your intake of digestive enzymes, or sauté or steam your produce.

❑ Phytochemicals should be used to enhance the healing process. For example, citrus bioflavonoids (from oranges, lemons, and grapefruits), grape seed extract, and carotenoids can all help fight capillary permeability and the red starburst appearance of the skin, characteristic of alcoholism.

❏ Detoxify every month for six months. Perform the Enzyme Kidney Flush every day for one week. During the following week, perform the Enzyme Toxin Flush. During the subsequent two weeks, perform the Vitamin C/Enzyme Flush. Follow directions in Part Three.

❏ During recovery, a proteolytic enzymatic skin salve and exfoliant can be used to assist in restoring vibrant skin and in decreasing capillary permeability.

❏ See the list of Resource Groups in the Appendix for a list of organizations that can provide you with more information about alcoholism.

Alkalosis

The maintenance of an acid-base balance in the blood is critical for life. A slight variation in pH level can be devastating, possibly causing shock, coma, and even death. When there is too much base in the blood as compared with the level of acid, the condition is called *alkalosis*. When there is too much acid in the blood compared with the level of base, the condition is *acidosis*. (*See* ACIDOSIS in Part Two.)

The body does all it can to maintain a constant blood pH range of 7.35 to 7.45 (See the discussion of pH in Chapter 5 of Part One.) Its first line of defense against changes in pH levels is the kidneys. The kidneys excrete excess acid from the bloodstream if the blood's pH level falls too low (acidic). This process, however, takes several days to rectify the problem. There are also pH buffers in the blood, such as bicarbonate and carbon dioxide. Buffers are substances that neutralize any excess acid or alkali in the blood to maintain a constant pH level. Bicarbonate is a basic substance and carbon dioxide is acidic. These two compounds exist in an equilibrium in the blood. However, if excess acid enters the bloodstream or excess base is lost from the bloodstream, more bicarbonate is produced. If more base enters the bloodstream or acid is lost, more carbon dioxide is produced. The third mechanism the body uses to control the blood pH level is the excretion of carbon dioxide (which is acidic) through the lungs. Excess levels of carbon dioxide in the blood cause the body to breathe deeper and faster. When respiration increases, the level of carbon dioxide in the blood decreases. Likewise, when respiration decreases, the level of carbon dioxide in the blood increases.

When illness causes any of these mechanisms to malfunction, acidosis or alkalosis will result. There are two types of alkalosis—metabolic and respiratory.

Metabolic alkalosis results from excessively high levels of bicarbonates in the blood. It usually develops when the body loses too much acid, such as occurs with prolonged vomiting. Rarely, it may occur as a result of ingesting too much of a basic substance, such as baking soda (a bicarbonate). Those suffering from metabolic alkolosis may experience no symptoms at all. Others may experience irritability and muscle twitching and cramping. As the condition gets more severe, the symptoms become more severe.

Respiratory alkalosis results from hyperventilation, which causes the the carbon dioxide level to decrease. Hyperventilation can be caused by anxiety, pain, fever, low blood levels of oxygen, or aspirin overdoses. Respiratory alkalosis produces a tingling feeling in the face and can cause more anxiety. As it gets worse, it can cause muscle spasms and mental fog.

Alkalosis is not a condition, but rather a symptom of an underlying problem. So, treatment involves removing the cause or treating the causative disease. However, alkalosis is a very serious symptom, so medical care is necessary.

Enzyme Therapy

Digestive enzyme therapy is used to strengthen the body as a whole and build general resistance. Enzymes improve the digestion and absorption of food nutrients, reduce stress on the gastrointestinal mucosa, help maintain normal pH levels, detoxify the body, and promote the growth of healthy intestinal flora. Enzymes help transport nutrients throughout the body, remove waste products, and improve health.

Enzyme Absorption System Enhancers (EASE) maximize enzyme activity. They can also improve the absorption and bioavailability of various nutrients and decrease the drain on the body's own digestive enzymes, thus prolonging their lives.

In addition to the standard multivitamin, multimineral, and multiglandular complexes recommended in the introduction to Part Two, the following supplements are recommended for the treatment of alkalosis.

ESSENTIAL ENZYMES

Enzyme	Suggested Dosage	Actions
A digestive enzyme product containing protease, amylase, and lipase. For a list of digestive enzyme products, see Appendix. *or*	Take as indicated on the label with meals.	Can normalize digestion by improving the breakdown and absorption of food nutrients; fights toxic build-up.
NESS Formula #1 *or*	Take 2 tablets with each meal.	Helps the body utilize food nutrients; replaces enzyme activity lost from food during processing and cooking.
Fiber Enzyme Formula from Prevail	Take as indicated on the label.	Contains high-potency alkalizing enzymes.

Acid-Forming Foods

If your body is too alkaline, increase your intake of acid-forming foods from the following list. Be aware that citrus fruits are not acid-forming foods, as they become alkaline in the body.

- Barley
- Beans (lima or white)
- Beef
- Chestnuts
- Corn and corn meal
- Cranberries
- Fish
- Lentils
- Millet
- Oatmeal

- Peanuts and peanut butter
- Peas
- Plums
- Pork
- Poultry
- Prunes
- Rice (brown)
- Rye flour
- Sweet potatoes
- Whole-grain breads, crackers, and cereals

SUPPORTIVE ENZYMES

Enzyme	Suggested Dosage	Actions
Formula Three—See page 38 of Part One for a list of products.	Take as indicated on the label.	Potent antioxidant enzymes.

ENZYME HELPERS

Nutrient	Suggested Dosage	Actions
Alfalfa tablets	Take as indicated on the label.	Important for the digestive tract; good source of vitamins, minerals, and phytochemicals important in maintaining health.
Amino acid complex	Take as indicated on the label.	Can help to acidify the body and normalize pH levels.
Betaine hydrochloric acid	Take as indicated on the label.	Aids digestion in those with insufficient hydrochloric acid levels in their digestive tracts.
L-Cysteine	Take 1,000 mg per day with juice or water between meals. For better absorption, take with vitamins B_6 and C.	Improves detoxification; helps make tissues more acidic.
Selenium	Take 100 mcg twice per day.	Potent free-radical fighter.
Sulfur	Take two 250 mg tablets per day.	Helps balance pH level; an acid-forming mineral.
Vitamin B complex	Take 250 mg per day.	Critical for maintaining normal pH levels.
Vitamin C	Take 1,000–2,000 mg three times per day.	Potent antioxidant; improves pH levels.

Comments

❑ In order to avoid pH imbalance, see the insets on pages 93

and 110 for acid- and alkaline-forming foods. Periodically review these lists, and adjust your diet accordingly.

❑ Grains, including whole-grain bread, crackers, cereals, and ricel; nuts; lentils; and beans, should constitute 75 percent of your diet. The remaining 25 percent of your diet should include fresh vegetables, fresh fruit, chicken, fish, natural cheese, and eggs.

❑ Do not eat sodium-containing products. Use sea salt, which is high in potassium, instead of table salt.

❑ Avoid mineral supplements, other than selenium and sulfur, for three weeks. Also avoid antacids completely, if possible.

❑ Knowing whether your blood pH level is too alkaline is a very important measure of health. Test strips for measuring urine pH levels are widely available at pharmacies, and are a quick and easy way to measure pH levels. Follow instructions on the container. Your blood pH level should be between 7.35 and 7.45. You can also check your body's pH levels using red and blue litmus paper. Blue litmus paper turns red in an acid environment, and red litmus paper turns blue in an alkaline environment.

❑ Exercise every day for twenty to thirty minutes. Try walking, bicycling, or swimming at a moderate pace.

Allergies

An allergy is a hypersensitivity to a normally harmless substance. Allergies can be produced by a seemingly endless list of apparently innocuous substances present in our food and drink, the air we breathe, the things we touch, and the clothes we wear. (See the inset "Allergens" on page 111.)

Allergies are very common. In fact, at least 30 percent of the

Allergens

Almost any substance has the potential to be an allergen (a substance that provokes an allergic reaction) for a given person. However, some substances are more likely than others to provoke an allergic reaction. Following is a list of just a few possible allergens.

Inhaled allergens	Molds, pollens, dust, fragrances (natural and bottled), animal dander.
Ingested allergens	Eggs, milk, peanuts, shellfish, chocolate, strawberries, pork and other foods (particularly proteins), food colorings and additives, drugs (including aspirin, antibiotics, and cortisone).
Contact allergens	Leather, fur, cosmetics, soaps, metals, chemicals.
Other allergens	Insect bites and stings.

population suffers from them. In addition, most allergic individuals are sensitive to a number of substances, not just one. Since allergies affect the health of so many Americans, their diagnosis, treatment, and prevention are very important.

It is the immune system's job to protect us from any and all foreign invaders, including bacteria, viruses, parasites, and fungi. In allergic reactions, for some reason, the body recognizes harmless substances as foreign invaders. The immune system produces an antibody called IgE (immunoglobulin E—nicknamed the "allergy antibody") in response to allergen exposure. When IgE is produced, it sets off a chain of events to fight the allergen, causing the symptoms collectively known as an allergic reaction.

The exact cause of allergies is still unknown. However, heredity appears to play an important role, since an individual is more likely to have allergies if someone else in the family suffers from allergies. This doesn't mean you'll have the same exact allergy that your mother or father has, but it does mean that you may have a tendency to develop an allergic response to something.

Many physicians encourage mothers to breast-feed their babies because of the antibodies that the mother can pass to the child through breast milk. One study found significantly higher levels of antibodies in breast-fed than in formula-fed babies. The researchers concluded that breast-feeding enhances the immune response, thus helping the body better fight antigens that cause allergies. (See the inset "The Importance of Breast-Feeding in the Prevention of Allergies" on page 112.)

Allergies can appear during any period of life, from infancy to the golden years of old age. In most cases, however, allergic reactions will assert their ugly heads before one reaches his or her forties. Exactly when an individual develops an allergic sensitivity is determined by a number of factors, such as the amount and length of exposure to an allergen. An allergy may develop slowly over a period of years until the individual becomes aware that he or she does indeed have an allergy and that some form of treatment is necessary. However, some people may lose their sensitivity to a substance and recover spontaneously.

Food Allergies

Any food can cause a food allergy. However, the most common food allergens include milk, eggs, fish, shellfish, wheat, peanuts, soy, and tree nuts (such as walnuts). The protein component of the food is what usually triggers the allergic reaction.

What's surprising is that just because one is allergic to milk does not necessarily mean that he or she will be allergic to other beef products, such as the meat itself. In fact, people with milk allergies can usually tolerate beef, and people who are allergic to eggs can usually still eat chicken or inhale particles from chicken feathers without any problem.

If you suspect that you have a food allergy, keep a food diary. The food diary consists of a detailed record by date and time of the foods and beverages you eat and drink and any symptoms that may occur. Over time, you may begin to see a pattern evolve between intake of a particular food and resulting symptoms.

After completing a food diary, if you suspect a hypersensitivity to a particular food, try eliminating that food for three weeks. If no improvement is seen, resume eating that food and try eliminating something else. If your symptoms disappear, you've probably found the culprit. However, this does not necessarily mean that you are allergic to that particular food. You may just be missing an enzyme and can't properly digest the food. For instance, people with lactase deficiency may think they are allergic to milk, when they actually are deficient in the enzyme necessary to digest lactose (a sugar in milk). This is a food intolerance, not an allergy. It is possible, however, to be allergic to milk proteins. See a physician for a proper diagnosis.

Physicians use several methods to test for allergies. In skin-prick testing, the patient's skin is punctured or scratched with a diluted extract of the suspected food. Any reaction indicates an allergy. A blood test can measure anti-

The Importance of Breast-Feeding in the Prevention of Allergies

Though scientists aren't sure exactly how breast-feeding protects infants from potential allergens and other threats to immunity, they do know that mother's milk contains immunologic compounds (including IgA) and other defense factors. Compared with formula-fed babies, breast-fed babies:

• Have fewer gastrointestinal illnesses.
• Have significantly higher antibody levels.
• Have a lower incidence of wheezing, prolonged colds, diarrhea, vomiting, eczema, and food and respiratory allergies.

bodies made against a substance. The double-blind, placebo-controlled food challenge test involves giving the patient a capsule containing either a sample of the allergen or a placebo (a harmless pill) and watching for a response. Neither the physician nor the patient knows which substance is in the capsule. This keeps the patient's and the doctor's expectations from affecting the test results.

Symptoms

Symptoms of an allergy can include sneezing, nasal congestion, wheezing, coughing, watery eyes, runny nose, headache, and fatigue. There may be nausea, vomiting, diarrhea, and cramps. Some allergies cause an itchy rash or hives or swelling and itching of the lips, mouth, or throat. Severe allergic reactions (called anaphylactic reactions) can cause such symptoms as dangerously low blood pressure, lung spasms, and shock. Anaphylactic reactions can be life-threatening.

Enzyme Therapy

Enzyme therapy is especially effective at fighting allergies because enzymes can break down the protein allergens and work to block the process that causes an allergic reaction. Digestive enzyme therapy is used to improve the digestion and assimilation of food nutrients, reduce stress on the gastrointestinal mucosa, help maintain normal pH levels, and detoxify the body. Enzymes can promote the growth of healthy intestinal flora and thereby bolster immune function. Digestive enzymes also serve as replacements for the body's pancreatic enzymes, leaving the pancreatic enzymes free to perform other functions in the body, such as boosting immunity and decreasing inflammation.

Systemic enzyme therapy is used to reduce inflammation, stimulate the immune system, improve circulation, transport nutrients throughout the body, and remove waste products.

Enzyme Absorption System Enhancers (EASE) maximize enzyme activity. They can also improve the absorption and bioavailability of various nutrients and decrease the drain on the body's own digestive enzymes, thus prolonging their lives.

In addition to the standard multivitamin, multimineral, and multiglandular complexes recommended in the introduction to Part Two, the following supplements are recommended for the treatment of allergies. The following recommended dosages are for adults. Children between the ages of two and five should take one-quarter the recommended dosage, those between the ages of six and twelve should take one-half the recommended dosage, and those between the ages of thirteen and seventeen should take three-quarters the recommended adult dosage.

ESSENTIAL ENZYMES

Enzyme	Suggested Dosage	Actions
A digestive enzyme product containing protease, amylase, and lipase enzymes. See the list of digestive enzyme products in Appendix. *or*	Take as indicated on the label with meals.	Improves the breakdown and absorption of food nutrients; fights toxic build-up.
A proteolytic enzyme or mixture *or*	Take as indicated on the label.	Stimulates immune function; breaks up circulating immune complexes; fights free-radical formation and inflammation.
Formula One— See page 38 of Part One for a list of products. *or*	Take 5 tablets twice per day.	Fights inflammation, pain, and swelling; speeds healing; stimulates the immune system; strengthens capillary walls.
Phlogenzym from Mucos	Take 4 tablets three times per day.	Stimulates the immune system; fights inflammation, toxins, and free radicals.

SUPPORTIVE ENZYMES

Enzyme	Suggested Dose	Actions
Formula Three—See page 38 of Part One for a list of products.	Take as indicated on the label.	Potent antioxidant enzymes.

ENZYME ABSORPTION SYSTEM ENHANCERS (EASE)

Combination	Suggested Dosage	Actions
Sinease from Prevail *or*	Take as indicated on the label.	Decongestant.
Combat from Life Plus *or*	Take as indicated on the label.	Helps support respiratory system; stimulates immune function.
Inflamzyme from PhytoPharmica *or*	Take as indicated on the label.	Immune system activator.
Connect-All from Nature's Plus *or*	Take as indicated on the label.	Supports tissue repair.
Mucous Dissolver Liquezyme from Enzyme Process Laboratories	Take as indicated on the label.	Dissolves mucus.

ENZYME HELPERS

Nutrient	Suggested Dosage	Actions
Acidophilus and other probiotics	Take as indicated on the label.	Stimulates enzyme activity; improves gastrointestinal function.
Adrenal glandular	Take 100–400 mg per day.	Helps fight stress and adrenal exhaustion; fights inflammation.
Coenzyme Q_{10}	Take 30–60 mg one to two times per day with meals.	Important for immune system function; has antioxidant activity.
Essential fatty acid complex	Take 100–200 mg two to three times per day with meals; or follow the advice of your health-care professional.	Strengthens immunity; helps strengthen cell membranes.
Kyolic Aged Garlic Liquid Extract with Vitamins B_1 and B_{12}	Take 10 drops in water or juice three times per day.	Stimulates immune system function.
Spleen glandular	Take 50 mg per day.	Supports the immune system; fights bacteria; inhibits inflammation.
Thymus glandular	Take 100 mg per day; or take as indicated on the label.	Assists the thymus gland in immune resistance; improves immune response; fights infections.
Vitamin A	Take up to 50,000 IU per day for no more than one month (for higher doses, see a physician). If pregnant, do not take more than 10,000 IU per day.	Needed for strong immune system and healthy mucous membranes; potent antioxidant.
Vitamin B complex	Take 100–150 mg three times per day.	B vitamins act as coenzymes.
Vitamin C	Take 10,000 mg per day; or 1–2 grams every hour until symptoms improve.	Fights infections, diseases, and allergies; stimulates the adrenal glands; promotes a healthy immune system; powerful antioxidant.
Vitamin E	Take 1,200 IU per day from all sources (including the multivitamin complex in the introduction to Part Two).	Acts as an intercellular antioxidant; improves circulation; helps resist diseases at the cellular level; aids in stimulating heart function; fights allergies.
Zinc	Take 50 mg per day. Do not take more than 100 mg per day from all sources.	Needed by over 300 enzymes; necessary for white blood cell immune function; improves healing.

Comments

❑ The best way to fight any allergy is to avoid the cause. Avoid any foods to which you know you are allergic. If you are allergic to pollen, try to stay indoors during the height of the pollen season, or wear a filtering mask. Vacuum your home regularly to keep it free from dust and pollen.

❑ Perform the Enzyme Kidney Flush every day for one week. During the following week, perform the Enzyme Toxin Flush. Perform these flushes only once per month. See Part Three for instructions for both flushes.

❑ See the list of Resource Groups in the Appendix for a list of organizations that can provide you with more information about allergies.

❑ *See also* ASTHMA; HAY FEVER.

Alzheimer's Disease

Alzheimer's disease (AD) is a degenerative type of dementia (decline in intellectual function) marked by atrophy of the brain. The parts of the brain that control memory, language, and thought are those most affected. Alzheimer's disease affects over 4 million Americans over 65 years of age; in fact, it is the most common type of dementia in older people.

Alzheimer's disease is named after the German physician Dr. Alois Alzheimer. In 1906, Dr. Alzheimer found tangled clumps of fibers (now known as neurofibrillary tangles and senile plaques) in the brain tissue of a woman who died from an unusual mental illness, now known as Alzheimer's disease. The tangles and plaques Dr. Alzheimer found are characteristic of the disease.

In addition to these tangles of nerve fibers and plaques, loss of nerve cells in those areas of the brain that are essential for several mental abilities, including memory, is also characteristic of the disease. AD patients also have lower levels of the chemicals responsible for the transmission of complex messages between billions of nerve cells in the brain, which disrupts normal memory and thinking. Alzheimer's disease can be either sporadic—with a weak genetic predisposition— or familial—with a strong genetic predisposition.

AD develops slowly, beginning with mild problems in memory and ending with severe mental damage. How fast the disease progresses varies from person to person with some having the disease as long as twenty years, and others only five.

The major risk factors for AD are family history and age. The amount one uses his or her brain to perform mental tasks throughout life also seems to play a role in the development of Alzheimer's disease. According to research from the National Institute on Aging, the more years of formal education one has, the less likely one is to develop Alzheimer's disease as they age. This draws a link between increased mental activity and decreased risk for Alzheimer's disease.

Environmental factors, viruses, serious head injury, and excessive exposure to aluminum, sulfur, calcium, silicon, and bromine may all play roles in the development of Alzheimer's disease; as may brain deficiencies of zinc, potassium, boron, vitamin B_{12}, selenium, and magnesium.

It is possible that a high intake of any neurotoxic metal, such as aluminum, may inhibit the activity of enzymes requiring magnesium. In fact, research indicates that dementia is associated with a relative insufficiency of magnesium in the brain. Such insufficiency may be due to low intake (inadequate nutritional intake is a common problem among the elderly), low retention, or impaired transport of the mineral.

The Centers for Disease Control consider Alzheimer's disease a leading cause of death in America; however, tracking the incidences of death due to Alzheimer's disease is difficult because, in its late stages, Alzheimer's disease disrupts normal body function, leading to a number of conditions to which the cause of death may be attributed.

The incidence of Alzheimer's disease is age-related, but it is not a normal part of aging. Even though it usually begins after age 65, younger people may also have AD, although it is much less common. The risk of AD increases with age. About 3 percent of men and women ages 65 to 74 have Alzheimer's disease, and nearly half of those age 85 and older may have the disease.

Symptoms

The primary symptoms of Alzheimer's disease are progressive loss of memory, language skills, and thought (dementia); personality changes; and difficulty performing intellectual tasks. Additional symptoms include agitation, sleeplessness, anxiety, depression, wandering, and disinterest in personal appearance.

Enzyme Therapy

Systemic enzyme therapy is used to help balance the body, eliminate toxins, and improve nerve communication. Enzymes can also improve circulation, strengthen the body as a whole, build general resistance, stimulate the immune system, and reduce any inflammatory processes.

Digestive enzyme therapy is used to improve metabolism, decrease any nutritional deficiencies and eliminate toxins. Enzyme therapy reduces stress on the gastrointestinal mucosa, helps maintain normal pH levels, detoxifies the body, and promotes the growth of healthy intestinal flora thereby improving the overall health of the body. Digestive enzymes also serve as replacements for the body's pancreatic enzymes, leaving the pancreatic enzymes free to perform other functions in the body, such as boosting immunity and improving circulation.

Enzyme Absorption System Enhancers (EASE) maximize enzyme activity. They can also improve the absorption and bioavailability of various nutrients and decrease the drain on the body's own digestive enzymes, thus prolonging their lives.

Autopsies of Alzheimer's disease patients revealed that, although catalase and glutathione peroxidase were similar in the brains of AD patients and control cadavers, the level of superoxide dismutase (SOD) in several parts of the brains of AD patients was up to 35-percent less than it was in the brains of those who did not die of Alzheimer's disease. In an article published in *The Journal of Neurochemistry*, authors Dr. W. Gsell, et al. measured the activities of the antioxidant enzymes catalase and superoxide dismutase in the brains of patients who had died from dementias like Alzheimer's disease and compared them with the brains of those of the same age who did not die of Alzheimer's disease. They found a significant reduction in the activity of catalase in parts of the brains of Alzheimer's disease patients. Therefore, supplemental antioxidant enzymes (as well as other enzymes) are essential in controlling this condition.

In addition to the standard multivitamin, multimineral, and multiglandular complexes recommended in the introduction to Part Two, the following supplements are recommended for the treatment of Alzheimer's disease.

ESSENTIAL ENZYMES

Enzyme	Suggested Dosage	Actions
Formula Three—See page 38 of Part One for a list of products.	Take 200 mcg three times per day between meals.	Potent antioxidant enzymes.

AND SELECT ONE OF THE FOLLOWING:

Inflazyme Forte from American Biologics	Take as indicated on the label between meals.	Improves circulation and strengthens capillary walls, thus improving delivery of oxygen and other nutrients to the brain; fights free radicals, which have been implicated as possible causes of Alzheimer's disease; improves immunity; reduces inflammation.
Formula One (particularly Wobenzym N from Naturally	Take 8–10 tablets three times per day between meals for six months or	Improves circulation and strengthens capillary walls, thus improving delivery of oxy-

Vitamin Supplements) —See page 38 of Part One for a list of products.	until symptoms improve. Thereafter, take 6 tablets three times per day between meals.	gen and other nutrients to the brain; fights free radicals, which have been implicated in causing Alzheimer's disease; improves immunity; reduces inflammation.
Bromelain	Take four to six 230–250 mg capsules per day.	Improves circulation and strengthens capillary walls, thus improving delivery of oxygen and other nutrients to the brain; fights free radicals, which have been implicated in Alzheimer's disease; improves immunity; reduces inflammation.
Enzyme complexes rich in proteases, plus amylases and lipases	Take as indicated on the label between meals.	Improves circulation and strengthens capillary walls, thus improving delivery of oxygen and other nutrients to the brain; fights free radicals, which have been implicated in Alzheimer's disease; improves immunity; reduces inflammation.

SUPPORTIVE ENZYMES

Enzyme	Suggested Dosage	Actions
A digestive enzyme formula containing protease, amylase, and lipase enzymes. See the list of digestive enzyme products in Appendix.	Take as indicated on the label with meals.	Improves the breakdown and absorption of proteins, carbohydrates, and fats; fights toxic build-up.

ENZYME ABSORPTION SYSTEM ENHANCERS (EASE)

Combination	Suggested Dosage	Actions
Food for Thought from Life Plus or	Take as indicated on the label.	Improves mental health; contains free-radical fighters and anti-aging nutrients.
Endocryn DHEA from Life Plus or	Take one 30-mg tablet once or twice per day as a dietary supplement; or take as indicated on the label.	Assists prolongation of life span; DHEA is a metabolic precursor of many significant hormones.
Somazyme from Life Plus or	Take 1–5 tablets one to three times per day, preferably on an empty stomach between meals.	Fights free radicals; supports healthy pancreatic function; supports immune system; fosters a healthy bloodstream; fights the aging process.
Food 4 Life from Rainbow Light	Take as indicated on the label.	Increases energy.

ENZYME HELPERS

Nutrient	Suggested Dosage	Actions
Aged garlic extract	Take as indicated on the label.	Lowers cholesterol and blood pressure; fights free radicals.
Bioflavonoids, such as rutin, quercetin, pycnogenol	Take 500 mg three times per day.	Fights atherosclerosis; potent antioxidant; reduces capillary fragility.
Brain glandular	Take as indicated on the label.	Improves brain function.
Calcium	Take 1,000 mg per day.	Plays important roles in muscle contraction and nerve transmission.
Carnitine	Take 500 mg two times per day.	Increases energy; lowers cholesterol; reduces risk of heart disease.
Coenzyme Q_{10}	Take 75 mg one to two times per day with meals.	Potent antioxidant; prevents heart attacks; decreases blood pressure and total serum cholesterol.
Gamma-linolenic acids	Take as indicated on the label.	Lowers blood pressure and high serum cholesterol levels; inhibits platelet aggregation; decreases vascular obstruction.
Grape seed extract	Take 100 mg one to three times per day with meals.	Free-radical fighter; protects blood vessels.
Heart glandular	Take 140 mg per day; or take as indicated on the label.	Improves heart function to improve circulation.
Lecithin	Take 3 capsules with meals; or take as indicated on the label.	Emulsifies fat.
Magnesium	Take 500 mg per day.	Plays important roles in protein building, muscle contraction, nerve function, energy production, calcium absorption, and blood clotting.
Niacin-bound chromium	Take up to 20 mg per day; or take as indicated on the label.	Improves circulation by dilating blood vessels.
Vitamin A	Take 5,000 IU per day.	Necessary for skin and cell membrane structure and strong immune system; potent antioxidant.
Vitamin C	Take 8,000 mg per day.	Essential for vascular function; potent antioxidant; improves the function of many enzymes; enhances absorption of iron; assists in converting cholesterol to bile salt.
Vitamin E	Take 1,200 IU per day from all sources (including the multivitamin complex in the introduction to Part Two).	Potent antioxidant; helps prevent cardiovascular disease; decreases risk of heart attack; stimulates circulation; strengthens capillary walls; prevents blood clots; maintains cell membranes.
Zinc	Take 50 mg per day. Do not take more than 100 mg per day from all sources.	Needed by over 300 enzymes; potent antioxidant.

Comments

❑ Several herbs enhance blood circulation, including astragalus, bilberry, ginkgo biloba, gotu kola, hawthorn, and wheat grass powder.

❑ Bilberry extract, raspberries, cranberries, red wine, grapes, hawthorn, black currants, and red cabbage contain anthocyanidins, known to fight free radicals, reduce blood vessel plaque formation, maintain capillary blood flow, and fight inflammation.

❑ Helpful enzyme-rich foods include all fruits, especially bananas and raisins, vegetables (such as broccoli, carrots, green leafy vegetables, sweet potatoes), bran, whole wheat, and fish foods.

❑ Vitamin A, carotenoids, and vitamin C help fight free radicals, so increase your intake of carrots, spinach, parsley, and kale; rich sources of these nutrients.

❑ Garlic and turnips contain selenium, a mineral critical for proper function of the antioxidant enzyme superoxide dismutase, so you should increase your intake of these vegetables.

❑ Eat more grapes, cherries, and citrus fruits. They're rich in bioflavonoids, which are necessary for capillary strength and integrity.

❑ Avoid aluminum cooking utensils, antacids, or any foods high in aluminum (processed cheese usually contains aluminum).

❑ Detoxification is critical for those with Alzheimer's disease. It is important to eliminate heavy metals and toxins from the body. Follow instructions in Part Three to perform the Enzyme Kidney Flush and the Enzyme Toxin Flush.

❑ See the list of Resource Groups in the Appendix for a list of organizations that can provide you with more information about Alzheimer's disease.

❑ *See also* AGING.

Anemia

Anemia is a blood disorder in which there are too few red blood cells in the blood or too little hemoglobin in them. Hemoglobin is the oxygen-bearing protein in the red blood cells. Consequently, if there is not enough hemoglobin in the blood, there is not enough oxygen getting to all of the cells of the body. All of our bodies' cells need oxygen in order to function. Anemia is actually a symptom of some underlying illness that is causing the decreased hemoglobin production, rather than a disease.

Anything that causes a deficiency of red blood cells, and thus hemoglobin, can lead to anemia. Such causes include poor bone marrow function or production; infections; medications; pregnancy; heavy menstruation; a poor diet lacking such vitamins and minerals as vitamin B_{12}, vitamin C, folic acid, and iron; excessive bleeding; alcoholism; illness; malabsorption; and inflammation in the body.

There are many different types of anemia. *Iron deficiency anemia* is the most common type. Iron is required for the production of hemoglobin, and thus red blood cells, so if there is a deficiency of iron, anemia results. Iron deficiency anemia is found most frequently in children and pregnant women. Like iron, folic acid and vitamin B_{12} are also important in red blood cell production. Deficiency of folic acid can lead to *folic acid anemia*, and deficiency of vitamin B_{12} can result in *vitamin B_{12} anemia*, also called *pernicious anemia*.

Aplastic anemia is characterized by decreased bone marrow production of red and white blood cells and platelets (another component of blood). It is usually caused by certain drugs. *Hemolytic anemia* occurs due to the abnormal and premature destruction of red blood cells. Hemolytic anemia is often congenital, but it can be caused by the use of certain drugs or other unknown causes. *Sickle-cell anemia* is a blood disorder characterized by sickle-shaped, rather than round, red blood cells that contain an abnormal type of hemoglobin. It leads to anemia. *Anemia of pregnancy* is also common.

Symptoms

Depending on the severity and type of anemia, symptoms may vary. The person may be pale (especially pale lips, fingernails, palms, and lining of the eyelids). The individual may feel tired and weak all the time. In more serious cases, there may be a pounding heartbeat, depression, loss of appetite, insomnia, shortness of breath, irritability, decreased resistance to infection, and dizziness. A deficiency of hemoglobin can be confirmed by a blood test.

Enzyme Therapy

Currently, enzyme therapy is recommended only for iron deficiency anemia. Digestive enzyme therapy is used to improve the digestion and absorption of food nutrients, including iron. Soy and other legumes, vegetable fibers, phosphates, phytates, bran, and antacids decrease the absorption of iron by binding with it. While vitamin C can increase the absorption of iron, it can be further increased with digestive enzymes, particularly amylases, which can help break down those fibers and other factors that may inhibit absorption of iron. Digestive enzymes can also reduce stress on the gastrointestinal mucosa, help maintain normal pH levels, detoxify the body, promote the growth of healthy intestinal flora, and build general resistance, all of which can help the body function more efficiently. Systemic enzyme therapy is used to reduce inflammation, stimulate the immune system, improve circulation, help speed tissue repair, bring nutrients to the damaged area, remove waste products, enhance wellness, and strengthen the body as a whole.

Enzyme Absorption System Enhancers (EASE) maximize enzyme activity. They can also improve the absorption and bioavailability of various nutrients and decrease the drain on the body's own digestive enzymes, thus prolonging their lives.

In addition to the standard multivitamin, multimineral, and multiglandular complexes recommended in the introduction to Part Two, the following supplements are recommended for the treatment of iron-deficiency anemia. The following recommended dosages are for adults. Children between the ages of two and five should take one-quarter the recommended dosage, those between the ages of six and twelve should take one-half the recommended dosage, and those between the ages of thirteen and seventeen should take three-quarters the recommended adult dosage.

ESSENTIAL ENZYMES

Enzyme	Suggested Dosage	Actions
A digestive enzyme formula containing protease, amylase, and lipase enzymes. For a list of digestive enzyme products, see Appendix. or	Take as indicated on the label.	Improves digestion by breaking down proteins, carbohydrates, and fats.
Pancreatin	Take as indicated on the label.	Contains proteases, amylases, and lipases.

SUPPORTIVE ENZYMES

Enzyme	Suggested Dosage	Actions
Formula Three—See page 38 of Part One for a list of products.	Take as indicated on the label.	Potent antioxidant enzymes.

ENZYME ABSORPTION SYSTEM ENHANCERS (EASE)

Combination	Suggested Dosage	Actions
Metabolic Liver Formula from Prevail or	Take as indicated on the label.	Improves oxygen function.
Bio-Immunozyme Forte from Biotics or	Take as indicated on the label.	Stimulates immune function; has antioxidant action.
Antioxidant Complex from Tyler or	Take 2–4 tablets between meals.	Potent free-radical fighter and immune system stimulator.
Hepazyme from Enzymatic Therapy or	Take as indicated on the label.	Stimulates immune function.
Detox Enzyme Formula from Prevail	Take 1 capsule between meals one or two times per day with	Helps rid the body of harmful toxins.

	a 6- to 8-ounce glass of water or juice.	
or		
Oxy-5000 Forte from American Biologics	Take as indicated on the label.	Helps the body excrete heavy metals; helps relieve problems caused by biochemical stressors, common infections, and degenerative processes.

ENZYME HELPERS

Nutrient	Suggested Dosage	Actions
Acidophilus or other probiotics, such as bifidobacterium; or Acidophilasé from Wakunaga, which contains *L. acidophilus*, *B. bifidum*, and proteolytic, lipolytic, and amylolytic enzymes.	Take as indicated on the label.	Improves gastrointestinal function; stimulates enzyme activity: improves iron absorption.
Aged garlic extract	Take two 300 mg capsules three times per day; or take as indicated on the label.	Potent antioxidant; improves digestion; stimulates healthy intestinal bacteria; improves iron absorption.
Amino-acid complex, such as AminoLogic from PhysioLogics	Take as indicated on the label.	Amino acids serve as building blocks for the body's enzymes.
A green food product, such as Green Kamut, Kyo-Green, or Green Magma	Take as indicated on the label.	Green foods are rich in enzymes, vitamins, minerals, chlorophyll, and amino acids.
Iron	Take as indicated on the label, or at the direction of your physician.	Important for protein metabolism, phospholipid synthesis, and wound healing; necessary for the formation of hemoglobin; works with manganese in immune response and with zinc to keep the arteries flexible.
Liver glandular	Take 1,000 mg per day.	Improves liver function; improves iron function.
Ox bile extract	Take 120 mg per day.	Aids digestion; improves iron absorption.
Vitamin A	Take 8,000–12,000 IU per day. If pregnant, take no more than 10,000 IU per day.	Needed by digestive glands to produce enzymes; required for healthy mucous linings, cell membrane structure, and immune system; potent antioxidant.
Vitamin B complex	Take 150 mg three times per day.	B vitamins act as coenzymes.
Vitamin C	Take 1,000–2,000 mg per day.	Stimulates the adrenal glands to produce hormones (which also act as coenzymes); aids in detoxification; improves the function of many enzymes; enhances absorption of iron.
Vitamin E	Take 400 IU per day.	Acts as an intercellular antioxidant; improves circulation; helps resist diseases at the cellular level.

Zinc	Take 50 mg per day. Do not take more than 100 mg per day from all sources.	Needed by over 300 enzymes, many of which work to metabolize carbohydrates, alcohol, and essential fatty acids; synthesize proteins; dispose of free radicals; and manufacture heme (a constituent of hemoglobin).

Comments

❏ Since iron is an important component in hemoglobin, foods containing iron are important. Therefore, increase your intake of heart, liver, kidney, and lean meat, whole wheat bread, green leafy vegetables, dried beans and peas, blackstrap molasses, raisins, apricots, beans, and almonds.

❏ Detoxification (cleansing, fasting, and juicing) is important in fighting anemia because toxins may interfere with the bioavailability and absorption of iron. See Part Three for further information on detoxification.

❏ If your iron intake is adequate, but you are anemic due to decreased iron absorption, stimulating organ function through such alternative therapies as meridian therapy, acupressure, acupuncture, and applied kinesiology may be benficial.

Angina Pectoris

Angina is pain in the chest, usually caused by an inadequate supply of oxygen to the tissues of the heart. It is actually a symptom, not a disease, and is an indication of atherosclerosis in the heart's coronary arteries or of a possible impending heart attack. Angina can occur when there is an increased demand for blood by the heart muscle (myocardium), but the coronary arteries are unable to adequately supply that blood. This is why angina victims may experience symptoms while engaged in some physical activity, such as jogging, lifting a heavy object, walking up a flight of stairs; while excited; or even after eating a heavy meal.

In the United States alone, it is estimated that there are 6,750,000 people who have angina pectoris, and an estimated 350,000 new cases occur every year. Men experience angina attacks more often than do women. In 1993, some 930 Americans died from the condition.

The blood vessel linings are covered by a thin film of fibrin. Normally, an equilibrium exists between fibrin synthesis and decomposition in the body—the body does not produce more fibrin than is needed to replace the fibrin that is destroyed. However, in angina, this equilibrium is disturbed, and excessive deposit of lipids and fibrin form. As the deposits build up, the blood vessels narrow. This causes a disturbance of blood supply, eventually resulting in angina pectoris. If angina is suspected, see a physician immediately.

Symptoms

The symptoms of angina may vary. Angina pain is usually (but not always) experienced as a heavy pressure or a crushing sensation in the chest. Sometimes, it may be a vague and only slightly noticeable ache. The pain generally feels like a constriction in the chest, rather than a sharp pain. It can be located in the neck or shoulders and may radiate down the left arm and even into the fingers. It may also radiate into the throat, the teeth, the jaws, or straight through to the back. Sometimes, the pain may even radiate down the right arm or be felt in the upper abdomen. Palpitations and dizziness may also occur. Attacks usually last only a few minutes.

Variant angina differs from typical angina in that it is characterized by pain at rest, rather than upon exertion, and by changes seen on an electrocardiogram during an attack of angina pain. It is caused by spasms of the large coronary arteries.

In *unstable angina*, the symptom pattern changes (for example pain becomes more severe or more frequent). Unstable angina is very serious. It is usually a result of a worsening of coronary artery obstruction. This is a medical emergency, as the risk for heart attack is very high.

Enzyme Therapy

Enzymes help to remove the causes of angina pectoris, that is, the excess deposition of fibrin and lipids in the coronary arteries. One of the causes of angina pectoris is a reduction in the level of plasmin (a natural proteolytic enzyme in the body) over time. This decrease means that there are no longer enough enzymes in the blood to dissolve the ongoing deposition of fibrin and lipids in the blood vessels. Certain enzymes and enzyme mixtures (such as bromelain, papain, pancreatin, trypsin, and chymotrypsin) stimulate fibrinolysis (fibrin breakdown) and increase blood flow. As a result, symptoms decrease, fibrin deposits are dissolved, and swelling is reduced.

Systemic enzyme therapy is used to reduce inflammation, break up fibrin, stimulate the immune system, and improve circulation. Enzymes decrease plaquing along the blood vessel walls, help speed tissue repair, bring nutrients to the damaged area, remove waste products, build general resistance, and enhance wellness.

Digestive enzyme therapy is used to improve the digestion of food, reduce stress on the gastrointestinal mucosa, help maintain normal pH levels, detoxify the body, promote the growth of healthy intestinal flora, and strengthen the body as a whole, thereby improving metabolism and helping the body to function optimally. Digestive enzymes also serve as replacements for the body's pancreatic enzymes, leaving the pancreatic enzymes free to perform other functions in the body, such as decreasing inflammation and improving circulation.

Enzyme Absorption System Enhancers (EASE) maximize enzyme activity. They can also improve the absorption and bioavailability of various nutrients and decrease the drain on the body's own digestive enzymes, thus prolonging their lives.

In addition to the standard multivitamin, multimineral, and multiglandular complexes recommended in the introduction to Part Two, the following supplements are recommended for the treatment of angina pectoris.

ESSENTIAL ENZYMES

Enzyme	Suggested Dosage	Actions
Formula Three—See page 38 of Part One for a list of products.	Take 200 mcg three times per day between meals; or take as indicated on the label.	Potent antioxidant enzymes.

AND SELECT ONE OF THE FOLLOWING:

Formula One (particularly Wobenzym N from Naturally Vitamin Supplements)—See page 38 of Part One for a list of products.	Take 8–10 tablets three times per day between meals for six months or until symptoms improve. Thereafter, take 6 tablets three times per day between meals.	Dissolves blood clots; fights inflammation, pain, and swelling; speeds healing; strengthens capillary walls.
Inflazyme Forte from American Biologics	Take as indicated on the label between meals.	Dissolves blood clots; fights inflammation, pain, and swelling; speeds healing; strengthens capillary walls.
Bromelain	Take four to six 230-mg capsules or tablets per day.	Fights inflammation, pain, and swelling; speeds healing; stimulates the immune system.
Enzyme complexes rich in proteases, plus amylases and lipases	Take as indicated on label between meals.	Dissolves blood clots; fights inflammation, pain, and swelling; speeds healing; strengthens capillary walls.

SUPPORTIVE ENZYMES

Enzyme	Suggested Dosage	Actions
A digestive enzyme formula containing protease, amylase, and lipase enzymes. See the list of digestive enzyme products in Appendix.	Take as indicated on the label with meals.	Improves the breakdown and absorption of proteins, carbohydrates, and fats; frees the body's enzymes to improve circulation; fights toxic build-up.

ENZYME ABSORPTION SYSTEM ENHANCERS (EASE)

Combination	Suggested Dosage	Actions
Protease Concentrate from Tyler *or*	Take as indicated on the label.	Breaks up fibrin; fights inflammation.
Bromelain Complex from PhytoPharmica *or*	Take as indicated on the label.	Decreases swelling; improves circulation; helps break up circulating immune complexes.
Cardio-Chelex from Life Plus	If under 40 and in good health, take 5 tablets once a day—	Supports healthy circulation; improves energy.

first thing in the morning. If you have a history of heart problems, take 5 tablets twice per day— first thing in the morning and midday.

ENZYME HELPERS

Nutrient	Suggested Dosage	Actions
Aged garlic extract	Take three 300 mg capsules three times per day; or take as indicated on the label.	Lowers cholesterol and blood pressure; fights free radicals.
Aorta glandular	Take as indicated on the label.	Helps improve blood vessel function and peripheral and cerebral circulation.
Bioflavonoids, such as rutin or quercetin	Take 500 mg three times per day.	Fight atherosclerosis; potent antioxidants; reduce capillary fragility.
Calcium	Take 1,000 mg per day.	Plays important roles in muscle contraction; helps maintain the heartbeat.
Carnitine	Take 500 mg two times per day.	Increases energy; lowers cholesterol; reduces risk of heart disease.
Coenzyme Q_{10}	Take 75 mg one to four times per day with meals.	Potent antioxidant; prevents heart attacks; decreases blood pressure and total serum cholesterol.
DHEA	Take as indicated on the label with meals.	Supports cardiovascular function.
Grape seed extract	Take 100 mg one to three times a day with meals.	Free-radical fighter; protects blood vessels.
A green food product, such as Kyo-Green, Green Kamut, or Green Magma	Take 1–2 teaspoons in water or juice three times per day.	Reduces blood pressure; effective against heart disease; fights free radicals.
Heart glandular	Take 140 mg per day; or take as indicated on the label.	Improves heart function.
Lecithin	Take one 1,200-mg capsule three times per day between meals; or take as indicated on the label.	Emulsifies fat.
Magnesium	Take 500 mg per day.	Plays important roles in protein building, muscle contraction, nerve function, energy production, calcium absorption, and blood clotting.
Niacin-bound chromium	Take up to 20 mg per day, or as indicated on the label.	Improves circulation by dilating blood vessels.
Vitamin A	Take 5,000 IU per day; or take as per your doctor's directions.	Necessary for skin and cell membrane structure and a strong immune system; potent antioxidant.

Vitamin B complex	Take 50 mg three times per day; or take as indicated on the label.	B vitamins act as coenzymes; support neuromuscular function in circulation.
Vitamin C	Take 5,000–10,000 mg per day.	Essential to vascular function; potent antioxidant; improves the function of many enzymes; enhances absorption of iron; assists in converting cholesterol to bile salt.
Vitamin E	Take 1,200 IU per day from all sources (including the multivitamin complex in the introduction to Part Two).	Potent antioxidant; helps prevent cardiovascular disease; stimulates circulation; strengthens capillaries; prevents blood clots; maintains cell membranes.
Zinc	Take 50 mg per day. Do not take more than 100 mg per day from all sources.	Needed for white blood cell immune function; interacts with platelets in blood clotting.

Comments

❏ A number of foods contain phytonutrients that can help circulatory conditions. For instance, garlic, onions, shallots, chives, leeks, and scallions contain allium compounds, which are potent free-radical fighters, help lower cholesterol, and fight circulatory disorders. Other foods, including wheat germ, oats, olives, nuts, nut and seed oils, organ meats, and eggs, contain alpha-tocopherol (vitamin E), known to help prevent cardiovascular disease and prevent blood clots. Several fruits contain anthocyanidins, which fight blood vessel plaque formation and maintain blood flow in small vessels. These foods include raspberries, cranberries, grapes, red wine, and black currants. Oolong and green teas contain epicatechin, which is a potent antioxidant that lowers blood pressure and strengthens capillaries.

❏ Angina pectoris can be a fatal condition, so you must make a commitment in order to improve. Angina is a warning to change your lifestyle and seek health care.

❏ If you are overweight, go on a low-fat diet.

❏ Reduce or eliminate your intake of coffee and of all alcohol. A little red wine, however, may be beneficial. Research shows that one to two glasses of red wine per day can decrease cholesterol levels.

❏ Smokers should stop smoking.

❏ Drink eight ounces of juice from fresh vegetables and fruits at least twice per day. Use live, enzyme-rich, antioxidant-rich foods, such as pineapple, broccoli, and parsley. See Part Three for instructions for the Enzyme Toxin Flush and other detoxification methods.

❏ See the list of Resource Groups in the Appendix for an organization that can provide you with more information about angina pectoris.

Ankylosing Spondylitis

Ankylosing spondylitis (AS) is a chronic rheumatic disorder characterized by inflammation and stiffening of the joints. AS patients often exhibit a bent over posture due to the stiffening of the spine and hips. The condition may progress to a point where the connective tissue between the vertebrae turn to bone, fusing the bones of the spine together. This is referred to as bamboo spine. This problem can also affect the hips, shoulders, and rib cage.

Ankylosing spondylitis occurs three times more frequently in men than in women. Although experts are unsure of its cause, the disease appears to have a genetic basis, as it is ten to twenty times more commonly found in the immediate family of an AS patient than it is in the general population. It may also be caused by an autoimmune disorder.

Symptoms

Recurrent back pain is the most common symptom of ankylosing spondylitis. Increased pain at night and early morning stiffness can also occur. Early symptoms and signs may also include loss of appetite, weight loss, fatigue, anemia, and fever. When the disease affects the rib cage the patient experiences diminished rib cage expansion, and thus difficulty breathing.

Because many of us have back pain, a major complicating factor in successfully treating ankylosing spondylitis is delayed diagnosis. In fact, generally, from three to ten years may pass before a correct diagnosis is finally made. If left untreated, the patient eventually develops a hunchback curvature of the spine, called a kyphosis. As the curvature progressively worsens, organs surrounded by the rib cage become increasingly pressured and confined. Serious problems in digestion, circulation, nerve supply, and organ function can result. Preventing, delaying, or correcting the deformity requires a combination of enzyme and nutritional therapies along with the care of a back specialist.

Enzyme Therapy

Systemic enzyme therapy is used to relieve joint discomfort, reduce inflammation, and stimulate the immune system. Enzymes improve circulation, help speed tissue repair, bring nutrients to the damaged area, remove waste products, and enhance wellness.

Digestive enzyme therapy is used to improve the breakdown and assimilation of food nutrients, reduce stress on the gastrointestinal mucosa, help maintain normal body pH levels, detoxify the body, promote the growth of healthy intestinal flora, and strengthen the body as a whole. In this way, the body's immune system is strengthened and better able to fight ankylosing spondylitis. Digestive enzymes also serve as replacements for the body's pancreatic enzymes, leaving the pancreatic enzymes free to perform other functions in the body, such as decreasing inflammation.

Past research indicates that the best results can be obtained by taking eight to ten coated tablets of an enzyme combination containing bromelain, papain, trypsin, chymotrypsin, and pancreatin between meals for six months (with significant improvement after three months). After six months, six coated enzyme tablets should be taken three times per day between meals.

Enzyme Absorption System Enhancers (EASE) maximize enzyme activity. They can also improve the absorption and bioavailability of various nutrients, and decrease the drain on the body's own digestive enzymes, thus prolonging their lives.

In addition to the standard multivitamin, multimineral, and multiglandular complexes recommended in the introduction to Part Two, the following supplements are recommended for the treatment of ankylosing spondylitis. The following recommended dosages are for adults. Children between the ages of two and five should take one-quarter the recommended dosage, those between the ages of six and twelve should take one-half the recommended dosage, and those between the ages of thirteen and seventeen should take three-quarters the recommended adult dosage.

ESSENTIAL ENZYMES

Enzyme	Suggested Dosage	Actions
Formula One (particularly Wobenzym N from Naturally Vitamin Supplements) —See page 38 of Part One for a list of products. *or*	Take 8–10 tablets between meals for six months or until symptoms improve. Thereafter, take 6 tablets three times per day between meals.	Decreases inflammation, pain, and swelling; increases rate of healing; strengthens capillary walls.
Inflazyme Forte from American Biologics *or*	Take as indicated on the label between meals.	Decreases inflammation, pain, and swelling; increases rate of healing; strengthens capillary walls.
Enzyme complexes rich in proteases, plus amylases and lipases	Take as indicated on the label between meals.	Decreases inflammation, pain, and swelling; increases rate of healing; strengthens capillary walls.

SUPPORTIVE ENZYMES

Enzyme	Suggested Dosage	Actions
A digestive enzyme formula containing protease, amylase, and lipase enzymes. See the list of digestive enzyme products in Appendix.	Take as indicated on the label with meals.	Improves the breakdown and absorption of proteins, carbohydrates, and fats; fights toxic build-up, thereby strengthening the body's immune system.
Formula Three—See page 38 of Part	Take 200 mcg three times per day between meals.	Potent antioxidant enzymes.

One for a list of products.

ENZYME ABSORPTION SYSTEM ENHANCERS (EASE)

Combination	Suggested Dosage	Actions
Protease Concentrate from Tyler *or*	Take as indicated on the label.	Fights inflammation.
Bromelain Complex from PhytoPharmica *or*	Take as indicated on the label.	Decreases swelling; improves circulation; helps break up circulating immune complexes.
Joint-Ease from Nature's Life *or*	Take as indicated on the label.	Decreases joint inflammation and swelling.
Mobil-Ease from Prevail	Take one capsule two to three times daily between meals. As with any product, discontinue use immediately if adverse reactions occur.	Decreases joint pain and inflammation.

ENZYME HELPERS

Nutrient	Suggested Dosage	Actions
Adrenal glandular	Take 100–400 mg per day.	Fights inflammation and adrenal exhaustion.
Calcium	Take 1,000 mg per day.	Maintains strong, healthy bones.
Chondroitin sulfate	Take 250 mg two to three times per day with meals; or follow advice of your health-care professional.	Supports connective tissue; nourishes cartilage, ligaments, and tendons; promotes cell hydration.
Citrus bioflavonoids	Take 500 mg three times per day.	Strengthen capillaries; improves delivery of nutrients to affected area; antioxidants.
Coenzyme Q_{10}	Take 60 mg one to two times per day with meals; or follow the advice of your health-care professional.	Potent antioxidant.
Glucosamine hydrochloride *or*	Take 500 mg two to four times per day with meals.	Stimulates cartilage repair.
Glucosamine sulfate	Take 500 mg one to three times per day with meals.	
Kyolic Aged Garlic Liquid Extract with Vitamins B_1 and B_{12}	Take as indicated on the label.	Fights free radicals; improves immune systemic function; eases inflammation.
Magnesium	Take 500 mg per day.	Required for strong, healthy bones and calcium absorption.
Niacin-bound chromium	Take up to 20 mg per day; or take as indicated on the label.	Improves circulation, thereby improving delivery of nutrients to the affected area.
Vitamin A	Take 5,000 IU per day.	Important for a strong immune system and for bone development.

Vitamin B complex	Take 50 mg three times per day; or take as indicated on the label.	B vitamins act as coenzymes and support neuromusculoskeletal function.
Vitamin C	Take 2,000 mg per day. If pain is severe, take an additional 5,000–6,000 mg per day until pain subsides.	Helps produce collagen and normal connective tissue; promotes healthy immune system; powerful antioxidant.
Whey to Go from Solgar (a combination of free-form amino acids and protein)	Take as indicated on the label.	Provides immunoglobulins, which help improve the function of the immune system.
Zinc	Take 50 mg per day. Do not take more than 100 mg per day from from all soruces.	Needed by over 300 enzymes; potent antioxidant.

Comments

❏ Raw, fresh, enzyme-rich fruits and vegetables (particularly green vegetables) should comprise over 60 percent of your diet. If your body cannot tolerate raw fruits and vegetables, increase your intake of digestive enzymes, or sauté or steam your produce.

❏ Improve overall health by avoiding alcohol, refined white flour, and sugar.

❏ Follow the dietary recommendations in Part Three.

❏ Gargle with a combination of four ounces of fresh juiced pineapple and four ounces of lemon or grapefruit juice.

❏ Follow the detoxification and juicing plans in Part Three.

❏ A well-planned exercise program is essential, as exercise can decrease the severity of pain. The goal is to build up the muscle groups responsible for spinal strength and flexibility.

❏ An individual with ankylosing spondylitis should be under the care of a spinal specialist (such as a chiropractor, medical doctor, naturopath, or osteopath) who will develop a treatment program. Your treatment program might include certain chiropractic techniques, massage, physiotherapy (such as electrical stimulation), plus home therapy (including heat), and a nutritional program. Your specialist may also train you in postural techniques, such as the proper ways to lift, bend, stand, walk, drive, sit, and sleep.

❏ Cortisone should be avoided if possible, and cytostatics or gold D-penicillamine should only be used in the most severe cases. For long-term therapy, the use of enzymes has proven to be superior to nonsteroidal anti-inflammatory drugs (NSAIDs), which are often recommended. Pain at rest, during exercise, and at night should improve after one to three months. Any nutritional enzyme intervention requires a great deal of compliance and has to be a long-term effort. Therefore, patience is required. In contrast to NSAIDs, there are few (if any) adverse effects along with definite improvement when using enzyme therapy.

Arteriosclerosis

Arteriosclerosis is a generic term that refers to a group of diseases that involve hardening and thickening of the blood vessel walls. The most common type of arteriosclerosis is *atherosclerosis*, which is characterized by fatty deposits in the medium to large arteries. It occurs most frequently in men, until the menopausal years of women—then the incidence is about the same between the genders. *Arteriolosclerosis* refers to a thickening of the small arteries due to age. *Mönckeberg's arteriosclerosis* is a narrowing of the arteries due to calcium deposits. *Arteriosclerosis obliterans* is a narrowing of the arteries supplying the extremities.

Most of us are born with healthy blood vessels. Over time and for a variety of reasons, soft, fatty material and fibrin begin to accumulate on the blood vessel walls. This material increases, hardens, and forms plaques that stiffen and narrow the vessel openings (called the lumen). Blood pressure rises because the heart has to pump harder to get blood through the constricted arteries. Oxygen supply to the heart is decreased. Exertion causes a sensation of tightness and pain in the chest (angina pectoris). In addition, clots can easily develop in the narrowed vessels. The formation of such clots in a constricted area sometimes causes complete closure of a coronary artery. This will cause the heart to suffocate, which leads to myocardial infarction (death of heart muscle). If the arteriosclerosis affects arteries going to the brain, a stroke can result. Arteriosclerosis is a very serious condition—it can even be fatal. Anyone suffering from arteriosclerosis should be under a physician's care.

Over 60 percent of all deaths in the United States each year are from cardiovascular disease—caused by atherosclerosis and its resulting damage. Although virtually epidemic in industrialized countries, economically deprived areas, such as India and countries in Africa, have lower occurrences. This could be because many of us in industrialized countries are sedentary, eat too much fat, eat too many calories, and smoke cigarettes. Many of us are also under a lot of stress, which may not cause arteriosclerosis, but can accelerate its progress. These are some risk factors for arteriosclerosis, others include hypertension, elevated serum cholesterol levels, having diabetes, being overweight, being male, having a family history of the disease, aging, and having a chronic inflammatory disease.

It is possible for a doctor to estimate your chance of arteriosclerosis by measuring a number of elements in the blood (including lipid levels, fibrinogen concentration, and elevation of lipoprotein A concentrations). These indicators, along with the presence of known risk factors, can serve as your "wake up call."

Symptoms

Arteriosclerosis is silent at first and usually has no symptoms until the disease reaches advanced stages. Only laboratory tests indicating elevated blood lipid levels will signal the problem in early stages. Symptoms of later stages include muscle cramping if the arteriosclerosis affects the arms and legs, angina pectoris or heart attack if the disease affects the vessels to the heart, and stroke if the vessels to the brain are affected.

Enzyme Therapy

Systemic enzyme therapy is used to break up fibrin, remove any blood clots, and decrease swelling and inflammation. Enzymes help decrease plaquing along the arterial walls and improve circulation. Enzymes also reduce the tendency of blood clot formation, stimulate the immune system, help speed tissue repair, bring nutrients to the damaged area, remove waste products, and enhance wellness.

Digestive enzyme therapy is used to improve the digestion of food, reduce stress on the gastrointestinal mucosa, help maintain normal pH levels, detoxify the body, and promote the growth of healthy intestinal flora, thereby improving metabolism and helping the body to function optimally. Digestive enzymes also serve as replacements for the body's pancreatic enzymes, leaving the pancreatic enzymes free to perform other functions in the body, such as improving circulation.

Enzymes can have a dramatic effect on cholesterol, as some researchers accidentally discovered. While studying the effect of enzyme therapy on rheumatism, researchers observed an unexpected benefit. After several weeks of systemic enzyme therapy, the blood lipid levels of the study participants tended to improve substantially. The protective high-density lipoproteins (HDLs) increased markedly, while the bad cholesterol (low-density lipoproteins—LDLs) and triglyceride values were reduced by as much as 25 percent.

Studies repeatedly show that pancreatic enzymes, such as pancreatin, trypsin, and chymotrypsin, and other proteolytic enzymes (including papain and bromelain) improve angina pain, inhibit platelet aggregation, break down atherosclerotic plaques and reduce blood pressure.

Enzyme Absorption System Enhancers (EASE) maximize enzyme activity. They can also improve the absorption and bioavailability of various nutrients and decrease the drain on the body's own digestive enzymes, thus prolonging their lives.

In addition to the standard multivitamin, multimineral, and multiglandular complexes recommended in the introduction to Part Two, the following supplements are recommended for the treatment of arteriosclerosis.

ESSENTIAL ENZYMES

Enzyme	Suggested Dosage	Actions
Formula Three—See page 38 of Part One for a list of products.	Take 200 mcg three times per day between meals.	Potent antioxidant enzymes.

AND SELECT ONE OF THE FOLLOWING:

Enzyme	Suggested Dosage	Actions
Formula One (particularly Wobenzym N from Naturally Vitamin Supplements)—See page 38 of Part One for a list of products.	Take 8–10 tablets three times per day between meals for six months or until symptoms improve. Thereafter, take 6 tablets three times per day between meals.	Dissolves blood clots; fights inflammation, pain, and swelling; speeds healing; strengthens capillary walls.
Inflazyme Forte from American Biologics	Take as indicated on the label between meals.	Dissolves blood clots; fights inflammation, pain, and swelling; speeds healing; strengthens capillary walls.
Bromelain	Take three 230–250 mg capsules or tablets twice per day between meals for two to three weeks, then continue taking 2 capsules or tablets twice per day between meals as a maintenance dose.	Decreases platelet aggregation.
Enzyme complexes rich in proteases, plus amylases and lipases	Take as indicated on the label between meals.	Dissolves blood clots; fights inflammation, pain, and swelling; speeds healing; strengthens capillary walls.

SUPPORTIVE ENZYMES

Enzyme	Suggested Dosage	Actions
A digestive enzyme formula containing protease, amylase, and lipase enzymes. See the list of digestive enzyme products in Appendix.	Take as indicated on the label with meals.	Improves the breakdown and absorption of proteins, carbohydrates, and fats. This aiding of digestion fights toxic build-up.

ENZYME ABSORPTION SYSTEM ENHANCERS (EASE)

Combination	Suggested Dosage	Actions
Protease Concentrate from Tyler or	Take as indicated on the label.	Breaks up fibrin; fights inflammation.
Bromelain Complex from PhytoPharmica or	Take as indicated on the label.	Decreases swelling; improves circulation; helps break up circulating immune complexes.
Cardio-Chelex from Life Plus	If under 40 and in good health, take 5 tablets once a day—first thing in the morning. With a history of heart problems, take	Supports healthy circulation; improves energy.

5 tablets twice per day—first thing in the morning and midday.

ENZYME HELPERS

Nutrient	Suggested Dosage	Actions
Adrenal glandular	Take as indicated on the label.	Helps fight stress and adrenal exhaustion.
Bioflavonoids, such as rutin and quercetin	Take 500 mg three times per day.	Fight atherosclerosis; potent antioxidants; reduce capillary fragility.
Calcium	Take 1,000 mg per day.	Helps maintain heartbeat; may lower blood pressure; important in proper blood clotting.
Carnitine	Take 500 mg two times per day.	Increases energy; lowers cholesterol; reduces risk of heart disease.
Coenzyme Q_{10}	Take 75 mg one to two times per day with meals.	Potent antioxidant; prevents heart attacks; decreases blood pressure and total serum cholesterol.
Essential fatty acid complex	Take 75–100 mg two to three times per day with meals.	Lowers serum cholesterol level; strengthens the immune system; helps strengthen cell membranes.
Heart glandular	Take 140 mg; or take as indicated on the label.	Improves heart function.
A green food product, such as Kyo-Green, Green Kamut, or Green Magma	Take 1–2 teaspoons in water or juice three times per day.	Reduces blood pressure; effective against heart disease; fights free radicals.
Kyolic Reserve	Take 3 capsules per day.	Lowers cholesterol; improves circulation.
Lecithin	Take as indicated on the label.	Emulsifies fat.
Magnesium	Take 500 mg per day.	Plays important roles in calcium absorption; important for the function of many enzymes.
Niacin *or* Niacin-bound chromium	Take 50 mg the first day, then increase dosage daily by 50 mg to 1,000 mg per day. Before taking doses higher than 1,000 mg, consult a trained metabolic physician. Take 20 mg per day; or take as indicated on the label.	Improves circulation by dilating blood vessels.
Vitamin A	Take 15,000 IU per day; or take as per your doctor's directions. If pregnant, take no more than 10,000 IU per day.	Necessary for epithelial tissue and cell membrane structure and a strong immune system; potent antioxidant; guards against heart disease and stroke.
Vitamin B complex	Take 50 mg three times per day; or take as indicated on the label.	B vitamins act as coenzymes and support neuromuscular function in circulation.
Vitamin C	Take 5,000–10,000 mg per day.	Essential to vascular function; potent antioxidant; improves the function of many enzymes; enhances absorption of iron; assists in converting cholesterol to bile salt.
Vitamin E	Take 1,200 IU per day from all sources (including the multivitamin complex in the introduction to Part Two).	Potent antioxidant; helps prevent cardiovascular disease; decreases risk of heart attack; stimulates circulation; strengthens capillary walls; prevents blood clots; maintains cell membranes.
Zinc	Take 50 mg per day. Do not take more than 100 mg per day from all sources.	Needed by over 300 enzymes; potent antioxidant.

Comments

❑ Garlic, onions, shallots, chives, leeks, and scallions contain allium compounds, which are potent free-radical fighters that help lower cholesterol, and fight circulatory disorders.

❑ Oats, wheat germ, olives, nuts, nut and seed oils, organ meats, and eggs contain alpha-tocopherol (vitamin E), known to help prevent cardiovascular disease and blood clots.

❑ Raspberries, cranberries, grapes, red wine, black currants, and red cabbage contain anthocyanidins, which reduce blood vessel plaque formation and maintain blood flow in small vessels.

❑ Oolong and green tea contain epicatechin, which is a potent antioxidant that lowers blood pressure, and strengthens capillaries. Other functional foods and phytonutrient concentrates are also excellent for fighting arteriosclerosis.

❑ Avoid toxins such as those formed by smoking or consuming alcoholic beverages. These toxins destroy enzymes and cause free-radical formation. (See Part Three for information on detoxification.)

❑ It is important to eliminate all changeable risk factors for heart disease, avoid all fatty foods, exercise regularly, and change your dietary habits.

❑ Many people take aspirin and other salicylate preparations to thin the blood and guard against arteriosclerotic development, thereby decreasing their risk of a myocardial infarction. But many people cannot tolerate the long-term administration of aspirin because it can cause ulcers. Enzymes are effective—without the side effects.

❑ *See also* CARDIOVASCULAR DISORDERS.

Arthritis

See ANKYLOSING SPONDYLITIS; GOUT; OSTEO-ARTHRI-TIS; REITER'S SYNDROME; RHEUMATOID ARTHRITIS.

Arthritis, Reactive

See REITER'S SYNDROME.

Asthma

Asthma is a common lung disorder that interferes with normal breathing. It is characterized by an increased responsiveness of the trachea and bronchial tubes to a variety of stimuli, causing inflammation and narrowing of the airways. The majority of asthma attacks are mild; however, if left untreated, attacks may occur closer together and last longer, so that breathing becomes a constant effort.

There are two types of asthma. *Allergic asthma* is often hereditary. It occurs after inhaling an allergen, such as dust, molds, pollen, or animal dander; or after exposure to nonairborne allergens, such as certain foods or drugs. The allergic reaction causes antibodies to form in the lungs' cells to fight the allergen. Histamine is released in an effort to fight the allergen, causing the bronchial muscles to contract, which causes the coughing and wheezing of an asthma attack. *Intrinsic asthma* usually begins later in life than allergic asthma. It can be triggered by inhaling irritants such as dust particles, smoke, or fumes. It can also be triggered by cold, damp weather; exercise; laughing; coughing; stress; or extreme emotions. Asthma caused by infection (such as a cold or bronchitis) is another type of intrinsic asthma.

Asthma can easily be confused with heart disease and other breathing disorders, such as emphysema. If asthma is suspected, a physician should be consulted. Approximately 10 million Americans are affected by this serious chronic condition, with nearly 20 percent suffering from limitation of their daily activities. Among children, it is the leading cause of school absenteeism. The medical costs for the treatment of asthma exceed $4 billion each year. Asthma can begin at any age but is much more common among children than adults. And, unfortunately, the prevalence of asthma is on the increase. In the years between 1979 and 1987, the number of people suffering from asthma increased by about one-third.

Increases in asthma have been reported in all age, race, and sex groups. This increase may be due to environmental factors, including air pollution and ozone.

Nutrition can have a dramatic effect on the onset, as well as the treatment, of asthma. For instance, a selenium deficiency will affect blood levels of the antioxidant enzyme glutathione peroxidase and is associated with an increased risk for asthma (as is low dietary intake of vitamins C and E). A recent study found that antioxidants can improve asthma symptoms. High body iron stores may increase asthma risk by increasing free-radical production. Exposure to environmental lead reduces the activities of several enzymes that influence immune function and therefore may increase the risk of asthma.

Symptoms

Symptoms of asthma include shortness of breath, a "tightness" in the chest, difficulty breathing, wheezing, coughing (either dry or with expectoration of thick mucus). An acute, severe attack of asthma is called *status asthmaticus*. The patient may experience an inability to speak or breathe, and his or her skin may turn blue. Unconsciousness may soon follow. This is a medical emergency, and the patient should seek immediate medical care.

Enzyme Therapy

Systemic enzyme therapy is used to reduce inflammation and stimulate the immune system. Enzymes can improve circulation, help speed tissue repair (by bringing nutrients to the damaged area and removing waste products), strengthen the body as a whole, and enhance wellness.

Digestive enzyme therapy is used to improve the digestion and absorption of food nutrients, reduce stress on the gastrointestinal mucosa, help maintain normal pH levels, detoxify the body, promote the growth of healthy intestinal flora, and build general resistance. Digestive enzymes also serve as replacements for the body's pancreatic enzymes, leaving the pancreatic enzymes free to perform other functions in the body, such as decreasing inflammation.

Enzyme Absorption System Enhancers (EASE) maximize enzyme activity. They can also improve the absorption and bioavailability of various nutrients and decrease the drain on the body's own digestive enzymes, thus prolonging their lives.

In addition to the standard multivitamin, multimineral, and multiglandular complexes recommended in the introduction to Part Two, the following supplements are recommended for the treatment of asthma. The following recommended dosages are for adults. Children between the ages of two and five should take one-quarter the recommended dosage, those between the ages of six and twelve should take one-half the recommended dosage, and those between the ages of thirteen and seventeen should take three-quarters the recommended adult dosage.

ESSENTIAL ENZYMES

Enzyme	Suggested Dosage	Actions
Formula One—See page 38 of Part One for a list of products. *or*	Take 25–30 tablets per day (spread intake throughout the day) between meals for six months. Then decrease dosage to 5 tablets three times per day between meals.	Bolsters the immune system; breaks up antigen-antibody complexes; fights free-radical formation; fights inflammation.
Papain *or*	Take 500 mg three times per day for six months between meals. Then decrease dosage to 300 mg three times per day between meals.	Stimulates immune function; breaks up circulating immune complexes; fights free radicals; stimulates immune function at the cellular level; fights inflammation; speeds recovery.
Bromelain *or*	Take four to six 250-mg capsules per day. If an attack occurs, take eight to ten 250-mg capsules within one to two hours or all at once, then, starting the following day, begin taking 4–6 capsules per day again as a maintenance dose.	Bolsters immunity; fights free-radical formation and inflammation; improves respiratory conditions.
Trypsin with chymotrypsin *or*	Take 200 mg of trypsin and 10 mg of chymotrypsin three times per day between meals for six months. Then decrease dosage to 125 mg of trypsin and 5 mg of chymotrypsin three times per day between meals.	Stimulates immunity; breaks up antigen-antibody complexes; fights free-radical formation and inflammation; speeds recovery.
Serratiopeptidase *or*	Take 5 mg three times per day between meals.	Fights inflammation; treats respiratory conditions; liquefies sputum.
Any proteolytic enzyme	Take 25–30 tablets per day (spread throughout the day) for six months. Then decrease dosage to 5 tablets three times per day.	Stimulates immune function; breaks up antigen-antibody complexes; fights free-radical formation; fights inflammation.

SUPPORTIVE ENZYMES

Enzyme	Suggested Dosage	Actions
Formula Three—See page 38 of Part One for a list of products.	Take as indicated on the label.	Potent antioxidant enzymes.
A digestive enzyme product containing protease, amylase, and lipase enzymes. See the list of digestive enzyme products in Appendix.	Take as indicated on the label with meals.	Improves the breakdown and absorption of food nutrients; fights toxic build-up.

ENZYME ABSORPTION SYSTEM ENHANCERS (EASE)

Combination	Suggested Dosage	Actions
Sinease from Prevail	Take as indicated	Decongestant.
or	on the label.	
Combat from Life Plus *or*	Take as indicated on the label.	Helps support respiratory system; stimulates immune function; helps control infections.
Inflamzyme from PhytoPharmica *or*	Take as indicated on the label.	Supports tissue repair; activates immune system.
Connect-All from Nature's Plus *or*	Take as indicated on the label.	Supports tissue function.
Mucous Dissolver Liquezyme from Enzyme Process Laboratories	Take as indicated on the label.	Dissolves mucus.

ENZYME HELPERS

Nutrient	Suggested Dosage	Actions
Acidophilus and other probiotics	Take as indicated on the label.	Stimulates enzyme activity; improves gastrointestinal and immune function to help the body better fight off allergens.
Adrenal glandular	Take 100–400 mg per day.	Helps fight stress and adrenal exhaustion; fights inflammation.
Arginine	Take as indicated on the label.	Accelerates healing; enhances thymus activity.
Bioflavonoids, such as rutin, quercetin, or pycnogenol	Take 200–400 mg three times per day.	Potent antioxidants; strengthen capillaries.
Catechins	Take as indicated on the label.	Strengthen the immune system; reduce capillary permeability.
Coenzyme Q_{10}	Take 30–60 mg one to two times per day with meals.	Important to immune system function; has antioxidant activity.
DHEA	Take 25 mg per day.	Supports immune, cardiovascular, and brain function.
Kyolic Aged Garlic Liquid Extract with Vitamins B_1 and B_{12}	Take 10 drops in liquid three times per day.	Stimulates immune system function.
Selenium	Take 50–100 mcg one or two times per day.	Acts as an enzyme activator; required by glutathione peroxidase.
Thymus glandular	Take 100 mg; or take as indicated on the label.	Assists the thymus gland in immune resistance; improves immune response; fights infections.
Vitamin A	Take up to 50,000 IU for no more than one month (for higher doses, see a physician). If pregnant, do not take more than 10,000 IU.	Needed for strong immune system and healthy mucous membranes; potent antioxidant.
Vitamin B complex	Take 50–100 mg three times per day.	B vitamins act as coenzymes.
Vitamin C	Take 10,000–15,000 mg per day; or 1–2 grams every hour until symptoms improve.	Stimulates the adrenal glands; promotes immune system; powerful antioxidant.

Comments

❑ A number of herbs are helpful in treating asthma, including alfalfa concentrate, astragalus, chlorella, echinacea, ginseng, and goldenseal.

❑ Raw, fresh, enzyme-rich fruits and vegetables (particularly green vegetables) should comprise over 60 percent of your diet. If your body cannot tolerate raw fruits and vegetables, increase your intake of digestive enzymes, or sauté or steam your produce.

❑ Follow the instructions for the Enzyme Toxin Flush and Kidney Flush in Part Three. Perform no more than once per month.

❑ Phytochemicals should be used to enhance the healing process. For example, citrus bioflavonoids, grape seed extract, and carotenoids fight capillary permeability and free radicals and are important enzyme helpers.

❑ Minimize exposure to irritants and allergens by altering the environment or moving away from the irritants.

❑ Hospitalization may be necessary if the asthma becomes resistant to conventional treatment.

❑ See the list of Resource Groups in the Appendix for a list of organizations that can provide you with more information about asthma.

Atherosclerosis

See ARTERIOSCLEROSIS.

Athlete's Foot

See FUNGAL SKIN INFECTIONS.

Backache

If your back hurts, you're not alone. Chronic back conditions are common conditions that are debilitating for many Americans. Annually, 5 to 14 percent of Americans suffer low back pain, and from 60 to 90 percent of us will suffer low back pain at some time in our lives. As a major cause of activity limitation, chronic back pain rivals heart disease and arthritis. It is the leading cause of time lost from work or play. Back pain can affect your lifestyle, job, state of mind, and happiness.

Usually, acute back pain is the result of an injury. Chronic back pain can be caused by weight gain or a series of small injuries that continually reaggravate the problem, or it may be a sign of a more serious chronic disorder. Most cases of chronic back pain are due to soft tissue injury rather than to broken bones and are a direct result of a sedentary lifestyle. We sit at school, at work, at home, while driving the car, and while watching television. No longer are we an agrarian society, where physical activity is part of everyday life. We have essentially become "couch potatoes."

Some individuals are particularly prone to suffering back injuries, including those who are overweight or whose jobs or other activities require frequent lifting of heavy objects or bending. At special risk are those individuals who frequently lift objects while in a twisted or forward-bent position, or those whose jobs include exposure to vibration (from industrial equipment or prolonged vehicle driving). Back pain is also associated with increased age. See the inset on page 128 for some conditions that can result in back pain.

The spine (sometimes referred to as the backbone) is composed of thirty-one small bones called vertebrae, twenty-four of which are moveable. Like building blocks, one vertebra sits on top of another. The bones become increasingly larger as they descend from the neck to the tailbone to carry the greater weight of the lower body. Discs between the vertebrae, like cushions, serve as shock absorbers.

An intricate network of nerves responsible for carrying messages to and receiving messages from various muscles, tissues, and organs of the body originate from the spinal cord. Besides giving the body structure, the vertebrae function to protect these important nerves.

Viewed from the side, the spine has four curves—it bends forward in the neck, backward in the midback, forward in the lower back, and backward at the tailbone. Back pain can occur anywhere along the spine.

Acute or chronic back pain is a symptom of inflammation. At the moment of injury, the body begins a series of defense measures, collectively called the inflammatory reaction. Blood flow to the affected area initially increases. Then, as blood becomes more viscous, blood flow decreases, the blood clotting mechanism is "turned on," and all channels for fluid exchange are blocked. This causes swelling (edema), which is partly responsible for the other signs of inflammation which include heat, redness, and pain.

Symptoms

Backache can be a constant aching pain or a sharp, stabbing, boring, or burning pain that increases when twisting, turning, or bending (like a "catch" in your back). Sometimes back pain is so severe that you are unable to get out of bed or a chair, or even walk.

Conditions That Can Result in Back Pain

Most of us will suffer from back pain at some point in our lives. The cause may be something as simple and reversible as obesity or pregnancy, or as serious as cancer. The following are a few conditions that can result in back pain.

CONDITION	CHARACTERISTICS
Aging	A slowdown in the body's enzymes and metabolic processes as a result of illness, injuries, or the degenerative process often lead to inflammation, circulatory disorders, immune deficiencies, and digestive disorders, all of which can result in back pain.
Ankylosing spondylitis	Ossification of the joints and ligaments of the spine create a bamboo appearance; chronic stiffness and pain that seem worse in the mornings result.
Bone spurs	An abnormal bony projection from a vertebra or other bone.
Cancer	Tumorous growths can cause constant back pain, unrelieved when lying down.
Disk protrusion or rupture	Causes decreased space between the vertebrae leading to nerve root irritation and inflammation; marked by sudden onset of a shooting pain down the back of one leg that may increase after strenuous exercise, sneezing, twisting, or coughing; lying on one side in a fetal position (with legs drawn up) usually eases pain.
Kidney infection	Can cause lower back pain, fever of 100° F or higher, nausea, vomiting, or painful urination.
Muscle spasms	An involuntary muscle contraction can cause twinges of pain.
Neuritis	Nerve inflammation, usually degenerative, that can cause pain anywhere in the neck and back.
Osteoarthritis	Noninfectious degeneration of freely moveable joints, usually in the spine and the weight-bearing joints, marked by protective muscle spasms, tenderness, pain on movement, and decreased range of motion.
Osteoporosis	A generalized, gradual, but progressive loss of bone tissue mass. This decrease in mass causes thinning and weakening of bony strength and development of small holes in the bone. Symptoms include sharp, stabbing pain in the spine and hips.
Obesity	Can cause a decrease in the space between the vertebrae through which the spinal nerves pass; often causes low-back pain.
Poor posture	Puts stress on the nerves coming from the spine, causing them to be stretched or irritated, resulting in back pain.
Pregnancy	Increased weight pulls the abdomen forward and causes pressure on the spinal nerves, resulting in lower back pain, usually after the fourth month of pregnancy.
Rheumatoid arthritis	An autoimmune disorder that primarily affects the joints of the body causing pain.
Sciatica	A shooting pain following the sciatic nerve down the back of the leg, which is caused by a pinched vertebral nerve.
Scoliosis	Curvature of the spine causes localized pain and fatigue in the area of the scoliosis.
Spondylolysthesis	A forward slipping of a vertebra, most commonly associated with defects and separation of the vertebral bony arch; marked by low-back pain and pain in the legs.
Sprains, strains	Sprains are overstretching or tearing of the ligaments that support and limit the mobility of a joint. A strain is an overstretching of the muscles.
Subluxations	Occur when a spinal vertebra is rotated beyond its normal range of movement and is out of alignment in relationship to the vertebrae just above and below it. This causes a decrease in the intervertebral opening, which results in swelling. The swelling, which presses or pinches the nerve and blood vessels going through the opening, results in pain.

Back Posture

Body balance is essential for a healthy back and a healthy body. This is achieved through proper body mechanics, muscle strengthening, muscle stretching, management of stress and tension, and sufficient rest.

Poor posture, caused by many factors, such as obesity, pregnancy, and lack of exercise, can cause the stomach to extend forward, resulting in the pelvis tilting down and forward. This causes an accentuated forward curvature in the back that some call *swayback*. The result is low-back pain. Hunching the shoulders forward can result in a "hunchback" posture, which places added stress on the back and neck, causing pain. Here are some postural techniques to avoid this pain.

- When lifting and bending, keep your knees bent, stomach tucked in, back straight, and buttocks pulled under. Don't lock your knees or bend your back.

- Don't twist, instead turn, making sure your shoulders, hips, and feet are in line and turn together. Don't plant your feet.

- When reaching, tuck your stomach in, keep your back straight, and pull your buttocks under. Do not stand in a swayback position.

- When pulling and/or pushing an object, keep knees bent, stomach in, back straight, and buttocks pulled under. Don't lock your knees or bend your back.

- When lying on your back, keep your knees bent and your back straight and flattened against the surface. Try to remove the arch in your back as much as possible. When sitting, sit straight and support the lower back, don't bend or hunch your back.

Enzyme Therapy

Enzymes are effective in fighting back pain because they treat inflammation by increasing the rate at which the body repairs itself. They do this by breaking up the debris in the injured area and allowing blood vessels to "unclog." This reduces swelling and pain and leads to a more rapid repair of the injured area. Enzymes often used to fight inflammation include trypsin, chymotrypsin, papain, bromelain, and microbial proteases.

Systemic enzyme therapy is used to decrease pain, swelling, and inflammation; stimulate the immune system; improve circulation; help speed tissue repair; bring nutrients to the damaged area; remove waste products; and strengthen the body as a whole.

Digestive enzyme therapy is used to improve the digestion of food, reduce stress on the gastrointestinal mucosa, help maintain normal pH levels, help eliminate the toxic by-products of inflammation, promote the growth of healthy intestinal flora, and build general resistance. In this way, digestive enzymes improve the absorption of nutrients, allowing them to reach areas of inflammation. Digestive enzymes also serve as replacements for the body's pancreatic enzymes, leaving the pancreatic enzymes free to perform other functions in the body, such decreasing inflammation.

Enzyme Absorption System Enhancers (EASE) maximize enzyme activity. They can also improve the absorption and bioavailability of various nutrients and decrease the drain on the body's own digestive enzymes, thus prolonging their lives.

In addition to the standard multivitamin, multimineral, and multiglandular complexes recommended in the introduction to Part Two, the following supplements are recommended for the treatment of back pain. The following recommended dosages are for adults. Children between the ages of two and five should take one-quarter the recommended dosage, those between the ages of six and twelve should take one-half the recommended dosage, and those between the ages of thirteen and seventeen should take three-quarters the recommended adult dosage.

ESSENTIAL ENZYMES

Enzyme	Suggested Dosage	Actions
Phlogenzym from Mucos	Take 2–3 tablets three times per day between meals.	Decreases inflammation, pain, and swelling; increases rate of healing; stimulates the immune system; strengthens capillary walls.
or		
Formula One (particularly Wobenzym N from Naturally Vitamin Supplements)—See page 38 of Part One for a list of products.	Take 10 tablets three times per day between meals. As symptoms subside (which should occur in about one week) decrease dosage to 3–5 tablets three times per day between meals until pain disappears.	Decreases inflammation, pain, and swelling; increases rate of healing; stimulates the immune system; strengthens capillary walls.
or		
Bromelain	Take one to two 250 mg tablets three times per day until symptoms subside. If pain is severe, take two to three 250 mg tablets three times per day between meals.	Decreases inflammation, pain, and swelling; increases rate of healing; stimulates the immune system; strengthens capillary walls.
or		
Microbial protease	Take as indicated on the label between meals.	Decreases inflammation, pain, and swelling; increases rate of healing; stimulates the immune system; strengthens capillary walls.
or		

| Inflazyme Forte from American Biologics | Take as indicated on the label between meals. | Decreases inflammation, pain, and swelling; increases rate of healing; stimulates the immune system; strengthens capillary walls. |

SUPPORTIVE ENZYMES

Enzyme	Suggested Dosage	Actions
A digestive enzyme formula containing proteases, amylases, and lipases. For a list of digestive enzyme products, see Appendix.	Take as indicated on the label with meals.	Improves the breakdown and absorption of food nutrients that can control inflammation; fights toxic by-products of inflammation.
Formula Three—See page 38 of Part One for a list of products.	Take 50 mcg three times per day between meals.	Potent antioxidant enzymes.

ENZYME ABSORPTION SYSTEM ENHANCERS (EASE)

Combination	Suggested Dosage	Actions
Inflamzyme from PhytoPharmica *or*	Take as indicated on the label.	Promotes tissue repair; decreases inflammation.
Connect-All from Nature's Plus *or*	Take as indicated on the label.	Supports connective tissue.
CTR Support from PhysioLogics *or*	Take as indicated on the label.	Nourishes the body during tissue recovery and healing from injury; helps maintain healthy connective tissue; diminishes damage caused by swelling and inflammation.
Joint-Ease from Nature's Life *or*	Take as indicated on the label.	Decreases inflammation and swelling.
Mobil-Ease from Prevail	Take one capsule between meals two to three times per day. As with any product, discontinue immediately if adverse effects occur.	Has anti-inflammatory activity.

ENZYME HELPERS

Nutrient	Suggested Dosage	Actions
Adrenal glandular	Take 100–400 mg per day.	Fights inflammation and adrenal exhaustion.
Amino-acid complex, which should include L-ornithine, L-glutamine, L-arginine, L-proline, and L-lysine	Take as indicated on the label on an empty stomach.	Amino acids are the building blocks of proteins and enzymes and are required for the growth and maintenance of all body tissues.
Bioflavonoids	Take 150–500 mg three times per day.	Strengthen capillaries; potent antioxidants.

Branched-chain amino acids (isoleucine, leucine, and valine)	Take as indicated on the label between meals.	Often found together in a supplement, branched-chain amino acids aid in healing of muscle, bone, and skin and act as fuel for muscles.
Calcium	Take 1,000 mg two times per day.	Plays important roles in muscle contraction and nerve transmission.
Chondroitin sulfate	Take 250 mg two to three times per day with meals; or follow the advice of your health-care professional.	Supports connective tissue; promotes cell hydration.
Coenzyme Q_{10}	Take 30–60 mg one to two times per day with meals; or follow the advice of your health-care professional.	Potent antioxidant.
Glucosamine hydrochloride *or* Glucosamine sulfate	Take 500 mg two to four times per day. Take 500 mg one to three times per day.	Stimulates cartilage repair.
Magnesium	Take 500 mg two times per day.	Plays important role in muscle contraction, nerve function, and calcium absorption.
Mucopolysaccharides, found in bovine cartilage extract and shark cartilage extract	Take as indicated on the label.	Support healthy, flexible joints; control inflammation; help keep joint membranes fluid; speed healing and recovery from inflammation.
Vitamin B complex	Take 50 mg three times per day; or take as indicated on the label.	B vitamins act as coenzymes and support neuromusculoskeletal function.
Vitamin C	Take 5,000–10,000 mg per day.	Helps to produce collagen and normal connective tissue; promotes immune system; powerful antioxidant.
Zinc	Take 15–50 mg per day. Do not take more than 100 mg per day from all sources.	Needed by over 300 enzymes; important for cell and protein synthesis.

Comments

❏ Eat enzyme-rich fresh fruits and vegetables, which are also high in antioxidants, including pineapples, papaya, sprouted seeds, broccoli, sunflower seeds, peanuts, and whole grapes (including the seeds and skins). Fresh fruits and vegetables should comprise 60 percent of your diet. See Part Three for dietary recommendations.

❏ Avoid food additives, preservatives, refined foods, and foods high in saturated fat.

❏ Drink at least two 8-ounce glasses of freshly juiced, enzyme-rich fruits and vegetables every day.

❏ Perform the Enzyme Kidney Flush every day for one week. During the following week, perform the Enzyme Toxin Flush.

Use these flushes only once per month. See Part Three for instructions on both flushes.

❏ Do no heavy lifting or repeated lifting or bending. See the inset "Back Posture" on page 129.

❏ Initial treatment is usually conservative. If pain is severe, you should stay off your feet and rest in bed, lying in a fetal position (on your side with your knees drawn up) as often as possible. You may find it more comfortable to lie on your back with your knees bent and your head and legs propped up on two pillows.

❏ Ice should be used for the first one to three days. Fill two or three small paper cups with water and freeze. When frozen, take one out and tear off the top half inch of paper revealing the ice. Place a paper towel or dish towel over the ice and apply to the painful area. Rub the affected area in a constant circulatory motion with the ice. Do not apply the ice directly to the skin. Apply the ice for five minutes, then keep the ice off the area for five minutes. Repeat this every one to two hours. The purpose is to stimulate blood flow to the area, remove waste from the area, and reduce inflammation.

❏ After three days, alternate ice with heat packs every fifteen minutes until symptoms improve. Be sure to begin and end with ice application. "Blue ice" or a frozen vegetable packet works better for this application. Wrap it with a dish towel and apply to the painful area. For heat, use a hot water bottle or heating pad. If a hot water bottle is used, wrap in a towel and apply. If a heating pad is used, put a warm, moist towel on the affected area, then place a sheet of wax paper over it, then the heating pad, and finally, a large bath towel. Make sure the wax paper is wide and long enough to cover the entire heating pad with at least one inch more extending out on every side. The heating pad should never touch the moist towel or the patient's skin. Great care should be exercised to assure this does not happen.

❏ You may require the services of a back specialist, such as a chiropractor, medical doctor, naturopath, or osteopath. Your physician will develop a treatment program for you that might include chiropractic adjustments, manipulation, massage, physiotherapy (including electrical stimulation, ultrasound, and ice and heat applications), exercises, and postural training (including techniques for proper lifting, bending, standing, walking, driving, sitting, and sleeping positions). If you are overweight, a weight-loss program will be instituted.

❏ Chiropractic spinal manipulation has been recognized by the United States Agency for Health Care Policy and Research as an effective therapy for acute low-back pain.

❏ An elastic back support (approximately 8 inches wide, with a lumbar support) should be used at all times. For neck pain, a cervical collar is sometimes fitted to give support, restrict motion, and reduce the chance of reaggravation.

❏ If pain persists, see a doctor. Back pain can be a symptom of many serious conditions, such as cancer, circulatory problems, infections, and autoimmune diseases.

❏ *See also* SCOLIOSIS; SLIPPED DISC.; SUBLUXATION.

Bacterial Infections

Bacteria are microorganisms found both inside and outside our bodies. We eat, drink, and breathe microbes. They are in our water, our air, and our food. Many of these organisms normally live on our skin, in the mouth, in the gastrointestinal and respiratory tracts, and even in the genitalia. However, only a relatively few bacteria have the ability to cause disease.

A bacterial infection can occur anywhere in or on the body when bacteria gain entrance through a wound or an opening, or when they are ingested or inhaled. A bacterial infection can also occur when the balance between "good" and "bad" bacteria in the body is disturbed.

The body has a number of defense mechanisms to inhibit infection, including the skin and mucous membranes, natural chemicals that kill harmful bacteria (such as hydrochloric acid in the stomach), beneficial bacteria (such as those in the small and large intestine), and the various components of the immune system (including phagocytes and antibodies).

In a healthy body, foreign cells (called antigens), such as bacteria, are recognized as being foreign by the body and are attacked by antibodies. The antibodies recognize and attach to the antigens, forming an immune complex. These immune complexes send out signals and the body reacts by sending in macrophages, the soldiers of the immune system, which seem to engulf the immune complexes and break them up through enzymatic action.

Unfortunately, sometimes the body isn't able to eliminate enough of the harmful bacteria, and they multiply in the tissues causing infection. If left unattended, a bacterial infection can enter the bloodstream causing a potentially fatal condition called bacteremia.

Enzyme Therapy

The orthodox medical approach to the control of bacterial diseases generally requires the use of antibiotics. Unfortunately, more and more bacteria are becoming resistant to antibiotics. Systemic enzyme therapy is a viable alternative and is used to fight bacteria and strengthen the body as a whole. Enzymes also reduce inflammation, stimulate the immune system, improve circulation, help speed tissue repair, bring nutrients to the damaged area, remove waste products, and enhance wellness. Orally administered enzymes synergistically activate macrophages and other immune system components. Enzymes also assist in breaking up and cleaning out the circulating immune complexes.

Digestive enzyme therapy is used to improve digestion of food, reduce stress on the gastrointestinal mucosa, help maintain normal pH levels, detoxify the body and promote the growth of healthy intestinal flora. Digestive enzymes serve as replacements for the body's pancreatic enzymes. This frees the pancreatic enzymes to perform other functions in the body, such as boosting immune function, decreasing inflammation, and improving circulation.

Enzyme Absorption System Enhancers (EASE) maximize enzyme activity. They can also improve the absorption and bioavailability of various nutrients and decrease the drain on the body's own digestive enzymes, thus prolonging their lives.

This book includes several bacterial infections with their own treatment plans, including ABSCESS; ACNE; ADENOIDITIS; ADNEXITIS; BLADDER INFECTIONS; BOILS; CONJUNCTIVITIS; DIARRHEA; EAR INFECTIONS; EMPYEMAS; EPIDIDYMITIS; GINGIVITIS; KIDNEY DISORDERS; LARYNGITIS; PNEUMONIA; RHEUMATIC FEVER; SINUSITIS; STAPHYLOCOCCAL INFECTIONS; and TONSILLITIS. However, the following is a general enzyme treatment plan for bacterial infections, which can be used in addition to the standard multivitamin, multimineral, and multiglandular complexes recommended in the introduction to Part Two.

ESSENTIAL ENZYMES

Enzyme	Suggested Dosage	Actions
Superoxide dismutase	Take 150 mcg per day.	Fights inflammation; potent antioxidant.

AND SELECT ONE OF THE FOLLOWING:

Bromelain	Take 750 mg two times per day between meals.	Fights inflammation, pain, and swelling; speeds healing; stimulates the immune system; strengthens capillary walls.
Serratiopeptidase	Take 10 mg two times per day between meals.	Fights inflammation; reduces pain and swelling.
Trypsin with chymotrypsin	Take 250 mg of trypsin and 10 mg of chymotrypsin three times per day between meals.	Fights inflammation, pain, and swelling; increases rate of healing; stimulates the immune system.
Pancreatin	Take 500 mg six times per day on an empty stomach.	Fights inflammation, pain, and swelling; increases rate of healing; stimulates the immune system; strengthens capillary walls.
Papain or a microbial protease	Take 300 mg two times per day between meals.	Fights inflammation; speeds healing.
Formula One—See page 38 of Part One for a list of products.	Take 5 tablets twice per day between meals.	Fights inflammation, pain and swelling; speeds healing; stimulates the immune system.

SUPPORTIVE ENZYMES

Enzyme	Suggested Dosage	Actions
Catalase and glutathione peroxidase	Take as indicated on the labels.	Potent antioxidant enzymes.
A digestive enzyme product containing protease, amylase, and lipase enzymes. See the list of digestive enzyme products in Appendix.	Take as indicated on the label with meals.	Improves the breakdown and absorption of food nutrients; fights toxic build-up.

ENZYME ABSORPTION SYSTEM ENHANCERS (EASE)

Combination	Suggested Dosage	Actions
Detox-Zyme from Rainbow Light or	Take as indicated on the label.	Helps cleanse the body and eliminate toxins.
Catimune from Life Plus or	Take 1 tablet twice daily between meals.	Helps detoxify the kidney and liver; flushes out toxins; supports blood and circulatory system.
Hepatic Complex from Tyler Encapsulations or	Take as indicated on the label.	Helps detoxify the liver.
Bioprotect from Biotics or	Take as indicated on the label.	Antioxidant combination with free radical fighting capabilities.
Metabolic Liver Formula from Prevail or	Take as indicated on the label.	Improves liver function.
Bio-Immunozyme Forte from Biotics or	Take as indicated on the label.	Stimulates immune function.
Advanced Nutritional System from Rainbow Light or	Take as indicated on the label.	Potent free radical fighter and immune system stimulator.
NESS Formula #416 or	Take as indicated on the label.	Breaks down fibrin; fights inflammation.
Protease Concentrate from Tyler and	Take as indicated on the label.	Breaks down fibrin; fights inflammation.
Bromelain Complex from Phyto Pharmica and	Take as indicated on the label.	Decreases swelling; improves circulation; helps break up circulating immune complexes.
Cardio-Chelex from Life Plus	Take 5 tablets once a day first thing in the morning.	Supports healthy circulation; improves removal of toxins.

ENZYME HELPERS

Nutrient	Suggested Dosage	Actions
Acidophilus and other probiotics	Take as indicated on the label.	Improve gastrointestinal function; help produce antibacterial substances; stimulate enzyme activity.

Adrenal glandular	Take 100–400 mg per day.	Fights inflammation and adrenal exhaustion.
Aged garlic extract	Take two 300 mg capsules three times per day.	Fights inflammation; stimulates the immune system; potent antioxidant.
Amino-acid complex	Take as indicated on the label on an empty stomach.	Required for enzyme production; helps the body maintain proper fluid and acid/base balance—directly related to overall health.
Bioflavonoids, such as rutin or quercetin	Take 500 mg three times per day between meals.	Strengthen capillaries, which improves the delivery of nutrients to the affected area and the removal of toxic by-products; potent antioxidant.
Coenzyme Q₁₀	Take 30–60 mg one to two times per day with meals; or follow the advice of your health-care professional.	Potent antioxidant.
Green Barley Extract from Green Foods Corporation or Kyo-Green from Wakunaga	Take as indicated on the label.	Potent free-radical fighter; rich in vitamins, minerals, and enzyme activity.
Thymus glandular	Take 250 mg twice per day.	Assists the thymus gland in immune resistance; improves immune response; fights infections.
Vitamin A	Take 30,000–35,000 IU per day until the infection is resolved. Take dosages over 50,000 IU only under the supervision of a physician. Pregnant women should take no more than 10,000 IU per day.	Needed for strong immune system and healthy mucous membranes; potent antioxidant.
Vitamin B complex	Take 150 mg three times per day; or take as indicated on the label.	B vitamins act as coenzymes.
Vitamin C	Take 10,000–15,000 mg per day.	Fights infections; improves wound healing; stimulates the adrenal glands; promotes immune system; powerful antioxidant.
Vitamin E	Take 4,000 IU per day.	Potent antioxidant and immune stimulator.
Zinc	Take 50 mg per day. Do not take more than 100 mg per day from all sources.	Needed by over 300 enzymes; necessary for white blood cell immune function; improves wound healing.

Comments

❏ To prevent a bacterial infection:

• Thoroughly wash and disinfect every wound.

• Wash your hands often and thoroughly. Many bacterial infections can be prevented with proper hygiene.

• Strengthen your immune system by eating right and taking nutritional supplements. The onset and severity of any illness caused by a bacterial infection is greatly influenced by the ability of the body's immune system to resist attack.

❏ To help reduce stress on the liver, combinations of microbial enzymes and other nutrients can be used. The liver is a major organ for detoxification.

❏ Eat plenty of garlic; it has antibacterial activity.

❏ Foods high in zinc, such as meat, liver, eggs, and oysters will stimulate and enhance immune system activity.

❏ Enzyme-rich fruits and vegetables (particularly green vegetables) should comprise 60 percent of your diet. If your body cannot tolerate raw fruits and vegetables, increase your intake of digestive enzyme supplements, or sauté or steam your produce.

❏ A number of herbs, including pau d'arco, alfalfa, echinacea, ginger root, ginkgo biloba, licorice, and white willow bark can help control inflammation. These herbs are taken orally.

❏ Perform the Enzyme Kidney Flush and Enzyme Toxin Flush as outlined in Part Three. These flushes can help restore the normal functional capacity of the kidney, gallbladder, and liver.

❏ Exercising helps to increase antioxidant and enzymatic activity.

Bamboo Spine

See ANKYLOSING SPONDYLITIS.

Bed Sores

See SKIN ULCERS.

Bee Stings

See INSECT BITES AND STINGS.

Bell's Palsy

Bell's palsy is the sudden paralysis of muscles on one side of the face because of impaired facial nerve conduction. This may be a result of nerve or nerve root irritation or injury. Luckily, the majority of those affected by Bell's palsy improve and completely recover. The likelihood of complete recovery depends upon the extent of nerve damage. The disorder is named after the physician who first described it.

What triggers the nerve impairment is not always known, but the condition is presumed to involve swelling of the nerve due to trauma, extended exposure to cold (such as the wind coming in the car window hitting the left side of the face continuously), a virus, an immune condition, a deficiency of blood (due to functional constriction or actual obstruction of a blood vessel), or pressure on the facial nerve as it courses through the narrow confines of the temporal bone. Bell's palsy is also called truck driver's syndrome, since it can occur after driving with the left car window down, allowing the wind to continually hit one side of the face.

Symptoms

Pain behind the ear on the affected side of the face often appears first. Facial weakness soon develops, which may lead to sudden paralysis on one side of the face. That side of the face becomes flat and expressionless, and the other side may appear a bit twisted or distorted. Often the patient cannot close the eye on the affected side of the face. Because of paralysis, chewing may become difficult on the affected side of the face. Taste, salivation, and tear formation may also be affected.

Enzyme Therapy

Systemic enzyme therapy is used to reduce the facial pain, swelling, and inflammation and to stimulate the immune system. Enzymes improve circulation, help speed tissue repair, bring nutrients to the damaged area, remove waste products, and enhance wellness.

Digestive enzyme therapy is used to improve the digestion of food, reduce stress on the gastrointestinal mucosa, help maintain normal pH levels, detoxify the body (which decreases the drain on our body's enzymes), and promote the growth of healthy intestinal flora, thereby strengthening the body's immune system.

Enzyme Absorption System Enhancers (EASE) maximize enzyme activity. They can also improve the absorption and bioavailability of various nutrients and decrease the drain on the body's own digestive enzymes, thus prolonging their lives.

In addition to the standard multivitamin, multimineral, and multiglandular complexes recommended in the introduction to Part Two, the following supplements are recommended for the treatment of Bell's palsy. The following recommended dosages are for adults. Children between the ages of two and five should take one-quarter the recommended dosage, those between the ages of six and twelve should take one-half the recommended dosage, and those between the ages of thirteen and seventeen should take three-quarters the recommended adult dosage.

ESSENTIAL ENZYMES

Enzyme	Suggested Dosage	Actions
Superoxide dismutase	Take 150 mcg per day.	Fights inflammation; potent antioxidant.

AND SELECT ONE OF THE FOLLOWING:

Enzyme	Suggested Dosage	Actions
Bromelain	Take 750 mg two times per day between meals. Continue until symptoms subside.	Fights inflammation, pain, and swelling; speeds recovery after injury or surgery; stimulates the immune system; strengthens capillary walls.
Formula One—See page 38 of Part One for a list of products.	Take 20 tablets per day between meals. Gradually decrease dosage to 6 tablets per day as the condition improves.	Speeds recovery from inflammation; stimulates the immune system; fights viruses.
Serratiopeptidase	Take 10 mg two times per day between meals.	Fights inflammation; reduces pain and swelling.
Trypsin with chymotrypsin	Take 250 mg of trypsin and 10 mg of chymotrypsin three times per day between meals.	Fights inflammation, pain, and swelling; increases rate of healing; stimulates the immune system.
Pancreatin	Take 500 mg six times per day on an empty stomach.	Fights inflammation, pain, and swelling; increases rate of healing; stimulates the immune system; strengthens capillary walls.
Papain	Take 300 mg two times per day between meals.	Fights inflammation; speeds healing.
A microbial protease	Take 300 mg two times per day between meals.	Fights inflammation; speeds healing.

SUPPORTIVE ENZYMES

Enzyme	Suggested Dosage	Actions
Catalase and glutathione peroxidase	Take as indicated on the label.	Potent antioxidant enzymes.
A digestive enzyme product containing protease, amylase, and lipase enzymes. See the list of digestive enzyme products in Appendix.	Take as indicated on the label with meals.	Improves the breakdown and absorption of food nutrients; fights toxic build-up.

ENZYME ABSORPTION SYSTEM ENHANCERS (EASE)

Combination	Suggested Dosage	Actions
Oxy-5000 Forte from American Biologics or	Take as indicated on the label.	Aids detoxification.

Inflamzyme from PhytoPharmica *or*	Take as indicated on the label.	Promotes tissue repair; improves immune functions and enzymatic activity.
Mobil-Ease from Prevail	Take as indicated on the label.	Has anti-inflammatory activity.

ENZYME HELPERS

Nutrient	Suggested Dosage	Actions
Adrenal glandular	Take 100–400 mg per day.	Fights inflammation and adrenal exhaustion.
Bioflavonoids, like rutin or quercetin	Take 500 mg three times per day between meals.	Strengthen capillaries; potent antioxidants.
Calcium	Take 1,000 mg per day.	Plays important roles in muscle contraction and nerve transmission.
Coenzyme Q_{10}	Take 30–60 mg one to two times per day with meals; or follow the advice of your health-care professional.	Potent antioxidant.
Magnesium	Take 500 mg per day.	Plays important role in muscle contraction, nerve function, and calcium absorption.
Niacin-bound chromium	Take 20 mg per day; or take as indicated on the label.	Dilates blood vessels; improves circulation.
Thymus glandular	Take 100 mg per day.	Assists the thymus gland in immune resistance; improves immune response; fights infections.
Vitamin B complex	Take 50 mg three times per day; or take as indicated on the label.	B vitamins act as coenzymes and support neuromusculoskeletal function.
Vitamin B_1 (thiamin)	Take 50 mg three times per day; or take as indicated on the label.	Essential for nerve message conveyance.
Vitamin C	Take 10,000–15,000 mg per day.	Fights infections, diseases, and allergies; stimulates the adrenal glands; promotes immune system; powerful antioxidant.
Zinc	Take 50 mg per day. Do not take more than 100 mg per day from all sources.	Needed by over 300 enzymes; potent antioxidant.

Comments

❑ Follow the dietary guidelines as outlined in Part Three.

❑ Detoxification through cleansing, fasting, and juicing may be of value. Follow directions in Part Three.

❑ A positive mental attitude can help you deal with the condition until it resolves. See Part Three.

Benign Prostate Hyperplasia

See PROSTATE DISORDERS.

Bladder Infection

A bladder infection, or cystitis, occurs when bacteria enter the bladder causing infection and inflammation. It is very common in women. The bladder is the reservoir for urine. Liquid wastes are filtered from the bloodstream by the kidneys and held in the bladder, which can hold about one pint of liquid. The waste is then discharged from the bladder, through the urethra, and out the body. Bacteria from the vagina can travel up the urethra and enter the bladder. But this bacteria is usually voided from the bladder with the urine. Infection occurs, however, when the bacteria is not expelled from the bladder. Bacteria can also reach the bladder by traveling from another part of the body through the bloodstream to the bladder. Though common and usually very treatable, bladder infections can sometimes be a symptom of a more serious problem in the body.

Bladder infections are so common in women, presumably, because the urethra is shorter in women, making the anus, vagina, and urethra so close together. Women frequently get bladder infections after engaging in sexual intercourse, which may cause bruising of the urethra. A bacterial infection of the prostate can cause a bladder infection in men.

The risk of cystitis increases with extended bed rest and immobility (such as after major surgery), untreated diabetes mellitus, with infection in other parts of the genitourinary system, systemic illness, obstruction of the urinary tract, and decreased immune resistance. In women, risk increases with increased sexual activity, with the use of a diaphragm or contraceptive sponge, during pregnancy and after childbirth, and during menopause. In men, risk increases with prostate infection and use of a catheter. Bladder infections can happen at any age, even in infancy.

Symptoms

Symptoms of a bladder infection can be mild or extremely troublesome and may be chronic or acute. The primary symptom is a frequent urge to urinate even though the bladder may be empty. Other symptoms include frequent urination, especially at night, and dark or cloudy urine with an unpleasant or strong odor. Lower abdominal pain, lower back pain, painful sexual intercourse, and a painful, burning sensation upon urination may accompany the infection. There may be blood in the urine, and one may suffer chills, fever, loss of appetite, nausea, vomiting, and back pain.

Enzyme Therapy

Systemic enzyme therapy is used to reduce bladder inflammation and infection. Enzymes can stimulate the immune system, improve circulation, help speed tissue repair, bring nutrients to the damaged area, remove waste products, and enhance wellness.

Digestive enzyme therapy is used to improve the digestion of food, reduce stress on the gastrointestinal mucosa, help maintain normal pH levels, detoxify the body, and promote the growth of healthy intestinal flora. Digestive enzymes also serve as replacements for the body's pancreatic enzymes, leaving the pancreatic enzymes free to perform other functions in the body, such as boosting immunity and decreasing inflammation.

Enzyme Absorption System Enhancers (EASE) maximize enzyme activity. They can also improve the absorption and bioavailability of various nutrients and decrease the drain on the body's own digestive enzymes, thus prolonging their lives.

In addition to the standard multivitamin, multimineral, and multiglandular complexes recommended in the introduction to Part Two, the following supplements are recommended for the treatment of cystitis. The following recommended dosages are for adults. Children between the ages of two and five should take one-quarter the recommended dosage, those between the ages of six and twelve should take one-half the recommended dosage, and those between the ages of thirteen and seventeen should take three-quarters the recommended adult dosage.

ESSENTIAL ENZYMES

Enzyme	Suggested Dosage	Actions
Superoxide dismutase	Take 150 mcg per day.	Fights inflammation; potent antioxidant.

AND SELECT ONE OF THE FOLLOWING:

Papain	Take 300 mg two times per day between meals.	Fights inflammation; speeds recovery from injuries and surgery.
Bromelain	Take 750 mg two times per day between meals.	Fights inflammation, pain, and swelling; speeds recovery after injury or surgery; stimulates the immune system.
Formula One—See page 38 of Part One for a list of products.	Take 10 tablets three times per day between meals.	Figths inflammation, pain, and swelling; speeds recovery after injury or surgery; stimulates the immune system.

Serratiopeptidase	Take 10 mg two times per day between meals.	Fights inflammation; reduces pain and swelling.
Trypsin with chymotrypsin	Take 250 mg of trypsin and 10 mg of chymotrypsin three times per day between meals.	Fights inflammation, pain, and swelling; increases rate of healing; stimulates the immune system.
Pancreatin	Take 500 mg six times per day on an empty stomach.	Fights inflammation, pain, and swelling; increases rate of healing; stimulates the immune system; strengthens capillary walls.
Microbial protease	Take 300 mg two times per day between meals.	Fights inflammation; speeds recovery from injuries and surgery.

SUPPORTIVE ENZYMES

Enzyme	Suggested Dosage	Actions
Catalase and glutathione peroxidase	Take as indicated on the label.	Potent antioxidant enzymes.
A digestive enzyme product containing protease, amylase, and lipase enzymes. See the list of digestive enzyme products in Appendix.	Take as indicated on the label with meals.	Improves the breakdown and absorption of food nutrients; fights toxic build-up.

ENZYME ABSORPTION SYSTEM ENHANCERS (EASE)

Combination	Suggested Dosage	Actions
Diuplex from Life Plus or	Take 2–3 tablets once or twice per day, preferably first thing in the morning and no later than midday.	Eliminates excess fluid; beneficial for urinary system health; decreases water retention.
Catimune from Life Plus or	Take 1 tablet twice daily between meals.	Helps detoxify the kidneys and bladder and flushes out toxins; supports circulatory system.
Detox-Zyme from Rainbow Light or	Take as indicated on the label.	Helps cleanse the body and eliminate toxins.
Bromelain Complex from PhytoPharmica or	Take as indicated on the label.	Reduces swelling and inflammation.
Bioprotect from Biotics or	Take as indicated on the label.	Has varied free-radical fighting capabilities.
Bio-Immunozyme Forte from Biotics	Take as indicated on the label.	Stimulates immune function.

ENZYME HELPERS

Nutrient	Suggested Dosage	Actions
Acidophilus and other probiotics, such as bifidobacterium	Take as indicated on the label.	Improves gastrointestinal function; helps produce antibacterial substances; stimulates enzyme activity.

Adrenal glandular	Take 100–400 mg per day.	Fights inflammation and adrenal exhaustion.
Aged garlic extract	Take two 300 mg capsules three times per day.	Fights inflammation; stimulates the immune system; potent antioxidant.
Bioflavonoids, such as rutin, quercetin, or Pycnogenol	Take 500 mg three times per day between meals.	Potent antioxidants; improve detoxification.
Coenzyme Q$_{10}$	Take 30–60 mg one to two times per day with meals; or follow the advice of your health-care professional.	Potent antioxidant.
Cranberry extract	Take as indicated on the label; or take 2 tablets four times per day. Take the last dose just before retiring.	Shown to fight bladder and other urinary tract infections; contains vitamin C and anthocyanidins, which fight free radicals and inflammation.
Green Barley Extract from Green Foods Corporation or Kyo-Green from Wakunaga	Take as indicated on the label.	Potent free-radical fighter; rich in vitamins, minerals, and enzyme activity.
Spirulina	Take as indicated on the label.	Rich in minerals, vitamins, enzymes, proteins, essential amino acids, and essential fatty acids, especially gamma-linolenic acid; good source of beta-carotene; improves immune function.
Thymus glandular	Take 100 mg per day.	Assists the thymus gland in immune resistance; improves immune response; fights infections.
Vitamin B complex	Take 50 mg three times per day; or take as indicated on the label.	B vitamins act as coenzymes.
Vitamin C	Take 10,000–15,000 mg per day.	Fights infections, diseases, and allergies; improves wound healing; stimulates the adrenal glands; promotes immune system; powerful antioxidant.
Zinc	Take 50 mg per day. Do not take more than 100 mg per day from all sources.	Needed by over 300 enzymes; potent antioxidant.

Comments

❏ Fruits and vegetables (particularly green vegetables) should comprise over 60 percent of your diet. If your body cannot tolerate raw fruits and vegetables, increase your intake of digestive enzymes, or sauté or steam your produce. Especially helpful foods include alfalfa juice concentrate, parsley and parsley leaves, horseradish, citrus fruits, cranberries, and other acidic foods.

❏ Avoid alcohol, refined white flour, and refined sugar. Small amounts of honey are allowed.

❏ Natural spices are fine, but chemical preservatives, particularly nitrites, should be avoided.

❏ Avoid chlorinated water. Distilled or bottled spring water or even well water is recommended.

❏ Limit protein intake. Protein places an enormous strain on the kidneys, weakening their function, changing the pH of the urine, and ultimately affecting the bladder.

❏ Instead of ordinary salt, use sea salt because it contains minerals that are important to nutrition and are advantageous in the treatment of bladder infections.

❏ Increase fluid intake. You should drink six to eight 8-ounce glasses of water or juice per day.

❏ Drink a citrus juice blend. To make: Place two peeled oranges, one peeled grapefruit, and one peeled lemon in a blender and liquefy, adding distilled water to produce the desired consistency. Drink throughout each day. If heartburn or diarrhea occurs, decrease consumption to three times per week. This is an excellent way to supplement with high doses of vitamin C.

❏ Perform the Enzyme Kidney Flush as outlined in Part Three.

❏ Health-care practitioners are trained to work with bladder conditions. They can provide a consultation, an examination, and laboratory tests to help with your specific needs.

Blood Clots

See EMBOLISM; THROMBOSIS.

Boils

A boil or furuncle is a tender nodule formed in the skin caused by a staphylococcal bacterial infection that enters through the hair follicles. Typically, boils occur on the face, neck, buttocks, and breasts. If several boils occur in a cluster, the cluster is referred to as a *carbuncle.* Furunculosis is a condition marked by recurrent and troublesome boils. The risk for boils increases with poor digestion, poor overall health, and local irritation. A *sty* is similar to a boil but occurs on the eyelid. A sty is caused by infection of the glands along the edge of the lid.

Symptoms

Pain and inflammation under the skin of the affected area is the first stage of a boil. There may be throbbing and intense pain, especially if the boil occurs on the ear, fingers, or nose. The boil then develops into a raised nodule that has a yellow or white center with a red rim (it looks like a very bad pimple). This stage occurs several days after the initial signs appear. Because pus builds up under the skin, putting pressure on nerves, this stage may be extremely painful.

Enzyme Therapy

Systemic enzyme therapy is used to reduce pain and inflammation, and bring the boil to a head. Enzymes stimulate the immune system to fight any infection, improve circulation, help speed tissue repair, bring nutrients to the damaged area, remove waste products, and enhance wellness.

Digestive enzyme therapy is used to improve the digestion of food, reduce stress on the gastrointestinal mucosa, help maintain normal pH levels, detoxify, and promote the growth of healthy intestinal flora, thereby strengthening the body's immune system.

Enzyme Absorption System Enhancers (EASE) maximize enzyme activity. They can also improve the absorption and bioavailability of various nutrients and decrease the drain on the body's own digestive enzymes, thus prolonging their lives.

In addition to the standard multivitamin, multimineral, and multiglandular complexes recommended in the introduction to Part Two, the following supplements are recommended for the treatment of boils, carbuncles, and sties. The following recommended dosages are for adults. Children between the ages of two and five should take one-quarter the recommended dosage, those between the ages of six and twelve should take one-half the recommended dosage, and those between the ages of thirteen and seventeen should take three-quarters the recommended adult dosage.

ESSENTIAL ENZYMES

Enzyme	Suggested Dosage	Actions
Superoxide dismutase	Take 150 mcg per day.	Fights inflammation; potent antioxidant.

AND SELECT ONE OF THE FOLLOWING:

Bromelain	Take 750 mg two times per day between meals.	Fights inflammation, pain, and swelling; speeds recovery after injury or surgery; stimulates the immune system; strengthens capillary walls.
Serratiopeptidase	Take 10 mg two times per day between meals.	Fights inflammation; reduces pain and swelling.
Trypsin with chymotrypsin	Take 250 mg of trypsin and 10 mg of chymotrypsin three times per day between meals.	Fights inflammation, pain, and swelling; increases rate of healing; stimulates the immune system.
Pancreatin	Take 500 mg six times per day on an empty stomach.	Fights inflammation, pain, and swelling; increases rate of healing; stimulates the immune system; strengthens capillary walls.

Papain or a microbial protease	Take 300 mg two times per day between meals.	Fights inflammation; speeds recovery from injuries and surgery.
Formula One—See page 38 of Part One for a list of products.	Take 5 tablets twice per day between meals.	Fights inflammation, pain, and swelling; speeds healing; stimulates the immune system.

SUPPORTIVE ENZYMES

Enzyme	Suggested Dosage	Actions
Catalase and glutathione peroxidase	Take as indicated on the label.	Potent antioxidant enzymes.
A digestive enzyme product containing protease, amylase, and lipase enzymes. See the list of digestive enzyme products in Appendix.	Take as indicated on the label with meals.	Improves the breakdown and absorption of food nutrients; fights toxic build-up.

ENZYME ABSORPTION SYSTEM ENHANCERS (EASE)

Combination	Suggested Dosage	Actions
Detox-Zyme from Rainbow Light *or*	Take as indicated on the label.	Helps cleanse the body and eliminate toxins.
Catimune from Life Plus *or*	Take 1 tablet twice daily between meals.	Flushes out toxins; supports circulatory system.
Hepatic Complex from Tyler *or*	Take as indicated on the label.	Helps detoxify the liver and thus detoxify the body as a whole.
Bioprotect from Biotics *or*	Take as indicated on the label.	Antioxidant combination.
Metabolic Liver Formula from Prevail *or*	Take as indicated on the label.	Improves liver function, thus helping to detoxify the body.
Bio-Immunozyme Forte from Biotics *or*	Take as indicated on the label.	Stimulates immune function.
Advanced Nutritional System from Rainbow Light *or*	Take as indicated on the label.	Potent free-radical fighter and immune system stimulator.
NESS Formula #416 *or* Protease Concentrate from Tyler *and*	Take as indicated on the label.	Breaks down fibrin; fights inflammation.
Bromelain Complex from PhytoPharmica *and*	Take as indicated on the label.	Decreases swelling; improves circulation; helps break up circulating immune complexes.
Cardio-Chelex from Life Plus	If under the age of 40 and in good health, take 5 tablets once a day first thing in the morning. With a history of heart problems or if over 40, take	Supports healthy circulation; improves removal of toxins.

5 tablets twice per day, first thing in the morning and at midday.

ENZYME HELPERS

Nutrient	Suggested Dosage	Actions
Acidophilus and other probiotics	Take as indicated on the label.	Improves gastrointestinal function, which helps produce antibacterial substances; stimulates enzyme activity.
Adrenal glandular	Take 100–400 mg per day.	Fights inflammation and adrenal exhaustion.
Aged garlic extract	Take two 300 mg capsules three times per day.	Fights inflammation; stimulates the immune system; potent antioxidant.
Amino-acid complex	Take as indicated on the label on an empty stomach.	Required for enzyme, peptide, and hormone production; helps the body maintain proper fluid and acid/base balance, which are directly related to overall health.
Bioflavonoids, such as rutin or quercetin	Take 500 mg three times per day between meals.	Strengthen capillaries, which improves the delivery of nutrients to the affected area and the removal of toxic by-products of inflammation; potent antioxidant.
Coenzyme Q_{10}	Take 30–60 mg one to two times per day with meals; or follow the advice of your health-care professional.	Potent antioxidant.
Green Barley Extract from Green Foods Corporation or Kyo-Green from Wakunaga	Take as indicated on the label.	Potent free-radical fighter; rich in vitamins, minerals, and enzyme activity.
Thymus glandular	Take 250 mg twice per day.	Assists the thymus gland in immune resistance; fights infections.
Vitamin A	Take 30,000–35,000 IU per day until the boil is healed. Dosages of 50,000 IU and higher should be taken only under the supervision of a physician. Pregnant women should take no more than 10,000 IU per day.	Needed for strong immune system and healthy mucous membranes; potent antioxidant.
Vitamin B complex	Take as indicated on the label.	B vitamins act as coenzymes.
Vitamin C	Take 10,000–15,000 mg per day.	Fights infections, diseases, allergies; improves wound healing; stimulates the adrenal glands; promotes immune system; powerful antioxidant.
Vitamin E	Take 1,200 IU per day from all sources (including	Potent antioxidant and immune system stimulator.
	the multivitamin complex in the introduction to Part Two).	
Zinc	Take 50 mg per day. Do not take more than 100 mg per day from all sources.	Needed by over 300 enzymes; necessary for white blood cell immune function; improves wound healing.

Comments

❑ To help reduce stress on the liver (which is a major organ for detoxification), combinations of microbial enzymes and other nutrients can be used. NESS makes a product containing microbial amylase, protease, lipase, and cellulase enzymes, plus acidophilus. Another product contains microbial amylase and lipase, plus safflower petals. A digestive enzyme product combines betaine hydrochloride and ox bile (required for digestion to decrease the drain on the liver). Ox bile extract helps assist digestive enzymes.

❑ Eat plenty of garlic; it has antibacterial activity.

❑ Foods high in zinc, such as meat, liver, eggs, and oysters will stimulate and enhance immune system activity.

❑ Enzyme-rich fruits and vegetables (particularly green vegetables) should comprise 60 percent of your diet. If your body cannot tolerate raw fruits and vegetables, increase your intake of digestive enzymes, or sauté or steam your produce.

❑ Reports indicate that drinking a tea made from pau d'arco is effective at treating boils. Other herbs taken orally, including alfalfa, echinacea, ginger root, ginkgo biloba, licorice, and white willow bark can help control inflammation.

❑ There are a number of ways of treating boils topically:

• Apply an enzymatic skin salve twice daily.

• Place ten drops of liquid aged garlic extract on a cotton swab or gauze. Dab gently along the affected area.

• Break open and spread the contents of two to four vitamin A and vitamin E capsules on the affected area.

• Zinc oxide salve can be spread over the affected area once daily.

• Spread tea tree oil on the affected area five times a day.

• Epsom salt may help draw the pus from the boil. Follow directions on the label.

• Apply ichthammol ointment (also called ammonium ichthyosulfonate and ammonium sulfoichthyolate) directly to the boil. Cover the area with a bandage. Change twice daily.

❑ Moist heat will help the boil come to a point and drain spontaneously. Place a warm, moist washcloth over the boil and cover with wax paper. Place a heating pad over the wax paper. Apply for fifteen to thirty minutes three to four times per day.

❏ Perform the Enzyme Kidney Flush and Enzyme Toxin Flush as outlined in Part Three. These flushes can help restore the normal functional capacity of the kidneys, gallbladder, and liver.

❏ Exercising helps to increase antioxidant and enzymatic activity.

Bone Fractures

A broken bone can be either a simple (closed) fracture—the bone breaks, but the skin stays intact—or a compound (open) fracture—the bone breaks through the skin, creating an opening for bacteria to enter. When a bone breaks into two or more fragments, the tissue within and around the area is torn. Blood vessels are also often torn at the site of the fracture. This interferes with the flow of blood and delivery of nutrients to the damaged area.

Blood infiltrates the surrounding muscles, and within only a few hours, the inflammatory process develops (swelling, pain, heat, and possibly redness). Immediate care is important, because if left unattended, the bone can heal misaligned and/or the surrounding tissue can develop chronic inflammation. In a compound fracture, bacteria can enter through the break in the skin resulting in infection and deterioration of the bone.

Symptoms

Usually, an individual knows immediately if a bone is fractured. However, in some cases, as in toe, finger, and rib fractures, a fracture may not be as obvious. Symptoms include pain, limited function, limb deformities and shortened limb length (in fractures of the long bones), swelling, and discoloration of the skin in the area of the fracture.

After a fracture, it is important to return the bones to their normal position to reduce the pain and swelling, fight any potential infection, and treat for possible effects of shock. The patient should see a physician to set the fractured bone.

Enzyme Therapy

The use of proteolytic enzymes for treating a fracture (and many soft tissue injuries, such as strains, sprains, hematomas, and bruises) causes rapid removal of inflammatory debris; the break up of microthrombi (blood clots); the return of circulation to the small blood vessels; an increase of tissue permeability; a decrease in swelling, pain, and inflammation; speedier healing; and a marked decrease in the duration of the inflammatory process.

Digestive enzyme therapy is used to improve the digestion and assimilation of food nutrients essential for bone repair, reduce stress on the gastrointestinal mucosa, and help maintain normal pH levels. Enzymes can help detoxify the body and promote the growth of healthy intestinal flora, thus reducing the possibility of bacterial invasion and infection. Digestive enzymes also serve as replacements for the body's pancreatic enzymes, leaving the pancreatic enzymes free to perform other functions in the body, such as decreasing inflammation.

Enzyme Absorption System Enhancers (EASE) maximize enzyme activity. They can also improve the absorption and bioavailability of various nutrients and decrease the drain on the body's own digestive enzymes, thus prolonging their lives.

Sometimes, surgery is required to reset a broken bone. Enzymes have been administered from one to five days before surgery as prophylactic therapy to reduce the swelling (which can be a major complication) and the inflammatory reaction. Therefore, the bone placement is easier and there are fewer complications. Research shows this can also substantially decrease postoperative infection, since the skin can heal more quickly, reducing the danger of infection. Recovery is more rapid and the healing occurs sooner with reduced scar tissue development. This also decreases rehabilitation time and the cost of therapy.

Enzymes can also be helpful when used before a fracture or injury occurs. In a double-blind study that I conducted at Portland State University with the director of sports medicine there, we found that of football players who sustained fractures, those who had been taking enzymes before the injury healed about twice as fast as those who did not receive enzyme therapy.

In addition to the standard multivitamin, multimineral, and multiglandular complexes recommended in the introduction to Part Two, the following supplements are recommended for the treatment of bone fractures. The following recommended dosages are for adults. Children between the ages of two and five should take one-quarter the recommended dosage, those between the ages of six and twelve should take one-half the recommended dosage, and those between the ages of thirteen and seventeen should take three-quarters the recommended adult dosage.

ESSENTIAL ENZYMES

Enzyme	Suggested Dosage	Actions
Formula One—See page 38 of Part One for a list of products.	Take 10 tablets three times per day between meals. As symptoms subside (which should occur in approximately one week) decrease dosage to 3–5 tablets three times per day between meals. When free of symptoms, begin taking a maintenance dose of 2–3 tablets after each meal.	Decreases inflammation, pain, and swelling; increases rate of healing; stimulates the immune system; strengthens capillary walls.
or		

Bromelain	Take three to four 230 mg capsules per day until symptoms subside. If pain is severe, take 6–8 capsules per day.	Fights inflammation; speeds recovery after injury or surgery; bolsters the immune system.
or		
Serratiopeptidase	Take as indicated on the label between meals.	Decreases inflammation, pain, and swelling; increases rate of healing; stimulates the immune system.
or		
A proteolytic enzyme	Take as indicated on the label between meals.	Decreases inflammation, pain, and swelling; increases rate of healing; stimulates the immune system.

SUPPORTIVE ENZYMES

Enzyme	Suggested Dosage	Actions
Formula three—See page 38 of Part One for a list of products.	Take as indicated on the labels.	Potent antioxidant enzymes.
A digestive enzyme product containing protease, amylase, and lipase enzymes. See the list of digestive enzyme products in Appendix.	Take as indicated on the label with meals.	Improves digestion thereby improving the overall health of the body.

ENZYME ABSORPTION SYSTEM ENHANCERS (EASE)

Combination	Suggested Dosage	Actions
Osteo Formula from Prevail	Take as indicated on the label.	Nutritional support for individuals at risk for bone loss.
or		
Magnesium Plus from Life Plus	Take 1 or 2 tablets one to three times per day.	Important for calcium utilization and many enzyme reactions.
or		
Protease Concentrate from Tyler	Take as indicated on the label.	Fights inflammation.
or		
Bromelain Complex from PhytoPharmica	Take as indicated on the label.	Decreases swelling; improves circulation; helps break up circulating immune complexes.
or		
Joint-Ease from Nature's Life	Take as indicated on the label.	Decreases inflammation and swelling.
or		
Mobil-Ease from Prevail	Take one capsule between meals two to three times per day.	Has anti-inflammatory activity.
or		
Inflamzyme from PhytoPharmica	Take as indicated on the label.	Promotes tissue repair.
or		
Connect-All from Nature's Plus	Take as indicated on the label.	Supports connective tissue.
or		
CTR Support from PhysioLogics	Take as indicated on the label.	Nourishes the body during tissue recovery and healing from injury; helps maintain healthy connective tissue; diminishes damage caused by swelling and inflammation.

ENZYME HELPERS

Nutrient	Suggested Dosage	Actions
Adrenal glandular	Take 100–400 mg per day.	Fights inflammation and adrenal exhaustion.
Alpha-linolenic acid	Take as indicated on the label.	Stimulates immune system; decreases inflammation.
Amino-acid complex	Take as indicated on the label on an empty stomach.	Amino acids are the building blocks of protein and enzymes and are essential for normal bone formation.
Bioflavonoids	Take 150–500 mg three times per day.	Strengthen capillaries; potent antioxidants.
Boron	Take 4 mg per day.	Improves mineral absorption and affects hormone metabolism, both important in bone repair and growth.
Branched-chain amino acids (isoleucine, leucine, and valine)	Take as indicated on the label between meals.	Often found together in a supplement, branched-chain amino acids aid in healing of muscle, bone, and skin and act as fuel for muscles.
Calcium	Take 1,000 mg per day.	Plays important roles in bone formation; essential for healthy cells.
Coenzyme Q_{10}	Take 30–60 mg one to two times per day with meals; or follow the advice of your health-care professional.	Potent antioxidant.
Magnesium	Take 500 mg per day.	Plays important role in calcium absorption, helpful in building bone; is also a cofactor for many enzymes involved in bone building.
Vitamin A	Take 50,000 IU per day for 6 weeks, then decrease dosage to 20,000 IU per day. If pregnant, take no more than 5,000 to 10,000 IU per day.	Potent antioxidant; required for healthy skin and cell membrane structure, a strong immune system, growth, and bone development; needed by digestive glands to produce enzymes.
Vitamin B complex	Take 50 mg three times per day, or take as indicated on the label.	B vitamins act as coenzymes and support neuromusculoskeletal function.
Vitamin C	Take 5,000–10,000 mg per day.	Helps to produce collagen and normal connective and bony tissue; promotes immune system; powerful antioxidant.
Zinc	Take 15–50 mg per day. Do not take more than 100 mg per day from all sources.	Needed by over 300 enzymes; antioxidant; essential for protein metabolism and wound healing.

Comments

❏ Eat plenty of fresh fruits and vegetables high in enzymatic and antioxidative activity, including pineapples, carrots, raw citrus fruits, papaya, figs, spinach, sweet potatoes, and turnip greens.

141

❏ Eat like a grazing animal; that is, eat frequent small meals (four or five meals per day) rather than two or three large meals. Eliminate all food additives, preservatives, refined foods, pesticides, heavy metals, and fried foods (see Part Three for dietary recommendations).

❏ Drink at least two 8-ounce glasses of freshly juiced fruits and vegetables per day. Use plenty of pineapple (about half of a pineapple per day). See Part Three for further information on juicing, cleansing, and fasting.

❏ As your bone repairs, stretching exercises, then swimming, walking, or bicycling (if possible) can help keep your muscles in shape and create a positive mental attitude.

❏ Alternative therapy for fractures may include mild pressure point therapy, massage, acupressure or acupuncture, electrical stimulation, ice and heat applications, and exercises. This should be done in cooperation with your physician.

Brain Tumors

Brain tumors are clumps of abnormal cells growing in the brain. They may be benign or malignant (cancerous). According to the Brain Tumor Society, nearly 44 percent of all primary brain tumors are benign. A primary brain tumor is one that originates in the brain, while a secondary tumor occurs when cancer cells spread to the brain from other areas in the body.

Because the brain controls every movement and thought, a benign tumor, which would normally be harmless in any other area of the body, can still interfere with normal brain activity. A tumor in any area of the brain can obstruct blood flow or put pressure on tissue and interfere with the job normally conducted by that particular area of the brain.

To diagnose a brain tumor, physicians use neurological examination (testing eye movement, hearing, balance, coordination, and sensation) and special imaging techniques, including computed tomography (CT), magnetic resonance imaging (MRI), and positron emission tomography (PET). When it is possible to access the tumor, a biopsy may be performed to analyze the cells under a microscope.

Symptoms

Symptoms of a brain tumor can include headaches, seizures, nausea and vomiting, drowsiness, and vision problems (including loss of vision and double vision). The patient may be unsteady, and there may be hearing loss or speech difficulties. Sometimes the person's personality changes markedly.

Enzyme Therapy

Standard treatment for brain tumors includes surgery, radiation, and chemotherapy. The concurrent use of enzymes as adjunctive therapy can lead to earlier response to chemotherapy and a decrease in the adverse side effects normally associated with chemotherapy. When used before and after surgery, enzymes can help improve wound healing and reduce the risk of blood clots. They can reduce the adverse effects and delayed damages resulting from radiotherapy (including radiation sickness). When used long-term, they can help prevent the spread of the cancer. With systemic enzyme therapy, patients demonstrate a significant improvement in quality of life.

Digestive enzyme therapy can help improve the digestion and assimilation of food nutrients, reduce stress on the gastrointestinal mucosa, and help maintain normal pH levels. They can help detoxify the body (which is particularly important in any cancer), promote the growth of healthy intestinal flora, and build general resistance. All of these actions improve the overall health of the body.

Enzyme Absorption System Enhancers (EASE) maximize enzyme activity. They can also improve the absorption and bioavailability of various nutrients and decrease the drain on the body's own digestive enzymes, thus prolonging their lives.

In addition to the standard multivitamin, multimineral, and multiglandular complexes recommended in the introduction to Part Two, the following supplements are recommended for the treatment of brain tumors. The following recommended dosages are for adults. Children between the ages of two and five should take one-quarter the recommended dosage, those between the ages of six and twelve should take one-half the recommended dosage, and those between the ages of thirteen and seventeen should take three-quarters the recommended adult dosage.

ESSENTIAL ENZYMES

Enzyme	Suggested Dosage	Actions
Before and after surgery:		
WobeMugos E from Mucos	Take 2–3 tablets three times per day.	Speeds wound healing; decreases inflammation.
Before and after chemotherapy or radiotherapy, and as long-term therapy:		
WobeMugos E from Mucos	Take 2–4 tablets three times per day.	Leads to earlier response to chemotherapy and reduced side effects; prevents metastasis; reduces appetite loss, depression, and pain.

SUPPORTIVE ENZYMES

Enzyme	Suggested Dosage	Actions
Bromelain	Take eight to ten	Fights inflammation, pain, and

	250 mg capsules per day. As soon as tumor growth is under control, the dose can be reduced gradually to a maintenance level of approximately 4 capsules per day.	swelling; stimulates the immune system; helps detoxify the body; improves digestion, thereby strengthening the immune system.
Formula One—See page 38 of Part One for a list of products.	Take 10 tablets three times per day between meals	Fights inflammation, pain, and swelling; stimulates the immune system; helps detoxify the body.
A digestive enzyme formula containing protease, amylase, and lipase enzymes. For a list of digestive enzyme products, see Appendix.	Take as indicated on the label.	Improves digestion by improving the breakdown of proteins, carbohydrates, and fats, thus improving the overall health of the body; helps detoxify the body.
Formula Three—See page 38 of Part One for a list of products.	Take as indicated on the label.	Potent antioxidant enzymes.

ENZYME ABSORPTION SYSTEM ENHANCERS (EASE)

Combination	Suggested Dosage	Actions
Vita-C-1000 from Life Plus *or*	Take as indicated on the label.	Helps fight cancer; boosts immune function.
Hepazyme from Enzymatic Therapy *or*	Take as indicated on the label.	Stimulates immune function.
Detox Enzyme Formula from Prevail *or*	Take 1 capsule between meals one or two times per a day. Take with 6- to 8-ounce glass of water or juice.	Helps rid the body of harmful toxins.
Antioxidant Complex from Tyler *or*	Take as indicated on the label.	Potent free-radical fighter and immune system stimulator.
Oxy-5000 Forte from American Biologics *or*	Take as indicated on the label.	Helps the body excrete heavy metals; helps rid the body of biochemical stressors, common infections, and degenerative processes.
Kyolic Formula 102 (with enzymes)	Take 2 capsules or tablets with a meal two to three times per day.	Strengthens the immune system; helps alleviate stress; helps fight free radicals; improves circulation; helps fight inflammation.

ENZYME HELPERS

Nutrient	Suggested Dosage	Actions
Acidophilus or other probiotics	Take as indicated on the label.	Stimulates enzyme activity; improves gastrointestinal function, thereby strengthening the immune system.
Aged garlic extract	Take as indicated on the label.	Helps fight cancer.

Bioflavonoid, such as rutin or quercetin	Take 150–500 mg three times per day.	Strengthens capillaries; potent antioxidant.
Magnesium	Take 250 mg per day.	Activates many of the body's enzymes; essential for proper cell function.
Thymus glandular	Take 30 mg per day.	Assists the thymus gland (which is important in immune function); helps improve immune response; fights cancer.
Vitamin A	Take 50,000 IU per day. (Do not take more than 10,000 IU if you are pregnant.) Take higher doses only under the supervision of a physician.	Stabilizes the cell membrane; improves response to cancer therapies; needed for strong immune system and healthy mucous membranes; potent antioxidant.
Vitamin C	Take 2–10 grams per day. Take more than 10 grams per day only under the supervision of a physician.	Prevents capillary weakness; fights infection; improves wound healing; stimulates the adrenal glands; promotes immune system; powerful antioxidant.
Vitamin D	Take 3,000–6,000 IU per day.	Supports immune system function; hinders tumor growth; promotes the break-up and removal of degenerated cells.
Vitamin E	Take 1,200 IU per day from all sources (including the multivitamin complex in the introduction to Part Two).	Causes more efficient consumption of oxygen; acts as an intercellular antioxidant; improves circulation; helps resist diseases at the cellular level.
Zinc	Take 50 mg per day. Do not take more than 100 mg per day from all sources.	Needed for white blood cell immune function; required by hundreds of enzymes in the body; helps protect the body from heavy metal poisoning, which can interfere with normal metabolism, cause toxic build-up, and lead to cancer.

Comments

❑ See Part Three for dietary recommendations.

❑ Avoid pesticides, food preservatives, additives, and colors.

❑ Detoxification is extremely important in treating cancer. See Part Three for detoxification methods.

❑ Maintain a positive mental attitude. Exercises and meditation can help. In addition, exercise can increase enzymatic and antioxidant activity.

❑ Physicians sometimes use enzyme retention enemas if oral administration of enzymes is difficult, or when greater absorption and concentration of enzymes is desired. An enzyme retention enema is a means of instilling some enzymes into the colon to be retained. This allows the enzymes to be absorbed into the body's systems. Physicians use enzyme products formulated specifically for this use.

❑ See the list of Resource Groups in the Appendix for an

143

organization that can provide you with more information about brain tumors.

❏ *See also* CANCER.

Breast Cancer

Breast cancer is the most common type of cancer in American women. The chance of developing cancer increases with age, with the incidence climbing dramatically after a women reaches 40. According to Y-Me National Breast Cancer Organization, three-quarters of all breast cancers occur in women over the age of 50. Fortunately, breast cancer can be beaten, especially if detected and treated early.

Though malignancy of the breast is often given the generic label of breast cancer, there are actually many types of breast cancer. *Ductal carcinoma* forms in the walls of the milk ducts of the breast, which secrete the milk produced in the milk glands. It may be localized, or it may spread to other breast tissue (invasive). *Lobular carcinoma* begins in the milk glands. This form of cancer, which usually develops before menopause, is less common than are ductal carcinomas. It usually becomes invasive at some point, and may even occur in both breasts at the same time. *Paget's disease of the nipple* is a type of breast cancer that causes discharge from the nipple and a crusty sore to form. It is a signal of an underlying malignancy.

Although no one particular cause of cancer has been pinpointed, it seems that the breasts' prolonged exposure to estrogen plays a large role. Risk factors include age—risk is greatest after the age of 75; prolonged exposure to oral contraceptives (which contain estrogen) or estrogen replacement therapy; menstruation onset before the age of 12; menopause onset after the age of 55; and having a first child after the age of 30, or having no pregnancies at all. All of these factors lead to prolonged exposure to estrogen. Other risk factors include a family history of breast cancer; obesity (which can cause an increase in estrogen production); use of breast implants, which may release carcinogens into the body; and exposure to radiation or certain pesticides.

Men can get breast cancer too. Though only 1 percent of breast-cancer cases are men, these situations are usually more serious. Because the disease is so rare in men, it is not normally detected early. Men generally do not examine themselves regularly for breast lumps, and once they begin to feel ill, breast cancer is not usually considered until all other possibilities are ruled out. Breast cancer in men is usually not detected until the disease has progressed to a more serious stage.

Symptoms

Breast cancer symptoms include any changes in the breast, such as dimpling, swelling, thickening, a lump, or a change in size.

There may be pain in the breast and tenderness in, or a discharge from, the nipple, or there may be changes in the nipple, such as its turning inward. The skin may appear irritated and scaly. A mammogram (an X-ray of the breast) is extremely valuable in detecting lumps before they are large enough to be felt.

Enzyme Therapy

For decades, enzyme therapy has been used effectively in Europe to treat breast cancer. Systemic enzyme therapy is used as part of an overall therapy program. Enzyme therapy is used to stimulate and stabilize the immune system, thus increasing the opportunities to decrease the rate of tumor growth and destroy residual tumor cells; to help differentiate the healthy tissue from the cancerous tissue, and better identify the tumor for surgical removal; to reduce inflammation; to help speed tissue repair by bringing nutrients to the area and removing waste products; to improve the patient's quality of life; to decrease the adverse side effects of the primary therapy; to reduce the cost of medical therapy by minimizing the necessity for its use (if enzyme therapy is initiated early enough); and to effectively treat lymphedema (swelling in the lymphatic vessels in the arm pit) which can follow breast cancer surgery. (*See* LYMPHEDEMA in Part Two to learn more about this condition.)

Digestive enzymes are essential for the gastrointestinal tract (and the body as a whole) in helping to prevent the onset of cancer because they aid in the breakdown, digestion, and absorption of nutrients; decrease the enzyme production load on the pancreas, thereby helping to reduce stress on the pancreas and increase its productive life. This helps prevent cancer because it helps prevent toxin build-up, especially in the small and large intestine; stimulates the production of friendly microflora in the intestine, thereby strengthening the body's immune system; and facilitates the elimination of potential cancer-causing waste from food and the cells of the body.

Enzyme Absorption System Enhancers (EASE) maximize enzyme activity. They can also improve the absorption and bioavailability of various nutrients and decrease the drain on the body's own digestive enzymes, thus prolonging their lives.

In addition to the standard multivitamin, multimineral, and multiglandular complexes recommended in the introduction to Part Two, the following supplements are recommended for the treatment of breast cancer.

ESSENTIAL ENZYMES

Enzyme	Suggested Dosage	Actions
Before and after surgery:		
WobeMugos E from Mucos	Take 4 tablets per day.	Speeds wound healing; reduces inflammation; stimulates immune activity.

Before and after chemotherapy or radiotherapy, and as long-term therapy:

WobeMugos E from Mucos	Take 4 tablets three times per day.	Leads to earlier response to chemotherapy and reduced side effects; prevents metastasis; reduces appetite loss, depression, and pain.

SUPPORTIVE ENZYMES

Enzyme	Suggested Dosage	Actions
Bromelain	Take eight to ten 230–250 mg capsules per day. As soon as tumor growth is under control, gradually reduce dose to a maintenance level of 4 capsules per day.	Fights inflammation, pain, and swelling; stimulates the immune system; improves digestion, which can improve overall health.
Formula One—See page 38 of Part One for a list of products.	Take 10 tablets three times per day between meals.	Stimulates the immune system; fights inflammation, pain, and swelling; helps detoxify the body; breaks up circulating immune complexes.
A digestive enzyme formula containing protease, amylase, and lipase enzymes. For a list of digestive enzyme products, see Appendix.	Take as indicated on the label.	Improves digestion by improving the breakdown of proteins, carbohydrates, and fats, which can improve overall health.
Formula Three—See page 38 of Part One for a list of products.	Take as indicated on the label.	Potent antioxidant enzymes.

ENZYME ABSORPTION SYSTEM ENHANCERS (EASE)

Combination	Suggested Dosage	Actions
Vita-C-1000 from Life Plus *or*	Take as indicated on the label.	Helps fight cancer; boosts immune function.
Hepazyme from EnzymaticTherapy *or*	Take as indicated on the label.	Stimulates immune function.
Detox Enzyme Formula from Prevail *or*	Take 1 capsule between meals one or two times per day. Take with a 6- to 8-ounce glass of water or juice.	Helps rid the body of harmful toxins.
Antioxidant Complex from Tyler *or*	Take as indicated on the label.	Potent free-radical fighter and immune system stimulator.
Kyolic Formula 102 (with enzymes) *or*	Take 2 capsules or tablets with a meal two to three times per day.	Strengthens the immune system; helps alleviate stress; helps fight free radicals; helps fight inflammation.
Oxy-5000 Forte from American Biologics	Take as indicated on the label.	Helps the body excrete heavy metals; helps rid the body of biochemical stressors, common infections, and degenerative processes.

ENZYME HELPERS

Nutrient	Suggested Dosage	Actions
Acidophilus or other probiotics	Take as indicated on the label.	Stimulates enzyme activity; improves gastrointestinal function, which reduces the potential for toxin build-up.
Bioflavonoids, such as rutin or quercetin	Take 150–500 mg three times per day.	Strengthens capillaries, so improves delivery of nutrients; potent antioxidant.
Magnesium	Take 250 mg per day.	Plays important roles in protein building and energy production; necessary for the synthesis of fats, proteins, and nucleic acid, which are helpful in improving overall health.
Thymus glandular	Take 30 mg per day.	Helps stimulate the thymus gland, which is important in immune resistance; fights cancer.
Vitamin A	Take 50,000 IU (Do not take more than 10,000 IU if you are pregnant.) Take higher doses only under the supervision of a physician.	Stabilizes the cell membrane; improves response to therapy; needed for strong immune system and healthy mucous membranes; potent antioxidant.
Vitamin C	Take 2–10 grams per day. For amounts over 10 grams, consult a physician.	Prevents capillary weakness; fights infection; improves wound healing; stimulates the adrenal glands; promotes immune system; powerful antioxidant.
Vitamin D	Take 3,000–6,000 IU per day.	Supports immune system function; hinders tumor growth; promotes break-up and removal of degenerated cells.
Vitamin E	Take 1,200 IU per day from all sources (including the multivitamin complex in the introduction to Part Two).	Causes more efficient consumption of oxygen; acts as an intercellular antioxidant; improves circulation; helps resist diseases at the cellular level; fights tumor cell development.
Zinc	Take 50 mg per day. Do not take more than 100 mg from all sources per day.	Potent free-radical fighter that combines with the antioxidant superoxide dismutase to fight breast cancer; helps protect the body from heavy metal poisoning; needed for white blood cell immune function.

Comments

❏ The risk of breast cancer is greater in those with a family history of breast cancer and also increases with age. It is important to conduct a breast self-examination monthly. If you are not familiar with the proper technique for breast self-examination, have your physician show you. Early detection and treatment can greatly increase your chance of survival.

❏ If you feel a lump in your breast, see a physician immediately for a mammogram and/or biopsy. Although over 80 percent of all lumps are benign (harmless), any delay in diagnosis and treatment is not worth the risk.

❏ The American Cancer Society urges annual mammograms for all women over the age of 40.

❏ Follow the dietary guidelines in Part Three. Although researchers don't know exactly what causes breast cancer, there seems to be an increased risk in women who drink alcohol or have a high intake of dietary fat.

❏ Increasing your intake of enzyme-rich fresh vegetables may help. A recent study found that women who eat lots of vegetables (including tomatoes, spinach, greens, corn, carrots, summer squash, and cucumbers) appear to have a decreased risk of developing premenopausal breast cancer.

❏ Increase your intake of phytochemical enzymes and antioxidants found in food extracts.

❏ Perform the Enzyme Kidney Flush and the Enzyme Toxin Flush. See instructions in Part Three.

❏ Physicians may use enzyme retention enemas if oral administration of enzymes is difficult or when greater absorption and concentration of enzymes is desired. An enzyme retention enema is a means of instilling some enzymes into the colon to be retained. This allows the enzymes to be absorbed into the body's systems. Physicians use enzyme products that have been formulated specifically for this use.

❏ See the list of Resource Groups in the Appendix for a list of organizations that can provide you with more information about breast cancer.

❏ *See also* CANCER.

Bronchitis

Bronchitis is an inflammation of the bronchial tubes. It may seem like a severe chest cold in its mild form, or it may lead to pneumonia in its severe forms.

Acute bronchitis is usually caused by some type of infection, such as a common cold, influenza, or other viral infection of the nose, throat, trachea, or bronchial tree. It is often associated with measles, whooping cough, allergies, and chickenpox. Acute bronchitis may also be caused by inhaling such irritants as dust, smoke, and fumes from acids, solvents, and other poisons.

Chronic bronchitis is caused by repeated inflammation of the airways, causing the walls of the bronchial tubes to enlarge, narrowing the airways. It is a frequently disabling widespread condition characterized by chronic and excessive mucus production and coughing. Chronic bronchitis is usually caused by allergies, smoking, or prolonged exposure to polluted air or other airborne irritants. It is usually not caused by infection. Those living in the city are more frequently affected than suburban people, and men are affected more frequently than are women. In mild cases, the only symptom may be a persistent cough, but lung function impairment and difficulty breathing may occur in advanced cases. If the cause is removed, the condition may be at least partially reversible; however, if the inflammation is present for a long time, permanent damage may result.

Although a typical case of bronchitis usually lasts a week or two, sometimes longer, it can be particularly dangerous in the elderly and those who suffer from chronic heart or lung disorders. Terminally ill patients are also at risk because their immune systems are often compromised.

Symptoms

For a day or two there may be a fever without any other symptoms. After this, there is increased coughing, which may begin as a dry cough but rapidly develops into a deep, rasping cough with thick sputum. Other symptoms of bronchitis include difficulty breathing, sore throat, and chills. There may also be malaise and pain in the muscles and back. Bronchitis can develop into pneumonia.

Enzyme Therapy

Systemic enzyme therapy is used to decrease swelling of the bronchial mucosa, as well as to liquefy the phlegm and other secretions, improving expectoration and the ability to breathe. Enzymes reduce inflammation, relieve chest and back pain, and stimulate the immune system by eliminating any present antigen-antibody complexes. Enzymes can improve circulation (thereby improving oxygen supply), helping speed tissue repair by bringing nutrients to the damaged area and removing waste products.

Digestive enzyme therapy is used to improve the digestion and absorption of food nutrients, reduce stress on the gastrointestinal mucosa, help maintain normal pH levels, detoxify the body, promote the growth of healthy intestinal flora, and strengthen the body as a whole, thereby strengthening the immune system. Digestive enzymes also serve as replacements for the body's pancreatic enzymes, leaving the pancreatic enzymes free to perform other functions in the body, such as decreasing inflammation.

Enzyme Absorption System Enhancers (EASE) maximize enzyme activity. They can also improve the absorption and bioavailability of various nutrients and decrease the drain on the body's own digestive enzymes, thus prolonging their lives.

Enzymes can also help transport antibiotics to the areas where they are needed and increase their concentration in the tissues. In fact, in cases of bronchitis, enzymes and antibiotic combinations result in astonishingly rapid improvement. Several bronchial conditions can be helped with enzyme therapy.

In addition to the standard multivitamin, multimineral, and multiglandular complexes recommended in the introduction to Part Two, the following supplements are recommended for the treatment of bronchitis. The following recommended dosages are for adults. Children between the ages of two and five should take one-quarter the recommended dosage, those between the ages of six and twelve should take one-half the recommended dosage, and those between the ages of thirteen and seventeen should take three-quarters the recommended adult dosage.

ESSENTIAL ENZYMES

Enzyme	Suggested Dosage	Actions
Formula Three—See page 38 of Part One for a list of products.	Take as indicated on the label.	Potent antioxidant enzymes.

AND SELECT ONE OF THE FOLLOWING:

Lysozyme	Take as indicated on the label.	Fights inflammation; improves respiratory conditions; bolsters the immune system; breaks up mucus.
Formula One—See page 38 of Part One for a list of products.	Take 3 tablets three times per day between meals as long as symptoms persist. In severe cases, increase dosage to 4–10 tablets three times per day.	Fights inflammation, pain, and swelling; speeds healing; strengthens capillary walls; bolsters the immune system.
Bromelain	Take three to four 230–250 mg capsules per day until symptoms subside. If severe, take 6–8 capsules per day.	Fights inflammation, pain, and swelling; stimulates the immune system.
Serratiopeptidase	Take 5 mg three times per day between meals.	Fights inflammation; improves respiratory conditions; bolsters the immune system; breaks up mucus.
Trypsin	Take 500 mg three times per day between meals.	Fights inflammation; improves respiratory conditions; bolsters the immune system; breaks up mucus.
A microbial protease	Take as indicated on the label between meals.	Fights inflammation, pain, and swelling; speeds healing; strengthens capillary walls.

SUPPORTIVE ENZYMES

Enzyme	Suggested Dosage	Actions
A digestive enzyme formula containing protease, amylase, and lipase enzymes. See the list of digestive enzyme products in Appendix.	Take as indicated on the label with meals.	Improves the breakdown and absorption of proteins, carbohydrates, and fats, thus improving the overall health of the body; fights toxic build-up; reduces stress.

ENZYME ABSORPTION SYSTEM ENHANCERS (EASE)

Combination	Suggested Dosage	Actions
Sinease from Prevail or	Take as indicated on the label.	Decongestant.
Combat from Life Plus or	Take as indicated on the label.	Helps support the respiratory system; stimulates immune system function; helps control infections.
Kyolic Formula 102 (with enzymes) or	Take 2 capsules or tablets with each meal two or three times per day.	Strengthens the immune system; helps alleviate stress; improves circulation; helps fight free radicals; helps fight inflammation.
Inflamzyme from PhytoPharmica or	Take as indicated on the label.	Supports tissues; an immune system activator.
Connect-All from Nature's Plus or	Take as indicated on the label.	Supports tissue repair.
Protease Concentrate from Tyler or	Take as indicated on the label.	Breaks up fibrin; fights inflammation.
Bromelain Plus from Enzyme Therapy or	Take as indicated on the label.	Breaks up circulating immune complexes.
Mucous Dissolver Liquezyme from Enzyme Process Laboratories	Take as indicated on the label.	Dissolves mucus.

ENZYME HELPERS

Nutrient	Suggested Dosage	Actions
Adrenal glandular	Take 100–400 mg per day.	Fights inflammation and adrenal exhaustion.
Amino-acid complex	Take on an empty stomach as indicated on the label.	Required for enzyme, peptide, and hormone production; helps the body maintain proper fluid and acid/base balance to improve overall health.
Bioflavonoids	Take 100–300 mg twice per day.	Strengthen capillaries; potent antioxidant.
Green Barley Extract or Kyo-Green	Take 1 heaping teaspoon in 8 ounces of liquid three times per day.	Fights free radicals.
Kyolic Aged Garlic Liquid Extract with Vitamins B$_1$ and B$_{12}$	Take 1–3 capsules every hour; or take as indicated on the label until symptoms are relieved.	Stimulates antioxidant activity and immune system function.
Lung glandular	Take as indicated on the label.	Improves lung function.
Pantothenic acid (vitamin B$_5$)	Take 25–50 mg per day.	Important in the formation of antibodies; used by the adrenal glands to fight stress.
Thymus glandular	Take 100 mg per day.	Immune system enhancer.

Vitamin A	Take 10,000–50,000 IU for no more than one month (for higher doses see a physician). If pregnant, take no more than 10,000 IU per day.	Necessary for mucous linings, skin and cell membrane structure, and strong immune system; potent antioxidant.
Vitamin B complex	Take 100–150 mg three times per day.	B vitamins act as coenzymes.
Vitamin C	Take 1–2 grams every hour until symptoms improve; or use a powder or crystalline form: Mix a heaping teaspoon with 6–8 ounces of water or orange or lemon juice; if lemon juice is used, you may sweeten with honey if you desire. Gargle a mouthful and swallow; repeat every hour.	Fights infections, diseases, allergies, and common cold; potent antioxidant.
Zinc or zinc gluconate lozenges	Take 50 mg per day. Take no more than 100 mg per day from all sources.	Needed for white blood cell immune function; needed by more than 300 enzymes.

Comments

❏ Hot broth or chicken or vegetable soup is helpful for cleaning out the body and sweating out toxins. Before making the soup, be sure to remove the skin from the chicken (chicken skins contain many toxins). Sip the broth or soup as frequently as possible.

❏ Milk and milk products are mucus-formers and should be strictly limited. They can cause increased congestion.

❏ Gargle twice daily with eight ounces of a combination made from four ounces of fresh juiced pineapple and four ounces of lemon or grapefruit, or gargle twice daily with the Enzymatic Gargle as described in Part Three. Swallow after gargling.

❏ Echinacea enhances immunity and fights inflammation, viral and bacterial infections, and upper respiratory tract infections. Slippery elm soothes sore throats and can help alleviate mucus build-up.

❏ Put five to six drops of eucalyptus in a sink or bowl filled with very hot water. Cover your head and shoulders with a large bath towel over the sink or bowl and inhale the vapors.

❏ If bronchitis is chronic, get tested for allergies, thyroid dysfunction, and immune system deficiencies.

❏ Liver and kidney flushes are of value in detoxification. Toxic build-up can weaken the immune system and the body as a whole so that it cannot fight off a viral or bacterial invasion. See Part Three for instructions on the Enzyme Kidney Flush and the Enzyme Toxin Flush.

❏ Until the fever subsides, rest is recommended. During the course of rehabilitation, a high fluid intake is suggested. Drink at least six to eight 8-ounce glasses of water per day.

❏ A hot water bottle on the chest or standing in a hot shower can help break up the mucus in the chest and give some relief.

❏ Manual lymph drainage is a form of massage that helps drain the lymph nodes and stimulates immune function.

❏ If a troublesome cough interferes with your sleep, use a cough suppressant (there are many natural forms at your health-food store). However, if you also have chronic obstructive pulmonary disease, see your physician.

❏ Expectorants may help loosen mucus.

❏ Steam inhalations might help, and a vaporizer is useful for an irritating cough. Check with your physician.

Bruises and Hematomas

A *bruise* is an injury to tissue under the skin following pressure, a blow, or trauma of some sort without breaking the skin. The injury causes the small blood vessels, called capillaries, under the skin to rupture, causing bleeding under the skin. This superficial bleeding under the skin is what causes the black-and-blue appearance so typical of bruises.

A *hematoma* is very similar to a bruise. It is a collection of blood that gets trapped in an organ, tissue, or tissue space because of a break in the blood vessel wall. This can occur due to injury, surgery, or a disorder of the blood vessels. The blood clots there causing swelling and often pain. While bruises usually resolve themselves, hematomas usually have to be drained to bring relief.

Spontaneous bruising (for no apparent reason) can be a symptom of a more serious problem, such as capillary fragility (which can result from scurvy, ascorbic acid deficiency, alcoholism, heavy smoking, or overuse of antacids), diabetes, aging, certain skin allergies, leukemia, and obesity.

Since the major problem of bruising is capillary fragility with possible inflammation, the main purpose of treatment should be to strengthen the blood vessel walls, end the inflammatory process as quickly as possible, clean out the waste products, and bring nutrients to the area. Enzymes and other nutrients (such as bioflavonoids) should be chosen to help achieve these goals. In the controlled, double-blind research study at Portland State University that I conducted with the director of sports medicine there, we found that bruising of injured football players healed twice as fast with enzyme therapy as without. Research in Germany agrees with my research and shows bruised athletes (karate and ice hockey participants) healed in half the average healing time with proteolytic enzymes.

Symptoms

Bruising can cause discoloration, pain, and possible heat and swelling of the skin. Tenderness can be local or widespread.

Symptoms of a hematoma include the characteristic black and blue marks on the skin and swelling, although hematomas are not limited to tissue immediately under the skin. Fractures are almost always accompanied by hematomas in the surrounding tissue. Though a rather benign condition when it occurs superficially, a hematoma can be life-threatening when it occurs in the heart, head, or around the spinal cord.

Enzyme Therapy

Systemic enzyme therapy is used to reduce inflammation, improve circulation, help speed tissue repair, fight free-radical formation, bring nutrients to the damaged area, remove waste products, stimulate the immune system, and improve wellness. The proteolytic enzymes work to strengthen the capillary walls, stimulate fibrinolysis (the break-up of fibrin and cholesterol), and increase blood flow.

Digestive enzyme therapy is used to improve the digestion of food, reduce stress on the gastrointestinal mucosa, help maintain normal pH levels, detoxify the body, and promote the growth of healthy intestinal flora. This allows the nutrients to reach and help repair the bruised area. Digestive enzymes also serve as replacements for the body's pancreatic enzymes, leaving the pancreatic enzymes free to perform other functions in the body, such as decreasing inflammation.

Enzyme Absorption System Enhancers (EASE) maximize enzyme activity. They can also improve the absorption and bioavailability of various nutrients and decrease the drain on the body's own digestive enzymes, thus prolonging their lives.

In addition to the standard multivitamin, multimineral, and multiglandular complexes recommended in the introduction to Part Two, the following supplements are recommended for the treatment of superficial bruises and hematomas. Hematomas in the head, heart, or spine are very serious conditions that should receive prompt medical care. The following recommended dosages are for adults. Children between the ages of two and five should take one-quarter the recommended dosage, those between the ages of six and twelve should take one-half the recommended dosage, and those between the ages of thirteen and seventeen should take three-quarters the recommended adult dosage.

ESSENTIAL ENZYMES

Enzyme	Suggested Dosage	Actions
Formula One (particularly Wobenzym N from Naturally Vitamin Supplements)—See page 38 of Part One	Take 10 tablets three times per day between meals. As symptoms subside (which should occur in approximately one week) decrease dosage to 3–5	Decreases inflammation, pain, and swelling; increases rate of healing; stimulates the immune system.
for a list of products.	tablets three times per day between meals until bruise disappears.	
Inflazyme Forte from American Biologics	Take as indicated on the label between meals.	Decreases inflammation, pain, and swelling; increases rate of healing; stimulates the immune system.
Bromelain	Take three to four 230–250 mg capsules per day until symptoms subside. If bruising is severe, take 6–8 capsules per day.	Decreases inflammation, pain, and swelling; increases rate of healing; stimulates the immune system.
Enzyme complexes rich in proteases, plus amylases and lipases	Take as indicated on the label between meals.	Decreases inflammation, pain, and swelling; increases rate of healing; stimulates the immune system.

SUPPORTIVE ENZYMES

Enzyme	Suggested Dosage	Actions
A digestive enzyme product containing protease, amylase, and lipase enzymes. For a list of digestive enzyme products, see Appendix.	Take as indicated on the label with meals.	Breaks down proteins, carbohydrates, and fats, improving digestion and helping to flush toxins from the body.
Formula Three—See page 38 of Part One for a list of products.	Take 50 mcg three times per day between meals.	Fights free radicals.

ENZYME ABSORPTION SYSTEM ENHANCERS (EASE)

Combination	Suggested Dosage	Actions
Inflamzyme from PhytoPharmica or	Take as indicated on the label.	Promotes tissue repair.
Connect-All from Nature's Plus or	Take as indicated on the label.	Supports connective tissue.
CTR Support from PhysioLogics	Take as indicated on the label.	Nourishes the body during tissue recovery and healing from injury; helps maintain healthy connective tissue; diminishes damage caused by swelling and inflammation.

ENZYME HELPERS

Nutrient	Suggested Dosage	Actions
Adrenal glandular	Take 100–400 mg per day.	Fights inflammation and adrenal exhaustion.
Amino-acid complex	Take as indicated on the label on an empty stomach.	Amino acids are the building blocks of protein and enzymes and are required for tissue repair.
Bioflavonoids	Take 150–500 mg three times per day.	Strengthen capillaries; potent antioxidants.

Coenzyme Q$_{10}$	Take 30–60 mg one to two times per day with meals; or follow the advice of your health-care professional.	Potent antioxidant.
Essential fatty acid complex	Take 75–100 mg two to three times per day with meals; or follow the advice of your health-care professional.	Helps strengthen cell membranes.
Green Barley Extract from Green Foods Corporation or Kyo-Green from Wakunaga	Take as indicated on the label.	Potent free-radical fighter; rich in vitamins, minerals, and enzyme activity.
Iron	Take as indicated on the label.	Needed for the formation of hemoglobin; essential component of many enzymes required for proper tissue repair; improves cell function.
Mucopolysaccharides found in bovine cartilage extract and shark cartilage extract	Take as indicated on the label.	Enhances healing through the body's normal inflammatory response.
Vitamin B complex	Take 50 mg three to four times per day; or take as indicated on the label.	B vitamins act as coenzymes and help build new cells to deliver nutrients to damaged tissue.
Vitamin C	Take 5,000–10,000 mg per day.	Helps to produce collagen and normal connective tissue; essential for wound healing; promotes immune system; powerful antioxidant.
Vitamin E	Take 1,200 IU per day from all sources (including the multivitamin complex in the introduction to Part Two).	Potent antioxidant; supplies oxygen to cells, thus speeding recovery from injury; necessary to maintain cell membrane integrity.
Whey to Go from Solgar (a combination of free-form amino acids and protein)	Take as indicated on the label.	Maintains muscles; improves muscle metabolism.
Zinc	Take 50 mg per day. Do not take more than 100 mg per day from all sources.	Needed by over 300 enzymes; potent antioxidant.

Comments

❏ Enzymes combined with herbs and other nutrients are used to treat bruises. One product by NESS includes kelp and Irish moss with protease. Another contains kelp, Irish moss, protease, amylase, and lipase with calcium lactate. Tyler produces a protease concentrate that contains protease, amylase, lipase, and calcium citrate.

❏ A number of foods, including raspberries, cranberries, red wine, grapes, black currants, and red cabbage contain phytochemicals known as anthocyanidins. These beneficial nutrients maintain flow in the small blood vessels and fight inflammation and free radicals. Try juicing them and drink one 8-ounce glass twice per day (see Part Three for instructions on juicing).

❏ Eat plenty of enzyme-rich fresh fruits and vegetables high in antioxidant activity, such as kelp, pineapple, soybeans, spinach, whole wheat, sprouted seeds, figs, sunflower seeds, rosehips, and alfalfa.

❏ Eat no fast foods, preservatives, or food additives. For further information on dietary recommendations, see Part Three.

❏ Decrease or stop smoking and drinking alcoholic beverages and other toxins.

❏ Ice should be used for the first one to three days to stimulate blood flow to, and remove waste from, the area. Fill two or three small paper cups with water and freeze. When frozen, take one out, tear off the top half inch of paper revealing the ice. Rub the affected area in a constant circulatory motion with the ice. Apply the ice for five minutes, then keep the ice off the area for five minutes. Repeat this every one to two hours. This will reduce swelling. Do not apply ice directly to skin to minimize risk of frostbite. If skin becomes irritated, stop therapy.

❏ After three days, alternate ice with heat packs every 15 minutes for about an hour. "Blue ice" or a frozen vegetable packet works better for this application. Wrap it with a dish towel and apply to the painful area. For heat, use a hot water bottle or heating pad. If a hot water bottle is used, wrap in a towel and apply. If a heating pad is used, put a warm, moist towel on the patient, then place a sheet of wax paper over it, then the heating pad, and finally, a large bath towel. Make sure the wax paper is wide and long enough to cover the entire heating pad with approximately one inch more extending on every side. The heating pad should never touch the moist towel or the patient's skin. Great care should be exercised to ensure that this does not happen.

❏ *See also* CARDIOVASCULAR DISORDERS.

Burns

A burn is an injury to the skin and other tissues caused by chemicals, electricity, heat, or radiation. Burns are classified according to their severity as first-degree, second-degree, or third-degree. First-degree burns are the most minor burns. Only the epidermis (the outermost layer of skin) is affected. Second-degree burns are a little worse than first-degree burns and may sometimes affect deeper layers of skin. Third-degree burns are the most severe and cause the most damage. They affect all layers of skin and sometimes even tissue under the skin, such as muscles.

Internal organs can also be burned, even when the skin is

not. For instance, by drinking a very hot liquid or ingesting a caustic substance, one can burn the esophagus and/or stomach. By inhaling very hot air when in the midst of a fire, one can burn the lungs. When these burns are mild, they heal rather quickly; however, more severe burns can lead to scarring, which leads to a narrowing of passageways. This can block the passage of food in the esophagus and air in the lungs.

Symptoms

A first-degree burn will have no blisters, but the skin will turn red and swell. When touched, the skin whitens. Only the epidermis is injured. In a few days, the burn usually heals completely with minimal or no scarring. The damage from a second-degree burns goes much deeper. The skin turns very red and there is blistering. The area is very painful and may turn white when touched. Superficial second-degree burns usually heal with little to no scarring; however, deeper second-degree burns that affect the dermis (deeper layers of skin) heal very slowly with much scarring. The surface of third-degree burns may be soft and white or black and leathery. There is sometimes blistering. Third-degree burns destroy all layers of the skin. Since the dermis does not regenerate like the epidermis does, healing is very slow, and significant scarring results. There is generally little or no pain with a third-degree burn because nerve endings in the skin are destroyed. Some burn tissue can become necrotic, and surrounding tissue can develop a serious infection. Skin elasticity can be destroyed.

A severe burn can cause dangerous systemic damage, such as respiratory tract injury, infection, and shock. Anyone suffering from a severe burn should seek immediate medical attention to counter these potentially life-threatening effects.

Enzyme Therapy

The major component of skin is collagen. It composes 75 percent of skin tissue. In fact, about one-third of all the protein in humans is collagen. Collagen fibers interlace, forming a network with elastin (an essential segment of connective tissue) that lends skin its elasticity, smoothness, and strength.

In undamaged young skin, collagen molecules apparently slide over each other. With scarring, however, collagen molecules fuse together (cross-link) making the damaged tissue rigid and leathery.

When dead tissue and surface debris (such as from a burn) are present in and around a wound, tissue repair is delayed. This dead tissue is often attached to the surface of the wound by strands of collagen. These collagen strands must be broken to get rid of the dead tissue and debris. This process, called debridement, is necessary to prevent infection. Collagen can be decomposed with the use of enzymes known as collagenases.

Since collagenase is effective in the debridement of wounds and because it reduces the extent of scar tissue, its most widespread use to date has probably been in the treatment of scars, ulcers, and burns. The collagen in a third-degree burn is often destroyed. Completely destroyed collagen can be decomposed by any proteolytic enzyme (such as pancreatin, trypsin, chymotrypsin, bromelain, and papain), but the less denatured collagen can most easily be decomposed by collagenase.

A series of successful animal experiments with burn therapy using collagenase was reported by Dr. E.L. Howes. This was followed by clinical trials in which small burns were treated with collagenase in a sterile ointment. The first three cases involved burns of the forearm, which responded very well to the therapy. The debris and dead skin came off within twenty-four hours (sooner than expected) with no complications.

When antiseptics and cleaning solutions were used as first aid and in cleaning the burned area, they caused cross-linking of the surface of the burn. This cross-linking gives collagen a leatherlike quality and makes it resistant to enzymatic action.

The individual plant enzymes, such as papain and bromelain, are successfully being used as exfoliants (see Part Three), as are salves made from enzyme combinations (including papain, bromelain, trypsin, and chymotrypsin, with rutin). Using these enzyme salves can reduce scar tissue formation.

Systemic enzyme therapy is used to encourage normal collagen growth, break up free-radical formation, strengthen the body as a whole, and build general resistance. Enzymes stimulate the immune system, reduce inflammation, improve circulation, help speed tissue repair, bring nutrients to the damaged area, remove waste products, and enhance wellness.

Digestive enzyme therapy is used to improve the digestion of food, reduce stress on the gastrointestinal mucosa, help maintain normal pH levels, detoxify the body, and promote the growth of healthy intestinal flora, all factors that can affect enzyme activity.

Enzyme Absorption System Enhancers (EASE) maximize enzyme activity. They can also improve the absorption and bioavailability of various nutrients and decrease the drain on the body's own digestive enzymes, thus prolonging their lives.

Topical enzymes are used to dissolve dead and damaged skin cells and exfoliate the skin while leaving healthy skin cells intact. These topical enzymes are used as salves, creams, and exfoliates. Enzymes fight free radicals and heal burns from the outside in and the inside out.

In addition to the standard multivitamin, multimineral, and multiglandular complexes recommended in the introduction to Part Two, the following supplements are recommended for the treatment of burns. The following recommended dosages are for adults. Children between the ages of two and five should take one-quarter the recommended dosage, those between the ages of six and twelve should take one-half the recommended dosage, and those between the ages of thirteen and seventeen should take three-quarters the recommended adult dosage.

ESSENTIAL ENZYMES

Enzyme	Suggested Dosage	Actions
Papain salve	Apply at least three times per day.	Fights inflammation; stimulates healing.
Superoxide dismutase	Take 150 mcg per day.	Fights inflammation; potent antioxidant.

AND SELECT ONE OF THE FOLLOWING:

Enzyme	Suggested Dosage	Actions
Formula One—See page 38 of Part One for a list of products.	Take 10 tablets three times per day between meals.	Stimulates the immune system; fights inflammation, pain, and swelling; helps detoxify the body; breaks up circulating immune complexes.
Bromelain	Take 750 mg two times per day between meals.	Fights inflammation, pain, and swelling; speeds recovery after injury or surgery; stimulates the immune system; strengthens capillary walls.
Trypsin with chymotrypsin	Take 250 mg of trypsin and 10 mg of chymotrypsin three times per day between meals.	Fights inflammation, pain, and swelling; increases rate of healing; stimulates the immune system.
Pancreatin	Take 500 mg six times per day on an empty stomach.	Fights inflammation, pain, and swelling; increases rate of healing; stimulates the immune system; strengthens capillary walls.
Papain or a microbial protease	Take 300 mg two times per day between meals.	Fights inflammation; speeds recovery from injuries and surgery.

SUPPORTIVE ENZYMES

Enzyme	Suggested Dosage	Actions
Catalase and glutathione peroxidase	Take as indicated on the labels.	Potent antioxidant enzymes.
A digestive enzyme product containing protease, amylase, and lipase enzymes. See the list of digestive enzyme products in Appendix.	Take as indicated on the label with meals.	Improves the breakdown and absorption of food nutrients; fights toxic build-up.

ENZYME ABSORPTION SYSTEM ENHANCERS (EASE)

Combination	Suggested Dosage	Actions
CTR Support from PhysioLogics	Take 2–3 capsules per day, preferably on an empty stomach; or follow the advice of your health-care professional.	Nourishes the body during tissue recovery and healing; contains a combination of herbs that support and maintain healthy connective tissue.
or		

Bio-Musculoskeletal Pak from Biotics	Take as indicated on the label.	Decreases inflammation; fights free-radical formation; stimulates immune function.
or		
Bioprotect from Biotics	Take 1–3 capsules each day.	Potent antioxidant combination; improves immune function and tissue repair.
or		
Detox Enzyme Formula from Prevail	Take 1 capsule between meals one or two times per day. Take with a 6- to 8-ounce glass of water or juice.	Helps rid the body of harmful toxins.
or		
Inflamzyme from PhytoPharmica	Take as indicated on the label.	Improves tissue repair.
or		
Connect-All from Nature's Plus	Take as indicated on the label.	Supports connective tissue.

ENZYME HELPERS

Nutrient	Suggested Dosage	Actions
Acidophilus and other probiotics	Take as indicated on the label.	Improve gastrointestinal function, so improve absorption of beneficial nutrients; needed for production of biotin, niacin, folic acid, and pyridoxine, which are required by many of the body's enzymes; help produce antibacterial substances; stimulate enzyme activity.
Adrenal glandular	Take 100–400 mg per day.	Fights inflammation and adrenal exhaustion.
Aged garlic extract	Take 2 capsules three times per day.	Fights inflammation; stimulates the immune system; potent antioxidant.
Amino-acid complex	Take as indicated on the label on an empty stomach.	Required for enzyme production; helps the body maintain proper fluid and acid/base balance, both of which are essential to health; helps build new tissue.
Bioflavonoids, such as rutin or quercetin	Take 500 mg three times per day between meals.	Strengthen capillaries; potent antioxidants.
Branched-chain amino acids (isoleucine, leucine, and valine)	Take as indicated on the label between meals.	Often found together in a supplement, branched-chain amino acids aid in healing of muscle, bone, and skin and act as fuel for muscles.
Coenzyme Q$_{10}$	Take 30–60 mg one to two times per day with meals; or follow the advice of your health-care professional.	Potent antioxidant.
Essential fatty acid complex	Take 100–200 mg two to three times per day with meals; or follow the advice of your health-care professional.	Help strengthen cell membranes; support immune function.
Green Barley Extract from Green Foods Corporation or Kyo-Green from Wakunaga	Take as indicated on the label.	Potent free-radical fighter; rich in vitamins, minerals, and enzyme activity.

Thymus glandular	Take 100 mg per day.	Assists the thymus gland in immune resistance; improves immune response; fights infections.
Vitamin A	Take 10,000–45,000 IU for four weeks. Any amount over 50,000 IU should be be taken only under the direction of a physician. Pregnant women should not exceed 10,000 IU per day.	Potent antioxidant; enhances immunity.
or		
Carotenoids	Take 10,000–30,000 IU per day; or take as indicated on the label.	Needed for strong immune system and healthy mucous membranes; potent antioxidant.
Vitamin B complex	Take 150 mg three times per day; or take as indicated on the label.	B vitamins act as coenzymes and support neuromuscu-loskeletal function.
Vitamin C	Take 10,000–15,000 mg per day.	Fights infections, diseases, allergies; improves wound healing; stimulates the adrenal glands; promotes immune system; powerful antioxidant.
Vitamin E	Take 1,200 IU per day from all sources (including the multivitamin complex in the introduction to Part Two).	Potent antioxidant; supplies oxygen to cells; improves circulation and level of energy.
Zinc	Take 50 mg per day. Do not take more than 100 mg per day from all sources.	Needed by over 300 enzymes; necessary for white blood cell immune function; improves wound healing.

Comments

❏ Enzyme gels and salves can be used to break up free-radical formation and cross-linking of collagen.

• As scar tissue begins to form (on approximately the third day after burn injury), apply papain enzyme cream. Do not apply to fresh burns, blisters, or open wounds. Papain enzyme salve helps break up free-radical formation. It dissolves dead, damaged skin cells, while leaving healthy cells intact. Once the dead cells are dissolved, the body manufactures new collagen and elastin, thus rebuilding firmer, more youthful skin.

• Enzyme-rich aloe vera gel or products like BurnGel from Aerobic Life Industries assist in fighting burns.

• Enzyme combination salves (including papain, bromelain, trypsin, and chymotrypsin, with the bioflavonoid rutin) bathe the burn area as the fluids weeping from the burn subside and fibrin formation begins. Never apply salve to an open wound.

❏ Fresh, raw, enzyme-rich fruits and vegetables (particularly green vegetables) should comprise over 60 percent of your diet. If your body cannot tolerate raw fruits and vegetables, increase your intake of digestive enzymes, or sauté or steam your produce.

❏ Eat foods and plant nutrients high in enzymes and potent antioxidants, such as pineapples, grapes, broccoli, and citrus fruits. Follow dietary recommendations in Part Three.

❏ One's diet should be higher in protein while the burn is healing. Approximately 4,000 to 7,000 calories per day should come from protein.

❏ Detoxification is particularly important after a burn injury. Gargle twice daily with the Enzymatic Gargle as described in Part Three.

❏ The pain from a minor burn can be relieved by applying aloe vera gel. With third-degree burns, go to the hospital's emergency room or see a doctor immediately.

❏ A wet, cool compress can help alleviate the pain from a minor burn. Apply for ten to fifteen minutes, then leave off for ten to fifteen minutes. Repeat until pain subsides.

❏ Break open vitamin A and E capsules and spread the contents over the burned area two to three times per day or as frequently as possible until the tissue heals. Emulsified A and E seem to have higher potencies.

❏ See the list of Resource Groups in the Appendix for an organization that can provide you with more information about burns.

Bursitis and Synovitis

Bursitis is a painful inflammation of the flat sacs called bursae, which protect certain tissues (such as bones and ligaments) from friction and promote muscular movement. Normally, bursae contain little fluid; however, when they become inflamed, fluid accumulates, which causes the bursal sac around the joint to become swollen, reducing the joint's range of motion. Bursitis is most common in the shoulder, but it can occur in any bursa, including the hips, elbows, knees, toes, pelvis, and heels. In synovitis, the synovial membranes (similar to bursae) that line the joints to reduce friction and facilitate movement become inflamed

Bursitis or synovitis can be caused by injury, overuse, extraordinary strenuous exercise, calcium deposits that cause degeneration of the tendon, or acute or chronic infection. Bursitis and synovitis can also be caused by the degenerative process in arthritis or gout, or it can flare up for no known cause.

Symptoms

Bursitis and synovitis usually causes severe pain in the affected joints that increases with movement. There may also be swelling, heat, redness, and loss of function.

Acute bursitis occurs suddenly. The affected area becomes inflamed and painful. If the affected bursa is near the skin surface, the area may appear red and swollen.

Chronic bursitis may follow repeated injuries or several bouts of acute bursitis. Symptoms are similar to those of acute bursitis; however, chronic pain and swelling may restrict movement of the affected area, causing atrophy of the muscles in that area. Flare-ups of chronic bursitis may last from a few days to several weeks.

Enzyme Therapy

Systemic enzyme therapy is used to reduce swelling, pain, and inflammation in the joints. Enzymes can stimulate the immune system, improve circulation, help speed tissue repair, bring nutrients to the damaged area, remove waste products, and improve health—all without serious long-term side effects. Digestive enzyme therapy is used to improve the digestion of food, reduce stress on the gastrointestinal mucosa, help maintain normal pH levels, detoxify the body, and promote the growth of healthy intestinal flora. They also serve as replacements for the body's pancreatic enzymes. This frees the pancreatic enzymes to perform other functions in the body, such as boosting immune function, decreasing inflammation, and improving circulation. Topical applications of enzymes are also helpful in penetrating the skin and attacking the inflamed bursae and synovial membranes.

Enzyme Absorption System Enhancers (EASE) maximize enzyme activity. They can also improve the absorption and bioavailability of various nutrients and decrease the drain on the body's own digestive enzymes, thus prolonging their lives.

In addition to the standard multivitamin, multimineral, and multiglandular complexes recommended in the introduction to Part Two, the following supplements are recommended for the treatment of bursitis and synovitis. The following recommended dosages are for adults. Children between the ages of two and five should take one-quarter the recommended dosage, those between the ages of six and twelve should take one-half the recommended dosage, and those between the ages of thirteen and seventeen should take three-quarters the recommended adult dosage.

ESSENTIAL ENZYMES

Enzyme	Suggested Dosage	Actions
A proteolytic salve, such as WobeMugos E Ointment	Apply to the affected area two or three times per day; or use as instructed on the label.	Reduces pain and inflammation.

AND SELECT ONE OF THE FOLLOWING:

Formula One (particularly Wobenzym N from Naturally Vitamin Supplements) —See page 38 of Part One for a list of products.	Take 10 tablets three times per day between meals. As symptoms subside (which should occur in approximately one week) decrease dosage to 3–5 tablets three times per day between meals.	Decreases inflammation, pain, and swelling; increases rate of healing; stimulates the immune system.
Phlogenzym from Mucos	Take 4 tablets three times per day between meals.	Decreases inflammation, pain, and swelling; increases rate of healing; stimulates the immune system; strengthens capillary walls.
Inflazyme Forte from American Biologics	Take as indicated on the label between meals.	Decreases inflammation, pain, and swelling; increases rate of healing; stimulates the immune system; strengthens capillary walls.
Papain	Take 600 mg three times per day between meals.	Decreases inflammation, pain, and swelling; increases rate of healing; stimulates the immune system.
Microbial protease	Take as indicated on the label.	Decreases inflammation, pain, and swelling; increases rate of healing; stimulates the immune system.
Bromelain	Take three to four 250 mg capsules per day until symptoms subside. If pain is severe, take 6–8 capsules per day.	Decreases inflammation, pain, and swelling; increases rate of healing; stimulates the immune system.

SUPPORTIVE ENZYMES

Enzyme	Suggested Dosage	Actions
Formula Three—See page 38 of Part One for a list of products.	Take 50 mcg three times per day between meals.	Fights free radicals.
A digestive enzyme formula containing protease, amylase, and lipase. See the list of digestive enzyme products in Appendix.	Take as indicated on the label with meals.	Improves the breakdown and absorption of food nutrients; fights toxic build-up.

ENZYME ABSORPTION SYSTEM ENHANCERS (EASE)

Combination	Suggested Dosage	Actions
Inflamzyme from PhytoPharmica or	Take as indicated on the label.	Promotes tissue repair.
Connect-All from Nature's Plus or	Take as indicated on the label	Supports connective tissue.
CTR Support from PhysioLogics	Take as indicated on the label.	Nourishes the body during tissue recovery and healing from injury; helps maintain healthy connective tissue; diminishes damage caused by swelling and inflammation.

ENZYME HELPERS

Nutrient	Suggested Dosage	Actions
Bioflavonoids	Take 150–500 mg three times per day.	Strengthen capillaries; potent antioxidants.
Branched-chain amino acids	Take as indicated on the label between	Often found together in a supplement, branched-chain

(isoleucine, leucine, and valine)	meals.	amino acids aid in healing of muscle, bone, and skin and act as fuel for muscles.
Calcium	Take 400 mg three times per day.	Necessary for proper nerve transmission; a reduced intake and metabolism of calcium can contribute to bursitis development.
Chondroitin sulfate	Take 250 mg two to three times per day with meals; or follow advice of your health-care professional.	Supports connective tissue; promotes cell hydration.
Coenzyme Q$_{10}$	Take 30–60 mg one to two times per day with meals; or follow the advice of your health-care professional.	Potent antioxidant.
Essential fatty acid complex	Take 75–100 mg two to three times per day with meals; or follow the advice of your health-care professional.	Helps strengthen cell membranes.
Glucosamine hydrochloride *or* Glucosamine sulfate	Take 500 mg two to four times per day. Take 500 mg one to three times per day.	Stimulates cartilage repair.
Green Barley Extract or Kyo-Green	Take as indicated on the label.	Potent free-radical fighters; rich in vitamins, minerals, and enzyme activity.
Kyolic Aged Garlic Liquid Extract with Vitamins B$_1$ and B$_{12}$	Take as indicated on the label.	Fights inflammation; improves immune function.
Magnesium	Take 200 mg per day.	Increases calcium absorption.
Mucopolysaccharides found in bovine cartilage extract and shark cartilage extract	Take as indicated on the label.	Supports healthy, flexible joints; controls inflammation; helps keep joint membranes fluid; improves the body's normal response to inflammation.
Vitamin B$_{12}$	Take 1,000 mcg per day for 2 weeks.	Needed for activation of the coenzyme folic acid (and vice versa); essential for healthy cells.
Vitamin C	Take 1,000–8,000 mg per day.	Helps to produce collagen and normal connective tissue; promotes immune system; powerful antioxidant.
Zinc	Take 50 mg per day. Do not take more than 100 mg per day from all sources.	Needed by over 300 enzymes; potent antioxidant.

Comments

❑ Herbs, including white willow bark and ginkgo biloba can help decrease inflammation. Hawthorn and gotu kola can help support collagen. Other beneficial nutrients would include rose hips, acerola cherries, and curcumin.

❑ X-rays often reveal a calcium build-up in the bursal sac in bursitis. This can actually be due to too little calcium in your body, as calcium is being pulled from other areas of the body. Increasing your intake of calcium will improve blood calcium levels and decrease the necessity to draw from other areas. Therefore, foods rich in calcium are critical for improvement. (See Part Three for dietary recommendations.)

❑ Eat enzyme-rich fresh vegetables and fruits, including pineapple, citrus fruits, papaya, apple, cherries, apricots, spinach, and sweet potatoes.

❑ Avoid toxins, pesticides, herbicides, food additives, preservatives, and food colors.

❑ Drink at least one 8-ounce glass of freshly juiced fruits and vegetables, twice per day. (See Part Three for additional information on juicing.)

❑ Aspirin, ibuprofen, acetaminophen, and other NSAIDs may help decrease pain, but too much of anti-inflammatory drugs can have serious, long-term side effects. Proteolytic enzymes can reduce the pain and swelling safely and effectively.

❑ Ice should be used for the first one to three days. Fill two or three small paper cups with water and freeze. When frozen, take one out, tear off the top half inch of paper revealing the ice. Place a paper towel or dish towel over the ice and apply to the painful area. Rub the affected area in a constant circulatory motion with the ice. Do not apply the ice directly to the skin. Apply the ice for five minutes, then keep the ice off the area for five minutes. Repeat this every one to two hours. The purpose is to stimulate blood flow to the area, remove waste from the area, and reduce inflammation.

❑ After three days, alternate ice with heat packs every fifteen minutes for thirty to sixty minutes. "Blue ice" or a frozen vegetable packet works better for this application. Wrap it with a dish towel and apply to the painful area. For heat, use a hot water bottle or heating pad. If a hot water bottle is used, wrap in a towel and apply. If a heating pad is used, put a warm, moist towel on the patient, then place a sheet of wax paper over it, then the heating pad, and finally, a large bath towel. Make sure the wax paper is wide and long enough to cover the entire heating pad with approximately one inch more extending out on every side. The heating pad should never touch the moist towel or the patient's skin. Great care should be exercised to ensure that this does not happen.

❑ Exercising increases antioxidant and enzymatic activity in the body. Try low-impact exercises such as walking, bicycling, and swimming; but avoid stress and strain of the affected area. For example, if you have bursitis in the shoulder, swimming could aggravate it. Try walking or bicycling instead.

❑ A positive mental attitude and meditation help give inner peace and relax the blood vessels and musculature.

❏ If your bursitis or synovitis continues for two weeks or longer, see a musculoskeletal expert, such as a chiropractor, medical doctor, naturopath, or osteopath who will develop a treatment program. Your treatment program might include chiropractic adjustments; manipulation; massage; and physiotherapy, including electrical stimulation, ice and heat, exercises, and postural training (which includes proper lifting, bending, standing, walking, driving, sitting, and sleeping techniques). Iontophoresis is used quite frequently. In this process, an enzyme salve is applied to the affected joint. Then ultrasound is used over the area to increase absorption of the enzymes.

❏ See also TENDONITIS.

Cancer

Cancer is the uncontrolled growth and spread of abnormal cells. Some cells in the body multiply regularly. Children and infants produce new cells for repair and to complete their growth and development. However, if cells multiply uncontrolled, a tumor develops. As the tumor grows, it can interfere with organ function, leading to death. Tumors can be either benign, meaning that they do not metastasize (spread) and are unlikely to return after removal, or malignant (cancerous).

One out of every five deaths (about 520,000) in the United States is caused by cancer. In 1992, over one million Americans were diagnosed with this dreaded disease. In fact, one in three Americans now living (about 75 million) will eventually have cancer. Although cancer can occur at any age, it strikes most frequently with advancing age. The good news is that many forms of cancer can be prevented, and others, if detected and treated early, can be cured.

There are five basic types of cancer.

- *Carcinomas* are the most common type of cancer. They form in epithelial cells, which are in tissues that cover surfaces or line internal organs. Carcinomas will usually metastasize through lymphatic channels. Carcinomas include cancers of the breast, colon, lung and bronchi, pancreas, skin, kidneys, ovaries, prostate, uterus, and squamous cell carcinomas of the head and neck.
- *Leukemias* are cancers of the tissues of the bone marrow, spleen, and lymph—those tissues that form blood.
- *Lymphomas* are cancers of the lymphatic system.
- *Myelomas* are rare cancers that form in the plasma of the bone marrow.
- *Sarcomas* are the rarest type of cancer and the most deadly. They develop in connective tissue. Because organs are made of connective tissue, sarcomas can occur in almost any part of the body. Sarcomas include cancer of the lymph and soft tissue.

The body's immune system does not normally react to its own cells. However, when a cell becomes cancerous, antigens form on the surface with which the body is unfamiliar. The immune system may then identify these cells as foreign and destroy them; however, even a healthy immune system cannot always destroy all cancer cells. Sometimes the "tumor antigen" is covered by a fibrin coating on the cell, making it difficult for the body to identify it.

There are two processes necessary for the development of cancer. The first is called *initiation*. Initiation is the process in which genetic material within a cell changes, making it possible for the cell to become cancerous. This change is brought about by carcinogens, such as chemicals or radiation, or by viruses. Initiation is a common occurrence that happens even in healthy bodies. Cancer cell production in our bodies is a normal phenomenon. With billions of cells produced every minute, every healthy person has hundreds or thousands of cancer cells at any given moment. This does not mean that everyone is suffering from cancer. Under normal conditions, our body's defense systems recognize, engulf, and eliminate the cells. Cancer does not grow in the body until the next step, *promotion*, occurs.

Promoters are agents (such as cell mutations, immune disorders, hormones, or carcinogens) that allow the cancer to grow. Promoters can create further cellular damage, allowing the cells to spread; they can inhibit the removal of the cancerous cells by damaging the immune system's function; or they can alter body tissues to make them more favorable for cancer growth. Often initiation cannot happen without the help of promotion, and promotion is ineffective on cells that have not been initiated.

Tumors crowd out healthy tissue and ultimately interfere with the essential functioning of involved organs. Further, cells from the primary (or original) cancer site tend to enter the bloodstream and the lymph vessels. In this way, malignant cells are carried to distant parts of the body where secondary malignant neoplasms or tumors develop. This spreading of cancer is called metastasis.

Symptoms

Symptoms of cancer depend upon the location and type of cancer. See the inset "Cancer Warning Signs" on page 157 for different types of cancer and their warning signs.

Enzyme Therapy

In a healthy body, certain proteolytic enzymes strip away the fibrin that protects cancer cells from detection. This paves the way for cancer cell destruction by the immune system. The more cancer cells the body produces, the more enzymes the body needs. The absence of certain enzymes allows the cancer cells to grow.

Cancer Warning Signs

To be forewarned is to be forearmed! Knowing the symptoms of cancer can lead to early treatment of, and victory over, the disease. See a doctor at the first warning sign of cancer and help reduce the possible consequences of this deadly disease.

SITE	SYMPTOMS
Brain	Headaches, seizures, nausea and vomiting, drowsiness, and vision problems (including loss of vision and double vision). The patient may be unsteady and there may be hearing loss or speech difficulties. Sometimes the person's personality changes markedly.
Breast	Changes in the breast, including a lump, swelling, thickening, tenderness, dimpling, pain, or a discharge from the nipple.
Colon/rectum	Bowel habit changes, bleeding from the rectum, or blood in the stool.
Lung	Persistent cough, chest pain, blood-streaked sputum, recurring bronchitis or pneumonia.
Pancreas	Usually occurs without any symptoms until in advanced stages, and so is considered a "silent" disease.
Prostate	Difficulty starting or stopping the flow of urine; weak or interrupted urine flow; inability to urinate; frequent need to urinate (especially at night); painful or burning urination; blood in the urine; pain in the lower back, pelvis, or upper thigh.
Skin	Skin changes, such as a change in the color or size of a mole; bleeding; scaliness; oozing; the spread of pigmentation beyond the borders of a mole; a change in sensation; itchiness; pain or tenderness.

Information obtained from the American Cancer Society.

In a healthy individual, there is a balance between cancer cell production and cancer cell destruction. However, as we age, injuries and illnesses place added stress on our bodies, and we cannot produce sufficient proteolytic enzymes to help break down and eliminate cancer cells. The number of cancer cells increases. Therefore, enzyme supplementation is essential. Enzyme therapy can help fight cancer. Proteolytic enzyme mixtures (including pancreatin, trypsin, chymotrypsin, bromelain, and papain) are able to attack cancer cells by breaking down the cancer cell antigens and removing the "glue" by which cancer cells attach themselves to vessel walls and tissues. The more enzymes present in the body, the greater the chances that the immune defense systems will be able to identify individual cancer cells. This means that people exposed to an increased risk of cancer could possibly reduce that risk by taking certain enzyme mixtures.

Enzyme therapy pioneers, Max Wolf, M.D., and Karl Ransberger, Ph.D., developed a multiple enzyme program using enzymes, vitamins, minerals, and the bioflavonoid rutin. Wolf and Ransberger found that enzyme therapy seems to inhibit metastasis and prolong patient life; that enzyme therapy must be continued indefinitely for the best results in inhibiting metastasis; and that higher doses of enzymes, particularly of enzyme combinations from plant and animal sources, give the best results.

Stephen J. Taussig has researched bromelain for several decades and has found that it can produce certain anticancer effects. In his research, he has found that he can retard the growth and metastasis of several malignant cell lines in vitro and in vivo. Extracted from the pineapple, bromelain is actually a mixture of a number of proteolytic enzymes.

Further, proteolytic enzymes could be taken before and after a cancer operation. This (along with other immune-stimulating products) could compensate for the weakened immune system that always follows surgical intervention.

Enzyme therapy during the treatment of cancer is gaining popularity in hospitals. Numerous oncologists in Germany, Italy, France, and the United States know the value of enzyme therapy. In addition to the use of enzyme tablets administered orally, enzymes and enzyme mixtures are applied as topical ointments, used as enemas, or injected directly into accessible malignant tumors. Enzyme therapy is capable of initiating complete tumor disintegration. (See the inset "Cancers That Have Been Helped by Systemic Enzyme Therapy" on page 158.)

Enzyme therapy has been used concurrently with medical

therapy to increase the effectiveness of radiation and chemotherapy when their use is necessary. Further, it is felt that an increased enzyme intake reduces the potential of radiation hangover and other unpleasant radiation or chemotherapeutic side effects (*see* RADIATION SICKNESS in Part Two). The administration of an enzyme mixture is often accompanied by an improvement in the patient's morale because he or she tends to regain appetite, put on lost weight, cease to be depressed, and feel considerably more energetic—both physically and mentally—with enzyme therapy.

In the treatment of cancer, systemic enzyme therapy is used as a complementary therapy with medical treatment. It is used to stimulate and stabilize the immune system, thus increasing the opportunities to destroy the residual tumor cells and decrease the rate of tumor growth. It helps differentiate the healthy tissue from the cancerous tissue and better identify the tumor for surgical removal. Enzyme therapy reduces inflammation, improves wound healing, helps speed tissue repair by bringing nutrients to the damaged area and removing waste products, improves the long-term prognosis, and improves the patient's quality of life. Enzyme therapy can also decrease the adverse side effects of the primary therapy (radiation, chemotherapy, and/or surgery), reduce the cost of therapy by minimizing its use (if enzyme therapy is initiated early enough), reduce the risk of complications, such as blood clots, and decrease immune suppression caused by surgery and anesthesia.

Digestive enzymes are essential for the gastrointestinal tract (and the body as a whole) in helping to prevent the onset of cancer and to improve the cancer patient's health because they aid the breakdown, digestion, and absorption of nutrients; and decrease the enzyme production load on the pancreas, which allows the pancreas to produce proteolytic enzymes that help the body fight cancer by breaking up the protective fibrin coat surrounding the cancer cells, thus allowing the immune cells to attack the cancer. This also helps to reduce stress on the pancreas increasing its productive life; help prevent toxin build-up, especially in the small and large intestine; stimulate friendly microflora production in the intestine; and facilitate elimination of potential cancer-causing waste from the food we eat and from cells of the body. Digestive enzymes also aid normal digestion by helping to replace enzymes that are lost when food is processed or cooked.

Enzyme Absorption System Enhancers (EASE) maximize enzyme activity. They can also improve the absorption and bioavailability of various nutrients and decrease the drain on the body's own digestive enzymes, thus prolonging their lives.

Enzyme therapy has a definite place in the prevention, treatment, and follow-up care of cancer patients. An individual with cancer should work with his or her physician to plan a treatment plan including enzymes. The treatment plan recommended here should serve as a guideline.

In addition to the standard multivitamin, multimineral, and multiglandular complexes recommended in the introduction to Part Two, the following supplements are recommended for the treatment of cancer, including before and after surgery; before, during, and after chemotherapy or radiation therapy; and as long-term and palliative therapy. The following recommended dosages are for adults. Children between the ages of two and five should take one-quarter the recommended dosage, those between the ages of six and twelve should take one-half the recommended dosage, and those between the ages of thirteen and seventeen should take three-quarters the recommended adult dosage.

The following enzyme therapy plan is effective in the treatment and prevention of most types of cancers. *See also* BRAIN TUMORS; BREAST CANCER; COLORECTAL CANCER; LUNG CANCER; PANCREATIC CANCER; PROSTATE CANCER; and SKIN CANCER for specific treatment plans.

Cancers That Have Been Helped by Systemic Enzyme Therapy

Because of the way enzymes work, it stands to reason that they could have a positive effect on every type of cancer; however, research has shown systemic enzyme therapy to be particularly helpful with certain cancers as a complementary therapy used with other treatments.

Research has supported the efficacy of enzyme therapy in the treatment of the following types of cancer:

- Brain tumors
- Breast cancer
- Colon cancer
- Kidney cancer
- Leukemia
- Lung and bronchopulmonary cancer
- Ovarian cancer
- Pancreatic cancer
- Prostate cancer
- Skin cancer
- Stomach cancer
- Uterine cancer

ESSENTIAL ENZYMES

Enzyme	Suggested Dosage	Actions
WobeMugos E from Mucos *or* Formula One—See page 38 of Part One for a list of products.	Take 2–4 tablets orally three times per day between meals, before, during, and after chemotherapy or radiation therapy, and as a long-term therapy. Take 10 tablets three times per day between meals·	Decreases the possibility of suppressed immune function after surgery or anesthesia; breaks down fibrin; decreases inflammation; improves wound healing; decreases risk of blood clot complications following surgery; improves effectiveness of chemotherapy or radiation; decreases radiation sickness and nausea; stimulates immune function; reduces possibility of malnutrition, weakness, and muscle wasting following medical treatment; prevents metastasis; prevents disturbances of lymphatic drainage.

SUPPORTIVE ENZYMES

Enzyme	Suggested Dosage	Actions
Bromelain	Take eight to ten 230–250 mg capsules per day. When tumor growth is under control, the dose can be reduced gradually to a maintenance level of approximately 4 capsules per day.	Fights inflammation, pain, and swelling; stimulates the immune system; improves digestion.
A digestive enzyme product containing protease, amylase, and lipase enzymes. See the list of digestive enzyme products in Appendix.	Take as indicated on the label with meals.	Improves the breakdown and absorption of food nutrients; fights toxic build-up.
Formula Three—See page 38 of Part One for a list of products.	Take as indicated on the label.	Potent antioxidant enzymes. The cellular activity of both catalase and superoxide dismutase is decreased by carcinogens and tumor promoters.

ENZYME ABSORPTION SYSTEM ENHANCERS (EASE)

Combination	Suggested Dosage	Actions
Antioxidant Complex from Tyler *or*	Take as indicated on the label.	Potent free-radical fighter and immune system stimulator.
Detox Enzyme Formula from Prevail *or*	Take 1 capsule between meals one or two times per day. Take with a 6- to 8-ounce glass of water or juice.	Helps rid the body of harmful toxins.
Hepazyme from Enzymatic Therapy *or*	Take as indicated on the label.	Stimulates immune function.
Kyolic Formula 102	Take 2 capsules or	Strengthens the immune sys-

(with enzymes)	tablets per day as desired.	tem; helps alleviate stress; improves circulation; helps fight inflammation.
or Oxy-5000 Forte from American Biologics *or*	Take as indicated on the label.	Helps the body excrete toxins; fights biochemical stressors, infections, and degenerative processes.
Vita-C-1000 from Life Plus	Take as indicated on the label.	Helps fight cancer; boosts immune function.

ENZYME HELPERS

Nutrient	Suggested Dosage	Actions
Acidophilus or other probiotics, such as Kyo-Dophilus from Wakunaga	Take as indicated on the label.	Stimulates enzyme activity; improves gastrointestinal function.
Adrenal glandular	Take 100–400 mg per day.	Improves body's response to stress.
Amino-acid complex, such as AminoLogic from PhysioLogics	Take as indicated on the label.	Amino acids are the building blocks of protein and enzymes and are required for the proper growth and maintenance of all body tissues.
Bioflavonoids, such as rutin and quercetin	Take 500 mg three times per day.	Potent antioxidants.
Coenzyme Q_{10}	Take 30–60 mg one to two times per day with meals.	Important to immune system function; has antioxidant activity.
Kidney glandular	Take as indicated on the label.	Helps fight toxicity; fights hypertension—sometimes a complication of cancer.
Kyolic Aged Garlic Liquid Extract with Vitamins B_1 and B_{12}	Put 10 drops on the tongue or in an 8-ounce glass of juice and drink three times per day.	Stimulates immune system function.
Liver glandular	Take 100 mg one to three times per day.	Improves immune defense and liver function.
Pancreas glandular	Take 325 mg twice per day with large meals; or take as indicated on the label.	Aids digestion; fights viral problems; assists the pancreatic enzymes trypsin and chymotrypsin.
Selenium	Take 50 mcg one or two times per day.	Acts as an enzyme activator; required by glutathione peroxidase.
Thymus glandular *or* T-Cell Formula from Ecological Formula	Take 100 mg per day; or take as indicated on the label. Take 2 tablets 3 times per day.	Assists the thymus gland in immune resistance; improves immune response; fights infections.
Vitamin A	Take up to 50,000 IU for no more than one month. If pregnant, do not take more than 10,000 IU. Doses higher than 50,000 IU should be taken only under the supervision of	Needed for strong immune system and healthy mucous membranes; potent antioxidant; consumption of alcohol may lead to severe headaches when taking high doses of vitamin A as long as 14 days after treat-

159

	a physician.	ment. Alcohol should be strictly avoided. If peeling skin occurs with high doses of vitamin A, see your doctor.
or Carotenoids, including alpha- and beta-carotene.	Take 5,000–25,000 IU per day; or take as indicated on label.	Potent antioxidants; block DNA damage.
Vitamin B complex	Take 100–150 mg three times per day.	B vitamins act as coenzymes.
Vitamin C	Take 10,000–15,000 mg per day; or take 1–2 grams every hour until symptoms improve.	Fights infections and diseases such as cancer; improves healing; stimulates the adrenal glands; promotes immune system; powerful antioxidant; acts as a natural chelator, which helps detoxify the body.
Vitamin D	Take 3,000 IU per day. Doses over 1,000 IU per day may cause nausea, loss of appetite, headache, diarrhea, fatigue, restlessness, and hypercalcemia. Take this high dose only under the supervision of a physician.	Supports immune system function; hinders tumor growth; promotes removal of degenerated cells; activates cancer-fighting monocytes (a type of white blood cell) in the body.
Vitamin E	Take 1,200 IU per day.	Acts as an intercellular antioxidant; improves circulation; helps resist diseases at the cellular level.
Zinc	Take 50 mg per day. Do not take more than 100 mg from all sources per day.	Potent free-radical fighter; combines with the antioxidant enzyme superoxide dismutase to fight cancer; needed by over 300 enzymes; necessary for white blood cell immune function; improves healing.

Comments

❑ A body with excessive tumor formation is not a healthy body and must be treated accordingly. An adequate intake of enzymes, vitamins, and minerals for a healthy, active 16-year-old athlete would not be sufficient for a 55-year-old man or woman with cancer. This is why such high dosages are recommended.

❑ It is estimated that approximately 35 percent of all cancers are related to diet. Decrease your intake of fats, as high-fat diets have been associated with colon, rectal, breast, and prostate cancers and possibly cancers of the ovary, uterus, and pancreas. Enzyme-rich fresh fruits and vegetables should comprise over 65 percent of your diet. If your body cannot tolerate raw fruits and vegetables, increase your intake of digestive enzymes, or sauté or steam your produce. Increase your intake of fiber-rich foods, which are associated with a lower risk for rectal and colon cancers. See the dietary recommendations in Part Three and the inset "Reducing your Risk of Cancer" below.

❑ Several herbs are helpful in the treatment of cancer, including alfalfa, aloe vera, astragalus, buckthorn, burdock, echinacea, ginger, ginseng, licorice, maitake mushrooms, pau d'arco, and wheat grass.

❑ Decrease your alcohol intake. High alcohol consumption has been associated with cancer of the mouth, larynx, pharynx, liver, esophagus, breast, neck, head, and large bowel. Alcohol and other refined carbohydrates produce undesirable fermentation in the bowel and stimulate improper metabolism of insulin, chromium, and zinc.

❑ Eat lots of garlic. For an odor-free garlic, try aged garlic extract.

Reducing Your Risk of Cancer

To reduce your risk of cancer, the American Cancer Society recommends the following:

• Don't smoke. Cigarette smoking is responsible for 90 percent of lung cancers among men and 79 percent among women—about 87 percent overall. Use of chewing tobacco or snuff increases the risk of cancers of the mouth, larynx, throat, and esophagus, and it is a highly addictive habit.

• Eat a varied diet.

• Include a variety of vegetables and fruits in the daily diet. Studies have shown that daily consumption of fresh vegetables and fruits is associated with a decreased risk of lung, prostate, bladder, esophagus, colorectal, and stomach cancers.

• Maintain a desirable weight.

• Eat more high-fiber foods, such as whole-grain cereals, breads, pasta, vegetables, and fruits. High-fiber diets are a healthy substitute for fatty foods and may reduce the risk of colon cancer.

• Cut down on total fat intake. A diet high in fat may be a factor in the development of certain cancers, particularly breast, colon, and prostate.

• Limit consumption of alcohol.

• Limit consumption of salt-cured, smoked, and nitrite-cured foods.

• Limit exposure to sunlight. Almost all of the more than 800,000 cases of basal and squamous cell skin cancer diagnosed each year in the United States are sun-related.

❑ Avoid chlorinated water. Distilled, bottled spring water, or well water is recommended.

❑ To successfully fight cancer, the continual elimination of toxins by keeping the excretory systems flushed is essential. Therefore, the Enzyme Kidney Flush and the Enzyme Toxin Flush should be performed on a continuing basis. Perform the Enzyme Kidney Flush for one week. The following week, perform the Enzyme Toxin Flush. Each flush should be repeated once every month. See Part Three for instructions for these two flushes.

❑ Vitamin B$_{15}$ (pangamic acid) is often used in the treatment of cancer. While not a true vitamin, pangamic acid is a natural chelator to assist in the removal of toxins from the body. Try taking 500 milligrams every day.

❑ During the weeks that you are not using the flushes, use the coffee enema once weekly as described in Part Three. Coffee enemas are used to stimulate the release of bile from the liver and facilitate the removal of waste metabolites from the body and into the intestines for excretion.

❑ Physicians sometimes use enzyme injections or enzyme retention enemas in the treatment of cancer. Both methods allow a greater concentration of the enzymes to be absorbed than might be accomplished if the enzymes were taken orally. An enzyme retention enema is a means of instilling some enzymes into the colon to be retained. This allows the enzymes to be absorbed into the body's systems. Physicians use products specially formulated for these uses.

❑ PAP tests and mammography screening can reduce the risk of cervical cancer and breast cancer, respectively, in women. Other procedures for the early detection of cancer include sigmoidoscopy; fecal occult blood testing; and skin, oral, and rectal examinations.

❑ A positive mental attitude is essential in combating cancer. See Part Three for more information.

❑ See the list of Resource Groups in the Appendix for a list of organization that can provide you with more information about cancer.

❑ *See also* BRAIN TUMORS; BREAST CANCER; COLORECTAL CANCER; LUNG CANCER; PANCREATIC CANCER; PROSTATE CANCER; SKIN CANCER.

Candidiasis

See FUNGAL SKIN INFECTIONS.

Canker Sores

Canker sores are small, painful ulcers found alone or in clusters on the inside of the mouth. Although the exact cause of canker sores is unknown, they may be triggered by stress, irritation or injury of the mouth lining, a viral infection, or poor dental hygiene. Deficiencies of certain nutrients, including vitamin B$_{12}$, folic acid, lysine, and iron increase one's susceptibility to canker sores. Women are usually more likely to suffer from canker sores than are men.

Symptoms

The first noticeable symptom of a canker sore may be a burning or tingling sensation that may occur a few hours before the sores develop. The sores may appear on the lips, on the soft palate, inside the cheeks, on the tongue, or in the throat. They may have red borders and be light yellow or pale white in color, but they do not form blisters as cold sores do. Canker sores may be extremely painful for three to four days, but symptoms usually diminish until the sore heals (usually without scarring) in seven to ten days. Severe attacks may be marked by fever, swollen lymph glands, and malaise. Many people get canker sores repeatedly. Subsequent attacks may vary from one lesion two or three times a year, to a sequence of uninterrupted multiple lesions.

Enzyme Therapy

Systemic enzyme therapy is used to relieve pain until the canker sore heals, reduce swelling and inflammation, and stimulate the immune system. Enzymes can improve circulation, help speed tissue repair, bring nutrients to the damaged area, remove waste products, and enhance wellness.

Digestive enzyme therapy is used to improve the digestion and assimilation of food nutrients, reduce stress on the gastrointestinal mucosa, help maintain normal pH levels, detoxify the body, promote the growth of healthy intestinal flora, strengthen the body as a whole, and thus build general resistance.

Topical enzyme exfoliants and creams are used to dissolve dead and damaged skin cells and exfoliate the skin, while leaving healthy skin cells intact.

Enzyme Absorption System Enhancers (EASE) maximize enzyme activity. They can also improve the absorption and bioavailability of various nutrients and decrease the drain on the body's own digestive enzymes, thus prolonging their lives.

In addition to the standard multivitamin, multimineral, and multiglandular complexes recommended in the introduction to Part Two, the following supplements are recommended for the treatment of canker sores. The following recommended dosages are for adults. Children between the

161

ages of two and five should take one-quarter the recommended dosage, those between the ages of six and twelve should take one-half the recommended dosage, and those between the ages of thirteen and seventeen should take three-quarters the recommended adult dosage.

ESSENTIAL ENZYMES

Enzyme	Suggested Dosage	Actions
Superoxide dismutase	Take 150 mcg per day.	Fights inflammation; potent antioxidant.

AND SELECT ONE OF THE FOLLOWING:

Bromelain	Tale 750 mg two times per day between meals.	Fights inflammation, pain, and swelling; speeds recovery; stimulates the immune system; strengthens capillary walls.
Formula One—See page 38 of Part One for a list of products.	Take as indicated on the label.	Fights inflammation, pain, and swelling.
A microbial protease	Take as indicated on the label.	Fights inflammation, pain, and swelling.
Pancreatin	Take 500 mg six times per day on an empty stomach.	Fights inflammation, pain, and swelling; increases rate of healing; stimulates the immune system; strengthens capillary walls.
Papain	Take 300 mg two times per day between meals.	Fights inflammation; speeds recovery.
Trypsin with chymotrypsin	Take 250 mg of trypsin and 10 mg of chymotrypsin three times per day between meals.	Fights inflammation, pain, and swelling; increases rate of healing; stimulates the immune system.

SUPPORTIVE ENZYMES

Enzyme	Suggested Dosage	Actions
Catalase and glutathione peroxidase	Take as indicated on the labels.	Potent antioxidant enzymes.
A digestive enzyme product containing protease, amylase, and lipase enzymes. See the list of digestive enzyme products in Appendix.	Take as indicated on the label with meals.	Improves the breakdown and absorption of food nutrients; fights toxic build-up.

ENZYME ABSORPTION SYSTEM ENHANCERS (EASE)

Combination	Suggested Dosage	Actions
AminoLogic from PhysioLogics	Take 1 capsule three times per day between meals; or follow the advice of your health-care professional.	Helps maintain normal tissue function and repair; furnishes lysine, which may be lacking; contains proteolytic enzymes to fight inflammation, stimulate immune function, and aid improved nutrient absorption.
or		

Detox Enzyme Formula from Prevail	Take 1 capsule between meals one or two times per day with a 6- to 8-ounce glass of water or juice.	Helps rid the body of harmful toxins.
or		
Hepazyme from Enzymatic Therapy	Take as indicated on the label.	Stimulates immune function.
or		
Kyolic Formula 102 (with enzymes)	Take 2 capsules or tablets per day as desired.	Strengthens the immune system; helps alleviate stress; fights free radicals; fights viral infections.
or		
Oxy-5000 Forte from American Biologics	Take as indicated on the label.	Helps rid the body of biochemical stressors and infections.

ENZYME HELPERS

Nutrient	Suggested Dosage	Actions
Acidophilus and other probiotics	Take as indicated on the label.	Improves gastrointestinal function, which improves immunity; needed for production of biotin, niacin, folic acid, and pyridoxine, which are important for the function of many enzymes; helps produce antibacterial substances; stimulates enzyme activity; fights toxin build-up.
Adrenal glandular	Take 100–400 mg per day.	Fights inflammation and adrenal exhaustion.
Amino-acid complex containing lysine	Take as indicated on the label on an empty stomach.	Required for enzyme production; helps the body maintain proper fluid and acid/base balance, which are critical for maintaining health.
Bioflavonoids, including pycnogenol, rutin, or quercetin	Take 500 mg three times per day between meals.	Strengthen capillaries and improves circulation, which speeds healing; potent antioxidants.
Branched-chain amino acids (isoleucine, leucine, and valine)	Take as indicated on the label between meals.	Often found together in a supplement, branched-chain amino acids aid in healing of tissue and skin.
Brewer's yeast	Take 1 tablespoon in food or juice three times per day.	Natural source for B vitamins; rich in minerals, important for healing and immunity.
Coenzyme Q_{10}	Take 30–60 mg one to two times per day with meals; or follow the advice of your health-care professional.	Potent antioxidant.
Essential fatty acid complex	Take 100–200 mg two to three times per day with meals; or follow the advice of your health-care professional.	Helps strengthen cell membranes; supports immune function.
Green Barley Extract from Green Foods Corporation or Kyo-Green from Wakunaga	Take as indicated on the label.	Potent free-radical fighter; rich in vitamins, minerals, and enzyme activity.

162

Kyolic Aged Garlic Liquid Extract with Vitamins B₁ and B₁₂	Take 10 drops in liquid or place directly on tongue two times per day.	Fights inflammation and free radicals; improves immune response.
Spirulina	Take as indicated on the label.	Rich in minerals, vitamins, enzymes, proteins, essential amino acids, and essential fatty acids, especially gamma-linolenic acid; good source of beta-carotene; improves immune function.
Thymus glandular	Take 100 mg per day.	Assists the thymus gland in immune resistance; improves immune response; fights infections.
Vitamin A	Take 5,000–40,000 IU per day. Begin with the lowest dosage and increase gradually as needed. Any dose over 50,000 IU should be monitored by a physician. If pregnant, take no more than 10,000 IU.	Needed for strong immune system and healthy mucous membranes, such as those of the mouth; potent antioxidant.
Vitamin B complex	Take 150 mg three times per day; or take as indicated on the label.	B vitamins act as coenzymes and support healthy tissues.
Vitamin B₁₂	Take 3,000 mcg twice per day for no more than three days, then decrease dosage to 500 mcg for the next three days.	Required by many enzymes; needed for the synthesis of nerve cells in tissue repair.
Vitamin C	Take 10,000–15,000 mg per day.	Fights infections, diseases, allergies; improves wound healing; stimulates the adrenal glands; promotes immune system; powerful antioxidant.
Whey	Take 1 heaping tablespoon in food or liquid three times per day.	Improves digestion so improves the absorption and availability of nutrients needed to fight canker sores; helps detoxify the body.
Zinc	Take 50 mg per day. Do not exceed 100 mg per day from all sources.	Needed for white blood cell immune function, and healing.

Comments

❑ Treat canker sores topically with an enzyme exfoliant or an enzyme salve (containing individual enzymes, such as papain, or enzyme mixtures with such enzymes as bromelain, papain, trypsin, chymotrypsin, pancreatin, lipase, and amylase). Some enzyme salves are combined in a base of vitamin A and vitamin E to improve antioxidant activity and prolong enzyme activity. Apply one to three times per day with cotton or gauze.

❑ Maintaining the body's alkaline/acid balance is essential. Add fermented foods (such as kefir, yogurt, and sauerkraut) to your diet.

❑ A number of enzyme-rich foods, including raspberries,

cranberries, red wine, grapes, black currants, and red cabbage, contain anthocyanidins, a phytochemical known to fight inflammation.

❑ To improve the overall health of your body, avoid alcohol and refined white flour and sugar. Small amounts of honey are allowed.

❑ Avoid chlorinated water. Distilled or bottled spring water, or even well water is recommended.

❑ In place of ordinary salt, use sea salt because it contains important minerals.

❑ To flush your body, drink at least eight 8-ounce glasses of distilled or bottled water per day. Also include plenty of freshly juiced vegetables and fruits (including carrots, citrus fruits, grapes, and pineapples).

❑ Rinse your mouth with a rinse made from two ounces of water, two ounces of hydrogen peroxide, a teaspoon of salt and a teaspoon of baking soda. Do not swallow. Repeat four times per day.

❑ Place ten drops of liquid aged garlic extract on a cotton swab or gauze. Dab gently on any canker sores. Repeat at least twice per day.

❑ Break open and spread the contents of two to four vitamin A and vitamin E capsules on the affected area twice per day.

Carbohydrate Intolerance

When we eat foods containing carbohydrates, such as breads, cereals, fruits, and vegetables, enzymes in our digestive tracts break them down into simple sugars that can be absorbed by the bloodstream (see Chapter 2 of Part One for a more complete discussion of digestion). The body then uses these sugars as fuel. Some individuals, however, don't make enough, or any, of one or more of the enzymes necessary to break down carbohydrates.

Carbohydrates are composed of mono-, di-, poly-, and oligosaccharides. Normally, enzymes in the small intestine (including maltase, lactase, isomaltase, or sucrase) break disaccharides down into monosaccharides. Disaccharides that haven't been broken down remain in the intestine and ferment, causing diarrhea and flatulence. Secondary enzyme deficiencies can occur because of conditions such as tropical sprue, acute intestinal infections, celiac disease and neomycin toxicity. These and other conditions can alter the mucosal lining of the jejunal portion of the small intestine and can decrease the numbers of several important digestive enzymes that are produced in the mucosal lining's brush border. Abdominal surgery or internal infections (especially in infants) can sometimes cause temporary secondary disaccharidase deficiencies, which resolve when the patient's health improves.

Some individuals suffering from carbohydrate intolerance may have sufficient enzymes but lack the transport system in the intestine for the absorption of monosaccharides such as galactose and glucose. These individuals will develop symptoms of carbohydrate intolerance after eating most kinds of sugars. Fortunately, glucose/galactose intolerance is extremely rare. At present, there has been no research conducted on enzymes in the treatment of glucose/galactose transport disorders.

Symptoms

Symptoms of carbohydrate intolerance may include diarrhea, bloating, flatulence, and abdominal cramps after eating carbohydrates. There also may be weight loss and nausea. Nutrient deficiencies may develop since the individual does not have the proper enzymes to digest his or her foods. Serious chronic intestinal infections can be a complication.

Enzyme Therapy

Digestive enzyme therapy is used to replace the missing enzymes, improve digestion of food, reduce stress on the gastrointestinal mucosa, help maintain normal pH levels, detoxify the body, promote the growth of healthy intestinal flora, strengthen the body as a whole, and build general resistance.

Systemic enzyme therapy is used as supportive care to reduce any inflammation, stimulate the immune system, improve circulation, help speed tissue repair, transport nutrients, remove waste products, and enhance wellness.

Enzyme Absorption System Enhancers (EASE) maximize enzyme activity. They can also improve the absorption and bioavailability of various nutrients and decrease the drain on the body's own digestive enzymes, thus prolonging their lives.

In addition to the standard multivitamin, multimineral, and multiglandular complexes recommended in the introduction to Part Two, the following supplements are recommended for the treatment of carbohydrate intolerance. The following recommended dosages are for adults. Children between the ages of two and five should take one-quarter the recommended dosage, those between the ages of six and twelve should take one-half the recommended dosage, and those between the ages of thirteen and seventeen should take three-quarters the recommended adult dosage.

ESSENTIAL ENZYMES

Enzyme	Suggested Dosage	Actions
A digestive enzyme product containing amylase	Take 50 mg three times per day with meals.	Breaks down carbohydrates.
NESS Formula #21	Take 2 tablets with each meal.	Provides multiple enzymes, including disaccharidases, to help the body utilize foods,

especially starch, sugar, and fiber.

SUPPORTIVE ENZYMES

Enzyme	Suggested Dosage	Actions
Formula Three—See page 38 of Part One for a list of products.	Take as indicated on the label.	Potent antioxidant enzymes.
Bromelain	Start with one 230–250 mg capsule per day and increase dosage by one capsule per day, if necessary, until symptoms begin to gradually disappear. Take no more than 3 capsules per day. If diarrhea develops, decrease dose.	Improves digestion; fights inflammation, pain, and swelling; stimulates the immune system.

ENZYME ABSORPTION SYSTEM ENHANCERS (EASE)

Combination	Suggested Dosage	Actions
Bean & Vegi Enzyme Formula from Prevail *or*	Take as indicated on the label.	Facilitates carbohydrate digestion and absorption.
Detox Enzyme Formula from Prevail *or*	Take 1 capsule between meals one or two times per day. Take with a 6- to 8-ounce glass of water or juice.	Helps rid the body of harmful toxins that accumulate when digestion is impaired.
Kyolic Formula 102 (with enzymes) *or*	Take 2 capsules or tablets per day as desired.	Has anti-inflammatory properties; improves carbohydrate digestion and absorption.
NESS Formula #10 *or*	Take 2 capsules immediately after each meal.	Helps the body to absorb carbohydrates and sugars.
NESS Formula #601 *or*	Take 2 capsules with each meal.	For gastric discomfort; an anti-inflammatory agent; helps protect the mucosal tissue of the intestine.
Oxy-5000 Forte from American Biologics *or*	Take as indicated on the label.	Helps the body excrete biochemical stressors, infections, and degenerative processes.
Sugar Check from Enzymatic Therapy	Take as indicated on the label.	Contains nutrients for improved sugar and carbohydrate metabolism and absorption.

ENZYME HELPERS

Nutrient	Suggested Dosage	Actions
Acidophilus or other probiotics, such as bifidobacterium or Acidophilasé from Wakunaga, which contains protease, lipase, and amylase	Take as indicated on the label.	Stimulates production of digestive enzymes in the mucosal membrane of the small intestine.

Adrenal glandular	Take 100–400 mg per day.	Fights adrenal exhaustion; restores normal carbohydrate balance.
Betaine Hydrochloric acid	Take 150 mg three times per day with meals.	Provides hydrochloric acid to improve digestion. If gastric upset occurs, discontinue use.
Fiber	Take as indicated on the label.	Provides bulk; needed for proper elimination.
Fructooligo-saccharides	Take as indicated on the label.	Increases peristaltic action of intestines; improves liver function.
Lecithin	Take 1,000 mg three times per day with meals.	Emulsifies fat.
Ox bile extract	Take 120 mg per day.	Aids in digestion.
Vitamin B$_1$ (thiamin)	Take 50 mg three times per day.	Essential for carbohydrate metabolism and proper energy production; assists enzymes that break down carbohydrates.
Vitamin B$_6$ (pyridoxine)	Take 50 mg three times per day.	Assists enzymes that metabolize amino acids and fats; assists in transport of potassium into cells; needed for healthy cell function.
Whey	Take 1 heaping tablespoon in food or liquid three times per day with meals.	Improves digestion.

Comments

❏ Eat a diet high in fresh, enzyme-rich fruits and vegetables. Carrots, broccoli, spinach, sweet peppers, kale, and turnip greens are high in vitamins, minerals, and enzymes.

❏ Garlic and onions should be eaten as often as possible.

❏ Eliminate all refined sugars and other refined products. Follow the Caveman's Enzyme Diet in Part Three.

❏ Cleansing, fasting, and juicing are very important, but drink no fruit juices. Instead, rely on vegetable juices. See Part Three for more information on juicing.

❏ Exercise, a positive mental attitude, and meditation stimulate antioxidant and enzyme activity in the body. Meditate daily and do some form of exercise every day, such as brisk walking, swimming, or bicycling, for a minimum of thirty minutes.

❏ See also LACTASE DEFICIENCY and CELIAC DISEASE.

Carbuncles

See BOILS.

Cardiovascular Disorders

Atherosclerosis, heart disease, cerebrovascular disease, and hypertension are all conditions falling under the umbrella term of cardiovascular disease ("cardio" relates to the heart and "vascular" pertains to the vessels—the arteries, veins, and capillaries, which provide channels for the flow of blood). In 1990, almost one million Americans died from the major cardiovascular diseases (which collectively have the deadly distinction of being the number-one killer in America).

Enzyme therapy is used to treat those cardiovascular disorders that result from "sticky" blood and blood clots. (See the inset "Cardiovascular Disorders Helped With Enzyme Therapy" on page 166.) If you are to remain healthy, your blood must maintain a constant dynamic balance between its ability to remain liquid and free-flowing and its ability to form blood clots to keep you from bleeding to death in the event of an injury. If blood can't clot, tissues can't heal. But excessive bleeding can cause death. This is what can happen to people suffering from hemophilia because their blood lacks the ability to coagulate.

On the other hand, if your blood clots too easily, life-threatening clots can form in your blood vessels, inhibiting circulation, and sometimes breaking free and causing heart attacks (when they reach the heart) or strokes (when they reach the brain). Cholesterol and other fatty materials may collect around the clotted deposits in the blood vessels. When this pathological condition develops, it is called hardening of the arteries, or atherosclerosis.

It is critical to maintain a proper equilibrium between blood clotting and blood liquefaction. Unfortunately, a disturbance in this equilibrium is probably the most common source of illness and death in this country. In fact, one American dies every 33 seconds from cardiovascular disease, a condition that is largely preventable.

Every day, the body produces about two grams of fibrin—the glue needed for coagulation. The body uses fibrin to seal wounds and to coat the internal walls of blood vessels. This coating protects the delicate vessel wall linings from damage by any passing particles and also smoothes any rough spots in the vessel walls. This assures an even, rather than a sporadic or disturbed, blood flow. The body constantly produces a steady supply of fibrin so it will have enough on hand in the event of an emergency (for instance, if you were to cut your finger).

For safety reasons, fibrin is present in the bloodstream in its inactive form, as fibrinogen. If a need arises, a series of reactions (triggered by enzymes) is put into play. First prothrombin (the precursor of the enzyme thrombin) is activated and changes to thrombin. Then the thrombin converts fib-

Cardiovascular Disorders Helped With Enzyme Therapy

According to past research, enzyme therapy can help many cardiovascular disorders, including the following conditions, many of which are discussed individually in this book.

- Angina pectoris
- Arteriosclerosis/atherosclerosis
- Coronary heart disease
- Degenerative inflammatory processes (both acute and chronic) resulting from arterial and venous diseases
- Eczema (as a result of a cardiovascular disorder)
- Embolism
- Endarteritis obliterans
- Hematomas
- High blood pressure
- Intermittent claudication
- Lymphatic edema
- Occlusive arterial disease
- Phlebitis

- Post-infarct healing process stimulation
- Post-thrombotic syndrome
- Raynaud's disease
- Scleroderma
- Smoker's and diabetic circulatory diseases
- Stroke prevention
- Swelling (edema)
- Thromboangitis obliterans
- Thromboembolic complications
- Thrombosis (both superficial and deep)
- Varicose ulcers
- Varicose veins
- Vasculitis
- Venous disorders

rinogen to fibrin, which forms a clot. When the clot is exposed to air, it dries out and forms a hard protective scab.

Fortunately, the body also has a protective mechanism to keep all of this fibrin from accumulating on the vessel walls (interrupting blood flow) or from being deposited in the wrong places. This "safety catch" is the enzyme *plasmin* (which exists in the blood as the proenzyme plasminogen). Plasmin dissolves clots through a process called fibrinolysis (the lysing or cutting of fibrin), and thus maintains blood flow equilibrium.

Even with all the built-in protective mechanisms, sometimes something goes wrong. Cardiovascular diseases occur when the blood becomes too sticky (thick blood is the most frequent accompanying symptom of fatal heart and vascular system disorders). This sticky blood can occur from excessive fibrin production. But it can also occur if there is an error in plasmin activation and an insufficient amount of fibrin is broken down. This can lead to blood flow problems in the arteries and veins, as well as other areas of the body (including the tear duct and the mammary and salivary glands).

Unfortunately, as we age, our plasmin levels decrease. By the age of 60, we only possess a fraction of the plasmin we had when we were young. This means that blood flow is more sluggish, toxic debris remains in the vessels, and the vessels become narrower and harden, increasing the risk of a heart attack or stroke.

Symptoms

Symptoms of a cardiovascular disease will vary depending on the specific condition. However, some of the most frequent symptoms are high blood pressure, elevated cholesterol, fatigue, and other effects of reduced circulation.

Enzyme Therapy

Numerous studies have demonstrated that one of the best ways to dissolve large clots is with the help of enzymes. A number of enzymes are currently used (primarily in a hospital setting) as thrombolytic and fibrinolytic agents, including brinase (a protease from *Aspergillus oryzae*), streptokinase (from bacteria), and urokinase (from human urine). Recently, an Oregon man suffered a stroke and had no signs of brain activity. Treatment with the enzyme urokinase (administered directly into the area of the blood clot), was started nine hours after his stroke began (as a "last-ditch" effort). Not only did he regain consciousness, he also regained the ability to move his arms and legs (remember, stroke is one of the leading causes of adult disability). He subsequently went through six weeks of rehabilitation and is now home but is continuing physical and speech therapies.

Maintaining proper enzyme levels in the blood can also help *prevent* cardiovascular disease. Research has shown that proteolytic enzymes can improve circulatory imbalances and help normalize the fibrinolytic equilibrium. Proteolytic enzymes increase the fibrinolytic activity of the blood and therefore normalize the equilibrium between blood clotting and the break-up of blood clots. The result is that deposits of fibrin are dissolved. The edema, swelling, and the pathologic condition is reduced or eliminated. Proteolytic enzymes improve the blood fluidity and, therefore, improve blood flow. This improvement in circulation improves the supply of nutrients to tissues. Proteolytic enzymes are also natural inhibitors of inflammation (which can occur in the vessels because of clot formation).

Systemic enzyme therapy is used as supportive care to improve circulation and reduce any pain, swelling, and inflammation. Enzymes stimulate the immune system, help speed tissue repair, bring nutrients to the damaged area, remove waste products, and improve health.

Digestive enzyme therapy is used to improve the digestion of food, reduce stress on the gastrointestinal mucosa, help maintain normal pH levels, detoxify the body, and promote the growth of healthy intestinal flora, thus relieving the stress on the body's own enzymes. Digestive enzymes also serve as replacements for the body's pancreatic enzymes, leaving the pancreatic enzymes free to perform other functions in the body, such as improving circulation.

A prolonged regular intake of enzyme supplement preparations with the main meals may have a beneficial effect on the function of the pancreas, according to Drs. Max Wolf and Karl Ransberger in their book *Enzyme Therapy* (LA: Regent House, 1972). Pancreatic enzymes can then be freed to reduce vascular inflammation.

Antioxidant enzymes are used to help eliminate free radicals. Normally, exceptionally large amounts of extracellular superoxide dismutase (SOD) are found in the blood vessel walls, especially the arteries. The arterial smooth muscle cells secrete large amounts of SOD and are probably the main source of the enzyme in the vascular wall. The SOD concentration is high enough to subdue the damaging effects of the superoxide free radicals. However, because of acute injury, chronic disease, or aging, sufficient amounts of SOD might not be available at the cellular level. Therefore, increased amounts (through supplementation) seem necessary to fight cardiovascular disorders.

Enzyme Absorption System Enhancers (EASE) maximize enzyme activity. They can also improve the absorption and bioavailability of various nutrients and decrease the drain on the body's own digestive enzymes, thus prolonging their lives.

In addition to the standard multivitamin, multimineral, and multiglandular complexes recommended in the introduction to Part Two, the following supplements are recommended for the treatment, as well as the prevention, of cardiovascular disorders caused by clotting blood. The following recommended dosages are for adults. Children between the ages of two and five should take one-quarter the recommended dosage, those between the ages of six and twelve should take one-half the recommended dosage, and those between the ages of thirteen and seventeen should take three-quarters the recommended adult dosage.

ESSENTIAL ENZYMES

Enzyme	Suggested Dosage	Actions
Formula Three—See page 38 of Part One for a list of products.	Take 200 mcg three times per day between meals.	Free-radical fighter.

AND SELECT ONE OF THE FOLLOWING:

Enzyme	Suggested Dosage	Actions
Formula One (particularly Wobenzym N from Naturally Vitamin Supplements)—See page 38 of Part One for a list of products.	Take 8–10 tablets three times per day between meals for six months or until symptoms improve. Thereafter, take 6 tablets three times per day between meals.	Dissolves blood clots; fights inflammation, pain, and swelling; speeds healing; strengthens capillary walls.
Inflazyme Forte from American Biologics	Take as indicated on the label between meals.	Dissolves blood clots; fights inflammation, pain, and swelling; speeds healing; strengthens capillary walls.
Phlogenzym from Mucos	Take 8–10 tablets three times per day between meals for six months or until symptoms improve. Thereafter, take 6 tablets three times per day between meals.	Dissolves blood clots; fights inflammation, pain, and swelling; speeds healing; strengthens capillary walls.
Bromelain	Take four to six 230–250 mg capsules per day.	Fights inflammation, pain, and swelling; speeds healing; stimulates the immune system.
Proteolytic enzymes	Take as indicated on label between meals.	Dissolves blood clots; fights inflammation, pain, and swelling; speeds healing; strengthens capillary walls.

SUPPORTIVE ENZYMES

Enzyme	Suggested Dosage	Actions
A digestive enzyme formula containing protease, amylase, and lipase enzymes. See the list of digestive enzyme products in Appendix.	Take as indicated on the label with meals.	Improves the breakdown and absorption of proteins, carbohydrates, and fats; fights toxic build-up.

ENZYME ABSORPTION SYSTEM ENHANCERS (EASE)

Combination	Suggested Dosage	Actions
Bromelain Complex from PhytoPharmica *or*	Take as indicated on the label.	Reduces swelling; improves circulation; helps break up circulating immune complexes.
Cardio-Chelex from Life Plus *or*	Take 5 tablets twice per day—first thing in the morning and at midday.	Supports healthy circulation; improves energy.
Kyolic Formula 102 (with enzymes) *or*	Take 2 capsules or tablets per day as desired.	Strengthens the immune system; helps alleviate stress; helps reverse circulatory disorders; helps fight free radicals; lowers cholesterol; helps fight inflammation; improves digestion.

NESS Formula #14	Take 2 tablets twice per day between meals.	Provides herbs and cofactors that work with enzymes to improve circulation and decrease inflammation and swelling.
or		
Protease Concentrate from Tyler Encapsulations	Take as indicated on the label.	Breaks up fibrin; fights inflammation.
or		
Quercezyme-Plus from Enzymatic Therapy	Take 1–2 capsules between meals.	Decreases inflammation; improves circulation; strengthens capillaries.

ENZYME HELPERS

Nutrient	Suggested Dosage	Actions
Adrenal glandular	Take as indicated on the label.	Helps fight stress and adrenal exhaustion.
Aged garlic extract	Take three 300 mg capsules three times per day.	Lowers cholesterol and blood pressure; fights free radicals.
Aorta glandular	Take as indicated on the label.	Improves aortic function; helps improve blood vessel function; improves peripheral and cerebral circulation; helps protect against atherosclerosis.
Bioflavonoids	Take 500 mg three times per day between meals.	Potent antioxidants; protect against heart disease; normalize blood pressure; help lower cholesterol; fight capillary fragility.
Calcium	Take 1,000 mg per day.	Plays important roles in muscle contraction and nerve transmission.
Coenzyme Q$_{10}$	Take 75 mg one to two times per day with meals.	Potent antioxidant; prevents heart attacks; decreases blood pressure and total serum cholesterol.
Essential fatty acid complex	Take 75–100 mg two to three times per day with meals.	Lowers cholesterol level; strengthens the immune system; helps strengthen cell membranes.
Green barley extract	Take 1–2 teaspoons in water or juice three times per day.	Reduces blood pressure; effective against heart disease; fights free radicals.
Heart glandular	Take 140 mg; or take as indicated on the label.	Improves heart function.
Lecithin	Take as indicated on the label.	Emulsifies fat.
Magnesium	Take 500 mg per day.	Plays important roles in protein building, muscle contraction, nerve function, energy production, calcium absorption, and blood clotting.
Niacin or niacin-bound chromium	Take as indicated on the label.	Improves circulation by dilating blood vessels.
Pycnogenol	Take 1,000 mcg per day.	Potent antioxidant; fights heart disease.
Vitamin A	Take 5,000 IU per day; or take as per your doctor's	Necessary for skin and cell membrane structure and

	directions.	strong immune system; potent antioxidant.
or		
Carotenoids (such as alpha-and beta-carotene, xanthophylls, cryptoxanthin, lutein, or zeaxanthin)	Take as indicated on the label.	Potent antioxidants; reduce arterial plaque.
Vitamin B complex	Take 50 mg three times per day; or take as indicated on the label.	B vitamins act as coenzymes and support neuromusculoskeletal function.
Vitamin C	Take 5,000–10,000 mg per day.	Essential to vascular function; potent antioxidant; improves the function of many enzymes; enhances absorption of iron; assists in converting cholesterol to bile salt.
Vitamin E	Take 1,200 IU per day from all sources (including the multivitamin complex in the introduction to Part Two).	Potent antioxidant; helps prevent cardiovascular disease; decreases risk of heart attack; stimulates circulation; strengthens capillary walls; prevents blood clots; maintains cell membranes.
Zinc	Take 50 mg per day. Do not take more than 100 mg per day.	Needed for white blood cell immune function; interacts with platelets in blood clotting.

Comments

❑ Eat more fresh, enzyme-rich fruits, vegetables, and sprouts. Buy fruits and vegetables in season.

❑ Garlic, onions, shallots, chives, leeks, and scallions contain allium compounds that are potent free-radical fighters, help lower cholesterol, and fight circulatory disorders.

❑ Oats, wheat germ, olives, nuts, nut and seed oils, and organ meats, contain alpha-tocopherol (vitamin E), known to help prevent cardiovascular disease and prevent blood clots.

❑ Raspberries, cranberries, grapes, red wine, black currants, and red cabbage contain anthocyanidins, which reduce blood-vessel plaque formation and maintain blood flow in small vessels.

❑ Oolong and green tea contain epicatechin, which is a potent antioxidant that lowers blood pressure and strengthens capillaries.

❑ A number of herbs have a long history of use in the treatment of heart and circulatory problems, including astragalus, bilberry, ginkgo biloba, ginseng, gotu kola, and hawthorn.

❑ Decrease your intake of fatty animal foods, hydrogenated vegetable oils, and cholesterol in your diet. Just visualize the fat in that marbled steak clinging to and clogging your blood vessels and shutting the blood supply to your heart.

❑ One to two glasses of red wine per day can be beneficial. It contains tannins and quercetin, bioflavonoids instrumental in fighting cardiovascular disease.

❏ Eat more fresh fish, fish oils, and flaxseed oil to increase your intake of omega-3 and omega-6 fatty acids.

❏ Avoid all refined foods, food additives, preservatives, toxins, and pesticides in your foods.

❏ Eat four or five times per day, rather than once per day. This provides your body's enzymes and your digestive enzyme supplements with an optimal environment to digest and absorb food nutrients.

❏ Drink plenty of fresh juices made with fresh, enzymatically active antioxidant foods, including green peppers, kale, green vegetables (such as spinach, broccoli, beet greens, and parsley), turnips, carrots, sweet peppers, ginger root, garlic, pineapple, and citrus fruits. Eat the white pulp on the inner lining of the skin of citrus fruits with the rest of the fruit. It is loaded with health-giving citrus bioflavonoids and enzymes.

❏ Since physical activity is directly related to cholesterol levels in the body, a daily program of exercise is critical. Simple brisk walking for thirty to forty minutes a day is of great benefit.

❏ Once each month use the Enzyme Kidney Flush and the Enzyme Toxin Flush. Perform the Enzyme Kidney Flush for one week. The following week, perform the Enzyme Toxin Flush. See Part Three for directions.

❏ A positive mental attitude and meditation can help fight free-radical formation and stimulate enzymatic activity in the body.

❏ See the list of Resource Groups in the Appendix for a list of organizations that can provide you with more information about cardiovascular disorders.

❏ *See also* ANGINA; ARTERIOSCLEROSIS; EMBOLISM; HIGH BLOOD PRESSURE; RAYNAUD'S DISEASE; STROKE; THROMBOSIS.

Carpal Tunnel Syndrome

Carpal tunnel syndrome (CTS) is a painful condition of the wrists and hands, resulting from nerve compression in the wrist and between the tendons of the forearm muscles. The eight bones of the wrist (carpus) form three sides of a tunnel. The fourth side is covered with soft tissue. Repeated irritation from the compression causes the soft tissue surrounding the carpal tunnel nerves and tendons to swell and become inflamed and painful.

CTS may occur in one or both wrists and is found more often in women than in men. This condition occurs most frequently in people who do repetitive wrist and hand movements, including typists, computer operators, grocery store clerks, assembly line workers, waiters and waitresses, musicians, carpenters and other laborers who routinely use hand tools, and athletes (such as tennis players). Symptoms of

carpal tunnel syndrome also appear in such conditions as rheumatoid arthritis, Raynaud's disease, hypothyroidism, menopause, and diabetes mellitus, and as a result of fluid changes due to pregnancy.

A number of nutritional deficiencies may cause carpal tunnel syndrome. For instance, a symptom of a vitamin B_1 deficiency is "wrist drop"—a lack of strength in the wrist. Low vitamin B_6 levels have also been associated with carpal tunnel syndrome. Certain drugs block the action of vitamin B_6, including hydralazine, isoniazid, dopamine, oral contraceptives, and penicillamine, as can excessive protein intake and the intake of yellow dyes.

Symptoms

The symptoms of carpal tunnel syndrome include pain, tenderness, soreness, numbness, a burning sensation, wrist inflammation, and swelling of the carpal tunnel. There may be a weakness in the wrist and muscles of the fingers and hand, plus burning, numbness, and tingling of the first three fingers of the affected hand. The thumb and the first three fingers are affected by carpal tunnel syndrome. The pinky finger is not affected. Symptoms may worsen at night.

The Carpal Tunnel Syndrome Self-Test

Electromyography is the only true confirmatory test for carpal tunnel syndrome. This involves sending electrical impulses through the nerves of your hand. If the nerves are "pinched" (entrapped) or damaged, the message can't get through the nerve fibers, and you probably have carpal tunnel syndrome.

You may want to try a self-test first to determine if you have carpal tunnel syndrome before undergoing electromyography. To perform the "T" self-test, bend both wrists so that the fingers are facing downward. Place the backs of both hands against each other or against a flat surface (such as a wall or table). This position tends to resemble a "T." Hold this position. If pain and other CTS symptoms appear, it is quite possible you have carpal tunnel syndrome. See a physician well trained in its treatment.

Enzyme Therapy

Systemic enzyme therapy is used to reduce the inflammation and swelling of the soft tissue in the carpal tunnel, thus decreasing the pressure on the tendons, nerves, and blood vessels passing through it. Enzymes stimulate the immune system, improve circulation, help speed tissue repair by bringing nutrients to the damaged area and removing waste products, and enhance wellness. Proteolytic enzymes, including trypsin, chymotrypsin, bromelain, and papain have well-documented effects in virtually all inflammatory conditions, regardless of their cause. Repeated clinical studies have shown that proteolytic enzymes are effective at reducing swelling, bruising, healing time, and pain.

Digestive enzyme therapy is used to improve the digestion of food, reduce stress on the gastrointestinal mucosa, help maintain normal pH levels, detoxify the body, promote the growth of healthy intestinal flora, strengthen the body as a whole, and building general resistance. All of these actions help reduce stress on the pancreas, which can free the body's enzymes to help reduce inflammation, swelling, and pain.

Enzyme Absorption System Enhancers (EASE) maximize enzyme activity. They can also improve the absorption and bioavailability of various nutrients and decrease the drain on the body's own digestive enzymes, thus prolonging their lives.

Topical enzyme salves can be used. Papain, bromelain, or enzyme combinations (such as papain, bromelain, pancreatin, trypsin, and chymotrypsin, with the bioflavonoid rutin) can be rubbed onto the wrist. They penetrate and help break up the fibrin formation and reduce swelling. Using ultrasound with the cream will improve the absorption.

In addition to the standard multivitamin, multimineral, and multiglandular complexes recommended in the introduction to Part Two, the following supplements are recommended for the treatment of carpal tunnel syndrome. The following recommended dosages are for adults. Children between the ages of two and five should take one-quarter the recommended dosage, those between the ages of six and twelve should take one-half the recommended dosage, and those between the ages of thirteen and seventeen should take three-quarters the recommended adult dosage.

ESSENTIAL ENZYMES

Enzyme	Suggested Dosage	Actions
Superoxide dismutase	Take 150 mcg per day.	Fights inflammation; potent antioxidant.

AND SELECT ONE OF THE FOLLOWING:

Formula One (particularly Wobenzym N from Naturally Vitamin Supplements) —See page 38 of Part One for a list of products.	Take 5 tablets twice a day.	Fights inflammation, pain, and swelling; speeds healing; stimulates the immune system; strengthens capillary walls.
Bromelain	Take three to four 230–250 mg capsules per day until symptoms subside. If pain is severe, take 6–8 capsules per day.	Fights inflammation, pain, and swelling; speeds healing; stimulates the immune system.
Trypsin with chymotrypsin	Take 250 mg of trypsin and 10 mg of chymotrypsin three times per day between meals.	Fights inflammation, pain, and swelling; increases rate of healing; stimulates the immune system.
Pancreatin	Take 500 mg six times per day on an empty stomach.	Fights inflammation, pain, and swelling; increases rate of healing; stimulates the immune

		system; strengthens capillary walls.
Papain	Take 300 mg two times per day between meals.	Fights inflammation; speeds healing.
Microbial protease	Take 300 mg two times per day between meals.	Fights inflammation; speeds recovery from injuries and surgery.

SUPPORTIVE ENZYMES

Enzyme	Suggested Dosage	Actions
Catalase and glutathione peroxidase	Take as indicated on the labels.	Potent antioxidant enzymes.
A digestive enzyme product containing protease, amylase, and lipase enzymes. See the list of digestive enzyme products in Appendix.	Take as indicated on the label with meals.	Improves the breakdown and absorption of food nutrients; fights toxic build-up.

ENZYME ABSORPTION SYSTEM ENHANCERS (EASE)

Combination	Suggested Dosage	Actions
Bio-Musculoskeletal Pak from Biotics *or*	Take as indicated on the label.	Decreases inflammation; fights free-radical formation.
Connect-All from Nature's Plus *or*	Take as indicated on the label.	Supports tissue maintenance and repair.
CTR Support from PhysioLogics *or*	Take as indicated on the label.	Nourishes the body during tissue recovery and healing from inflammation; diminishes damage caused by swelling and inflammation.
Inflamzyme from PhytoPharmica *or*	Take as indicated on the label.	Speeds tissue repair.
Joint-Ease from Nature's Life *or*	Take as indicated on the label.	Decreases inflammation and swelling.
Mobil-Ease from Prevail *or*	Take 1 capsule between meals two to three times per day.	Has anti-inflammatory activity.
Protease Concentrate from Tyler Encapsulations *or*	Take as indicated on the label.	Fights inflammation.
Quercezyme-Plus from Enzymatic Therapy	Take 1–2 capsules between meals.	Decreases inflammation; improves circulation by strengthening capillaries.

ENZYME HELPERS

Nutrient	Suggested Dosage	Actions
Acidophilus and other probiotics	Take as indicated on the label.	Improve gastrointestinal function to aid detoxification; needed for production of biotin, niacin, folic acid, and pyridoxine, which

		are important to many enzymes; stimulates enzyme activity.
Adrenal glandular	Take 100–400 mg per day.	Fights inflammation and adrenal exhaustion.
Bioflavonoids, including pycnogenol, rutin, quercetin	Take 500 mg three times per day between meals.	Potent antioxidants; improve circulation and fight inflammation.
Calcium	Take 1,000 mg per day.	Plays important roles in muscle contraction and nerve transmission.
Green Barley Extract from Green Foods Corporation or Kyo-Green from Wakunaga	Take as indicated on the label.	Potent free-radical fighter; rich in vitamins, minerals, and enzyme activity.
Kyolic EPA	Take 3 capsules per day.	Fights inflammation; improves immune system.
Magnesium	Take 500 mg per day.	Plays important role in muscle contraction, nerve function, and calcium absorption.
Niacin-bound chromium	Take up to 20 mg per day; or take as indicated on the label.	Dilates blood vessels; improves circulation.
Vitamin B₁	Take 500 mg three times per day for eight to ten weeks.	Increases activity at the nerves' synaptic junctions.
Vitamin B₆	Take 500 mg three times per day for eight to ten weeks.	Involved in the synthesis of several neurotransmitters and neurohormones; assists in transport of potassium into cell; needed for healthy nerves, muscle, blood, and skin.
Vitamin C	Take 10,000–15,000 mg per day.	Improves healing; stimulates the adrenal glands; promotes immune system; powerful antioxidant.
Vitamin E	Take 1,200 IU per day from all sources (including the multivitamin complex in the introduction to Part Two).	Potent antioxidant; supplies oxygen to cells; improves circulation and level of energy.

Comments

❑ Eat plenty of enzyme-rich fresh fruits and vegetables and other foods high in enzymatic activity, particularly proteolytic enzymes found in pineapple, papayas, and figs. These enzymes help fight inflammation and reduce swelling.

❑ A number of foods are rich in anthocyanidins enzyme helpers. Some foods include raspberries, cranberries, red wine, grapes, black currants, red cabbage, grape seed extract, and bilberry extract. Anthocyanidins maintain blood flow in the small vessels, protect blood vessels, inhibit edema, and fight inflammation and free radicals.

❑ Avoid ham, potato chips, salami, French fries, pretzels, and other foods high in salt or sodium. Use sea salt instead of table salt. Increase your potassium intake.

❑ Cleansing, fasting, and juicing are of benefit in assisting

your body in optimal function. Use juices from apples, pineapple, parsley, broccoli, and carrots. See Part Three for more information on juicing and detoxification.

❑ Weight control is important. Excess weight causes greater stress on the wrist and carpal tunnel. Therefore, weight loss helps decrease CTS in many people.

❑ Splints or wrist supports may be necessary to support and immobilize the wrist. These can be obtained at a health-food store, drug store, or from a health-care professional, such as a chiropractor.

❑ If possible, remove the cause of the irritation. Review your job tasks, your work at home, and your recreational activities. Avoid any movements, postures, or activities that place stress on, or tend to aggravate, your wrist action. It may be necessary to seek a different line of work.

❑ If your job requires repetitive hand movements or repeated twisting of the wrists, try altering your work place or periodically changing your job task to give your wrists a rest. For example, if you are a secretary or computer operator, check the height of your chair and its distance from your computer, the size and location of your computer screen, the angle you must place your fingers and wrist when typing, and so on. Wrist support pads placed in front of your keyboard are sometimes of help. Egghead Software and other computer retailers carry "natural" keyboards, engineered to more closely align with the natural position of your arms and hands. Talk to a number of experts and get their opinions. Employers are usually very cooperative and will work with you. With today's skyrocketing health-care costs, any successful employer wants to keep on-the-job injury rates as low as possible.

❑ Ice should be used for the first one to three days of treatment. Fill two or three small paper cups with water and freeze. When frozen, take one out and tear off the top half-inch of paper, revealing the ice. Rub the affected area in a circular motion with the ice for one minute. Remove for one minute, then reapply for one minute. Do this five times. The ice should be applied in this manner a minimum of three times a day for the first three days. The purpose is to stimulate blood flow to the area, remove waste from the area, and reduce inflammation.

❑ After three days, alternate ice with heat packs every 15 minutes until pain subsides. "Blue ice" or a frozen vegetable packet works better for this application. Wrap it with a dish towel and apply to the painful area. For heat, use a hot water bottle or heating pad. If a hot water bottle is used, wrap in a towel and apply. If a heating pad is used, put a warm, moist towel on the affected area, then place a sheet of wax paper over it, then the heating pad, and finally, a large bath towel. Make sure the wax paper is wide and long enough to cover the entire heating pad with at least one inch more extending out on every side. The heating pad should never touch the moist towel or the patient's skin. Great care should be exercised to ensure that this does not happen.

❏ After following these recommendations, if you see no improvement, visit a chiropractor, naturopath, or other healthcare provider. They can devise a treatment program that may include physiotherapy, such as ultrasound, acupressure, acupuncture, or applied kinesiology.

Cataracts

Vision requires the teamwork of many parts of the eye. Light enters through the iris and is received by the lens, which focuses the light on the retina, which senses the light and sends impulses to the brain through the optic nerve. The brain then interprets these impulses as a visual image. So, you see, if just one of these structures of the eye is not working correctly, vision is affected.

Cataracts are a clouding of the lens of the eye. If the lens is clouded, all light will not pass through the lens. This will cause distorted vision. The exact cause of cataracts is unknown, although they often result from the use of certain drugs; exposure to radiation, such as X-rays and sunlight; certain eye diseases; and as a complication of other diseases, such as diabetes. Senile cataracts occur with aging. Rarely, cataracts are present at birth (congenital cataracts) or develop in young people. Cataracts are the leading cause of blindness in the world.

When cataracts are still in the early stages, eyeglasses may improve vision. But as they advance and impair vision more and more, surgery is necessary to restore vision. This procedure involves the removal of the lens of the eye and replacement with an artificial lens. The key is to prevent cataract formation by improving circulation to the eyes.

Symptoms

Cataracts progressively blur vision. The degree of vision loss depends on the location of the cataract and its thickness. A cataract at the back of the lens (posterior subcapsular cataract) affects vision the greatest because it is located at the point where the light is supposed to pass on to the retina. Also, the thicker a cataract is, the more it will affect vision. Cataracts are generally painless, although rarely, swelling of the lens may result, or they may lead to glaucoma, which can be painful.

Enzyme Therapy

Systemic enzyme therapy is used to improve circulation to the eye. Enzymes help speed tissue repair, bring nutrients to the damaged area, remove waste products, and enhance wellness.

Digestive enzyme therapy is used to improve the digestion and assimilation of food nutrients, reduce stress on the gastrointestinal mucosa, help maintain normal pH levels, detoxify the body, and promote the growth of healthy intestinal flora, thus making the body healthier so that it is better able to combat cataracts. Digestive enzymes also serve as replacements for the body's pancreatic enzymes, leaving the pancreatic enzymes free to perform other functions in the body, such as improving circulation and decreasing inflammation.

Enzyme Absorption System Enhancers (EASE) maximize enzyme activity. They can also improve the absorption and bioavailability of various nutrients and decrease the drain on the body's own digestive enzymes, thus prolonging their lives.

In addition to the standard multivitamin, multimineral, and multiglandular complexes recommended in the introduction to Part Two, the following supplements are recommended for the treatment of cataracts. The following recommended dosages are for adults. Children between the ages of two and five should take one-quarter the recommended dosage, those between the ages of six and twelve should take one-half the recommended dosage, and those between the ages of thirteen and seventeen should take three-quarters the recommended adult dosage.

ESSENTIAL ENZYMES

Enzyme	Suggested Dosage	Actions
Superoxide dismutase	Take 150 mcg per day.	Fights inflammation; potent antioxidant.

AND SELECT ONE OF THE FOLLOWING:

Enzyme	Suggested Dosage	Actions
Formula One—See page 38 of Part One for a list of products.	Take 5 tablets twice per day.	Fights inflammation, pain, and swelling; speeds recovery after injury or surgery; stimulates the immune system; strengthens capillary walls.
Bromelain	Take three to four 250 mg capsules per day until symptoms subside. With advanced cataracts, take 6–8 capsules per day.	Fights inflammation and swelling; speeds healing; stimulates the immune system.
Serratiopeptidase	Take 10 mg two times per day between meals.	Fights inflammation; reduces swelling.
Trypsin with chymotrypsin	Take 250 mg of trypsin and 10 mg of chymotrypsin three times per day between meals.	Fights inflammation and swelling; increases rate of healing; stimulates the immune system.
Pancreatin	Take 500 mg six times per day on an empty stomach.	Fights inflammation and swelling; increases rate of healing; stimulates the immune system; strengthens capillary walls.
Papain	Take 300 mg two times per day between meals.	Fights inflammation; speeds recovery from surgery.
Microbial protease	Take 300 mg two times per day between meals.	Fights inflammation; speeds recovery from surgery.

SUPPORTIVE ENZYMES

Enzyme	Suggested Dosage	Actions
Catalase and glutathione peroxidase	Take as indicated on the labels.	Potent antioxidant enzymes.
A digestive enzyme product containing protease, amylase, and lipase enzymes. See the list of digestive enzyme products in Appendix.	Take as indicated on the label with meals.	Improves the breakdown and absorption of food nutrients; fights toxic build-up.

ENZYME ABSORPTION SYSTEM ENHANCERS (EASE)

Combination	Suggested Dosage	Actions
Antioxidant Complex from Tyler or	Take as indicated on the label.	Potent free-radical fighter and immune system stimulator.
Detox Enzyme Formula from Prevail or	Take 1 capsule between meals one or two times per day with a 6- to 8-ounce glass of water or juice.	Helps rid the body of harmful toxins.
Hepazyme from Enzymatic Therapy or	Take as indicated on the label.	Stimulates immune function.
Kyolic Super Formula 102 (with enzyme complex) or	Take 2 capsules or tablets with meals two to three times per day.	Strengthens the immune system; helps alleviate stress; helps fight free radicals; helps fight inflammation.
Oxy-5000 Forte from American Biologics or	Take as indicated on the label.	Helps the body excrete toxins; fights biochemical stressors, infections, and degenerative processes.
Vita-C-1000 from Life Plus	Take as indicated on the label.	Helps maintain vision; boosts immune function.

ENZYME HELPERS

Nutrient	Suggested Dosage	Actions
Acidophilus or other probiotic	Take as indicated on the label.	Improves gastrointestinal function and aids in the digestion and absorption of nutrients, which improves the health of the body; stimulates enzyme activity.
Adrenal glandular	Take 100–400 mg per day.	Fights inflammation and adrenal exhaustion.
Aged garlic extract	Take 2 capsules three times per day.	Fights inflammation; stimulates the immune system; potent antioxidant.
Bioflavonoids, such as rutin or quercetin	Take 500 mg three times per day between meals.	Strengthen capillaries; potent antioxidant.
Coenzyme Q$_{10}$	Take 30–60 mg one to two times per day with meals; or follow the advice	Potent antioxidant.

Nutrient	Suggested Dosage	Actions
	of your health-care professional.	
Essential fatty acid complex	Take 100–200 mg two to three times per day with meals; or follow the advice of your health-care professional.	Help strengthen cell membranes; support immune function.
Niacin or	Take 50 mg the first day, then increase dosage daily by 50 mg to 1,000 mg. If adverse effects occur, decrease dosage. Take doses higher than 1,000 mg only under the supervision of a physician.	Dilates blood vessels; improves circulation.
Niacin-bound chromium	Take 20 mg per day or as indicated on the label.	
Pycnogenol	Take as indicated on the label.	Potent antioxidant; improves capillary wall strength.
Vitamin A or	Take 10,000–15,000 IU per day. If pregnant, take no more than 10,000 IU per day.	Needed for strong immune system and healthy mucous membranes; potent antioxidant.
Carotenoids	Take 5,000–10,000 IU per day, or as indicated on the label.	Potent antioxidant; enhances immunity.
Vitamin B complex	Take 150 mg three times per day, or as indicated on the label.	B vitamins act as coenzymes.
Vitamin C	Take 10,000–15,000 mg per day.	Promotes immune system; powerful antioxidant.
Vitamin E	Take 1,200 IU per day from all sources (including the multivitamin complex in the introduction to Part Two).	Potent antioxidant; supplies oxygen to cells; improves circulation and level of energy.
Zinc	Take 15–50 mg per day. Do not take more than 100 mg per day from all sources.	Needed by over 300 enzymes; necessary for white blood cell immune function.

Comments

❏ Bilberry protects and strengthens the blood vessels that feed the eyes and supports arteries and capillaries.

❏ Garlic, onions, shallots, chives, leeks, and scallions contain allium compounds that are potent free-radical fighters, help lower cholesterol, and fight circulatory disorders.

❏ Raspberries, cranberries, grapes, red wine, black currants, and red cabbage contain anthocyanidins, which reduce blood vessel plaque formation and maintain blood flow in small vessels.

❏ See the list of Resource Groups in the Appendix for a list of organizations that can provide you with more information about cataracts.

Celiac Disease

Celiac disease is an inherited chronic digestive malabsorption disorder. Individuals with this disease have an allergic reaction to foods containing gluten, a protein found in some grains, such as rye, barley, and wheat. When celiac disease patients ingest gluten, their bodies see it as an antigen and releases antibodies. The immune complexes that are formed damage the lining of the small intestine, making the digestion and absorption of food difficult. Frequently, celiac disease patients lack some digestive enzymes, which normally break down toxins and help flush them from the gastrointestinal tract (see "Percentage of Celiac Patients With Enzyme Deficiencies" below). This allows toxins to accumulate in the epithelium, resulting in further damage to the intestine.

According to the Celiac Disease Foundation, only about 1 in every 2,500 Americans suffers from celiac disease. However, some populations have higher rates, such as in Ireland, where the incidence is 1 in every 100 to 300 people, and in their neighboring country England, the incidence is 1 in 500 to 2,000. Celiac disease primarily affects Caucasians. It is rarely found in the Chinese, Japanese, or Africans.

Celiac disease is genetic and is also associated with other immune disorders, including asthma, rheumatoid arthritis, thyroid disease, pulmonary diseases, serum IgA deficiency, IgA nephropathy, Sjögren's syndrome, glomerulonephritis, systemic lupus erythematosus (SLE), Crohn's disease, insulin-dependent diabetes mellitus, vasculitis, sarcoidosis, liver disorders, scleroderma, polymyositis, and ulcerative colitis.

Trauma can sometimes trigger the onset of celiac disease. Factors such as pregnancy, environmental conditions, emotional stress, viral infections, physical trauma, and surgery can promote celiac disease in susceptible individuals.

Symptoms

The symptoms of celiac disease can occur at any age. In infants, celiac disease can cause vomiting and growth failure. According to the Celiac Disease Foundation, the most common symptoms are intestinal gas, abdominal cramping, bloating, and distention. There can be constipation or diarrhea, and stools may be bulky, pale, and very foul-smelling. The individual may be anemic and suffer from weakness, lack of energy, fatigue, headaches, or mental confusion. Steatorrhea can occur, as can weight loss, bone pain, and "burning" feet. The individual may be depressed and irritable. Dermatitis herpetiformis, a skin condition marked by severe itching and blistering of the skin, can also occur.

Because the symptoms of celiac disease are varied and often vague, patients can sometimes suffer from the condition for many years before a correct diagnosis is made. Dental problems and osteoporosis may result because of poor mineral absorption (calcium, magnesium, and phosphorus are critical for healthy teeth and bones). Difficulties absorbing vitamins and other minerals, such as iron, can lead to a variety of anemias and nutritional deficiencies. But celiac disease can also cause severe intestinal damage, which is probably its most serious consequence.

Enzyme Therapy

Digestive enzyme therapy is used to improve the digestion and assimilation of food nutrients and reduce stress on the gastrointestinal mucosa. Enzymes help maintain normal pH levels, detoxify the body, promote the growth of healthy intestinal flora, strengthen the body as a whole, and build general resistance. Digestive enzymes also serve as replacements for the body's pancreatic enzymes, leaving the pancreatic enzymes free to perform other functions in the body, such as boosting immunity and decreasing inflammation.

Percentage of Celiac Patients With Enzyme Deficiencies

Because celiac disease can damage the lining of the small intestine, it can interfere with the production of important digestive enzymes, such as lactase, sucrase, and maltase. The following chart indicates the occurrence of deficiencies of these enzymes in celiac disease patients.

SEVERITY OF CELIAC DISEASE	PERCENT DEFICIENT IN LACTASE	PERCENT DEFICIENT IN SUCRASE	PERCENT DEFICIENT IN MALTASE
Mild	25	0	0
Moderate	100	80	20
Severe	100	100	75

Adapted from Gray, G.M. "Disaccharidases of the Small Intestine in Selected Diseases," In: Katz, D.D., ed., Human Health and Disease, Bethesda: FASEB, 1980: p 124.

Enzyme Absorption System Enhancers (EASE) maximize enzyme activity. They can also improve the absorption and bioavailability of various nutrients and decrease the drain on the body's own digestive enzymes, thus prolonging their lives.

According to Drs. Michael Murray and Joseph Pizzorno in their book *An Encyclopedia of Natural Medicine* (Rocklin, CA: Prima Publishing, 1991), taking a papain supplement (500 to 1,000 mg) with meals may enable some individuals to tolerate gluten.

Oral papain was used in an adult suffering from celiac disease with partial atrophy of the intestinal villi (small, fingerlike projections in the intestinal wall) and intestinal malabsorption. He was placed on a gluten-free diet, and he gained some weight. Although his symptoms improved, his steatorrhea continued and after six months, he continued to have absorption difficulties. The patient began taking 1,800 mg of enterically coated papain tablets with each meal (papain is able to digest wheat gluten). He reported that he did not suffer from loose stools as frequently, and there were no side effects from the papain. A month after beginning the papain therapy, absorption had improved to normal. The patient abandoned his gluten-free diet and expected a relapse. However, no problems developed.

In addition to digestive enzyme therapy, systemic enzyme therapy is used as supportive care to reduce inflammation, stimulate the immune system, and improve circulation. Because celiac disease patients frequently suffer damage to the intestinal lining, systemic enzyme therapy can help speed tissue repair, bring nutrients to the damaged area, remove waste products, and enhance wellness.

In addition to the standard multivitamin, multimineral, and multiglandular complexes recommended in the introduction to Part Two, the following supplements are recommended for the treatment of celiac disease. The following recommended dosages are for adults. Children between the ages of two and five should take one-quarter the recommended dosage, those between the ages of six and twelve should take one-half the recommended dosage, and those between the ages of thirteen and seventeen should take three-quarters the recommended adult dosage.

ESSENTIAL ENZYMES

Enzyme	Suggested Dosage	Actions
Papain *or*	Take 1,800–2,000 mg three times per day with meals.	Potent proteolytic enzyme; breaks down proteins.
A digestive enzyme product rich in proteases *or*	Take as indicated on the label.	Improves digestion and breaks down proteins.
Bromelain	Start with one 230–250 mg capsule per day and increase dosage gradually, if necessary.	Improves digestion; fights inflammation, pain, and swelling; stimulates the immune system.
or	Take no more than 3 capsules per day. If diarrhea develops, decrease dose.	
Pancreatin *or*	Take 300 mg three times per day with meals.	Contains proteases, amylases, and lipases, so is able to digest proteins, carbohydrates, and fat.
Wobenzym N from Naturally Vitamin Supplements	Take 2–3 tablets between meals.	Improves digestion; contains proteases, amylases, and lipases.

SUPPORTIVE ENZYMES

Enzyme	Suggested Dosage	Actions
Formula Three—See page 38 of Part One for a list of products.	Take as indicated on the labels.	Potent antioxidant enzymes.

ENZYME ABSORPTION SYSTEM ENHANCERS (EASE)

Combination	Suggested Dosage	Actions
Detox Enzyme Formula from Prevail *or*	Take 1 capsule between meals one or two times per day with a 6- to 8-ounce glass of water or juice.	Helps rid the body of harmful toxins.
Hepazyme from Enzymatic Therapy *or*	Take as indicated on the label.	Stimulates immune function.
Kyolic Formula 102 (with enzymes) *or*	Take 2 capsules or tablets between meals two to three times per day.	Strengthens the immune system; helps alleviate gastrointestinal stress; helps fight free radicals; improves circulation; helps fight inflammation; improves digestion and absorption of fats, carbohydrates, and proteins.
Oxy-5000 Forte from American Biologics *or*	Take as indicated on the label.	Helps the body excrete toxins; fights biochemical stressors, infections, and degenerative processes.
Pro-Flora from Tyler Encapsulations *or*	Take as indicated on the label.	Helps restore normal digestion and gastrointestinal flora.
Similase from Tyler Encapsulations *or*	Take as indicated on the label.	Assists the body in digestion and absorbing protein and other nutrients.
Vita-C-1000 from Life Plus	Take as indicated on the label.	Boosts immune function; fights infection.

ENZYME HELPERS

Nutrient	Suggested Dosage	Actions
Acidophilus or other probiotics, such as bifidobacterium or Acidophilasé from Wakunaga, which	Take as indicated on the label.	Improves gastrointestinal function; needed for production of biotin, niacin, folic acid, and pyridoxine, which are required by many enzymes; helps pro-

contains protease, amylase, and lipase enzymes.		duce antibacterial substances; stimulates enzyme activity.
Aged garlic extract	Take two 300 mg capsules three times per day.	Potent antioxidant; improves digestion; stimulates healthy intestinal bacteria.
Betaine hydrochloric acid	Take 150 mg three times per day with meals.	Provides hydrochloric acid to improve digestion. If gastric upset occurs, discontinue use.
Calcium	Take 500 mg per day.	Those with celiac disease often have low levels of calcium in their bodies.
Fiber	Take as indicated on the label.	Provides bulk; needed for proper elimination.
Folic acid	Take 50 mcg three times per day.	Required for proper cell division and synthesis of nucleotides; maintains good appetite; important for red blood cell formation, body growth, protein metabolism, reproduction, cell division, and hydrochloric acid production.
Fructooligo- saccharides	Take as indicated on the label.	Increases peristaltic action of intestines; improves liver function.
Lecithin	Take 1,000 mg three times per day with meals.	Emulsifies fat.
Magnesium	Take 250 mg per day.	Plays important roles in energy production and calcium absorption; necessary for glucose metabolism, blood clotting, and the synthesis of fats, proteins, and nucleic acid.
Niacin-bound chromium	Take 10 mg three times per day.	Stimulates gastric juices and hydrochloric acid; important in energy metabolism; supports healthy gastrointestinal tract.
Ox bile extract	Take 120 mg per day.	Aids digestion.
Vitamin B$_1$ (thiamin)	Take 150 mg three times per day.	Essential for carbohydrate metabolism, proper energy production, nerve function, and appetite.
Vitamin B$_6$ (pyridoxine)	Take 50 mg three times per day.	Assists enzymes that metabolize amino acids and fats; necessary for proper immune function; needed for healthy nerves, muscle, blood, and skin.
Vitamin C	Take 5,000 mg three times per day.	Stimulates the adrenal glands to produce hormones (which also act as coenzymes); aids in detoxification; is a potent antioxidant; improves the function of many enzymes; necessary for a healthy digestive tract.
Whey	Take 1 heaping tablespoon three times per day with meals; or take as indicated on the label.	Improves digestion.

Zinc	Take 10–50 mg per day. Do not exceed 100 mg per day from all sources.	Needed by over 300 enzymes. Many zinc enzymes work to metabolize carbohydrates, alcohol, and essential fatty acids; synthesize proteins; dispose of free radicals; and manufacture heme.

Comments

❏ Eliminate all sources of gluten from your diet, such as wheat (including kamut, einkorn, and spelt), oats, barley, rye, and triticale. This also includes anything made from these grains, such as soy sauce, malt, textured or hydrolyzed vegetable and plant proteins, grain alcohols, grain vinegars, and some natural flavorings. A number of prescription medicines and vitamins use grains as binders and fillers. Read labels carefully.

❏ Even small amounts of gluten can cause problems. Contact a doctor, nutritionist, or dietitian for a list of gluten-free foods. It is difficult to avoid gluten because it is widely used in processing so many foods (including hot dogs, soups, ice cream, and sauces).

❏ Substitute rice or corn for wheat, rye, barley, or oats. Other grasses, such as millet, Job's tears, ragi, teff grass, and sorghum may be safe for celiac disease patients, but their effects on patients have not yet been studied adequately.

❏ A number of herbs can improve general digestion, including ginger root and licorice root; as can foods rich in chlorophyll, such as chlorella, spirulina, alfalfa concentrate, and kelp.

❏ Cleansing the body helps remove the toxins that can aggravate gluten intolerance. The Enzyme Toxin Flush and Enzyme Kidney Flush further help to remove toxins from the body. See Part Three for directions.

❏ See the list of Resource Groups in the Appendix for a list of organizations that can provide you with more information about celiac disease.

Chickenpox

Chickenpox is an acute viral infection marked by a skin rash. This highly contagious disease usually lasts about two weeks after symptoms appear, and it is not considered serious for most; however, the disease is usually more severe in adults and can be very serious in newborns. Chickenpox is spread by contact with someone who has the disease, or by breathing infected droplets in the air. It is one of the most common childhood diseases, but is rare among adults, probably because most of us are exposed to and develop immunity to it in childhood. Once one contracts chickenpox, one generally has a lifetime immunity to the illness; however, the virus that causes chickenpox (varicella zoster virus) remains dormant in the body. Sometimes the virus is reactivated many years later, usually after the age of fifty, in the form of *shingles*. The virus can be reactivated by decreased immunity, an illness, or a drug.

Symptoms

Fever, headache, and loss of appetite usually occur seven to twenty-one days after exposure. About a day later, a rash of small, red marks develops. This rash can appear anywhere on the body. These marks become blisters in a day or so. The fluid leaks from the blisters, forming a crust. New crops of blisters appear every three to four days. The blisters are no longer infectious once they are completely dry.

The rash and subsequent scabs that form in this disease can be very itchy, but avoid scratching, as the scabs can become infected and leave scars. Symptoms and their complications are worse in adults and newborns than they are in children. Pneumonia may be a complication in these high risk groups.

Enzyme Therapy

Systemic enzyme therapy is used to fight the viral attack, stimulate the immune system, and reduce inflammation. Enzymes improve circulation, help speed tissue repair (by bringing nutrients to the damaged area and removing waste products), and enhance wellness.

Digestive enzyme therapy is used to improve the digestion of food, reduce stress on the gastrointestinal mucosa, help maintain normal pH levels, detoxify the body, promote the growth of healthy intestinal flora, strengthen the body as a whole, and build general resistance, all of which are essential for boosting immune function. Digestive enzymes also serve as replacements for the body's pancreatic enzymes, leaving the pancreatic enzymes free to perform other functions in the body, such as decreasing inflammation.

Enzyme Absorption System Enhancers (EASE) maximize enzyme activity. They can also improve the absorption and bioavailability of various nutrients and decrease the drain on the body's own digestive enzymes, thus prolonging their lives.

In addition to the standard multivitamin, multimineral, and multiglandular complexes recommended in the introduction to Part Two, the following supplements are recommended for the treatment of chickenpox. The following recommended dosages are for adults. Children between the ages of two and five should take one-quarter the recommended dosage, those between the ages of six and twelve should take one-half the recommended dosage, and those between the ages of thirteen and seventeen should take three-quarters the recommended adult dosage.

ESSENTIAL ENZYMES

Enzyme	Suggested Dosage	Actions
Formula One—See page 38 of Part One for a list of products.	Take 10 tablets three times per day between meals for six months. Then decrease dosage to 5 tablets three times per day between meals.	Bolsters the immune system; fights free-radical formation; fights inflammation.
Papain	Take 500 mg three times per day between meals for six months. Then decrease dosage to 300 mg three times per day between meals.	Stimulates immune function; fights free-radical formation; stimulates immune function at the cellular level; fights inflammation.
Bromelain	Take 400 mg three times per day between meals for six months. Then decrease dosage to 250 mg three times per day between meals.	Bolsters immunity; fights free-radical formation; fights inflammation; fights viruses.
Any proteolytic enzyme	Take 10 tablets three times per day for six months. Then decrease dosage to 5 tablets three times per day.	Bolsters the immune system; fights free-radical formation; fights inflammation.
Trypsin with chymotrypsin	Take 200 mg of trypsin and 10 mg of chymotrypsin three times per day between meals for six months. Then decrease dosage to 125 mg of trypsin and 5 mg of chymotrypsin three times per day between meals.	Stimulates immune function; fights free radicals; fights inflammation.

SUPPORTIVE ENZYMES

Enzyme	Suggested Dosage	Actions
Formula Three—See page 38 of Part One for a list of products	Take as indicated on the label.	Free-radical fighters.
A digestive enzyme product containing protease, amylase, and lipase enzymes. See the list of digestive enzyme products in Appendix.	Take as indicated on the label with meals.	Improves the breakdown and absorption of food nutrients; fights toxic build-up.

ENZYME ABSORPTION SYSTEM ENHANCERS (EASE)

Combination	Suggested Dosage	Actions
Antioxidant Complex from Tyler *or*	Take as indicated on the label.	Potent free-radical fighter and immune system stimulator.
Detox Enzyme Formula from Prevail *or*	Take 1 capsule between meals one or two times per day with a 6- to 8-ounce glass of water or juice.	Helps rid the body of harmful toxins.

177

Hepazyme from Enzymatic Therapy *or*	Take as indicated on the label.	Stimulates immune function.
Kyolic Formula 102 (with enzymes) *or*	Take 2 capsules or tablets between meals two to three times per day.	Strengthens the immune system; improves circulation; fight inflammation.
NESS Formula #7 *or*	Take 2 capsules immediately after each meal; or take 4 capsules upon rising and at bedtime.	Helps the body utilize essential fatty acids.
Oxy-5000 Forte from American Biologics *or*	Take as indicated on the label.	Helps the body excrete toxins; fights biochemical stressors, infections, and degenerative processes.
Vita-C-1000 from Life Plus	Take as indicated on the label.	Boosts immune function.

ENZYME HELPERS

Nutrient	Suggested Dosage	Actions
Acidophilus	Take as indicated on the label.	Stimulates enzyme activity; improves gastrointestinal function, which helps detoxify the gastrointestinal tract.
Adrenal glandular	Take 100–400 mg per day.	Helps fight stress and adrenal exhaustion; fights inflammation.
Arginine	Take as indicated on the label.	Accelerates healing; enhances thymus activity.
Coenzyme Q_{10}	Take 30–60 mg one to two times per day with meals.	Important to immune system function; has antioxidant activity.
DHEA	Take 25 mg per day.	Supports immune, cardiovascular, and brain function.
Essential fatty acid complex	Take 100–200 mg two to three times per day with meals; or follow the advice of your health-care professional.	Strengthens immunity; helps strengthen cell membranes.
Kidney glandular	Take as indicated on the label.	Improves kidney function to improve the detoxification of the body.
Kyolic Aged Garlic Liquid Extract with Vitamins B_1 and B_{12}	Take 10 drops in food or liquid, or place directly on tongue three times per day.	Stimulates immune system function.
Selenium	Take 50 mcg one or two times per day.	Acts as an enzyme activator; required by glutathione peroxidase.
Spirulina	Take as indicated on the label.	Rich in minerals, vitamins, enzymes, proteins, essential amino acids, and essential fatty acids; improves immune function.
Thymus glandular	Take 100 mg per day; or take as indicated on the label.	Assists the thymus gland in immune resistance; improves immune response; fights infections.

Vitamin A *or*	Take up to 50,000 IU for no more than one month (for higher doses, see a physician). If pregnant, do not take more than 10,000 IU per day.	Needed for strong immune system and healthy mucous membranes; potent antioxidant.
Carotenoids	Take 5,000–25,000 IU per day; or take as indicated on the label.	Potent antioxidants and immune system stimulants.
Vitamin B complex	Take 100–150 mg three times per day.	B vitamins act as coenzymes.
Vitamin C	Take 10,000 mg per day; or 1–2 grams every hour until symptoms improve.	Fights infections, diseases, and allergies; improves wound healing; stimulates the adrenal glands; promotes immune system; powerful antioxidant.
Vitamin E	Take 1,200 IU per day from all sources (including the multivitamin complex in the introduction to Part Two).	Acts as an intercellular antioxidant; improves circulation; helps resist diseases at the cellular level.
Zinc	Take 50 mg per day. Do not take more than 100 mg per day from all sources.	Needed by over 300 enzymes; necessary for white blood cell immune function; improves healing.

Comments

❑ Since chickenpox is a viral disease, a patient's diet should stimulate and enrich the immune system, help resist infection, and reduce fever. Foods should include enzyme-rich, raw, fresh vegetables (such as cabbage, broccoli, spinach), citrus fruits, vegetable oils, fish-liver oils, brewer's yeast, whole grains, and unpolished rice (rice from which the hulls have not been removed).

❑ Detoxify to cleanse the body of toxins and decrease taxation of the body's enzymes. Perform the Enzyme Kidney Flush for one week. The following week, perform the Enzyme Toxin Flush. See Part Three for instructions for both detoxification flushes.

❑ Relieve the itching of chickenpox by taking a cold bath (throw in a handful of oatmeal or baking soda) or by applying calamine or witch-hazel lotion. Herbs, such as calendula and rosemary, can also help relieve the itching. Make a tea from about one ounce of either in four cups of water. Apply the strained liquid to the rash with a washcloth.

❑ *Warning:* Do not give aspirin to a child with chickenpox, as it has been connected to the development of Reye's syndrome, a potentially fatal condition.

❑ Anyone with chickenpox should be isolated and kept away from contact with others.

Cholesterol, Elevated

Cholesterol is a waxy substance produced primarily by the liver and absorbed from foods of animal origin in the diet. Though high levels of blood cholesterol have been implicated as a factor in arteriosclerosis and other cardiovascular problems, the body actually needs a certain amount of cholesterol to form adrenal and sex hormones, bile salts, and vitamin D. It is also required for nerve and brain function.

Some 60 million American adults have blood cholesterol levels high enough to require medical intervention. A change in diet is the primary treatment. Treatment is important because the chance of a heart attack is greatly increased by high blood cholesterol levels. In fact, research results from the well-known Framingham Study indicate that as blood cholesterol levels increase, coronary heart disease and its death rates also increase. This is because fatty deposits (composed of calcium and cholesterol) gather in the heart and arteries, decreasing the flow of blood.

Cholesterol must attach itself to a protein in order to travel through the bloodstream. This combination is called a *lipoprotein*. One's total serum cholesterol (the level of cholesterol in the blood) is composed of *low-density lipoproteins* (*LDLs*—the so-called bad cholesterol) and *high-density lipoproteins* (*HDLs*—the so-called good cholesterol). When the levels of LDLs become too high, they have a tendency to stick to the blood vessel walls, causing atherosclerosis. LDLs are actually the component of serum cholesterol that lead to cardiovascular disease. This is why they are referred to as "bad" cholesterol. HDLs, on the other hand, actually act as cholesterol scavengers, attracting cholesterol and bringing it back to the liver to be removed from the body. This is why they are termed "good" cholesterol.

The danger caused by high cholesterol in the blood is not only the fatty substance that attaches to our blood vessel walls. According to Lopez, Williams, and Miehlke in their book *Enzymes: The Foundation of Life* (Charleston, SC: Neville Press, Inc., 1994), cholesterol may take the form of pointed crystals that press into the blood vessel walls when blood pressure is high. The affected portion of the blood vessel loses its elasticity and cholesterol plaques form. This is the beginning of arteriosclerosis.

Dr. Wilhelm Glenk and Sven Neu postulate in their book *Enzyme* (Munich, Germany: Wilhelm Heyne Verlag, 1990) that the damaged blood vessel walls are mistakenly seen by the immune system as not part of the body but rather foreign matter that must be attacked and eliminated. Antibodies in the vessel walls mobilize against foreign bodies, which may result in immunologically caused vascular inflammation.

Symptoms

Like a thief in the night, symptoms of high cholesterol may not be noticeable until it is too late. Inner ear problems and hearing loss may be an indication. So may shortness of breath when climbing stairs. Because these symptoms may indicate other health problems, see a physician for laboratory blood tests to confirm the problem. Often, obesity, lack of exercise, or a family history of high blood cholesterol are indications that your blood cholesterol levels should be checked.

Enzyme Therapy

At present, there is a great deal of research in Germany on arteriosclerosis, low-density-lipoprotein cholesterol, and enzymes. Researchers have found that combinations of enzymes (bromelain and trypsin) with the bioflavonoid rutin reduce oxidized LDL cholesterol and increase nonoxidized high-density lipoprotein (HDL). According to Dr. Karl Ransberger, world-famous expert in systemic enzyme therapy, there is substantial evidence that enzyme treatment reduces arteriosclerosis and myocarditis.

In addition, Professor Heidland from the University of Würzburg, Germany demonstrated that a combination of trypsin, bromelain, and rutin reduces arteriosclerosis in kidney arterioles of kidney transplant patients. Arteriosclerosis is a very severe, life-threatening complication with kidney transplant patients. The rate of success using enzymes is substantial. Enzyme therapy has proven to be a significant discovery in the treatment of arteriosclerosis.

Elimination of elevated cholesterol, decreased inflammation, and an improvement in circulation and immune function can all occur by assisting the body's own enzymes and by taking enzyme supplements. Systemic enzyme therapy is used to decrease cholesterol levels (including low-density lipoproteins), help remove antigen-antibody complexes (which sometimes result), and reduce inflammation. Enzymes stimulate the immune system, improve circulation, help speed repair of any tissue damaged by cholesterol deposition, bring nutrients to the damaged area, remove waste products, and improve health. Digestive enzyme therapy is used to improve the digestion of food, reduce stress on the gastrointestinal mucosa, and help maintain normal pH levels. Digestive enzymes can release the nutrients in your foods that are effective at lowering blood cholesterol levels.

Enzyme Absorption System Enhancers (EASE) maximize enzyme activity. They can also improve the absorption and bioavailability of various nutrients and decrease the drain on the body's own digestive enzymes, thus prolonging their lives.

In addition to the standard multivitamin, multimineral, and multiglandular complexes recommended in the introduction to Part Two, the following supplements are recommended for the treatment of hypercholesterolemia. The following recommended dosages are for adults. Children between the ages of two and five should take one-quarter the recommended dosage, those between the ages of six and twelve should take one-half the recommended dosage, and those between the ages of thirteen and seventeen should take three-quarters the recommended adult dosage.

ESSENTIAL ENZYMES

Enzyme	Suggested Dosage	Actions
Formula Three—See page 38 of Part One for a list of products.	Take 200 mcg three times per day between meals.	Free-radical fighters.

AND SELECT ONE OF THE FOLLOWING:

Enzyme	Suggested Dosage	Actions
Phlogenzym from Mucos	Take 4–5 tablets three times per day between meals.	Can lower cholesterol; dissolves blood clots; fights inflammation, pain, and swelling; speeds healing; strengthens capillary walls.
Formula One (particularly Wobenzym N from Naturally Vitamin Supplements)—See page 38 of Part One for a list of products.	Take 8–10 tablets three times per day between meals for six months or until symptoms improve. Thereafter, take 6 tablets three times per day between meals.	Dissolves blood clots; fights inflammation, pain, and swelling; speeds healing; strengthens capillary walls.
Inflazyme Forte from American Biologics	Take as indicated on the label between meals.	Dissolves blood clots; fights inflammation, pain, and swelling; speeds healing; strengthens capillary walls.
Bromelain	Start with one 250 mg capsule per day. Increase dosage, if necessary. Take no more than 2 or 3 capsules per day. If diarrhea develops, decrease dosage.	Improves digestion; fights inflammation, pain, and swelling; stimulates the immune system.
A proteolytic enzyme or mixture	Take as indicated on the label between meals.	Dissolves blood clots; fights inflammation, pain, and swelling; speeds healing; strengthens capillary walls.

SUPPORTIVE ENZYMES

Enzyme	Suggested Dosage	Actions
A digestive enzyme formula containing lipase, amylase, and protease enzymes. See the list of digestive enzyme products in Appendix.	Take 2–3 tablets three times daily; or take as indicated on the label. Take with meals or 30 minutes before meals.	Improves the breakdown and absorption of proteins, carbohydrates, and fats; fights toxic build-up. Lipase enhances lipid metabolism and fat digestion.

ENZYME ABSORPTION SYSTEM ENHANCERS (EASE)

Combination	Suggested Dosage	Actions
AminoLogic from PhysioLogics	Take 1 capsule three times per day between meals; or follow the advice of your health-care professional.	Helps maintain normal cholesterol levels; contains enzymes to stimulate immune function; aids improved nutrient absorption.
or		
Kyolic Formula 102 (with enzymes)	Take 2 capsules or tablets per day as desired.	Lowers cholesterol; strengthens the immune system; helps alleviate stress; helps reverse circulatory disorders; helps fight free radicals; helps fight inflammation; improves digestion.
or		
NESS Formula #13	Take 2 capsules immediately after each meal; or take 3 capsules between meals.	Aids in fat metabolism and maintaining healthy cell function.
or		
Vita-C-1000 from Life Plus	Take as indicated on the label.	Lowers cholesterol.

ENZYME HELPERS

Nutrient	Suggested Dosage	Actions
Adrenal glandular	Take as indicated on the label.	Helps fight stress and adrenal exhaustion.
Aged garlic extract	Take 3 capsules three times per day.	Lowers cholesterol and blood pressure; fights free radicals.
Amino-acid complex including carnitine	Take 500 mg two times per day.	Carnitine increases energy; lowers cholesterol; reduces risk of heart disease.
Bioflavonoids, including pycnogenol, quercetin, grape seed extract, and rutin	Take 500 mg three times per day between meals.	Fight atherosclerosis; potent antioxidants; reduce capillary fragility.
Carotenoids (such as alpha- or beta-carotene, xanthophylls, cryptoxanthin, lutein, zeaxanthin)	Take 15,000–30,000 IU per day; or take as indicated on the label.	Potent antioxidants; reduce arterial plaque.
Coenzyme Q_{10}	Take 75 mg one to two times per day with meals.	Potent antioxidant; prevents heart attacks; decreases blood pressure and total serum cholesterol.
Gamma-linolenic acids	Take as indicated on the label.	Lowers blood pressure and high serum cholesterol levels; inhibits platelet aggregation; decreases vascular obstruction.
Green barley extract	Take 1–2 teaspoons in water or juice three times per day.	Reduces blood pressure; effective against heart disease; fights free radicals.
Heart glandular	Take 140 mg; or take as indicated on the label.	Improves heart function.
Lecithin	Take 3 capsules with meals; or take as indicated on the label.	Emulsifies fat.
Niacin-bound chromium	Take 20 mg per day; or take as indicated on the label.	Improves circulation by dilating blood vessels; can lower cholesterol levels in certain individuals.
Vitamin B complex	Take 50 mg three times per day; or take as indicated on the label.	B vitamins act as coenzymes to cholesterol-lowering enzymes.

Vitamin C	Take 5,000–10,000 mg per day; increase dosage as your doctor indicates.	Essential to vascular function; potent antioxidant; improves the function of many enzymes; assists in converting cholesterol to bile salt.
Vitamin E	Take 1,200 IU per day from all sources (including the multivitamin complex in the introduction to Part Two).	Potent antioxidant; helps prevent cardiovascular disease; stimulates circulation; strengthens capillary walls; prevents blood clots; maintains cell membranes.

Comments

❑ Eat foods high in proteolytic enzymes, such as pineapples, figs, and papaya.

❑ Apples contain pectin, which can lower cholesterol and decrease the risk of heart disease.

❑ Garlic, onions, shallots, chives, leeks, and scallions contain allium compounds, which are potent free-radical fighters that help lower cholesterol and fight circulatory disorders.

❑ Raspberries, cranberries, grapes, red wine, black currants, and red cabbage contain anthocyanidins, which reduce blood vessel plaque formation and maintain blood flow in small vessels.

❑ Oolong and green tea contain epicatechin, which is a potent antioxidant that lowers blood pressure and strengthens capillaries.

❑ One's diet should be low in fats and refined sugars, but high in complex carbohydrates and fiber (noted for its ability to lower cholesterol levels). One study indicated that individuals following such a cholesterol-reducing diet experienced a 9-percent greater reduction in blood cholesterol levels and had a 19-percent lower incidence of coronary heart disease than those who were not on a cholesterol-lowering diet. So diet does make a difference. Remember, cholesterol does not occur in vegetable fat. It is only found in animal fat. For further information on dietary recommendations, see Part Three.

❑ Drink at least two 8-ounce glasses of enzyme-rich fresh fruit and vegetable juice per day, such as pineapple, papaya, grape, parsley, and broccoli juices. For further information about juicing, turn to Part Three.

❑ A number of herbs are noted for their ability to lower cholesterol, including chlorella, ginseng, spirulina, and maitake, reishi, and shiitake mushrooms.

❑ Exercising thirty to forty minutes per day, having a positive mental attitude, and meditating fifteen minutes per day can stimulate enzymatic activity in the body.

❑ Many physicians prescribe cholesterol-lowering drugs for their patients who suffer from high cholesterol levels. In fact, there has been more than a tenfold increase in prescriptions for lipid-lowering drugs in the past decade (more than 26 million prescriptions were written for these drugs in the United States during 1992). There is some evidence that most of these cholesterol-lowering drugs either cause or promote cancer in test animals. Patients for whom these drugs are prescribed are exposed over time to doses that have been shown to cause cancer in animals. Therefore, use of these drugs should be under the supervision of a well-trained physician and restricted to individuals who are at high risk of imminent death from cardiovascular disease.

Chronic Fatigue Syndrome

Chronic fatigue syndrome (CFS) is a baffling and complex condition marked by persistent fatigue. More than one million Americans suffer from this disease and most of them (about two-thirds) are young, middle-class women.

Though CFS seems to involve an immune system dysfunction, the exact nature of this dysfunction is not known, although it may relate to the central nervous system. CFS may begin after an acute viral infection of some type, such as with the Epstein-Barr virus, herpes virus, cytomegalovirus, or infectious mononucleosis. In some individuals, symptoms began after a bout of bronchitis, hepatitis, or other infection. Most likely, CFS develops as a combination of a number of factors.

Since the cause of CFS is probably a combination of factors, if any of the following disorders are present, see their sections for their enzyme therapy programs as well: ALLERGIES; CARDIOVASCULAR DISORDERS; INDIGESTION; and MULTIPLE SCLEROSIS.

Symptoms

Probably, the most striking symptom of chronic fatigue syndrome is a debilitating, persistent, and profound exhaustion, along with poor stamina. Other symptoms include low-grade fever; swollen, painful lymph nodes; sore throat; muscle aches and weakness; headaches; gastrointestinal upset; joint pain; sleep disturbances; and neurological complaints, such as depression, irritability, forgetfulness, and confusion. Diagnosing CFS is difficult because these symptoms mimic so many other diseases. In fact, CFS is often misdiagnosed. Complicating the problem is the fact that there is no definitive diagnostic test for CFS.

According to the CFIDS Association of America, Inc., a patient is considered to be suffering from CFS if he or she meets both of the following criteria:

1. Clinically evaluated, unexplained, persistent or relapsing chronic fatigue that is of definite or new onset (not lifelong), is not substantially alleviated by rest, is not the result of

ongoing exertion, and results in substantial reduction in previous levels of educational, occupational, personal, or social activities.

2. The concurrent occurrence of four or more of the following symptoms: substantial impairment in concentration or short-term memory: tender lymph nodes; sore throat; multi-joint pain with redness or swelling; muscle pain; headache of a new type, severity, or pattern; malaise after exercise that lasts more than twenty-four hours; and unrefreshing sleep. These symptoms must not predate the fatigue and must have recurred or persisted during six or more consecutive months of illness. Also, all other causes must have first been ruled out.

Enzyme Therapy

The purpose of enzyme treatment for chronic fatigue syndrome is primarily to relieve specific symptoms. However, since enzymes have a broad spectrum of applications, enzymes can naturally adapt to a person's needs. Pain, allergies, and gastrointestinal difficulties are some of the symptoms that can be relieved through the use of enzymes, enzyme helpers, and diet changes.

Since proteolytic enzymes are extremely effective in treating viral, immune, and inflammatory conditions and in breaking up antigen-antibody complexes, enzymes are used in the treatment of chronic fatigue syndrome. Systemic enzyme therapy is used as supportive care to reduce pain and inflammation, stimulate the immune system, and improve circulation. Enzymes transport nutrients throughout the body, remove waste products, and enhance wellness.

Digestive enzyme therapy is used to improve the digestion of food and alleviate some of the gastrointestinal difficulties that often accompany CFS. Enzymes can reduce stress on the gastrointestinal mucosa, help maintain normal pH levels, detoxify and promote the growth of healthy intestinal flora. Digestive enzymes also serve as replacements for the body's pancreatic enzymes, leaving the pancreatic enzymes free to perform other functions in the body, such as boosting immunity, decreasing inflammation, and improving circulation.

Enzyme Absorption System Enhancers (EASE) maximize enzyme activity. They can also improve the absorption and bioavailability of various nutrients and decrease the drain on the body's own digestive enzymes, thus prolonging their lives.

In addition to the standard multivitamin, multimineral, and multiglandular complexes recommended in the introduction to Part Two, the following supplements are recommended for the treatment of chronic fatigue syndrome. The following recommended dosages are for adults. Children between the ages of two and five should take one-quarter the recommended dosage, those between the ages of six and twelve should take one-half the recommended dosage, and those between the ages of thirteen and seventeen should take three-quarters the recommended adult dosage.

ESSENTIAL ENZYMES

Enzyme	Suggested Dosage	Actions
Formula One—See page 38 of Part One for a list of products. *or*	Take 10 tablets three times per day between meals for six months. Then decrease dosage to 5 tablets three times per day between meals.	Bolsters the immune system; breaks up antigen-antibody complexes; fights free-radical formation; fights inflammation.
Papain *or*	Take 500 mg three times per day for six months between meals. Then decrease dosage to 300 mg three times per day between meals.	Stimulates immune function; breaks up circulating immune complexes; fights free-radical formation; fights inflammation.
Bromelain *or*	Take 400 mg three times per day between meals for six months. Then decrease dosage to 250 mg three times per day between meals.	Bolsters the immune system; breaks up circulating immune complexes; fights free-radical formation; fights inflammation.
Trypsin with chymotrypsin *or*	Take 200 mg of trypsin and 10 mg of trypsin three times per day between meals for six months. Then decrease dosage to 125 mg of trypsin and 10 mg of chymotrypsin three times per day between meals.	Stimulates immune function; breaks up antigen-antibody complexes; fights free-radical formation; fights inflammation.
Any proteolytic enzyme	Take 10 tablets three times per day for six months. Then decrease dosage to 5 tablets three times per day.	Bolsters the immune system; breaks up antigen-antibody complexes; fights free-radical formation; fights inflammation.

SUPPORTIVE ENZYMES

Enzyme	Suggested Dosage	Actions
Formula Three—See page 38 of Part One for a list of products.	Take as indicated on the label.	Free radical fighters.
A digestive enzyme product containing protease, amylase, and lipase enzymes. See the list of digestive enzyme products in Appendix.	Take as indicated on the label with meals.	Assists in the breakdown and absorption of food nutrients; fights toxic build-up.

ENZYME ABSORPTION SYSTEM ENHANCERS (EASE)

Combination	Suggested Dosage	Actions
Advanced Nutritional System from Rainbow Light *or*	Take as indicated on the label.	Potent free-radical fighter and immune system stimulator.
Bromelain Complex from PhytoPharmica *or*	Take as indicated on the label.	Reduces swelling and inflammation.
AminoLogic from PhysioLogics	Take 1 capsule three times per day between	Helps maintain normal tissue function and repair; furnishes

	meals; or follow the advice of your health-care professional.	lysine, which may be lacking; contains proteolytic enzymes to fight inflammation, stimulate immune function, and aid improved nutrient absorption.
or		
Immuzyme from Tyler Encapsulations	Take as indicated on the label.	Helps stimulate immune function.
or		
NESS Formula #11	Take 2 capsules immediately after each meal; or take 3 capsules upon rising and at bedtime.	Helps support the immune system.

ENZYME HELPERS

Nutrient	Suggested Dosage	Actions
Acidophilus or other probiotic	Take as indicated on the label.	Stimulates enzyme activity; improves gastrointestinal function.
Adrenal glandular	Take 100–400 mg per day.	Helps fight stress and adrenal exhaustion; fights inflammation.
Calcium	Take 1,000 mg per day.	Plays important roles in muscle contraction and nerve transmission.
Coenzyme Q_{10}	Take 30–60 mg one to two times per day with meals.	Important to immune system function; has antioxidant activity.
DHEA	Take 25 mg per day.	Supports immune, cardiovascular, and brain function.
Essential fatty acid complex	Take 100–200 mg two to three times per day with meals; or follow the advice of your health-care professional.	Strengthens immunity; helps strengthen cell membranes.
Kyolic Aged Garlic Extract with vitamins B_1 and B_{12}	Take 10 drops in food or a beverage, or place directly on tongue, three times per day.	Stimulates immune system function.
Magnesium	Take 250 mg per day.	Plays important roles in protein building, muscle contraction, nerve function, energy production, and calcium absorption; necessary for glucose metabolism, blood clotting, and the synthesis of fats, proteins, and nucleic acid.
Selenium	Take 50 mcg one or two times per day.	Acts as an enzyme activator; required by glutathione peroxidase.
Spirulina	Take as indicated on the label.	Rich in minerals, vitamins, enzymes, proteins, essential amino acids, and essential fatty acids; improves immune function.
Thymus glandular	Take 100 mg per day; or take as indicated on the label.	Assists the thymus gland in immune resistance; improves immune response; fights infections.

Vitamin A	Take up to 50,000 IU for no more than one month (for higher doses, see a physician). If pregnant, do not take more than 10,000 IU per day.	Needed for strong immune system; potent antioxidant.
Vitamin B complex with vitamin B_{12}	Take 100–150 mg three times per day.	B vitamins act as coenzymes; vitamin B_{12} fights chronic fatigue.
Vitamin C	Take 10,000 mg per day; or 1–2 grams every hour until symptoms improve.	Fights infections, diseases, and allergies; improves healing; stimulates the adrenal glands; promotes immune system; powerful antioxidant.
Vitamin E	Take 1,200 IU per day from all sources (including the multivitamin complex in the introduction to Part Two).	Acts as an intercellular antioxidant; improves circulation; helps resist diseases at the cellular level; aids in stimulating heart function.
Zinc	Take 50 mg per day. Do not take more than 100 mg per day from all sources.	Needed by over 300 enzymes; necessary for white blood cell immune function; improves healing.

Comments

❏ To effectively combat CFS, it is important to make certain lifestyle changes, including getting more rest and reducing sources of stress in your life.

❏ Eat a well-balanced diet with plenty of fresh fruits and vegetables. See dietary recommendations in Part Three.

❏ Perform the Enzyme Toxin Flush and the Enzyme Kidney Flush as outlined in Part Three.

❏ Counseling and other supportive therapy may be helpful in teaching you to cope with this condition.

❏ See the list of Resource Groups in the Appendix for a list of organizations that can provide you with more information about chronic fatigue syndrome.

Cirrhosis

Cirrhosis is a chronic disease of the liver, characterized by destruction of normal tissue, causing scar tissue that interferes with liver function. In the United States, three-quarters of cirrhosis cases are due to alcoholism. Chronic hepatitis is the leading cause of cirrhosis in those parts of the world where viral hepatitis is common. In children, cirrhosis may be caused by a number of inherited disorders (such as alpha$_1$-antitrypsin deficiency, cystic fibrosis, glycogen storage disease, biliary atresia, or other rare diseases). Cirrhosis may also be caused by severe reactions to prescription drugs, malnutrition, obstruction of bile ducts, prolonged exposure to environmental toxins, chronic inflammation, and abnormal storage of metals by the body.

According to the American Liver Foundation, cirrhosis and other liver diseases rank eighth as a cause of death in America, taking the lives of over 25,000 Americans each year. In the United States, cirrhosis is exceeded only by cancer and cardiovascular disease as a major cause of death in individuals aged 45 to 65.

With early alcoholic cirrhosis, if the person stops drinking, the spread of liver scarring stops; however, any scarred tissue will remain. Liver cancer is common in people with cirrhosis.

Symptoms

Surprisingly, many patients with mild cirrhosis are well nourished and don't have any apparent symptoms, although some may feel weak and sick, have a poor appetite, and lose weight. Unfortunately, the disease may not cause obvious symptoms until far advanced, making diagnosis difficult. Symptoms of later stages of cirrhosis may include fever, jaundice, constipation or diarrhea, nausea, loss of appetite, vomiting, liver enlargement, ascites (abdominal swelling due to fluid accumulation), itching, increased drug sensitivity due to the liver's inability to inactivate drugs, vomiting of blood, malnutrition, anemia, hair loss, dark yellow or brown urine, spider veins, breast enlargement in men (gynecomastia), muscle wasting, red palms, edema, enlarged spleen, bleeding and bruising, liver encephalopathy, abnormal nerve function, kidney failure, and coma.

Enzyme Therapy

Oral enzymes can restore immune function and break down antigen-antibody complexes, allowing the cells of the immune system to destroy and expel them. According to one study, patients treated with an enzyme combination containing pancreatin, trypsin, chymotrypsin, bromelain, and papain with rutin showed rapid relief from itching, jaundice, and indigestion. In addition, liver and spleen size decreased, ascites disappeared, cholesterol levels decreased, and liver cell function improved—all without side effects, according to Dr. A.M. Vasilenko and Dr. S.V. Svec of the State Medical Academy in Dnepropetrovsk, Russia.

Digestive enzyme therapy is used to fight cirrhosis of the liver. Enzymes improve the digestion of food, reduce stress on the gastrointestinal mucosa, and decrease the burden on the liver. Enzymes help maintain normal pH levels; detoxify the body; promote the growth of healthy intestinal flora, thus reducing strain on the liver; strengthen the body as a whole; help fight constipation and diarrhea and loss of appetite; and build general resistance.

Systemic enzyme therapy is used to reduce inflammation, stimulate the immune system, improve circulation, help speed tissue repair, bring nutrients to the damaged area, remove waste products, and enhance wellness.

Enzyme Absorption System Enhancers (EASE) maximize enzyme activity. They can also improve the absorption and bioavailability of various nutrients and decrease the drain on the body's own digestive enzymes, thus prolonging their lives.

In addition to the standard multivitamin, multimineral, and multiglandular complexes recommended in the introduction to Part Two, the following supplements are recommended for the treatment of cirrhosis. The following recommended dosages are for adults. Children between the ages of two and five should take one-quarter the recommended dosage, those between the ages of six and twelve should take one-half the recommended dosage, and those between the ages of thirteen and seventeen should take three-quarters the recommended adult dosage.

ESSENTIAL ENZYMES

Enzyme	Suggested Dosage	Actions
Superoxide dismutase	Take 150 mcg per day.	Fights inflammation; potent antioxidant.

AND SELECT ONE OF THE FOLLOWING:

Enzyme	Suggested Dosage	Actions
Formula One—See page 38 of Part One for a list of products.	Take 10 tablets three times per day for three months or until the symptoms disappear.	Fights inflammation, pain and swelling; speeds recovery; stimulates immune function; strengthens capillary walls.
Bromelain	Take 750 mg two times per day between meals.	Fights inflammation, pain, and swelling; speeds recovery; stimulates the immune system.
Serratiopeptidase	Take 10 mg two times per day between meals.	Fights inflammation; reduces pain and swelling.
Trypsin with chymotrypsin	Take 250 mg of trypsin and 10 mg of chymotrypsin three times per day between meals.	Fights inflammation, pain, and swelling; increases rate of healing; stimulates the immune system.
Pancreatin	Take 500 mg six times per day on an empty stomach.	Fights inflammation, pain, and swelling; increases rate of healing; stimulates the immune system.
Papain or a microbial protease	Take 300 mg two times per day between meals.	Fights inflammation; speeds recovery.

SUPPORTIVE ENZYMES

Enzyme	Suggested Dosage	Actions
Catalase and glutathione peroxidase	Take as indicated on the labels.	Potent antioxidant enzymes.
A digestive enzyme product containing protease, amylase, and lipase enzymes. See the list of digestive enzyme products in Appendix.	Take as indicated on the label with meals.	Improves the breakdown and absorption of food nutrients; fights toxic build-up.

ENZYME ABSORPTION SYSTEM ENHANCERS (EASE)

Combination	Suggested Dosage	Actions
Bioprotect from Biotics or	Take as indicated on the label.	Contains antioxidants; has varied free-radical fighting capabilities.
Catimune from Life Plus or	Take 1 tablet twice daily between meals.	Helps detoxify the kidneys and liver; flushes out toxins; supports circulatory system.
Detox-Zyme from Rainbow Light or	Take as indicated on the label.	Helps cleanse the body and eliminate toxins.
Hepatic Complex from Tyler or	Take as indicated on the label.	Helps detoxify the liver.
Metabolic Liver Formula from Prevail or	Take as indicated on the label.	Improves liver function.
NESS Formula #7	Take 2 capsules immediately after each meal; or take 4 capsules upon rising and at bedtime.	Helps the body utilize essential fatty acids.

ENZYME HELPERS

Nutrient	Suggested Dosage	Actions
Acidophilus and other probiotics	Take as indicated on the label.	Improves gastrointestinal function so improves digestion; needed for production of biotin, niacin, folic acid, and pyridoxine, all of which serve as coenzymes.
Adrenal glandular	Take 100–400 mg per day.	Fights inflammation and adrenal exhaustion.
Amino-acid complex, such as Amino Forte from Nutrisupplies, Inc.	Take as indicated on the label on an empty stomach.	Required for enzyme production; helps the body maintain proper fluid and acid/base balance, which are essential for overall health.
Bioflavonoids, especially hesperidin	Take as indicated on the label.	Improves liver function; fights liver disease.
Choline	Take as indicated on the label.	Needed for cholesterol and fat utilization; a basic component of lecithin, which helps emulsify fat; essential for the health of the liver.
Coenzyme Q$_{10}$	Take 30–60 mg one to two times per day with meals; or follow the advice of your health-care professional.	Potent antioxidant.
Inositol	Take as indicated on the label.	Works closely with pantothenic acid, PABA, folic acid, and vitamin B$_6$ to encourage proper liver function; plays a role in cell membrane integrity;
Kyolic Aged Garlic Liquid Extract with Vitamins B$_1$ and B$_{12}$	Take 3 teaspoons per day either in 6–8 ounces of water or juice, or placed directly on the tongue.	Fights inflammation and free radicals; stimulates immunity.
Lecithin	Take 1,500 mg two to three times per day.	Emulsifies fat.
Liver glandular	Take 100 mg per day.	Helps stimulate liver function.
Milk thistle extract	Take as indicated on the label.	Potent antioxidant; protects the liver; supports new liver cell growth.
Thymus glandular	Take 100 mg per day.	Assists the thymus gland in immune resistance; improves immune response; fights infections.
Vitamin A	Take 15,000–40,000 IU per day. Daily intakes of more than 50,000 IU for over a month should be monitored by a physician. Pregnant women should take no more than 10,000 IU per day.	Needed for strong immune system; potent antioxidant.
Vitamin B complex	Take 50 mg three times per day; or take as indicated on the label.	B vitamins act as coenzymes.
Vitamin B$_{12}$	Take 500 mg three times per day.	Essential for the synthesis of choline and methionine, which can prevent fat from accumulating in the liver, and red blood cell function.
Vitamin C	Take 10,000–15,000 mg per day.	Fights infections and diseases; improves healing; stimulates the adrenal glands; promotes healthy immune system; powerful antioxidant.
Zinc	Take 50 mg per day. Do not take more than 100 mg per day from all sources.	Needed by over 300 enzymes; potent antioxidant.

Comments

❑ Avoid alcohol and drugs of all kinds (unless prescribed by your physician).

❑ Detoxification through fasting and juicing is important to cleanse the liver. See Part Three for information about the Enzyme Toxin Flush.

❑ A "caveman" diet, high in fresh, enzyme-rich fruits and vegetables should be eaten. See Part Three for further dietary recommendations.

❑ Eat four to five times per day instead of three. This takes stress off the liver.

❑ Exercise, positive mental attitude, and meditation are important. See Part Three for further information.

❑ See the list of Resource Groups in the Appendix for a list of organizations that can provide you with more information about cirrhosis.

Cold Sores

See HERPES SIMPLEX VIRUS.

Colds

A common cold is a contagious viral infection of the upper respiratory tract. Any number of viruses can cause colds, with new cold-causing viruses cropping up constantly. The symptoms of a cold usually appear eighteen to forty-eight hours after infection and may last from a few days to a few weeks, although there is no way to predict how long a cold will last. A physician is rarely needed to care for a cold, unless complications develop.

The cough that often accompanies a cold is the body's attempt to clear the air passage of mucus, dust, or other substances that cause irritation. Coughing is a symptom of a number of conditions, from the common cold to more serious diseases, such as tuberculosis, lung cancer, and emphysema. If you have a persistent cough for more than a few weeks, or if you cough up bloody sputum, see a doctor immediately. If an infant has a cough, seek medical attention without delay.

Cough medicine may relieve, but not cure, a cough. Most cough medicines reduce or suppress the cough by suppressing the cough reflex response. Because a cough is a reflex mechanism that helps clear the respiratory tract, it serves a useful purpose and should not be suppressed.

Sometimes colds are accompanied by a mild fever. A slightly elevated body temperature is the body's way of increasing enzyme activity in order to fight the inflammation and infection and eliminate viruses and bacteria. One study showed that elderly people who suffered from an annual cold, severe enough to require bed rest, with a fever developed cancer less frequently than those who didn't suffer from a fever, according to Dr. Ulrich Abel of the Cancer Research Institute in Heidelberg, Germany. Therefore, fever and enzymes are helpful in fighting colds, flu, and cancer. However, a too-elevated fever can be dangerous. See a physician if you have a prolonged high fever.

Symptoms

Symptoms of a common cold include watery eyes, runny and/or stuffy nose (rhinitis), head congestion (with a mild, moderate, or severe headache), fatigue, coughing, and sneezing. Your sense of taste and smell may be decreased, and you may run a low-grade fever or suffer from chills. A general feeling of discomfort and listlessness (malaise) may be present. There may be a sore throat, ranging from mild to severe as the cold develops. Your cold may show all or only some of these symptoms.

Even the common cold may pose a dangerous problem for some individuals. Those at high risk include small infants, the aged, and individuals with heart disease, asthma, tuberculosis, kidney disease, and chronic bronchitis. The viruses that cause a common cold can migrate to the lungs and develop into pneumonia in serious cases.

Enzyme Therapy

Systemic enzyme therapy is used to fight head colds and chest colds and to reduce inflammation and stimulate the immune system. Enzymes can also improve circulation, bring nutrients to the infected area, remove waste products, strengthen the body as a whole, build general resistance, and enhance wellness.

Digestive enzyme therapy is used to improve the digestion of food, cleanse the gastrointestinal tract, reduce stress on the gastrointestinal mucosa, help maintain normal body pH levels, detoxify the body, and promote the growth of healthy intestinal flora, thereby making the body healthier so that it can fight off a cold. Digestive enzymes also serve as replacements for the body's pancreatic enzymes, leaving the pancreatic enzymes free to perform other functions in the body, such as decreasing inflammation.

Enzyme Absorption System Enhancers (EASE) maximize enzyme activity. They can also improve the absorption and bioavailability of various nutrients and decrease the drain on the body's own digestive enzymes, thus prolonging their lives.

In addition to the standard multivitamin, multimineral, and multiglandular complexes recommended in the introduction to Part Two, the following supplements are recommended for the treatment of colds. The following recommended dosages are for adults. Children between the ages of two and five should take one-quarter the recommended dosage, those between the ages of six and twelve should take one-half the recommended dosage, and those between the ages of thirteen and seventeen should take three-quarters the recommended adult dosage.

ESSENTIAL ENZYMES

Enzyme	Suggested Dosage	Actions
Formula Three—See page 38 of Part One for a list of products.	Take 200 mcg three times per day between meals.	Free-radical fighters.

AND SELECT ONE OF THE FOLLOWING:

Lysozyme	Take as indicated on the label.	Fights inflammation; improves respiratory conditions; bolsters the immune system; breaks up mucus.

Formula One—See page 38 of Part One for a list of products.	Take 3 capsules three times per day between meals.	Reduces inflammation and congestion; stimulates immune function.
Bromelain	Take 500 mg three times per day between meals.	Fights inflammation; improves respiratory conditions; bolsters the immune system; breaks up mucus.
Serratiopeptidase	Take 5 mg three times per day between meals.	Fights inflammation; improves respiratory conditions; bolsters the immune system; breaks up mucus.
Trypsin	Take 500 mg three times per day between meals.	Fights inflammation; improves respiratory conditions; bolsters the immune system; breaks up mucus.
Sfericase	Take 500 mg three times per day between meals.	Fights inflammation; improves respiratory conditions; bolsters the immune system; breaks up mucus.
Enzyme complexes rich in proteases, plus amylases and lipases, working synergistically	Take as indicated on the label between meals.	Fights inflammation, pain, and swelling; speeds healing; strengthens capillary walls.

SUPPORTIVE ENZYMES

Enzyme	Suggested Dosage	Actions
A digestive enzyme formula containing protease, amylase, and lipase enzymes. See the list of digestive enzyme products in Appendix.	Take as indicated on the label with meals.	Improves the breakdown and absorption of proteins, carbohydrates, and fats; fights toxic build-up.

ENZYME ABSORPTION SYSTEM ENHANCERS (EASE)

Combination	Suggested Dosage	Actions
Cold Zzap from Naturally Vitamin Supplements	Take as indicated on the label.	Supports healthy immunity; shortens the duration of a cold; has antibacterial and antiviral effects.
or		
Combat from Life Plus	Take as indicated on the label.	Helps support the respiratory system; stimulates immune system function; helps control infections.
or		
Kyolic Formula 102 (with enzymes)	Take 2 capsules or tablets with a meal two to three times per day.	Strengthens the immune system; helps alleviate stress; helps fight free radicals; helps fight inflammation; improves digestion.
or		
Mucous Dissolver Liquezyme from Enzyme Process Laboratories	Take as indicated on the label.	Dissolves mucus.
or		
Sinease from Prevail	Take as indicated on the label.	Decongestant; helps promote a healthy digestive system.
or		
Sinus Ease from	Take as indicated on	Stimulates immune function;

Nature's Life	the label.	reduces inflammation.
or		
Zym-Eeze Lysozyme Lozenges from Future Foods	Take as indicated on the label.	Natural enzymatic antimicrobial agent; fights sore throat, colds, and strep throat.

ENZYME HELPERS

Nutrient	Suggested Dosage	Actions
Adrenal glandular	Take 300 mg per day.	Supports adrenal function.
Aged garlic extract	Take one to three 300 mg capsules with meals twice per day.	Detoxifies and provides energy.
Bioflavonoids, such as quercetin and rutin	Take 100–300 mg twice per day.	Fights inflammation; reduces capillary fragility; has antihistamine activity.
DHEA	Take 25 mg per day.	Supports immune function.
Grape seed extract	Take 100 mg per day.	Contains anthocyanidins, which fight free radicals and inflammation.
Green Barley Extract or Kyo-Green	Take 1 heaping teaspoon in 8 ounces of liquid three times per day.	Fights free radicals.
Lung glandular	Take as indicated on the label.	Improves lung function.
Melatonin	Take 3 mg at bedtime.	Sleep aid.
Vitamin A	Take 10,000–40,000 IU for no more than one month (for higher doses see a physician). Pregnant women should not take more than 10,000 IU per day.	Necessary for mucous linings, skin and cell membrane structure, and a strong immune system; potent antioxidant.
Vitamin C	Take 1–2 grams every hour until symptoms improve. If you use a powder or crystalline vitamin C, mix a heaping teaspoon with 6 to 8 ounces of water or orange or lemon juice. If lemon juice is used, you may sweeten the mixture with honey. Gargle a mouthful and swallow. Repeat every hour.	Fights infections, diseases, allergies, and the common cold; potent antioxidant.
Vitamin E	Take 400 IU per day.	Potent antioxidant; resists diseases at the cellular level; effective against allergies and sinusitis.
Zinc lozenge	Suck up to 50 mg of lozenges per day. Do not take more than 100 mg per day from all sources.	Needed for white blood cell immune function; can help reduce cold symptoms.

Comments

❏ Fresh, enzyme-rich fruits and vegetables (particularly green vegetables) should comprise over 60 percent of your diet. If your body cannot tolerate raw fruits and vegetables, increase your intake of digestive enzymes, or sauté or steam your produce.

❑ During a cold, increase your intake of vitamin C in such foods as oranges, limes, lemons, grapefruit, tangerines, apples, cantaloupe, strawberries, papayas, guavas, mangoes, black currants, persimmons, dark green vegetables (such as broccoli and turnip greens), red and green peppers, lettuce, tomatoes, potatoes, and cauliflower.

❑ Hot broth or chicken or vegetable soup is helpful for cleaning out the body and sweating out toxins. Add plenty of onions and garlic (known for their ability to stimulate immunity and fight free radicals). Before making the soup, be sure to remove the skin from the chicken (chicken skins contain many toxins). Sip the broth or soup as frequently as possible.

❑ A number of herbs can help strengthen the immune system and fight off viral infections, including echinacea, ginger, ginkgo biloba, goldenseal, licorice, and pau d'arco. Maitake and reishi mushrooms can be used to treat respiratory tract conditions.

❑ Gargle three times per day with the Enzymatic Gargle. See Part Three for instructions for its preparation.

❑ Put five to six drops of eucalyptus in a sink or bowl of very hot water. Cover your head and shoulders with a large bath towel over the sink or bowl and inhale the vapors.

❑ With flu and colds, ask your doctor to check for the presence of allergies, thyroid dysfunction, and immune system deficiencies.

❑ The Enzyme Toxin Flush and the Enzyme Kidney Flush are of great value in detoxification. See Part Three for instructions.

❑ How can colds be avoided? One way is to stay away from people who have colds or to persuade them to cover their sneezes. Do not touch their hands, as the virus will probably be on their hands due to their covering their noses and mouths when they sneeze and wiping their noses. Stay away from crowds. A single sneeze can propel the infected droplets of virus several feet. Avoid excessive fatigue and maintain a good diet. Colds are more frequent during the cold weather; however, only a virus can cause a cold. Also, avoid sources of throat and nose irritation, such as smoking, stress, etc.

Colitis, Ulcerative

See INFLAMMATORY BOWEL DISEASE.

Colorectal Cancer

Colorectal cancer is cancer of the the large intestine and rectum. The large intestine is about 5-feet long and two-and-one-half inches in diameter and is divided into several subsections (the cecum, appendix, colon, rectum, and anal canal). The large intestine's job is to eliminate food wastes, that is, those food components that the body couldn't break down, absorb, and utilize. If you suffer from constipation, and waste is allowed to sit in the colon, toxins can build up, which can lead to cancer.

Colon cancer (cancer of the large intestine) is more common in women than in men. Rectal cancer is more common in men than in women. Overall, in Western countries, colorectal cancer is the second most common type of cancer.

Risk for colorectal cancer increases with age, a family history of the disease, other disorders of the colon, and Crohn's disease. Diet also appears to play a role in the development of colon cancer. Diets high in fat, animal protein, and refined carbohydrates, and low in fiber seem to increase risk. Risk for colorectal cancer seems to be reduced with a diet high in calcium, vitamin D, and cruciferous vegetables.

Symptoms

In the early stages of colorectal cancer, there may be no obvious signs or symptoms. Symptoms depend on the type, location, and extent of the cancer. Bloody stools, continuous diarrhea or constipation, a change in stools, and gastrointestinal discomfort can all signal the disease. However, many of these symptoms could be signs of other disease or even simple indigestion. If you believe you may have colorectal cancer (see "Warning Signs of Colorectal Cancer" on page 189), immediately seek medical care. Even though colorectal cancer is the second leading cause of cancer death in this country, if it is detected and treated early, the cure rate can often exceed 80 percent.

Enzyme Therapy

In the treatment of cancer, systemic enzyme therapy acts as a complementary therapy and should be integrated with the overall therapy program. The primary therapy for colorectal cancer includes surgery, chemotherapy, and radiation. Enzyme therapy is used to stimulate and stabilize the immune system, thus increasing the opportunities to destroy the residual tumor cells and decrease the rate of tumor growth; help differentiate the healthy from the cancerous tissue and better identify the tumor for surgical removal; reduce inflammation; improve wound healing and help speed tissue repair by bringing nutrients to the damaged area and removing waste products; improve the long-term prognosis; improve the patient's quali-

Warning Signs of Colorectal Cancer

Knowing the early warning signs of colorectal cancer might save your life. The American Cancer Society estimated in 1997 that over 46,000 Americans would die from colon cancer that year. Many of these deaths might have been prevented if the cancer had been detected and treated early. If you notice any of the following warning signs of colorectal cancer, see your doctor immediately.

- Blood in the stool.
- Diarrhea or constipation that lasts longer than two weeks.
- Stools that are smaller than usual in width.

- Chronic fatigue.
- Bloating, fullness, cramping, frequent gas pains or other stomach discomfort.
- Unexplained weight loss.

ty of life; decrease the adverse side effects of the primary therapy (radiation, chemotherapy, and/or surgery); reduce the cost of primary therapy by minimizing its use if enzyme therapy is begun early enough; reduce the risk of complications, such as blood clots; and decrease immune suppression caused by surgery and anesthesia.

Digestive enzymes are essential for the gastrointestinal tract and the body as a whole in helping to prevent the onset of cancer and to improve the cancer patient's health because they aid the breakdown, digestion, and absorption of nutrients; decrease the enzyme production load on the pancreas, thereby helping to reduce its stress and increase its productive life; help prevent toxin build-up especially in the small and large intestine; stimulate friendly microflora production in the intestine; and facilitate elimination of potential cancer-causing waste from food components and body cells.

Enzyme Absorption System Enhancers (EASE) maximize enzyme activity. They can also improve the absorption and bioavailability of various nutrients and decrease the drain on the body's own digestive enzymes, thus prolonging their lives.

In addition to the standard multivitamin, multimineral, and multiglandular complexes recommended in the introduction to Part Two, the following supplements are recommended for the treatment and prevention of colorectal cancer.

ESSENTIAL ENZYMES

Enzyme	Suggested Dosage	Actions
Before and after surgery; before, during, and after chemotherapy or radiation therapy; and as long-term and palliative therapy:		
WobeMugos E From Mucos	Take 2–4 tablets three times per day.	Speeds wound healing; leads to earlier response to chemotherapy and radiotherapy and reduced side effects; Reduces anorexia, muscle wasting, malnutrition, and weakness, depression, and pain following medical treatment; prevents metastasis; eliminates or prevents disturbances of lymphatic drainage.
and		
Formula One—See page 38 of Part One for a list of products.	Take 10 tablets three times per day between meals.	

SUPPORTIVE ENZYMES

Enzyme	Suggested Dosage	Actions
Bromelain	Take eight to ten 250 mg capsules per day. When tumor growth is under control, gradually reduce dosage to a maintenance dose of approximately 4 capsules per day.	Fights inflammation, pain, and swelling; stimulates the immune system; improves digestion.
A digestive enzyme formula containing protease, amylase, and lipase enzymes. For a list of digestive enzyme products, see Appendix.	Take as indicated on the label.	Improves digestion by improving the breakdown of proteins, carbohydrates, and fats.
Formula Three—See page 38 of Part One for a list of products.	Take as indicated on the label.	Free-radical fighters.

ENZYME ABSORPTION SYSTEM ENHANCERS (EASE)

Combination	Suggested Dosage	Actions
Antioxidant Complex from Tyler	Take as indicated on the label.	Potent free-radical fighter and immune system stimulator.
or		
Detox Enzyme Formula from Prevail	Take 1 capsule between meals one or two times per day with a 6- to 8-ounce glass of water or juice.	Helps rid the body of harmful toxins.
or		
Hepazyme from Enzymatic Therapy	Take as indicated on the label.	Stimulates immune function.
or		
Kyolic Formula 102 (with enzymes)	Take 2 capsules or tablets with meals two to three times per day.	Strengthens the immune system; helps alleviate stress; helps fight free radicals; helps fight inflammation.
or		
Oxy-5000 Forte from American Biologics	Take as indicated on the label.	Helps the body excrete toxins; fights biochemical stressors, common infections, and degenerative processes.
or		
Vita-C-1000 from Life Plus	Take as indicated on the label.	Helps fight cancer; boosts immune function.

ENZYME HELPERS

Nutrient	Suggested Dosage	Actions
Acidophilus or other probiotics	Take as indicated on the label.	Stimulates enzyme activity; improves gastrointestinal function.
Bioflavonoids, such as rutin or quercetin	Take 150–500 mg three times per day.	Strengthen capillaries; potent antioxidants.
Magnesium	Take 250 mg per day.	Plays important roles in protein building and energy production; necessary for the synthesis of fats, proteins, and nucleic acid, thus reducing irritation of the gastrointestinal tract.
Thymus glandular	Take 30 mg per day.	Assists thymus gland (important in immune resistance); fights cancer.
Vitamin A	Take 50,000 IU per day. Do not take more than 10,000 IU if you are pregnant. Dosages over 50,000 IU should be taken only under the supervision of a physician.	Stabilizes the cell membranes; improves response to therapy; needed for strong immune system and healthy mucous membranes, including those of the large intestine; potent antioxidant.
Vitamin C	Take 2–10 grams per day. For amounts over 10 grams, consult a physician.	Especially effective in fighting tumors of the gastrointestinal tract; potent free-radical scavenger and immune system stimulator.
Vitamin D	Take 3,000–6,000 IU per day. These high dosages should be taken only under the supervision of a physician.	Has an immune system stimulating effect; hinders the advance of tumors; activates cancer-fighting monocytes (a type of white blood cell).
Vitamin E	Take 1,200 IU per day from all sources (including the multivitamin complex in the introduction to Part Two).	Causes more efficient consumption of oxygen; acts as an intercellular antioxidant; improves circulation; helps resist diseases at the cellular level.
Zinc	Take 50 mg per day. Do not take more than 100 mg per day from all sources.	Needed for white blood cell immune function, wound healing, growth, development, and thyroid hormone function; a cofactor for hundreds of enzymes; potent antioxidant.

Comments

❑ Surgery, radiation, and chemotherapy are the standard treatments for colorectal cancer. But prevention is always easier than treatment. Recent research indicates that exercise, such as a daily hourlong walk, can reduce the risk of developing colon cancer by 46 percent. Exercising may also help you maintain proper weight. Increased weight can actually double your risk of developing colon cancer.

❑ Eat a diet low in fats but high in fiber and complex carbohydrates. See Part Three for further information on diet.

❑ A stool blood test is an easy way to test for blood in the stool—an indication of colorectal cancer. Ask your doctor to have this test done if you suspect cancer.

❑ To help detoxify the body, perform the Enzyme Toxin Flush and the Enzyme Kidney Flush once per month, or as per your physician's instructions. See Part Three for further information on both flushes.

❑ During the weeks that you are not using the flushes, use the coffee enema once per week as described in Part Three. Coffee enemas are used to stimulate the release of bile from the liver and facilitate the removal of waste metabolites from the body and into the intestines for excretion.

❑ Physicians sometimes use enzyme injections or enzyme retention enemas in the treatment of cancer. Both methods allow a greater concentration of the enzymes to be absorbed than might be accomplished if the enzymes were taken orally. An enzyme retention enema is a means of instilling some enzymes into the colon to be retained. This allows the enzymes to be absorbed into the body's systems. Physicians use products specially formulated for these uses.

❑ See the list of Resource Groups in the Appendix for a list of organizations that can provide you with more information about colorectal cancer.

❑ *See also* CANCER.

Conjunctivitis

Conjunctivitis is an inflammation of the membrane that lines the underside of the eyelids and covers the eyeball. Conjunctivitis is usually caused by a virus (such as the ones that cause the common cold or measles), bacteria, or an allergy. Irritants, including smoke, dust, wind, and air pollution, can also cause the condition, as can damage from intense ultraviolet light.

Symptoms

Conjunctivitis is characterized by a discharge from the eye, as well as swelling, redness, and pain. *Bacterial conjunctivitis* may cause moderate lid swelling and a purulent (sticky and heavy) discharge. The eyes may burn and itch. *Viral conjunctivitis* may cause only minimal swelling of the eyelid. In contrast to bacterial conjunctivitis, the discharge may be light and clear. The lymph nodes may be swollen. *Allergic conjunctivitis* may cause moderate to severe lid swelling, intense itching, and a clear discharge. The eyes may be red.

Enzyme Therapy

Systemic enzyme therapy is used to reduce inflammation, stimulate the immune system, improve circulation, and help speed tissue repair by bringing nutrients to the damaged area and removing waste products. Enzymes enhance wellness, strengthen the body as a whole, and build general resistance.

Digestive enzyme therapy is used to improve digestion of food, reduce stress on the gastrointestinal mucosa, help maintain normal pH levels, detoxify the body and promote the growth of healthy intestinal flora. All of these actions make the body healthier so that it is better able to fight conjunctivitis. Digestive enzymes also serve as replacements for the body's pancreatic enzymes, leaving the pancreatic enzymes free to perform other functions in the body, such as decreasing inflammation.

Enzyme Absorption System Enhancers (EASE) maximize enzyme activity. They can also improve the absorption and bioavailability of various nutrients and decrease the drain on the body's own digestive enzymes, thus prolonging their lives.

In addition to the standard multivitamin, multimineral, and multiglandular complexes recommended in the introduction to Part Two, the following supplements are recommended for the treatment of most types of conjunctivitis, although I would not recommend this therapy for the treatment of infants born with conjunctivitis. The following recommended dosages are for adults. Children between the ages of two and five should take one-quarter the recommended dosage, those between the ages of six and twelve should take one-half the recommended dosage, and those between the ages of thirteen and seventeen should take three-quarters the recommended adult dosage.

ESSENTIAL ENZYMES

Enzyme	Suggested Dosage	Actions
Superoxide dismutase	Take 150 mcg per day.	Fights inflammation; potent antioxidant.

AND SELECT ONE OF THE FOLLOWING:

Bromelain	Take three to four 230–250 mg capsules per day until symptoms subside. If severe, take 6 to 8 capsules per day.	Fights inflammation, pain, and swelling; speeds healing; stimulates the immune system.
Formula One—See page 38 of Part One for a list of products.	Take 5 tablets twice per day beteen meals.	Fights inflammation, pain, and swelling; speeds healing; stimulates the immune system; strengthens capillary walls.
Microbial protease	Take 300 mg two times per day between meals.	Fights inflammation; speeds healing.
Pancreatin	Take 500 mg six times per day on an empty stomach.	Fights inflammation, pain, and swelling; increases rate of healing; stimulates the immune system.
Papain	Take 300 mg two times per day between meals.	Fights inflammation; speeds healing.

Serratiopeptidase	Take 10 mg two times per day between meals.	Fights inflammation; reduces pain and swelling.
Trypsin with chymotrypsin	Take 250 mg of trypsin and 10 mg of chymotrypsin three times per day between meals.	Fights inflammation, pain, and swelling; increases rate of healing; stimulates the immune system.

SUPPORTIVE ENZYMES

Enzyme	Suggested Dosage	Actions
Catalase and glutathione peroxidase	Take as indicated on the label.	Potent antioxidant enzymes.
A digestive enzyme product containing protease, amylase, and lipase enzymes. See the list of digestive enzyme products in Appendix.	Take as indicated on the label with meals.	Improves the breakdown and absorption of food nutrients; fights toxic build-up.

ENZYME ABSORPTION SYSTEM ENHANCERS (EASE)

Combination	Suggested Dosage	Actions
Antioxidant Complex from Tyler or	Take as indicated on the label.	Potent free-radical fighter and immune system stimulator.
Bioprotect from Biotics or	Take as indicated on the label.	Contains antioxidants; has varied free-radical fighting capabilities.
Bromelain Complex from PhytoPharmica or	Take as indicated on the label.	Reduces swelling and inflammation; breaks up circulating immune complexes.
Detox-Zyme from Rainbow Light or	Take as indicated on the label.	Helps cleanse the body and eliminate toxins.
Kyolic Formula 102 (with enzymes) or	Take 2 capsules or tablets with meals two to three times per day.	Strengthen the immune system; helps alleviate stress; helps fight free radicals; helps fight inflammation.
NESS Formula #301	Take 3 capsules between meals three times per day.	Provides cleansing herbs with enzymes to deliver nutrients to the lymphatic system.

ENZYME HELPERS

Nutrient	Suggested Dosage	Actions
Adrenal glandular	Take 100–400 mg per day.	Fights inflammation and adrenal exhaustion.
Bioflavonoids, such as rutin or quercetin	Take 500 mg three times per day between meals.	Strengthen capillaries; potent antioxidants.
Coenzyme Q_{10}	Take 30–60 mg one to two times per day with meals; or follow the advice of your health-care professional.	Potent antioxidant.
Niacin	Take 50 mg the first day, then increase daily by 50 mg to dosage	Dilates blood vessels; improves circulation to inflamed areas, thus allowing

191

or	1,000 mg per day. If adverse effects occur, reduce dosage. Take doses higher than 1,000 mg only under the supervision of a physician.	nutrients to get in and waste products to be removed.
Niacin-bound chromium	Take 20 mg per day; or take as indicated on the label.	
Thymus glandular	Take 100 mg per day.	Assists the thymus gland in immune resistance; improves immune response; fights infections.
Vitamin A *or*	Take 5,000 IU per day.	Needed for strong immune system and healthy mucous membranes; potent antioxidant.
Carotenoid complex	Take 5,000–10,000 IU per day.	Potent antioxidant; stimulates immune function; precursor of vitamin A.
Vitamin B complex	Take 150 mg three times per day; or take as indicated on the label.	B vitamins act as coenzymes.
Vitamin C	Take 10,000–15,000 mg per day.	Fights infections, diseases, and allergies; stimulates the adrenal glands; promotes immune system; powerful antioxidant.
Zinc	Take 50 mg per day. Do not take more than 100 mg per day from all sources.	Needed by over 300 enzymes; necessary for white blood cell immune function; improves wound healing.

Comments

❏ Conjunctivitis is easily spread, so wash your hands often and avoid touching the affected eye.

❏ Rinse the affected eye with an herbal wash several times a day. You can make your own by steeping three teaspoons of chamomile or one teaspoon of eyebright in a pint of boiling water; strain; cool; then use to rinse the eye.

❏ Detoxification is essential for a healthy body and healthy eyes. See Part Three for further information on detoxification.

Constipation

The term constipation has come to indicate any absence or irregularity in bowel movements. However, the quantity of feces evacuated, the interval between bowel movements, and the character of the stool (if it is dry, hard, compact, and difficult to pass) are all important in determining whether or not one is constipated.

After the digestive tract has digested and extracted the nutrients from your food, the resulting waste passes through the colon pushed by the peristaltic action of the intestinal muscles. Although the waste begins as a thick liquid, water is constantly pulled from the waste and absorbed through the colon walls. The longer the waste stays in the colon, the more water is absorbed and the drier and more compact the waste becomes. If, for whatever reason, the colon muscles become temporarily inactive, waste will not promptly pass out of the body, and you will suffer from constipation.

The frequency of bowel movements depends on your physical make-up and physical and dietary habits. Most people have one movement every twenty-four hours, but some individuals have a movement every thirty-six or forty-eight hours (or at even greater intervals) and do not suffer from constipation. There is nothing to worry about if you have only occasional minor discomfort or irregularity.

Constipation can be caused by a number of factors. Enzyme-poor diets with too many refined carbohydrates; too much fat, meat, caffeine, and alcohol; and too little fiber (see "The Importance of Fiber" on page 193) can wreak havoc on your intestinal tract. Inadequate water intake, nervous tension, insufficient exercise, poor or inconsistent toilet habits, and laxative overuse can also cause constipation. A number of diseases can interfere with bowel function, including hyper- and hypothyroidism, colon disturbances (such as fistulas, inflammation, polyps, obstructions, and tumors), and circulatory disturbances. Many drugs, such as aluminum hydroxide (found commonly in antacids), bismuth and iron salts, antihypertensives, and many sedatives, can cause constipation.

Symptoms

Symptoms of constipation include difficulty emptying the bowel or straining at the stool. Often, the waste material becomes hard and compact making evacuation quite painful. Additional symptoms may include gas and flatulence, nausea, fatigue, a coated tongue, nervous irritability, body odor, bad breath, headaches, sallow skin and mental dullness. Lower intestinal tract discomfort may also be associated with the condition.

Unless constipation is the result of an organic disease, it is rarely serious in itself, but it can lead to diverticulitis and diverticulosis. Toxin build-up in the colon can lead to colon cancer.

Enzyme Therapy

Digestive enzyme therapy is used to relieve constipation by improving the digestion of food and improving and normalizing elimination. Enzyme therapy reduces stress on the gastrointestinal mucosa, helps maintain normal pH levels, detoxifies the body, and promotes the growth of healthy intestinal flora. Digestive enzymes also serve as replacements for the body's pancreatic enzymes. This frees the pancreatic enzymes to perform other health functions in the

The Importance of Fiber

Dietary fiber refers to a group of components in plant foods (fruits, vegetables, and the outer coverings of grains, seeds, nuts, and legumes) that are resistant to the digestive enzymes of the human gastrointestinal tract. Soluble fibers can be broken down by water, insoluble fiber cannot.

The soluble fibers include gums, mucilages, and some pectins and hemicelluloses. Oat bran and beans, for example, contain relatively large proportions of soluble fibers. Some soluble fibers have been found to reduce blood cholesterol, enhance glucose tolerance, and increase insulin sensitivity.

Insoluble fibers include cellulose, lignin, and other pectins and hemicelluloses. Wheat bran is a good source of insoluble fiber. Insoluble fibers that retain water increase stool weights. Insoluble fibers are the best for the treatment of constipation.

Though the average intake of fiber in the United States is 12 grams a day, it is recommended that the average adult consume at least 40 to 60 grams of fiber per day. This can be achieved by eating lots of fruits and vegetables high in fiber, high-fiber cereals, and whole grains. You can tell if your dietary intake of fiber is sufficient, if:

- Stools have minimal odor and are light enough to float.
- Bowel movements are daily, effortless, and regular.
- There is no gas or flatulence.

---

body, such as boosting immune function, decreasing inflammation, and improving circulation.

Systemic enzyme therapy is used to reduce any inflammation, stimulate the immune system, improve circulation, transport nutrients throughout the body, remove waste products, strengthen the body as a whole, build general resistance, and enhance wellness.

Enzyme Absorption System Enhancers (EASE) maximize enzyme activity. They can also improve the absorption and bioavailability of various nutrients and decrease the drain on the body's own digestive enzymes, thus prolonging their lives.

In addition to the standard multivitamin, multimineral, and multiglandular complexes recommended in the introduction to Part Two, the following supplements are recommended for the treatment of constipation. The following recommended dosages are for adults. Children between the ages of two and five should take one-quarter the recommended dosage, those between the ages of six and twelve should take one-half the recommended dosage, and those between the ages of thirteen and seventeen should take three-quarters the recommended adult dosage.

ESSENTIAL ENZYMES

Enzyme	Suggested Dosage	Actions
A digestive enzyme formula containing protease, amylase, and lipase enzymes. See the list of digestive enzyme products in Appendix.	Take as indicated on the label.	Improves digestion by breaking down proteins, carbohydrates, and fats.
Formula One—See page 38 of Part One for a list of products.	Take 2 tablets with each meal; increase as needed.	Improves digestion by breaking down protein, carbohydrates, and fats.
Bromelain	Start with one 250 mg capsule per day, and increase dosage gradually, if necessary. Take no more than 3 capsules per day. If diarrhea develops, decrease dosage.	Improves digestion; fights inflammation, pain, and swelling; stimulates the immune system.
Pancreatin	Take 300 mg three times per day with meals.	Contains proteases, amylases, and lipases, so is able to digest proteins, carbohydrates, and fat.

SUPPORTIVE ENZYMES

Enzyme	Suggested Dosage	Actions
Formula Three—See page 38 for a list of products.	Take as indicated on the labels.	Potent antioxidant enzymes.

ENZYME ABSORPTION SYSTEM ENHANCERS (EASE)

Combination	Suggested Dosage	Actions
Biodias A Granules from Amano Pharmaceutical Co., Ltd.

or | Adults should take one packet in 8 ounces of water between or after meals three times per day. Children ages 11 to 14 should take two-thirds of a packet; children between the ages of 8 and 10 should take half of the packet; and children between the ages of 5 and 7 should take one-third of a packet. | Helps improve digestion of nutrients; relieves inflammation; promotes regeneration of mucous membranes; decreases stress on the stomach and intestines; reduces gastric pain. |

Bromelain Complex from PhytoPharmica *or*	Take as indicated on the label.	Reduces constipation.
Metabolic Liver Formula from Prevail *or*	Take as indicated on the label.	Improves digestion.
Pro-Flora from Tyler Encapsulations	Take as indicated on the label.	Helps restore normal gastrointestinal flora, digestion and elimination of waste.

ENZYME HELPERS

Nutrient	Suggested Dosage	Actions
Acidophilus or other probiotics, such as bifidobacterium or Acidophilasé from Wakunaga, which contains lipolytic, proteolytic, and amylolytic enzymes	Take as indicated on the label.	Improves gastrointestinal function; stimulates enzyme activity; improves elimination.
Aged garlic extract	Take 2 capsules three times per day.	Potent antioxidant; improves digestion; stimulates healthy intestinal bacteria and destroys harmful bacteria.
Amino-acid complex, such as AminoLogic from PhysioLogics	Take as indicated on the label.	Amino acids serve as building blocks for the body's enzymes to improve digestion.
Betaine Hydrochloric acid	Take 150 mg three times per day with meals.	Provides hydrochloric acid, which improves digestion; if gastric upset occurs, discontinue use.
Calcium	Take 500 mg per day.	Essential for proper digestion and elimination.
Choline	Take 100 mg three times per day.	Needed for cholesterol and fat metabolism; essential for the health of the kidneys, liver, and gallbladder.
Fiber	Take as indicated on the label.	Provides bulk; needed for proper elimination.
Fructooligo-saccharides	Take as indicated on the label.	Increases peristaltic action of intestines; improves liver function.
Green food product, such as Green Kamut, Kyo-Green, or Green Barley Extract	Take as indicated on the label.	Green foods are rich in enzymes, vitamins, minerals, chlorophyll, and amino acids.
Inositol	Take 150 mg three times per day.	An inositol deficiency can lead to constipation.
Lecithin	Take 1,000 mg three times per day with meals.	Emulsifies fat.
Magnesium	Take 250 mg per day.	Plays important roles in protein building, muscle contraction, nerve function, energy production, and calcium absorption; necessary for synthesis of fats, proteins, and nucleic acid.

Niacin-bound chromium	Take up to 10 mg three times per day.	Stimulates gastric juices and hydrochloric acid, essential for digestion; important in energy metabolism; supports healthy gastrointestinal tract; important for the metabolism of fat, carbohydrates, and protein.
Vitamin B$_1$ (thiamin)	Take 50 mg three times per day.	Essential for carbohydrate metabolism, proper energy production, nerve function, appetite, and digestion.
Vitamin B$_6$ (pyridoxine)	Take 50 mg three times per day.	Assists enzymes that metabolize amino acids and fats; necessary for proper immune function; aids digestion.
Whey	Take 1 heaping tablespoon three times per day with meals.	Improves digestion.

Comments

❑ Eat plenty of fiber-rich foods, including fruits, whole-grain breads and cereals, and leafy green vegetables. Figs, dates, prunes, and raisins can improve constipation.

❑ Lubricants and fluids are essential in the treatment of constipation. Drink plenty of water or fruit juice, and use oily salad dressings, butter, and oils with your food.

❑ Eat yogurt to help establish friendly intestinal bacteria.

❑ Eat frequent, small meals, rather than one or two large meals a day.

❑ Several herbs can improve digestion, including alfalfa concentrate, aloe vera, chlorella, ginger, and spirulina.

❑ To rid the bowel of wastes, drink enzyme-rich, colon-cleansing juices. See directions for juicing in Part Three. Also perform the Enzyme Kidney Flush and the Enzyme Toxin Flush. See Part Three for directions.

❑ For an easy-to-make fiber drink, mix equal parts of flaxseed and oat bran in water. Let it sit refrigerated overnight. In the morning, add 2 tablespoons of this mixture to a an 8-ounce glass of juice, and drink.

❑ The following are some effective superfoods: Bioflavonoid, Fiber & C Support Drink or Cho-Lo-Fiber Tone Drink from Crystal Star; Aloe Juice With Ginger from Aloe Falls; Pro Greens With Flax from Nutricology; and Whey to Go protein drink from Solgar.

❑ Short-term use of laxatives is usually effective, but avoid long-term use because it can interfere with the normal action of the intestinal muscles. Laxatives containing mineral oil also interfere with the absorption of numerous vitamins (including vitamins A, D, E, and K), as well as the action of a number of drugs. Habitual laxative or enema use will eventually lead to an inability of the bowels to function normally.

❑ Exercise thirty minutes every day. Although walking, bicycling, and swimming are excellent ways to relieve constipation, any type of exercise will increase enzymatic activity in your body and encourage elimination.

❑ See the list of Resource Groups in the Appendix for a list of organizations that can provide you with more information about constipation.

Cramps, Menstrual

See DYSMENORRHEA.

Cramps, Muscle

See MUSCLE CRAMPING.

Crohn's Disease

See INFLAMMATORY BOWEL DISEASE.

Cuts

See SCARS; SKIN ULCERS.

Cystic Fibrosis

Cystic fibrosis (CF) is a hereditary disease of the exocrine glands (those glands that secrete fluids into a duct), affecting primarily the gastrointestinal tract and the lungs. It affects Caucasians more than other races—in fact, it is the most common inherited disease that leads to death among Caucasians in the United States. It usually becomes evident shortly after birth. This disease causes thick mucus to be produced, which obstructs glands and ducts in various organs. The mucus clogs the air passages and allows bacteria to multiply, causing severe coughing spells and leaving the patient particularly susceptible to pneumonia, bronchitis, emphyse-

ma, and repeated lung infections (cystic fibrosis is among the most serious lung problems of American children). The mucus also prevents pancreatic enzymes from reaching the intestine and assisting the digestion and absorption of food (nearly 95 percent of CF patients have pancreatic insufficiency). The result can be weight loss or poor weight gain, malnutrition, steatorrhea (fatty stools), or intestinal obstruction. In fact, up to 10 percent of CF babies require immediate surgery to relieve intestinal obstructions.

Symptoms

The first symptoms of cystic fibrosis usually appear in infancy. The first symptom is often *meconium ileus*—the infant's first stool is very thick, passing more slowly than normal, sometimes blocking the intestines. Other early symptoms in babies include poor weight gain despite good appetite; frequent, bulky, foul-smelling, oily stools; a distended abdomen. Coughing is usually the most noticeable symptom. It is often accompanied by wheezing, respiratory tract infections, difficulty breathing, gagging, vomiting, and disturbed sleep. Cystic fibrosis also causes excessive sweating, with high concentrations of salt in the sweat. As the disease progresses, lack of oxygen may turn the skin bluish and cause clubbing of the fingers. The chest may also become barrel-shaped. As the patient gets older, his or her growth is slowed, and puberty is delayed. The patient experiences continued respiratory and gastrointestinal difficulties. Complications include the coughing up of blood, a collapsed lung, and heart failure. Some CF patients develop diabetes mellitus and liver problems.

Enzyme Therapy

Steatorrhea (fatty stools due to fat malabsorption) can lead to deficiencies in essential fatty acids and vitamins, since the body requires a certain amount of fat for proper function. Oral administration of pancreatic enzyme preparations from pork sources has historically been the treatment of choice to replace any deficient digestive enzymes. Most cystic fibrosis patients use enterically coated enzyme preparations that release only in the alkaline environment of the small intestine, protecting the enzymes from being killed by the acid in the stomach. Research shows that enterically coated enzymes not only decrease steatorrhea, but also improve the absorption of beneficial nutrients, thus decreasing the risk of nutritional deficiencies. Further, patients with cystic fibrosis often experience delays in normal growth, but enzymes seem to improve the growth patterns and lead to increases in weight. Recent research indicates that acid-stable lipases taken from microbial sources may also improve fat absorption.

Digestive enzyme therapy is used in the treatment of cystic fibrosis to improve the digestion and absorption of food nutrients and to improve fat absorption. Enzymes reduce stress on the gastrointestinal mucosa, help maintain normal pH levels, detoxify the body, promote the growth of healthy intestinal flora, and strengthen the body as a whole.

Systemic enzyme therapy is used as supportive care to reduce any inflammation, stimulate the immune system, improve circulation, help transport nutrients throughout the body, remove waste products, and improve overall health.

Enzyme Absorption System Enhancers (EASE) maximize enzyme activity. They can also improve the absorption and bioavailability of various nutrients and decrease the drain on the body's own digestive enzymes, thus prolonging their lives.

In addition to the standard multivitamin, multimineral, and multiglandular complexes recommended in the introduction to Part Two, the following supplements are recommended for the treatment of cystic fibrosis. The following recommended dosages are for adults. Children between the ages of two and six should take one-quarter of the recommended dosage, those between the ages of seven and twelve should take one-half of the recommended dosage, and those between the ages of thirteen and sixteen should take three-quarters of the recommended dosage. Cystic fibrosis is a very serious, potentially deadly disease. Therefore, any program should be reviewed by a well-trained physician or health-care practitioner.

ESSENTIAL ENZYMES

Enzyme	Suggested Dosage	Actions
Pancrelipase or pancreatin *or*	Take as directed by your physician.	Contains proteases, amylases, and lipases, so is able to digest proteins, carbohydrates, and fat.
A digestive enzyme formula containing protease, amylase, and lipase enzymes. See the list of digestive enzyme products in Appendix. *or*	Take as indicated on the label or as directed by your physician.	Improves digestion by breaking down proteins, carbohydrates, and fats.
Formula One—See page 38 of Part One for a list of products.	Take 5–10 tablets with every meal until steatorrhea resolves.	Contains proteases, amylases, and lipases, so is able to digest proteins, carbohydrates, and fat.

SUPPORTIVE ENZYMES

Enzyme	Suggested Dosage	Actions
Formula Three—See page 38 of Part One for a list of products.	Take as indicated on the label.	Potent antioxidant enzymes.

ENZYME ABSORPTION SYSTEM ENHANCERS (EASE)

Combination	Suggested Dosage	Actions
Detox Enzyme Formula from Prevail	Take 1 capsule between meals one or two	Helps rid the body of harmful toxins.

	times per day with a 6- to 8- ounce glass of water or juice.	
or		
PTE Support from PhysioLogics	Take 1–2 capsules three times per day on an empty stomach (one hour before or three hours after a meal); or follow the advice of your health-care professional.	Eliminates protein waste; controls inflammatory response.

ENZYME HELPERS

Nutrient	Suggested Dosage	Actions
Acidophilus or other probiotics, such as bifidobacterium; or Acidophilasé from Wakunaga, which contains lipolytic, proteolytic, and amylolytic enzymes	Take as indicated on the label.	Improves gastrointestinal function; needed for production of biotin, niacin, folic acid, and pyridoxine; helps produce antibacterial substances; and stimulates enzyme activity, all of which improve digestion and immune function.
Adrenal glandular	Take 100–400 mg per day.	Fights adrenal exhaustion; restores normal carbohydrate balance, thus assisting in the digestion and absorption of food nutrients.
Amino-acid complex such as AminoLogic from PhysioLogics	Take as indicated on the label.	Amino acids serve as building blocks for the body's enzymes that improve digestion.
Fiber	Take as indicated on the label.	Provides bulk; needed for proper elimination.
Fructooligo-saccharides	Take as indicated on the label.	Increases peristaltic action of intestines; improves liver function.
Green food product, such as Green Kamut, Kyo-Green, or Green Barley Extract	Take as indicated on the label.	Green foods are rich in enzymes, vitamins, minerals, chlorophyll, and amino acids.
Lecithin	Take 1,000 mg three times per day with meals.	Emulsifies fat.
Lung glandular	Take as indicated on the label.	Improves lung function.
Ox bile extract	Take 120 mg per day.	Aids digestion.
Pancreas glandular	Take 300 mg per day.	Aids digestion; decreases food allergies; fights steatorrhea; assists the pancreatic enzymes trypsin and chymotrypsin.
Vitamin A	Take 5,000–10,000 IU per day.	Needed by digestive glands to produce enzymes; required for healthy mucous linings and skin and cell membrane structure, and a strong immune system; potent antioxidant.
Vitamin B complex	Take 50 mg three times per day.	B vitamins act as potent coenzymes.
Vitamin C	Take 1,000–2,000 mg per day.	Stimulates the adrenal glands to produce hormones (which

		also act as coenzymes); aids in detoxification; is a potent antioxidant; improves the function of many enzymes; assists in converting cholesterol to bile salts.
Whey	Take 1 heaping tablespoon three times per day with meals.	Improves digestion.
Zinc	Take 10–50 mg per day. Do not take more than 100 mg from all sources per day.	Needed by over 300 enzymes (including pancreatic enzymes and superoxide dismutase). Many zinc enzymes work to metabolize carbohydrates, alcohol, and essential fatty acids; synthesize proteins; and dispose of free radicals.

Comments

❑ Standard treatment for cystic fibrosis includes enzyme therapy, antibiotics, good nutrition, and chest physiotherapy (where the chest is "thumped" to help break up mucus). This is a serious condition and should be monitored by a healthcare professional well trained in the treatment of cystic fibrosis.

❑ Some CF patients are able to get relief by using Pulmozyme (rhDNase). Pulmozyme decreases the viscoelasticity and adhesiveness of the mucus, breaking it up and improving the transport and elimination of sputum. The drug is administered through an aerosol and is safe and usually well-tolerated. It improves pulmonary function and the patient's well-being and reduces the risk of infections and exacerbations, which might require antibiotics.

❑ See the list of Resource Groups in the Appendix for a list of organizations that can provide you with more information about cystic fibrosis.

❑ *See also* Steatorrhea.

Cystic Mastitis

See FIBROCYSTIC BREAST DISEASE.

Cystitis

See BLADDER INFECTION.

Dermatitis

Dermatitis, also called eczema, is inflammation of the skin that may cause itchiness, blisters, redness, scaling, scabbing, and thickening of the skin. It may have many different causes, including allergies, chemical or physical irritation, and infection. Sometimes the cause cannot be determined. In many cases, dermatitis may be an allergic reaction.

There are many types of dermatitis. *Atopic dermatitis* is a chronic type of dermatitis. People with atopic dermatitis often have hay fever or asthma or some other type of allergic disorder. It is characterized by very itchy rashes in areas where heat and moisture are retained, such as the creases behind the elbows and knees, the neck, face, groin, and genitals. It most commonly affects children. Atopic dermatitis is aggravated by stress, temperature or humidity changes, bacterial skin infections, contact with irritating clothing, and, in infants, food allergies.

Contact dermatitis is caused by contact with an irritating substance. The irritant can be anything—soaps, detergents, lotions and creams, cosmetics, certain metals, plants (such as poison ivy), or chemicals used in the manufacture of clothing. Removal of the offending agent usually causes the redness and blistering to disappear in a few days, but the scaling, thickening of the skin, and itching may last for several days to several weeks.

Seborrheic dermatitis most often affects the scalp and face, although it may also affect other areas of the body. It causes scaling of the skin, which manifests itself as dandruff when it affects the scalp. In more severe cases, yellowish or reddish pimples may appear. Newborns may experience a thick, crusted rash on the scalp, called cradle rash, accompanied by diaper rash.

Stasis dermatitis affects the lower legs. It is caused by blood and fluid pooling under the skin, causing the skin to turn dark brown. It is characterized by chronic inflammation of the skin of the lower legs, causing, redness, scaling, warmth, and swelling. It occurs most often in those with edema and varicose veins.

Enzyme Therapy

Enzyme therapy can work from the outside in and from the inside out. Systemic enzyme therapy is used to reduce the signs of inflammation, including pain, swelling, and redness. Enzymes can stimulate the immune system, improve circulation, help speed tissue repair, bring nutrients to the damaged area, remove waste products, and enhance wellness.

Digestive enzyme therapy is used to improve digestion of food, reduce stress on the gastrointestinal mucosa, help maintain normal pH levels, detoxify, and promote the growth of healthy intestinal flora. Digestive enzyme therapy can also help replace the enzymes lost when foods are cooked or processed. All of these actions make the body healthier so that it can fight dermatitis.

Enzyme Absorption System Enhancers (EASE) maximize enzyme activity. They can also improve the absorption and bioavailability of various nutrients and decrease the drain on the body's own digestive enzymes, thus prolonging their lives.

In contrast to systemic and digestive enzymes, which work from the inside out, enzymes can also be applied topically. When applied directly to the affected area, they work from the outside in to reduce inflammation, relieve pain, and enhance healing. Topical enzyme creams are used to dissolve dead and damaged skin cells and exfoliate the skin, while leaving healthy skin cells intact.

In addition to the standard multivitamin, multimineral, and multiglandular complexes recommended in the introduction to Part Two, the following supplements are recommended for the treatment of dermatitis. The following recommended dosages are for adults. Children between the ages of two and five should take one-quarter the recommended dosage, those between the ages of six and twelve should take one-half the recommended dosage, and those between the ages of thirteen and seventeen should take three-quarters the recommended adult dosage.

ESSENTIAL ENZYMES

Enzyme	Suggested Dosage	Actions
Superoxide dismutase	Take 150 mcg per day.	Fights inflammation; potent antioxidant.

AND SELECT ONE OF THE FOLLOWING:

Formula One—See page 38 of Part One for a list of products.	Take 3 tablets twice per day between meals.	Fights inflammation, pain, swelling, and free-radical formation; speeds healing; stimulates the immune system.
Bromelain	Take three to four 230–250 mg capsules per day until symptoms subside. If severe, take 6–8 capsules per day.	Fights inflammation, pain, and swelling; speeds healing; stimulates the immune system.
Serratiopeptidase	Take 10 mg two times per day between meals.	Fights inflammation; reduces pain and swelling.
Trypsin with chymotrypsin	Take 250 mg of trypsin and 10 mg of chymotrypsin three times per day between meals.	Fights inflammation, pain, and swelling; increases rate of healing; stimulates the immune system.
Pancreatin	Take 500 mg six times per day on an empty stomach.	Fights inflammation, pain, and swelling; increases rate of healing; stimulates the immune system; strengthens capillary walls.
Papain or a microbial protease	Take 300 mg two times per day between meals.	Fights inflammation; speeds healing.

SUPPORTIVE ENZYMES

Enzyme	Suggested Dosage	Actions
Catalase and glutathione peroxidase	Take as indicated on the label.	Potent antioxidant enzymes.
A digestive enzyme product containing protease, amylase, and lipase enzymes. See the list of digestive enzyme products in Appendix.	Take as indicated on the label with meals.	Improves the breakdown and absorption of food nutrients; fights toxic build-up.

ENZYME ABSORPTION SYSTEM ENHANCERS (EASE)

Combination	Suggested Dosage	Actions
Advanced Nutritional System from Rainbow Light *or*	Take as indicated on the label.	Potent free-radical fighter and immune system stimulator.
Akne-Zyme from Enzymatic Therapy *or*	Take 1 or 2 capsules daily as a dietary supplement.	Maintains skin integrity.
Bio-Musculoskeletal Pak from Biotics *or*	Take as indicated on the label.	Decreases inflammation; fights free-radical formation; stimulates immune function.
Bromelain Complex from PhytoPharmica *or*	Take as indicated on the label.	Reduces swelling and inflammation; improves circulation; helps break up any circulating immune complexes.
Connect-All from Nature's Plus *or*	Take as indicated on the label.	Supports connective tissue.
CTR Support from PhysioLogics *or*	Take 2–3 capsules per day, preferably on an empty stomach; or follow the advice of your health-care professional.	Nourishes the body during tissue recovery and healing; helps maintain healthy tissue; diminishes damage caused by swelling and inflammation.
Detox Enzyme Formula from Prevail *or*	Take 1 capsule between meals one or two times per day with a 6- to 8-ounce glass of water or juice.	Helps rid the body of harmful toxins.
Kyolic Formula 102 (with enzymes) *or*	Take 2 capsules or tablets with meals two to three times per day.	Strengthens skin integrity; strengthens the immune system; helps alleviate stress; helps fight free radicals; helps fight inflammation.
NESS Formula #416 *or*	Take as indicated on the label.	Fights inflammation.
Protease Concentrate from Tyler Encapsulations	Take as indicated on the label.	Fights inflammation.

ENZYME HELPERS

Nutrient	Suggested Dosage	Actions
Acidophilus and other probiotics	Take as indicated on the label.	Improves gastrointestinal function, thus detoxifying the body;

		needed for production of biotin, niacin, folic acid, and pyridoxine, all of which are required by many enzymes; helps produce antibacterial substances; stimulates enzyme activity.
Adrenal glandular	Take 100–400 mg per day.	Fights inflammation and adrenal exhaustion.
Aged garlic extract	Take two 300 mg capsules three times per day.	Fights inflammation; stimulates the immune system; potent antioxidant.
Bioflavonoids, such as rutin or quercetin	Take 500 mg three times per day between meals.	Strengthen capillaries; potent antioxidants.
Coenzyme Q$_{10}$	Take 30–60 mg one to two times per day with meals; or follow the advice of your health-care professional.	Potent antioxidant.
Essential fatty acid complex	Take 100–200 mg two to three times per day with meals; or follow the advice of your health-care professional.	Helps strengthen cell membranes; supports immune function.
Green Barley Extract from Green Foods Corporation or Kyo-Green from Wakunaga	Take as indicated on the label.	Potent free-radical fighter; rich in vitamins, minerals, and enzyme activity.
Niacin *or*	Take 50 mg the first day, then increase dosage daily by 50 mg to 1,000 mg. Decrease dosage if adverse effects begin to appear. Take dosages over 1,000 mg only under the supervision of a physician.	Dilates blood vessels; improves circulation.
Niacin-bound chromium	Take 20 mg per day; or take as indicated on the label.	
Pituitary glandular	Take 100 mg per day.	Helps form new blood vessels; aids tissue repair and healing.
Thymus glandular	Take 100 mg per day.	Assists the thymus gland in immune resistance; fights infections.
Vitamin A	Take 10,000–40,000 IU for 4 weeks, then decrease dosage to 10,000 IU. Any amount over 50,000 should be taken only at the direction of a physician. Do not take more than 10,000 IU if you are pregnant.	Needed for strong immune system and healthy tissue; potent antioxidant.
Vitamin B$_6$	Take 500 mg per day between meals.	Activates many enzymes; aids in immune system function and antibody production; inhibits formation of toxic chemicals.
Vitamin C	Take 10,000–15,000 mg per day.	Fights infections, diseases, and allergies; improves wound healing; stimulates the adrenal glands; promotes immune system; powerful antioxidant.
Vitamin E	Take 1,200 IU per day from all sources (including the multivitamin complex in the introduction to Part Two).	Potent antioxidant; supplies oxygen to cells; improves circulation and level of energy.
Zinc	Take 50 mg per day. Do not take more than 100 mg per day from all sources.	Needed by over 300 enzymes; necessary for white blood cell immune function; improves wound healing.

Comments

❏ Eat plenty of fresh, enzyme-rich fruits and vegetables, such as pineapple, papaya, and figs. Follow the dietary guidelines in Part Three.

❏ Several herbs can improve digestion, including alfalfa concentrate, chlorella, and ginger root extract.

❏ Detoxify the body with juicing and fasting and other detoxification techniques described in Part Three, including the Enzyme Kidney Flush and the Enzyme Toxin Flush.

❏ Put ten drops of liquid aged garlic extract on a cotton swab or gauze. Dab gently on the affected area.

❏ Break open and spread the contents of two to four vitamin A and vitamin E capsules on the affected area.

❏ Zinc oxide salve can be spread over the affected area once daily.

❏ A positive mental attitude can help stimulate antioxidants, which fight free-radical formation. See Part Three for more information.

❏ Papain and other enzyme exfoliants and salves (such as Arve's Poi and Papaya Enzyme Exfoliating Gel) can be helpful. Follow directions on the label, and see SKIN CARE in Part Three for more information.

❏ Apply a vitamin C paste topically twice per day. See directions for its preparation in Part Three.

❏ Dermatitis may be caused by a vitamin B$_6$ deficiency. Eat foods high in vitamin B$_6$, including wheat germ and brown rice.

❏ Avoid any foods that seem to worsen your skin condition. In many people this may include seafood, dairy products (especially milk), eggs, and nuts. Keep a food diary to help you identify any food allergies.

❏ Avoid any clothing that may irritate the skin or cause an allergic reaction, such as wool.

❏ Make a poultice from slippery elm, white oak leaves and bark, and comfrey to reduce pain and encourage healing.

❏ To boost immune function, prepare a tea from goldenseal and/or echinacea. Drink two to three cups per day.

Dermatomyositis

Dermatomyositis is an inflammatory condition of the connective tissue, skin, and muscles, causing muscle degeneration. It is a rare, chronic disease, possibly caused by an autoimmune reaction, that progresses slowly and, in adults, often accompanies cancer. It usually occurs in children between five and fifteen years of age, and adults between forty and sixty years old. It affects women twice as frequently as it does men.

Symptoms

Symptoms include muscle weakness (which may occur suddenly or over time), muscle and joint pain, rash, fever, fatigue, Raynaud's phenomenon (a circulatory disorder that causes the fingers and toes to turn white, especially when exposed to cold), and weight loss. When dermatomyositis affects the esophagus, the patient may experience difficulty swallowing and even regurgitation of food.

Enzyme Therapy

Systemic enzyme therapy is used to fight pain and swelling, reduce inflammation, break up circulating antigen-antibody complexes, and stimulate the immune system. Enzymes improve circulation, help speed tissue repair by bringing nutrients to the damaged area and removing waste products, and enhance wellness.

Digestive enzyme therapy is used to improve the digestion of food, reduce stress on the gastrointestinal mucosa, help maintain normal pH levels, detoxify the body, promote the growth of healthy intestinal flora, and strengthen the body as a whole. All of these actions make the body healthier so that it can fight dermatomyositis. Digestive enzymes also serve as replacements for the body's pancreatic enzymes, leaving the pancreatic enzymes free to perform other functions in the body, such as decreasing inflammation.

Enzyme Absorption System Enhancers (EASE) maximize enzyme activity. They can also improve the absorption and bioavailability of various nutrients and decrease the drain on the body's own digestive enzymes, thus prolonging their lives.

In addition to the standard multivitamin, multimineral, and multiglandular complexes recommended in the introduction to Part Two, the following supplements are recommended for the treatment of dermatomyositis. The following recommended dosages are for adults. Children between the ages of two and five should take one-quarter the recommended dosage, those between the ages of six and twelve should take one-half the recommended dosage, and those between the ages of thirteen and seventeen should take three-quarters the recommended adult dosage.

ESSENTIAL ENZYMES

Enzyme	Suggested Dosage	Actions
Superoxide dismutase	Take 150 mcg per day.	Fights inflammation; potent antioxidant.

AND SELECT ONE OF THE FOLLOWING:

Enzyme	Suggested Dosage	Actions
Formula One—See page 38 of Part One for a list of products.	Take 10 tablets three times per day until symptoms subside.	Fights inflammation, pain, and swelling; stimulates the immune system.
Bromelain	Take three to four 250 mg capsules per day until symptoms subside. If severe, take 6–8 capsules per day.	Fights inflammation, pain, and swelling; speeds healing; stimulates the immune system.
Trypsin with chymotrypsin	Take 250 mg of trypsin and 10 mg of chymotrypsin three times per day between meals.	Fights inflammation, pain, and swelling; increases rate of healing; stimulates the immune system.
Pancreatin	Take 500 mg six times per day on an empty stomach.	Fights inflammation, pain, and swelling; increases rate of healing; stimulates the immune system; strengthens capillary walls.
Papain or a microbial protease	Take 300 mg two times per day between meals.	Fights inflammation; speeds healing; stimulates the immune system.

SUPPORTIVE ENZYMES

Enzyme	Suggested Dosage	Actions
Catalase and glutathione peroxidase	Take as indicated on the labels.	Potent antioxidant enzymes.
A digestive enzyme product containing protease, amylase, and lipase enzymes. See the list of digestive enzyme products in Appendix.	Take as indicated on the label with meals.	Improves the breakdown and absorption of food nutrients; fights toxic build-up.

ENZYME ABSORPTION SYSTEM ENHANCERS (EASE)

Combination	Suggested Dosage	Actions
Bromelain Complex from PhytoPharmica or	Take as indicated on the label.	Reduces swelling and inflammation; improves circulation; helps eliminate any circulating immune complexes.
Inflamzyme from PhytoPharmica or	Take as indicated on the label.	Promotes tissue repair, immune function, and enzymatic activity.
Mobil-Ease from Prevail or	Take as indicated on the label.	Has anti-inflammatory activity.
Oxy-5000 Forte from American Biologics or	Take as indicated on the label.	Helps the body excrete toxins.
Protease Concentrate from Tyler Encapsulations	Take as indicated on the label.	Fights inflammation.

200

ENZYME HELPERS

Nutrient	Suggested Dosage	Actions
Acidophilus and other probiotics, such as bifido-bacteria	Take as indicated on the label on an empty stomach.	Improve gastrointestinal function; aid in the digestion and absorption of nutrients; stimulate enzyme activity.
Adrenal glandular	Take 100–400 mg per day.	Fights inflammation and adrenal exhaustion.
Aged garlic extract	Take two 300 mg capsules three times per day.	Fights inflammation; stimulates the immune system; potent antioxidant.
Amino-acid complex that contains L-ornithine, L-glutamine, L-arginine, L-proline, and L-lysine	Take as indicated on the label on an empty stomach.	Required for enzyme production; helps the body maintain proper fluid and acid/base balance, necessary for a healthy body; promotes healthy tissue.
Bioflavonoid, such as rutin or quercetin	Take 500 mg three times per day between meals.	Strengthen capillaries; potent antioxidants.
Coenzyme Q$_{10}$	Take 30–60 mg one to two times per day with meals; or follow the advice of your health-care professional.	Potent antioxidant.
Essential fatty acid complex	Take 100–200 mg two to three times per day with meals; or follow the advice of your health-care professional.	Helps strengthen cell membranes; support immune function.
Niacin-bound chromium	Take up to 20 mg per day; or take as indicated on the label.	Dilates blood vessels; improves circulation.
Thymus glandular	Take 100 mg per day.	Assists the thymus gland in immune resistance; improves immune response.
Vitamin A	Take 5,000 IU per day.	Needed for strong immune system and healthy tissue; potent antioxidant.
Vitamin B complex	Take 50 mg three times per day; or take as indicated on the label.	B vitamins act as coenzymes and support neuromusculoskeletal function.
Vitamin C	Take 10,000–15,000 mg per day.	Fights infections, diseases, and allergies; improves wound healing; stimulates the adrenal glands; promotes immune system; powerful antioxidant.
Vitamin E	Take 400 IU per day.	Potent antioxidant; supplies oxygen to cells; improves circulation and level of energy.
Zinc	Take 15–50 mg per day. Do not take more than 100 mg per day from all sources.	Needed by over 300 enzymes; necessary for white blood cell immune function; improves wound healing.

Comments

❏ Since this condition may be triggered by an autoimmune reaction, treatment should be geared toward stimulating the immune system and breaking up antigen-antibody complexes.

❏ A number of herbs can enhance immunity, including echinacea, ginseng, licorice, and maitake, reishi, and shiitake mushrooms.

❏ Raw, fresh, enzyme-rich fruits and vegetables (particularly green vegetables) should comprise over 60 percent of your diet. If your body cannot tolerate raw fruits and vegetables, increase your intake of digestive enzymes, or sauté or steam your produce.

❏ Increase your intake of foods high in enzymes (such as pineapple), beta-carotene (such as carrots), and bioflavonoids (oranges, lemons, and grapefruit).

❏ Avoid foods that will weaken the immune system, such as alcohol, refined white flour, and sugar. These foods make your body's enzymes work harder, thus decreasing their ability to enhance immunity.

❏ Natural spices are fine, but chemical preservatives, particularly nitrites, should be avoided.

❏ Avoid chlorinated water. Distilled or bottled spring water, or even well water is recommended.

❏ Use the Enzymatic Gargle twice a day. See directions in Part Three.

❏ Perform the Enzyme Kidney Flush as outlined in Part Three.

❏ An enzymatic salve, may be of help in treating the skin problems associated with dermatomyositis. See Part Three for information.

❏ Break open and spread the contents of two to four vitamin A and E capsules on the skin twice a day.

❏ Make a paste by mixing a heaping teaspoon of powdered or crystalline vitamin C with distilled or bottled water. Use a tongue depressor to spread the paste across the skin or dab with a cloth. Allow the paste to remain on the skin for no more than 30 minutes. Repeat the process at least two times per day.

❏ *See also* POLYMYOSITIS.

Diabetes

Diabetes mellitus is a condition marked by high blood sugar (glucose) levels because the body either does not produce insulin or does not properly use it. Insulin is a hormone whose chief role is to promote the entry of glucose into cells. We need insulin to convert the starches, sugars, and other carbohydrates we eat into energy. When glucose cannot enter and nourish the cells, body tissues are subjected to the equivalent of starvation.

There are two forms of diabetes mellitus: *Type I* or *insulin-dependent diabetes mellitus* (IDDM, which used to be called juvenile-onset diabetes) and *Type II* or *non-insulin dependent diabetes mellitus* (NIDDM, which used to be called adult-onset diabetes). In his book *Dr. Bernstein's Diabetes Solution* (Boston: Little, Brown and Co., 1997), Richard K. Bernstein states that more accurate terms for these disorders may be "autoimmune diabetes" for Type I and "insulin-resistant diabetes" for Type II. The causes and treatment vary depending on the type of diabetes.

Type I diabetes usually develops before the age of 30, although it may develop at any time in one's life. For some reason, the body's antibodies attack and destroy the beta cells of the pancreas (the cells responsible for insulin production). Some believe that a viral infection or some nutritional factor in childhood or early adulthood may be a factor in the cause of this type of diabetes. There may be a genetic tendency necessary to develop IDDM. Whatever its cause, Type I diabetes is serious because the body needs insulin. Therefore, Type I diabetics require insulin injections. Fortunately, only 10 to 15 percent of all diabetics suffer from Type I diabetes.

Although Type II diabetes can occur in children, it most often occurs in those 30 years and older, hence it is often referred to as adult-onset diabetes. This type of diabetes is marked by either impaired insulin production or a decrease in insulin's effectiveness—a condition called insulin resistance. Obesity is a high risk factor for diabetes—80 to 90 percent of those with diabetes are obese. Certain racial groups, such as Blacks and Hispanics, are at increased risk. Risk also increases with aging. Type II diabetes also appears to be genetic. Unlike Type I diabetes, those suffering from Type II often can control their condition with proper diet or medications, although insulin injections may be necessary.

Do you have diabetes and don't know it? In the United States, about 50 percent of the 16 million people with diabetes are unaware that they have the condition, according to the American Diabetes Association. This means that half—or about 8 million people—have the disease and don't know it! Unfortunately, diabetes is the fourth leading cause of death by disease in the United States, and more than 178,000 will die this year from the disease and its related complications. (See the inset "Complications of Diabetes" on page 203.)

Symptoms

Symptoms of diabetes include frequent urination, excessive thirst, extreme hunger, unusual weight loss, extreme fatigue, and irritability. Other symptoms may include frequent infections (women may suffer from recurrent vaginal infections), blurred vision, slow healing of cuts and bruises, and tingling or numbness in the hands or feet.

Enzyme Therapy

There is a substantial amount of research and clinical data on the use of enzymes in diabetes, according to Dr. Karl Ransberger, a world-renowned enzyme expert. According to Ransberger, the German Society for Diabetes is currently conducting a clinical study to assess the ability of an enzyme combination called Phlogenzym to prevent the development of Type I diabetes in children. The study involves children of diabetic parents and children with other diabetic relatives. The researchers are screening the children and studying those who have antibodies to the enzyme glutamic acid decarboxylase (an enzyme found in the beta cells of the pancreas, which are destroyed in those with Type I diabetes) and insulin and, therefore, run a 50-percent risk of becoming diabetic within two years. Researchers are treating the children with Phlogenzym (which contains bromelain and trypsin and the bioflavonoid rutin) or a placebo to see if Phlogenzym prevents the development of Type I diabetes.

A number of animal experiments have been conducted showing that Phlogenzym does inhibit the onset of diabetes in non-obese diabetic mice (which are frequently used in diabetes research). It also prevents chemically induced diabetes. In these investigations, this combination proved to inhibit diabetes, which, according to Ransberger, is why the Germany Society for Diabetes has chosen Phlogenzym as the product with the highest potential to show inhibitory effects. Once the insulin-producing cells of the pancreas are destroyed and Type I diabetes is established, it cannot be reversed, according to Dr. Ransberger. These patients must be treated with insulin in order to stay alive.

Diabetics (whether Type I or Type II) frequently suffer from poor digestion. Digestive enzyme therapy can improve digestion of food, improving absorption of nutrients essential for improving immune function and circulation; reduce stress on the gastrointestinal mucosa; help maintain normal pH levels; detoxify the body; and promote the growth of healthy intestinal flora. Digestive enzymes serve as replacements for the body's pancreatic enzymes, leaving the pancreatic enzymes free to perform other functions in the body, such as boosting immune function, decreasing inflammation, and improving circulation.

Digestive enzyme therapy is particularly effective at treating a condition that often occurs in diabetics called *gastroparesis diabeticorum* (delayed stomach-emptying). Gastroparesis occurs because prolonged elevated blood sugar levels can damage the nerves responsible for enzyme secretion, hormone production, and muscular activity in the stomach and intestines. Symptoms of this condition will vary by individual, but usually include heartburn, bloating, belching, constipation (which may alternate with diarrhea), nausea, and vomiting. Because this condition delays stomach emptying, it may also affect blood sugar levels because it delays the

---

Complications of Diabetes

It is important to carefully monitor and control diabetes because of the serious long-term complications associated with the disease. According to the Centers for Disease Control and Prevention, diabetics have an increased risk of suffering from a number of conditions and complications.

PROBLEM	COMMENT
Amputations	Every year, nearly 54,000 people with diabetes have some portion of their lower extremities (legs or feet) amputated due to poor circulation that can lead to gangrene. In fact, those with diabetes constitute half of all the patients requiring amputations.
Blindness	Every year, between 12,000 and 24,000 diabetics lose their eyesight because of diabetic retinopathy. In fact, diabetes is the primary cause of blindness in those aged 20 to 74.
Cardiovascular/ Cerebrovascular disease	Diabetics have a two to three times higher risk of developing cardiovascular disease and a two to four times higher risk of suffering a stroke than those without diabetes. Among diabetics, nearly 48 percent of all deaths are due to cardiovascular disease.
Dental disease	Individuals with diabetes have a greater risk of developing periodontal disease (a leading cause of tooth loss). One study found that 30 percent of Type I diabetics 19 years of age and older suffered from periodontal disease.
High blood pressure	The majority of diabetics (sixty to sixty-five percent) suffer from elevated blood pressure levels.
Kidney disease	Diabetes is the leading cause of kidney failure. In 1992 alone, over 56,000 diabetics either received a kidney transplant or were forced to undergo dialysis.
Nerve disease	Diabetes can cause mild to severe nerve damage. In fact, between 60 and 70 percent of those with the disease have suffered nerve damage to some degree. This can lead to delayed stomach emptying (gastroparesis), peripheral neuropathy, carpal tunnel syndrome, and impaired sensation in the hands and/or feet .

---

normal rise in blood sugar that occurs after eating a meal. Digestive enzymes can help relieve some of the symptoms of gastroparesis. Chewable enzymes are very effective in treating this condition.

In addition to helping to prevent Type I diabetes and improving digestion in both types, enzymes can successfully inhibit many of the diseases that occur as a complication of diabetes. Diabetes can cause kidney failure, sometimes making dialysis necessary. A substantial number of animal experiments and clinical studies demonstrate that enzymes can inhibit diabetic nephropathy, a kidney disease caused by diabetes. Here, the enzymes are highly successful in preventing autoimmune destruction and fibrosis of the kidney. Dr. Ransberger states that his research shows that proteolytic enzyme mixtures can inhibit the gradual development of kidney insufficiency in most patients. He believes this is a very important finding because it can keep thousands of diabetic patients from developing kidney disease.

Systemic enzyme therapy is used to improve circulation (which is often poor in diabetics), stimulate the immune system, reduce inflammation, transport nutrients throughout the body, remove waste products, and enhance wellness. Enzyme therapy can also be used to stimulate antioxidant activity and fight free radical formation using antioxidant and proteolytic enzymes.

Enzyme Absorption System Enhancers (EASE) maximize enzyme activity. They can also improve the absorption and bioavailability of various nutrients, and decrease the drain on the body's own digestive enzymes, thus prolonging their lives.

In addition to standard multivitamin, multimineral, and multiglandular complexes recommended in the introduction to Part Two, the following supplements are recommended for the treatment of diabetes. The following recommended dosages are for adults. Children between the ages of two and five should take one-quarter the recommended dosage, those between the ages of six and twelve should take one-half the recommended dosage, and those between the ages of thirteen and seventeen should take three-quarters the recommended adult dosage.

ESSENTIAL ENZYMES

Enzyme	Suggested Dosage	Actions
To prevent the onset of Type I diabetes:		
Phlogenzym from Mucos	Take 3 tablets three times per day between meals.	Helps to inhibit the onset of Type I diabetes; stimulates immune function; helps break-up antigen-antibody complexes and free radicals.
Formula One—See page 38 of Part One for a list of products.	Take 5 tablets three times per day between meals.	Helps to inhibit the onset of Type I diabetes; stimulates immune function; helps break-up antigen-antibody complexes and free radicals.
For treatment of Type I and Type II diabetes:		
A digestive enzyme product. For a list of digestion enzyme products, see Appendix.	Take as indicated on the label.	Improves digestion and absorption; helps to break down fats, carbohydrates, and proteins.
If you suffer from gastroparesis, also take a chewable digestive enzyme product.	Take as indicated on the label.	Improves digestion and absorption; helps to break down fats, carbohydrates, and proteins; improves gastroparesis.

AND SELECT ONE OF THE FOLLOWING:

Enzyme	Suggested Dosage	Actions
Wobenzym N from Naturally Vitamin Supplements	Take 10 tablets three times per day between meals.	Improves immunity, circulation, and nerve function; reduces inflammation.
Inflazyme Forte from American Biologics	Take as indicated on the label.	Improves immunity, circulation, and nerve function; reduces inflammation.
Mega-Zyme from Enzymatic Therapy	Take 10 tablets three times per day between meals.	Improves immunity, circulation, and nerve function; reduces inflammation.

SUPPORTIVE ENZYMES

Enzyme	Suggested Dosage	Actions
Formula Three—See page 38 of Part One for a list of products.	Take as indicated on the labels.	Potent antioxidant enzymes. SOD levels are typically decreased in diabetic patients.

ENZYME ABSORPTION SYSTEM ENHANCERS (EASE)

Combination	Suggested Dosage	Actions
DiaBest from PhysioLogics *or*	Take 1–3 tablets daily with each meal; or follow the advice of your health-care professional.	Supports healthy blood glucose levels; aids circulation to the feet; nourishes a healthy cardiovascular system; helps maintain a healthy immune system.
Sugar Check from Enzymatic Therapy *or*	Take as indicated on the label.	Contains nutrients important for sugar and carbohydrate metabolism.
NESS Formula #13	Take as indicated on the label.	Helps fight diabetes; plays a role in protein, fat, and carbohydrate metabolism.

ENZYME HELPERS

Nutrient	Suggested Dosage	Actions
Acidophilus or other probiotics, such as bifidobacterium or Acidophilasé from Wakunaga, which also include enzymes.	Take as indicated on the label.	Improves gastrointestinal function; stimulates enzyme activity.
Adrenal extract	Take 100–400 mg per day.	Fights adrenal exhaustion; restores normal carbohydrate balance.
Aged garlic extract	Take two capsules three times per day.	Potent antioxidant; improves digestion; stimulates immune function and healthy intestinal bacteria.
Amino-acid complex containing L-carnitine and L-glutamine, such as AminoLogic from PhysioLogics	Take as indicated on the label.	Amino acids serve as building blocks for the body's enzymes.
Betaine Hydrochloric acid	Take 150 mg three times per day with meals.	Provides Hydrochloric acid to improve digestion. If gastric upset occurs, discontinue use.
Brewer's yeast	Take as indicated on the label.	Contains chromium, which is important for insulin production.
Calcium	Take 750 mg twice per day.	Essential for the activity of several enzymes.
Chromium polynicotinate or chromium picolinate	Take as indicated on the label.	Essential to human nutrition and health; required for normal carbohydrate, protein, and fat metabolism and blood sugar regulation; a cofactor for insulin; plays a vital role as glucose tolerance factor.
Coenzyme Q_{10}	Take as indicated on the label.	Improves circulation.
Copper	Take as indicated on the label.	Essential cofactor necessary for superoxide dismutase activity.
Fiber	Take as indicated on the label.	Provides bulk; needed for proper elimination.
Green food product, such as Green Kamut, Kyo-Green, or Green Barley Extract	Take as indicated on the label.	Green foods are rich in enzymes, vitamins, minerals, chlorophyll, and amino acids.
Liver extract	Take as indicated on the label.	Improves liver function; helps regulate sugar metabolism.
Magnesium	Take 250 mg per day.	Necessary for glucose metabolism; plays important roles in protein building, muscle contraction, nerve function, and energy production; necessary for synthesis of fats, proteins, and nucleic acid.
Maitake mushrooms	Take as indicated on the label.	High in chromium; assists in glucose metabolism.

Manganese	Take as indicated on the label.	Necessary for normal protein and carbohydrate breakdown; activates several enzymes.
Ox bile extract	Take 120 mg per day.	Improves digestion.
Pancreas extract	Take as indicated on the label.	Aids digestion; assists the pancreatic enzymes trypsin and chymotrypsin.
Potassium	Take 100 mg three times per day.	Maintains proper water balance between body fluids and cells; important in nerve transmission, muscle activity, and protein synthesis; necessary for cellular enzymes to work properly.
Vitamin A	Take 5,000 IU per day.	Needed by digestive glands to produce enzymes; required for healthy mucous linings and a strong immune system; potent antioxidant.
Vitamin B complex	Take 150 mg three times per day.	B vitamins act as coenzymes.
Vitamin B_1 (thiamin)	Take 50 mg three times per day.	Essential for carbohydrate metabolism and proper nerve function.
Vitamin B_3 (niacin) or Niacin-bound chromium	Take 50 mg three times per day. Take 10 mg three times per day.	Stimulates gastric juices and hydrochloric acid; important in energy metabolism; maintains nervous system and brain; supports healthy gastrointestinal tract.
Vitamin B_5 (pantothenic acid)	Take 50 mg per day.	Coenzyme needed by many enzymes.
Vitamin B_6 (pyridoxine)	Take 50 mg three times per day.	Assists enzymes that metabolize amino acids and fats; necessary for proper immune function; involved in sugar metabolism; helps to support adrenal glands often exhausted in diabetics.
Vitamin C	Take 2,000 mg three times per day.	Stimulates the adrenal glands to produce hormones (which also act as coenzymes); helps normalize sugar metabolism; aids in detoxification; is a potent antioxidant; improves the function of many enzymes. This vitamin is often depleted in diabetics.
Vitamin D	Take 400–800 IU per day.	Involved in the process of insulin secretion and in regulating the transport and absorption of calcium and phosphorus.
Vitamin E	Take 1,200 IU per day from all sources (including the multivitamin complex in the introduction to Part Two).	Acts as an intercellular antioxidant; improves circulation. This vitamin is often depleted in diabetics.
Zinc	Take 10–50 mg per day. Do not exceed 100 mg per day from all sources.	Needed by over 300 enzymes. Many zinc enzymes work to metabolize carbohydrates, synthesize proteins, and help dispose of free radicals.

Comments

❑ Anyone with diabetes should be under the care of a physician well trained in its treatment who will instruct the patient on how to monitor blood glucose levels and to plan meals, and who may also prescribe oral medicines or insulin injections. Treatment of diabetes is an ongoing process that requires close and regular monitoring.

❑ Eat small, frequent meals rather than two or three large meals. Snack five or six times per day.

❑ Diet is an extremely important part of diabetic therapy. Your diet should include complex carbohydrates that are high in fiber, such as fresh vegetables, rather than simple carbohydrates, such as breads and pastries. Complex carbohydrates take longer for the body to break down and absorb and, therefore, provide for a slower and more gradual increase in blood sugar levels.

❑ Eat foods with a high chromium content to improve insulin's effect on body tissues. High chromium foods include whole-grain products, bran, mushrooms, brewer's yeast, and wheat germ.

❑ Avoid sugar and other refined carbohydrates; coffee and other caffeine-containing beverages and drugs; tobacco; alcohol; salt; food additives, preservatives, and other chemicals that cause allergies and may affect blood sugar levels; and physical and emotional stress.

❑ Herbs beneficial in treating blood sugar problems include licorice root (*Glycyrrhiza glabra*); juniper cedar berries (*Juniperus sabina pinaceae*), Mexican wild yam, copalquin or capalachi (also known as quina blanca in Mexico), goldenseal (*Hydrastis*), lobelia (*Lobelia inflata*), garlic, and dandelion.

❑ Losing weight can sometimes normalize blood sugar levels. If you are overweight, begin a weight-loss program.

❑ Thirty minutes or more of exercise every day is critical in maintaining healthy circulation and metabolism and can help regulate blood sugar levels. Walking is especially helpful in combating diabetes. The body can improve the use of its own insulin and fight off diabetes when you walk as little as thirty minutes per day, according to a recent study conducted by Dr. Elizabeth Mayer-Davis, University of South Carolina and published in the *Journal of the American Medical Association*, March, 1998. Her findings are based on a study of 1,400 women ranging in age from 40 to 69 years. Some women had a mild form of diabetes; others had normal blood sugar levels. In addition to walking, other moderate activity that may improve insulin use would include climbing stairs, swimming, bicycling, doing many household chores, and gardening. This moderate physical activity should be performed most days of the week—if not every day.

❑ Work to reduce stress and to increase relaxation. See Part Three for information on relaxation and meditation.

❑ Detoxification can be beneficial in helping to improve overall body health. Detoxify only under the supervision of a well-trained physician. See Part Three for information on detoxification. Note: Although juice fasts are an important part of most standard biological treatment programs for many diseases, juice fasts are *not* recommended in the treatment of diabetes except with the approval of, and under the supervision of, an experienced physician.

❑ The future of enzyme therapy for the treatment of Type I diabetes may include injections of enzymes. Stanford University and UCLA scientists have shown that injecting mice that are genetically prone to diabetes with the enzyme glutamic acid decarboxylase can prevent the immune system errors that lead to this disease. Researchers note that any possible use in humans is years away as trials need to be conducted on animals and subsequently on humans.

❑ See the list of Resource Groups in the Appendix for a list of organizations that can provide you with more information about diabetes.

Diarrhea

Diarrhea is the excessive and frequent discharge of watery material from the bowel. Water is removed from stools in the large intestine. When the stools leave the large intestine too quickly, or if something prohibits the water from being removed from the stools, diarrhea results. There are many causes of diarrhea.

Osmotic diarrhea occurs when substances are not absorbed into the bloodstream and remain in the intestines, causing too much water to remain in the stool. It can be caused by certain foods and sugar substitutes and enzyme deficiencies. Osmotic diarrhea stops soon after the osmotic substance stops being consumed.

Secretory diarrhea is usually the result of some type of infection. It occurs when toxins produced by the infection or by certain tumors cause the intestines to secrete salt and water into the stool. Certain laxatives may also cause secretory diarrhea.

Exudative diarrhea occurs when the lining of the large intestine becomes inflamed, causing it to release blood, mucus, and other fluids into the stools. It can be caused by Crohn's disease, ulcerative colitis, cancer, and tuberculosis.

Diarrhea may also be caused by malabsorption syndromes and overgrowth of normal "friendly" bacteria in the intestines or the presence of bacteria not normally found in the intestines. An overactive thyroid, gastrointestinal surgery, certain drugs, and caffeine may also cause diarrhea.

Symptoms

Symptoms of diarrhea include frequent watery and loose stools. Accompanying symptoms might include abdominal pain and cramping, fever, thirst, and vomiting. Excessive or prolonged diarrhea can cause dehydration due to the water lost from the body. It can also interfere with nutrient absorption, since food is rushed through the intestines before nutrients can be extracted and absorbed. Diarrhea can also throw your body's pH level out of balance. Diarrhea in young children may be especially dangerous because they cannot tolerate much fluid loss.

See Allergies, Carbohydrate Intolerance, Celiac Disease, Inflammatory Bowel Disease, Influenza, Irritable Bowel Syndrome, Lactase Deficiency, or Steatorrhea in Part Two, if diarrhea is a symptom of any of these conditions, for the enzyme therapy program for its treatment.

Enzyme Therapy

Digestive enzyme therapy is used to restore a fluid balance to the body, strengthen the body as a whole, and build general resistance. Enzymes improve the digestion and assimilation of food nutrients, reduce stress on the gastrointestinal mucosa, help maintain normal pH levels, detoxify the body, and promote the growth of healthy intestinal flora.

Systemic enzyme therapy is used to reduce inflammation, stimulate the immune system, and improve circulation. Enzymes help transport nutrients through the body, remove waste products, and enhance wellness.

Enzyme Absorption System Enhancers (EASE) maximize enzyme activity. They can also improve the absorption and bioavailability of various nutrients and decrease the drain on the body's own digestive enzymes, thus prolonging their lives.

In one study, researchers treated diarrhea caused by enterotoxigenic *Escherichia coli* (ETEC) with an enterically coated protease preparation used to modify the intestinal surface. Oral administration of protease was successful in treating the diarrhea. There was a 99.5-percent success rate using the protease treatment. Results of the study indicated the use of protease to prevent ETEC-caused diarrhea has considerable potential.

Another study examined the effect of bromelain on *Escherichia coli* bacteria in the small intestine. Bromelain was administered orally and was found to inhibit the bacterial attachment to the small intestine. There were no adverse side effects from the bromelain treatment. It is concluded that administration of bromelain can inhibit *Escherichia coli* bacterial activity and may, therefore, be useful for the prevention of E. coli-caused diarrhea.

In addition to the standard multivitamin, multimineral, and multiglandular complexes recommended in the introduction to Part Two, the following supplements are recommended for the treatment of diarrhea. The following recommended dosages are for adults. Children between the ages of two and five should take one-quarter the recommended dosage, those between the ages of six and twelve should take one-half the recommended dosage, and those between the ages of thirteen and seventeen should take three-quarters the recommended adult dosage.

ESSENTIAL ENZYMES

Enzyme	Suggested Dosage	Actions
Formula One—See page 38 of Part One for a list of products. *or*	Take as indicated on the label.	Improves digestion.
Pancreatin *or*	Take 300 mg three times per day with meals.	Contains proteases, amylases, and lipases so is able to digest proteins, carbohydrates, and fats.
Bromelain *or*	Take 230–250 mg three times per day. Take no more than 750 mg per day. If diarrhea worsens, decrease the dosage until diarrhea stops.	Inhibits bacterial attachment to the small intestine.
A digestive enzyme formula containing protease, amylase, and lipase enzymes. For a list of digestive enzyme products, see Appendix. *or*	Take as indicated on the label.	Improves digestion by breaking down proteins, carbohydrates, and fats.
If swallowing tablets or pills is difficult, try chewable digestive enzymes, such as those taken from bromelain, papain, or papaya.	Take as indicated on the label.	Improves digestion.

SUPPORTIVE ENZYMES

Enzyme	Suggested Dosage	Actions
Fromula Three—See page 38 of Part One for a list of products.	Take as indicated on the labels.	Potent antioxidant enzymes.

ENZYME ABSORPTION SYSTEM ENHANCERS (EASE)

Combination	Suggested Dosage	Actions
Biodias A Granules from Amano Pharmaceutical Co., Ltd. *or*	Adults should take one packet in eight ounces of water between or after meals three times per day. Children ages 11 to 14 should take two-thirds of a packet; children between the ages of 8 and 10 should take half of the packet; and children between the ages of 5 and 7 should take one-third of a packet.	Helps improve digestion of nutrients; relieves inflammation; promotes regeneration of mucous membranes; decreases stress on the stomach and intestines; reduces gastric pain.

Bromelain Complex from PhytoPharmica *or*	Take as indicated on the label.	Reduced diarrhea by fighting unhealthy bacteria.
Digest-eze RX from PhytoTherapy *or*	Take as indicated on the label.	Helps prevent gas, bloating, and diarrhea; relieves gastrointestinal discomfort.
Inner Ecology from Prevail *or*	Take as indicated on the label.	Stimulates the growth of beneficial intestinal bacteria.
Pro-Flora from Tyler Encapsulations	Take as indicated on the label.	Helps restore normal digestion and gastrointestinal flora.

ENZYME HELPERS

Nutrient	Suggested Dosage	Actions
Acidophilus or other probiotics, such as bifidobacterium; or Acidophilasé from Wakunaga, which contains amylase, lipase, and protease	Take as indicated on the label.	Improves gastrointestinal function; replaces lost friendly bacteria; needed for production of biotin, niacin, folic acid, and pyridoxine, essential to many of the body's enzymes; helps produce antibacterial substances; stimulates enzyme activity.
Aged garlic extract	Take two 300 mg capsules three times per day.	Potent antioxidant; improves digestion; stimulates healthy intestinal bacteria; kills unfriendly bacteria.
Betaine hydrochloric acid	Take 150 mg three times per day with meals.	Hydrochloric acid improves digestion. If gastric upset occurs, discontinue use.
Fiber	Take as indicated on the label.	Provides bulk; needed for proper elimination.
Fructooligo-saccharides	Take as indicated on the label.	Help normalize peristaltic action of intestines; improve liver function.
Lecithin	Take 1,000 mg three times per day with meals.	Emulsifies fat.
Niacin-bound chromium	Take 10 mg three times per day.	Stimulates gastric juices and hydrochloric acid; important in energy metabolism; supports healthy gastrointestinal tract; necessary for normal digestion and absorption of nutrients.
Ox bile extract	Take 120 mg per day.	Aids digestion.
Vitamin B$_1$ (thiamin)	Take 150 mg three times per day.	Essential for carbohydrate metabolism, proper energy production and appetite; necessary for normal digestion and absorption of nutrients.
Vitamin B$_6$ (pyridoxine)	Take 50 mg three times per day.	Assists enzymes that metabolize amino acids and fats; necessary for proper immune function; assists in the transport of potassium (an electrolyte, which are often lost during bouts of diarrhea) into cells.
Vitamin C (Use abuf-buffered form, such as calcium ascorbate or magnesium	Take three 250 mg tablets twice per day.	Essential for immunity.

ascorbate.)		
Whey	Take 1 heaping tablespoon three times per day with meals.	Improves digestion.
Zinc (Zinc monome-thionine and zinc gluconate are good choices.)	Take 40–50 mg per day. Do not exceed 100 mg per day for total zinc intake.	Helps repair tissue damage in the digestive tract; improves immune function; potent antioxidant.

Comments

❑ Several herbs can improve digestion, including alfalfa, blackberry root, chlorella, ginger root extract, raspberry leaves, and green foods, such as Kyo-Green from Wakunaga or Green Magma from Green Foods Corp.

❑ Identify and eliminate the cause of the diarrhea (if possible).

❑ Drink three eight-ounce glasses of juiced or blended pineapple, papaya, banana, or apples.

❑ Replace lost fluids by drinking plenty of water, tea, and light soups.

❑ Replace lost electrolytes with such electrolyte replacement fluids as Pedialyte from Abbot Laboratories.

❑ Take apple pectin tablets or activated charcoal (on a short-term basis).

❑ For bloating and gas, try Indigestion Relief from Bioforce.

❑ For burping and belching, try Hyland's Indigestion after meals, Beano from AKPharma, or Digestive Formula from Country Life before meals.

❑ To improve digestion try a tea made from any of the following:

- A pinch of nutmeg, cloves, cinnamon, and ginger.
- Thyme.
- Slippery elm.
- Mint and alfalfa.
- Fennel and catnip.
- Wild yam.

❑ Eat cultured foods (such as kefir, yogurt, miso), high-fiber foods (like whole grains), enzyme-rich foods (including pineapple and papaya), and plenty of fresh fruits and vegetables.

❑ Eat four to five small meals per day. Chew food longer to put less stress on your gastrointestinal tract.

❑ Avoid spicy and fatty foods, red meats, fried foods, dairy products, sugary foods, and caffeine (coffee or cola drinks). Decrease acid-forming food intake.

❑ To cleanse your gastrointestinal tract: On day one, eat no foods, but drink six to eight 8-ounce glasses of fresh apple juice. Add Source of Life Energy Shake from Nature's Plus or a similar product to the first glass. On days two through four, drink six to eight 8-ounce glasses of enzyme-rich juices, such as papaya, pineapple, kiwi, carrot, and parsley. Follow this program for four days while gradually incorporating your normal diet. See Part Three for further information on juicing.

❑ See Part Three for the Enzyme Kidney Flush and the Enzyme Toxin Flush.

❑ Consult a physician if diarrhea lasts more than a day or two.

Digestive Disorders

See CARBOHYDRATE INTOLERANCE, CELIAC DISEASE; CONSTIPATION; DIARRHEA; DIVERTICULAR DISEASE; GAS, ABDOMINAL; GASTRITIS; HEARTBURN; HYPO-CHLORHYDRIA; INDIGESTION; INFLAMMATORY BOWEL DISEASE; IRRITABLE BOWEL SYNDROME; LACTASE DEFICIENCY; LEAKY GUT SYNDROME; NAUSEA; PANCREATIC INSUFFICIENCY; PANCREATITIS; PEPTIC ULCERS; STEATORRHEA.

Diverticular Disease

Diverticulosis is the presence of saclike outpouchings in the large intestine. They may appear anywhere in the large intestine, but are most common in the colon. These small pouches (called diverticula) form along the mucous membrane that lines the colon and can protrude through the colon's muscular outer wall. It is usually found where the blood vessels penetrate the muscular wall to supply the mucosa (typically near the rectum).

Diverticulosis is fairly common after the age of 40. In fact, most people in the United States over the age of 60 have the condition and almost everyone who reaches the age of 90 has it. Diverticulosis in and of itself is a rather benign condition; in fact it rarely even causes symptoms. It is when the condition develops into diverticulitis that problems occur.

Diverticulitis is inflammation of one or more diverticula. Fortunately, only about 20 percent of diverticulosis patients develop complications such as diverticulitis, which can lead to perforation, bleeding, or obstruction of the colon. Preventing the condition by maintaining a healthy colon is critical.

One of the leading causes of the diseases is a diet low in fiber. Insufficient fiber intake can cause spasms of the colon muscles, especially in the last part of the colon near the rectum, which is called the sigmoid colon. The mucosa can eventually push through the muscular coat at weak points when pressure in the colon builds up. Usually, this occurs when waste products from the body build up in the colon and press outward. Poor bowel habits and constipation can cause the waste to build up pressure in the colon. The disease worsens with age.

Symptoms

Diverticulosis usually produces no symptoms, and the diverticula are discovered only after an intestinal examination or a barium X-ray is performed. Those who suffer from diverticulitis might complain of abdominal cramps, tenderness, elevated temperature, constipation or diarrhea, nonspecific abdominal distress (including flatulence and pain), muscle spasms and pains (often in the lower left side of the abdomen), and disturbed bowel function. Other possible causes for the symptoms should be eliminated before symptoms are attributed to diverticulitis.

Enzyme Therapy

Digestive enzyme therapy is used to soften the stool, improve the digestion of food, increase regularity, and reduce stress on the gastrointestinal mucosa. Enzymes can help maintain normal pH levels, detoxify the body, promote the growth of healthy intestinal flora, and decrease pressure in the colon. Systemic enzyme therapy is used to reduce inflammation in the colon, stimulate the immune system, improve circulation, help speed tissue repair by bringing nutrients to the damaged area and removing waste products, strengthen the body as a whole, build general resistance, and improve wellness.

Enzyme Absorption System Enhancers (EASE) maximize enzyme activity. They can also improve the absorption and bioavailability of various nutrients and decrease the drain on the body's own digestive enzymes, thus prolonging their lives.

In addition to the standard multivitamin, multimineral, and multiglandular complexes recommended in the introduction to Part Two, the following supplements are recommended for the treatment of diverticular disease.

ESSENTIAL ENZYMES

Enzyme	Suggested Dosage	Actions
Superoxide dismutase	Take 150 mcg per day.	Fights inflammation; potent antioxidant.

AND SELECT ONE OF THE FOLLOWING:

Bromelain	Take three to four 250 mg capsules per day until symptoms subside. If pain is severe, take 6–8 capsules per day.	Fights inflammation, pain, and swelling; speeds healing; stimulates the immune system.
Trypsin with chymotrypsin	Take 230–250 mg of trypsin and 10 mg of chymo-	Fights inflammation, pain, and swelling; increases rate of heal-

	trypsin three times per day between meals.	ing; stimulates proper digestion.
Pancreatin	Take 500 mg six times per day on an empty stomach.	Fights inflammation, pain, and swelling; increases rate of healing; stimulates proper digestion.
Papain or a microbial protease	Take 300 mg two times per day between meals.	Fights inflammation; speeds recovery.
Formula One—See page 38 of Part One for a list of products.	Take 3 tablets with each meal.	Fights inflammation, pain, and swelling; increases rate of healing; stimulates proper digestion.

SUPPORTIVE ENZYMES

Enzyme	Suggested Dosage	Actions
Catalase and glutathione peroxidase	Take as indicated on the labels.	Potent antioxidant enzymes.
A digestive enzyme product containing protease, amylase, and lipase enzymes. See the list of digestive enzyme products in Appendix.	Take as indicated on the label with meals.	Improves the breakdown and absorption of food nutrients; fights toxic build-up.

ENZYME ABSORPTION SYSTEM ENHANCERS (EASE)

Combination	Suggested Dosage	Actions
Detox-Zyme from Rainbow Light or	Take as indicated on the label.	Helps cleanse the body and eliminate toxins.
Inner Act from Life Plus or	Take as indicated on the label.	Cleanses the body and detoxifies the colon.
NESS #13 or	Take as indicated on the label.	Will help break down various food groups and normalize the body.
Zyme Dophilus from Enzymatic Therapy	Take as indicated on the label.	Promotes the growth of friendly gastrointestinal bacteria and enzymes to better break down proteins, fats, and carbohydrates.

ENZYME HELPERS

Nutrient	Suggested Dosage	Actions
Acidophilus and other probiotics, or Acidophilasé from Wakunaga, which contains lipase, amylase, and protease	Take as indicated on the label.	Improve gastrointestinal function; needed for production of biotin, niacin, folic acid, and pyridoxine, which are needed by many enzymes; helps produce antibacterial substances in the colon; stimulates enzyme activity.
Adrenal glandular	Take 100–400 mg per day.	Fights inflammation and adrenal exhaustion.
Aged garlic extract	Take two 300 mg capsules three times per day.	Fights inflammation; stimulates the immune system; potent antioxidant.

Amino-acid complex, such as AminoLogic from PhysioLogics	Take as indicated on the label on an empty stomach.	Required for enzyme, peptide, and hormone production; helps the body maintain proper fluid and acid/base balance.
Bioflavonoids, such as rutin or quercetin	Take 500 mg three times per day between meals.	Potent antioxidant; improves capillary wall strength.
Coenzyme Q_{10}	Take 30–60 mg one to two times per day with meals; or follow the advice of your health-care professional.	Potent antioxidant.
Essential fatty acid complex	Take 100–200 mg two to three times per day with meals; or follow the advice of your health-care professional.	Helps strengthen cell membranes; support immune function.
Fiber	Take as indicated on the label.	A source of bulk to improve elimination.
Fructooligo-saccharides (FOS)	Take as indicated on the label.	Work in the body as a fiber source and as a "prebiotic," stimulating the growth of probiotics in the gut, thus improving digestion of nutrients and elimination of wastes.
Green Barley Extract from Green Foods Corporation or Kyo-Green from Wakunaga	Take as indicated on the label.	Potent free-radical fighter; rich in vitamins, minerals, and enzyme activity.
Thymus glandular	Take 100 mg per day.	Assists the thymus gland in immune resistance; improves immune response; fights infections.
Vitamin A	Take 5,000 IU twice per day. Take no more than 10,000 IU from all sources per day if pregnant.	Needed for strong immune system and healthy mucous membranes; potent antioxidant.
Vitamin B complex	Take 50 mg three times per day; or take as indicated on the label.	B vitamins act as coenzymes and are involved in energy production and gastrointestinal function.
Vitamin C	Take 8,000–13,000 mg per day.	Fights infections; improves healing; stimulates the adrenal glands; promotes immune system; powerful antioxidant.
Whey	Take as indicated on the label.	Improves digestion.
Zinc	Take 15–50 mg per day. Do not take more than 100 mg per day from all sources.	Needed by over 300 enzymes; necessary for white blood cell immune function; improves wound healing.

Comments

❏ Eat foods high in complex carbohydrates, including whole grains and nuts in all forms. Try grain cereals, whole wheat bread (check the label to ensure that whole wheat is the first ingredient) or other whole grain breads, millet, oats, and granolas with no sugar added. Add one to two tablespoons

of wheat bran to foods, or put in cold or hot liquid and drink with each meal.

❏ Raw, enzyme-rich fruits and vegetables (particularly green vegetables) should comprise over 60 percent of your enzyme-rich diet. If your body cannot tolerate raw fruits and vegetables, increase your intake of digestive enzymes, or sauté or steam your produce. Augment your enzyme-rich diet with enzyme supplements.

❏ Drink at least eight 8-ounce glasses of distilled or bottled water every day.

❏ Avoid alcohol, refined white flour, and sugar. These produce undesirable fermentation in the bowel and stimulate improper metabolism of insulin, chromium, and zinc.

❏ Periodically use a coffee enema and other detoxification methods, such as fasting, the Enzyme Kidney Flush, and the Enzyme Toxin Flush, as outlined in Part Three.

❏ Regular exercise is essential for proper regularity, health, and fitness. Exercise stimulates antioxidant activity in the body.

❏ See the list of Resource Groups in the Appendix for a list of organizations that can provide you with more information about diverticular disease.

Diverticulitis

See DIVERTICULAR DISEASE.

Diverticulosis

See DIVERTICULAR DISEASE.

Dysmenorrhea

Dysmenorrhea simply means painful menstruation. There are two types of dysmenorrhea—primary and secondary. Dysmenorrhea is referred to as primary when no disorder is found that may be causing the pain. Uterine contractions that occur when blood supply to the uterus is reduced are thought to cause the pain. Contributing factors may also include malposition of the uterus, a narrow cervical opening (interfering with the passage of tissue through the cervix), anxiety about menses, and lack of exercise. Dysmenorrhea is a common disorder of women (it affects more than 50 percent of women) and causes significant time loss from work or school. It usually appears during adolescence and seems to lessen with age and/or following pregnancy.

Secondary or acquired dysmenorrhea is pain with menstruation that is caused by some gynecological disorder. Endometriosis is one of the most common causes of dysmenorrhea. (See the inset "Endometriosis" on page 212.) Adenomyosis (ingrowth of the endometrium into the uterine musculature), fibroids, a pelvic lesion, and inflammation of the fallopian tubes may also cause this condition.

Symptoms

Dysmenorrhea may cause pain in the lower abdomen and sometimes the lower back and thighs that may begin shortly before menstruation and that usually subsides after about two days. A woman may also experience nausea and vomiting, constipation or diarrhea, and headache.

Enzyme Therapy

Systemic enzyme therapy is used to reduce menstrual pain and inflammation in the lower abdomen. Enzymes stimulate the immune system, improve circulation, help speed tissue repair, bring nutrients to the damaged area, remove waste products, enhance wellness, strengthen the body as a whole, and build general resistance.

Digestive enzyme therapy is used to improve the digestion of food, reduce stress on the gastrointestinal mucosa, help maintain normal pH levels, detoxify the body, and promote the growth of healthy intestinal flora. Digestive enzymes also serve as replacements for the body's pancreatic enzymes, leaving the pancreatic enzymes free to perform other functions in the body, such as decreasing inflammation.

Enzyme Absorption System Enhancers (EASE) maximize enzyme activity. They can also improve the absorption and bioavailability of various nutrients and decrease the drain on the body's own digestive enzymes, thus prolonging their lives.

In addition to the standard multivitamin, multimineral, and multiglandular complexes recommended in the introduction to Part Two, the following supplements are recommended for the treatment of both primary and secondary dysmenorrhea. The following recommended dosages are for adults. Girls between the ages of twelve and seventeen should take three-quarters the recommended adult dosage.

ESSENTIAL ENZYMES

Enzyme	Suggested Dosage	Actions
Superoxide dismutase	Take 150 mcg per day.	Fights inflammation; potent antioxidant.

AND SELECT ONE OF THE FOLLOWING:

Bromelain	Take 750 mg two times per day between meals.	Fights inflammation, pain, and swelling; speeds healing; stimulates the immune system; strengthens capillary walls.
Serratiopeptidase	Take 10 mg two times per day between meals.	Fights inflammation; reduces pain and swelling.
Trypsin with chymotrypsin	Take 250 mg of trypsin and 10 mg of chymotrypsin three times per day between meals.	Fights inflammation, pain, and swelling; increases rate of healing; stimulates the immune system.
Pancreatin	Take 500 mg six times per day on an empty stomach.	Fights inflammation, pain, and swelling; increases rate of healing; stimulates the immune system; strengthens capillary walls.
Papain	Take 300 mg two times per day between meals.	Fights inflammation; speeds healing.
Microbial protease	Take 300 mg two times per day between meals.	Fights inflammation; speeds healing.
Formula One—See page 38 of Part One for a list of products.	Take 5 tablets twice per day.	Fights inflammation, pain, and swelling; speeds healing; stimulates the immune system; strengthens capillary walls.

SUPPORTIVE ENZYMES

Enzyme	Suggested Dosage	Actions
Catalase and glutathione peroxidase	Take as indicated on the label.	Potent antioxidant enzymes.
A digestive enzyme product containing protease, amylase, and lipase enzymes. See the list of digestive enzyme products in Appendix.	Take as indicated on the label with meals.	Improves the breakdown and absorption of food nutrients; fights toxic build-up.

ENZYME ABSORPTION SYSTEM ENHANCERS (EASE)

Combination	Suggested Dosage	Actions
Bioprotect from Biotics or	Take as indicated on the label.	Contains antioxidants; has varied free-radical fighting capabilities.
Bromelain Complex from PhytoPharmica or	Take as indicated on the label.	Reduces swelling and inflammation; improves circulation.
Cardio-Chelex from Life Plus or	Take 5 tablets first thing in the morning.	Supports healthy circulation; improves energy.
Connect-All from Nature's Plus or	Take as indicated on the label.	Supports healing of tissue.
CTR Support from PhysioLogics	Take as indicated on the label.	Nourishes the body during tissue recovery and healing; diminishes damage caused by

Endometriosis

The endometrium is the tissue lining the inside of the uterus. This lining builds up and sheds during the monthly menstrual cycle. Endometriosis occurs when cells like those lining the uterus grow abnormally on the surfaces of other structures in the abdominal cavity, causing pain, infertility, and other problems. For women of reproductive age, the presence of this endometrial tissue in abnormal locations is relatively common.

Common symptoms of endometriosis include pain before and during periods (usually worse than "normal" menstrual cramps), heavy or irregular bleeding, infertility, and pain during or after sexual activity. Additional symptoms experienced during menstruation may include painful bowel movements, fatigue, diarrhea and/or constipation, lower back pain, and intestinal upset. Some women with endometriosis seem to have no noticeable symptoms.

Endometrial growths most commonly occur in the abdomen and involve the fallopian tubes, the pelvic lining, the ovaries, the outer surface of the uterus, the ligaments that support the uterus, and the area between the rectum and the vagina. But these growths can also be found in the rectum, bladder, cervix, vagina, intestines, lung, or other locations.

There are a number of theories regarding the cause of endometriosis, but they remain just that—theories. One theory states that menstrual tissue may go back up through the fallopian tubes, implanting and growing in the abdomen. However, because endometriosis has been found at distant locations (such as the lungs and nasal mucosa), perhaps the endometrial particles travel through the blood vessels or lymphatic system. It may be that tissue backs up in all women, and the tissue is only able to implant and grow in those with hormone problems or immune deficiencies. Endometriosis may also be hereditary.

Infertility is a common result of the disease's progression and affects 30 to 40 percent of women with the condition. Women with endometriosis are often advised not to postpone pregnancy. This is because the risk of infertility is greater the longer the disease is present. Also, pregnancy can give the body a break from menstruation, sometimes forcing the endometriosis into remission. Endometriosis should be suspected in any woman with infertility, as 25 to 50 percent of infertile women have the condition. Sometimes, symptoms will improve only to return after several years. At that time, dysmenorrhea and dyspareunia (painful or difficult sexual intercourse) may occur. Rectal pain or pain above the pubic region may also occur. Although its cause is unknown, abnormal bleeding is frequent.

Minor laparoscopic surgery is sometimes used to remove a growth. This can relieve symptoms and, in some cases, make pregnancy possible. However, recurrences are common. At present, there is no cure for endometriosis, although removal of the ovaries or hysterectomy can eliminate the symptoms (but this, in turn, causes other side effects). For further information, contact: Endometriosis Association/ 8585 N. 76th Place/ Milwaukee, Wisconsin 53223/ Tel: (414) 355-2200, Fax: (414) 355-6065.

Nutrient	Suggested Dosage	Actions
or		swelling and inflammation.
Detox-Zyme from Rainbow Light *or*	Take as indicated on the label.	Helps cleanse the body and eliminate toxins.
Inflamzyme from PhytoPharmica *or*	Take as indicated on the label.	Promotes tissue repair.
Protease Concentrate from Tyler Encapsulations	Take as indicated on the label.	Breaks up fibrin; fights inflammation.

ENZYME HELPERS

Nutrient	Suggested Dosage	Actions
Acidophilus and other probiotics	Take as indicated on the label.	Improves gastrointestinal function; needed for production of biotin, niacin, folic acid, and pyridoxine; stimulates enzyme activity.
Adrenal glandular	Take 100–400 mg per day.	Fights inflammation and adrenal exhaustion.
Aged garlic extract	Take two 300 mg capsules three times per day.	Fights inflammation; stimulates the immune system; potent antioxidant.
Bioflavonoids, such as rutin and quercetin	Take 500 mg three times per day between meals.	Strengthen capillaries; potent antioxidant.
Calcium	Take 1,000 mg per day.	Plays important roles in muscle contraction and nerve transmission.
Gamma-linolenic acids (GLA)	Take as indicated on the label.	Inhibit platelet aggregation; decrease vascular obstruction; help fight menstrual cycle disorders and inflammation; stimulate immune system.
Magnesium	Take 500 mg per day.	Plays important role in muscle contraction, nerve function, and calcium absorption.

Raw ovary glandular	Take 50 mg per day.	Improves ovarian function.
Raw uterus glandular	Take 25 mg per day.	Improves uterine function.
Thymus glandular	Take 100 mg per day.	Assists the thymus gland in immune resistance; improves immune response; fights infections.
Vitamin A	Take 5,000 IU per day.	Needed for strong immune system and healthy mucous membranes; potent antioxidant.
Vitamin B complex	Take 150 mg three times per day; or take as indicated on the label.	B vitamins act as coenzymes and support enzymes that reduce swelling and inflammation; increase circulation.
Vitamin B$_3$ (niacin)	Take 50 mg twice per day between meals. Increase dosage by 50 mg per day to 1,000 mg, or until symptoms disappear, whichever occurs first.	Needed for proper circulation; helps fight inflammation and reduces swelling.
Vitamin C	Take 10,000–15,000 mg per day.	Improves tissue healing; promotes immune system; powerful antioxidant.

Comments

❑ Black haw, dong quai, passion flower, and wild yam are herbs that are effective in the treatment of dysmenorrhea.

❑ The discomfort of primary dysmenorrhea may be relieved with the application of local heat (a hot water bottle or heating pad applied to the abdomen).

❑ Secondary dysmenorrhea can be relieved by relieving the cause(s) of the symptoms, sometimes with surgery.

❑ Exercise can help reduce the discomfort of dysmenorrhea. See Part Three for general recommendations and exercise and dietary guidelines.

Ear Infection

The ear is composed of three sections—the outer ear, which consists of the external part of the ear and the ear canal; the middle ear, which consists of the eardrum and a chain of three small bones that runs from the eardrum to the inner ear, called the ossicular chain; and the inner ear, filled with fluid and consisting of the seashell-shaped cochlea, which converts sound waves into nerve impulses and sends them on to the brain, and the semicircular canals, which provide balance. An ear infection can occur in any of the three sections.

An infection of the outer ear (external otitis) is often called swimmer's ear because it occurs most often during the summer swimming season. It can be caused by bacteria or, rarely, fungus. Injury to the ear canal often leads to external otitis.

Also, when debris and ear wax accumulate in the ear canal, they tend to trap water that gets into the ear while washing hair or swimming, making the canal more prone to infection.

A middle ear infection (otitis media) is caused by bacteria or viruses. This type of ear infection is most common in young children. It usually develops as a result of a cold when viruses or bacteria from the throat infect the middle ear. When a middle ear infection is left untreated or is not adequately treated, secretory otitis media can develop, in which fluid accumulates in the middle ear. Acute mastoiditis is another complication that can result from an untreated or improperly treated middle ear infection. Acute mastoiditis is a bacterial infection of the prominent bone behind the ear. Chronic otitis media results from perforation in the eardrum caused by acute otitis media, blockage of the eustachian tube, or injury. If an ear infection is left untreated, it can result in complete hearing loss, and often, facial paralysis. It can also spread to the inner ear and then to the brain, where it can cause meningitis (inflammation of the membrane surrounding the brain).

Symptoms

Symptoms of an ear infection will vary depending on the area of the ear that is affected. An outer ear infection causes itching, pain, a foul-smelling discharge from the ear, and perhaps a slight fever and a temporary loss of hearing in the affected ear. If not treated, an outer ear infection can progress to an infection of the middle ear. Reduced hearing, severe earache, a "full" feeling in the ear, nausea, vomiting, and fever are all symptoms of a middle ear infection. The eardrum becomes inflamed and may swell. If it ruptures, there may be a bloody, then clear, discharge from the ear that may change to pus. An inner ear infection can cause hearing loss, nausea, severe vertigo, and fever. Tinnitus may also result from an ear infection. (See "Tinnitus" on page 214.)

Enzyme Therapy

Systemic enzyme therapy is used to decrease pain, swelling, inflammation, and infection in the ear. Enzymes stimulate the immune system, improve circulation, help speed tissue repair, bring nutrients to the damaged area, remove waste products, enhance wellness, and build general resistance.

Digestive enzyme therapy is used to improve the digestion of food, reduce stress on the gastrointestinal mucosa, help maintain normal pH levels, detoxify the body, promote the growth of healthy intestinal flora, and strengthen the body as a whole, thus making the body healthier so that it can fight off ear infections. Digestive enzymes also serve as replacements for the body's pancreatic enzymes, leaving the pancreatic enzymes free to perform other functions in the body, such as boosting immunity and decreasing inflammation.

Enzyme Absorption System Enhancers (EASE) maximize enzyme activity. They can also improve the absorption and

bioavailability of various nutrients and decrease the drain on the body's own digestive enzymes, thus prolonging their lives.

In addition to the standard multivitamin, multimineral, and multiglandular complexes recommended in the introduction to Part Two, the following supplements are recommended for the treatment of ear infections. The following recommended dosages are for adults. Children between the ages of two and five should take one-quarter the recommended dosage, those between the ages of six and twelve should take one-half the recommended dosage, and those between the ages of thirteen and seventeen should take three-quarters the recommended adult dosage.

ESSENTIAL ENZYMES

Enzyme	Suggested Dosage	Actions
Superoxide dismutase	Take 150 mcg per day.	Fights inflammation; potent antioxidant.

AND SELECT ONE OF THE FOLLOWING:

Formula One—See page 38 of Part One for a list of products.	Take 6 tablets three times per day between meals.	Reduces inflammation and swelling; stimulates immune function; enhances tissue repair.
Bromelain	Take three to four 230–250 mg capsules per day until symptoms subside. If severe, take 6–8 capsules per day.	Fights inflammation, pain, and swelling; speeds healing; stimulates the immune system.
Trypsin with chymotrypsin	Take 250 mg of trypsin and 10 mg of chymotrypsin three times per day between meals.	Fights inflammation, pain, and swelling; increases rate of healing; stimulates the immune system.
Pancreatin	Take 500 mg six times per day on an empty stomach.	Fights inflammation, pain, and swelling; increases rate of heal- ing; stimulates the immune system; strengthens capillary walls.
Papain or a microbial protease	Take 300 mg two times per day between meals.	Fights inflammation; speeds healing.

SUPPORTIVE ENZYMES

Enzyme	Suggested Dosage	Actions
Catalase and glutathione peroxidase	Take as indicated on the label.	Potent antioxidant enzymes.
A digestive enzyme product containing protease, amylase, and lipase enzymes. See the list of digestive enzyme products in Appendix.	Take as indicated on the label with meals.	Improves the breakdown and absorption of food nutrients; fights toxic build-up.

ENZYME ABSORPTION SYSTEM ENHANCERS (EASE)

Combination	Suggested Dosage	Actions
Cold Zzap from Naturally Vitamin Supplements *or*	Take as indicated on the label.	Fights infection from viruses or bacteria; boosts immunity; has antioxidant activity.
Hepazyme from Enzymatic Therapy *or*	Take as indicated on the label.	Stimulates immune function.
Inflamzyme from PhytoPharmica *or*	Take as indicated on the label.	Immune system activator; reduces inflammation; fights free radicals.
Kyolic Formula 102 (with enzymes)	Take 2 capsules or tablets with meals two to three times per day.	Strengthens the immune system; helps alleviate stress; helps fight free radicals; helps fight inflammation; improves digestion and absorption of

---

Tinnitus

Tinnitus is noise heard in the ear that does not originate in the environment. It is usually experienced as a ringing (the word tinnitus is latin for "to ring"), but the noise may also be a buzzing, hissing, roaring, or whistling in the ears. The sounds may be continuous, intermittent, or pulsating (usually synchronized with the heartbeat). Hearing loss is often associated with this nerve-wracking condition.

Almost everything that can go wrong with the ear has tinnitus as an associated symptom. Problems ranging from cancer in the ear to an overproduction of wax or an infection in the ear can produce tinnitus. Tinnitus can also be a symptom of an obstruction, infection, otosclerosis (a condition marked by the formation of spongy bone in the ear), or Meniere's disease.

Sometimes tinnitus results from allergies, antibiotic or other drug use (including salicylates and quinine), inhalation of carbon monoxide, cardiovascular disease (including arteriosclerosis, aneurysms, and hypertension), and ear trauma caused by loud noise. Certain diuretics can cause the condition, as can alcohol, heavy metal intoxication, hypothyroidism, anemia, head trauma, and temporomandibular joint problems. As many as 200 drugs can cause tinnitus as a side effect.

Some 50 million American adults have tinnitus to some degree or another. Of those, 12 million suffer severely enough to seek the help of a physician or a hearing specialist. For further information contact the American Tinnitus Association/ P.O. Box 5/ Portland, OR 97207-0005/ Tel: (503) 248-9985/ Fax: (503) 248-0024.

or		nutrients that fight infections.
Vita-C-1000 from Life Plus	Take as indicated on the label.	Boosts immune function; fights infection.

ENZYME HELPERS

Nutrient	Suggested Dosage	Actions
Acidophilus (and other probiotics)	Take as indicated on the label.	Has antiviral, antifungal, and antibacterial activity; stimulates enzyme activity, which decreases infection and inflammation and increases immune function.
Aged garlic extract	Take two 300 mg capsules three times per day.	Fights inflammation; stimulates the immune system; potent antioxidant.
Amino-acid complex or individual amino acids, such as L-ornithine, L-glutamine, L-arginine, L-proline, and L-lysine	Take as indicated on the label on an empty stomach.	Required for enzyme and hormone production; helps the body maintain proper fluid and acid/base balance; helps fight infection and inflammation; promotes formation of healthy tissue; stimulates immune function.
Bioflavonoids, such as rutin or quercetin	Take 500 mg three times per day between meals.	Strengthen capillaries; potent antioxidants.
Coenzyme Q$_{10}$	Take 30–60 mg one to two times per day with meals; or follow the advice of your health-care professional.	Potent antioxidant.
Essential fatty acid complex	Take 100–200 mg two to three times per day with meals; or follow the advice of your health-care professional.	Helps strengthen cell membranes; supports immune function.
Green Barley Extract from Green Foods Corporation or Kyo-Green from Wakunaga	Take as indicated on the label.	Potent free-radical fighters; rich in vitamins, minerals, and enzyme activity.
Thymus glandular or ThymuPlex # 398 from Enzymatic Therapy	Take 90 mg per day; or take as directed on the label.	Assists the thymus gland in immune resistance; improves immune response; fights infections.
Vitamin A	Take 10,000–15,000 IU per day until the infection resolves.	Needed for strong immune system and healthy mucous membranes; potent antioxidant.
or A-mulsin from Naturally Vitamin Supplements	Take two capsules per day until the infection resolves.	
or Beta-carotene	Take 50,000 IU per day until the infection resolves.	
Vitamin C	Take 8,000–13,000 mg per day.	Fights infections, diseases, and allergies; improves wound healing; stimulates the adrenal glands; promotes immune system; powerful antioxidant.
Vitamin E	Take 1,200 IU per day from all sources (including the multivitamin complex	Potent antioxidant; supplies oxygen to cells; improves circulation and level of energy.

	in the introduction to Part Two).	
Zinc	Take 50 mg per day. Do not take more than 100 mg per day from all sources.	Needed by more than 300 enzymes; potent antioxidant.

Comments

❑ Several herbs can help reduce the inflammation that accompanies an ear infection, including aged garlic extract, echinacea, ginger root extract, ginkgo biloba, licorice, and white willow bark.

❑ Rest is very helpful in the treatment of an ear infection.

❑ Apply warm moist packs to the area. To prepare the warm moist pack, place a heating pad on the bed, place a piece of wax paper over the heating pad, with a one-inch overlap around all sides of the heating pad, then place a warm, moist folded washcloth on top of the heating pad. Lie down with the affected ear facing down on the warm moist pack.

❑ Place a ball of cotton containing warm garlic or olive oil in the external ear.

❑ During a fever, the body requires more enzymes and vitamins A and C.

❑ A well-balanced diet can help the body fight infection. Also, increase your consumption of protein, as proteins help produce antibodies to help the body fight infection.

❑ Use the Enzymatic Gargle, the Garlic/Enzyme Gargle, or the Echinacea/Enzyme Gargle at least three times per day to fight inflammation and infection and stimulate immune function. See Part Three for instructions for their preparation.

❑ Maitake, reishi, and shiitake mushrooms are valuable in fighting inflammation and infection.

❑ Take one 500 mg capsule of evening primrose oil twice per day. Children under the age of 8 should take the dosage once per day.

❑ For dietary recommendations and detoxification methods (cleansing, fasting, and juicing), see Part Three.

❑ Exercises (including walking and bicycling), positive mental attitude, and meditation are helpful in reducing stress, and stimulating enzymatic, immune, and antioxidant activity in the body. When exercising outdoors in a cold or windy environment, be sure to cover your ears during an ear infection.

Eczema

See DERMATITIS.

Embolism

An embolism is a sudden blocking of a blood vessel by an embolus (any foreign material, such as a blood clot). The embolus may be part of a blood clot that has broken off, a fat globule, a gas bubble, a mass of bacteria or parasites, or similar foreign object. As the embolus is carried through the circulatory system from where it originated, it may arrive at a place in a blood vessel that is too narrow for it to pass through. At this point it lodges, causing an obstruction. If a blood clot causes blockage at its point of origin, it is called a thrombus. The resulting condition is called a thrombosis (*see* THROMBOSIS in Part Two).

An embolism is one of the main causes of stroke. Usually, the obstruction is caused by a segment of a blood clot that lodges in one of the carotid arteries (the pair of major blood vessels in the neck that lead to the brain), or in an artery of the brain. This causes the blood supply to an area of the brain to be cut off, and, therefore, brain cells are deprived of oxygen and nutrients, and starve and die. An embolus obstructing an artery in a leg or arm can result in gangrene. When a clot lodges in the lungs, it is called a pulmonary embolism. Where there has been damage to vein walls, clots can sometimes arise in the legs as a result of inflammation (phlebitis). The blood vessels of the heart or the legs frequently form blood clots that produce blockages elsewhere. Poor circulation can also lead to an embolism.

Symptoms

Symptoms of an embolism will vary depending on where the vessel obstruction occurs. If an embolism occurs in the brain, it may cause a stroke, hemiplegia (paralysis on one side of the body), numbness, and aphasia (difficulties in speaking and understanding), often followed by permanent nerve damage. The victim may be unable to speak; to move his or her mouth, arms, or legs; or even to coordinate movements. An embolism in an artery of the lung can result in breathlessness, which might be the only symptom; however, a large clot might quickly cause death. Additional symptoms of a pulmonary embolism could include rapid, shallow breathing, restlessness and anxiety, sharp pain upon breathing, lightheadedness, fainting, and/or convulsions. An embolism in the vessels to the heart can cause a heart attack. In the arms or legs, it can cause pain and cramping.

An infarct may result when an embolus completely obstructs a vessel. An infarct is an area of dead or dying tissue resulting from obstruction of the blood vessels normally supplying that part.

Enzyme Therapy

Enzymes break up the clot (or other foreign particle) without causing a blockage of the blood vessel. Anticoagulants are ordinarily used and are the basic treatment for embolism. These medicines reduce the tendency of the blood to clot. Blood flow can also be increased by drugs that dilate blood vessels. Sometimes, surgery is recommended if the blockage is within the surgeon's reach (such as the carotid artery). If this is the case, proteolytic enzymes play a role in recovery from surgery, reduced inflammatory response, wellness, and the prevention of small thrombi and fibrin formation.

Systemic enzyme therapy is used to break up the fibrin formation, the LDL cholesterol, and the embolism. It is not helpful, however, in the treatment of gas bubble embolisms. Enzymes also reduce inflammation, stimulate the immune system, improve circulation, help speed tissue repair, bring nutrients to the damaged area, remove waste products, enhance wellness, strengthen the body as a whole and build general resistance.

Digestive enzyme therapy is used to improve the digestion of food, reduce stress on the gastrointestinal mucosa, help maintain normal pH levels, detoxify, and promote the growth of healthy intestinal flora, thus making the body healthier so that it can deal with the effects of an embolism. Digestive enzymes also serve as replacements for the body's pancreatic enzymes, leaving the pancreatic enzymes free to perform other functions in the body, such as improving circulation.

Enzyme Absorption System Enhancers (EASE) maximize enzyme activity. They can also improve the absorption and bioavailability of various nutrients and decrease the drain on the body's own digestive enzymes, thus prolonging their lives.

In addition to the standard multivitamin, multimineral, and multiglandular complexes recommended in the introduction to Part Two, the following supplements are recommended for the treatment of embolisms. The following recommended dosages are for adults. Children between the ages of two and five should take one-quarter the recommended dosage, those between the ages of six and twelve should take one-half the recommended dosage, and those between the ages of thirteen and seventeen should take three-quarters the recommended adult dosage.

ESSENTIAL ENZYMES

Enzyme	Suggested Dosage	Actions
Formula Three—See page 38 of Part One for a list of products.	Take 200 mcg of each three times per day between meals.	Potent antioxidant enzymes.

AND SELECT ONE OF THE FOLLOWING:

Formula One (particularly Wobenzym	Take 8–10 tablets three times per day between	Dissolves blood clots; fights inflammation, pain, and

N from Naturally Vitamin Supplements)—See page 38 of Part One for a list of products.	meals for six months or until symptoms improve. Thereafter, take 6 tablets three times per day between meals.	swelling; speeds healing; strengthens capillary walls.
Inflazyme Forte from American Biologics	Take as indicated on the label between meals.	Dissolves blood clots; fights inflammation, pain, and swelling; speeds healing; strengthens capillary walls.
Bromelain	Take four to six 230–250 mg capsules per day.	Fights inflammation, pain, and swelling; speeds healing; stimulates the immune system.
Enzyme complexes rich in proteases, plus amylases and lipases, working synergistically	Take as indicated on label between meals.	Dissolves blood clots; fights inflammation, pain, and swelling; speeds healing; strengthens capillary walls.

SUPPORTIVE ENZYMES

Enzyme	Suggested Dosage	Actions
A digestive enzyme formula containing protease, amylase, and lipase enzymes. See the list of digestive enzyme products in Appendix.	Take as indicated on the label with meals.	Improves the breakdown and absorption of proteins, carbohydrates, and fats; fights toxic build-up.

ENZYME ABSORPTION SYSTEM ENHANCERS (EASE)

Combination	Suggested Dosage	Actions
Bromelain Complex from PhytoPharmica or	Take as indicated on the label.	Reduces swelling and inflammation; improves circulation.
Cardio-Chelex from Life Plus or	Take 5 tablets twice per day—first thing in the morning and at midday.	Supports healthy circulation; improves energy.
Cardio-Protector from Phyto-Therapy, Inc. or	Take as indicated on the label.	Contains nutrients essential for heart and vascular health with enzymes to improve absorption of nutrients.
Kyolic Formula 102 (with enzymes) from Wakunaga or	Take 2 capsules or tablets between meals two to three times per day.	Helps reverse cardiovascular/circulatory disorders; strengthens the immune system; helps alleviate stress; helps fight free radicals; helps fight inflammation.
Protease Concentrate from Tyler Encapsulations	Take as indicated on the label.	Breaks up fibrin; fights inflammation.

ENZYME HELPERS

Nutrient	Suggested Dosage	Actions
Adrenal glandular	Take as indicated on the label.	Helps fight stress and adrenal exhaustion.
Aged garlic extract	Take 3 capsules three times per day.	Lowers cholesterol and blood pressure; fights free radicals.
Bioflavonoids, such as rutin and quercetin	Take 500 mg three times per day.	Fights atherosclerosis; potent antioxidant; reduces capillary fragility.
Calcium	Take 1,000 mg per day.	Plays important roles in muscle contraction; involved in the activation of several enzymes, including lipase.
Coenzyme Q_{10}	Take 75 mg one to two times per day with meals.	Potent antioxidant; prevents heart attacks; decreases blood pressure and total serum cholesterol.
Heart glandular	Take 140 mg; or take as indicated on the label.	Improves heart function.
Lecithin	Take 3 capsules with meals; or take as indicated on the label.	Emulsifies fat.
Magnesium	Take 500 mg per day.	Plays important roles in protein building, muscle contraction, calcium absorption, and blood clotting; vital catalyst to enzyme activity; protects arterial lining from such stressors as inflammation and edema.
Niacin or	Take 50 mg the first day, then increase dosage daily by 50 mg to 1,000 mg. If adverse effects occur, decrease dosage. Take doses higher than 1,000 mg only under the supervision of a physician.	Improves circulation by dilating blood vessels, which can help fight edema; lowers cholesterol.
Niacin-bound chromium	Take 20 mg per day, or as indicated on the label.	
Vitamin A	Take 15,000 IU per day; or take as per your doctor's directions. If pregnant, take no more than 10,000 IU per day.	Necessary for skin and cell membrane structure and strong immune system; potent antioxidant.
Vitamin B complex	Take 50 mg three times per day, or as indicated on the label.	B vitamins act as coenzymes to enzymes that reduce inflammation and edema.
Vitamin C	Take 5,000–10,000 mg per day.	Essential to vascular function; potent antioxidant; improves the function of many enzymes; essential in reducing edema and inflammation; protects against excessive blood clotting; reduces cholesterol; assists in converting cholesterol to bile salt.
Vitamin E	Take 1,200 IU per day from all sources (including the multivitamin complex in the introduction to Part Two).	Potent antioxidant; helps prevent cardiovascular disease; decreases risk of heart attack; stimulates circulation; strengthens capillary walls; prevents blood clots; maintains cell membranes.
Wheat germ oil	Take 100 mg per day.	Improves oxygen utilization; a good source of vitamin E and other nutrients.

| Zinc (zinc monomethionine is an excellent form) | Take 50 mg per day. Do not take more than 100 mg per day. | Needed for white blood cell immune function; potent antioxidant; cofactor for more than 300 enzymes including superoxide dismutase. |

Comments

❏ Raw, enzyme-rich vegetables and fruits should comprise 60 percent of your diet. Eat lots of foods high in enzymatic activity, including pineapples, papayas, and figs.

❏ Garlic, onions, shallots, chives, leeks, and scallions contain allium compounds that are potent free-radical fighters, help lower cholesterol, and fight circulatory disorders.

❏ Oats, wheat germ, olives, nuts, nut and seed oils, organ meats, and eggs contain alpha-tocopherol (vitamin E), known to help prevent cardiovascular disease and prevent blood clots.

❏ Raspberries, cranberries, grapes, red wine, black currants, and red cabbage contain anthocyanidins, which reduce blood vessel plaque formation and maintain blood flow in small vessels.

❏ Oolong and green tea contain epicatechin, which is a potent antioxidant that lowers blood pressure and strengthens capillaries.

❏ Try the Enzyme Kidney Flush for a week. The following week, perform the Enzyme Toxin Flush. Perform both only once per month. For instructions, see Part Three.

❏ An embolism can be dangerous. See a doctor for examination and consultation.

❏ *See also* THROMBOSIS.

Empyemas

An empyema is a condition in which pus forms in a body cavity. It usually occurs in the pleural cavity—the cavity formed by a thin membrane that covers the lungs and lines the chest wall. Pneumonia is the most common cause of a pleural empyema. It can also be caused by a lung abscess that spreads into the pleural cavity. A subdural empyema occurs when pus collects between the brain and its surrounding tissue. It can result from a sinus infection, severe ear infection, head injury, surgery, or a blood infection.

Symptoms

The primary symptom of a pleural empyema is difficulty breathing and pain in the chest. A fever may also be present and there may be frequent coughing. A subdural empyema

may cause headaches, sleepiness, seizures, and other complications of the brain. Without treatment, it may be fatal.

Enzyme Therapy

Systemic enzyme therapy is used to break down the fibrin formation, reduce the inflammation and pus in the chest, stimulate the immune system, improve circulation, help speed tissue repair, bring nutrients to the damaged area, and remove waste products. Enzymes enhance wellness, strengthen the body as a whole and build general resistance.

Digestive enzyme therapy is used to improve the digestion of food, reduce stress on the gastrointestinal mucosa, help maintain normal pH levels, detoxify the body, and promote the growth of healthy intestinal flora, thus making the body healthier so that it can rid itself of empyemas.

Enzyme Absorption System Enhancers (EASE) maximize enzyme activity. They can also improve the absorption and bioavailability of various nutrients and decrease the drain on the body's own digestive enzymes, thus prolonging their lives.

In addition to the standard multivitamin, multimineral, and multiglandular complexes recommended in the introduction to Part Two, the following supplements are recommended for the treatment of empyemas. The following recommended dosages are for adults. Children between the ages of two and five should take one-quarter the recommended dosage, those between the ages of six and twelve should take one-half the recommended dosage, and those between the ages of thirteen and seventeen should take three-quarters the recommended adult dosage.

ESSENTIAL ENZYMES

Enzyme	Suggested Dosage	Actions
Superoxide dismutase	Take 150 mcg per day.	Fights inflammation; potent antioxidant.

AND SELECT ONE OF THE FOLLOWING:

Enzyme	Suggested Dosage	Actions
Formula One—See page 38 of Part One for a list of products.	Take 5 tablets twice per day.	Fights inflammation, pain, and swelling; speeds recovery after injury or surgery; stimulates the immune system; strengthens capillary walls.
Bromelain	Take 690–750 mg two times per day between meals.	Fights inflammation, pain, and swelling; speeds healing; stimulates the immune system.
Serratiopeptidase	Take 10 mg two times per day between meals.	Fights inflammation; reduces pain and swelling.
Pancreatin	Take 500 mg six times per day on an empty stomach.	Fights inflammation, pain, and swelling; increases rate of healing; stimulates the immune system; strengthens capillary walls.
Trypsin with chymotrypsin	Take 230–250 mg of trypsin and 10 mg of	Fights inflammation, pain, and swelling; increases rate of

	chymotrypsin three times per day between meals.	healing; stimulates the immune system.
Papain	Take 300 mg two times per day between meals.	Fights inflammation; speeds recovery from injuries and surgery.
Microbial protease	Take 300 mg two times per day between meals.	Fights inflammation; speeds recovery from injuries and surgery.

SUPPORTIVE ENZYMES

Enzyme	Suggested Dosage	Actions
Catalase and glutathione peroxidase	Take as indicated on the labels.	Potent antioxidant enzymes.
A digestive enzyme product containing protease, amylase, and lipase enzymes. See the list of digestive enzyme products in Appendix.	Take as indicated on the label with meals.	Improves the breakdown and absorption of food nutrients; fights toxic build-up.

ENZYME ABSORPTION SYSTEM ENHANCERS (EASE)

Combination	Suggested Dosage	Actions
Advanced Nutritional System from Rainbow Light *or*	Take as indicated on the label.	Potent free-radical fighter and immune system stimulator.
Cold Zzap from Naturally Vitamin Supplements *or*	Take as indicated on the label.	Fights infection from viruses or bacteria; boosts immunity; has antioxidant activity.
Kyolic Formula 102 (with enzymes) *or*	Take 2 capsules or tablets with meals two to three times per day.	Strengthens the immune system; helps alleviate stress; helps fight free radicals; helps fight inflammation.
NESS Formula #416 *or*	Take as indicated on the label.	Breaks up fibrin; fights inflammation.
Protease Concentrate from Tyler Encapsulations *or*	Take as indicated on the label.	Breaks up fibrin; fights inflammation.
Quercezyme-Plus from Enzymatic Therapy *or*	Take as indicated on the label.	Stimulates immune system function and body tissue repair.
Vita-C-1000 from Life Plus	Take as indicated on the label.	Boosts immune function; fights infection; helps detoxify the body.

ENZYME HELPERS

Nutrient	Suggested Dosage	Actions
Acidophilus and other probiotics	Take as indicated on the label.	Improves gastrointestinal function; aids in the digestion and absorption of nutrients; helps produce antibacterial substances; stimulates enzyme activity, which decreases infection and inflammation and stimulates immune function.

Adrenal glandular	Take 100–400 mg per day.	Fights inflammation and adrenal exhaustion.
Aged garlic extract	Take two 300 mg capsules three times per day.	Fights inflammation; stimulates the immune system; potent antioxidant.
Amino-acid complex	Take as indicated on the label on an empty stomach.	Required by enzymes that increase immune function and fight infection and inflammation; helps the body maintain proper fluid and acid/base balance, essential for body health.
Bioflavonoids, including rutin or quercetin	Take 500 mg three times per day between meals.	Strengthen capillaries; potent antioxidants.
Essential fatty acid complex	Take 100–200 mg two to three times per day with meals; or follow the advice of your health-care professional.	Helps strengthen cell membranes; support immune function.
Green Barley Extract from Green Foods Corporation or Kyo-Green from Wakunaga	Take as indicated on the label.	Potent free-radical fighter; rich in vitamins, minerals, and enzyme activity.
Lung glandular	Take as indicated on the label.	Improves lung function and respiratory distress.
Thymus glandular	Take 100 mg per day.	Assists the thymus gland in immune resistance; improves immune response; fights infections.
Vitamin A	Take 5,000–10,000 IU per day.	Needed for strong immune system and healthy mucous membranes; potent antioxidant.
Vitamin B complex	Take 50 mg three times per day; or take as indicated on the label.	B vitamins act as coenzymes.
Vitamin C	Take 10,000–15,000 mg per day.	Fights infections, diseases, and allergies; improves wound healing; stimulates the adrenal glands; promotes immune system; powerful antioxidant.
Zinc	Take 15–50 mg per day. Do not take more than 100 mg per day.	Needed by over 300 enzymes; necessary for white blood cell immune function; improves wound healing; potent antioxidant.

Comments

❑ Eat plenty of foods rich in anthocyanidins, including raspberries, cranberries, grapes, and red cabbage. These phytochemicals fight inflammation, maintain blood flow in the capillaries, and fight free radicals.

❑ Several herbs, including astragalus, echinacea, ginseng, licorice, and maitake, reishi, and shiitake mushrooms can bolster immunity.

❑ Alfalfa, echinacea, ginger, ginkgo biloba, licorice, and white willow bark can help fight the inflammation that accompanies empyemas.

❑ The Enzymatic Gargle, the Echinacea/Enzyme Gargle, and the Garlic/Enzyme Gargle are of benefit in fighting inflammation and infection and stimulating the immune system. See Part Three for instructions.

❑ Medical treatment includes removal of the pus through a hollow needle inserted in the chest cavity (aspiration) and high doses of antibiotics. One or two daily aspirations may be enough for a small amount of thin pus. But thick pus may require open drainage over weeks or months.

Epididymitis

The epididymis is an almost 20-foot long coiled tube that is attached to the upper part of each testicle from which it collects sperm and allows them to mature there. Epididymitis is an inflammation or an infection of the epididymis. Urinary tract infections, prostatitis, sexually transmitted diseases (such as gonorrhea or chlamydia), or prostate surgery or other genitourinary procedures can cause the condition. It can affect any man between puberty and old age. Any case of epididymitis that lasts more than one to two months is considered chronic.

Symptoms

Epididymitis is marked by pain, swelling, and hardening of the testicles; tenderness and fluid in the scrotal sac; and fever. The pain can be mild or severe and can affect one or both testicles.

Enzyme Therapy

Systemic enzyme therapy is used to reduce the inflammation, pain, and swelling and stimulate the immune system. Enzymes improve circulation, help speed tissue repair, bring nutrients to the damaged area, remove waste products, and improve health.

Digestive enzyme therapy is used to improve the digestion of food, reduce stress on the gastrointestinal mucosa, help maintain normal pH levels, detoxify the body, and promote the growth of healthy intestinal flora, thus making the body healthier so that it can fight epididymitis. Digestive enzymes also serve as replacements for the body's pancreatic enzymes, leaving the pancreatic enzymes free to perform other functions in the body, such as boosting immunity and decreasing inflammation.

Enzyme Absorption System Enhancers (EASE) maximize enzyme activity. They can also improve the absorption and bioavailability of various nutrients and decrease the drain on the body's own digestive enzymes, thus prolonging their lives.

In addition to the standard multivitamin, multimineral, and multiglandular complexes recommended in the introduction to Part Two, the following supplements are recommended for the treatment of epididymitis. The following recommended dosages are for adults. Boys between the ages of twelve and seventeen should take three-quarters the recommended adult dosage.

ESSENTIAL ENZYMES

Enzyme	Suggested Dosage	Actions
Formula One—See page 38 of Part One for a list of products.	Take 10 tablets three times per day or 6 tablets five times per day between meals for six months. Then decrease dosage to 5 tablets three times per day between meals.	Fights free-radical formation and inflammation; bolsters the immune system.
Bromelain	Take three to four 250 mg capsules per day until symptoms subside. If pain is severe, take 6–8 capsules per day.	Fights inflammation, pain, and swelling; speeds healing; stimulates the immune system.
A proteolytic enzyme	Take 10 tablets three times per day or 6 tablets five times per day for six months. Then decrease dosage to 5 tablets three times per day between meals.	Bolsters the immune system; breaks up antigen-antibody complexes; fights free-radical formation; fights inflammation.

SUPPORTIVE ENZYMES

Enzyme	Suggested Dosage	Actions
Formula Three—See page 38 of Part One for a list of products.	Take as indicated on the labels.	Potent antioxidant enzymes.
A digestive enzyme product containing protease, amylase, and lipase. See the list of digestive enzyme products in Appendix.	Take as indicated on the label with meals.	Improves the breakdown and absorption of food nutrients; fights toxic build-up.

ENZYME ABSORPTION SYSTEM ENHANCERS (EASE)

Combination	Suggested Dosage	Actions
Advanced Nutritional System from Rainbow Light or	Take as indicated on the label.	Potent free-radical fighter and immune system stimulator.
Biovital Plus from Enzymatic Therapy or	Take as indicated on the label.	Anti-inflammatory agent; decreases swelling; fights infection; contains many antioxidants; contains many enzymes that aid in the absorption of glandulars, herbs, and other nutrients.
NESS Formula #416 or	Take as indicated on the label.	Fights inflammation.

| Protease Concentrate from Tyler Encapsulations *or* | Take as indicated on the label. | Fights inflammation. |
| Vita-C-1000 from Life Plus | Take as indicated on the label. | Boosts immune function; fights infection. |

ENZYME HELPERS

Nutrient	Suggested Dosage	Actions
Acidophilus and other probiotics	Take as indicated on the labels.	Stimulates enzyme activity; improves gastrointestinal function.
Adrenal glandular	Take 100–400 mg per day.	Helps fight stress and adrenal exhaustion; fights inflammation.
Kyolic Aged Garlic Liquid Extract with Vitamins B_1 and B_{12}	Take 10 drops in water or juice three times per day.	Stimulates immune system function.
Thymus glandular	Take 100 mg per day, or as indicated on the label.	Assists the thymus gland in immune resistance; improves immune response; fights infections.
Vitamin A	Take up to 50,000 IU for no more than one month (for higher doses, see a physician). If pregnant, do not take more than 10,000 IU.	Needed for strong immune system and healthy mucous membranes; potent antioxidant.
Vitamin C	Take 10,000–15,000 mg per day; or 1–2 grams every hour until symptoms improve.	Fights infections, diseases, and allergies; improves wound healing; stimulates the adrenal glands; promotes immune system; powerful antioxidant.
Zinc	Take 50 mg per day. Do not take more than 100 mg per day.	Needed by over 300 enzymes; necessary for white blood cell immune function; improves healing.

Comments

❑ Epididymitis is usually treated with antibiotics to combat infection and reduce the risk of sterility. Enzymes can improve the efficacy of antibiotics and control inflammation and pain.

❑ Apply ice packs to help relieve the pain and reduce swelling.

❑ An athletic supporter can help support the weight of the testicles and thereby relieve some discomfort.

❑ Bed rest is usually advised until symptoms improve.

❑ Help support the urinary system by drinking plenty of fresh water and avoiding alcohol and coffee and other acidic beverages that irritate the urinary tract.

❑ The Enzyme Toxin Flush and the Enzyme Kidney Flush can be of benefit. See Part Three for instructions.

Eye Problems

See CATARACTS; CONJUNCTIVITIS; MACULAR DEGENERATION; RETINOPATHY.

Fever Blisters

See HERPES SIMPLEX VIRUS.

Fibrocystic Breast Disease

A benign condition, fibrocystic breast disease is marked by the formation of tender noncancerous cysts in the breasts. It is the most common female breast disorder and is found in some 50 percent of premenopausal women. Ovarian hormones seem to be involved in the development of fibrocystic breast disease, since new cysts do not usually appear after menopause. When cysts occur, they are usually found in both breasts, but solitary cysts are also common. When the nodular cysts are relatively large and near the surface of the breast, they may be moved freely.

Symptoms

Symptoms of fibrocystic breast disease are tender nodular cysts, pain, and premenstrual breast discomfort. However, the condition can often be asymptomatic (without symptoms), only discovered when pressure is accidentally placed on the cysts. Most women with fibrocysts do not have an increased risk for breast cancer.

Enzyme Therapy

Systemic enzyme therapy is used to help break up the fibrosis in the breast, reduce any inflammation that may be present, and stimulate the immune system. Enzymes improve circulation, help speed tissue repair, bring nutrients to the damaged area and remove waste products, enhance wellness, strengthen the body as a whole, and build general resistance.

Digestive enzyme therapy is used to improve the digestion of food, reduce stress on the gastrointestinal mucosa, help maintain normal pH levels, detoxify, and promote the growth of healthy intestinal flora, thereby making the body healthier so that it is better able to deal with the effects of fibrocysts. Digestive enzymes also serve as replacements for the body's pancreatic enzymes, leaving the pancreatic enzymes free to perform other functions in the body, such as improving circulation and decreasing inflammation.

Enzyme Absorption System Enhancers (EASE) maximize enzyme activity. They can also improve the absorption and bioavailability of various nutrients and decrease the drain on the body's own digestive enzymes, thus prolonging their lives.

In addition to the standard multivitamin, multimineral, and multiglandular complexes recommended in the introduction to Part Two, the following supplements are recommended for the treatment of fibrocystic breast disease. The following recommended dosages are for adults. Girls between the ages of twelve and seventeen should take three-quarters the recommended adult dosage.

ESSENTIAL ENZYMES

Enzyme	Suggested Dosage	Actions
Superoxide dismutase	Take 150 mcg per day.	Fights inflammation; potent antioxidant.

AND SELECT ONE OF THE FOLLOWING:

Formula One—See page 38 of Part One for a list of products.	Take 5 tablets twice per day.	Fights inflammation, pain, and swelling; speeds recovery after injury or surgery; stimulates the immune system; strengthens capillary walls.
Bromelain	Take three to four 230–250 mg capsules per day until symptoms subside. If pain is severe, take 6–8 capsules per day.	Fights inflammation, pain, and swelling; speeds recovery after injury or surgery; stimulates the immune system.
Serratiopeptidase	Take 10 mg two times per day between meals.	Fights inflammation; reduces pain and swelling.
Trypsin with chymotrypsin	Take 250 mg of trypsin and 10 mg of chymotrypsin three times per day between meals.	Fights inflammation, pain, and swelling; increases rate of healing; stimulates the immune system.
Pancreatin	Take 500 mg six times per day on an empty stomach.	Fights inflammation, pain, and swelling; increases rate of healing; stimulates the immune system; strengthens capillary walls.
Papain	Take 300 mg two times per day between meals.	Fights inflammation; speeds recovery from injuries and surgery.
Microbial protease	Take 300 mg two times per day between meals.	Fights inflammation; speeds recovery from injuries and surgery.

SUPPORTIVE ENZYMES

Enzyme	Suggested Dosage	Actions
Catalase and glutathione peroxidase	Take as indicated on the label.	Potent antioxidant enzymes.
A digestive enzyme product containing protease, amylase, and lipase. See the list of digestive enzyme products in Appendix.	Take as indicated on the label with meals.	Improves the breakdown and absorption of food nutrients; fights toxic build-up.

ENZYME ABSORPTION SYSTEM ENHANCERS (EASE)

Combination	Suggested Dosage	Actions
Antioxidant Complex from Tyler *or*	Take as indicated on the label.	Potent free-radical fighter and immune system stimulator.
Detox Enzyme Formula from Prevail *or*	Take 1 capsule between meals one or two times per day with a 6- to 8-ounce glass of water or juice.	Helps rid the body of harmful toxins.
Hepazyme from Enzymatic Therapy *or*	Take as indicated on the label.	Stimulates immune function.
Oxy-5000 Forte from American Biologics *or*	Take as indicated on the label.	Helps the body excrete toxins; fights biochemical stressors, common infections, and degenerative processes.
Vita-C-1000 from Life Plus	Take as indicated on the label.	Boosts immune function; fights infection.

ENZYME HELPERS

Nutrient	Suggested Dosage	Actions
Acidophilus and other probiotics	Take as indicated on the label.	Improves gastrointestinal function; aids the digestion and absorption of nutrients; helps produce antibacterial substances; stimulates enzyme activity, which decreases inflammation and stimulates immune function.
Adrenal glandular	Take 100–400 mg per day.	Fights inflammation and adrenal exhaustion.
Aged garlic extract	Take 2 capsules three times per day.	Fights inflammation; stimulates the immune system; potent antioxidant.
Bioflavonoids, such as rutin or quercetin	Take 500 mg three times per day between meals.	Strengthen capillaries; potent antioxidants.
Coenzyme Q$_{10}$	Take 30–60 mg one to two times per day with meals; or follow the advice of your health-care professional.	Potent antioxidant.
Essential fatty acids	Take 100–200 mg two to three times per day with meals; or follow the advice of your health-care professional.	Help strengthen cell membranes; support immune function.
Green Barley Extract from Green Foods Corporation or Kyo-Green from Wakunaga	Take as indicated on the label.	Potent free-radical fighter; rich in vitamins, minerals, and enzyme activity.
Niacin	Take 50 mg the first day,	Dilates blood vessels; improves

or	then increase dosage daily by 50 mg to 1,000 mg. If adverse effects occur, decrease dosage. Take doses higher than 1,000 mg only under the supervision of a physician.	circulation.
Niacin-bound chromium	Take 20 mg per day or as indicated on the label.	
Spirulina	Take as indicated on the label.	Rich in minerals, vitamins, enzymes, proteins, essential amino acids, and essential fatty acids, especially gamma-linolenic acid; good source of beta-carotene, an antioxidant that improves immune function.
Thymus glandular	Take 100 mg per day.	Assists the thymus gland in immune resistance; improves immune response; fights infections.
Vitamin A	Take 5,000 IU per day.	Needed for strong immune system and healthy tissues; potent antioxidant.
Vitamin B complex	Take 150 mg three times per day, or as indicated on the label.	B vitamins act as coenzymes; support soft tissue function; stimulate enzymes, which break up fibrin and stimulate immune function.
Vitamin C	Take 10,000–15,000 mg per day.	Fights infections, diseases, and allergies; improves wound healing; stimulates the adrenal glands; promotes immune system; powerful antioxidant.
Vitamin E	Take 1,200 IU per day from all sources (including the multivitamin complex in the introduction to Part Two).	Potent antioxidant; supplies oxygen to cells; improves circulation and level of energy; can relieve many of the symptoms of fibrocystic disease.
Zinc	Take 50 mg per day. Do not take more than 100 mg per day from all sources.	Needed by over 300 enzymes; necessary for white blood cell immune function; improves tissue healing; potent antioxidant.

Comments

❑ Eliminate coffee, tea, chocolate, and cola from your diet. These substances contain methylxanthine, which causes increased cell growth and formation of fibrous tissue because it interferes with the action of the enzyme phosphodiesterase.

❑ Follow the diet plans and Enzyme Detoxification Program found in Part Three.

❑ A program of exercise (walking, bicycling, and swimming) is essential to increase circulation and enzyme and antioxidant activity.

❑ Meditate for fifteen minutes every day. Maintain a positive mental attitude.

Fibroids

A fibroid is a noncancerous growth composed of muscle and fibrous tissue that occurs in the uterus. These firm, gray-white tumors occur in the uterus of as many as 40 percent of women over the age of 40. The cause of fibroids is unknown, but it is apparent that estrogen is required for their growth, as they often grow larger during pregnancy and shrink after menopause, and they are quite rare in girls that have not yet reached puberty.

Symptoms

Fibroids usually have no symptoms and are often found only during a gynecological exam. Often, however, when they are numerous or large, they may cause pain, pressure, and/or heaviness in the pelvic region; heavy, prolonged, and/or more frequent menstrual periods; bleeding between periods; a frequent urge to urinate; painful sexual intercourse; and anemia. Rarely, fibroids may cause infertility if they block the fallopian tubes.

Enzyme Therapy

Systemic enzyme therapy is used to help break up fibrin formation, reduce inflammation, stimulate the immune system, and improve circulation. Enzymes help speed tissue repair, bring nutrients to the damaged area, remove waste products, enhance wellness, strengthen the body as a whole, and build general resistance.

Digestive enzyme therapy is used to improve digestion of food, reduce stress on the gastrointestinal mucosa, help maintain normal pH levels, detoxify, and promote the growth of healthy intestinal flora, thereby making the body healthier so that it can fight the formation of fibroids. Digestive enzymes also serve as replacements for the body's pancreatic enzymes, leaving the pancreatic enzymes free to perform other functions in the body, such as boosting immunity, decreasing inflammation, and improving circulation.

Enzyme Absorption System Enhancers (EASE) maximize enzyme activity. They can also improve the absorption and bioavailability of various nutrients and decrease the drain on the body's own digestive enzymes, thus prolonging their lives.

In addition to the standard multivitamin, multimineral, and multiglandular complexes recommended in the introduction to Part Two, the following supplements are recommended for the treatment of fibroids. The following recommended dosages are for adults. Girls between the ages of twelve and seventeen should take three-quarters the recommended adult dosage.

ESSENTIAL ENZYMES

Enzyme	Suggested Dosage	Actions
Superoxide dismutase	Take 150 mcg per day.	Fights inflammation; potent antioxidant.

AND SELECT ONE OF THE FOLLOWING:

Enzyme	Suggested Dosage	Actions
Formula One—See page 38 of Part One for a list of products.	Take 5 tablets twice per day.	Fights inflammation, pain, and swelling; speeds healing; stimulates the immune system; strengthens capillary walls.
Bromelain	Take 690–750 mg two times per day between meals.	Fights inflammation, pain, and swelling; speeds healing; stimulates the immune system; strengthens capillary walls.
Trypsin with chymotrypsin	Take 250 mg of trypsin and 10 mg of chymotrypsin three times per day between meals.	Fights inflammation, pain, and swelling; increases rate of healing; stimulates the immune system.
Pancreatin	Take 500 mg six times per day on an empty stomach.	Fights inflammation, pain, and swelling; increases rate of healing; stimulates the immune system; strengthens capillary walls.
Papain	Take 300 mg two times per day between meals.	Fights inflammation; speeds healing.
Microbial protease	Take 300 mg two times per day between meals.	Fights inflammation; speeds healing.

SUPPORTIVE ENZYMES

Enzyme	Suggested Dosage	Actions
Catalase and glutathione peroxidase	Take as indicated on the labels.	Potent antioxidant enzymes.
A digestive enzyme product containing protease, amylase, and lipase enzymes. See the list of digestive enzyme products in Appendix.	Take as indicated on the label with meals.	Improves the breakdown and absorption of food nutrients; fights toxic build-up.

ENZYME ABSORPTION SYSTEM ENHANCERS (EASE)

Combination	Suggested Dosage	Actions
Bromelain Complex from Enzymatic Therapy *or*	Take as indicated on the label.	Decreases swelling; improves circulation; helps break up any circulating immune complexes.
Connect-All from Nature's Plus *or*	Take as indicated on the label.	Supports connective tissue.
CTR Support from PhysioLogics *or*	Take as indicated on the label.	Nourishes the body during tissue recovery; diminishes damage caused by swelling and inflammation.
Inflamzyme from PhytoPharmica *or*	Take as indicated on the label.	Promotes tissue repair and immune function.

Protease Concentrate from Tyler Encapsulations	Take as indicated on the label.	Breaks up fibrin; fights inflammation.

ENZYME HELPERS

Nutrient	Suggested Dosage	Actions
Acidophilus and other probiotics	Take as indicated on the label.	Improve gastrointestinal function and aid digestion and absorption of nutrients; stimulate enzyme activity, which decreases inflammation and stimulates immune system function; help break up fibrin.
Adrenal glandular	Take 100–400 mg per day.	Fights inflammation and adrenal exhaustion.
Aged garlic extract	Take two 300 mg capsules three times per day.	Fights inflammation; stimulates the immune system; potent antioxidant.
Bioflavonoids, such as rutin or quercetin	Take 500 mg three times per day between meals.	Strengthen capillaries; potent antioxidant.
Calcium	Take 1,000 mg per day.	Involved in the activation of several enzymes.
Coenzyme Q_{10}	Take 30–60 mg one to two times per day with meals; or follow the advice of your healthcare professional.	Potent antioxidant.
Essential fatty acids	Take as indicated on the label.	Help strengthen cell membranes; support immune function.
Green Barley Extract from Green Foods Corporation or Kyo-Green from Wakunaga	Take as indicated on the label.	Potent free-radical fighter; rich in vitamins, minerals, and enzyme activity.
Magnesium	Take 500 mg per day.	Plays important role in calcium absorption and blood clotting; catalyst of enzyme activity.
Pituitary glandular	Take 100 mg per day.	Helps form new blood vessels; fights tumor growth; aids wound healing.
Raw ovary glandular	Take 50 mg per day.	Improves ovarian function.
Raw uterus glandular	Take 25 mg per day.	Improves uterine function.
Spirulina	Take as indicated on the label.	Rich in minerals, vitamins, enzymes, proteins, essential amino acids, and essential fatty acids, especially gamma-linolenic acid; good source of beta-carotene—an antioxidant; improves immune function.
Thymus glandular	Take 100 mg per day.	Assists the thymus gland in immune resistance; improves immune response; fights infections.
Vitamin A	Take 10,000–15,000 IU per day. If pregnant, take no more than 10,000 IU per day.	Needed for strong immune system and healthy tissue; potent antioxidant.
Vitamin B complex	Take 150 mg three times per day, or as indicated	B vitamins act as coenzymes to enzymes that break up fibrin;

	on the label.	stimulate the immune system; fight free radicals; support tissue function.
Vitamin C	Take 10,000–15,000 mg per day.	Improves healing; stimulates the adrenal glands; promotes immune system; powerful antioxidant.
Vitamin E	Take 1,200 IU per day from all sources (including the multivitamin complex in the introduction to Part Two).	Potent antioxidant; supplies oxygen to cells; improves circulation and level of energy.

Comments

❑ A number of herbs can help control inflammation (if present), stimulate immune function, and fight free radicals. Some of these include alfalfa, echinacea, ginger root, ginkgo biloba, licorice, and white willow bark.

❑ Eat fresh, enzyme-rich foods high in proteolytic activity, such as pineapple, papaya, and figs. See Part Three for dietary recommendations.

❑ Keep the body detoxified with cleansing, juicing, and fasting. See Part Three for further information on detoxification.

❑ Meditate about fifteen minutes per day. Void your mind and think of nothing. Allow your body to re-energize itself, increasing your enzymatic and antioxidant activity.

Fibromyalgia

See MYOFASCIAL PAIN SYNDROME.

Flatulence

See GAS, ABDOMINAL.

Flu

See INFLUENZA.

Fungal Skin Infections

Fungal skin infections usually occur in moist areas of the body, such as between toes, in the groin area, and under the breasts. Fungi grow in the superficial layers of the skin only. Often patients develop allergic reactions to fungal infections on one part of the body and develop a rash in another part of the body.

There are several types of fungal skin infections. *Ringworm* is one common type. Ringworm is usually caused by one or more of five types of fungus—often *Trichophyton*. General ringworm of the body (*tinea corporis*) is marked by flat, pink to red, scaling circular lesions with clearly defined borders. The lesions tend to be clear in the center.

Athlete's foot is ringworm of the foot (*tinea pedis*) and usually occurs in warm weather. It generally appears between the toes and sometimes on the soles of the feet. Mild athlete's foot infections may produce only scaling. However, worse infections may produce severe scaling and an itchy, painful rash. In severe cases, blisters may even form on the feet. It is usually caused by either *Trichophyton* or *Epidermophyton*.

Nail ringworm (*tinea unguium*) is marked by lusterless, deformed, and thickened nails. There may be redness and swelling around the nail. The nail may become separated from the nailbed and ultimately crumble or fall off. Nail ringworm is generally painless and not contagious.

Scalp ringworm (*tinea capitis*) is common in children and is very contagious. It is caused by either *Trichophyton* or *Microsporum*. It may cause a red, scaly, itchy rash and patchy hair loss.

Jock itch is groin ringworm (*tinea cruris*). This infection is most common in men but may occur in women as well. It is marked by red, itchy, painful ringed lesions on the skin of the groin area and of the inner thighs. Blisters may also occur. Jock itch occurs most often in the warm weather, and recurrence is common.

Candidiasis is an infection of the yeast *Candida albicans*. *Candida* are normally found in the digestive tract and vagina but, under normal circumstances, are harmless. However, when one is taking antibiotics, which kill bacteria that keep *Candida* under control; when one is immunosuppressed; or often in warm, humid weather, *Candida* can cause skin and mucous membrane infections. Rarely, *Candida* can infect deeper tissues and the blood, causing life-threatening infections. This happens usually in those who are immunosuppressed. Symptoms of candidiasis depend upon the location of the infection. (See the Inset "Candidiasis" on page 226.)

Infection of body skin produce a red rash that may itch or burn. Small pustules may appear in or around the rash.

Vaginal infections (also called yeast infections) are a common affliction of women, particularly pregnant women and women with diabetes. Such infections may cause a white or yellowish discharge from the vagina and itching, burning, and redness in and around the vagina. Vaginal infections may produce no symptoms at all. Penile infections are often contracted sexually from a woman with a vaginal infection,

Candidiasis

Candidiasis (also called candidosis, moniliasis, and oilomycosis) is a fungal infection that usually occurs on the moist superficial areas of the body, such as in the mouth or vagina. However, it can also affect the skin (dermatocandidiasis), respiratory tract and lungs (bronchocandidiasis), central nervous system, and other tissues and organs of the body. Individuals with compromised immune systems, including those on immunosuppressive drugs or suffering from HIV/AIDS, are at particular risk of developing candidiasis. It can also occur as a result of excessive refined sugar and refined carbohydrate intake; the use of certain drugs, including oral contraceptives and antibiotics; digestive disorders; or a stressful lifestyle, which decreases immune function and increases free-radical formation.

Because candidiasis can affect many systems, the symptoms may vary by location. Generally, however, they include low energy, chronic fatigue, malaise, and loss of sex drive. Symptoms in the gastrointestinal tract include abdominal pain, bloating, changes in bowel function, colitis, constipation, diarrhea, gas, intestinal cramps, persistent heartburn, and rectal itching.

If candidiasis infects the bloodstream, effects can be felt throughout the body. Some include depression, extreme fatigue, headaches, hyperactivity, inability to concentrate, irritability, memory loss, mood swings, muscle and joint pain, a numb or tingling feeling in the face or extremities, adrenal problems, diabetes, hypothyroidism, premenstrual syndrome (PMS) and other menstrual problems, allergies, chemical sensitivities, and decreased immunity.

Genitourinary symptoms may include burning, frequent bladder or kidney infections, impotence, intense itching, prostatitis, or vaginal yeast infections (vaginitis).

Respiratory signs include clogged sinuses, congestion, a nagging cough, and a sore throat. Other symptoms include acne, bad breath, a burning tongue, canker sores, or thrush (a white carpet appearance on the tongue or inside the cheeks).

Usually the term candidiasis is used to refer to a vaginal infection. Vaginal candidiasis can occur when *Candida*, fungi that normally live in the vagina, grow unchecked. This can happen due to pregnancy, antibiotic treatment, menstruation, diabetes, the use of oral contraceptives or feminine hygiene sprays, frequent douching, and the use of spermicidal jelly or creams.

In all forms of candidiasis, the best defense is a strong, healthy immune system. There is a growing amount of evidence that women who suffer from recurrent *Candida* infections have weakened immune systems, perhaps due to inadequate nutrient intake or fluctuations in female hormones.

The overgrowth of *Candida* is usually controlled or prevented by digestive secretions, including pancreatic enzymes, bile, and hydrochloric acid. A deficiency of any of these essential digestive ingredients can permit excessive growth of yeast fungus. Therefore, it is critical to restore the gastrointestinal tract's normal digestive balance and secretions by furnishing assistance to the body with supplementation of enzymes, hydrochloric acid, and other nutrients. Adequate enzyme production in the body and supplementation is essential to maintain and restore normal pH levels to the gastrointestinal tract.

Try Candi-Stat from Tyler Encapsulations. It contains amylase, lipase, protease, lactase, sucrase, maltase, and cellulase enzymes. Use it along with Pro Flora, pau d'arco, garlic, barberry, goldenseal, milk thistle seed, and silybum marianum extract.

---

or occur in men with diabetes. They produce a red, scaling rash on the underside of the penis that may hurt, or it may produce no symptoms at all. An anal infection may produce a red, itchy, raw rash around the anus.

Thrush is candidiasis inside the mouth. It is very common in infants, and may occur in older children and adults. When it occurs in adults, it is usually due to immunosuppression or the use of antibiotics. It produces creamy white patches in the mouth or on the tongue that are usually not painful unless scraped off. Then small, painful ulcers may occur. When candidiasis affects the corners of the mouth, it is called *perlèche*.

When *Candida* infect the nail, it produces symptoms similar to nail ringworm.

Tinea versicolor is a common infection among young adults.

In dark-skinned people, it causes light patches to form, and in fair-skinned individuals, it causes dark patches to form. The patches may scale and they do not tan. This condition usually does not cause pain or itching.

Enzyme Therapy

Systemic enzyme therapy is used to reduce inflammation, stimulate the immune system, improve circulation, help speed tissue repair, bring nutrients to the damaged area, remove waste products, improve health, strengthen the body as a whole, and build general resistance.

Digestive enzyme therapy is used to improve the digestion of food, reduce stress on the gastrointestinal mucosa, help maintain normal pH levels, detoxify the body, and promote the growth of healthy intestinal flora. Digestive enzymes also serve as replacements for the body's pancreatic enzymes, leaving the pancreatic enzymes free to perform other functions in the body, such as boosting immunity.

Topical enzyme creams are used to dissolve dead and damaged skin cells and exfoliate the skin while leaving healthy skin cells intact. These creams or salves should contain proteolytic enzymes. The plant enzymes bromelain and papain are used most frequently; however, microbial proteases can be quite helpful. In Germany, a combined enzyme product containing bromelain, papain, trypsin, and chymotrypsin is used in the treatment of fungal infections. Treatment combines topical enzyme creams or salves with enzymes taken orally.

Enzyme Absorption System Enhancers (EASE) maximize enzyme activity. They can also improve the absorption and bioavailability of various nutrients and decrease the drain on the body's own digestive enzymes, thus prolonging their lives.

In addition to the standard multivitamin, multimineral, and multiglandular complexes recommended in the introduction to Part Two, the following supplements are recommended for the treatment of fungal infections. The following recommended dosages are for adults. Children between the ages of two and five should take one-quarter the recommended dosage, those between the ages of six and twelve should take one-half the recommended dosage, and those between the ages of thirteen and seventeen should take three-quarters the recommended adult dosage.

ESSENTIAL ENZYMES

Enzyme	Suggested Dosage	Actions
WobeMugos E Ointment from Mucos	Follow directions on the label. Do not apply to open wounds.	Helps fight fungal infections.

AND SELECT ONE OF THE FOLLOWING:

Formula One (particularly Wobenzym N from Naturally Vitamin Supplements)—See page 38 of Part One for a list of products.	Take 10 tablets three times per day between meals for six months or until symptoms improve. Then decrease dosage to 5 tablets three times per day between meals.	Stimulates immune function; breaks up circulating immune complexes; fights inflammation.
WobeMugos E from Mucos	Take 2 tablets three times per day between meals for six months. Then take as needed.	Bolsters the immune system; breaks up antigen-antibody complexes; fights free radicals and inflammation.
Bromelain	Take 360–400 mg three times per day between meals for six months. Then decrease dosage to 250 mg three times per day between meals.	Stimulates immune function; breaks up circulating immune complexes; stimulates immune function at the cellular level; fights inflammation.
A microbial	Take as indicated	Bolsters the immune system;

| proteolytic enzyme combination | on the label. | breaks up antigen-antibody complexes; fights free radicals and inflammation. |

SUPPORTIVE ENZYMES

Enzyme	Suggested Dosage	Actions
Formula Three—See page 38 of Part One for a list of products.	Take as indicated on the labels.	Potent antioxidant enzymes.
A digestive enzyme product containing protease, amylase, and lipase. See the list of digestive enzyme products in Appendix.	Take as indicated on the label with meals.	Improves the breakdown and absorption of food nutrients; fights toxic build-up.

ENZYME ABSORPTION SYSTEM ENHANCERS (EASE)

Combination	Suggested Dosage	Actions
Ecology Pak from Life Plus

or | Take 2 or 3 tablets once or twice per day, preferably in the morning and no later than midday. | Assists with detoxification; stimulates immune function. |
| Hepazyme from Enzymatic Therapy
or | Take as indicated on the label. | Stimulates immune function. |
| Inflamzyme from PhytoPharmica
or | Take as indicated on the label. | Immune system activator; reduces inflammation; fights free radicals. |
| Kyolic Formula 102 (with enzymes)

or | Take 2 capsules or tablets with meals two to three times per day. | Strengthens the immune system; helps alleviate stress; helps fight free radicals; helps fight inflammation; improves digestion. |
| Vita-C-1000 from Life Plus | Take as indicated on the label. | Boosts immune function; fights infection. |

ENZYME HELPERS

Nutrient	Suggested Dosage	Actions
Acidophilus or other probiotics	Take as indicated on the label.	Stimulates enzyme activity; improves gastrointestinal function.
Adrenal glandular	Take 100–400 mg per day.	Helps fight stress and adrenal exhaustion; fights inflammation.
Kyolic Aged Garlic Liquid Extract with Vitamins B_1 and B_{12}	Take 10 drops in water or juice three times per day.	Stimulates immune system function.
Selenium	Take 50 mcg one or two times per day.	Acts as an enzyme activator; required by glutathione peroxidase.
Thymus glandular	Take 100 mg per day or as indicated on the label.	Assists the thymus gland in immune resistance; improves immune response; fights infections.
Vitamin A	Take up to 50,000 IU for no more than one month (for higher doses, see a	Needed for strong immune system and healthy mucous membranes; potent antioxidant.

or	physician). If pregnant, do not take more than 10,000 IU per day.	
Carotenoids	Take 25,000 IU per day.	Potent antioxidants; block tissue damage; improve immune function.
Vitamin B complex	Take 100–150 mg three times per day.	B vitamins act as coenzymes.
Vitamin C	Take 10,000–15,000 mg per day, or 1–2 grams every hour until symptoms improve.	Fights infections; improves healing; stimulates the adrenal glands; promotes immune system; powerful antioxidant.
Vitamin E	Take 1,200 IU per day from all sources (including the multivitamin complex in the introduction to Part Two).	Acts as an intercellular antioxidant; improves circulation; promotes immune function.
Zinc	Take 50 mg. Do not take more than 100 mg per day from all sources.	Needed by over 300 enzymes; necessary for white blood cell immune function; improves healing.

Comments

❑ Garlic and other allium vegetables contain allylic sulfides, which are effective at combating fungi.

❑ An external application of tea tree oil is effective as a natural antifungal agent, while drinking pau d'arco tea fights candida infection.

❑ Avoid all refined sugars; food additives, preservatives, and colors; canned juices; and alcoholic beverages.

❑ Avoid fried, smoked, or pickled foods.

❑ Avoid all dairy products except plain yogurt. Prepare your own yogurt (see directions in Chapter 4 of Part One), or buy it fresh.

❑ Cleansing the body internally and externally is essential in treating fungal skin infections. Perform the Enzyme Kidney Flush every day for one week. The following week, perform the Enzyme Toxin Flush. Use these flushes no more than once per month. See instructions for both flushes in Part Three.

Furuncles

See BOILS.

Gas, Abdominal

Gas is a typical symptom of indigestion and may be the result of eating spoiled food; eating or drinking too much or too fast, which causes us to swallow air; inadequate digestion; food allergies; or enzyme deficiencies. Certain foods can cause gas in susceptible individuals, including foods high in fiber (especially if you're not used to fiber), fried foods, and sugary foods. Even eating while emotionally upset can cause gas.

Gases such as hydrogen, methane, and carbon dioxide are produced in the colon by the bacterial metabolism of fermentable nutrients (including amino acids and carbohydrates). They are produced in especially large quantities after we eat certain vegetables (such as cabbage or baked beans) or fruits containing indigestible carbohydrates. Large quantities of disaccharides may pass into the colon in those people with malabsorption syndromes and disaccharidase deficiencies. The disaccharides then ferment, producing hydrogen, which causes abdominal gas. Therefore, a number of conditions, including sprue, lactase deficiency, carbohydrate malabsorption, and pancreatic insufficiency, should be considered as possible causes of excessive gas. But even those without these digestive disorders sometimes have trouble digesting carbohydrates.

Symptoms

Abdominal gas can create a bloated feeling, pain, pressure, discomfort, heartburn, and nausea. It can be accompanied by constipation or diarrhea, belching, flatulence, or a distended stomach. Abdominal gas is a symptom of many disorders, including indigestion. Excessive gas retention can contribute to the development of irritable bowel syndrome.

Enzyme Therapy

Digestive enzyme therapy is used to provide any missing enzymes and thereby improve the digestion of food. This will reduce stress on the gastrointestinal mucosa, help maintain normal pH levels, detoxify the body, and promote the growth of healthy intestinal flora.

Systemic enzyme therapy is used to reduce any inflammation, stimulate the immune system, improve circulation, help transport nutrients throughout the body, remove waste products, and improve health.

Enzyme Absorption System Enhancers (EASE) maximize enzyme activity. They can also improve the absorption and bioavailability of various nutrients and decrease the drain on the body's own digestive enzymes, thus prolonging their lives.

Alpha-galactosidase, cellulase, and other enzyme liquid drops and tablets help prevent gas from developing after you eat many high-fiber foods. These enzymes break down the complex sugars and fibers in many foods, including beans, lentils, pasta, onions, broccoli, cabbage, and other vegetables, fruits, and whole grains. This makes the foods more digestible and helps prevent gas formation.

In addition to the standard multivitamin, multimineral, and multiglandular complexes recommended in the introduction to Part Two, the following supplements are recommended for the relief of abdominal gas. The following recommended dosages are for adults. Children between the ages of two and five should take one-quarter the recommended dosage, those between the ages of six and twelve should take one-half the recommended dosage, and those between the ages of thirteen and seventeen should take three-quarters the recommended adult dosage.

ESSENTIAL ENZYMES

Enzyme	Suggested Dosage	Actions
A digestive enzyme formula containing protease, amylase, and lipase enzymes. For a list of digestive enzyme products, see Appendix. *or*	Take as indicated on the label with meals.	Improves digestion by breaking down proteins, carbohydrates, and fats.
Pancreatin	Take 300 mg three times per day with meals.	
If you have trouble digesting protein, try a protease, such as bromelain, carboxypeptidase, microbial protease, papain, pepsin, peptidase, rennin, or trypsin and chymotrypsin.	Take as indicated on the label with meals.	Breaks down proteins.
If you have trouble digesting carbohydrates, try an amylase, such as alpha-galactosidase, beta-glucosidase, carbohydrase, cellulase, diastase, glucoamylase, hemicellulase, invertase, lactase, maltase, phytase, or sucrase.	Take as indicated on the label with meals.	Breaks down carbohydrates.
If you have trouble digesting fats, try an enzyme supplement containing lipase.	Take as indicated on the label with meals.	Breaks down lipids.

SUPPORTIVE ENZYMES

Enzyme	Suggested Dosage	Actions
Bromelain	Start with one 230–250 mg capsule per day and increase dosage gradually, if necessary. Take no more than 3 capsules per day. If diarrhea develops, decrease dose.	Improves digestion; fights inflammation, pain, and swelling; stimulates the immune system.
Formula Three—See page 38 of Part One for a list of products.	Take 150 mcg per day.	Free-radical fighters.

ENZYME ABSORPTION SYSTEM ENHANCERS (EASE)

Combination	Suggested Dosage	Actions
Biodias A Granules from Amano Pharmaceutical Co., Ltd. *or*	Adults should take one packet in 8 ounces of water between or after meals three times per day. Children ages 11 to 14 should take two-thirds of a packet; children between the ages of 8 and 10 should take half of the packet; and children between the ages of 5 and 7 should take one-third of a packet.	Helps improve digestion of nutrients; relieves inflammation; promotes regeneration of mucous membranes; decreases stress on the stomach and intestines; reduces gastric pain.
Source of Life Vibra-Gest from Nature's Plus *or*	Take as indicated on the label.	Contains microbial enzymes with probiotics to improve digestion.
Sugar Check from Enzymatic Therapy *or*	Take as indicated on the label.	Contains nutrients for improved sugar and carbohydrate metabolism and absorption.
Zygest from PhysioLogics	Take 1 or 2 capsules before meals to aid digestion; or take as directed by your health-care professional.	Helps the body digest proteins, carbohydrates, fats, and plant fibers.

ENZYME HELPERS

Nutrient	Suggested Dosage	Actions
Acidophilus or other probiotics, such as bifidobacterium; or Acidophilasé from Wakunaga, which also contains amylolytic, lipolytic, and proteolytic enzymes	Take as indicated on the label.	Improves gastrointestinal function; stimulates enzyme activity.
Aged garlic extract	Take 2 capsules three times per day.	Potent antioxidant; improves digestion; stimulates healthy intestinal bacteria.
Betaine hydrochloric acid	Take 150 mg three times per day with meals.	Provides hydrochloric acid, which helps improve digestion. If gastric upset occurs, discontinue use.
Fructooligo-saccharides	Take as indicated on the label.	Increases peristaltic action of intestines; improves liver function.
Lecithin	Take 1,000 mg three times per day with meals.	Emulsifies fat.
Ox bile extract	Take 120 mg per day.	Aids digestion.
Vitamin A	Take 5,000 IU per day.	Needed by digestive glands to produce enzymes; required for healthy mucous linings, such as those in the intestine.
Vitamin B complex	Take 50 mg three times per day.	B vitamins act as potent coenzymes.

Vitamin B₁ (thiamin)	Take 150 mg three times per day.	Essential for carbohydrate metabolism.
Vitamin C	Take 1,000–2,000 mg per day.	Stimulates the adrenal glands to produce hormones (which also act as coenzymes); aids in detoxification; improves the function of many enzymes.
Whey	Take 1 heaping tablespoon three times per day with meals.	Improves digestion.
Zinc	Take 10–50 mg per day.	Needed by over 300 enzymes. Many zinc enzymes work to metabolize carbohydrates, alcohol, and essential fatty acids; synthesize proteins; dispose of free radicals.

Comments

❏ Foods rich in chlorophyll, including chlorella, spirulina, Kyo-Green, Green Kamut, Green Barley Extract, and Green Magma, can regulate cholesterol, improve bowel function, stimulate the growth of healthful bacteria, absorb intestinal poisons, and promote normal peristalsis.

❏ Avoid food additives, preservatives, refined foods (especially those with refined sugars and starches), and fast foods.

❏ Increase your intake of fiber foods to improve elimination. But until your system gets used to more fiber, also increase your intake of enzyme supplements, plus foods high in enzymes, particularly amylase and cellulase.

❏ Toxin build-up and poor elimination can cause gas. Therefore, fasting and juicing for cleansing the system are extremely valuable. See Part Three for further information.

❏ Exercising (walking, swimming, and bicycling) increases circulation, improves peristaltic activity, and promotes elimination and a healthy intestine.

❏ A positive mental attitude and meditation can help reduce stress levels and improve digestion.

❏ At times, expert advice may be necessary. See a well-trained health-care professional or nutritionist.

❏ See the list of Resource Groups in the Appendix for a list of organizations that can provide you with more information about abdominal gas.

❏ *See also* CELIAC DISEASE; HEARTBURN; LACTASE DEFICIENCY; PANCREATIC INSUFFICIENCY.

Gastritis

Gastritis is inflammation of the stomach lining. There are many types of gastritis, which are identified by their causes.

Acute stress gastritis is the most severe type. It is caused by a severe sudden illness or injury. *Atrophic gastritis* is caused by antibodies attacking the stomach lining. It usually affects elderly people or those who have had part of their stomach surgically removed. *Bacterial gastritis* is caused by an infection of *Helicobacter pylori* bacteria. *Chronic erosive gastritis* is caused by irritants, such as drugs; Crohn's disease; and infections. Ulcers may also develop. It most commonly affects alcoholics. In *eosinophilic gastritis*, eosinophils, a type of white blood cell, accumulate in the stomach wall, probably as a result of an allergic reaction to a roundworm infection. In *plasma cell gastritis*, plasma cells accumulate in the stomach wall for some unknown reason. *Viral* and *fungal gastritis* are also pretty common. They usually develop in those with compromised immune systems.

In 1988, 2.7 million cases of gastritis were reported. Data from the 1990s show more than 3 million doctor visits in addition to about 621,000 hospital visits for the treatment of gastritis. More than 2,000,000 outpatient prescriptions were written in 1985 to treat gastritis.

Symptoms

Symptoms of gastritis vary depending on the cause, but generally include indigestion and abdominal pain. Other symptoms may include appetite loss, diarrhea, nausea, and vomiting. Ulcers are common with many types of gastritis. Black or bloody stools or vomiting blood indicates internal bleeding, and immediate medical care should be obtained.

Acute gastritis comes on suddenly and can last for a short time (a few days). Chronic gastritis may have no pain, but individuals may develop nausea and/or appetite loss. From chronic gastritis, pernicious anemia (vitamin B₁₂ deficiency) may develop.

Enzyme Therapy

Digestive enzyme therapy is used to improve the digestion of food, help normalize the acid and pepsin levels in the stomach, reduce stress on the gastrointestinal mucosa, help maintain normal pH levels, detoxify the body, and promote the growth of healthy intestinal flora. Digestive enzymes also serve as replacements for the body's pancreatic enzymes, leaving the pancreatic enzymes free to perform other functions in the body, such as boosting immunity and decreasing inflammation.

Systemic enzyme therapy is used to help restore normal body function, reduce inflammation, and stimulate the immune system. Enzymes can improve circulation, help speed tissue repair, transport nutrients throughout the body, remove waste products, and improve health.

Enzyme Absorption System Enhancers (EASE) maximize enzyme activity. They can also improve the absorption and bioavailability of various nutrients and decrease the drain on the body's own digestive enzymes, thus prolonging their lives.

In addition to the standard multivitamin, multimineral, and multiglandular complexes recommended in the introduction to Part Two, the following supplements are recommended for the treatment of gastritis. The following recommended dosages are for adults. Children between the ages of two and five should take one-quarter the recommended dosage, those between the ages of six and twelve should take one-half the recommended dosage, and those between the ages of thirteen and seventeen should take three-quarters the recommended adult dosage.

ESSENTIAL ENZYMES

Enzyme	Suggested Dosage	Actions
A digestive enzyme product containing protease, amylase, and lipase enzymes. For a list of digestive enzyme products, see Appendix.	Take as indicated on the label.	Improves digestion by breaking down proteins, carbohydrates, and fats.
If you have trouble digesting protein, try a protease, such as bromelain, carboxy-peptidase, a microbial protease, papain, pepsin, peptidase, rennin, or trypsin and chymotrypsin.	Take as indicated on the label with meals.	Breaks down proteins.
If you have trouble digesting carbohydrates, try an amylase, such as alpha-galactosidase, beta-glucosidase, carbohydrase, cellulase, diastase, glucoamylase, hemicellulase, invertase, lactase, maltase, phytase, or sucrase.	Take as indicated on the label with meals.	Breaks down carbohydrates.
If you have trouble digesting fats, try an enzyme supplement containing lipase.	Take as indicated on the label with meals.	Breaks down fats.
Formula One—See page 38 of Part One for a list of products.	Take 5 coated tablets or one tablespoon of a granulated enzyme mixture three times per day between meals.	Fights inflammation, swelling, and pain; speeds healing; stimulates immune function; fights free-radical formation; improves digestion.
If you have difficulty swallowing pills or tablets, try chewable digestive enzymes, such as bromelain or papain (papaya).	Take as indicated on the label.	Improves digestion.

SUPPORTIVE ENZYMES

Enzyme	Suggested Dosage	Actions
Superoxide dismutase, catalase, and glutathione peroxidase	Take as indicated on the labels.	Potent antioxidant enzymes.

ENZYME ABSORPTION SYSTEM ENHANCERS (EASE)

Combination	Suggested Dosage	Actions
AminoZyme from Nature's Plus	Take as indicated on the label.	Metabolizes protein and frees amino acids.
or		
Anti-Flam from Crystal Star Herbal Nutrition	Take as indicated on the label.	Soothing combination of anti-inflammatory herbs with enzymes.
or		
Biodias A Granules from Amano Pharmaceutical Co., Ltd.	Adults should take one packet in 8 ounces of water between or after meals three times per day. Children ages 11 to 14 should take two-thirds of a packet; children between the ages of 8 and 10 should take half of the packet; and children between the ages of 5 and 7 should take one-third of a packet.	Helps improve digestion of nutrients; relieves inflammation; promotes regeneration of mucous membranes; decreases stress on the stomach and intestines; reduces gastric pain.
or		
Pro-Gest-Ade from Enzymatic Therapy	Take as indicated on the label.	Contains nutrients essential for proper digestive function, plus enzymes to enhance digestion and absorption of proteins, fats, and carbohydrates.
or		
Ultra-Zyme from Nature's Plus	Take as indicated on the label.	High potency digestive enzyme combination.
or		
Zygest from PhysioLogics	Take 1 or 2 capsules before meals to aid digestion; or take as directed by your health-care professional.	Helps the body digest proteins, carbohydrates, fats, and plant fibers; can improve the digestion of milk sugars, so is helpful for gastritis caused by lactose intolerance.

ENZYME HELPERS

Nutrient	Suggested Dosage	Actions
Acidophilus or other probiotics, such as bifidobacterium; or Acidophilasé from Wakunaga, which also contains enzymes.	Take as indicated on the label.	Improves gastrointestinal function; needed for production of biotin, niacin, folic acid, and pyridoxine; helps produce antibacterial substances; stimulates enzyme activity.
Aged garlic extract	Take 2 capsules three times per day.	Potent antioxidant; improves digestion; stimulates healthy intestinal bacteria.

Fructooligo-saccharides	Take as indicated on the label.	Increases peristaltic action of intestines; improves liver function.
Lecithin	Take 1,000 mg three times per day with meals.	Emulsifies fat.
Ox bile extract	Take 120 mg per day.	Aids digestion.
Vitamin A	Take 5,000 IU per day.	Needed by digestive glands to produce enzymes; promotes normal taste and appetite; required for healthy mucous linings and skin and cell membrane structure and a strong immune system; potent antioxidant.
Vitamin B complex	Take 50 mg three times per day.	B vitamins act as coenzymes.
Vitamin B$_1$ (thiamin)	Take 150 mg three times per day.	Essential for carbohydrate metabolism, proper energy production, and appetite.
A basic (rather than acidic) form of vitamin C, such as calcium ascorbate or magnesium ascorbate.	Take 1,000–2,000 mg per day.	Stimulates the adrenal glands to produce hormones (which also act as coenzymes); aids in detoxification; is a potent antioxidant; improves the function of many enzymes; required for converting folic acid (a coenzyme) into its active form; improves digestion; free-radical fighter.
Whey	Take 1 heaping table-spoon three times per day with meals.	Improves digestion.
Zinc	Take 10–50 mg per day. Take no more than 100 mg per day from all sources.	Needed by over 300 enzymes. Many zinc enzymes work to metabolize carbohydrates, alcohol, and essential fatty acids; synthesize proteins; and dispose of free radicals.

Comments

❑ For acute indigestion, try raw pancreas glandular extract, DGL tablets from Enzymatic Therapy, or Alfa Juice caps by Solaray.

❑ For diarrhea, try apple pectin tablets or activated charcoal (on a short-term basis).

❑ For bloating and gas, try Indigestion Relief from Bioforce.

❑ For burping and belching, try Hyland's Indigestion after meals, Beano from AKPharma, or Digestive Formula from Country Life.

❑ To improve digestion, try teas made from the following:

- A pinch of nutmeg, cloves, cinnamon, and ginger.
- Thyme.
- Slippery elm.
- Mint and alfalfa.
- Fennel and catnip.
- Chamomile.

❑ Eat cultured foods (such as kefir, yogurt, miso), high-fiber foods (like whole grains), enzyme-rich foods (including pineapple and papaya), plus fresh fruits and vegetables.

❑ Eat four to five small meals per day. Chew food longer to decrease stress on your gastrointestinal tract.

❑ Avoid spicy foods, red meats, fatty foods, fried foods, dairy products, sugary foods, and caffeine. Decrease acid-forming food intake.

❑ To cleanse your gastrointestinal tract: On day one, eat no foods, but drink six to eight 8-ounce glasses of fresh apple juice. Add Source of Life Energy Shake from Nature's Plus or a similar product to the first glass. On days two through four, drink six to eight 8-ounce glasses of enzyme-rich juices, such as papaya, pineapple, kiwi, carrot, and parsley. Follow this program for four days while gradually incorporating your normal diet. See Part Three for further information on juicing.

❑ See Part Three for directions for the Enzyme Kidney Flush and the Enzyme Toxin Flush.

Gastroesophageal Reflux Disease

See HEARTBURN.

Gingivitis

Gingivitis is an inflammation of the gums and is an early stage of periodontal disease. Plaque is the main cause of gingivitis. When teeth are not brushed and flossed properly, plaque, which is made up primarily of bacteria, builds up along the gum line, causing the gums to become inflamed. Though plaque is the principal cause of gingivitis, there are many factors that can aggravate the condition, including pregnancy and other factors that affect the hormones, certain drugs, and vitamin deficiencies.

Symptoms

Gingivitis is marked by swollen, red, and painful gums. The gums are especially painful near and around the teeth, and they often bleed during eating and brushing. In advanced stages, halitosis and increased bleeding can occur. If gingivitis is left untreated, it can progress to periodontitis (pyorrhea). In adults, pyorrhea is the primary cause of tooth loss. In this condition, the pockets between the gums and the teeth deepen, and debris collects in the pockets, which allows bacterial toxins to spread (promoting the disease). Destruction of the supporting bony tissue is the earliest evidence on X-ray. Following progressive bone loss, there is tooth loss and the gums recede. Tooth movement is common in later stages. Unless an acute infection occurs, pain is usually absent.

Enzyme Therapy

Systemic enzyme therapy is used to decrease pain, swelling, and inflammation in the gums and stimulate the immune system. Enzymes improve circulation, help speed tissue repair, bring nutrients to the damaged area, remove waste products, improve health, strengthen the body as a whole, and build general resistance.

Digestive enzyme therapy is used to improve the digestion of food, reduce stress on the gastrointestinal mucosa, help maintain normal pH levels, detoxify the body, and promote the growth of healthy intestinal flora, thus improving the overall health of the body. Digestive enzymes also serve as replacements for the body's pancreatic enzymes, leaving the pancreatic enzymes free to perform other functions in the body, such as decreasing inflammation. Enzymes taken orally as chewables are very effective.

Enzyme Absorption System Enhancers (EASE) maximize enzyme activity. They can also improve the absorption and bioavailability of various nutrients and decrease the drain on the body's own digestive enzymes, thus prolonging their lives.

In addition to the standard multivitamin, multimineral, and multiglandular complexes recommended in the introduction to Part Two, the following supplements are recommended for the treatment of gingivitis. The following recommended dosages are for adults. Children between the ages of six and twelve should take one-half the recommended dosage, and those between the ages of thirteen and seventeen should take three-quarters the recommended adult dosage.

ESSENTIAL ENZYMES

Enzyme	Suggested Dosage	Actions
Superoxide dismutase	Take 150 mcg per day.	Fights inflammation; potent antioxidant.
A chewable papain or bromelain tablet, such as those from Nature's Plus	Chew one tablet 3 times per day between meals.	Reduces inflammation.

AND SELECT ONE OF THE FOLLOWING:

Bromelain	Take three to four 230–250 mg capsules per day until symptoms subside. If severe, take 6–8 capsules per day.	Fights inflammation, pain, and swelling; speeds healing; stimulates the immune system.
Serratiopeptidase	Take 10 mg two times per day between meals.	Fights inflammation; reduces pain and swelling.
Trypsin with chymotrypsin	Take 250 mg of trypsin and 10 mg of chymotrypsin three times per day between meals.	Fights inflammation, pain, and swelling; increases rate of healing; stimulates the immune system.
Pancreatin	Take 500 mg six times per day on an empty stomach.	Fights inflammation, pain, and swelling; increases rate of healing; stimulates the immune system; strengthens capillary walls.
Papain	Take 300 mg two times per day between meals.	Fights inflammation; speeds healing.
A microbial protease	Take 300 mg two times per day between meals.	Fights inflammation; speeds healing.

SUPPORTIVE ENZYMES

Enzyme	Suggested Dosage	Actions
Catalase and glutathione peroxidase	Take as indicated on the label.	Potent antioxidant enzymes.
A digestive enzyme product containing protease, amylase, and lipase enzymes. See the list of digestive enzyme products in Appendix.	Take as indicated on the label with meals.	Improves the breakdown and absorption of food nutrients; fights toxic build-up.

ENZYME ABSORPTION SYSTEM ENHANCERS (EASE)

Combination	Suggested Dosage	Actions
Anti-Flam Caps from Crystal Star Herbal Nutritional Products or	Take as indicated on the label.	Fights inflammation; decreases swelling.
Connect-All from Nature's Plus or	Take as indicated on the label.	Supports tissue repair.
CTR Support from PhysioLogics or	Take as indicated on the label.	Nourishes the body during tissue recovery and healing; helps maintain healthy connective tissue; diminishes damage caused by swelling and inflammation.
Curazyme from Enzymatic Therapy or	Take as indicated on the label.	Strengthens gum integrity.
Enzimmune from Enzyme Process Laboratories or	Take one tablet before each meal.	Stimulates immune function to better fight bacteria.
Inflamzyme from PhytoPharmica or	Take as indicated on the label.	Promotes tissue repair; decreases inflammation.
Kyolic Formula 102 (with enzymes)	Take 2 capsules or tablets with meals two to three times per day.	Strengthens the immune system; helps fight free radicals; helps fight inflammation; strengthens collagen integrity.

ENZYME HELPERS

Nutrient	Suggested Dosage	Actions
Acidophilus and other probiotics	Take as indicated on the label.	Improve gastrointestinal function, which improves the digestion and absorption of nutrients necessary to fight gingivitis; needed for production of biotin, niacin, folic acid, and pyridoxine, all needed as coenzymes; help produce antibacterial substances; stimulate enzyme activity.
Adrenal glandular	Take 100–400 mg per day.	Fights inflammation.
Aged garlic extract	Take 2 capsules three times per day.	Fights inflammation; stimulates the immune system; potent antioxidant.
Amino-acid complex	Take as indicated on the label on an empty stomach.	Required for enzyme production; helps the body maintain proper fluid and acid/base balance, necessary for overall health.
Bioflavonoids, such as rutin or quercetin	Take 500 mg three times per day between meals.	Strengthen capillaries; potent antioxidants.
Coenzyme Q$_{10}$	Take 30–60 mg one to two times per day with meals; or follow the advice of your health-care professional.	Potent antioxidant.
Pituitary glandular	Take 100 mg per day.	Helps form new blood vessels; aids healing.
Thymus glandular	Take 100 mg per day.	Assists the thymus gland in immune resistance; improves immune response; fights infections.
Vitamin A	Take 5,000 IU per day.	Needed for strong immune system to better fight bacteria and for healthy mucous membranes, such as those found in the mouth; potent antioxidant.
Vitamin B complex	Take 150 mg three times per day, or as indicated on the label.	B vitamins act as coenzymes.
Vitamin C	Take 10,000–15,000 mg per day.	Fights infections, diseases, and allergies; improves wound healing; stimulates the adrenal glands; promotes immune system; powerful antioxidant.
Vitamin E	Take 1,200 IU per day from all sources (including the multivitamin complex in the introduction to Part Two).	Potent antioxidant; supplies oxygen to cells; improves circulation and level of energy.

Comments

❑ Enzyme-rich fresh fruits and vegetables (particularly green vegetables) should comprise 60 percent of your diet. If your body cannot tolerate raw fruits and vegetables, increase your intake of digestive enzymes, or sauté or steam your produce.

❑ Whole grains stimulate and clean the gums and teeth. Eat plenty of high-fiber foods, legumes, and vegetables. The teeth and gums require exercise. Fiber foods clean and stimulate the gums and help to remove plaque from the teeth. Further, chewing helps to stimulate enzymatic activity in the mouth.

❑ Hawthorn berries, rose hips, and echinacea help decrease inflammation and stimulate the immune system.

❑ Phytochemicals should be used to enhance the healing process. For example, citrus bioflavonoids, grape seed extract, and carotenoids fight capillary permeability.

❑ Gargle three times per day with the Enzymatic Gargle as outlined in Part Three.

❑ Follow the Enzyme Kidney Flush program for one week. The following week, use the Enzyme Toxin Flush program. Use both flushes no more often than once per month. See Part Three for instructions.

❑ Topical applications of certain nutrients can be very beneficial when fighting gingivitis. A few examples:

• Topical applications of coenzyme Q$_{10}$ can help cure gingivitis. Break open a capsule and rub on the gums.

• Break open vitamin A and E capsules and rub the contents on the healing gums.

• Rub clove oil on the gums and teeth to decrease pain.

• Rub aloe vera gel directly on the gums to soothe and help relieve pain.

❑ Take care of your teeth. Brush and floss after every meal (use a soft toothbrush). Make a toothpaste of baking soda and sea salt or use a natural toothpaste found in health-food stores, such as Nature's Gate (contains vitamin C, baking soda, and sea salt), Tom's of Maine Natural Toothpaste (contains calcium, propolis, and myrrh), Nature de France (clay-based), Vicco Pure Herbal Toothpaste (contains bark, flower, and roots), or Weleda Soft Toothpaste (contains baking soda, salt, herbs, and silica).

Glaucoma

Glaucoma is a serious eye condition in which pressure builds in the eyeball, causing optic nerve damage and vision loss. Glaucoma is one of the most serious and most common disorders of the eye in the United States. It is the leading cause of blindness and threatens the sight of 67 million people worldwide. Those at highest risk to develop glaucoma are over 50 years old, very nearsighted, diabetic, relatives of people with glaucoma, or are African-Americans. While the risk for glaucoma increases with age, it is not just a disease of aging. Approximately 10,000 babies are born every year in the United States with glaucoma.

Glaucoma occurs when intraocular pressure builds up because the usual outflow of eye fluid is obstructed. If untreated, blindness can occur. There are two main types of glaucoma—open-angle glaucoma and narrow-angle, or closed-angle, glaucoma.

Open-angle glaucoma is a chronic type of glaucoma. It is often called primary open-angle glaucoma when there appears to be no underlying cause of the condition. In open-angle glaucoma, pressure builds slowly because fluid from the eye drains too slowly. This causes slow, progressive vision loss. Open-angle glaucoma is the most common type of glaucoma.

Initially, open-angle glaucoma produces no symptoms. Symptoms may not appear until permanent damage occurs in the eyes. Such symptoms may include a gradual narrowing of vision until tunnel vision develops; vision problems, such as seeing halos when looking at light and difficulty adapting to the dark; and headaches.

Narrow-angle, or *closed-angle glaucoma* occurs because the space between the iris and the cornea, where the eye fluid normally drains, is too small for the fluid to get through. Anything that causes the pupil to dilate can cause the iris to completely block the drainage area causing an increase in eye pressure. This type of glaucoma is often referred to as acute glaucoma because it results in sudden attacks of increased intraocular pressure.

Narrow-angle glaucoma produces a sudden slight decrease in vision, headaches and eye pain, and vision problems, such as seeing halos around lights. Nausea and vomiting may also occur. Attacks go away as the pupil closes, but they can recur. With each attack, the field of vision reduces more and more.

Pigmentary-dispersion glaucoma is a type of open-angle glaucoma in which the iris is concave-shaped rather than convex-shaped. This causes the iris to rub against the lens, which scrapes pigment off the iris. The pigment particles ultimately wind up clogging the drainage area of the eye, causing intraocular pressure to increase.

Secondary glaucoma occurs as a result of systemic illnesses, such as diabetes, cardiovascular disorders, or infection; of disorders of the eye, such as inflammation, infection, cataracts, or any other eye disorder that may interfere with fluid drainage; or the use of some medications, including steroids.

Enzyme Therapy

Systemic enzyme therapy is used to reduce pain, swelling, and inflammation in the eye, to stimulate the immune system, and to improve circulation to the eye. Enzymes help speed tissue repair, bring nutrients to the damaged area, remove waste products, and enhance wellness.

Digestive enzyme therapy is used to improve the digestion and assimilation of food nutrients, reduce stress on the gastrointestinal mucosa, help maintain normal pH levels, detoxify the body, and promote the growth of healthy intestinal flora, thus making the body healthier so that it is better able to combat glaucoma. Digestive enzymes also serve as replacements for the body's pancreatic enzymes, leaving the pancreatic enzymes free to perform other functions in the body, such as decreasing inflammation and improving circulation.

Enzyme Absorption System Enhancers (EASE) maximize enzyme activity. They can also improve the absorption and bioavailability of various nutrients and decrease the drain on the body's own digestive enzymes, thus prolonging their lives.

In addition to the standard multivitamin, multimineral, and multiglandular complexes recommended in the introduction to Part Two, the following supplements are recommended for the treatment of glaucoma. The following recommended dosages are for adults. Children between the ages of two and five should take one-quarter the recommended dosage, those between the ages of six and twelve should take one-half the recommended dosage, and those between the ages of thirteen and seventeen should take three-quarters the recommended adult dosage.

ESSENTIAL ENZYMES

Enzyme	Suggested Dosage	Actions
Superoxide dismutase	Take 150 mcg per day.	Fights inflammation; potent antioxidant.

AND SELECT ONE OF THE FOLLOWING:

Formula One—See page 38 of Part One for a list of products.	Take 5 tablets twice per day.	Fights inflammation, pain, and swelling; speeds healing; stimulates the immune system; strengthens capillary walls.
Bromelain	Take three to four 230–250 mg capsules per day until symptoms subside. If the condition is advanced, take 6–8 capsules per day.	Fights inflammation and swelling; speeds healing; stimulates the immune system.
Serratiopeptidase	Take 10 mg two times per day between meals.	Fights inflammation; reduces swelling.
Trypsin with chymotrypsin	Take 250 mg of trypsin and 10 mg of chymotrypsin three times per day between meals.	Fights inflammation and swelling; increases rate of healing; stimulates the immune system.
Pancreatin	Take 500 mg six times per day on an empty stomach.	Fights inflammation and swelling; increases rate of healing; stimulates the immune system; strengthens capillary walls.
Papain	Take 300 mg two times per day between meals.	Fights inflammation; speeds healing.
Microbial protease	Take 300 mg two times per day between meals.	Fights inflammation; speeds recovery from surgery.

SUPPORTIVE ENZYMES

Enzyme	Suggested Dosage	Actions
Catalase and glutathione peroxidase	Take as indicated on the label.	Potent antioxidant enzymes.
A digestive enzyme product containing protease, amylase, and lipase enzymes. See the list of digestive enzyme products in Appendix.	Take as indicated on the label with meals.	Improves the breakdown and absorption of food nutrients; fights toxic build-up.

ENZYME ABSORPTION SYSTEM ENHANCERS (EASE)

Combination	Suggested Dosage	Actions
Antioxidant Complex from Tyler *or*	Take as indicated on the label.	Potent free-radical fighter and immune system stimulator.
Detox Enzyme Formula from Prevail *or*	Take 1 capsule between meals one or two times per day with a 6- to 8-ounce glass of water or juice.	Helps rid the body of harmful toxins.
Hepazyme from Enzymatic Therapy *or*	Take as indicated on the label.	Stimulates immune function.
Kyolic Formula 102 (with enzymes) *or*	Take 2 capsules or tablets with meals two to three times per day.	Strengthens the immune system; helps alleviate stress; helps reverse circulatory disorders; helps fight free radicals; helps fight inflammation.
Oxy-5000 Forte from American Biologics *or*	Take as indicated on the label.	Helps the body excrete toxins; fights biochemical stressors, infections, and degenerative processes.
Vita-C-1000 from Life Plus	Take as indicated on the label.	Helps maintain vision; boosts immune function.

ENZYME HELPERS

Nutrient	Suggested Dosage	Actions
Acidophilus or other probiotic	Take as indicated on the label.	Improves gastrointestinal function and aids in the digestion and absorption of nutrients, which improves the health of the body; stimulates enzyme activity.
Adrenal glandular	Take 100–400 mg per day.	Fights inflammation and adrenal exhaustion.
Aged garlic extract	Take 2 capsules three times per day.	Fights inflammation; stimulates the immune system; potent antioxidant.
Bioflavonoids, such as rutin or quercetin	Take 500 mg three times per day between meals.	Strengthen capillaries; potent antioxidants.
Coenzyme Q_{10}	Take 30–60 mg one to two times per day with meals; or follow the advice of your health-care professional.	Potent antioxidant.

Essential fatty acids (found in flaxseed oil)	Take 100–200 mg two to three times per day with meals; or follow the advice of your health-care professional.	Help strengthen cell membranes; support immune function.
Niacin *or*	Take 50 mg the first day, then increase dosage daily by 50 mg to 1,000 mg. If adverse effects occur, decrease dosage. Take doses higher than 1,000 mg only under the supervision of a physician.	Dilates blood vessels; improves circulation.
Niacin-bound chromium	Take 20 mg per day or as indicated on the label.	
Pycnogenol	Take as indicated on the label.	Potent antioxidant; improves capillary wall strength.
Vitamin A *or*	Take 10,000–15,000 IU per day. If pregnant, take no more than 10,000 IU per day.	Needed for strong immune system and healthy mucous membranes; potent antioxidant.
Carotenoids	Take 5,000–10,000 IU per day, or as indicated on the label.	Potent antioxidant; enhances immunity.
Vitamin B complex	Take 150 mg three times per day, or as indicated on the label.	B vitamins act as coenzymes.
Vitamin C	Take 10,000–15,000 mg per day.	Promotes immune system; powerful antioxidant.
Vitamin E	Take 1,200 IU per day from all sources (including the multivitamin complex in the introduction to Part Two).	Potent antioxidant; supplies oxygen to cells; improves circulation and level of energy.
Zinc	Take 15–50 mg per day. Do not take more than 100 mg per day from all sources.	Needed by over 300 enzymes; necessary for white blood cell immune function.

Comments

❑ Bilberry protects and strengthens the blood vessels that feed the eyes and supports arteries and capillaries.

❑ Garlic, onions, shallots, chives, leeks, and scallions contain allium compounds, which are potent free-radical fighters that help lower cholesterol and fight circulatory disorders.

❑ Raspberries, cranberries, grapes, red wine, black currants, and red cabbage contain anthocyanidins, which reduce blood vessel plaque formation and maintain blood flow in small vessels.

❑ Researchers at the Weizmann Institute in Rehovot, Israel are treating glaucoma in laboratory animals with collagenase, an enzyme that thins the hard coating of the eye (the sclera). This allows the fluid to leak out, thus relieving intraocular pressure. The technique (developed by Dr. Arieh Yaron and Dr. Jacob Dan) involves encapsulating the enzyme in a plastic disk (the size of a contact lens), which is applied to the eye for four hours. The enzyme slowly leaks out, breaking up the hard protein coating. Collagenase thins the layer and evidently forms a filter allowing the fluid to drain. The enzyme is able to complete its bloodless "surgery" within a few hours. It is then removed and any remaining solution rinsed from the eye. The researchers found that this form of treatment significantly lowered intraocular pressure and had long-lasting effects.

❏ See the list of Resource Groups in the Appendix for a list of organizations that can provide you with more information about glaucoma.

Glomerulonephritis

Glomerulonephritis is a common kidney disorder characterized by inflammation of the glomeruli (the thin-walled clusters of capillaries in the kidney where blood filtration takes place). There are four types of glomerulonephritis—acute, rapidly progressive, chronic, and nephrotic syndrome.

Acute glomerulonephritis (acute nephritic syndrome, postinfectious glomerulonephritis) usually follows some type of infection. It is characterized by the sudden appearance of blood in the urine along with protein clumps of red blood cells found during urinalysis. Most commonly, acute glomerulonephritis follows a streptococcal infection, such as strep throat (Poststreptococcal glomerulonephritis). The immune complexes formed as the body's antibodies fight the infection coat the glomerulus and interfere with its function. This type of glomerulonephritis is most common in children.

Many people with acute glomerulonephritis experience no symptoms. Those who do experience edema, decreased urination, and dark urine that contains blood. Most people who suffer from acute glomerulonephritis recover completely; however, if there are large amounts of protein in the blood, kidney damage may result.

Rapidly progressive glomerulonephritis (rapidly progressive nephritic syndrome) is an uncommon disorder that may occur as part of a disorder that also affects other organs, may be caused by antibodies attacking the glomeruli, may be caused by immune complexes formed elsewhere that are deposited in the kidney, or may be due to unknown causes. It results in partial destruction of the glomeruli, which causes kidney failure. The production of antibodies against the kidney and the deposition of immune complexes in the kidney may be related to a viral infection or an autoimmune disorder, such as systemic lupus erythematosus.

Rapidly progressive glomerulonephritis can cause weakness, fatigue, fever, nausea and vomiting, loss of appetite, and blood in the urine. Left untreated, this condition can cause kidney failure.

In *chronic glomerulonephritis* (chronic nephritic syndrome) the glomeruli are damaged and kidney function degenerates over years. Its cause is unknown. It produces no symptoms for years, causing the disease to go undetected in many until kidney failure occurs.

Nephrotic syndrome results in severe loss of protein into the urine, decreased blood levels of protein, salt and water retention, and increased blood levels of lipids. It may be caused by many diseases affecting the kidneys or the use of drugs that are toxic to the kidneys. Symptoms include malaise, loss of appetite, edema, muscle wasting, abdominal pain and distention, and frothy urine. Nutritional deficiencies, blood clotting disorders, high blood pressure, and opportunistic infections may also result. Children may experience very low blood pressure and blood pressure that falls upon standing up.

Enzyme Therapy

Systemic enzyme therapy is used to reduce pain, swelling, and inflammation in the kidneys and stimulate the immune system. Enzymes improve circulation, bringing nutrients to the damaged area and removing waste products.

Digestive enzyme therapy is used to improve digestion of food, reduce stress on the gastrointestinal mucosa, help maintain normal pH levels, detoxify the body, promote the growth of healthy intestinal flora, build general resistance, and strengthen the body as a whole. Digestive enzymes also serve as replacements for the body's pancreatic enzymes, leaving the pancreatic enzymes free to perform other functions in the body, such as decreasing inflammation.

Enzyme Absorption System Enhancers (EASE) maximize enzyme activity. They can also improve the absorption and bioavailability of various nutrients and decrease the drain on the body's own digestive enzymes, thus prolonging their lives.

In addition to the standard multivitamin, multimineral, and multiglandular complexes recommended in the introduction to Part Two, the following supplements are recommended for the treatment of glomerulonephritis. The following recommended dosages are for adults. Children between the ages of two and five should take one-quarter the recommended dosage, those between the ages of six and twelve should take one-half the recommended dosage, and those between the ages of thirteen and seventeen should take three-quarters the recommended adult dosage. This therapy program will help improve the symptoms of all types of glomerulonephritis; however it is particularly effective in the treatment of rapidly progressive glomerulonephritis.

ESSENTIAL ENZYMES

Enzyme	Suggested Dosage	Actions
Superoxide dismutase	Take 150 mcg per day.	Fights inflammation; potent antioxidant.

AND SELECT ONE OF THE FOLLOWING:

Formula One—See page 38 of Part One for a list of products.	Take 10 tablets three times per day between meals.	Fights inflammation, pain, and swelling; speeds healing; stimulates immunity.
Bromelain	Take three to four 230–250 mg capsules per day until symptoms subside. If severe, take 6–8 capsules per day.	Fights inflammation, pain, and swelling; speeds healing; stimulates the immune system.
Serratiopeptidase	Take 10 mg two times per day between meals.	Fights inflammation; reduces pain and swelling.
Trypsin with chymotrypsin	Take 250 mg of trypsin and 10 mg of chymotrypsin three times per day between meals.	Fights inflammation, pain, and swelling; increases rate of healing; stimulates the immune system.
Pancreatin	Take 500 mg six times per day on an empty stomach.	Fights inflammation, pain, and swelling; increases rate of healing; stimulates the immune system; strengthens capillary walls.
Papain or a microbial protease	Take 300 mg two times per day between meals.	Fights inflammation; speeds recovery from injuries and surgery.

SUPPORTIVE ENZYMES

Enzyme	Suggested Dosage	Actions
Catalase and glutathione peroxidase	Take as indicated on the label.	Potent antioxidant enzymes.
A digestive enzyme product containing protease, amylase, and lipase. See the list of digestive enzyme products in Appendix.	Take as indicated on the label with meals.	Improves the breakdown and absorption of food nutrients; fights toxic build-up.

ENZYME ABSORPTION SYSTEM ENHANCERS (EASE)

Combination	Suggested Dosage	Actions
Advanced Nutritional System from Rainbow Light *or*	Take as indicated on the label.	Potent free-radical fighter and immune system stimulator.
Biovital Plus from Enzymatic Therapy *or*	Take as indicated on the label.	Anti-inflammatory agent; decreases swelling; fights infection; contains many antioxidants and many enzymes that aid in the absorption of glandulars, herbs, and other nutrients that assist in urinary function.
Kyolic Formula 102 (with enzymes) *or*	Take 2 capsules or tablets with meals two to three times per day.	Strengthens the immune system; helps fight free radicals; helps fight inflammation; improves digestion.
NESS Formula #416 *or*	Take as indicated on the label.	Breaks up fibrin; fights inflammation.

Protease Concentrate from Tyler Encapsulations *or*	Take as indicated on the label.	Breaks up fibrin; fights inflammation.
Vita-C-1000 from Life Plus	Take as indicated on the label.	Boosts immune function; fights infection.

ENZYME HELPERS

Nutrient	Suggested Dosage	Actions
Acidophilus and other probiotics	Take as indicated on the label.	Improves gastrointestinal function to detoxify the body; needed for production of biotin, niacin, folic acid, and pyridoxine, all important coenzymes; helps produce antibacterial substances; stimulates enzyme activity.
Adrenal glandular	Take 100–400 mg per day.	Fights inflammation and adrenal exhaustion.
Bioflavonoid, such as rutin or quercetin	Take 500 mg three times per day between meals.	Strengthen capillaries; potent antioxidants.
Coenzyme Q_{10}	Take 30–60 mg one to two times per day with meals; or follow the advice of your health-care professional.	Potent antioxidant.
Kidney glandular	Take as indicated on the label.	Promotes kidney function.
Thymus glandular	Take 100 mg per day.	Assists the thymus gland in immune resistance; improves immune response; fights infections.
Vitamin A	Take 10,000–15,000 IU per day. For higher doses, see a physician. If pregnant, take no more than 10,000 IU per day under the supervision of a physician.	Needed for strong immune system and healthy tissues; potent antioxidant.
Vitamin B complex	Take 50 mg three times per day, or as indicated on the label.	B vitamins act as coenzymes.
Vitamin C	Take 10,000–20,000 mg per day. Doses over 10,000 per day should be taken only under the supervision of a physician.	Fights infections, diseases, and allergies; stimulates the adrenal glands; promotes immune system; powerful antioxidant.
Zinc	Take 15–50 mg per day. Do not take more than 100 mg per day from all sources.	Needed by over 300 enzymes; necessary for white blood cell immune function.

Comments

❑ Fruits and vegetables (particularly green vegetables) should comprise over 60 percent of your diet. If your body cannot tolerate raw fruits and vegetables, increase your intake of digestive enzymes, or sauté or steam your produce.

❑ Restrict your salt intake to reduce the work of the kidneys. This will help lower blood pressure.

❏ Limit your protein intake. Protein puts an enormous amount of stress on the kidneys.

❏ Eat foods high in complex carbohydrates, including whole grains and nuts in all forms. Try grain cereals, whole wheat bread (check the label to ensure that whole wheat is the first ingredient) or other whole grain breads, millet, oats, and granolas with no sugar added. Add 1–2 tablespoons of wheat bran to foods, or put in cold or hot liquid and drink with each meal.

❏ Avoid alcohol, refined white flour, and sugar. These produce undesirable fermentation in the bowel and stimulate improper metabolism of insulin, chromium, and zinc, which weakens overall health. Small amounts of honey are allowed.

❏ Natural spices are fine (if salt-free), but chemical preservatives, particularly nitrites, should be avoided.

❏ Avoid chlorinated water. Distilled or bottled spring water, or even well water is recommended.

❏ Cold-pressed vegetable oil can be used liberally.

❏ Follow directions for the Enzyme Kidney Flush and the Enzyme Toxin Flush in Part Three.

❏ See the list of Resource Groups in the Appendix for a list of organizations that can provide you with more information about glomerulonephritis.

❏ *See also* KIDNEY DISORDERS.

Gluten Enteropathy/ Intolerance

See CELIAC DISEASE.

Gout

Gout is a metabolic disorder caused by excessively high levels of uric acid in the blood due to either excessive production in the body or decreased elimination by the kidneys. The uric acid then deposits in the joints and sometimes the kidneys and other organs. It most commonly affects the big toe (a condition called podagra) but may also affect the ankles, insteps, knees, wrists, elbows, and the cartilage in the ears. It rarely affects such centrally located joints as the hips, shoulders, and spine. Gout occurs in men twenty times more frequently than it does in women.

A diet rich in fats, overindulgence in alcohol or food, and emotional stress can all trigger an attack of gout. Other triggers include minor trauma, surgery, fatigue, a penicillin injection, illness, and renal insufficiency. Risk for gout also increases with the presence of thyroid disorders, diabetes, kidney disorders, hyperlipidemia, anemia, vascular disorders, and hypertension.

Symptoms

A gout attack occurs without warning. It is characterized by sudden, severe pain that gets progressively worse in one or more joint (particularly the big toe), although the first few attacks usually affect only one joint. The joint is also inflamed, warm, and red. The surrounding and overlying tissue and skin may be swollen, hot, tense, shiny, and discolored. Fever, chills, malaise, and a rapid heartbeat may also accompany a gout attack. After a few days, the attack usually subsides and the symptoms disappear completely until the next attack. Untreated, the attacks may last longer, occur more frequently, and affect more joints. Gout may also permanently damage affected joints.

The most reliable test for gout is to draw a fluid sample from the affected joint or joints. The fluid is then examined under a microscope and if uric acid crystals are present, the diagnosis is gout.

Enzyme Therapy

Systemic enzyme therapy is used to decrease pain, swelling, and inflammation and to stimulate the immune system. Enzymes improve circulation, help speed tissue repair, bring nutrients to the damaged area, remove waste products, enhance wellness, and build general resistance.

Digestive enzyme therapy is used to improve digestion of food, reduce stress on the gastrointestinal mucosa, help maintain normal pH levels, detoxify the body, promote the growth of healthy intestinal flora, and strengthen the body as a whole. Digestive enzymes also serve as replacements for the body's pancreatic enzymes, leaving the pancreatic enzymes free to perform other functions in the body, such as decreasing inflammation.

Topical enzyme skin salves and creams are used to penetrate the superficial layers of the skin and decrease pain, swelling, and inflammation. Papain and bromelain are helpful, particularly in combinations with trypsin and chymotrypsin.

Enzyme Absorption System Enhancers (EASE) maximize enzyme activity. They can also improve the absorption and bioavailability of various nutrients and decrease the drain on the body's own digestive enzymes, thus prolonging their lives.

In addition to the standard multivitamin, multimineral, and multiglandular complexes recommended in the introduction to Part Two, the following supplements are recommended for the treatment of gout. The following recommended dosages are for adults. Children between the ages of two and five should take one-quarter the recommended dosage, those between the ages of six and twelve should take one-half the recommended dosage, and those between the ages of thirteen and seventeen should take three-quarters the recommended adult dosage.

ESSENTIAL ENZYMES

Enzyme	Suggested Dosage	Actions
A proteolytic enzyme salve	Apply to the affected area three times per day, or as indicated on the label.	Reduces inflammation and pain.

AND SELECT ONE OF THE FOLLOWING:

Formula One (particularly Wobenzym N from Naturally Vitamin Supplements)—See page 38 of Part One for a list of products.	Take 10 tablets three times per day between meals. As symptoms subside (which should occur in approximately one week), decrease dosage to 3–5 tablets three times per day between meals. When symptoms are gone, continue taking a maintenance dose of 2–3 tablets after each meal.	Decreases inflammation, pain, and swelling; increases rate of healing; stimulates the immune system; strengthens capillary walls.
Inflazyme Forte from American Biologics	Take as indicated on the label between meals.	Decreases inflammation, pain, and swelling; increases rate of healing; stimulates the immune system; strengthens capillary walls.
Enzyme complexes rich in proteases, plus amylases and lipases working synergistically	Take as indicated on the label between meals.	Decreases inflammation, pain, and swelling; increases rate of healing; stimulates the immune system; strengthens capillary walls.
Bromelain	Take three to four 230–250 mg capsules per day until symptoms subside. If severe, take 6–8 capsules per day.	Fights inflammation, pain, and swelling; speeds recovery after injury or surgery; stimulates the immune system.

SUPPORTIVE ENZYMES

Enzyme	Suggested Dosage	Actions
Superoxide dismutase, catalase, and glutathione peroxidase	Take 50 mcg of each three times per day between meals.	Potent antioxidant enzymes.
A digestive enzyme product containing protease, amylase, and lipase. See the list of digestive enzyme products in Appendix.	Take as indicated on the label with meals.	Improves digestion by breaking down proteins, carbohydrates, and fats.

ENZYME ABSORPTION SYSTEM ENHANCERS (EASE)

Combination	Suggested Dosage	Actions
Bromelain Plus from	Take as indicated on	Decreases swelling; improves
Enzymatic Therapy or	the label.	circulation; helps break up any circulating immune complexes.
Joint-Ease from Nature's Life or	Take as indicated on the label.	Decreases inflammation and swelling.
Mobil-Ease from Prevail or	Take one capsule between meals two to three times per day.	Has anti-inflammatory activity.
Protease Concentrate from Tyler Encapsulations	Take as indicated on the label.	Breaks up fibrin; fights inflammation.

ENZYME HELPERS

Nutrient	Suggested Dosage	Actions
Acidophilus and other probiotics	Take as indicated on the label.	Improves digestion; stimulates enzyme activity.
Adrenal glandular	Take 100–400 mg per day.	Fights inflammation and adrenal exhaustion.
Bioflavonoids	Take 150–500 mg three times per day.	Strengthen capillaries, thus improving circulation and speeding healing; potent antioxidant.
Coenzyme Q_{10}	Take 30–60 mg one to two times per day with meals; or follow the advice of your health-care professional.	Potent antioxidant.
Essential fatty acids (found in flaxseed oil)	Take 75–100 mg two to three times per day with meals; or follow the advice of your health-care professional.	Help strengthen cell membranes.
Glucosamine hydrochloride	Take 500 mg two to four times per day.	Stimulates cartilage repair.
and/or Glucosamine sulfate	Take 500 mg one to three times per day.	
Kyolic 101 Formula	Take 1 teaspoon per day.	Stimulates the immune system; reduces free-radical formation.
Mucopolysaccharides found in bovine cartilage extract and shark cartilage extract	Take as indicated on the label.	Support healthy, flexible joints; control inflammation; help keep joint membranes fluid; fight inflammation.
Pantothenic acid (vitamin B_5)	Take 300–500 mg three times per day, or as indicated on the label.	Essential to break uric acid down into harmless compounds.
Vitamin A	Take 10,000–25,000 IU per day. If pregnant, take no more than 10,000 IU per day.	Works with proteolytic enzymes.
Vitamin B complex	Take 50 mg three times per day, or as indicated on the label.	B vitamins act as coenzymes and support joint function.
Vitamin C	Take 1,000–10,000 mg per day.	Helps to produce collagen and normal connective tissue; promotes immune system; powerful antioxidant.
Vitamin E	Take 400 IU per day.	Potent antioxidant; supplies

240

		oxygen to cells; improves level of energy.
Zinc	Take 15–50 mg per day. Do not take more than 100 mg per day from all sources.	Needed by over 300 enzymes.

Comments

❑ Fresh, enzyme-rich fruits and vegetables (particularly green vegetables) should comprise over 60 percent of your diet. They increase the body's excretion rate of uric acid. Be sure to include pineapples, papayas, figs, and sprouting seeds. If your body cannot tolerate raw fruits and vegetables, increase your intake of digestive enzymes, or sauté or steam your produce.

❑ Cherries, grapes, and pineapples stimulate antioxidant activity, enzyme activity, and immune function.

❑ Absolutely avoid alcohol, which can aggravate gout.

❑ Decrease your protein intake.

❑ A high intake of water and other fluids is important in the treatment of gout. Drink at least eight 8-ounce glasses per day.

❑ Gargle with the Enzymatic Gargle at least once daily. See Part Three for directions for its preparation.

❑ Exercises are very important for good health. Do some form of exercise five to seven times per week for 30 to 40 minutes. The exercises don't have to be elaborate to work— you can "fast walk" (swing your arms while walking briskly), ride a bicycle, or swim.

❑ Meditate about 15 minutes per day. Void your mind and think of nothing. Allow your body to re-energize itself. Increase your enzymatic and antioxidant activity through exercises and a positive mental attitude.

❑ Gout therapy could include mild chiropractic and naturopathic care, pressure point therapy, massage, acupressure or acupuncture, electrical stimulation, and alternating ice and warm moist packs.

Guillain-Barré Syndrome

Guillain-Barré syndrome (acute febrile polyneuritis, acute polyneuritis, infectious polyneuropathy, or acute polyradiculitis) is an inflammation of the nerve roots that causes rapidly worsening muscle weakness and loss of sensation that may lead to paralysis. It is believed to be caused by an autoimmune reaction and is usually triggered by an infection, surgery, or immunization.

Symptoms

Guillain-Barré syndrome is marked by tingling, weakness, and loss of sensation that begins in the feet and moves upward toward the arms. In some, the muscles that support breathing become so weak that a respirator is needed. In some, the muscles used for chewing and swallowing become too weak to eat, and they must be fed intravenously or through a feeding tube. Some lose normal reflexes. Paralysis may result.

Enzyme Therapy

Systemic enzyme therapy is used to reduce the pain and inflammation in the affected nerves. Enzymes stimulate the immune system, improve circulation, help speed tissue repair, bring nutrients to the damaged area, remove waste products, and enhance wellness.

Digestive enzyme therapy is used to improve digestion of food, reduce stress on the gastrointestinal mucosa, help maintain normal pH levels, detoxify the body, and promote the growth of healthy intestinal flora, thereby promoting the health of the entire body. Digestive enzymes also serve as replacements for the body's pancreatic enzymes, leaving the pancreatic enzymes free to perform other functions in the body, such as boosting immunity and decreasing inflammation.

Enzyme Absorption System Enhancers (EASE) maximize enzyme activity. They can also improve the absorption and bioavailability of various nutrients and decrease the drain on the body's own digestive enzymes, thus prolonging their lives.

In addition to the standard multivitamin, multimineral, and multiglandular complexes recommended in the introduction to Part Two, the following supplements are recommended for the treatment of Guillain-Barré syndrome. The following recommended dosages are for adults. Children between the ages of two and five should take one-quarter the recommended dosage, those between the ages of six and twelve should take one-half the recommended dosage, and those between the ages of thirteen and seventeen should take three-quarters the recommended adult dosage.

ESSENTIAL ENZYMES

Enzyme	Suggested Dosage	Actions
Superoxide dismutase	Take 150 mcg per day.	Fights inflammation; potent antioxidant.

AND SELECT ONE OF THE FOLLOWING;

Formula One—See page 38 of Part One for a list of products.	Take 10 tablets three times per day between meals.	Fights inflammation, pain, and swelling; speeds healing; stimulates the immune system.
Bromelain	Take three to four 250 mg capsules per day until symptoms subside. If severe, take 6–8 capsules per day.	Fights inflammation, pain, and swelling; speeds healing; stimulates the immune system.

241

Serratiopeptidase	Take 10 mg two times per day between meals.	Fights inflammation; reduces pain and swelling.
Trypsin with chymotrypsin	Take 250 mg of trypsin and 10 mg of chymotrypsin three times per day between meals.	Fights inflammation, pain, and swelling; increases rate of healing; stimulates the immune system.
Pancreatin	Take 500 mg six times per day on an empty stomach.	Fights inflammation, pain, and swelling; increases rate of healing; stimulates the immune system; strengthens capillary walls.
Papain or a microbial protease	Take 300 mg two times per day between meals.	Fights inflammation; speeds recovery from injuries and surgery.

SUPPORTIVE ENZYMES

Enzyme	Suggested Dosage	Actions
Catalase and glutathione peroxidase	Take as indicated on the labels.	Potent antioxidant enzymes.
A digestive enzyme product containing protease, amylase, and lipase. See the list of digestive enzyme products in Appendix.	Take as indicated on the label with meals.	Improves the breakdown and absorption of food nutrients; fights toxic build-up.

ENZYME ABSORPTION SYSTEM ENHANCERS (EASE)

Combination	Suggested Dosage	Actions
Hepazyme from Enzymatic Therapy or	Take as indicated on the label.	Stimulates immune function.
Inflamzyme from PhytoPharmica or	Take as indicated on the label.	Immune system activator; reduces inflammation; fights free radicals.
Kyolic Formula 102 (with enzymes) or	Take 2 capsules or tablets with meals two to three times per day.	Strengthens the immune system; helps alleviate stress; helps fight free radicals; helps fight inflammation.
Vita-C-1000 from Life Plus	Take as indicated on the label.	Boosts immune function.

ENZYME HELPERS

Nutrient	Suggested Dosage	Actions
Acidophilus and other probiotics	Take as indicated on the label.	Needed for production of biotin, niacin, folic acid, and pyridoxine, all important coenzymes; help produce antibacterial substances; stimulates enzyme activity to fight inflammation.
Adrenal glandular	Take 100–400 mg per day.	Fights inflammation and adrenal exhaustion.
Aged garlic extract	Take 2 capsules three times per day.	Fights inflammation; stimulates the immune system; potent antioxidant.

Bioflavonoids, such as rutin or quercetin	Take 500 mg three times per day between meals.	Strengthen capillaries; potent antioxidant.
Calcium	Take 1,000 mg per day.	Plays important roles in muscle contraction and nerve transmission.
Magnesium	Take 500 mg per day.	Plays important role in muscle contraction, nerve function, and calcium absorption.
Thymus glandular	Take 100 mg per day.	Assists the thymus gland in immune resistance; improves immune response; fights infections.
Vitamin B complex	Take 150 mg three times per day, or as indicated on the label.	B vitamins act as coenzymes; support neuromusculoskeletal function.
Vitamin C	Take 10,000–15,000 mg per day.	Fights infections, diseases, and allergies; improves wound healing; stimulates the adrenal glands; promotes immune system; powerful antioxidant.
Zinc	Take 50 mg per day. Do not take more than 100 per day from all sources.	Needed by more than 300 enzymes; potent antioxidant.

Comments

❑ Fruits and vegetables (particularly green vegetables) should comprise over 60 percent of your diet. Beets, celery, onions, cauliflower, artichoke, and turnips, among others should be eaten raw, blended, or juiced. If your body cannot tolerate raw fruits and vegetables, increase your intake of digestive enzymes, or sauté or steam your produce. Follow the dietary guidelines as outlined in Part Three.

❑ A number of herbs can be helpful in fighting Guillain-Barré syndrome, including astragalus, echinacea, and maitake, reishi, and shiitake mushrooms, which help enhance immunity. Ginger root, ginkgo biloba, white willow bark and licorice fight inflammation.

❑ Detoxification is important in any autoimmune condition. Follow the detoxification, juicing, and fasting plans as outlined in Part Three.

❑ Gargle with the Enzymatic Gargle daily (see Part Three for instructions for its preparation).

❑ If you suffer from Guillain-Barré syndrome, seek the services of a well-trained health professional who will develop a treatment program for you.

❑ *See also* NEURITIS.

Hangover

A hangover occurs after excessive alcohol consumption (usually the next morning). Too much alcohol can leave you dehydrated (thus the thirst), play havoc with your blood sugar (elevating and then lowering levels), and strip your cells of nutrients (especially the B vitamins). Alcohol (even just one night of heavy drinking) can cause fat to accumulate in the liver. This fat then interferes with the distribution of oxygen and nutrients to the liver cells.

There are other ways alcohol affects the metabolism of nutrients in the tissues. Alcohol causes too much histamine to be secreted by the stomach cells, causing inflammation. Histamine is the immune system's inflammation-producing agent. In addition, alcohol causes an oversecretion of acid in the stomach. The result is that the stomach becomes vulnerable to the formation of ulcers. The efficiency of vitamin D activation and production is lost by the liver, and bile production and excretion is altered. Vitamin B_{12}, thiamin, and folic acid are not absorbed by the intestinal cells and increased amounts of zinc, calcium, potassium, and magnesium are excreted by the kidneys after alcohol consumption. It's no wonder you feel awful.

Symptoms

Hangover symptoms may include a dry mouth, thirst, nausea or vomiting, and headache. Irritability, shakiness or dizziness, fatigue, loss of appetite, perhaps heartburn, paleness, and unstable gait may also be present. Chronic, excessive alcohol intake can cause nutritional deficiencies and liver changes. Cirrhosis of the liver and acute pancreatitis are two diseases known to be caused by prolonged and excessive alcohol intake.

Enzyme Therapy

Enzymes are very effective at decreasing the effects of a hangover. Digestive enzyme therapy is used to improve the digestion and assimilation of food nutrients, reduce stress on the gastrointestinal mucosa, help maintain normal pH levels, detoxify the body, and promote the growth of healthy intestinal flora. Systemic enzyme therapy is used to reduce inflammation that can result from chronic drinking, stimulate the immune system, improve circulation, help speed tissue repair, bring nutrients to the damaged area, remove waste products, and enhance wellness.

Enzyme Absorption System Enhancers (EASE) maximize enzyme activity. They can also improve the absorption and bioavailability of various nutrients and decrease the drain on the body's own digestive enzymes, thus prolonging their lives.

In addition to the standard multivitamin, multimineral, and multiglandular complexes recommended in the introduction to Part Two, the following supplements are recommended for the treatment of hangover.

ESSENTIAL ENZYMES

Enzyme	Suggested Dosage	Actions
A digestive enzyme formula containing protease, amylase, and lipase enzymes. See the list of digestive enzyme products in Appendix. *or*	Take as indicated on the label.	Improves the digestion and assimilation of food nutrients. Breaks down proteins, carbohydrates, and fats.
Bromelain *or*	Start with one 230–250 mg capsule per day and increase dosage gradually, if necessary. Take no more than 2 or 3 capsules per day. If diarrhea develops, decrease dose.	Improves digestion; fights inflammation, pain, and swelling; stimulates the immune system.
Pancreatin	Take as indicated on the label.	Contains proteases, amylases, and lipases.

SUPPORTIVE ENZYMES

Enzyme	Suggested Dosage	Actions
Formula Three—See page 38 of Part One for a list of products.	Take as indicated on the label.	Free-radical fighters.

ENZYME ABSORPTION SYSTEM ENHANCERS (EASE)

Combination	Suggested Dosage	Actions
Biodias A Granules from Amano Pharmaceutical Co., Ltd. *or*	Adults should take one packet in 8 ounces of water between or after meals three times per day. Children ages 11 to 14 should take two-thirds of a packet; children between the ages of 8 and 10 should take half of the packet; and children between the ages of 5 and 7 should take one-third of a packet.	Helps improve digestion of nutrients; relieves inflammation; promotes regeneration of mucous membranes; decreases stress on the stomach and intestines; reduces gastric pain.
Kyolic Formula 102 (with enzymes) *or*	Take 2 capsules or tablets with meals two to three times per day.	Strengthens the immune system; helps alleviate stress; helps fight free radicals; helps fight inflammation; improves digestion.
Sugar Check from Enzymatic Therapy	Take as indicated on the label.	Contains nutrients for improved sugar and carbohydrate metabolism and absorption.

ENZYME HELPERS

Nutrient	Suggested Dosage	Actions
Acidophilus or other probiotics, such as Acidophilase from Wakunaga or Mega-Dophilus from Natren	Take as indicated on the label.	Improves gastrointestinal function; needed for production of biotin, niacin, folic acid, and pyridoxine; stimulates enzyme activity.

Aged garlic extract	Take 2 capsules three times per day.	Potent antioxidant; improves digestion; stimulates healthy intestinal bacteria.
Amino-acid complex	Take as indicated on the label.	Amino acids serve as building blocks for the body's enzymes.
Ox bile extract	Take 120 mg per day.	Aids in digestion.
Vitamin A	Take 5,000 IU per day.	Needed by digestive glands to produce enzymes; required for healthy mucous linings and a strong immune system, which is necessary for removing toxins from the body; potent antioxidant.
Vitamin B complex	Take 50 mg three times per day.	B vitamins act as coenzymes.
Vitamin C	Take 1,000–2,000 mg per day.	Stimulates the adrenal glands to produce hormones (which also act as coenzymes); aids in detoxification; is a potent antioxidant; improves the function of many enzymes.
Whey	Take 1 heaping tablespoon three times per day with meals.	Improves digestion.
Zinc	Take 10–50 mg per day. Do not take more than 100 mg per day from all sources.	Needed by over 300 enzymes. Many zinc enzymes work to metabolize carbohydrates, alcohol, and essential fatty acids; synthesize proteins; dispose of free radicals.

Comments

❏ Raw fruits and vegetables (particularly green vegetables) should comprise over 70 percent of your diet for the next few days (then drop back down to 50 or 60 percent). If your body cannot tolerate raw fruits and vegetables, increase your intake of digestive enzymes, or sauté or steam your produce.

❏ To improve overall health, avoid alcohol, refined white flour, and sugar.

❏ Cleansing and juicing are important in relieving a hangover. Drink at least four to five 8-ounce glasses of juice per day for two days and plenty of water.

❏ For one week following the hangover, follow the Enzyme Kidney Flush program. The following week, use the Enzyme Toxin Flush. See directions for performing both flushes in Part Three.

❏ Gargle with the Enzymatic Gargle three times per day (see Part Three for directions for its preparation).

❏ Add ten drops of Wakunaga's Liquid Aged Garlic Extract to a 6-ounce glass of tomato juice or V8 juice. Add two to three drops of Tabasco sauce, stir, and drink.

❏ Sweat out the toxins. Exercise is very important for good body health. Do some form of exercise five to seven times per week for thirty to forty minutes. The exercises don't have to be elaborate to work—try fast walking (swing your arms while walking briskly), riding a bicycle, or swimming.

❏ If you must drink, do so in moderation. Some types of alcoholic beverages have the potential to cause worse hangovers than others. Alcoholic beverages contain *congeners*. Congeners are by-products of alcohol fermentation. It seems the more congeners in your alcoholic beverage, the worse the hangover. Some types of alcohol, such as vodka and gin, have few congeners, while others, including whiskey and champagne, have more, so they can cause a worse hangover.

Hay Fever

Hay fever is an allergy to airborne allergens, usually pollen. Tree pollens from elm, alder, oak, maple, juniper, olive, and birch can cause spring hay fever; the summer form is primarily due to weed or grass pollens, including timothy, Bermuda, or Johnson grass; while fall hay fever is often caused by weed pollens, including ragweed. Sometimes, airborne mold spores cause hay fever.

Symptoms

Hay fever symptoms may include itching in the throat, the roof of the mouth, nose, and/or eyes; sneezing; stuffy and runny nose; and a watery, clear discharge from the eyes. This may be accompanied by headache, nervous irritability, loss of appetite, insomnia, depression, swollen nasal membranes, coughing, and asthmatic wheezing.

Enzyme Therapy

Systemic enzyme therapy is used to reduce inflammation, stimulate the immune system, and break up mucus in the respiratory tract. Enzymes can improve circulation, help speed tissue repair, bring nutrients to the damaged area, remove waste products, and improve health.

Digestive enzyme therapy is used to improve digestion of food, reduce stress on the gastrointestinal mucosa, help maintain normal pH levels, detoxify the body, and promote the growth of healthy intestinal flora, thereby making the body healthier so that it can combat hay fever. Digestive enzymes also serve as a replacement for the body's pancreatic enzymes, leaving the pancreatic enzymes free to perform other functions in the body, such as boosting immunity and decreasing inflammation.

Enzyme Absorption System Enhancers (EASE) maximize enzyme activity. They can also improve the absorption and bioavailability of various nutrients and decrease the drain on the body's own digestive enzymes, thus prolonging their lives.

In addition to the standard multivitamin, multimineral, and multiglandular complexes recommended in the introduction to Part Two, the following supplements are recommended for the treatment of hay fever. The following recommended dosages are for adults. Children between the ages of two and five should take one-quarter the recommended dosage, those between the ages of six and twelve should take one-half the recommended dosage, and those between the ages of thirteen and seventeen should take three-quarters the recommended adult dosage.

ESSENTIAL ENZYMES

Enzyme	Suggested Dosage	Actions
Formula One—See page 38 of Part One for a list of products. *or*	Take 10 tablets three times per day between meals for for six months. Then decrease dosage to 5 tablets three times per day between meals.	Bolsters the immune system; breaks up antigen-antibody complexes; fights free-radical formation; fights inflammation.
Bromelain *or*	Take 400–500 mg three times per day between meals for six months. Then decrease dosage to 250 mg three times per day between meals.	Bolsters the immune system; breaks up antigen-antibody complexes; fights free-radical formation; fights inflammation.
Trypsin with chymotrypsin *or*	Take 200 mg three times per day between meals for six months. Then decrease dosage to 125 mg three times per day between meals.	Stimulates immune function; breaks up circulating immune complexes; fights free radicals and inflammation.
Serratiopeptidase *or*	Take 5 mg three times per day between meals.	Fights inflammation; treats respiratory conditions; liquefies sputum.
A proteolytic enzyme	Take as indicated on the label.	Stimulates immune function; breaks up circulating immune complexes; fights free radicals and inflammation.

SUPPORTIVE ENZYMES

Enzyme	Suggested Dosage	Actions
Formula Three—See page 38 of Part One for a list of products.	Take as indicated on the label.	Potent antioxidant enzymes.
A digestive enzyme product containing protease, amylase, and lipase enzymes. See the list of digestive enzyme products in Appendix.	Take as indicated on label with meals.	Improves the breakdown and absorption of food nutrients; fights toxic build-up.

ENZYME ABSORPTION SYSTEM ENHANCERS (EASE)

Combination	Suggested Dosage	Actions
Combat from Life Plus *or*	Take as indicated on the label.	Helps support the respiratory system; stimulates immune system function.
Connect-All from Nature's Plus *or*	Take as indicated on the label.	Supports tissue repair.
Inflamzyme from PhytoPharmica *or*	Take as indicated on the label.	Promotes tissue repair; immune system activator.
Kyolic Formula 102 (with enzymes) *or*	Take 2 capsules or tablets with meals two to three times per day.	Strengthens the immune system; helps alleviate stress; helps fight free radicals; helps fight inflammation; improves digestion.
Mucous Dissolver Liquezyme from Enzyme Process Laboratories *or*	Take as indicated on the label.	Dissolves mucus.
Protease Concentrate from Tyler Encapsulations *and*	Take as indicated on the label.	Breaks up fibrin; fights inflammation.
Bromelain Complex from PhytoPharmica *or*	Take as indicated on the label.	Reduces swelling and inflammation.
Sinease from Prevail	Take as indicated on the label.	Decongestant.

ENZYME HELPERS

Nutrient	Suggested Dosage	Actions
Acidophilus and other probiotics	Take as indicated on the label.	Stimulates enzyme activity; improves gastrointestinal function; bolsters immunity.
Adrenal glandular	Take 300 mg per day.	Helps fight stress and adrenal exhaustion; fights inflammation.
Arginine	Take as indicated on the label.	Accelerates healing; enhances thymus activity.
Coenzyme Q_{10}	Take 30–60 mg one to two times per day with meals.	Important to immune system function; has antioxidant activity.
DHEA	Take 25 mg per day.	Supports immune function.
Essential-fatty-acid complex	Take 100–200 mg two to three times per day with meals; or follow the advice of your health-care professional.	Strengthens immunity; helps strengthen cell membranes.
Kyolic Aged Garlic Liquid Extract with Vitamins B_1 and B_{12}	Take 10 drops in water or juice three times per day.	Stimulates immune system function.
Thymus glandular	Take 90 mg per day.	Assists the thymus gland in immune resistance; improves immune response; fights infections.
Vitamin A or carotenoids	Take up to 50,000 IU per day from all sources for no more than one month. Take doses higher than 50,000 IU only under the supervision of a physician. If pregnant, do not take more than 10,000 IU per day.	Potent antioxidants; needed for strong immune system and healthy mucous membranes.

245

Vitamin C	Take 1–2 grams every hour until symptoms improve.	Fights infections, diseases, and allergies; stimulates the adrenal glands; promotes immune system; powerful antioxidant.
Vitamin E	Take 1,200 IU per day from all sources (including the multivitamin complex in the introduction to Part Two).	Acts as an intercellular antioxidant; improves circulation; helps resist diseases at the cellular level; fights headaches, allergies, and sinusitis.

Comments

❑ A number of herbs can help fight allergies, including alfalfa, licorice, pau d'arco, and royal jelly.

❑ Fresh, enzyme-rich fruits and vegetables should comprise 60 percent of your diet. If you can't tolerate raw foods, increase your intake of digestive enzymes, or try light steaming or stir-frying your produce. See dietary recommendations in Part Three.

❑ Avoid milk and other mucus-forming dairy products. These foods can add to congestion in your sinuses and throat.

❑ Exercises are very important for good health. Do some form of exercise five to seven times per week for 30 to 40 minutes. The exercises don't have to be elaborate to work—try fast walking (swing your arms while walking briskly), riding a bicycle, or swimming.

❑ Meditate 15 minutes every day. Void your mind and think of nothing. Allow your body to re-energize. Increase your enzymatic, immune, and antioxidant activity through positive mental attitude.

❑ With hay fever, moving to a different climate and various methods of air filtration can be helpful.

❑ *See also* ALLERGIES.

Headache

Headaches are a very common medical problem—one of the most common. There are many different types of headaches with many different causes. Many headaches are caused by some benign problem of the eyes, ears, nose, teeth, or throat. They can also be caused by fatigue, stress, or depression. Temporomandibular joint disorder (TMJ) may also cause headaches. More serious but less common causes of headaches include hypertension; infections, such as meningitis; injury to the head; a brain abscess or tumor; glaucoma or other eye problem; cerebral hypoxia (lack of oxygen to the brain); and serious illnesses, such as syphilis or cancer. Many headaches are due to unknown causes. Most headaches, however, fall into one of four categories.

Tension headaches (or stress headaches) are the most common type of headache. They are caused by muscle tension in the head, neck, or shoulders, usually due to fatigue or stress. A tension headache is characterized by a steady, moderately severe pain above the eyes or in the back of the head that usually starts in the morning and worsens throughout the day. A feeling of tightness around the neck and scalp, fatigue, and weakness may also accompany the headache.

Migraine headaches are marked by intense, throbbing, recurrent, incapacitating pain that usually affects one side of the head but sometimes affects both sides. The pain is usually accompanied by nausea and vomiting, appetite loss, and irritability. Many experience what is known as an aura or a prodrome before the pain begins. This is often experienced as bright lights seen in one's head, a blind spot, or distortion of images that help warn the sufferer that the headache is coming. Migraines are caused when arteries to the brain constrict. They are more common in women than in men. (See the inset "Migraine" on page 247.)

A *cluster headache* is an extremely painful but rare type of migraine headache. One nostril usually itches and runs just before the pain begins. Shortly afterward, there is sudden and intense pain on that side of the head that spreads around the eye. The headache usually lasts about one or two hours. It may be accompanied by infected conjunctiva in the eyes, nausea, and perspiration. After the headache subsides, the eyelid on the affected side of the head may droop. The attacks come in groups, occurring at the same time or times each day for consecutive days, followed by attack-free weeks or months before the attacks start all over again. Of those affected by cluster headaches, 90 percent are men.

Sinus headaches accompany upper respiratory tract infections. This type of headache is usually located at the front of the head above and below the eyes. It is usually worse in the morning and during cold, damp weather. A fever may accompany a sinus headache.

According to the American Association for the Study of Headache, headache is the seventh leading reason for seeing a physician. Americans spend over $1 billion each year on headache medication. Some 50 million Americans experience severe headaches (marked by recurring or long-lasting head pain). This results in disability, disruption of daily life, and much suffering. Emotionally, head pain can be devastating, causing strained relationships at work and at home.

Enzyme Therapy

Systemic enzyme therapy is used to eliminate the headache, reduce any inflammation, and improve circulation. Enzymes stimulate the immune system, bring nutrients to the affected areas, remove waste products, and enhance wellness.

Digestive enzyme therapy is used to improve digestion of food, reduce stress on the gastrointestinal mucosa, help maintain normal pH levels, detoxify, and promote the growth of healthy intestinal flora. All of these actions make the body healthier so that it can be better able to rid itself of headaches. Digestive enzymes also serve as replacements for the body's pancreatic enzymes, leaving the pancreatic enzymes free to perform other functions in the body, such as decreasing inflammation and improving circulation. Treatments usually vary depending on the cause of the headache. However, whatever the program, enzymes can help.

Enzyme Absorption System Enhancers (EASE) maximize enzyme activity. They can also improve the absorption and bioavailability of various nutrients and decrease the drain on the body's own digestive enzymes, thus prolonging their lives.

In addition to the standard multivitamin, multimineral, and multiglandular complexes recommended in the introduction to Part Two, the following supplements are recommended for the treatment of headaches. The following recommended dosages are for adults. Children between the ages of two and five should take one-quarter the recommended dosage, those between the ages of six and twelve should take one-half the recommended dosage, and those

between the ages of thirteen and seventeen should take three-quarters the recommended adult dosage.

ESSENTIAL ENZYMES

Enzyme	Suggested Dosage	Actions
Formula One (particularly Wobenzym N from Naturally Vitamin Supplements—See page 38 of Part One for a list of products. *or*	Take 10 tablets three times per day between meals. As symptoms subside (which should occur in approximately one week) decrease dosage to 3–5 tablets three times per day between meals.	Decreases inflammation, pain, and swelling; increases rate of healing; stimulates the immune system; strengthens capillary walls.
Inflazyme Forte from American Biologics *or*	Take as indicated on the label between meals.	Decreases inflammation, pain, and swelling; increases rate of healing; stimulates the immune system; strengthens capillary walls.
Bromelain *or*	Take four to six 230–250 mg capsules per day.	Fights inflammation, pain, and swelling; speeds recovery after injury or surgery; stimulates the immune system.

Migraine

A migraine headache is a particularly painful headache. Some 26 million Americans experience migraine attacks; two million of these are children. Migraines occur equally between boys and girls in children, but in adulthood, they are more likely to affect women (75 percent of those affected are female).

A migraine headache is characterized by an intense, throbbing, recurrent pain on one or both sides of the head and is usually accompanied by vomiting, nausea, sensitivity to noise, and difficulty looking at bright lights. The headache may last from a few hours to several days. One to four attacks per month is the average for most migraine sufferers.

There are two types of migraine: the classic migraine (with aura) and the common migraine (without aura). Only about one in five migraine patients suffers a migraine with an aura. Aura refers to a group of symptoms experienced up to an hour before headache pain begins. An aura can last ten to twenty-five minutes and can affect the migraine sufferer in many ways. He or she may see flashing lights, bright spots, or colors and may go blind in the central field of vision or in one half of the visual field. Other aura manifestations include facial numbness on one side, as well as tongue or hand numbness. Sometimes, hallucinations occur with migraines.

Neurochemicals (chemicals in the nervous system) control the body's pain mechanisms. These same chemicals are also involved in sleep, mood changes, and

other biological functions. It seems that in migraine patients, some of these neurochemicals function abnormally. During a migraine attack, the blood vessels in the head dilate and the muscles of the scalp become tender and tight. Researchers aren't sure of what causes this, but biological disorders of the central nervous system (especially the brain), seem to be involved. Regulation of these neurochemicals is a primary goal of treatment.

A variety of factors and events can trigger the onset of migraine, including hormonal changes, stress, eating certain foods, fluctuations in the temperature or weather, and changes in eating or sleeping patterns. However, there are a number of ways to help combat migraines. Proper diet and detoxification are very important. See Part Three for instructions. Eliminate any foods containing phenylethylanines, tyramine, alcohol, caffeine, and monosodium glutamate (all of which may trigger a migraine). Changes in lifestyle, including changing sleep patterns, job environment, and eating habits; exercising regularly; and getting adequate rest, may be necessary. Making changes to your diet and lifestyle may take time, but are extremely cost-effective and necessary for success in the treatment of migraines. Exercise a minimum of thirty to forty minutes per day. This can include fast walking, swimming, or bicycling. Meditate at least 15 minutes per day. See Part Three for instructions. If migraines are frequent, seek the services of a physician or alternative-health practitioner well-trained in the treatment of migraines.

Enzyme complexes rich in proteases plus amylases and lipases working synergistically	Take as indicated on the label between meals.	Decreases inflammation, pain, and swelling; increases rate of healing; stimulates the immune system; strengthens capillary walls.

SUPPORTIVE ENZYMES

Enzyme	Suggested Dosage	Actions
Superoxide dismutase, catalase, and glutathione peroxidase	Take 50 mcg of each three times per day between meals.	Potent antioxidant enzymes.
A digestive enzyme product containing protease, amylase, and lipase enzymes. See the list of digestive enzyme products in Appendix.	Take as indicated on the label.	Improves digestion by breaking down proteins, carbohydrates, and fats.

ENZYME ABSORPTION SYSTEM ENHANCERS (EASE)

Combination	Suggested Dosage	Actions
Biodias A Granules by Amano Pharmaceutical Co., Ltd.	Adults should take one packet in 8 ounces of water between or after meals three times per day. Children ages 11 to 14 should take two-thirds of a packet; children between the ages of 8 and 10 should take half of the packet; and children between the ages of 5 and 7 should take one-third of a packet.	Helps improve digestion of nutrients; relieves inflammation; promotes regeneration of mucous membranes; decreases stress on the stomach and intestines; reduces gastric pain.

ENZYME HELPERS

Nutrient	Suggested Dosage	Actions
Adrenal glandular	Take 100–400 mg per day.	Fights inflammation and adrenal exhaustion.
Aged garlic extract	Take 2 capsules three times per day.	For detoxification and energy.
Bioflavonoids	Take 150–500 mg three times per day.	Strengthen capillaries; potent antioxidants.
Brain glandular	Take as indicated on the label.	Helps improve brain function.
Calcium	Take 1,000 mg per day.	Plays important roles in muscle contraction and nerve transmission.
Coenzyme Q$_{10}$	Take 30–60 mg one to two times per day with meals; or follow the advice of your health-care professional.	Potent antioxidant.
Essential fatty acid complex	Take 75–100 mg two to three times per day	Helps strengthen cell membranes.

	with meals; or follow the advice of your health-care professional.	
Green Barley Extract from Green Foods Corporation or Kyo-Green from Wakunaga	Take as indicated on the label.	Potent free-radical fighters; rich in vitamins, minerals, and enzyme activity.
Magnesium	Take 500 mg per day.	Plays important role in muscle contraction, nerve function, and calcium absorption.
Niacin or Niacin-bound chromium	Take 50 mg the first day, then increase dosage daily by 50 mg to 1,000 mg. If adverse effects occur, decrease dosage. Take doses higher than 1,000 mg only under the supervision of a physician. Take 20 mg per day, or as indicated on the label.	Helps dilate blood vessels, thus improving circulation and oxygen supply to the brain; maintains nervous system and brain.
Vitamin B complex	Take 50 mg three times per day, or as indicated on the label.	B vitamins act as coenzymes; support neuromusculo-skeletal function.
Vitamin C	Take 1,000–10,000 mg per day.	Promotes a healthy immune system, which is helpful in fighting headaches caused by allergies; powerful antioxidant.
Vitamin E	Take 1,200 IU per day from all sources (including the multivitamin complex in the introduction to Part Two).	Potent antioxidant; supplies oxygen to cells; improves level of energy.
Zinc	Take 15–50 mg per day. Do not take more than 100 mg per day.	Needed by over 300 enzymes.

Comments

❏ Tension headaches will resolve as the source of the tension disappears. Relaxation techniques will also help.

❏ Sinus headaches can often be relieved by decongestants, antibiotics, or sinus drainage (with the help of herbal and other remedies).

❏ Heat, massage, medication, biofeedback, meditation (and other relaxation techniques), and sometimes a bite plate will help relieve a TMJ-caused headache.

❏ For further information on foods that help, dietary modifications, detoxification methods (cleansing, fasting, and juicing), exercises, positive mental attitude, meditation, and general suggestions, see Part Three.

❏ If your headaches are frequent or severe, you may want to seek the services of an alternative health-care professional. After conducting a history, physical examination, and X-rays, your chiropractor, naturopath, or osteopath will develop a treatment program. Your treatment program might include chiropractic adjustments; manipulation; massage; or physiotherapy, including electrical stimulation, ice and heat applications, exercises, and postural training.

❏ See the list of Resource Groups in the Appendix for a list of organizations that may provide you with further information about headaches.

Heart Disease

See CARDIOVASCULAR DISEASE.

Heartburn

Heartburn (also called acid stomach) is a burning pain in the stomach that spreads across the chest. It is a result of a disorder called *gastroesophageal reflux disease* (GERD). Gastroesophageal reflux is a common disorder that occurs when the stomach's contents back up into the esophagus. As mentioned in Chapter 2 of Part One, the glands of the stomach produce a number of substances, including hydrochloric acid and various enzymes to digest food. The lining of the stomach produces mucus to protect it from the acid. The esophagus doesn't have this protection, so it can be easily damaged when the acids back up (a process called reflux). With GERD, the esophagus is under constant attack and can be damaged by the stomach's contents. The lungs and voice box (larynx) can also be damaged.

Apparently, part of the esophagus called the LES (the lower esophageal sphincter) fails to work properly in people with GERD. This muscle's job is to open when you swallow, allowing food to travel to the stomach. It is supposed to close if there is no food present. But because the LES malfunctions, the stomach's contents are allowed to back up into the esophagus.

Fatty or spicy foods can worsen heartburn in some sensitive individuals, as can drinking too much alcohol or caffeine or eating while under stress. Gulping down your food and eating too fast can cause you to swallow air—and what goes in must come out. Within the body, the trapped air warms, expands, and produces pressure on the digestive tract, causing belching and an acid stomach. When the acid backs up into the esophagus, you have heartburn.

Symptoms

The primary symptom of heartburn is a burning-type pain in the stomach that may spread up the chest and into the neck or throat. It usually occurs after a meal or when you lie down. There may be some regurgitation of stomach contents. Persistent heartburn can be the sign of a serious problem, such as an ulcer. Frequent regurgitation of stomach contents can cause damage to the delicate tissues of the esophagus.

Enzyme Therapy

Digestive enzyme therapy is used to improve the digestion of food, reduce stress on the gastrointestinal mucosa, help maintain normal pH levels, detoxify the body, and promote the growth of healthy intestinal flora. Systemic enzyme therapy is used as supportive care to reduce any inflammation, stimulate the immune system, improve circulation, help speed tissue repair, enhance wellness, strengthen the body as a whole, and build general resistance.

Enzyme Absorption System Enhancers (EASE) maximize enzyme activity. They can also improve the absorption and bioavailability of various nutrients and decrease the drain on the body's own digestive enzymes, thus prolonging their lives.

In addition to the standard multivitamin, multimineral, and multiglandular complexes recommended in the introduction to Part Two, the following supplements are recommended for the treatment of heartburn. The following recommended dosages are for adults. Children between the ages of two and five should take one-quarter the recommended dosage, those between the ages of six and twelve should take one-half the recommended dosage, and those between the ages of thirteen and seventeen should take three-quarters the recommended adult dosage.

ESSENTIAL ENZYMES

Enzyme	Suggested Dosage	Actions
A digestive enzyme formula containing protease, amylase, and lipase enzymes. See the list of digestive enzyme products in Appendix. *or*	Take as indicated on the label.	Improves digestion by breaking down proteins, carbohydrates, and fats.
Pancreatin *or*	Take as indicated on the label.	Contains proteases, amylases, and lipases.
If you have trouble digesting protein, try a protease, such as bromelain, carboxypeptidase, a microbial protease, papain, pepsin, peptidase, rennin, or trypsin and chymotrypsin. *or*	Take as indicated on the label with meals.	Breaks down proteins.
If you have trouble digesting carbohydrates, try an amylase, such as alpha-galactosidase, beta-glucosidase, carbohydrase, cellulase, diastase, glucoamylase, hemicellulase, invertase, lactase, maltase, phytase, or sucrase. *or*	Take as indicated on the label with meals.	Breaks down carbohydrates.

If you have trouble digesting fats, try an enzyme supplement containing lipase. *or*	Take as indicated on the label with meals.	Break down fats.
Formula One—See page 38 of Part One for a list of products. *or*	Take 5 coated tablets or one tablespoon of a granulated enzyme mixture three times per day between meals.	Fights inflammation and pain; speeds healing; stimulates immune function; fights free-radical and immune complex formation; improves digestion.
If you have trouble swallowing tablets or pills, try chewable enzyme supplements, such as bromelain or papain (papaya).	Take as indicated on the label.	Improves digestion.

SUPPORTIVE ENZYMES

Enzyme	Suggested Dosage	Actions
Formula Three—See page 38 of Part One for a list of products.	Take as indicated on the label.	Potent antioxidant enzymes.

ENZYME ABSORPTION SYSTEM ENHANCERS (EASE)

Combination	Suggested Dosage	Actions
Acid Ease from Prevail *or*	Take as indicated on the label.	Helps maintain proper pH levels in the colon.
Biodias A Granules from Amano Pharmaceutical Co., Ltd. *or*	Adults should take one packet in water between or after meals three times per day. Children ages 11 to 14 should take two-thirds of a packet; children between the ages of 8 and 10 should take half of the packet; and children between the ages of 5 and 7 should take one-third of a packet.	Helps improve digestion of nutrients; relieves inflammation; promotes regeneration of mucous membranes; decreases stress on the stomach and intestines; reduces gastric pain.
Kyolic Formula 102 (with enzymes)	Take 2 capsules or tablets with meals two to three times per day.	Improves digestion.

ENZYME HELPERS

Nutrient	Suggested Dosage	Actions
Acidophilus or other probiotics, such as bifidobacteria or Acidophilasé from Wakunaga, which contains amylolytic, lipolytic, and proteolytic enzymes	Take as indicated on the label.	Improves gastrointestinal function; needed for production of biotin, niacin, folic acid, and pyridoxine, important coenzymes; stimulates enzyme activity.
Fiber	Take as indicated on the label.	Provides bulk; needed for proper elimination.
Fructooligo-saccharides	Take as indicated on the label.	Increases peristaltic action of intestines; improves liver function.
Lecithin	Take 1,000 mg three times per day with meals.	Emulsifies fat.
Ox bile extract	Take 120 mg per day.	Aids digestion.
Vitamin A	Take 5,000 IU per day.	Needed by digestive glands to produce enzymes; required for healthy mucous linings, such as those in the stomach.
Vitamin B complex	Take 50 mg three times per day.	B vitamins act as coenzymes.
Vitamin C (use a buffered form, such as Ester-C)	Take 1,000–2,000 mg per day.	Stimulates the adrenal glands to produce hormones (which also act as coenzymes); aids in detoxification; is a potent antioxidant; improves the function of many enzymes.
Whey	Take 1 heaping tablespoon three times per day with meals.	Improves digestion.
Zinc	Take 10–50 mg per day. Do not take more than 100 mg per day from all sources.	Needed by over 300 enzymes. Many zinc enzymes work to metabolize carbohydrates, alcohol, and essential fatty acids; synthesize proteins.

Comments

❏ Eat yogurt and other foods rich in *Lactobacillus acidophilus* or other healthy bacteria daily.

❏ Eat enzyme-rich foods, including papayas, which improve the normal digestive process.

❏ Foods rich in chlorophyll, including chlorella, spirulina, alfalfa, and green foods such as Green Magma, Green Kamut, Kyo-Green, or Green Barley Extract, can help improve digestion.

❏ A number of herbs can encourage proper digestion, including alfalfa concentrate, ginger root, and licorice root.

❏ Limit alcohol and caffeine intake. Caffeine is found in coffee, tea, cola drinks, chocolate, certain cold remedies, weight-control aids, and pain relievers.

❏ Chew foods slowly and never eat when tense or hurried.

❏ The Enzyme Kidney Flush and the Enzyme Toxin Flush can help cleanse the body and restore a normally functioning stomach. See Part Three for instructions.

❏ A number of herbs, including ginger, chamomile, and licorice, can help alleviate heartburn when drunk as a soothing tea.

❏ Avoid foods that increase stomach acid, including citrus fruits and juices, chocolate, caffeine, tomato products of all kinds, foods high in fat, and spicy foods.

❏ Avoid smoking cigarettes and other tobacco products.

❏ Don't overeat, and if you are overweight, go on a weight-loss program. Overeating and obesity place extra pressure on the abdomen, pushing the stomach's contents into the esophagus. It is for this same reason that you should avoid wearing tight clothing (such as a girdle).

❏ Don't do any lifting, bending, or stooping after eating.

❏ Don't eat or drink within three hours of bedtime. An empty stomach means there will be less material to flow back into the esophagus.

❏ Some medicines can exacerbate heartburn, including aspirin.

❏ Elevating the head of your bed six to eight inches can help keep gastric acids in your stomach. The easiest way is to place blocks under the legs at the head of the bed or to place a foam wedge in between the mattress and box springs.

❏ *See also* GAS, ABDOMINAL; INDIGESTION.

Hematomas

See BRUISES AND HEMATOMAS.

Hemorrhoids

Hemorrhoids (piles) are varicose (or dilated) veins of the rectum. Those that are located inside the anal canal are called internal hemorrhoids, and those that protrude outside the anus are called external hemorrhoids. Hemorrhoids are caused by pressure on the veins in the anus or rectum, such as constant straining during bowel movements, prolonged sitting or standing, or liver disease that increases blood pressure in the portal vein (a main blood vessel to the liver). During pregnancy, hemorrhoids often develop because the fetus pushes against the abdominal organs, putting pressure on the rectum. Pressure may also be caused by a tumor or large cyst in the colon.

Symptoms

Hemorrhoids may bleed, particularly during bowel movements, often turning the toilet water red. The hemorrhoid may protrude, causing discomfort and itching. If a blood clot forms in the hemorrhoid or it is irritated, it may swell and hurt.

Note: Cancer of the rectum can also cause bright red bleeding from the rectum, but most cases of rectal bleeding are caused by hemorrhoids. A doctor should evaluate any bleeding from the rectum.

Enzyme Therapy

Systemic enzyme therapy is used to reduce swelling and inflammation, improve circulation, and help speed tissue repair. Enzymes stimulate the immune system, bring nutrients to the damaged area, remove waste products, and enhance wellness.

Digestive enzyme therapy is used to improve the digestion and assimilation of food nutrients, reduce stress in the rectum and gastrointestinal mucosa, help maintain normal pH levels, detoxify the body, promote the growth of healthy intestinal flora, and keep the stool soft for easy elimination.

Enzyme Absorption System Enhancers (EASE) maximize enzyme activity. They can also improve the absorption and bioavailability of various nutrients and decrease the drain on the body's own digestive enzymes, thus prolonging their lives.

In addition to the standard multivitamin, multimineral, and multiglandular complexes recommended in the introduction to Part Two, the following supplements are recommended for the treatment of hemorrhoids. The following recommended dosages are for adults. Children between the ages of two and five should take one-quarter the recommended dosage, those between the ages of six and twelve should take one-half the recommended dosage, and those between the ages of thirteen and seventeen should take three-quarters the recommended adult dosage.

ESSENTIAL ENZYMES

Enzyme	Suggested Dosage	Actions
A digestive enzyme combination including protease, amylase, and lipase enzymes. For a list of digestive enzyme products, see Appendix. *or*	Take as indicated on the label.	Improves digestion and elimination.
Pancreatin	Take as indicated on the label.	Contains proteases, amylases, and lipases.

SUPPORTIVE ENZYMES

Enzyme	Suggested Dosage	Actions
Bromelain	Take four to six 230–250 mg capsules per day.	Fights inflammation, pain, and swelling; speeds recovery after injury or surgery; stimulates the immune system.
Formula One—See page 38 of Part One for a list of products.	Take 2–3 tablets three times per day between meals.	Fights inflammation; stimulates immune function; fights free radicals.
Formula Three—See page 38 of Part One for a list of products.	Take as indicated on the label.	Potent antioxidant enzymes.

251

ENZYME ABSORPTION SYSTEM ENHANCERS (EASE)

Combination	Suggested Dosage	Actions
Bromelain Complex from PhytoPharmica *or*	Take as indicated on the label.	Reduces swelling and inflammation; improves circulation.
Bromelain Plus from Enzymatic Therapy *or*	Take as indicated on the label.	Decreases swelling; improves circulation; fights inflammation.
Detox Enzyme Formula from Prevail *or*	Take as indicated on the label.	Helps rid the body of harmful toxins.
Protease Concentrate from Tyler Encapsulations	Take as indicated on the label.	Breaks up fibrin; fights inflammation.

ENZYME HELPERS

Nutrient	Suggested Dosage	Actions
Acidophilus or other probiotics, such as bifidobacterium or Acidophilasé from Wakunaga, which contains amylolytic, lipolytic, and proteolytic enzymes	Take as indicated on the label.	Improves gastrointestinal function; needed for production of biotin, niacin, folic acid, and pyridoxine, which serve as coenzymes; stimulates enzyme activity.
Aged garlic extract	Take 2 capsules three times per day.	Potent antioxidant; improves digestion; stimulates healthy intestinal bacteria.
Aloe vera extract	Take as indicated on the label.	Relieves pain; excellent colon cleanser; effective laxative; treats hemorrhoids.
Amino-acid complex	Take as indicated on the label.	Amino acids serve as building blocks for the body's enzymes.
Betaine hydrochloric acid	Take 150 mg three times per day with meals.	Improves digestion and elimination. If gastric upset occurs, discontinue use.
Bioflavonoids, such as rutin or quercetin	Take 150 mg three times per day.	Strengthen capillaries.
Fiber	Take as indicated on the label.	Provides bulk; needed for proper elimination.
Fructooligo-saccharides	Take as indicated on the label.	Increases peristaltic action of intestines; improves liver function.
Green food product, such as Green Kamut, Kyo-Green, or Green Barley Extract	Take as indicated on the label.	Green foods are rich in enzymes, vitamins, minerals, chlorophyll, and amino acids.
Lecithin	Take 1,000 mg three times per day with meals.	Emulsifies fat.
Ox bile extract	Take 120 mg per day.	Aids digestion.
Spirulina	Take as indicated on the label.	Rich in minerals, vitamins, enzymes, proteins, essential amino acids, and essential fatty acids; good source of beta-carotene which is helpful in maintaining healthy tissues.
Vitamin A	Take 25,000 IU per day. If pregnant, do not take more than 10,000 IU per day.	Needed by digestive glands to produce enzymes; required for healthy mucous linings; improves digestion and elimination.
Vitamin B complex	Take 50 mg three times per day.	B vitamins act as potent coenzymes.
Vitamin C	Take 5,000–10,000 mg per day.	Stimulates the adrenal glands to produce hormones, which also act as coenzymes; aids in detoxification; is a potent antioxidant; improves the function of many enzymes.
Zinc	Take 10–50 mg per day.	Needed by over 300 enzymes. Many zinc enzymes work to improve digestion and metabolize carbohydrates and essential fatty acids, synthesize proteins, and dispose of free radicals.

Comments

❏ Fresh, enzyme-rich fruits and vegetables (particularly green vegetables) should comprise 60 percent of your diet. If your body cannot tolerate raw fruits and vegetable, increase your intake of digestive enzymes, or sauté or steam your produce.

❏ Increase your intake of fiber-rich whole grains and nuts in all forms.

❏ Use psyllium seed or stool softeners to relieve straining and constipation.

❏ Drink a minimum of eight 8-ounce glasses of water (distilled or bottled) per day.

❏ Essential detoxifying tools are the Enzyme Toxin Flush and the Enzyme Kidney Flush. Follow directions in Part Three.

❏ Topical treatments can help relieve the discomfort of hemorrhoids:

• Apply an enzyme salve twice daily. Proteolytic enzymes decrease inflammation and swelling.

• Applying very cold water or a cold cloth directly to the anal area for a few minutes can help relieve pain and discomfort.

❏ Exercise (brisk walking, swimming, or bicycling) at least five days per week for thirty to forty minutes. This can help improve digestion and elimination, decreasing pressure on the rectum.

❏ See the list of Resource Groups in the Appendix for an organization that can provide you with more information about hemorrhoids.

Hepatitis

Hepatitis is any inflammation in the liver. According to the American Liver Foundation, hepatitis is the most common serious contagious disease of the liver. Every year, some 70,000 cases are reported to the Centers for Disease Control and Prevention. However, as large as this number is, it represents only a fraction of the cases occurring annually in this country.

Hepatitis is usually caused by a virus, particularly a hepatitis virus (hepatitis A, B, C, D, or E—see the inset "Hepatitis Viruses" on page 254). Less commonly, it may be caused by such viruses as yellow fever, cytomegalovirus, and infectious mononucleosis. Nonviral hepatitis is usually caused by alcohol use or certain drugs. Hepatitis can be either acute or chronic.

Acute hepatitis is always caused by one of the five hepatitis viruses. The inflammation and its symptoms begin suddenly and last only a few weeks. Symptoms include flulike symptoms, such as malaise, fever, fatigue, nausea, vomiting, and loss of appetite. Often smokers experience a distaste for cigarettes. Several days later, jaundice (yellow eyes and skin) develops, the urine becomes very dark, and the stools become very pale. At this point, the other symptoms disappear as the jaundice gets worse. Acute viral hepatitis usually lasts from four to eight weeks, even without treatment.

Chronic hepatitis is any liver inflammation that lasts for more than six months. Though it is usually rather mild and produces no symptoms or damage, it may last for years. Sometimes, however, the liver is damaged by chronic hepatitis, leading to liver failure. Chronic hepatitis may be caused by viral infection (often hepatitis C); such drugs as isoniazid, methyldopa, nitrofurantoin, and sometimes even prolonged use of acetominophen; alcohol abuse; Wilson's disease; and unknown causes. The hepatitis A and E viruses do not cause chronic hepatitis. Many with chronic hepatitis experience no symptoms at all. Those who do usually experience symptoms similar to those of acute viral hepatitis, including flu-like symptoms and upper abdominal discomfort. Jaundice sometimes develops but not always.

Enzyme Therapy

Systemic enzyme therapy is used to reduce inflammation, swelling, and pain in the liver; stimulate the immune system; and improve circulation. Enzymes help speed tissue repair, bring nutrients to the damaged area, remove waste products, and enhance wellness.

Digestive enzyme therapy is used to improve digestion of food, reduce stress on the gastrointestinal mucosa, help maintain normal pH levels, detoxify, and promote the growth of healthy intestinal flora. Digestive enzymes also serve as replacements for the body's pancreatic enzymes, leaving the pancreatic enzymes free to perform other functions in the body, such as boosting immunity and decreasing inflammation.

Enzyme Absorption System Enhancers (EASE) maximize enzyme activity. They can also improve the absorption and bioavailability of various nutrients and decrease the drain on the body's own digestive enzymes, thus prolonging their lives.

In addition to the standard multivitamin, multimineral, and multiglandular complexes recommended in the introduction to Part Two, the following supplements are recommended for the treatment of hepatitis. The following recommended dosages are for adults. Children between the ages of two and five should take one-quarter the recommended dosage, those between the ages of six and twelve should take one-half the recommended dosage, and those between the ages of thirteen and seventeen should take three-quarters the recommended adult dosage.

ESSENTIAL ENZYMES

Enzyme	Suggested Dosage	Actions
Formula One—See page 38 of Part One for a list of products. *or*	Take 10 tablets three times per day between meals for six months. Then decrease dosage to 5 tablets three times per day between meals.	Bolsters the immune system; breaks up antigen-antibody complexes; fights free-radical formation; fights inflammation.
Papain *or*	Take 500 mg three times per day for six months between meals. Then decrease dosage to 300 mg three times per day between meals.	Stimulates immune function; breaks up circulating immune complexes; stimulates immune function at the cellular level; fights inflammation.
Bromelain *or*	Take three to four 230–250 mg capsules per day until symptoms subside. If severe, take 6–8 capsules per day.	Fights inflammation, pain, and swelling; speeds recovery after injury or surgery; stimulates the immune system.
Trypsin with chymotrypsin *or*	Take 200 mg of trypsin and 10 mg of chymotrypsin three times per day between meals for six months. Then decrease dosage to 125 mg of trypsin and 5 mg of chymotrypsin three times per day between meals.	Stimulates immune function; breaks up circulating immune complexes; stimulates immune function at the cellular level; fights inflammation.
Serratiopeptidase *or*	Take 5 mg three times per day between meals.	Fights inflammation; treats acute and chronic conditions.
Any proteolytic enzyme	Take 10 tablets three times per day for six months. Then decrease dosage to 5 tablets three times per day.	Stimulates immune function; breaks up any circulating immune complexes; stimulates immune function at the cellular level; fights inflammation.

SUPPORTIVE ENZYMES

Enzyme	Suggested Dose	Rationale
Formula Three—See page 38 of Part One for a list of products.	Take as indicated on the label.	Potent antioxidant enzymes.
A digestive enzyme product containing protease, amylase, and lipase enzymes. See the	Take as indicated on the label with meals.	Improves the breakdown and absorption of food nutrients; fights toxic build-up.

Hepatitis Viruses

There are many viruses that can cause hepatitis. Understanding the various ways that these viruses can be transmitted can help you avoid contracting the disease. Knowing the symptoms each virus produces will help you recognize the symptoms if you do contract hepatitis.

Hepatitis A virus (formerly called infectious hepatitis) is seen frequently in adults in the Western world but is most common in children of developing countries. It is spread primarily through contaminated feces—the virus passes from the stool of an infected person to the mouth of another person—usually due to poor hygiene. It can be spread through direct contact with the feces, or indirectly through water, food, utensils, or raw shellfish. Most hepatitis A infections cause no symptoms, and they rarely become chronic.

Hepatitis B virus (formerly called serum hepatitis) is the most common hepatitis virus, with an estimated 1.2 million carriers in the United States and 300 million carriers worldwide. It is also one of the most serious

hepatitis viruses, particularly in the elderly. It is sometimes fatal. It can be spread from mother to child at birth, or through blood transfusions, contaminated needles, or sexual contact. It rarely becomes chronic.

Hepatitis C virus (formerly called non-A, non-B hepatitis) is spread through contaminated needles or blood transfusions. In America, more than 3.5 million people carry this virus. While sexual transmission and transmission at birth of the hepatitis C virus are quite uncommon, they may occur. Hepatitis C infections often become chronic. They may also lead to cirrhosis and liver cancer.

Hepatitis D (formerly called delta hepatitis) is found primarily in intravenous drug users. It occurs only in those who are infected with hepatitis B virus, and it increases the severity of a hepatitis B infection.

Hepatitis E virus (formerly called enteric or epidemic non-A, non-B hepatitis) is found in underdeveloped countries. It is spread in the same manner as hepatitis A virus and produces similar symptoms.

---

list of digestive enzyme products in Appendix.

ENZYME ABSORPTION SYSTEM ENHANCERS (EASE)

Combination	Suggested Dosage	Actions
Ecology Pak from Life Plus or	Take 2 or 3 tablets once or twice per day, preferably in the morning and no later than midday.	Assists with natural detoxification; stimulates immune function.
Hepazyme from Enzymatic Therapy or	Take as indicated on the label.	Stimulates immune function.
Inflamzyme from PhytoPharmica or	Take as indicated on the label.	Immune system activator; reduces inflammation; fights free radicals.
Kyolic Formula 102 (with enzymes) or	Take 2 capsules or tablets with meals two to three times per day.	Strengthens the immune system; helps alleviate stress; helps fight free radicals; helps fight inflammation.
Metabolic Liver Formula from Prevail	Take as indicated on the label.	Improves liver function.

ENZYME HELPERS

Nutrient	Suggested Dosage	Actions
Acidophilus and other probiotics	Take as indicated on the label.	Stimulates enzyme activity; improves gastrointestinal function.
Bioflavonoids, such as rutin and quercetin	Take as indicated on the label.	Strengthen capillaries.
Coenzyme Q_{10}	Take 30–60 mg one to two times per day with meals.	Important to immune system function; has antioxidant activity.
Essential fatty acid complex	Take 100–200 mg two to three times per day with meals; or follow the advice of your health-care professional.	Strengthens immunity; helps strengthen cell membranes.
Kyolic Aged Garlic Liquid Extract with Vitamins B_1 and B_{12}	Take 10 drops in water or juice three times per day.	Stimulates immune system function.
Liver glandular	Take 50–100 mg per day.	Improves immune defense and liver function.
Thymus glandular	Take 100 mg per day, or as indicated on the label.	Assists the thymus gland in immune resistance; improves immune response; fights infections.
Vitamin A	Take up to 50,000 IU for no more than one month (for higher doses, see a physician). If pregnant, do not take more than 10,000 IU.	Needed for strong immune system and healthy tissues; potent antioxidant.
Vitamin B complex	Take 100–150 mg three times per day.	B vitamins act as coenzymes.
Vitamin C	Take 10,000–15,000 mg per day; or 1–2 grams every hour until	Fights infections, diseases, and allergies; improves wound healing; stimulates the adrenal

	symptoms improve.	glands; promotes immune system; powerful antioxidant.
Vitamin E	Take 1,200 IU per day from all sources (including the multivitamin complex in the introduction to Part Two).	Acts as an intercellular antioxidant; improves circulation; helps resist diseases at the cellular level.
Zinc	Take 50 mg per day. Do not take more than 100 mg per day.	Needed by over 300 enzymes; necessary for white blood cell immune function; improves healing.

Comments

❏ A number of herbs can stimulate immune function, including astragalus, echinacea, ginseng, licorice, and maitake, reishi, and shiitake mushrooms.

❏ Silymarin (extract of milk thistle weed) helps to repair the liver and is used to treat chronic liver necrosis, cirrhosis, and hepatitis A and B. It also decreases fat deposits.

❏ Raw, enzyme-rich fruits and vegetables (particularly green vegetables) should comprise 60 percent of your diet. If your body cannot tolerate raw fruits and vegetables, sauté or steam them.

❏ To improve overall health, avoid alcohol, white flour, and refined sugar, all of which produce undesirable fermentation in the bowel that can lead to a build-up of toxins.

❏ Avoid chlorinated water. Use distilled or bottled water instead.

❏ Detoxification is particularly important in hepatitis. Follow the Enzyme Kidney Flush program every day for one week. The following week, perform the Enzyme Toxin Flush. See guidelines for both flushes in Part Three.

❏ See the list of Resource Groups in the Appendix for a list of organizations that can provide you with more information about hepatitis.

Herniated Disc

See SLIPPED DISC.

Herpes Simplex Virus

A herpes simplex infection causes outbreaks of clusters of painful blisters on the skin or mucous membranes. Once the eruption subsides, the virus remains dormant in the body. It is periodically reactivated (by such factors as exposure to the sun, fever, physical or emotional stress, immune system suppression, certain foods or drugs, or unknown factors) causing outbreaks to recur, often in the same location.

There are two types of herpes simplex viruses that infect the skin—HSV-1 and HSV-2. HSV-1 usually affects the mouth and, less commonly, the corneas. When it affects the mouth, it produces what are known as cold sores or fever blisters. When it affects the cornea, the infection is referred to as *herpes simplex keratitis*, or *corneal herpes simplex*. HSV-2 usually affects the genital area and is known as *genital herpes*.

The herpes simplex viruses cause an initial tingling, itching, or burning sensation. This feeling is followed by the eruption of small, painful, fluid-filled blisters, which appear anywhere from a few hours to a few days after the tingling begins. After a few days, the blisters may break and begin to dry, forming a yellowish crust and painful shallow ulcers. They usually begin to heal about one or two weeks after their appearance, and are usually totally healed in about three weeks. Some scarring may remain.

Corneal herpes simplex virus may resemble a bacterial infection, causing red, painful, watery eyes that are sensitive to light. Repeated outbreaks may cause damage to the cornea.

The initial outbreak of either virus in whatever region of the body it infects is usually the most severe. Infants often contract oral herpes simplex from adults with cold sores. This first infection causes extensive gum inflammation and mouth soreness along with possible fever and swollen lymph nodes in the neck. Those who do not first contract the virus until adulthood usually have more severe symptoms. Subsequent outbreaks are usually much less severe and produce cold sores. In addition to the previously mentioned factors that can trigger an eruption, oral herpes flare-ups may be triggered by mouth injury or dental work.

Genital herpes is sexually transmitted. Outbreaks may make urination and walking painful. The lymph nodes in the groin area enlarge. The first outbreak is the worst—it lasts longer, is more widespread, and more painful than subsequent outbreaks. It may also cause fever and malaise.

Enzyme Therapy

Regardless of the type of herpes, systemic enzyme therapy is used to reduce the inflammation associated with herpes and to fight the viral infection. Enzymes stimulate the immune system, improve circulation, speed tissue repair, bring nutrients to the damaged area, remove waste products, enhance wellness, strengthen the body as a whole, and build general resistance.

Digestive enzyme therapy is used to improve digestion of food, reduce stress on the gastrointestinal mucosa, help maintain normal pH levels, detoxify the body, and promote the growth of healthy intestinal flora. In this way, the immune system is strengthened to help fight off herpes simplex infections. Digestive enzymes also serve as replacements for the body's pancreatic enzymes, leaving the pancreatic enzymes free to perform other functions in the body, such as boosting immunity and decreasing inflammation.

Enzyme Absorption System Enhancers (EASE) maximize enzyme activity. They can also improve the absorption and bioavailability of various nutrients and decrease the drain on the body's own digestive enzymes, thus prolonging their lives.

Topical enzyme creams are used to dissolve dead and damaged skin cells and exfoliate the skin while leaving healthy skin cells intact. These creams or salves usually include proteolytic enzymes. The plant enzymes bromelain and papain are used most frequently; however, microbial proteases can be quite helpful. In Germany, a combined enzyme product containing bromelain, papain, trypsin, and chymotrypsin is used. The most effective treatment combines topical enzyme creams or salves with enzymes taken orally.

In addition to the standard multivitamin, multimineral, and multiglandular complexes recommended in the introduction to Part Two, the following supplements are recommended for the treatment of herpes simplex virus infections. The following recommended dosages are for adults. Children between the ages of two and five should take one-quarter the recommended dosage, those between the ages of six and twelve should take one-half the recommended dosage, and those between the ages of thirteen and seventeen should take three-quarters the recommended adult dosage.

ESSENTIAL ENZYMES

Enzyme	Suggested Dosage	Actions
WobeMugos E Ointment from Mucos	Apply salve directly to the affected areas of the skin as often as possible.	Caution: use externally only. Do not apply enzyme ointment to open wounds or to mucous membranes.

AND SELECT ONE OF THE FOLLOWING:

WobeMugos E from Mucos	Take 5 tablets three times per day.	Bolsters immunity; breaks up immune complexes; fights inflammation and free radicals.
Formula One—See page 38 of Part One for a list of products.	Take 10 tablets three times per day between meals for six months. Then decrease dosage to 5 tablets three times per day between meals.	Bolsters the immune system; breaks up antigen-antibody complexes; fights free-radical formation; fights inflammation.
Bromelain	Take 400 mg three times per day between meals for six months. Then decrease dosage to 250 mg three times per day between meals.	Bolsters the immune system; breaks up antigen-antibody complexes; fights free-radical formation; fights inflammation.
A proteolytic enzyme	Take 10 tablets three times per day for six months. Then decrease dosage to 5 tablets three times per day.	Stimulates immune function; breaks up circulating immune complexes; fights free radicals and inflammation.

SUPPORTIVE ENZYMES

Enzyme	Suggested Dosage	Actions
Formula Three—See page 38 of Part One for a list of products.	Take as indicated on the label.	Potent antioxidant enzymes.
A digestive enzyme product containing protease, amylase, and lipase. See the list of digestive enzyme products in Appendix.	Take as indicated on the label with meals.	Improves the breakdown and absorption of food nutrients; fights toxic build-up.

ENZYME ABSORPTION SYSTEM ENHANCERS (EASE)

Combination	Suggested Dosage	Actions
Defense Formula from Prevail *or*	Take as indicated on the label.	Enhances the body's natural resistance to viral infections.
Ecology Pak from Life Plus *or*	Take 2 or 3 tablets once or twice per day, preferably in the morning and no later than midday.	Assists with natural detoxification; stimulates immune function.
Hepazyme from Enzymatic Therapy *or*	Take as indicated on the label.	Stimulates immune function.
Inflamzyme from PhytoPharmica *or*	Take as indicated on the label.	Immune system activator; reduces inflammation; fights free radicals.
Kyolic Formula 102 (with enzymes) *or*	Take 2 capsules or tablets with meals two to three times per day.	Strengthens the immune system; helps alleviate stress; helps fight free radicals; helps fight inflammation.
Vita-C-1000 from Life Plus	Take as indicated on the label.	Boosts immune function; fights infection.

ENZYME HELPERS

Nutrient	Suggested Dosage	Actions
Acidophilus and other probiotics	Take as indicated on the label.	Stimulates enzyme activity; improves the body's natural immunity.
Adrenal glandular	Take 100–400 mg per day.	Helps fight stress and adrenal exhaustion; fights inflammation.
Bioflavonoids, such as rutin and quercetin	Take 500 mg three times per day.	Fight viral infections; strengthen capillaries.
Coenzyme Q_{10}	Take 30–60 mg one to two times per day with meals.	Important to immune system function; has antioxidant activity.
Essential fatty acid complex	Take 100–200 mg two to three times per day with meals; or follow the advice of your health-care professional.	Strengthens immunity; helps strengthen cell membranes.
Kyolic Aged Garlic Liquid Extract with Vitamins B_1 and B_{12}	Take 10 drops three times per day.	Stimulates immune system function.

Lysine	Take as indicated on the label.	Fights herpes; aids tissue repair; plays a role in enzyme, hormone, and antibody production.
Thymus glandular	Take 100 mg per day, or as indicated on the label.	Assists the thymus gland in immune resistance; improves immune response; fights infections.
Vitamin A *or*	Take up to 50,000 IU for no more than one month (for higher doses, see a physician). If pregnant, do not take more than 10,000 IU.	Needed for strong immune system and healthy mucous membranes; potent antioxidant.
Carotenoids	Take 25,000 IU per day.	Potent antioxidants; block tissue damage.
Vitamin B complex	Take 100–150 mg three times per day.	B vitamins act as coenzymes.
Vitamin C	Take 10,000–15,000 mg per day; or 1–2 grams every hour until symptoms improve.	Fights infections, diseases, and allergies; improves wound healing; stimulates the adrenal glands; promotes immune system; powerful antioxidant.
Zinc	Take 50 mg per day. Do not take more than 100 mg per day from all sources.	Needed by over 300 enzymes; necessary for white blood cell immune function; improves healing.

Comments

❏ Prevention, by avoiding contact with the infectious blisters of anyone, is the best "cure."

❏ Eat foods high in the amino acid lysine (such as corn, kidney beans, and split green peas), and avoid foods rich in arginine (chocolate, seeds, and nuts), which seems to encourage the virus.

❏ Although there is no cure for herpes, treatment can help alleviate the pain and improve resistance to the virus. Enzymes and enzyme combinations should be used as soon as the tingling begins. This will greatly reduce the duration of the cold sores.

❏ Physicians sometimes use enzyme injections or enzyme retention enemas in the treatment of herpes. Both methods allow a greater concentration of the enzymes to be absorbed than might be accomplished if the enzymes were taken orally. An enzyme retention enema is a means of instilling some enzymes into the colon to be retained. This allows the enzymes to be absorbed into the body's systems. Physicians use products specially formulated for these uses.

❏ See the list of Resource Groups in the Appendix for an organization that can provide you with more information about herpes simplex.

❏ *See also* HERPES ZOSTER.

Herpes Zoster

Herpes zoster (also called shingles) is an infection of the nervous system caused by the varicella-zoster virus, the same virus that causes chickenpox. Although shingles can occur at any age, it is most common after the age of 50. It is estimated that up to 90 percent of the population of Europe and North America is already infected with at least one of the six basic forms of herpes virus, according to Professor Dr. Heinrich Wrba and Dr. Otto Pecher (*Enzyme.* Vienna: Verlag, 1993). These viruses remain in one's body for the rest of one's life and usually become active due to some physical, emotional, or biochemical stress.

After the primary infection of chickenpox heals, the varicella-zoster virus remains dormant in the spinal or cranial nerve cells. If the body becomes weakened because of severe disease, drugs, or stress, the virus may be reactivated. Most of the time the reason for reactivation is unknown. When the virus reactivates, it manifests as shingles. Once one has an attack of shingles, he or she should have a lifelong immunity to it.

Further, the body's antibodies recognize the antigens as foreign and multiply rapidly in an effort to attack and eliminate the virus. The antibodies join with the antigens (the virus) to form immune complexes. If the body is weakened because of stress or illness, it will not be able to completely degrade these immune complexes. In shingles, the undegraded immune complexes tend to settle on the nerve cells already altered by the virus. This causes inflammation, pain, nerve cell damage, and the typical symptoms of shingles.

Symptoms

Herpes zoster symptoms may begin with fever, chills, gastrointestinal disturbances, and general malaise. After three or four days, crops of small blisters (vesicles) appear on the skin. There is severe pain and reddening of the skin along the course of the affected nerves. Lymph nodes may be swollen and the blisters may itch or burn.

Should this disease begin in the intercostal nerves (those between the ribs), the chest, the back, and/or the extremities may be affected. This is the most common affected area and occurs in about 50 percent of the herpes zoster cases. If the virus affects the auditory or facial nerves, it can result in headaches, earaches, and eventually facial nerve paralysis, as well as disorders of hearing and equilibrium. If it affects the optic nerve, it can cause inflammation, conjunctivitis, and even blindness.

The blisters begin to dry up and scab over about five days after they appear. However, until this point, the blisters are quite infectious and can cause chickenpox in susceptible people. Most people recover after two or three weeks, but the pain along the nerve may persist for months or even years after the rash disappears. This chronic, debilitating pain is called *post-zoster neuralgia*.

Enzyme Therapy

Systemic enzyme therapy is used to reduce inflammation, stimulate the immune system, break down circulating immune complexes, improve circulation, help speed tissue repair, bring nutrients to the damaged area, remove waste products, improve health, strengthen the body as a whole, and build general resistance.

Digestive enzyme therapy is used to improve the digestion of food, reduce stress on the gastrointestinal mucosa, help maintain normal pH levels, detoxify, and promote the growth of healthy intestinal flora, thereby making the body healthier so that it can fight off the herpes zoster virus. Digestive enzymes also serve as replacements for the body's pancreatic enzymes, leaving the pancreatic enzymes free to perform other functions in the body, such as boosting immunity and decreasing inflammation.

Enzyme Absorption System Enhancers (EASE) maximize enzyme activity. They can also improve the absorption and bioavailability of various nutrients and decrease the drain on the body's own digestive enzymes, thus prolonging their lives.

In 1964, Dr. R. Dorrer (senior physician of the hospital in Prien am Chiemsee, Germany), was the first therapist to use enzyme combinations in the treatment of herpes zoster patients. Dr. Dorrer found that with enzyme therapy, pain decreased within three days and the vesicles became encrusted much faster than normal. When enzyme therapy is started immediately after the first vesicles appear, the possibility of developing post-zoster neuralgia can be greatly reduced.

Note: Systemic enzyme therapy should be initiated immediately after the initial signs of herpes zoster appear and continued for a few days after the symptoms have disappeared.

In addition to the standard multivitamin, multimineral, and multiglandular complexes recommended in the introduction to Part Two, the following supplements are recommended for the treatment of shingles. The following recommended dosages are for adults. Children between the ages of two and five should take one-quarter the recommended dosage, those between the ages of six and twelve should take one-half the recommended dosage, and those between the ages of thirteen and seventeen should take three-quarters the recommended adult dosage.

ESSENTIAL ENZYMES

Enzyme	Suggested Dosage	Actions
WobeMugos E from Mucos	Take 5 tablets three times per day.	Stimulates immune function; breaks up circulating immune complexes; fights free radicals and inflammation.
or		
Formula One—See page 38 of Part One for a list of products.	Take 10 tablets three times per day between meals for six months. Then decrease dosage to 5 tablets three times per day between meals.	Stimulates immune function; breaks up circulating immune complexes; fights free radicals and inflammation.
or		
Bromelain	Take 400 mg three times per day between meals for six months. Then decrease dosage to 250 mg three times per day between meals.	Stimulates immune function; breaks up circulating immune complexes; fights free radicals and inflammation.
or		
Pancreatin	Take 500 mg six times per day on an empty stomach.	Fights inflammation, pain, and swelling; increases rate of healing; stimulates the immune system; strengthens capillary walls.

SUPPORTIVE ENZYMES

Enzyme	Suggested Dosage	Actions
Formula Three—See page 38 of Part One for a list of products.	Take as indicated on the label.	Potent antioxidant.
A digestive enzyme product containing protease, amylase, and lipase. See the list of digestive enzyme products in Appendix.	Take as indicated on the label with meals.	Improves the breakdown and absorption of food nutrients; fights toxic build-up.

ENZYME ABSORPTION SYSTEM ENHANCERS (EASE)

Combination	Suggested Dosage	Actions
Defense Formula from Prevail	Take as indicated on the label.	Enhances the body's natural resistance to viral infections.
or		
Ecology Pak from Life Plus	Take 2 or 3 tablets once or twice per day, preferably in the morning and no later than midday.	Assists with natural detoxification; stimulates immune function.
or		
Hepazyme from Enzymatic Therapy	Take as indicated on the label.	Stimulates immune function.
or		
Inflamzyme from PhytoPharmica	Take as indicated on the label.	Immune system activator; reduces inflammation; fights free radicals.
or		
Kyolic Formula 102 (with enzymes)	Take 2 capsules or tablets with meals two to three times per day.	Strengthens the immune system; helps alleviate stress; helps fight free radicals; helps fight inflammation.
or		
Vita-C-1000 from Life Plus	Take as indicated on the label.	Boosts immune function; fights infection.

ENZYME HELPERS

Nutrient	Suggested Dosage	Actions
Acidophilus and other probiotics	Take as indicated on the label.	Stimulates enzyme activity; improves the body's natural immunity.
Adrenal glandular	Take 100–400 mg per day.	Helps fight stress and adrenal exhaustion; fights inflammation.
Bioflavonoids, such as rutin and quercetin	Take 500 mg three times per day.	Strengthen capillaries.
Coenzyme Q_{10}	Take 30–60 mg one to two times per day with meals.	Important to immune system function; has antioxidant activity.
Essential fatty acid complex	Take 100–200 mg two to three times per day with meals; or follow the advice of your health-care professional.	Strengthens immunity; helps strengthen cell membranes.
Kyolic Aged Garlic Liquid Extract with Vitamins B_1 and B_{12}	Take 10 drops in water or juice or place directly on the tongue three times per day.	Stimulates immune system function.
Thymus glandular	Take 100 mg per day, or as indicated on the label.	Assists the thymus gland in immune resistance; improves immune response; fights infections.
Vitamin A	Take up to 50,000 IU for no more than one month (for higher doses, see a physician). If pregnant, do not take more than 10,000 IU per day.	Needed for strong immune system and healthy mucous membranes; potent antioxidant.
Vitamin B complex	Take 100–150 mg three times per day.	B vitamins act as coenzymes.
Vitamin C	Take 10,000–15,000 mg per day; or 1–2 grams every hour until symptoms improve.	Fights infections, diseases, and allergies; improves wound healing; stimulates the adrenal glands; promotes immune system; powerful antioxidant.
Vitamin E	Take 1,200 IU per day from all sources (including the multivitamin complex in the introduction to Part Two).	Improves antioxidant activity and fights the herpes virus.
Zinc	Take 50 mg per day. Do not take more than 100 mg per day.	Needed by over 300 enzymes; necessary for white blood cell immune function; improves healing.

Comments

❑ To help ease the discomfort, apply wet cool compresses to the affected areas.

❑ Most individuals recover completely, although there may be occasional scarring of the skin. However, the nerve pain (neuralgia) following a herpes zoster attack, may continue for months or years, most commonly in the elderly. Dr. W.

Bartsch feels that long periods of illness and pain can result if therapy is delayed. Therefore, he recommends enzyme therapy begin as soon as the condition is diagnosed.

❑ Today's standard medical treatment of zoster is oral acyclovir. However, the use of oral enzyme combinations is as effective and less expensive. Corticosteroids are sometimes used to treat post-zoster neuralgia, but the use of corticosteroids is controversial and has serious side effects.

❑ Elderly individuals and those who are immunodeficient due to drugs or diseases are particularly at risk.

❑ Enzyme therapy represents a new method of therapeutic action.

❑ For additional information on diet, fasting, juicing, and the Enzyme Toxin Flush and the Enzyme Kidney Flush, please see Part Three.

High Blood Pressure

Circulating blood exerts pressure against the blood vessel walls. The force fluctuates with every heartbeat. During the period of heartbeat called systole, the heart contracts and forces blood into the arteries. Blood pressure is greatest during this period (called systolic pressure). During the period known as diastole, the pressure drops to a minimum as the heart muscle relaxes. This is called diastolic pressure.

A systolic pressure above 140 millimeters and/or a diastolic over 90 is usually considered to be high blood pressure, while low blood pressure (hypotension) is suspected if the systolic pressure is below 90 millimeters. Hypertension is often associated with nervous tension, kidney disease, hardening of the arteries, high blood cholesterol levels, and other disorders. If your hypertension is related to another known disorder, refer to that section in Part Two for its enzyme therapy program.

From person to person and from time to time in the same person, blood pressure varies. Factors such as heredity, age, general health, emotional state, and activity level can influence blood pressure. Two important factors in a person's health are indicated by blood pressure: the elasticity of the blood vessels and the volume of circulating blood. For example, if a person's blood pressure declines rapidly after a severe injury, this is a warning that the person is going into shock. This very serious condition is associated with decrease in the volume of circulating blood. As we age, the blood vessel walls lose their elasticity, this causes blood pressure to increase. Since the rate at which your heart beats can affect blood pressure, blood-pressure measurements should be taken when a person is in a relaxed condition.

Symptoms

Until complications develop, hypertension has no symptoms. If symptoms do arise, it is usually because the blood pressure has been high for an extended period of time and has damaged a vital organ. But uncomplicated hypertension does not cause fatigue, headache, nervousness, and nosebleed as is commonly believed.

Enzyme Therapy

Systemic enzyme therapy is used to lower blood pressure, lower LDL cholesterol, increase fibrin activity, reduce inflammation, stimulate the immune system, improve circulation, help speed tissue repair, help transport nutrients throughout the body, and remove waste products, and enhance wellness.

Digestive enzyme therapy is used to improve the digestion of food, reduce stress on the gastrointestinal mucosa, help maintain normal pH levels, detoxify, and promote the growth of healthy intestinal flora, thus improving the overall health of the body. Digestive enzymes also serve as replacements for the body's pancreatic enzymes, leaving the pancreatic enzymes free to perform other functions in the body, such as improving circulation. Both systemic enzyme therapy and digestive enzyme therapy can be helpful in lowering blood pressure.

Enzyme Absorption System Enhancers (EASE) maximize enzyme activity. They can also improve the absorption and bioavailability of various nutrients and decrease the drain on the body's own digestive enzymes, thus prolonging their lives.

At present, there is a great deal of research being conducted in Germany on arteriosclerosis (often associated with high blood pressure) and low-density-lipoprotein cholesterol. Combinations of bromelain and trypsin with the bioflavonoid rutin reduce oxidized LDL cholesterol ("bad cholesterol") and increase non-oxidized high-density lipoproteins (HDL—the "good cholesterol"). According to Dr. Karl Ransberger, world-famous expert in systemic enzyme therapy, there is substantial evidence that enzyme treatment reduces arteriosclerosis and myocarditis.

In addition, Professor Heidland, from the University of Wuerzburg demonstrated that a combination of trypsin, bromelain, and rutin reduces arteriosclerosis in kidney arterioles of kidney transplant patients. The rate of success is substantial. Arteriosclerosis is a very severe, life-threatening complication with kidney transplant patients. Therefore, this is an important discovery.

In addition to the standard multivitamin, multimineral, and multiglandular complexes recommended in the introduction to Part Two, the following supplements are recommended for the treatment of hypertension. The following recommended dosages are for adults. Children between the ages of two and five should take one-quarter the recommended dosage, those between the ages of six and twelve should take one-half the recommended dosage, and those between the ages of thirteen and seventeen should take three-quarters the recommended adult dosage.

ESSENTIAL ENZYMES

Enzyme	Suggested Dosage	Actions
Formula Three—See page 38 of Part One for a list of products.	Take 200 mcg of each three times per day between meals.	Potent antioxidant enzymes.

AND SELECT ONE OF THE FOLLOWING:

Enzyme	Suggested Dosage	Actions
Formula One (particularly Wobenzym N from Naturally Vitamin Supplements)—See page 38 of Part One for a list of products.	Take 8–10 tablets three times per day between meals for six months or until symptoms improve. Thereafter, take 6 tablets three times per day between meals.	Improves circulation and blood vessel integrity; strengthens capillary walls.
Inflazyme Forte from American Biologics	Take as indicated on the label between meals.	Improves circulation and blood vessel integrity; strengthens capillary walls.
Bromelain	Take four to six 230–250 mg capsules per day.	Improves circulation and blood vessel integrity; stimulates the immune system.
Enzyme complexes rich in proteases, plus amylases and lipases, working synergistically	Take as indicated on the label between meals.	Improves circulation and blood vessel integrity; strengthens capillary walls.

SUPPORTIVE ENZYMES

Enzyme	Suggested Dosage	Actions
A digestive enzyme formula containing protease, amylase, and lipase enzymes. See the list of digestive enzyme products in Appendix.	Take as indicated on the label with meals.	Improves the breakdown and absorption of proteins, carbohydrates, and fats; fights toxic build-up.

ENZYME ABSORPTION SYSTEM ENHANCERS (EASE)

Combination	Suggested Dosage	Actions
Bromelain Complex from PhytoPharmica or	Take as indicated on the label.	Improves circulation; helps break up circulating immune complexes.
Cardio-Chelex from Life Plus or	Take as indicated on the label.	Supports healthy circulation; improves energy.
Protease Concentrate from Tyler Encapsulations	Take as indicated on the label.	Breaks up fibrin; fights inflammation.

ENZYME HELPERS

Nutrient	Suggested Dosage	Actions
Bioflavonoids, such as rutin or quercetin	Take 500 mg three times per day between meals.	Strengthen capillaries and blood vessel walls; potent antioxidants.
Calcium	Take 1,000 mg per day.	Low calcium intake has been implicated in hypertension.
Carnitine	Take 500 mg two times per day.	Increases energy; lowers cholesterol; reduces risk of heart disease.
Coenzyme Q$_{10}$	Take 75 mg one to two times per day with meals.	Potent antioxidant; decreases blood pressure and total serum cholesterol.
Essential fatty acid complex	Take 75–100 mg two to three times per day with meals.	Lowers cholesterol level; strengthen the immune system; helps strengthen cell membranes.
Fiber	Take as indicated on the label.	Increase in fiber has been directly related to a decrease in blood cholesterol levels.
Green Barley Extract	Take 1–2 teaspoons in water or juice three times per day.	Reduces blood pressure; effective against heart disease; fights free radicals.
Heart glandular	Take 140 mg, or as indicated on the label.	Improves heart function.
Kyolic Reserve	Take 2 capsules twice per day.	Lowers cholesterol and blood pressure; fights free radicals.
Lecithin	Take 3 capsules with meals, or as indicated on the label.	Emulsifies fat.
Magnesium	Take 500 mg per day.	May lower blood pressure; plays important roles in calcium absorption.
Niacin	Take 50 mg the first day, then increase dosage daily by 50 mg to 1,000 mg. If adverse effects occur, decrease dosage. Take doses higher than 1,000 mg only under the supervision of a physician.	Improves circulation by dilating blood vessels.
or		
Niacin-bound chromium	Take 20 mg per day, or as indicated on the label.	
Vitamin A	Take 15,000 IU per day, or as per your doctor's directions. If pregnant, do not take more than 10,000 IU per day.	Necessary for skin and cell membrane structure and strong immune system; potent antioxidant.
Vitamin B complex	Take 50 mg three times per day, or as indicated on the label.	B vitamins act as coenzymes.
Vitamin C	Take 5,000–10,000 mg per day.	Essential to vascular function; helps prevent high blood pressure; reduces blood cholesterol levels; potent antioxidant; improves the function of many enzymes.
Vitamin E	Take 1,200 IU per day from all sources (including the multivitamin complex in the introduction to Part Two).	Potent antioxidant; helps prevent cardiovascular disease; decreases risk of heart attack; stimulates circulation; strengthens capillary walls; prevents blood clots; maintains cell membranes.
Zinc	Take 50 mg per day. Do not take more than 100 mg per day.	Needed for white blood cell immune function; interacts with platelets in blood clotting.

Comments

❑ Garlic, onions, shallots, chives, leeks, and scallions contain allium compounds, which are potent free-radical fighters that help lower cholesterol and fight circulatory disorders.

❑ Oats, wheat germ, olives, nuts, nut and seed oils, organ meats, and eggs contain alpha-tocopherol (vitamin E), which is known to help prevent cardiovascular disease and prevent blood clots.

❑ Raspberries, cranberries, grapes, red wine, black currants, and red cabbage contain anthocyanidins, which reduce blood vessel plaque formation and maintain blood flow in small vessels.

❑ Oolong and green tea contain epicatechin, which is a potent antioxidant that lowers blood pressure and strengthens capillaries.

❑ A number of herbs can improve circulation, including astragalus, ginkgo biloba, and maitake mushrooms (which are also high in enzymatic activity).

❑ Women taking oral contraceptives for five years or longer are two to three times more likely to have high blood pressure than women not taking contraceptives.

❑ For further information on necessary dietary modifications, detoxification methods (through cleansing, fasting, and juicing), exercises, positive mental attitude, meditation, and general suggestions, see Part Three of this book.

❑ *See also* CARDIOVASCULAR DISORDERS; STRESS; STROKE.

HIV

See AIDS.

Hives

Whenever you come in contact with an allergen, your body releases chemicals known as histamines. In some people, this release of histamine causes a rash called hives. Usually hives (also called urticaria) develop within a few hours of exposure to the allergen. However, the rash may appear several days later in unusual cases. When this happens, it can be much more difficult to identify the cause of the reaction. Medications are the most common cause of delayed reactions. If a rash develops while you are taking any drug, immediately report it to your doctor.

Hives can be caused by a number of different allergens, including insect bites and stings, certain foods (especially eggs, milk, shrimp and other shellfish, strawberries, nuts, tomatoes, and chocolate), food additives (colors, preservatives, and artificial flavors), and some drugs (such as aspirin and penicillin). Hives may also be the first symptom of certain diseases, including hepatitis, rubella, strep throat (in children), and infectious mononucleosis.

Hives is an annoying, but usually not serious, condition. However, seek immediate medical attention if you also have difficulty swallowing, speaking, or breathing; swelling of the throat; abdominal pain; fever; or nausea. These symptoms are an indication of anaphylactic shock—a potentially fatal condition caused by respiratory and/or vascular collapse.

Symptoms

Hives are smooth, red, and slightly elevated wheals on the body that are accompanied by itching. The hives may change in size or shape. They usually disappear within a few hours, or in severe cases, several days. Other allergic symptoms may accompany hives, such as swelling of the eyes or other body parts.

Enzyme Therapy

Since hives are usually an allergic reaction, systemic enzyme therapy is used to reduce inflammation and stimulate the immune system. Enzymes can improve circulation, help speed tissue repair, bring nutrients to the damaged area, remove waste products, and enhance wellness.

Digestive enzyme therapy is used to improve the digestion of food, reduce stress on the gastrointestinal mucosa, help maintain normal pH levels, detoxify the body, and promote the growth of healthy intestinal flora. Digestive enzymes also as replacements for the body's pancreatic enzymes, leaving the pancreatic enzymes free to perform other functions in the body, such as decreasing inflammation.

Topical enzyme creams are used to dissolve dead and damaged skin cells, while leaving healthy skin cells intact. Enzyme salves and creams applied directly to the skin can be very helpful in decreasing the allergic reaction.

Enzyme Absorption System Enhancers (EASE) maximize enzyme activity. They can also improve the absorption and bioavailability of various nutrients and decrease the drain on the body's own digestive enzymes, thus prolonging their lives.

In addition to the standard multivitamin, multimineral, and multiglandular complexes recommended in the introduction to Part Two, the following supplements are recommended for the treatment of hives. The following recommended dosages are for adults. Children between the ages of two and five should take one-quarter the recommended dosage, those between the ages of six and twelve should take one-half the recommended dosage, and those between the ages of thirteen and seventeen should take three-quarters the recommended adult dosage.

ESSENTIAL ENZYMES

Enzyme	Suggested Dosage	Actions
Superoxide dismutase	Take 150 mcg per day.	Fights inflammation; potent antioxidant.
WobeMugos E Ointment from Mucos	Rub a generous amount on the affected area.	Helps to remove dead skin; decreases inflammation and swelling; stimulates immune action.

AND SELECT ONE OF THE FOLLOWING:

Enzyme	Suggested Dosage	Actions
Bromelain	Take 750 mg two times per day between meals.	Fights inflammation, pain, and swelling; speeds healing; stimulates the immune system; strengthens capillary walls.
Formula One—See page 38 of Part One for a list of products.	Take 10 tablets three times per day between meals.	Fights inflammation, pain, and swelling; speeds healing; stimulates the immune system.
Serratiopeptidase	Take 10 mg two times per day between meals.	Fights inflammation; reduces pain and swelling.
Trypsin with chymotrypsin	Take 250 mg of trypsin and 10 mg of chymotrypsin three times per day between meals.	Fights inflammation, pain, and swelling; increases rate of healing; stimulates the immune system.
Pancreatin	Take 500 mg six times per day on an empty stomach.	Fights inflammation, pain, and swelling; increases rate of healing; stimulates the immune system; strengthens capillary walls.
Papain or a microbial protease	Take 300 mg two times per day between meals.	Fights inflammation; speeds healing.

SUPPORTIVE ENZYMES

Enzyme	Suggested Dosage	Actions
Catalase and glutathione peroxidase	Take as indicated on the label.	Potent antioxidant enzymes.
A digestive enzyme product containing protease, amylase, and lipase. See the list of digestive enzyme products in Appendix.	Take as indicated on the label with meals.	Improves the breakdown and absorption of food nutrients; fights toxic build-up.

ENZYME ABSORPTION SYSTEM ENHANCERS (EASE)

Combination	Suggested Dosage	Actions
Akne-Zyme from Enzymatic Therapy *or*	Take 1 or 2 capsules daily as a dietary supplement.	Maintains skin integrity.
Bio-Musculoskeletal Pak from Biotics *or*	Take as indicated on the label.	Decreases inflammation; fights free-radical formation; stimulates immune function.
Bioprotect from Biotics *or*	Take 1–3 capsules each day.	Contains antioxidants; improves immune function and tissue repair.
Connect-All from Nature's Plus *or*	Take as indicated on the label.	Supports connective tissue.
CTR Support from PhysioLogics *or*	Take 2–3 capsules per day, preferably on an empty stomach, or follow the advice of your health-care professional.	Nourishes the body during tissue recovery and healing; helps maintain healthy connective tissue.
Detox Enzyme Formula from Prevail *or*	Take 1 capsule between meals one or two times per day with a 6- to 8-ounce glass of water or juice.	Helps rid the body of harmful toxins.
Inflamzyme from PhytoPharmica *or*	Take as indicated on the label.	Promotes tissue repair.
Kyolic Formula 102 (with enzymes)	Take 2 capsules or tablets with meals two to three times per day.	Strengthens the immune system; helps alleviate stress; helps fight free radicals; helps fight inflammation; improves digestion.

ENZYME HELPERS

Nutrient	Suggested Dosage	Actions
Acidophilus (and other probiotics)	Take as indicated on the label.	Improves gastrointestinal function; needed for production of biotin, niacin, folic acid, and pyridoxine, all important coenzymes; helps detoxify the body; stimulates enzyme activity.
Adrenal glandular	Take 300 mg per day.	Fights inflammation and adrenal exhaustion.
Aged garlic extract	Take 2 capsules three times per day.	Fights inflammation; stimulates the immune system; potent antioxidant.
Amino-acid complex	Take as indicated on the label on an empty stomach.	Required for enzyme production; helps the body maintain proper fluid and acid/base balance, necessary for overall health.
Bioflavonoids, such as rutin or quercetin	Take 500 mg three times per day between meals.	Fight inflammation; have antihistamine activity.
Coenzyme Q$_{10}$	Take 30–60 mg one to two times per day with	Potent antioxidant.

	meals; or follow the advice of your health-care professional.	
Essential fatty acid complex	Take 100–200 mg two to three times per day with meals; or follow the advice of your health-care professional.	Helps strengthen cell membranes; support immune function.
Green Barley Extract from Green Foods Corporation or Kyo-Green from Wakunaga	Take as indicated on the label.	Potent free-radical fighter; rich in vitamins, minerals, and enzyme activity.
Niacin *or*	Take 50 mg the first day, then increase dosage daily to 1,000 mg. If adverse effects occur, decrease dosage. Take doses higher than 1,000 mg only under the supervision of a physician.	Dilates blood vessels; improves circulation.
Niacin-bound chromium	Take up to 20 mg per day, or as indicated on the label.	
Thymus glandular	Take 90 mg per day.	Assists the thymus gland in immune resistance; improves immune response; fights infections.
Vitamin A	Take 5,000 IU per day. Pregnant women should not exceed 10,000 IU per day.	Needed for strong immune system and healthy mucous membranes; potent antioxidant.
Vitamin C	Take 8,000–13,000 mg per day.	Fights infections, diseases, allergies; improves wound healing; stimulates the adrenal glands; promotes immune system; powerful antioxidant.
Vitamin E	Take 1,200 IU per day from all sources (including the multivitamin complex in the introduction to Part Two).	Potent antioxidant; supplies oxygen to cells; improves circulation and level of energy.

Comments

❑ Eat plenty of natural, whole-grain breads and cereals and other niacin-rich foods to inhibit histamine release.

❑ Enzyme-rich fruits and vegetables (particularly green vegetables) should comprise 60 percent of your diet. If your body cannot tolerate raw fruits and vegetables, increase your intake of digestive enzymes, or sauté or steam your produce. See dietary recommendations in Part Three.

❑ Alfalfa concentrate, echinacea, ginger root extract, ginkgo biloba, and white willow bark fight inflammation. Ginseng and licorice enhance immunity.

❑ If you react adversely to aspirin, avoid foods containing salicylates, such as grapes, raisins, apricots, dried foods, berries, tea, and foods made with vinegar (including pickles and relishes).

❏ Gargle twice daily with the Enzymatic Gargle (see Part Three for instructions). It contains enzymes and vitamin C, which help fight hives.

❏ Follow the guidelines in Part Three for the Enzyme Kidney Flush and Enzyme Toxin Flush programs.

❏ Topical applications can help the condition:

  • Apply aloe vera topically to help soothe discomfort.

  • Bathe in water to which 5 or 6 tablespoons of oatmeal, 3 tablespoons of cornstarch or baking soda, or a strong infusion of chickweed has been added.

  • Apply papaya cream to the affected area three times per day.

  • Place ten drops of liquid aged garlic extract on a cotton swab or gauze. Dab gently along the affected area.

  • Break open and spread the contents of two to four vitamin A and vitamin E capsules on the affected area.

  • Zinc oxide salve can be spread over the affected area once daily.

  • Apply calamine lotion or cold compresses to the affected area.

❏ *See also* ALLERGIES.

Hypercholesterolemia

See CHOLESTEROL, ELEVATED.

Hypertension

See HIGH BLOOD PRESSURE.

Hypochlorhydria

Hypochlorhydria is a decreased or insufficient production of hydrochloric acid (HCl). The ability of the body to secrete gastric acid decreases with age. This condition can result in re-duced ability to absorb vitamins, such as folic acid, B_6, and B_{12}, and minerals, including calcium. Over a period of time, hypochlorhydria can contribute to malnutrition, malabsorption of nutrients, and even inflammatory skin conditions.

One of hydrochloric acid's roles is to kill harmful bacteria. Therefore, an insufficient production can allow harmful bacteria to grow. Hydrochloric acid also triggers the conversion of the inactive enzyme pepsinogen into the active enzyme pepsin, which is necessary for protein digestion.

With age, the incidence of hypochlorhydria usually increases. In fact, over 50 percent of those over 60 years of age have low levels of hydrochloric acid in the stomach.

Symptoms

Some symptoms of hypochlorhydria include belching, bloating, burning in the abdomen, diarrhea or constipation, indigestion, nausea, flatulence immediately following meals, food allergies, a feeling of "fullness" after meals, and itching around the rectum. In addition, the stool may contain undigested food, and there may be gas in the upper digestive tract, abnormal flora or chronic intestinal parasites, and chronic candida infections. Fingernails may peel, weaken, and crack, and acne may be present.

Enzyme Therapy

Digestive enzyme therapy is used to balance the HCl levels in the stomach, improve the digestion of food, improve the absorption of vitamins and minerals, reduce stress on the gastrointestinal mucosa, help maintain normal pH levels, detoxify the body, and promote the growth of healthy intestinal flora. Digestive enzymes also serve as replacements for the body's pancreatic enzymes, leaving the pancreatic enzymes free to perform other functions in the body, such as boosting immunity, decreasing inflammation, and improving circulation.

Systemic enzyme therapy is used to reduce indigestion and stimulate the immune system. Enzymes can improve circulation, transport nutrients throughout the body, remove waste products, and improve health.

Enzyme Absorption System Enhancers (EASE) maximize enzyme activity. They can also improve the absorption and bioavailability of various nutrients and decrease the drain on the body's own digestive enzymes, thus prolonging their lives.

In addition to the standard multivitamin, multimineral, and multiglandular complexes recommended in the introduction to Part Two, the following supplements are recommended for the treatment of hypochlorhydria. The following recommended dosages are for adults. Children between the ages of two and five should take one-quarter the recommended dosage, those between the ages of six and twelve should take one-half the recommended dosage, and those between the ages of thirteen and seventeen should take three-quarters the recommended adult dosage.

ESSENTIAL ENZYMES

Enzyme	Suggested Dosage	Actions
Formula Two—See page 38 of Part One for a list of products.	Take 1–2 capsules three times per day after each meal.	Digests protein, fats, carbohydrates, and fiber.
or		

Pancreatin *or*	Take as indicated on the label.	Aids digestion and absorption.
Bromelain *or*	Take 250–500 mg with meals.	Aids digestion and absorption.
Papain *or*	Take 500–1,000 mg with meals.	Aids digestion and absorption.
A digestion enzyme product containing protease, amylase, and lipase enzymes. For a list of digestive enzyme products, see Appendix. *or*	Take as indicated on the label.	Improves digestion by breaking down proteins, carbohydrates, and fats.
Formula One—See page 38 of Part One for a list of products. *or*	Take 5 coated tablets or two heaping tablespoons of a granulated mixture three times per day between meals.	Improves digestion.
If you have difficulty swallowing tablets or pills, try chewable enzymes, such as bromelain or papain (papaya).	Take as indicated on the label.	Improves digestion.

SUPPORTIVE ENZYMES

Enzyme	Suggested Dosage	Actions
Formula Three—See page 38 of Part One for a list of products.	Take as indicated on the label.	Free-radical fighters.

ENZYME ABSORPTION SYSTEM ENHANCERS (EASE)

Combination	Suggested Dosage	Actions
Biodias A Granules from Amano Pharmaceutical Co., Ltd.	Adults should take one packet in 8 ounces of water between or after meals three times per day. Children ages 11 to 14 should take two-thirds of a packet; children between the ages of 8 and 10 should take half of the packet; and children between the ages of 5 and 7 should take one-third of a packet.	Helps improve digestion of nutrients; relieves inflammation; promotes regeneration of mucous membranes; decreases stress on the stomach and intestines; reduces gastric pain.

ENZYME HELPERS

Nutrient	Suggested Dosage	Actions
Acidophilus or other probiotics, such as bifidobacterium; or Acidophilasé from Wakunaga, which also includes enzymes.	Take as indicated on the label.	Improves gastrointestinal function; needed for production of the coenzymes biotin, niacin, folic acid, and pyridoxine; helps produce antibacterial substances; stimulates enzyme activity.

Aged garlic extract	Take 2 capsules three times per day.	Potent antioxidant; improves digestion; stimulates healthy intestinal bacteria.
Fructooligo-saccharides	Take as indicated on the label.	Increase peristaltic action of intestines and improve digestion; improve liver function to help detoxify the body.
Betaine hydrochloric acid	The first day, take one 10-grain capsule three times per day with meals. Increase dosage by one 10-grain capsule per day until a feeling of warmth is experienced in the stomach. Then reduce dosage by 10 grains.	Stimulates production of hydrochloric acid, a digestive aid.
Lecithin	Take 1,000 mg three times per day with meals.	Emulsifies fat.
Ox bile extract	Take 120 mg per day.	Aids digestion.
Vitamin A	Take 5,000 IU per day.	Needed by digestive glands to produce enzymes; required for healthy mucous linings and skin and cell membrane structure and a strong immune system; potent antioxidant.
Vitamin B complex	Take 50 mg three times per day.	B vitamins act as coenzymes.
Vitamin B_1 (thiamin)	Take 150 mg three times per day.	Essential for carbohydrate metabolism, proper energy production, and appetite.
Vitamin C	Take 1,000–2,000 mg per day.	Stimulates the adrenal glands to produce hormones (which also act as coenzymes); aids in detoxification; a potent antioxidant; improves the function of many enzymes; required for converting the coenzyme folic acid to its active form.
Whey	Take 1 heaping tablespoon three times per day with meals.	Improves digestion.
Zinc	Take 10–50 mg per day. Take no more than 100 mg per day from all sources.	Needed by over 300 enzymes. Many zinc enzymes work to metabolize carbohydrates, alcohol, and essential fatty acids; synthesize proteins; and dispose of free radicals.

Comments

❑ Hypochlorhydria can be treated by altering any dietary imbalances and supplementing with enzymes and hydrochloric acid.

❑ Eat raw, enzyme-rich fresh foods, which are easier to digest, to reduce the body's need to produce more digestive enzymes of its own.

❑ Follow the Enzyme Kidney Flush and the Enzyme Toxin Flush programs to help balance your digestive tract and your body. See Part Three for directions.

Hypoglycemia

Hypoglycemia is a condition marked by low blood sugar (glucose) levels. The brain requires glucose for proper function. When blood sugar levels fall, many systems of the body are affected, particularly the brain, and mental confusion and a host of other symptoms can result. Generally, glucose levels below 70 milligrams per deciliter of blood are considered to be low. However, according to Dr. Carlton Fredericks, "There is no number, no point, no range of blood which constitutes hypoglycemia." He believes that the speed at which the blood sugar level drops causes the hypoglycemic symptoms, not the blood sugar level itself.

Hypoglycemia is generally classified as drug-related or nondrug-related. Most cases of hypoglycemia occur in diabetics and are drug-related. It occurs when diabetics take too high doses of insulin. It can also occur as a result of the drug pentamidine, used to treat a form of AIDS-related pneumonia. Alcohol consumption on an empty stomach may also cause hypoglycemia.

Nondrug-related hypoglycemia may be caused by too much insulin secreted by the pancreas or a problem with the mechanisms in the body that normalize blood sugar levels when they begin to fall too low, such as with the adrenal glands and the liver. There are two types of nondrug-related hypoglycemia—fasting and reactive. *Fasting hypoglycemia* usually occurs as a result of some other disorder, such as a disorder of the pituitary or adrenal glands or the liver. With this condition, blood sugar levels fall after fasting or, rarely, after prolonged exercise. In *reactive hypoglycemia*, blood sugar levels drop after consumption of food, usually carbohydrates.

Rarely, a type of hypoglycemia can occur as a result of antibodies in the body being produced against insulin.

Symptoms

Symptoms of hypoglycemia can include nervousness, sweating, tremulousness, fatigue, palpitations, faintness, and hunger. In more severe cases, confusion, inappropriate behavior, headache, dizziness, insomnia, stupor, visual disturbances, seizures, and coma may result. If the blood sugar levels remain too low for a long period of time, permanent brain damage may result.

Enzyme Therapy

Treatment of hypoglycemia should emphasize control of blood glucose through meal planning, use of supplements, regular physical activity, and evaluating and improving any relevant psychosocial and medical factors. Treatment of hypoglycemia should be ongoing. An important part of the process is patient and family education.

Digestive enzyme therapy is used to improve the digestion of food and normalize carbohydrate metabolism, thereby normalizing blood sugar levels. Digestive enzymes can reduce stress on the gastrointestinal mucosa, help maintain normal pH levels, detoxify the body, and promote the growth of healthy intestinal flora. Digestive enzymes also serve as replacements for the body's pancreatic enzymes, leaving the pancreatic enzymes free to perform other functions in the body.

Systemic enzyme therapy works at the cellular level and in organs to improve overall health by stimulating the immune system, improving circulation, helping speed tissue repair, transporting nutrients throughout the body, removing waste products, and enhancing wellness.

Enzyme Absorption System Enhancers (EASE) maximize enzyme activity. They can also improve the absorption and bioavailability of various nutrients and decrease the drain on the body's own digestive enzymes, thus prolonging their lives.

Nutrient and enzyme deficiencies can affect organs that participate in blood sugar level maintenance, such as the liver, pancreas, thyroid, pituitary gland, and adrenal glands. Therefore, enzyme-rich glandulars and organ extracts should be part of a total therapy program.

In addition to the standard multivitamin, multimineral, and multiglandular complexes recommended in the introduction to Part Two, the following supplements are recommended for the treatment of hypoglycemia. The following recommended dosages are for adults. Children between the ages of two and five should take one-quarter the recommended dosage, those between the ages of six and twelve should take one-half the recommended dosage, and those between the ages of thirteen and seventeen should take three-quarters the recommended adult dosage.

ESSENTIAL ENZYMES

Enzyme	Suggested Dosage	Actions
Formula One—See page 38 of Part One for a list of products. *or*	Take 3 tablets three times per day with meals.	Helps fight abnormal blood sugar levels; plays a role in protein, fat, and carbohydrate metabolism.
A digestive enzyme combination containing microbial protease, amylase, lipase, and possibly, cellulase. For a list of digestive enzyme products, see Appendix. *or*	Take as indicated on the label.	Improves sugar metabolism.
Zymase from Tyler Encapsulations	Take 2–4 tablets three times per day before meals.	Improves sugar, carbohydrate, protein, and fat digestion and metabolism.

SUPPORTIVE ENZYMES

Enzyme	Suggested Dosage	Actions
Phlogenzym from Mucos	Take as indicated on the label, or as per your physician's instructions.	Helps digest protein; helps stimulate immune function; fights free radicals.
Formula Three—See page 38 of Part One for a lsit of products.	Take as indicated on the label.	Potent antioxidant enzymes.

ENZYME ABSORPTION SYSTEM ENHANCERS (EASE)

Combination	Suggested Dosage	Actions
NESS Formula #13 or	Take as indicated on the label.	Helps fight abnormal blood sugar levels; plays a role in protein, fat, and carbohydrate metabolism.
Diabest from PhysioLogics or	Take as indicated on the label, or as directed by your physician.	Supports healthy blood sugar levels; supports ability of the body to produce insulin within the cell; may support the body's ability to regenerate insulin-producing beta cells.
Sugar Check from Enzymatic Therapy or	Take as indicated on the label.	Contains nutrients for improved sugar and carbohydrate metabolism.
Dr. Wright's Blood Sugar Improvement Formula from The Rockland Corporation	Take as indicated on the label.	Improves blood sugar levels.

ENZYME HELPERS

Nutrient	Suggested Dosage	Actions
Acidophilus or other probiotics, such as bifidobacterium or Acidophilasé from Wakunaga, which contains amylolytic, lipolytic, and proteolytic enzymes.	Take as indicated on the label.	Improves gastrointestinal function; needed for production of biotin, niacin, folic acid, and pyridoxine—coenzymes important for normal metabolism; stimulates enzyme activity.
Adrenal extract, such as Drentex from General Research Laboratories	Take 100–400 mg per day.	Fights adrenal exhaustion; restores normal carbohydrate balance.
Aged garlic extract	Take 2 capsules three times per day.	Potent antioxidant; improves digestion; stimulates healthy intestinal bacteria.
Amino-acid complex containing L-cysteine, L-carnitine, and L-glutamine, such as Amino Acid 1000 from Nature's Life	Take as indicated on the label.	Amino acids serve as building blocks for the body's enzymes.
Betaine hydrochloric acid	Take as indicated on the label.	Provides hydrochloric acid to improve digestion. If gastric upset occurs, discontinue use.
Brewer's yeast	Take as indicated on the label.	A very important supplement for hypoglycemics, it contains chromium, important for insulin production.
Calcium	Take 750 mg twice per day.	Vital for all body functions, but especially in proper utilization of other minerals, as well as vitamins A, C, and D; needed to balance excess phosphorus in brewer's yeast.
Chromium polynicotinate or chromium picolinate	Take as indicated on the label.	Works synergistically with insulin to improve blood sugar levels.
Copper	Take as indicated on the label.	Essential cofactor necessary for superoxide dismutase activity; important in maintaining normal blood sugar levels.
Green food product, such as Green Kamut, Kyo-Green, or Green Barley Extract	Take as indicated on the label.	Green foods are rich in enzymes, vitamins, minerals, chlorophyll, and amino acids; aid in maintaining normal blood sugar levels.
Lecithin	Take 1,000 mg three times per day with meals.	Emulsifies fat; important for proper fat metabolism.
Liver extract	Take as indicated on the label.	Improves liver function; can detoxify the body (those with hypoglycemia are often toxic); promotes liver cell regeneration; helps regulate the mechanisms of sugar metabolism.
Magnesium	Take 250 mg per day.	Necessary for glucose metabolism.
Maitake mushroom supplements	Take as indicated on the label.	High in chromium; assist in glucose metabolism.
Pancreas extract	Take as indicated on the label.	Aids digestion; important in sugar metabolism; assists the pancreatic enzymes trypsin and chymotrypsin.
Potassium	Take 100 mg three times per day.	Essential to nerve transmission; necessary for cellular enzymes to work properly.
Thyroid extract	Take as indicated on the label.	Helps regulate mechanisms of sugar metabolism.
Vitamin A	Take 5,000 IU per day.	Needed by digestive glands to produce enzymes; assists mineral absorption; important for strong immune system; potent antioxidant; has synergistic action with other vitamins.
Vitamin B complex	Take 150 mg three times per day.	B vitamins act as potent coenzymes and are essential for carbohydrate metabolism; involved in sugar metabolism and normalizing sugar levels.
Vitamin B_1 (thiamin)	Take 150 mg three times per day.	Essential for carbohydrate metabolism and proper energy production.

Vitamin B₃ (niacin) *or*	Take 50 mg three times per day.	Stimulates gastric juices and hydrochloric acid, necessary to improve digestion; important in energy metabolism; supports healthy liver and gastrointestinal tract.
Niacin-bound chromium	Take 10 mg three times per day.	
Vitamin B₅	Take 50 mg per day.	Coenzyme for many enzymes.
Vitamin B₆	Take 50 mg three times per day.	Assists enzymes that metabolize amino acids and fats; necessary for proper immune function; involved in sugar metabolism; helps support adrenal glands, which are often exhausted in hypoglycemics.
Vitamin B₁₂	Take 125 to 150 mcg per day.	Potent coenzyme; helpful in regeneration of liver, which is often toxic or overstressed in hypoglycemics.
Vitamin C	Take 2,000 mg per day.	Stimulates the adrenal glands to produce hormones (which also act as coenzymes); improves the function of many enzymes; increases the body's tolerance to sugars and carbohydrates; helps normalize sugar metabolism; improves adrenal output.
Vitamin D	Take 400–800 IU per day.	Involved in the process of insulin secretion.
Vitamin E	Take 1,200 IU per day from all sources (including the multivitamin complex in the introduction to Part Two).	Acts as an intercellular antioxidant.
Whey	Take 1 heaping tablespoon three times per day with meals.	Improves digestion.
Zinc	Take 10–50 mg per day. Do not exceed 100 mg per day from all sources.	Needed by over 300 enzymes. Many zinc enzymes work to metabolize carbohydrates, alcohol, and essential fatty acids; synthesize proteins; and help dispose of free radicals.

Comments

❏ Herbs beneficial in treating blood sugar problems include licorice root (*Glycyrrhiza glabra*); juniper cedar berries (*Juniperus sabina pinaceae*), Mexican wild yam, copalquin or capalachi, goldenseal (*Hydrastis*), lobelia (*Lobelia inflata*), garlic, and dandelion.

❏ Avoid sugar and other refined carbohydrates, coffee and other caffeine-containing beverages and drugs, tobacco, alcohol, salt, food additives and preservatives, other chemicals that cause allergies and may affect blood sugar levels, and physical and emotional stress.

❏ Diet is an extremely important part of hypoglycemia therapy and should include high-fiber foods, such as fresh raw vegetables and fruits, grains, and nuts and seeds, all of which have been shown to help maintain normal blood sugar levels.

❏ Eat several small meals throughout the day, rather than two or three large meals. Snacking between meals or in the evening can help normalize blood sugar levels. Snacks could include a cup of kefir or yogurt (mixed with one tablespoon of brewer's yeast); the juice of two apples and four carrots; one plain whole-grain rye cracker with a piece of cheese and a glass of fresh, slightly warmed milk; the juice of two apples and one-quarter to one-half of a fresh pineapple; or one to two pieces of fresh fruit (choose from pears, pineapples, melons, apples, cherries, and papaya; or prepare a mixed, enzyme-rich fruit salad).

❏ Eat foods high in chromium content, such as whole-grain products, bran, mushrooms, brewer's yeast, and wheat germ. These foods seem to improve insulin's effect on body tissues.

❏ Although juice fasts are an important part of most standard treatment programs for a majority of diseases, juice fasts are not recommended in the treatment of hypoglycemia, except with the approval and supervision of a physician experienced in the treatment of hypoglycemia.

❏ Daily exercise of thirty to sixty minutes is extremely important for the hypoglycemic patient. Brisk walking, swimming, bicycling, and aerobics are a few examples. Exercise stimulates metabolic and enzymatic activity necessary for your body's engine to function properly. See Part Three for further information on exercising.

❏ A positive mental attitude and peace of mind are critical. Mental and emotional stresses can overtax the adrenal and other glands and organs involved in sugar metabolism. Meditation and relaxation exercises can be helpful. See Part Three for further information.

Indigestion

Indigestion is a general term used to describe disturbances in the gastrointestinal tract. Causes of indigestion vary widely. It can occur if you gulp down food without properly chewing, if you eat while emotionally stressed, if you eat too fast, or if you overeat. Fatty foods cause indigestion in some individuals, while others suffer because there are one or two foods that just do not agree with them. Allergies to foods (such as wheat, sugar, and dairy products), protein deficiency, and enzyme deficiencies can also be contributing factors.

Excessive stomach acid can cause indigestion. However, some individuals have indigestion because of too little, rather than too much, stomach acid. Several medicines (such as aspirin) can irritate the stomach, causing indigestion. Tobacco and alcohol can also cause indigestion, as can stress (physical, nutritional, and emotional). See the inset "Frequent Causes of Indigestion, and Enzyme Cures" on page 270.

Chronic indigestion may overstress the immune system, interfere with proper nutrition, cause toxic build-up, and lead to infections. As a result, constipation, diarrhea, and/or chronic fatigue may follow.

Improper or inadequate digestion can do more than interfere with the breakdown and absorption of the nutrients in foods. Any incompletely digested food molecules that remain in the intestine can also be absorbed into the circulatory system. This can lead to food allergies and several diseases.

Improving your diet can also cause temporary indigestion. Remember, it takes time for your body to adjust when you drastically change your diet to include the extra fiber consumed when you eat more fresh fruits and vegetables and decrease your intake of meats and fats. Fiber-rich foods can cause gas and bloating until your body adjusts.

Symptoms

Indigestion can include such symptoms as an upset stomach, abdominal pain, a burning sensation in the abdomen, gas, heartburn, nausea, vomiting, poor food absorption, excessive belching, diarrhea, a feeling of fullness or bloating in the abdomen, cramps, and constipation. These symptoms are also common symptoms of many illnesses.

Enzyme Therapy

Digestive enzyme therapy is used to improve the digestion of food, reduce stress on the gastrointestinal mucosa, help maintain normal pH levels, detoxify the body, and promote the growth of healthy intestinal flora. Digestive enzymes also serve as replacements for the body's pancreatic enzymes, leaving the pancreatic enzymes free to perform other functions in the body, such as boosting immunity, decreasing inflammation, and improving circulation.

Systemic enzyme therapy is used to reduce indigestion, decrease inflammation, and stimulate the immune system. Enzymes can improve circulation, help speed tissue repair, transport nutrients throughout the body, remove waste products, and improve health.

Enzyme Absorption System Enhancers (EASE) maximize enzyme activity. They can also improve the absorption and bioavailability of various nutrients and decrease the drain on the body's own digestive enzymes, thus prolonging their lives.

In addition to the standard multivitamin, multimineral, and multiglandular complexes recommended in the introduction to Part Two, the following supplements are recommended for the treatment of indigestion. The following recommended dosages are for adults. Children between the ages of two and five should take one-quarter the recommended dosage, those between the ages of six and twelve should take one-half the recommended dosage, and those between the ages of thirteen and seventeen should take three-quarters the recommended adult dosage.

ESSENTIAL ENZYMES

Enzyme	Suggested Dosage	Actions
If you have trouble digesting protein, try a protease, such as bromelain, carboxypeptidase, a microbial protease, papain, pepsin, peptidase, rennin, or trypsin and chymotrypsin. *or*	Take as indicated on the label with meals.	Breaks down proteins.
If you have trouble digesting carbohydrates, try an amylase, such as alpha-galactosidase, beta-glucosidase, carbohydrase, cellulase, diastase, glucoamylase, hemicellulase, invertase, lactase, maltase, phytase, or sucrase. *or*	Take as indicated on the label with meals.	Breaks down carbohydrates.
If you have trouble digesting fats, try an enzyme supplement containing lipase. *or*	Take as indicated on the label with meals.	Breaks down fats.
If you have trouble swallowing tablets or pills, try chewable enzyme tablets, such as bromelain or papain (papaya).	Take as indicated on the label.	Improves digestion.

AND SELECT ONE OF THE FOLLOWING:

Enzyme	Suggested Dosage	Actions
A digestive enzyme product containing protease, amylase, and lipase enzymes. For a list of digestive enzyme products, see Appendix.	Take as indicated on the label.	Improves digestion by breaking down proteins, carbohydrates, and fats.
Formula Two—See page 38 of Part One for a list of products.	Take 1–2 capsules three times per day after each meal.	Digests protein, fats, carbohydrates, and fiber.

SUPPORTIVE ENZYMES

Enzyme	Suggested Dosage	Actions
Formula Three—See page 38 of Part One for a list of products.	Take as indicated on the label.	Potent antioxidant enzymes.

ENZYME ABSORPTION SYSTEM ENHANCERS (EASE)

Combination	Suggested Dosage	Actions
Biodias A Granules from Amano Pharmaceutical Co., Ltd.	Adults should take one packet in 8 ounces of water between or after meals three times per day. Children ages 11 to 14 should take two-thirds of a packet; children between the ages of 8 and 10 should take half of the packet; and children between the ages of 5 and 7 should take one-third of a packet.	Helps improve digestion of nutrients; relieves inflammation; promotes regeneration of mucous membranes; decreases stress on the stomach and intestines; reduces gastric pain.
or		
Colo-Clen from Enzyme Process International	Take as indicated on the label.	Helps to detoxify the colon.
or		
Kyolic Formula 102 (with enzymes)	Take 2 capsules or tablets with meals two to three times per day.	Strengthens the immune system; helps alleviate stress; fights free radicals.
or		
Sugar Check from Enzymatic Therapy	Take as indicated on the label.	Contains nutrients for improved sugar and carbohydrate metabolism and absorption.

ENZYME HELPERS

Nutrient	Suggested Dosage	Actions
Acidophilus or other probiotics, such as bifidobacterium; or Acidophilasé from Wakunaga, which also contains enzymes.	Take as indicated on the label.	Improves gastrointestinal function; needed for production of the coenzymes biotin, niacin, folic acid, and pyridoxine; helps produce antibacterial substances; stimulates enzyme activity.
Aged garlic extract	Take two 300 mg capsules three times per day.	Potent antioxidant; improves digestion; stimulates healthy intestinal bacteria.
Betaine hydrochloric acid	Take 150 mg three times per day with meals.	Provides hydrochloric acid to improve digestion. If gastric upset occurs, discontinue use.
Fructooligo-saccharides	Take as indicated on the label.	Increases peristaltic action of intestines; improves liver function.
Lecithin	Take 1,000 mg three times per day with meals.	Emulsifies fat.
Ox bile extract	Take 120 mg per day.	Aids digestion.
Vitamin A	Take 5,000 IU per day.	Needed by digestive glands to produce enzymes; promotes normal taste and appetite; required for healthy mucous linings and skin and cell membrane structure and a strong immune system; potent antioxidant.

Frequent Causes of Indigestion, and Enzyme Cures

There are many causes of indigestion. Several of these causes are treatable with enzymes. Following are a few causes of indigestion and the necessary enzymes for treating the condition.

INDIGESTION CAUSES	ENZYME CURES
Declining digestive function with age	A combination containing proteases, amylases, and lipases.
Reduced digestive function because of gastric resection (removal of a portion of the gastrointestinal tract)	A combination containing proteases, amylases, and lipases.
Chronic pancreatitis and pancreatic insufficiency	A combination containing proteases, amylases, and lipases.
Chronic hyperacidity	A combination containing proteases, amylases, and lipases.
Problems digesting fats (such as steatorrhea or reduced bile secretion)	Lipases.
Problems digesting proteins	Proteases.
Problems digesting carbohydrates (including celiac disease or lactase deficiency)	Amylases (including lactase, sucrase, invertase).

Vitamin B complex	Take 50 mg three times per day.	B vitamins act as coenzymes.
Vitamin B₁ (thiamin)	Take 150 mg three times per day.	Essential for carbohydrate metabolism and proper energy production; coenzyme to many enzymes.
Vitamin C	Take 1,000–2,000 mg per day.	Stimulates the adrenal glands to produce hormones (which also act as coenzymes); aids in detoxification; a potent antioxidant; improves the function of many enzymes; enhances absorption of iron; assists in converting cholesterol to bile salts; required for converting folic acid to its active form.
Whey	Take 1 heaping tablespoon three times per day with meals.	Improves digestion.
Zinc	Take 10–50 mg per day. Take no more than 100 mg per day from all sources.	Needed by over 300 enzymes. Many zinc enzymes work to metabolize carbohydrates, alcohol, and essential fatty acids; synthesize proteins; and dispose of free radicals.

Comments

❑ For acute indigestion, try raw pancreas glandular extract, DGL tablets from Enzymatic Therapy, or Alfa Juice caps from Solaray.

❑ For diarrhea, try apple pectin tablets or activated charcoal (on a short-term basis).

❑ For bloating and gas, try Indigestion Relief from Bioforce.

❑ For burping and belching, try Hyland's Indigestion after meals, Beano from AKPharma, or Digestive Formula from Country Life.

❑ To improve digestion, teas made from the following herbs can be helpful:

- A pinch of nutmeg, cloves, cinnamon, and ginger.
- Thyme.
- Slippery elm.
- Mint and alfalfa.
- Fennel and catnip.

❑ Eat cultured foods (such as kefir, yogurt, miso), high-fiber foods (like whole grains), enzyme-rich foods (including pineapple and papaya), plus fresh fruits and vegetables.

❑ Eat four to five small meals per day. Chew food longer to put less stress on your gastrointestinal tract.

❑ Avoid spicy and fatty foods, red meats, fried foods, dairy products, sugary foods, and caffeine. Decrease acid-forming food intake.

❑ To cleanse your gastrointestinal tract: On day one, eat no foods, and drink six to eight 8-ounce glasses of fresh apple juice. Add Source of Life Energy Shake from Nature's Plus or a similar product to the first glass. On days two through four, drink six to eight 8-ounce glasses of enzyme-rich juices, such as papaya, pineapple, kiwi, carrot, and parsley. Follow this program for four days. On the second day, begin adding one meal per day. See Part Three for further information on juicing.

❑ If indigestion occurs because of a deficiency in hydrochloric acid, try one to two teaspoons of cider vinegar in a glass of water.

❑ See Part Three for directions for the Enzyme Kidney Flush and the Enzyme Toxin Flush programs.

Inflammatory Bowel Disease

There are two types of inflammatory bowel disease (IBD): ulcerative colitis and Crohn's disease. Both diseases involve inflammation of the intestines (bowel). The symptoms of the two disorders are so similar that it is often difficult to distinguish one from the other. IBD is a chronic condition. Some people suffer from it continuously, while others have flare-ups and remissions. As many as two million Americans may suffer from inflammatory bowel disease. There is no known cause, but treating the symptoms can help improve the quality of life for IBD sufferers.

Ulcerative colitis (granulomatous colitis) is characterized by inflammation and ulceration of the large intestine, causing episodes of bloody diarrhea with mucus, abdominal cramping, and fever. It usually begins in the rectum and spreads through the large intestine. When it affects only the rectum it is called ulcerative proctitis, which causes fewer complications than ulcerative colitis. Most cases of ulcerative colitis begin before 30 years of age, although it can occur later in life. The attacks may begin gradually, or they may occur suddenly. Those with ulcerative colitis are at an increased risk for colon cancer.

Crohn's disease (regional ileitis, granulomatous ileitis, or ileocolitis) is inflammation anywhere along the digestive tract, but it most commonly occurs in the last part of the small intestine and the large intestine. Crohn's disease has been reported with increasing frequency since it was first identified a few decades ago. Though researchers are not sure of the cause, it could have a genetic basis or be caused by environmental, infectious, or immune factors. It appears to have a familial tendency and is seen most commonly in those with Anglo-Saxon or Northern European ancestry, particularly in Jews. Most cases begin before the age of 30, particularly between the ages of 14 and 24. As mentioned previously,

Crohn's disease produces symptoms similar to those of ulcerative colitis—chronic diarrhea, abdominal cramping, and fever, along with appetite and weight loss and an abdominal mass that can be felt. Often, intestinal obstructions, fistulas (abnormal channels that may connect different parts of the intestines), and abscesses develop. Crohn's disease may also cause inflammation in other parts of the body, such as the joints, the eyes, and the skin. It may also cause an increased risk for cancer of the large intestine, if that area is affected.

Crohn's disease and ulcerative colitis are different from irritable bowel syndrome (IBS) and spastic colon. There is no direct relationship between IBS and Crohn's disease or ulcerative colitis.

Enzyme Therapy

The key to effective treatment is to suppress the inflammatory process, permit healing of the colon, and relieve rectal bleeding, abdominal pain, and diarrhea. The advantage of enzyme therapy is that it decreases pain, swelling, and inflammation without the side effects of drugs. In addition, constant inflammation of the small intestine (as can occur in Crohn's disease) can destroy the mucosa and decrease disaccharidase activity, interfering with digestion. This problem will require supplementation with digestive enzymes. Absorption of fat-soluble vitamins can also be impaired.

Systemic enzyme therapy is used to reduce inflammation, stimulate the immune system, improve circulation, help speed tissue repair, bring nutrients to the damaged area, remove waste products, and improve health. Digestive enzyme therapy is used to improve the digestion of food, replace any missing enzymes, reduce stress on the gastrointestinal mucosa, help maintain normal pH levels, detoxify the body, and promote the growth of healthy intestinal flora.

Enzyme Absorption System Enhancers (EASE) maximize enzyme activity. They can also improve the absorption and bioavailability of various nutrients and decrease the drain on the body's own digestive enzymes, thus prolonging their lives.

In addition to the standard multivitamin, multimineral, and multiglandular complexes recommended in the introduction to Part Two, the following supplements are recommended for the treatment of inflammatory bowel disease. The following recommended dosages are for adults. Children between the ages of two and five should take one-quarter the recommended dosage, those between the ages of six and twelve should take one-half the recommended dosage, and those between the ages of thirteen and seventeen should take three-quarters the recommended adult dosage.

ESSENTIAL ENZYMES

Enzyme	Suggested Dosage	Actions
Amylase enzymes (including lactase, sucrase, and maltase)	Take as indicated on the label.	Replaces disaccharidase enzymes that can be lost in IBD because of mucosal damage.
Superoxide dismutase	Take 150 mcg per day.	Fights inflammation; potent antioxidant.

AND SELECT ONE OF THE FOLLOWING:

Formula One—See page 38 of Part One for a list of products.	Take 10 tablets three times per day between meals.	Fights inflammation, pain, and swelling; speeds recovery after injury or surgery; stimulates the immune system.
Bromelain	Take three to four 230–250 mg capsules per day until symptoms subside. If severe, take 6–8 capsules per day.	Fights inflammation, pain, and swelling; speeds recovery after injury or surgery; stimulates the immune system.
WobeMugos E from Mucos	Take 4 tablets three times per day between meals.	Fights inflammation, pain, and swelling; speeds recovery after injury or surgery; stimulates the immune system.
A microbial protease	Take as indicated on the label.	Fights inflammation; speeds recovery.
Pancreatin	Take 500 mg six times per day on an empty stomach.	Fights inflammation, pain, and swelling; increases rate of healing; stimulates the immune system.

SUPPORTIVE ENZYMES

Enzyme	Suggested Dosage	Actions
Catalase and glutathione peroxidase	Take as indicated on the label.	Potent antioxidant enzymes.
A digestive enzyme product containing protease, amylase, and lipase. See the list of digestive enzyme products in Appendix.	Take as indicated on the label with meals.	Improves the breakdown and absorption of food nutrients; fights toxic build-up.

ENZYME ABSORPTION SYSTEM ENHANCERS (EASE)

Combination	Suggested Dosage	Actions
Biodias A Granules from Amano Pharmaceutical Co., Ltd.	Adults should take one packet in 8 ounces of water between or after meals three times per day. Children ages 11 to 14 should take two-thirds of a packet; children between the ages of 8 and 10 should take half of the packet; and children between the ages of 5 and 7 should take one-third of a packet.	Helps improve digestion of nutrients; relieves inflammation; promotes regeneration of mucous membranes; decreases stress on the stomach and intestines; reduces gastric pain.
or		
Detox Enzyme Formula from Prevail	Take as indicated on the label.	Helps rid the body of harmful toxins.
or		

Hepazyme from Enzymatic Therapy *or*	Take as indicated on the label.	Stimulates immune function.
Kyolic Formula 102 (with enzymes) *or*	Take 2 capsules or tablets with meals two to three times per day.	Strengthens the immune system; helps alleviate stress; helps fight free radicals and inflammation; improves digestion and absorption of nutrients.
Oxy-5000 Forte from American Biologics *or*	Take as indicated on the label.	Helps the body excrete toxins; and biochemical stressors, common infections, and degenerative processes.
Pro-Flora from Tyler Encapsulations *or*	Take as indicated on the label.	Helps restore normal digestion and gastrointestinal flora.
Vita-C-1000 from Life Plus	Take as indicated on the label.	Boosts immune function; fights infection.

ENZYME HELPERS

Nutrient	Suggested Dosage	Actions
Acidophilus and other probiotics	Take as indicated on the label.	Improve gastrointestinal function; needed for production of biotin, niacin, folic acid, and pyridoxine, important coenzymes; help produce antibacterial substances; stimulate enzyme activity.
Aged garlic extract	Take two 300 mg capsules three times per day.	Fights inflammation; stimulates the immune system; potent antioxidant.
Amino-acid complex	Take as indicated on the label on an empty stomach.	Required for enzyme, peptide, and hormone production; helps decrease inflammation in the intestines.
Betaine hydrochloric acid	Take 150 mg three times per day with meals.	Provides hydrochloric acid for better digestion.
Calcium	Take 500 mg two times per day.	Works with enzymes to digest protein and fats; aids nutrient absorption.
Coenzyme Q_{10}	Take 30–60 mg one to two times per day with meals; or follow the advice of your health-care professional.	Potent antioxidant.
Fructooligo-saccharides	Take as indicated on the label.	Increases peristaltic action of intestines; improves liver function.
Green Barley Extract or Kyo-Green	Take as indicated on the label.	Potent free-radical fighter; rich in vitamins, minerals, and enzyme activity.
Magnesium	Take 500 mg per day.	Plays important role in tissue function, and calcium absorption.
Ox bile extract	Take 120 mg per day.	Aids digestion.
Thymus glandular	Take 100 mg per day.	Assists the thymus gland in immune resistance; improves immune response; fights infections.

Vitamin A	Take 10,000–15,000 IU per day. If pregnant, take no more than 10,000 IU per day.	Needed for strong immune system and healthy mucous membranes; potent antioxidant.
Vitamin B complex	Take 150 mg three times per day, or as indicated on the label.	B vitamins act as coenzymes; support enzymatic function.
Vitamin C	Take 10,000–15,000 mg per day.	Fights infections, diseases, and allergies; improves wound healing; stimulates the adrenal glands; promotes immune system; powerful antioxidant.
Whey	Take 1 heaping tablespoon three times per day with meals.	Improves digestion.
Zinc	Take 15–50 mg per day. Do not take more than 100 mg per day.	Needed by over 300 enzymes; necessary for white blood cell immune function; improves wound healing.

Comments

❑ Spicy foods or those high in fiber may cause some discomfort; so bland, soft foods may be desired during IBD episodes.

❑ Good nutrition is particularly important in this condition because of nutrient losses due to diarrhea and poor absorption, as well as the loss of appetite that sometimes accompanies IBD.

❑ Cortisone is often administered in either pills or enemas, but cortisone's serious side effects should be noted. In extreme cases, colon surgery may be necessary to give the tissues a chance to heal.

❑ See the list of Resource Groups in the Appendix for a list of organizations that can provide you with more information about inflammatory bowel disease.

Influenza

Often called *flu* or *grippe*, influenza is an acute viral respiratory infection. Influenza spreads rapidly and has an incubation period of only one to four days. The virus may be transmitted through the inhalation of droplets from the sneeze or cough of an infected person, or through direct contact with an infected person. Influenza A, influenza B, and influenza C are the main viruses that cause influenza epidemics every year; but they have the ability to mutate into different forms each year. Note that the "intestinal flu" is a popular term for a number of intestinal upsets that have no relationship to influenza.

Flu is often named for the area where a particular epidemic started. Thus, Spanish flu (the serious and often fatal pandemic of 1918) is the name given to the flu believed to have been brought from Spain by returning American GIs.

Since the Spanish flu, the Asian flu (of 1957), the Hong Kong flu (of 1968), and subsequent strains have proven to be so virulent, the Centers for Disease Control and Prevention closely monitors emerging flu viruses.

Flu viruses weaken the body's defenses against bacteria and other viruses making one susceptible to secondary infection by another virus or a bacteria. This is why individuals suffering from the flu risk developing either viral or bacterial pneumonia. Usually, however, the acute symptoms in uncomplicated cases last only a few days, followed by gradual recovery of normal fitness and strength.

The elderly and those with an underlying health problem (such as cardiovascular or pulmonary disease) have an increased risk of developing complications from the flu. For this reason, immediate care is essential and enzymes might be of assistance, either by fighting the virus directly or by stimulating the body's defenses so that the body can fight the virus itself.

Symptoms

The symptoms of influenza include fever (often high fevers), headache, chills, malaise, loss of appetite, weakness, fatigue, sore throat, cough, sneezing, runny nose, and general aches and pains. There may also be inflammation of the mucous membranes of the throat and nose, nausea, and vomiting.

Enzyme Therapy

There is no medication available to treat influenza once it has been acquired, only its symptoms. Therefore, prevention by strengthening the body (and its immune system) are of prime importance. Once the virus has invaded the body, a healthy immune system is critical. Enzymes help in both instances.

Systemic enzyme therapy is used to strengthen the body as a whole, build general immune resistance, reduce inflammation, and fight off invading viruses. Enzymes improve circulation, help speed tissue repair, bring nutrients to the infected areas, remove waste products, and enhance wellness.

Digestive enzyme therapy is used to improve the digestion of food, reduce stress on the gastrointestinal mucosa, help maintain normal pH levels, detoxify, and promote the growth of healthy intestinal flora, thereby making the body healthier so that it can fight influenza. Digestive enzymes also serve as replacements for the body's pancreatic enzymes, leaving the pancreatic enzymes free to perform other functions in the body, such as boosting immunity.

Enzyme Absorption System Enhancers (EASE) maximize enzyme activity. They can also improve the absorption and bioavailability of various nutrients and decrease the drain on the body's own digestive enzymes, thus prolonging their lives.

In addition to the standard multivitamin, multimineral, and multiglandular complexes recommended in the intro-

duction to Part Two, the following supplements are recommended for the treatment of influenza. The following recommended dosages are for adults. Children between the ages of two and five should take one-quarter the recommended dosage, those between the ages of six and twelve should take one-half the recommended dosage, and those between the ages of thirteen and seventeen should take three-quarters the recommended adult dosage.

ESSENTIAL ENZYMES

Enzyme	Suggested Dosage	Actions
Formula Three—See page 38 of Part One for a list of products.	Take 200 mcg three times per day between meals.	Potent antioxidant enzymes.

AND SELECT ONE OF THE FOLLOWING:

Enzyme	Suggested Dosage	Actions
Lysozyme	Take 200 mcg three times per day between meals.	Fights inflammation; improves respiratory conditions; bolsters the immune system; breaks up mucus.
Formula One (particularly Wobenzym N from Naturally Vitamin Supplements) —See page 38 of Part One for a list of products.	Take 8–10 tablets three times per day between meals for six months or until symptoms improve. Thereafter, take 6 tablets three times per day between meals.	Fights inflammation, pain, and swelling; speeds healing; strengthens capillary walls.
Bromelain	Take 500 mg three times per day between meals.	Fights inflammation; improves respiratory conditions; bolsters the immune system; breaks up mucus.
Serratiopeptidase	Take 5 mg three times per day between meals.	Fights inflammation; improves respiratory conditions; bolsters the immune system; breaks up mucus.
Sfericase	Take 500 mg three times per day between meals.	Fights inflammation; improves respiratory conditions; bolsters the immune system; breaks up mucus.
Inflazyme Forte from American Biologics	Take as indicated on the label between meals.	Fights inflammation, pain, and swelling; speeds healing; strengthens capillary walls.
Enzyme complexes rich in proteases, plus amylases and lipases.	Take as indicated on the label between meals.	Fights inflammation, pain, and swelling; speeds healing; strengthens capillary walls.

SUPPORTIVE ENZYMES

Enzyme	Suggested Dosage	Actions
A digestive enzyme formula containing protease, amylase, and lipase enzymes. See the list of digestive enzyme products in Appendix.	Take as indicated on the label with meals.	Improves the breakdown and absorption of proteins, carbohydrates, and fats; fights toxic build-up.

ENZYME ABSORPTION SYSTEM ENHANCERS (EASE)

Combination	Suggested Dosage	Actions
Cold Zzap from Naturally Vitamin Supplements *or*	Take as indicated on the label.	Fights infection from viruses or bacteria; boosts immunity; has antioxidant activity.
Combat from Life Plus *or*	Take as indicated on the label.	Helps support the respiratory system; stimulates immune system function; helps control infections.
Connect-All from Nature's Plus *or*	Take as indicated on the label.	Supports connective tissue.
Inflamzyme from PhytoPharmica *or*	Take as indicated on the label.	Promotes tissue repair; immune system activator.
Kyolic Formula 102 (with enzymes) *or*	Take 2 capsules or tablets with meals two to three times per day.	Strengthens the immune system; helps alleviate stress; helps fight free radicals; helps fight inflammation.
Mucous Dissolver Liquezyme from Enzyme Process Laboratories *or*	Take as indicated on the label.	Dissolves mucus.
Protease Concentrate from Tyler Encapsulations *and*	Take as indicated on the label.	Breaks up fibrin; fights inflammation.
Bromelain Complex from PhytoPharmica *or*	Take as indicated on the label.	Decreases swelling; improves circulation; helps break up circulating immune complexes.
Sinease from Prevail	Take as indicated on the label.	Decongestant.

ENZYME HELPERS

Nutrient	Suggested Dosage	Actions
Bioflavonoids	Take 100–300 mg twice per day.	Strengthen capillaries; potent antioxidants.
Green Barley Extract or Kyo-Green	Take 1 heaping teaspoon in 8 ounces of liquid three times per day.	Fights free radicals.
Kyolic Aged Garlic Liquid Extract with Vitamins B$_1$ and B$_{12}$	Take as indicated on the label.	Stimulates antioxidant activity and immune system function.
Lung glandular	Take as indicated on the label.	Improves lung function.
Pantothenic acid (vitamin B$_5$)	Take 125 mg per day.	Important in protein, fat, and carbohydrate metabolism; needed for healthy digestive system and to reduce toxins.
Vitamin A	Take 10,000–50,000 IU for no more than one month. (For higher doses, see a physician.) If pregnant, take no more than 10,000 IU per day.	Necessary for healthy mucous linings and skin and cell membrane structure and strong immune system; potent antioxidant.
Vitamin B complex	Take 100–150 mg three times per day.	B vitamins act as coenzymes.
Vitamin C	Take 1–2 grams every hour until symptoms improve; or mix a heaping teaspoon of powder or crystalline vitamin C with 6 to 8 ounces of water or orange or lemon juice (if lemon juice is used, you may sweeten with honey) and gargle a mouthful and swallow. Repeat every hour.	Fights infections, diseases, allergies, and common cold; potent antioxidant.
Winterized C from Naturally Vitamin Supplements	Take as indicated on the label.	Fights colds and flu.
Zinc or zinc gluconate lozenges	Take up to 100 mg per day.	Needed for white blood cell immune function; involved with more than 300 enzymes.

Comments

❑ A number of herbs can help bolster immunity, including astragalus, echinacea, ginseng, licorice, and maitake, reishi, and shiitake mushrooms.

❑ If influenza is contracted, a person should stay in bed during the acute phase and should gradually return to normal activities while convalescing.

❑ Hot broth or chicken or vegetable soup is helpful for cleaning out the body and sweating out the toxins. Before making the soup, be sure to remove the skin from the chicken (chicken skins contain many toxins). Sip the broth or soup as frequently as possible during the waking hours.

❑ Gargle with the Enzymatic Gargle two to three times per day. Alternatively, you may try the Echinacea/Enzyme Gargle. See directions for the preparation of these gargles in Part Three.

❑ Put five to six drops of eucalyptus in a sink or bowl of very hot water. Cover your head and shoulders with a large bath towel over the sink or bowl, and inhale the vapors.

❑ The Enzyme Kidney Flush and the Enzyme Toxin Flush are of value in detoxification (see Part Three for instructions).

❑ NESS makes some products that help fight infection. One is an enzyme combination that includes microbial amylase and protease with marshmallow root and rose hips (NESS Formula #6). Another product contains kelp and Irish moss in addition to microbial protease (NESS Formula #14).

❑ The orthodox medical approach to flu prevention is an influenza vaccination. It takes about two weeks after injection for a person to develop immunity. Sometimes this works, sometimes it does not. Once symptoms have developed, however, it is useless to receive a flu shot.

Insect Bites and Stings

At some time in our lives, most of us will be bitten or stung by a mosquito, bee, wasp, ant, spider, or tick. Those minor wounds that we often call bites are really more like very small puncture wounds inflicted on us by insects. It is the substance they leave in the wound, and not the wound itself, that typically does the damage. When an insect stings, it injects venom into the victim. Bloodsucking insects, such as lice or mosquitoes, inject an anticoagulant into the victim to temporarily prevent the blood from clotting. This causes the body to release histamine at the site, which causes the itching.

About five out of 1,000 people are allergic to the stings of insects from the order *hymenoptera*, which includes bees, hornets, and wasps. This allergy can be dangerous, if not life-threatening. It may take over 100 bee stings to cause a lethal reaction in most of us; however, in a hypersensitive person, one bee sting can cause a fatal anaphylactic reaction.

If you are stung by a bee and feel weak, if there is swelling of the body, or if the sting is in a very sensitive area, such as the lip, tongue, or eyelid, call a doctor immediately. You may be experiencing an allergic reaction and will need medical care. If you know you are allergic to bees, you should carry a kit containing an antihistamine and epinephrine. Such kits should be carried at all times in potentially dangerous areas.

In most instances, mosquito, louse, tick, and flea bites have no serious consequences and the discomfort usually disappears in a few hours. However, many serious diseases, including malaria, dengue fever, yellow fever, typhus, and lyme disease can be carried and transmitted by these bloodsucking insects.

Symptoms

Bites and stings frequently cause localized itching, pain, swelling, and redness. The site may be warm to the touch. If untreated, any bite or sting can fester and become infected. Even though itching may be severe, resist the urge to scratch, as a secondary infection could result.

An allergic reaction to a sting can cause such symptoms as labored breathing; weakness; confusion; difficulty swallowing; diarrhea; and swelling of the neck, tongue, and/or joints. An even more severe reaction can result in closing of the airways, shock, and death.

Enzyme Therapy

Systemic enzyme therapy is used to decrease the pain, swelling, and inflammation caused by the bite or sting. Enzymes stimulate the immune system, improve circulation, help speed tissue repair, bring nutrients to the damaged area, remove waste products, fight allergic reactions, and enhance wellness.

Digestive enzyme therapy is used to improve the digestion of food, reduce stress on the gastrointestinal mucosa, help maintain normal pH levels, detoxify the body, and promote the growth of healthy intestinal flora. This improves the absorption and metabolism of nutrients that help fight the toxins from insect bites and stings. Digestive enzymes also serve as replacements for the body's pancreatic enzymes, leaving the pancreatic enzymes free to perfom other functions in the body, such as decreasing inflammation.

Enzyme Absorption System Enhancers (EASE) maximize enzyme activity. They can also improve the absorption and bioavailability of various nutrients and decrease the drain on the body's own digestive enzymes, thus prolonging their lives.

In addition to the standard multivitamin, multimineral, and multiglandular complexes recommended in the introduction to Part Two, the following supplements are recommended for the treatment of insect bites and stings. The following recommended dosages are for adults. Children between the ages of two and five should take one-quarter the recommended dosage, those between the ages of six and twelve should take one-half the recommended dosage, and those between the ages of thirteen and seventeen should take three-quarters the recommended adult dosage.

ESSENTIAL ENZYMES

Enzyme	Suggested Dosage	Actions
Superoxide dismutase	Take 150 mcg per day.	Fights inflammation; potent antioxidant.
Proteolytic enzyme salve	Apply to bite or sting 2–3 times per day.	Reduces pain and inflammation.

AND SELECT ONE OF THE FOLLOWING:

Enzyme	Suggested Dosage	Actions
Papain	Take 300 mg two times per day between meals.	Fights inflammation; speeds recovery from injuries.
Formula One—See page 38 of Part One for a list of products.	Take 5 tablets twice a day.	Fights inflammation, pain, and swelling; increases rate of healing; stimulates the immune system; strengthens capillary walls.
Bromelain	Take 750 mg two times per day between meals.	Fights inflammation, pain, and swelling; speeds recovery after injury; stimulates the immune system; strengthens capillary walls.
Serratiopeptidase	Take 10 mg two times per day between meals.	Fights inflammation; reduces pain and swelling.
Trypsin with chymotrypsin	Take 250 mg of trypsin and 10 mg of chymotrypsin three times per day between meals.	Fights inflammation, pain, and swelling; increases rate of healing; stimulates the immune system.
Pancreatin	Take 500 mg six times per day on an empty stomach.	Fights inflammation, pain, and swelling; increases rate of healing; stimulates the immune system; strengthens capillary walls.

Microbial protease	Take 300 mg two times per day between meals.	Fights inflammation; speeds recovery from injuries.

SUPPORTIVE ENZYMES

Enzyme	Suggested Dosage	Actions
Catalase and glutathione peroxidase	Take as indicated on the label.	Potent antioxidant enzymes.
A digestive enzyme product containing protease, amylase, and lipase enzymes. See the list of digestive enzyme products in Appendix.	Take as indicated on the label with meals.	Improves the breakdown and absorption of food nutrients; fights toxic build-up.

ENZYME ABSORPTION SYSTEM ENHANCERS (EASE)

Combination	Suggested Dosage	Actions
Oxy-5000 Forte from American Biologics *or*	Take as indicated on the label.	Aids detoxification.
Inflamzyme from PhytoPharmica *or*	Take as indicated on the label.	Immune system activator; fights free radicals; improves tissue repair and enzymatic activity.
Vita-C-1000 from Life Plus *or*	Take as indicated on the label.	Boosts immune function; fights infection.
Kyolic Formula 102 (with enzymes) *or*	Take 2 capsules or tablets per day as desired.	Strengthens the immune system; helps alleviate stress; improves circulation; helps fight inflammation.
Ecology Pak from Life Plus *or*	Take 2 or 3 tablets once or twice per day, preferably in the morning and no later than midday.	Assists with natural detoxification; supports the immune system.
Hepazyme from Enzymatic Therapy	Take as indicated on the label.	Stimulates immune function.

ENZYME HELPERS

Nutrient	Suggested Dosage	Actions
Adrenal glandular	Take 100–400 mg per day.	Fights inflammation and adrenal exhaustion.
Aged garlic extract	Take two 300 mg capsules three times per day in between meals.	Fights inflammation; stimulates the immune system; potent antioxidants.
Aloe vera extract	Take as indicated on the label.	Aids in healing.
Citrus bioflavonoids	Take as indicated on the label.	Strengthen capillaries; potent antioxidants.
Essential fatty acid complex	Take 100–200 mg two to three times per day with meals; or follow the advice of your health-care professional.	Helps strengthen cell membranes; supports immune function.

Niacin *or*	Take 50 mg the first day, then increase dosage daily by 50 mg until the swelling of the bite disappears. Before taking doses higher than 1,000 mg per day, consult your physician.	Dilates blood vessels and allows toxins to be flushed from the body more rapidly.
Niacin-bound chromium	Take 20 mg per day; or take as indicated on the label.	
Pantothenic acid (vitamin B_5)	Take 250 mg three times per day.	Improves healing; needed for healthy digestive system and skin and for the formation of antibodies.
Thymus glandular	Take 100 mg per day.	Assists the thymus gland in immune resistance; fights infections.
Vitamin A	Take 10,000–15,000 IU per day. If pregnant, take no more than 10,000 IU per day.	Needed for strong immune system and healthy mucous membranes; potent antioxidant.
Vitamin B complex	Take 150 mg three times per day; or take as indicated on the label.	B vitamins act as coenzymes.
Vitamin C	Take 2,000 mg for the first two hours, then 1,000 mg per hour, until 10,000–15,000 mg is reached. Doses above 10,000 mg per day may cause flatulence and loose stools. These symptoms will disappear when the dosage is decreased.	Fights infections, diseases, and allergies; improves wound healing; stimulates the adrenal glands; promotes immune system; powerful antioxidant.
Zinc	Take 15–50 mg per day. Do not take more than 100 mg per day from all sources.	Needed by over 300 enzymes; necessary for white blood cell immune function; improves wound healing.

Comments

❏ Stings of many bumblebees, honeybees, hornets, wasps, and yellow jackets should be removed by scraping the skin with a fingernail. Other stingers can be removed with tweezers.

❏ Herbs, including alfalfa, echinacea, ginger root, ginkgo biloba, licorice, and white willow bark can help decrease the inflammation of a bite or sting.

❏ Apply an enzymatic salve one to three times daily. See directions in Part Three.

❏ Apply a vitamin-C paste made by mixing one heaping tablespoon of vitamin C powder with a few drops of water to make a paste. Soak a piece of gauze in the paste and apply liberally. Repeat the process three times per day.

❏ Apply calamine lotion, a paste made with water and sodium bicarbonate, or household ammonia to the affected site. Don't use paste made with mud; it can lead to infection.

277

❑ An ice cube rubbed on the bite or sting can help reduce pain.

❑ Make a poultice from slippery elm, white oak leaves and bark, and comfrey to reduce pain and encourage healing.

❑ Place ten drops of liquid aged garlic extract on a cotton swab or gauze. Dab gently along the affected area. This decreases inflammation and fights free-radical formation.

❑ Break open and spread the contents of two to four vitamin A and vitamin E capsules on affected areas.

❑ Phytochemicals should be used to enhance the healing process. For example, citrus bioflavonoids, grape seed extract, and carotenoids fight capillary permeability, fight free radicals, and work as enzyme helpers.

❑ To boost immune function, prepare a tea made from goldenseal or echinacea.

❑ Take one to three capsules of yellow dock every hour until symptoms subside.

❑ Follow directions in Part Three for the Enzyme Kidney Flush.

❑ Gargle with a combination of fresh juiced pineapple and lemon or grapefruit in an 8-ounce glass.

Intermittent Claudication

Intermittent claudication (also called occlusive arterial disease of the limbs) is a deficiency in the amount of blood supplied to an exercising muscle. This is usually caused by atherosclerotic plaques (atheromas) that reduce or block the blood supply to the extremity. *Thromboangiitis obliterans* (also called Buerger's disease) is a serious example of intermittent claudication. In this condition, which affects primarily smokers, the blood vessels, particularly of the legs, become inflamed and circulation is obstructed. Tissues starve and gangrene results.

Symptoms

Intermittent claudication is marked by increasing pain, aches, cramps, or a tired feeling that occurs during walking (most commonly felt in the calf, but also may be felt in the hip, thigh, foot, or buttocks). Resting from one to five minutes will usually relieve the pain. The patient can then walk the same distance before this pain returns. Pain is increased when walking uphill or rapidly. When the distance a person can walk without pain begins to decrease, it is a clear indication that the disease is progressing. The arms can also be affected. In this case, similar symptoms can occur upon exertion. If not treated effectively, the occlusive disease may eventually close the blood vessel opening so that, ultimately, the pain occurs even at rest.

Enzyme Therapy

Systemic enzyme therapy is used to decrease the pain and reduce or remove the atherosclerotic plaques and prevent them from blocking or reducing the flow of blood to the body. In addition, the enzymes reduce fibrin activity and inflammation, stimulate the immune system, improve circulation, help speed tissue repair, bring nutrients to the damaged area, remove waste products, and improve health. Enzymes strengthen the body as a whole and build general resistance.

Digestive enzyme therapy is used to improve the digestion of food, reduce stress on the gastrointestinal mucosa, help maintain normal pH levels, detoxify the body, and promote the growth of healthy intestinal flora. Digestive enzymes also serve as replacements for the body's pancreatic enzymes, leaving the pancreatic enzymes free to perform other functions in the body, such as improving circulation and decreasing inflammation.

Enzyme Absorption System Enhancers (EASE) maximize enzyme activity. They can also improve the absorption and bioavailability of various nutrients and decrease the drain on the body's own digestive enzymes, thus prolonging their lives.

In addition to the standard multivitamin, multimineral, and multiglandular complexes recommended in the introduction to Part Two, the following supplements are recommended for the treatment of intermittent claudication. The following recommended dosages are for adults. Children between the ages of two and five should take one-quarter the recommended dosage, those between the ages of six and twelve should take one-half the recommended dosage, and those between the ages of thirteen and seventeen should take three-quarters the recommended adult dosage.

ESSENTIAL ENZYMES

Enzyme	Suggested Dosage	Actions
Formula Three—See page 38 of Part One for a list of products.	Take 200 mcg of each three times per day between meals.	Potent antioxidant enzymes.

AND SELECT ONE OF THE FOLLOWING:

Formula One (particularly Wobenzym N from Naturally Vitamin Supplements) —See page 38 of Part One for a list of products.	Take 8–10 tablets three times per day between meals for six months or until symptoms improve. Thereafter, take 6 tablets three times per day between meals.	Dissolves blood clots; fights inflammation, pain, and swelling; speeds healing; strengthens capillary walls.
Phlogenzym from Mucos	Take 4–5 tablets three times per day between meals for six months or until symptoms improve. Thereafter, take 3 tablets three times per day between meals.	Fights inflammation, swelling, and pain; dissolves blood clots; improves healing; strengthens capillaries.
Inflazyme Forte from	Take as indicated on the	Dissolves blood clots; fights

American Biologics	label between meals.	inflammation, pain, and swelling; speeds healing; strengthens capillary walls.
Bromelain	Take four to six 230–250 mg capsules per day.	Fights inflammation, pain, and swelling; speeds recovery after injury or surgery; stimulates the immune system.
A proteolytic enzyme	Take as indicated on the label between meals.	Dissolves blood clots; fights inflammation, pain, and swelling; speeds healing; strengthens capillary walls.

SUPPORTIVE ENZYMES

Enzyme	Suggested Dosage	Actions
A digestive enzyme formula containing protease, amylase, and lipase enzymes. See the list of digestive enzyme products in Appendix.	Take as indicated on the label with meals.	Improves the breakdown and absorption of proteins, carbohydrates, and fats; fights toxic build-up.

ENZYME ABSORPTION SYSTEM ENHANCERS (EASE)

Combination	Suggested Dosage	Actions
Bromelain Complex from PhytoPharmica or	Take as indicated on the label.	Reduces swelling; improves circulation; helps break up circulating immune complexes.
Cardio-Chelex from Life Plus or	Take as indicated on the label.	Supports healthy circulation; improves energy.
Cardio-Protector Rx from Phyto-Therapy, Inc. or	Take as indicated on the label.	Contains nutrients essential for heart and vascular health with enzymes to improve absorption of nutrients.
Kyolic Formula 102 (with enzymes) from Wakanuga or	Take 2 capsules or tablets with meals two to three times per day.	Strengthens the immune system; helps alleviate stress; helps reverse cardiovascular disorders; helps fight free radicals and inflammation and lowers cholesterol.
NESS Formula #14 or	Take 2 tablets twice per day between meals.	Provides herbs and cofactors that work with enzymes to improve circulation and decrease inflammation and swelling.
Protease Concentrate from Tyler Encapsulations	Take as indicated on the label.	Breaks up fibrin; fights inflammation.

ENZYME HELPERS

Nutrient	Suggested Dosage	Actions
Adrenal glandular	Take as indicated on the label.	Helps fight stress and adrenal exhaustion.
Aged garlic extract	Take three 300 mg capsules three times per day.	Lowers cholesterol and blood pressure; fights free radicals.
Aorta or heart glandular	Take 140 mg, or as indicated on the label.	Improves heart function.
Bioflavonoids, such as rutin and quercetin	Take 500 mg three times per day between meals.	Fight atherosclerosis; potent antioxidants; reduce capillary fragility.
Calcium	Take 1,000 mg per day.	Plays important roles in muscle contraction and nerve transmission.
Carotenoid complex	Take 5,000–10,000 IU per day.	Carotenoids are potent antioxidants; reduce arterial plaque.
Coenzyme Q_{10}	Take 75 mg one to two times per day with meals.	Potent antioxidant; decreases blood pressure and total serum cholesterol.
Essential fatty acids	Take 75–100 mg two to three times per day with meals.	Lower cholesterol level; strengthen the immune system; help strengthen cell membranes.
Lecithin	Take 3 capsules with meals, or as indicated on the label.	Emulsifies fat.
Magnesium	Take 500 mg per day.	Plays important roles in protein building, muscle contraction, nerve function, energy production, calcium absorption, and blood clotting.
Niacin or	Take 50 mg the first day, then increase dosage daily by 50 mg to 1,000 mg. If adverse effects occur, decrease dosage. Take doses higher than 1,000 mg only under the supervision of a physician.	Improves circulation by dilating blood vessels.
Niacin-bound chromium	Take 20 mg per day, or as indicated on the label.	
Vitamin B	Take 50 mg three times per day, or as indicated on the label.	B vitamins act as coenzymes and support neuromusculoskeletal function.
Vitamin C	Take 5,000–10,000 mg per day.	Essential to vascular function; potent antioxidant; improves the function of many enzymes; assists in converting cholesterol to bile salt.
Vitamin E	Take 1,200 IU per day from all sources (including the multivitamin complex in the introduction to Part Two).	Potent antioxidant; helps prevent cardiovascular disease; decreases risk of heart attack; stimulates circulation; strengthens capillary walls; prevents blood clots; maintains cell membranes.
Zinc	Take 50 mg per day. Do not take more than 100 mg per day.	Interacts with platelets in normal blood clotting; needed for white blood cell immune function, wound healing, growth, development, and thyroid hormone function.

Comments

❑ A number of herbs are well known for their ability to regulate cholesterol, including ginseng, chlorella, and hawthorn. Ginkgo biloba inhibits blood clot formation, while gotu kola and astragalus promote circulation. Bilberry and many other herbs contain anthocyanidins—phytochemicals that reduce blood vessel plaque formation, maintain blood flow in small vessels, protect blood vessels, help prevent cardiovascular disease and heart attacks, inhibit edema, and fight free radicals and inflammation. Raspberries, cranberries, grapes, red wine, black currants, and red cabbage also contain these helpful phytochemicals.

❑ Garlic, onions, shallots, chives, leeks, and scallions contain allium compounds, which are potent free-radical fighters that help lower cholesterol and fight circulatory disorders.

❑ Oolong and green tea contain epicatechin, which is a potent antioxidant that lowers blood pressure and strengthens capillaries.

❑ Green foods, such as Kyo-Green or Green Barley Extract, can help reduce blood pressure and fight free radicals and are effective against heart disease.

❑ For further information on dietary modifications, detoxification methods (such as cleansing, fasting, and juicing), exercises, positive mental attitude, meditation, and general suggestions, see Part Three.

Irritable Bowel Syndrome

Irritable bowel syndrome (IBS) is a common functional disorder that affects both the small and large intestine. The condition causes pain in the abdomen and diarrhea or constipation due to the gastrointestinal tract's sensitivity to stress, diet, drugs, and other irritants. A bout produces strong contractions of the gastrointestinal tract, causing cramping and diarrhea. This may be followed by a period of constipation. IBS represents about one-half of all gastrointestinal complaints or referrals to doctors and hospitals. Women are affected by IBS three times more often than are men.

Often no physical cause can be found for the condition, although hormones, diet, drugs, and emotional factors may aggravate gastrointestinal sensitivity. IBS patients are depressed, anxious, or neurotic more often than non-IBS patients. Onset and recurrence of IBS are frequently related to periods of emotional conflict and stress.

Symptoms

Erratic frequency of bowel movements, abdominal distress (which can include flatulence, bloating, and pain), and variation in stool consistency are the usual symptoms of irritable bowel syndrome. Headache, nausea, weakness, depression, fatigue, difficulty concentrating, anxiety, diarrhea, and constipation may also be associated with this condition.

Although IBS causes a great deal of discomfort, it does not usually lead to other gastrointestinal disorders, and there is no evidence that it is a precursor to cancer.

Enzyme Therapy

Systemic enzyme therapy is used to reduce inflammation, stimulate the immune system, improve circulation, help speed tissue repair, bring nutrients to the damaged area, and remove waste products. Enzymes strengthen the body as a whole, build general resistance, and enhance wellness.

Digestive enzyme therapy is used to improve the digestion of food, reduce stress on the gastrointestinal mucosa, help maintain normal pH levels, detoxify, and promote the growth of healthy intestinal flora. Digestive enzymes also serve as replacements for the body's pancreatic enzymes, leaving the pancreatic enzymes free to perform other functions in the body, such as decreasing inflammation.

Enzyme Absorption System Enhancers (EASE) maximize enzyme activity. They can also improve the absorption and bioavailability of various nutrients and decrease the drain on the body's own digestive enzymes, thus prolonging their lives.

In addition to the standard multivitamin, multimineral, and multiglandular complexes recommended in the introduction to Part Two, the following supplements are recommended for the treatment of irritable bowel syndrome. The following recommended dosages are for adults. Children between the ages of two and five should take one-quarter the recommended dosage, those between the ages of six and twelve should take one-half the recommended dosage, and those between the ages of thirteen and seventeen should take three-quarters the recommended adult dosage.

ESSENTIAL ENZYMES

Enzyme	Suggested Dosage	Actions
Formula One—See page 38 of Part One for a list of products. *or*	Take 10 tablets three times per day between meals for six months. Then decrease dosage to 5 tablets three times per day between meals.	Bolsters the immune system; fights free-radical formation; fights inflammation.
Papain *or*	Take 500 mg three times per day for six months between meals. Then decrease dosage to 300 mg three times per day between meals.	Stimulates immune function; fights free-radical formation; fights inflammation.
Bromelain	Take three to four 230–250 mg capsules per day until symptoms	Fights inflammation, pain, and swelling; speeds healing; stimulates the immune system.

or	subside. If severe, take 6–8 capsules per day.	
Trypsin with chymotrypsin	Take 200 mg of trypsin and and 10 mg of chymotrypsin three times per day between meals for six months. Then decrease dosage to 125 mg of trypsin and 5 mg of chymotrypsin three times per day between meals.	Stimulates immune function; fights free-radical formation; fights inflammation.
or		
A microbial protease	Take as indicated on the label.	Stimulates immune function; fights free-radical formation; fights inflammation.

SUPPORTIVE ENZYMES

Enzyme	Suggested Dosage	Actions
Formula Three—See page 38 of Part One for a list of products.	Take as indicated on the label.	Potent antioxidant enzymes.
A digestive enzyme product containing protease, amylase, and lipase enzymes. See the list of digestive enzyme products in Appendix.	Take as indicated on the label with meals.	Improves the breakdown and absorption of food nutrients; fights toxic build-up.

ENZYME ABSORPTION SYSTEM ENHANCERS (EASE)

Combination	Suggested Dosage	Actions
Biodias A Granules from Amano Pharmaceutical Co., Ltd.	Adults should take one packet in 8 ounces of water between or after meals three times per day. Children ages 11 to 14 should take two-thirds of a packet; children between the ages of 8 and 10 should take half of the packet; and children between the ages of 5 and 7 should take one-third of a packet.	Helps improve digestion of nutrients; relieves inflammation; promotes regeneration of mucous membranes; decreases stress on the stomach and intestines; reduces gastric pain.
or		
Inner Ecology from Prevail	Take as indicated on the label.	Helps to re-establish and maintain a healthy balance in your digestive system; stimulates the growth of helpful intestinal bacteria.
or		
Kyolic Formula 102 (with enzymes)	Take 2 capsules or tablets with meals two to three times per day.	Strengthens the immune system; helps alleviate stress; helps fight free radicals and inflammation; improves carbohydrate, protein, and fat digestion.
or		
Pro-Flora Plus from Tyler Encapsulations	Take as indicated on the label.	Helps to re-establish and maintain a healthy balance in your digestive system.
or		
Pro-Gest-Ade from	Take as indicated on	Contains nutrients essential for

Enzymatic Therapy	the label.	proper digestive function, plus enzymes to enhance digestion and absorption of proteins, fats, and carbohydrates.

ENZYME HELPERS

Nutrient	Suggested Dosage	Actions
Acidophilus and other probiotics	Take as indicated on the label.	Stimulates enzyme activity; improves gastrointestinal function.
Adrenal glandular	Take 100–400 mg per day.	Helps fight stress and adrenal exhaustion; fights inflammation.
Aged garlic extract	Take two 300 mg capsules three times per day or 1–3 capsules with meals twice per day.	Stimulates the immune system; fights free radicals.
Calcium	Take 1,000 mg per day.	Plays important roles in muscle contraction and nerve transmission.
Coenzyme Q_{10}	Take 30–60 mg one to two times per day with meals.	Important to immune system function; has antioxidant activity.
Essential fatty acid complex	Take 100–200 mg two to three times per day with meals; or follow the advice of your health-care professional.	Strengthens immunity; helps strengthen cell membranes.
Magnesium	Take 500 mg per day.	Plays important roles in protein building, muscle contraction, nerve function, energy production, and calcium absorption; necessary for glucose metabolism and synthesis of fats, proteins, and nucleic acid.
Thymus glandular	Take 100 mg per day, or as indicated on the label.	Assists the thymus gland in immune resistance; improves immune response; fights infections.
Vitamin A	Take up to 50,000 IU for no more than one month. (For higher doses, see a physician.) If pregnant, do not take more than 10,000 IU.	Needed for strong immune system and healthy mucous membranes; potent antioxidant.
Vitamin B complex	Take 100–150 mg three times per day.	B vitamins act as coenzymes.
Vitamin C	Take 10,000–15,000 mg per day; or 1–2 grams every hour until symptoms improve.	Fights infections, diseases, and allergies; improves wound healing; stimulates the adrenal glands; promotes immune system; powerful antioxidant.
Vitamin E	Take 1,200 IU per day from all sources (including the multivitamin complex in the introduction to Part Two).	Acts as an intercellular antioxidant; improves circulation; helps resist diseases at the cellular level.
Zinc	Take 15–50 mg per day. Do not take more than 100 mg per day.	Needed by more than 300 enzymes; necessary for white blood cell immune function; improves healing.

Comments

❏ A positive mental attitude, meditation, and relaxation techniques can help alleviate the symptoms of IBS.

❏ A number of herbs can fight inflammation, including alfalfa, echinacea, ginger root, ginkgo biloba, licorice, and white willow bark.

❏ See Part Three for information regarding dietary recommendations and detoxification methods (cleansing, fasting, and juicing).

❏ See the list of Resource Groups in the Appendix for an organization that can provide you with more information about irritable bowel syndrome.

Jock Itch

See FUNGAL SKIN INFECTIONS.

Kidney Disorders

Kidney disease is actually a catch-all term that includes diseases ranging from kidney stones and urinary tract infections to more serious disorders such as glomerulonephritis and polycystic kidney disease.

The main function of the kidneys is to remove excess water, sodium, and waste products from the blood. The kidneys regulate the body's potassium levels, as well as its pH balance. The kidneys also produce hormones that affect other organs, including erythropoietin, which stimulates the production of red blood cells, and renin (not to be confused with the enzyme rennin), which helps regulate blood pressure.

Within each kidney are about a million units called nephrons. These nephrons consist of filtering units, called glomeruli, which are attached to tubules. When blood enters each glomerulus, it is filtered and the remaining fluid, passes along the tubule, where water and chemicals are either removed from or added to the filtered fluid, depending on what the body needs. The final product is the urine, which we eliminate.

There are many different kidney disorders. *Nephrotic syndrome* is a collection of symptoms caused by diseases that affect the kidney. It includes prolonged loss of large amounts of protein into the urine, resulting in decreased levels of protein in the blood; increased levels of lipids in the blood, and retention of salt and water in the body. *Nephritis*

is inflammation of the kidneys. There are many different types of nephritis, depending on which part of the kidney is inflamed, how quickly it progresses, and how long it lasts. *Pyelonephritis* is a bacterial infection of one or both kidneys.

Renal vein thrombosis is a blockage of the renal vein, which carries blood from the kidneys. It results in the nephrotic syndrome. There are also conditions resulting from blockages of arteries to the kidney, such as *atheroembolic kidney disease* (in which emboli composed of fat clog the small renal arteries), *cortical necrosis* (in which the small arteries to the kidney cortex are blocked, causing tissue death), *kidney infarction* (in which the main artery to the kidney, the renal artery, is blocked, causing tissue death), and *malignant nephrosclerosis* (a condition associated with severe hypertension in which the small arteries in the kidneys are damaged). All of these conditions resulting from blockages of arteries can result in *kidney failure,* which is a very serious condition in which the kidneys fail to filter the blood of toxic substances.

There are several hereditary and congenital disorders of the kidneys. *Alport's syndrome* is a hereditary nephritis that may also cause deafness and eye problems. *Bartter's syndrome* is a hereditary disorder in which the kidneys excrete too much of the electrolytes (potassium, sodium, and chloride), leading to low blood levels of potassium, sodium chloride, and water, and high blood levels of the hormones aldosterone and renin. Children with this disorder grow very slowly. They may also have a mental disability and experience muscle weakness, excessive thirst, and excessive urination. *Fanconi's syndrome* is a rare hereditary disorder that results in excessive losses of glucose, bicarbonate, phosphates, and some amino acids in the urine. It may also be caused by exposure to heavy metals, a vitamin-D deficiency, the use of certain drugs, and certain disorders. Symptoms may include weakness and bone pain. *Hartnup disease* is a rare hereditary disorder in which amino acids are not absorbed well from the intestine, causing excessive amounts to be excreted in the urine. It causes skin rashes and brain abnormalities including mental disabilities. *Liddle's syndrome* is a rare hereditary disorder in which the kidneys excrete potassium but not enough sodium and water, resulting in hypertension. *Polycystic kidney disease* is a hereditary disorder in which cysts form in the kidneys, causing them to grow larger, but with less functioning tissue, ultimately causing kidney failure. In *medullary cystic disease,* a disease that can be either hereditary or due to a congenital defect, cysts develop deep within the kidneys, leading to kidney failure. *Renal glycosuria* is a disorder in which glucose is excreted in the urine, even though blood levels of glucose are normal or even low. *Cystinuria* is a hereditary disorder in which the kidneys excrete the amino acid cystine into the urine due to a defect of the kidney tubules. This often causes cystine stones to form in the urinary tract.

Symptoms

Although each kidney condition has its own unique characteristics, certain symptoms are associated with the majority of kidney problems. These include: frequent night urination, a reduced output of urine, fluid retention (puffiness in the face and limbs or resulting weight gain), pain in the side or small of the back, vomiting, fever, nausea, and loss of appetite. The urine may be cloudy or bloody. Both kidneys are usually affected by any kidney disease. If the ability of the kidneys to filter blood is seriously damaged, excess fluid and wastes may build up in the body. This causes kidney failure.

Enzyme Therapy

Systemic enzyme therapy is used to reduce pain, swelling, and inflammation in the kidneys and to stimulate the immune system, helping to fight circulating antigen-antibody (immune) complexes (a common problem in many kidney disorders). Enzymes improve circulation, help speed tissue repair, bring nutrients to the damaged area, remove waste products, improve health, strengthen the body as a whole, and build general resistance.

Digestive enzyme therapy is used to improve the digestion of food, reduce stress on the gastrointestinal mucosa, help maintain normal pH levels, detoxify the body, and promote the growth of healthy intestinal flora, thereby making the body healthier. Digestive enzymes also serve as replacements for the body's pancreatic enzymes, leaving them free to perform other functions in the body, such as boosting immunity and decreasing inflammation.

Enzyme Absorption System Enhancers (EASE) maximize enzyme activity. They can also improve the absorption and bioavailability of various nutrients and decrease the drain on the body's own digestive enzymes, thus prolonging their lives.

In addition to the standard multivitamin, multimineral, and multiglandular complexes recommended in the introduction to Part Two, the following supplements are recommended for the treatment of kidney disorders. The following recommended dosages are for adults. Children between the ages of two and five should take one-quarter the recommended dosage, those between the ages of six and twelve should take one-half the recommended dosage, and those between the ages of thirteen and seventeen should take three-quarters the recommended adult dosage.

ESSENTIAL ENZYMES

Enzyme	Suggested Dosage	Actions
Superoxide dismutase	Take 150 mcg per day.	Fights inflammation; potent antioxidant.

AND SELECT ONE OF THE FOLLOWING:

Phlogenzym from Mucos	Take 10 tablets three times per day between meals.	Fights inflammation, pain, and swelling; speeds recovery after injury or surgery; stimulates the immune system.
WobeMugos E from Mucos	Take 10 tablets three times per day between meals.	Fights inflammation, pain, and swelling; speeds recovery after injury or surgery; stimulates the immune system.
Serratiopeptidase	Take 10 mg two times per day between meals.	Fights inflammation; reduces pain and swelling.
Formula One— See page 38 of Part One for a list of products.	Take 10 tablets three times per day between meals.	Fights inflammation, pain, and swelling; speeds recovery after injury or surgery; stimulates the immune system.

SUPPORTIVE ENZYMES

Enzyme	Suggested Dosage	Actions
Catalase and glutathione peroxidase	Take as indicated on the label.	Potent antioxidant enzymes.
A digestive enzyme product containing protease, amylase, and lipase enzymes. See the list of digestive enzyme products in Appendix.	Take as indicated on the label with meals.	Improves the breakdown and absorption of food nutrients; fights toxic build-up.

ENZYME ABSORPTION SYSTEM ENHANCERS (EASE)

Combination	Suggested Dosage	Actions
Defense Formula from Prevail *or*	Take as indicated on the label.	Enhances the body's natural resistance; fights infections.
Renatone from Enzymatic Therapy	Take one or two tablets with each meal.	Contains nutrients essential for kidney function.

ENZYME HELPERS

Nutrient	Suggested Dosage	Actions
Acidophilus and other probiotics	Take as indicated on the label.	Improve gastrointestinal function; needed for production of biotin, niacin, folic acid, and pyridoxine, all important coenzymes; help fight harmful bacteria; stimulate enzyme activity.
Aged garlic extract	Take 2 capsules three times per day.	Fights inflammation; stimulates the immune system; potent antioxidant.
Green Barley Extract or Kyo-Green	Take as indicated on the label.	Potent free-radical fighters; rich in vitamins, minerals, and enzyme activity.
Kidney glandular	Take 100 mg per day.	Improves kidney function.
Niacin	Take 50 mg the first day, then increase dosage daily by 50 mg to 1,000 mg. If adverse effects occur, decrease dosage. Take	Dilates blood vessels; improves circulation.

or Niacin-bound chromium	doses higher than 1,000 mg only under the supervision of a physician. Take 20 mg per day, or as indicated on the label.	
Spirulina	Take as indicated on the label.	Rich in minerals, vitamins, enzymes, proteins, essential amino acids, and essential fatty acids, especially gamma-linolenic acid; good source of betacarotene; improves immune function.
Thymus glandular	Take 100 mg per day.	Assists the thymus gland in immune resistance; improves immune response; fights infections.
Vitamin A	Take 10,000–15,000 IU per day. If pregnant, take nore more than 10,000 IU per day.	Needed for strong immune system and healthy tissue; potent antioxidant.
Vitamin B complex	Take 150 mg three times per day, or as indicated on the label.	B vitamins act as coenzymes and support healthy tissue.
Vitamin C	Take 10,000–15,000 mg per day.	Fights infections, diseases, and allergies; improves wound healing; stimulates the adrenal glands; promotes immune system; powerful antioxidant.
Zinc	Take 50 mg per day. Do not take more than 100 mg per day from all sources.	Needed by over 300 enzymes; necessary for white blood cell wound healing.

Comments

❑ Fruits and vegetables (particularly green vegetables) should comprise over 60 percent of your diet. If your body cannot tolerate raw fruits and vegetables, increase your intake of digestive enzymes, or sauté or steam your produce.

❑ Cranberry juice can help acidify the urine, making the environment inhospitable to bacteria, and improve kidney disorders.

❑ Avoid alcohol, refined white flour, and sugar, which produce undesirable fermentation in the bowel and stimulate improper metabolism of insulin, chromium, and zinc, putting added strain on the kidneys. Small amounts of honey are allowed.

❑ Natural spices are fine, but chemical preservatives, particularly nitrites, should be avoided.

❑ Limit protein intake. Protein places a tremendous burden on the kidneys.

❑ The kidneys filter some 4,000 quarts of blood per day. In order to maintain the proper alkaline/acid kidney balance, the elimination of waste is critical. Sufficient amounts of liquid (such as spring or distilled water) should be consumed daily. (Avoid chlorinated water.) Fruit and vegetable juices (in addition to raw, whole, enzyme-rich foods, enzyme supplements, and helpers) should be drunk if you are nutritionally deficient.

❑ To detoxify the kidneys, follow the directions in Part Three for the Enzyme Kidney Flush.

❑ Gargle with the Enzymatic Gargle twice daily (see instructions in Part Three), or gargle with eight ounces of a combination of 50 percent fresh juiced pineapple and 50 percent lemon or grapefruit juice.

❑ If you notice blood in your urine, immediately seek medical care.

❑ See the list of Resource Groups in the Appendix for a list of organizations that can provide you with more information about kidney disorders.

❑ *See also* GLOMERULONEPHRITIS.

Lactase Deficiency

Lactase deficiency is a deficiency or total lack of the enzyme lactase. Normally produced in the small intestine, lactase breaks lactose, or milk sugar, down into simple sugars—glucose and galactose. Without sufficient lactase, when foods containing lactose are eaten, the lactose travels to the colon and sits undigested. Bacteria begin to work on it, fermenting it and causing characteristic symptoms of diarrhea and pain. Lactase-deficient individuals suffer symptoms anywhere from fifteen minutes to three hours after consuming milk or milk products. Many of these individuals are otherwise healthy, were able to drink milk as infants, have no known allergies, and are able to consume small amounts of milk in their coffee or cereal.

As infants, most of us have sufficient lactase to digest mother's milk. However, lactase levels begin to decline by the age of three (lactase levels decrease during childhood in nearly all populations). Except for humans, most animals drink milk only during nursing, with lactase levels decreasing after weaning. So low lactase levels may be the norm in adult animals, and humans, as well. However, the enzyme levels of some groups, such as Scandinavians and Northwestern Europeans, remain high through adulthood. (See "Lactase Deficiency by Population Groups" on page 285.)

Some people are born with an enzyme deficiency, but damage to the intestinal mucosa (caused by celiac disease, gastroenteritis, acute infectious diarrhea, or chronic bowel problems) can also create a deficiency state (this is called secondary lactase deficiency). Because the intestinal lining also contains maltase and sucrase, individuals suffering from intestinal damage may also have trouble digesting other disaccharides, such as maltose and sucrose. Maltase and sucrase levels in humans can be increased by following a diet high in fructose or sucrose, but, unfortunately, diet cannot alter the body's lactase levels.

Scientists aren't sure if lactase deficiency is a genetic trait or an adaptive response. Are some people lactase deficient because their ancestors didn't drink enough milk? Or did their ancestors avoid milk because they were lactase deficient? Which came first, the chicken or the egg?

Some people who believe they are allergic to milk may actually be deficient in lactase. An oral lactose tolerance test or the lactose breath hydrogen test may help to differentiate between a lactase deficiency and an actual allergy to the protein in cow's milk. Similar tests can be conducted to evaluate other disaccharide enzyme deficiencies.

Symptoms

Common symptoms of lactase deficiency include gas, intestinal bloating, cramps, diarrhea, and stomach distress. Many of these symptoms, often mistakenly attributed to allergies or food sensitivities, are actually due to enzyme deficiencies. In those without sufficient lactase activity, even a half cup of milk can produce symptoms (there are about 12 grams of lactose in a cup of milk).

Enzyme Therapy

Because lactose intolerance is caused by an absence or deficiency of lactase, it is possible to provide the missing lactase with tablets or liquid. The best known is probably LactAid, but there are many types on the market. In addition, most grocery stores carry milk with added lactase enzymes.

Digestive enzyme therapy is used to improve the digestion of lactose, reduce stress on the gastrointestinal mucosa, help maintain normal pH levels, detoxify the body, and promote the growth of healthy intestinal flora. Systemic enzyme therapy is used as supportive care to reduce any intestinal inflammation, stimulate the immune system, improve circulation, remove waste products, and improve health.

Enzyme Absorption System Enhancers (EASE) maximize enzyme activity. They can also improve the absorption and bioavailability of various nutrients and decrease the drain on the body's own digestive enzymes, thus prolonging their lives.

In addition to the standard multivitamin, multimineral, and multiglandular complexes recommended in the introduction to Part Two, the following supplements are recommended for the treatment of lactase deficiency. The following recommended dosages are for adults. Children between the ages of two and five should take one-quarter the recommended dosage, those between the ages of six and twelve should take one-half the recommended dosage, and those between the ages of thirteen and seventeen should take three-quarters the recommended adult dosage.

ESSENTIAL ENZYMES

Enzyme	Suggested Dosage	Actions
Lactase—See Appendix for digestive enzyme products that contain lactase. *or*	Take as indicated on the label.	Provides the missing enzyme to digest lactose.
A digestive enzyme product containing protease, amylase, and lipase enzymes. See the list of digestive enzyme products in Appendix.	Take as indicated on the label with meals.	Improves the breakdown and absorption of food nutrients; fights toxic build-up.

SUPPORTIVE ENZYMES

Enzyme	Suggested Dosage	Actions
Formula Three— See page 38 of Part One for a list of products.	Take 50 mcg three times per day between meals.	Potent antioxidant enzymes.
Formula One—See page 38 of Part One for a list of products.	Take as indicated on the label.	Stimulates immune function.

ENZYME ABSORPTION SYSTEM ENHANCERS (EASE)

Combination	Suggested Dosage	Actions
Biodias A Granules from Amano Pharmaceutical Co., Ltd.	Adults should take one packet in 8 ounce of water between or after meals three times per day. Children ages 11 to 14 should take two-thirds of a packet; children between the ages of 8 and 10	Helps improve digestion of nutrients; relieves inflammation; promotes regeneration of mucous membranes; decreases stress on the stomach and intestines; reduces gastric pain.

LACTASE DEFICIENCY BY POPULATION GROUPS

In the United States, there may be as many as 50 million people who cannot tolerate lactose. However, worldwide certain racial and ethnic groups are more affected than others. The following table shows the breakdown of lactase deficiency according to ethnicity.

POPULATION GROUP	PERCENT LACTASE DEFICIENT
African	50–80%
Asian	55–97%
Danish	3%
Mexican	71%
North American: African-American	70–80%
North American: Caucasian	5–20%
Puerto Rican	21%

or	should take half of the packet; and children between the ages of 5 and 7 should take one-third of a packet.	
Digest-eze RX from PhytoTherapy *or*	Take as indicated on the label.	Helps prevent gas, bloating, and diarrhea.
Lactazyme from Tyler Encapsulations *or*	Take as indicated on the label.	Aids in the digestion of dairy proteins and milk sugar.
Sugar Check from Enzymatic Therapy *or*	Take as indicated on the label.	Contains nutrients for improved sugar and carbohydrate metabolism and absorption.
Ultra-Zyme from Nature's Plus *or*	Take as indicated on the label.	Helpful fat, carbohydrate, and protein digestive aid.
Zygest from PhysioLogics	Take 1 or 2 capsules before meals to aid digestion; or take as directed by your health-care professional.	Digests milk sugars; helps the body digest proteins, carbohydrates, fats, and plant fibers.

ENZYME HELPERS

Nutrient	Suggested Dosage	Actions
Acidophilus	Take as indicated on the label.	Improves gastrointestinal function; needed for production of biotin, niacin, folic acid, and pyridoxine, important coenzymes; helps produce antibacterial substances, which reduces stress on the gastrointestinal tract; stimulates enzyme activity.
Fructooligo-saccharides	Take as indicated on the label.	Promotes growth of probiotics and thus a healthy gastrointestinal tract.
Vitamin B complex	Take 50 mg three times a day.	B vitamins are essential coenzymes to many enzymes.
Vitamin C	Take 1,000–8,000 mg per day.	Essential for tissue function; potent antioxidant; improves function of many enzymes.
Whey	Take 1 heaping tablespoon in food or water three times per day.	Improves digestion; works as a prebiotic.
Zinc	Take 50 mg per day.	Needed by many enzymes.

Comments

❑ Eat a primarily non-dairy diet rich in fruits and vegetables (they should comprise over 60 percent of your diet), particularly green, leafy vegetables containing calcium. Eat vegetables raw, blended, or juiced. If your body cannot tolerate raw fruits and vegetables, increase your intake of digestive enzymes, or sauté or steam your produce.

❑ Eat plenty of yogurt. The beta-galactosidase in yogurt is able to digest lactose in the intestine. This is why so many people who are lactose deficient are able to eat yogurt. It is also rich in calcium.

❑ See the list of Resource Groups in the Appendix for an organization that can provide you with more information about lactase deficiency.

Laryngitis

Laryngitis is an inflammation of the larynx (voice box) that causes you to "lose your voice." Although viral infection is the common cause, bacterial infection can also result in laryngitis. Laryngitis may occur during an infection of influenza, pneumonia, bronchitis, whooping cough, diphtheria, or measles; or due to allergic reactions, inhaling irritating substances, or overuse of the voice. Chronic laryngitis may be an indication of polyps or nodules on the vocal cords, or possibly cancer.

Symptoms

The most obvious sign of laryngitis is a hoarse voice. Other symptoms include pain, swelling, and inflammation of the larynx. Fever, sore throat, and malaise may accompany a severe infection.

Enzyme Therapy

Systemic enzyme therapy is used to help reduce inflammation, pain, and swelling of the larynx; stimulate the immune system; and improve circulation. Enzymes help speed tissue repair, bring nutrients to the damaged area, remove waste products, improve health, strengthen the body as a whole, and build general resistance.

Digestive enzyme therapy is used to improve the digestion of food, reduce stress on the gastrointestinal mucosa, help maintain normal pH levels, detoxify the body, and promote the growth of healthy intestinal flora, thereby making the body healthier so that it can combat laryngitis. Digestive enzymes also serve as replacements for the body's pancreatic enzymes, leaving the pancreatic enzymes free to perform other functions in the body, such as boosting immunity and decreasing inflammation.

Enzyme Absorption System Enhancers (EASE) maximize enzyme activity. They can also improve the absorption and bioavailability of various nutrients and decrease the drain on the body's own digestive enzymes, thus prolonging their lives.

In addition to the standard multivitamin, multimineral, and multiglandular complexes recommended in the introduction to Part Two, the following supplements are recommended for the treatment of laryngitis. The following recommended dosages are for adults. Children between the ages of two and five should take one-quarter the recommended dosage, those between the ages of six and twelve should take one-half the recommended dosage, and those between the ages of thirteen and seventeen should take three-quarters the recommended adult dosage.

ESSENTIAL ENZYMES

Enzyme	Suggested Dosage	Actions
Formula Three—See page 38 of Part One for a list of products.	Take 200 mcg of each three times per day between meals.	Potent antioxidant enzymes.

AND SELECT ONE OF THE FOLLOWING:

Bromelain	Take three to four 230–250 mg capsules per day until symptoms subside. If severe, take 6–8 capsules per day.	Fights inflammation, pain, and swelling; speeds recovery after injury or surgery; stimulates the immune system.
Formula One (particularly Wobenzym N from Naturally Vitamin Supplements)—See page 38 of Part One for a list of products.	Take 8–10 tablets three times per day between meals for six months or until symptoms improve. Thereafter, take 6 tablets three times per day between meals.	Fights inflammation, pain, and swelling; speeds healing; strengthens capillary walls.
Inflazyme Forte from American Biologics	Take as indicated on the label between meals.	Fights inflammation, pain, and swelling; speeds healing; strengthens capillary walls.
WobeMugos E from Mucos	Take 3 tablets three times per day between meals.	Fights inflammation, pain, and swelling; speeds healing; strengthens capillary walls.
A proteolytic enzyme	Take as indicated on the label between meals.	Fights inflammation, pain, and swelling; speeds healing; strengthens capillary walls.

SUPPORTIVE ENZYMES

Enzyme	Suggested Dosage	Actions
A digestive enzyme formula containing protease, amylase, and lipase. See the list of digestive enzyme products in Appendix.	Take as indicated on the label with meals.	Improves the breakdown and absorption of proteins, carbohydrates, and fats; fights toxic build-up.

ENZYME ABSORPTION SYSTEM ENHANCERS (EASE)

Combination	Suggested Dosage	Actions
Cold Zzap from Naturally Vitamin Supplements *or*	Take as indicated on the label.	Fights infection from viruses or bacteria; boosts immunity; has antioxidant activity.
Combat from Life Plus *or*	Take as indicated on the label.	Helps support the respiratory system; stimulates immune system function; helps control infections.
Connect-All from Nature's Plus *or*	Take as indicated on the label.	Supports tissue repair.
Inflamzyme from PhytoPharmica *or*	Take as indicated on the label.	Promotes tissue repair; immune system activator.
Kyolic Formula 102 (with enzymes) *or*	Take 2 capsules or tablets with meals two to three times per day.	Strengthens the immune system; helps alleviate stress; helps fight free radicals; and inflammation.
Protease Concentrate from Tyler Encapsulations *and*	Take as indicated on the label.	Breaks up fibrin; fights inflammation.
Bromelain Complex from PhytoPharmica *or*	Take as indicated on the label.	Reduces swelling and inflammation; improves circulation; helps break up any circulating immune complexes.
Sinease from Prevail	Take as indicated on the label.	Decongestant.

ENZYME HELPERS

Nutrient	Suggested Dosage	Actions
Acidophilus and other probiotics	Take as indicated on the label.	Improves gastrointestinal function, thereby improving the health of the entire body; needed for production of some B vitamins, which are coenzymes for many enzymes; helps produce antibacterial substances; stimulates enzyme activity.
Adrenal glandular	Take 100–400 mg per day.	Fights inflammation and adrenal exhaustion.
Amino-acid complex	Take as indicated on the label on an empty stomach.	Required for enzyme production.
Bioflavonoids	Take 100–300 mg twice per day.	Strengthen capillaries; potent antioxidants.
Green Barley Extract or Kyo-Green	Take 1 heaping teaspoon in 8 ounces of liquid three times per day.	Fights free radicals.
Kyolic Aged Garlic Liquid Extract with Vitamins B_1 and B_{12}	Take as indicated on the label.	Stimulates antioxidant activity and immune system function.
Lung glandular	Take as indicated on the label.	Improves lung function.
Vitamin A	Take 10,000–50,000 IU per day for no more than one month. (For higher doses, see a physician.)	Necessary for mucous linings, cell membrane structure of the throat and larynx, and strong immune system; potent antioxidant.
Vitamin B complex	Take 100–150 mg three times per day.	B vitamins act as coenzymes.
Vitamin C	Take 1–2 grams every hour hour until symptoms improve; or mix a heaping teaspoon of powder or crystalline vitamin C with	Fights infections, diseases, allergies, and common cold; essential for wound healing; potent antioxidant.

	6 to 8 ounces of water or orange or lemon juice (if lemon juice is used, you may sweeten with honey) and gargle a mouthful and swallow. Repeat every hour.	
Zinc or zinc gluconate lozenge	Take 50 mg per day. Do not exceed 100 mg per day from all sources.	Needed for white blood cell immune function; involved with more than 300 enzymes.

Comments

❏ Enzyme-rich fresh fruits and vegetables (particularly green vegetables) should comprise over 60 percent of your diet. If your body cannot tolerate raw fruits and vegetables, increase your intake of digestive enzymes, or sauté or steam your produce.

❏ Avoid alcohol, refined white flour, and sugar. These produce undesirable fermentation in the bowel and stimulate improper metabolism of insulin and chromium. This can overtax the body, leading to fatigue and the reduced ability to respond to stress and to fight inflammation and infection.

❏ Milk and milk products, which are mucus-formers, should be strictly limited. These foods cause increased congestion. Yogurt and cheeses are probably somewhat less harmful than whole milk because they produce less mucus than do other dairy products.

❏ Liver and kidney flushes are of value in detoxification (see Part Three for instructions for the Enzyme Kidney Flush and the Enzyme Toxin Flush).

❏ Twice daily, gargle with the Enzymatic Gargle (see Part Three for instructions) or with four ounces of fresh juiced pineapple blended with four ounces of lemon or grapefruit juice. Swallow after gargling.

❏ Hot broth or chicken or vegetable soup is helpful for cleaning out the body and sweating out toxins. Before making soup, be sure to remove the skin from the chicken (chicken skins contain many toxins). Sip the broth or soup as frequently as possible.

❏ Put five to six drops of eucalyptus in a sink or bowl of very hot water. Cover your head and shoulders with a large bath towel over the sink or bowl and inhale the vapors.

Leaky Gut Syndrome

Although its main purpose is to digest food and assist in nutrient absorption, the lining of the gastrointestinal tract also acts as a barrier to protect us and to limit the loss of microbes and microbial products from the gastrointestinal tract into the bloodstream. This barrier function is weakened in individuals stressed by hemorrhage, injury, immunosuppression, or sepsis. Many experts feel a problem in the gastrointestinal tract's barrier function can play a central role in the process of critical illness.

"Leaky gut syndrome refers to an abnormal increase in the permeability of the intestinal lining, which allows toxins and antigens to 'leak' from the intestines into the general circulation," according to Dr. Corey Resnick, well-known enzyme authority. This can result in faulty immune reactions or toxicity in practically any area of the body.

The large intestine contains waste products, bacteria, parasites, undigested protein, fungi, and other toxins. When the gut becomes "leaky," it allows these and other toxic materials that would normally be eliminated or repelled to sneak across the intestinal barrier and enter the bloodstream. An increase in intestinal permeability seems to be associated with a number of disorders, including allergies, autoimmune diseases, chronic conditions, and digestive problems. This type of "leakiness" can be caused by alcohol, caffeine, parasites, molds, fungi, food-borne bacteria, food additives, enzyme deficiencies, drugs (including aspirin, ibuprofen, cortisone, and prescription drugs), and a high intake of carbohydrates.

Not only do toxins and waste products escape from the large intestine into the bloodstream, but important microbes can as well. Factors such as bacterial overgrowth in the intestine, weakened immune function (due to antibiotic use), and burns and other injuries can promote a leaking of intestinal microbes into the bloodstream (a process called translocation).

Symptoms

The existence of a number of conditions may indicate the presence of leaky gut syndrome. There may be abdominal complaints, including gas, bloating, or cramping, as well as headache, poor concentration, fatigue, and memory loss. Some conditions that occur as a result of increased permeability include allergies, ankylosing spondylitis, celiac disease, chronic dermatological conditions, Crohn's disease (and other inflammatory bowel diseases), inflammatory joint disease, Reiter's syndrome, rheumatoid arthritis, and schizophrenia.

Enzyme Therapy

Systemic enzyme therapy is used to stimulate the immune system, reduce inflammation, improve circulation, help speed tissue repair, transport nutrients throughout the body, remove waste products, and improve health.

Digestive enzyme therapy is used to improve the digestion of food, reduce stress (which weakens the gut) on the gastrointestinal mucosa, help maintain normal pH levels, detoxify the body, promote the growth of healthy intestinal flora, and improve wellness. Dr. Corey Resnick states that dietary supplementation with enzymes can help to increase digestive function and correct or reduce the effects of leaky gut syndrome. Digestive enzymes also serve as replacements for the body's pancreatic enzymes, leaving the pancreatic enzymes free to perform other functions in the body, such as improving circulation and boosting immunity.

Enzyme Absorption System Enhancers (EASE) maximize enzyme activity. They can also improve the absorption and bioavailability of various nutrients and decrease the drain on the body's own digestive enzymes, thus prolonging their lives.

In addition to the standard multivitamin, multimineral, and multiglandular complexes recommended in the introduction to Part Two, the following supplements are recommended for the treatment of leaky gut syndrome. The following recommended dosages are for adults. Children between the ages of two and five should take one-quarter the recommended dosage, those between the ages of six and twelve should take one-half the recommended dosage, and those between the ages of thirteen and seventeen should take three-quarters the recommended adult dosage.

ESSENTIAL ENZYMES

Enzyme	Suggested Dosage	Actions
A digestive enzyme formula containing proteases, amylases, and lipases. For a list of digestive enzyme products, see Appendix. *or*	Take as indicated on the label.	Improves digestion by breaking down proteins, carbohydrates, and fats.
Pancreatin	Take as indicated on the label.	Contains proteases, amylases, and lipases.

SUPPORTIVE ENZYMES

Enzyme	Suggested Dosage	Actions
Formula One—See page 38 of Part One for a list of products.	Take 10 tablets three times per day between meals for six months. Then decrease dosage to 5 tablets three times per day between meals.	Bolsters the immune system; breaks up antigen-antibody complexes; fights free-radical formation; fights inflammation.
Formula Three—See page 38 of Part One for a list of products.	Take as indicated on the label.	Free-radical fighters.

ENZYME ABSORPTION SYSTEM ENHANCERS (EASE)

Combination	Suggested Dosage	Actions
Biodias A Granules from Amano Pharmaceutical Co., Ltd.	Adults should take one packet in 8 ounces of water between or after meals three times per day. Children ages 11 to 14 should take two-thirds of a packet; children between the ages of 8 and 10 should take half of the packet; and children between the ages of 5 and 7 should take one-third of a packet.	Helps improve digestion of nutrients; relieves inflammation; promotes regeneration of mucous membranes; decreases stress on the stomach and intestines; reduces gastric pain.
or		
Detox Enzyme Formula from Prevail	Take 1 capsule between meals one or two times per day with a 6- to 8-ounce glass of water or juice.	Helps rid the body of harmful toxins.
or		
Digest-eze RX from PhytoTherapy	Take as indicated on the label.	Helps prevent gas, bloating, and diarrhea.
Kyolic Formula 102 (with enzymes)	Take 2 capsules or tablets with meals two to three times per day.	Strengthens the immune system; helps alleviate stress; helps fight free radicals and inflammation; improves digestion and absorption of nutrients.
or		
Oxy-5000 Forte from American Biologics	Take as indicated on the label.	Helps the body excrete toxins; fights biochemical stressors, common infections, and degenerative processes.
or		
Pro-Flora from Tyler Encapsulations	Take as indicated on the label.	Helps maintain normal intestinal flora and enzymatic activity.
or		
Pro-Gest-Ade from Enzymatic Therapy	Take as indicated on the label.	Contains nutrients essential for proper digestive function, plus enzymes to enhance digestion and absorption of proteins, fats, and carbohydrates.
or		
Ultra-Zyme from Nature's Plus	Take as indicated on the label.	High potency digestive enzyme combination that improves digestion.
or		
Vita-C-1000 from Life Plus	Take as indicated on the label.	Boosts immune function.

ENZYME HELPERS

Nutrient	Suggested Dosage	Actions
Acidophilus or other probiotics	Take as indicated on the label.	Improves gastrointestinal function; needed for the production of B vitamins, coenzymes for many enzymes; helps produce antibacterial substances; stimulates enzyme activity.
Adrenal glandular	Take 100–400 mg per day.	Fights adrenal exhaustion; restores normal carbohydrate balance.
Aged garlic extract	Take two 300 mg capsules three times per day.	Potent antioxidant; improves digestion; stimulates healthy intestinal bacteria.
Amino-acid complex	Take as indicated on the label.	Amino acids serve as building blocks for the body's enzymes.
Betaine hydrochloric acid	Take 150 mg three times per day with meals.	Provides hydrochloric acid for digestion; if gastric upset occurs, discontinue use.

Fiber	Take as indicated on the label.	Provides bulk; needed for proper elimination.
Fructooligo-saccharides	Take as indicated on the label.	Increases peristaltic action of intestines; improves liver function and detoxification of the body; aids digestion.
Lecithin	Take 1,000 mg three times per day with meals.	Emulsifies fat.
Ox bile extract	Take 120 mg per day.	Aids digestion.
Vitamin A	Take 5,000 IU per day.	Needed by digestive glands to produce enzymes; required for healthy mucous linings and skin and cell membrane structure and a strong immune system; potent antioxidant.
Vitamin B complex	Take 50 mg three times per day.	B vitamins act as coenzymes.
Vitamin C	Take 1,000–2,000 mg per day.	Stimulates the adrenal glands to produce hormones (which also act as coenzymes); aids in detoxification; is a potent antioxidant; improves the function of many enzymes.
Whey	Take 1 heaping tablespoon in food or water three times per day with meals.	Improves digestion.
Zinc	Take 10–50 mg per day. Do not take more than 100 mg per day from all sources.	Needed by over 300 enzymes.

Comments

❑ Increase your intake of yogurt and other acidophilus-containing foods to help replace some of the intestine's beneficial bacteria.

❑ Eliminate all white flour, refined sugar, alcohol, coffee (and other sources of caffeine), and high fat foods.

❑ Eliminate any food to which you know are allergic. These could include wheat, milk, and eggs.

❑ A number of herbs can help bolster the immune system, including echinacea, garlic, astragalus, licorice, and maitake mushrooms.

❑ Foods rich in chlorophyll and green food products (including Kyo-Green and Green Magma) can help this condition. They are rich in enzymes, vitamins, minerals, and amino acids.

Leiomyomas

See FIBROIDS.

Lung Cancer

Lung cancer is the most common cancer among both men and women and is the leading cause of death from cancer. But it is largely preventable, since most lung cancer is caused by tobacco smoke (90 percent in men and 70 percent in women). In fact, cigarette smokers are ten times more likely to develop lung cancer than are nonsmokers. The World Health Organization estimates that nearly three million people die worldwide each and every year because of smoking. Other lung cancer risk factors include exposure to chemicals (arsenic and other substances), asbestos, or air pollution; certain conditions such as tuberculosis; and secondhand smoke inhaled by nonsmokers.

Most lung cancers (over 90 percent) start in the bronchi. This is called *bronchogenic carcinoma*. Lung cancer can also originate in the air sacs of the lungs (called alveoli). This type of cancer is called *alveolar cell carcinoma*. Cancers may also start in another part of the body, such as the breast, bone, skin, thyroid, colon, stomach, prostate, or cervix, and spread to the lungs.

Symptoms

The symptoms of lung cancer depend on the type and location, but one of the most obvious symptoms is usually a persistent cough. Other symptoms include chest pain, bloody sputum, shortness of breath, and recurring respiratory infections, such as bronchitis or pneumonia. Later symptoms of lung cancer include appetite and weight loss and weakness. Lung cancers may secrete hormones or other substances that may cause effects, such as metabolic, muscle, and nerve disorders, in places far removed from the lungs. Lung cancers may also spread to other parts of the body, such as the adrenal glands, bone, brain, or liver.

Enzyme Therapy

In the treatment of cancer, systemic enzyme therapy is a complementary therapy to medical therapy and should be integrated with the overall therapy program. The primary therapy for lung cancer includes surgery, chemotherapy, and radiation. Enzyme therapy is used to stimulate and stabilize the immune system, thus increasing the opportunities to destroy the residual tumor cells and decrease the rate of tumor growth; to help differentiate the healthy from the cancerous tissue, and better identify the tumor for surgical removal; to reduce inflammation; to improve wound healing and help speed tissue repair by bringing nutrients to the damaged area and removing waste products; to improve the long-term prognosis; to improve the patient's quality of life; to decrease the adverse side effects of the primary therapy; to reduce the cost of therapy (if enzyme therapy has begun early enough); to reduce the risk of complications, such as blood clots; and to decrease immune suppression caused by surgery and anesthesia.

Digestive enzymes are essential for the gastrointestinal tract and the body as a whole in helping to prevent the onset of cancer and improve the cancer patient's health because they aid the breakdown, digestion, and absorption of nutrients; decrease the enzyme production load of the pancreas, thereby helping to reduce its stress and increase its productive life; help prevent toxin build-up especially in the small and large intestine; stimulate friendly microflora production in the intestine; and facilitate elimination of potential cancer-causing waste from the food and cells of the body. Digestive enzymes also serve as replacements of the body's pancreatic enzymes, leaving the pancreatic enzymes free to perform other functions in the body, such as boosting immunity and improving circulation.

Enzyme Absorption System Enhancers (EASE) maximize enzyme activity. They can also improve the absorption and bioavailability of various nutrients and decrease the drain on the body's own digestive enzymes, thus prolonging their lives.

Studies using systemic enzyme therapy with radio- and chemotherapies have found that it results in a reduction in tumor size, weight gain, and a sense of well-being in the patient. This frequently leads to an improvement in the general state of health and quality of life and an increase in survival time for the patient.

In addition to the standard multivitamin, multimineral, and multiglandular complexes recommended in the introduction to Part Two, the following supplements are recommended for the treatment of lung cancer. The following recommended dosages are for adults. Children between the ages of two and five should take one-quarter the recommended dosage, those between the ages of six and twelve should take one-half the recommended dosage, and those between the ages of thirteen and seventeen should take three-quarters the recommended adult dosage.

ESSENTIAL ENZYMES

Enzyme	Suggested Dosage	Actions
WobeMugos E from Mucos	Take 2–4 tablets three times per day between meals.	When used before and after surgery, enzymes speed wound healing and decrease inflammation and immune complexes, while stimulating immune function. When used before, during, and after chemotherapy or radiotherapy, enzymes lead to earlier response to therapies and reduced side effects. When used for long-term and palliative therapy, enzymes reduce the anorexia, malnutrition, weakness, muscle wasting,
and		
Formula One— See page 38 of Part One for a list of products.	Take 10 tablets three times per day between meals.	

depression, and pain, commonly following treatment; prevent metastasis; and eliminate or prevent disturbances of lymphatic drainage.

SUPPORTIVE ENZYMES

Enzyme	Suggested Dosage	Actions
Bromelain	Take eight to ten 230–250 mg capsules per day. As soon as tumor growth is under control, the dose can be reduced gradually to a maintenance level of approximately 4 capsules per day.	Fights inflammation, pain, and swelling; stimulates the immune system; improves digestion.
A digestive enzyme formula containing protease, amylase, and lipase enzymes. For a list of digestive enzyme products, see Appendix.	Take as indicated on the label.	Improves digestion by improving the breakdown of proteins, carbohydrates, and fats.
Formula Three—See page 38 of Part One for a list of products.	Take as indicated on the label.	Potent antioxidant enzymes.

ENZYME ABSORPTION SYSTEM ENHANCERS (EASE)

Combination	Suggested Dosage	Actions
Antioxidant Complex from Tyler or	Take as indicated on the label.	Potent free-radical fighter and immune system stimulator.
Detox Enzyme Formula from Prevail or	Take 1 capsule between meals one or two times per day with a 6- to 8-ounce glass of water or juice.	Helps rid the body of harmful toxins.
Hepazyme from Enzymatic Therapy or	Take as indicated on the label.	Stimulates immune function.
Kyolic Formula 102 (with enzymes) or	Take 2 capsules or tablets with meals two to three times per day.	Strengthens the immune system; helps alleviate stress; helps fight free radicals and inflammation.
Oxy-5000 Forte from American Biologics or	Take as indicated on the label.	Helps the body excrete toxins; fights biochemical stressors, common infections, and degenerative processes.
Vita-C-1000 from Life Plus	Take as indicated on the label.	Helps fight cancer; boosts immune function.

ENZYME HELPERS

Nutrient	Suggested Dosage	Actions
Acidophilus or other probiotics	Take as indicated on the label.	Stimulates enzyme activity; improves gastrointestinal function, thus freeing pancreatic enzymes to boost immunity.

Bioflavonoids, such as rutin or quercetin	Take 150–500 mg three times per day.	Strengthen capillaries; potent antioxidants.
Magnesium	Take 250 mg per day.	Plays important roles in protein building, energy production, and synthesis of nutrients.
Thymus glandular	Take 30 mg per day.	Assists thymus gland (important in immune resistance); helps improve immune response fights cancer.
Vitamin A	Take 50,000–100,000 IU per day. Do not take more than 10,000 IU if you are pregnant). For amounts over 50,000 IU, consult a physician.	Stabilizes the cell membrane; improves response to therapy; needed for strong immune system and healthy tissues; potent antioxidant.
Vitamin C	Take 2–10 grams per day. For amounts over 10 grams, consult a physician.	Prevents capillary weakness; fights infection; improves wound healing; stimulates the adrenal glands; promotes immune system; powerful antioxidant.
Vitamin D	Take 3,000–6,000 IU per day.	Involved in regulating the transport and absorption of calcium and phosphorus.
Vitamin E	Take 1,200 IU per day from all sources (including the multivitamin complex in the introduction to Part Two).	Causes more efficient consumption of oxygen; acts as an intercellular antioxidant; improves circulation; helps resist diseases at the cellular level.
Zinc	Take 50 mg per day. Do not take more than 100 mg from all sources per day.	Needed for white blood cell immune function, wound healing, growth, and development; helps protect the body from heavy metal poisoning; needed by over 300 enzymes; fights free radicals.

Comments

❏ If you smoke, stop immediately. According to the American Cancer Society, those smokers showing evidence of precancerous lung tissue damage can return that tissue to normal if they stop smoking. (See "Help for Smokers" below.)

❏ Lung cancer treatment often involves high doses of vitamin A. A highly concentrated retinol palmitate is used with doses ranging from 300,000 IU (and higher) for a period of two years. These high doses can be used only under a doctor's close supervision.

❏ Cut down on your intake of dietary fat. A low-fat diet can slow the progression of cancer.

❏ See Part Three for information on diet, exercise, and detoxification methods.

❏ Coffee enemas are often used with cancer patients to stimulate the release of bile from the liver and to facilitate the removal of waste metabolites from the body and into the intestines for excretion. See Part Three for instructions for using the coffee enema.

❏ Physicians sometimes use enzyme injections or enzyme retention enemas in the treatment of cancer. Both methods allow a greater concentration of the enzymes to be absorbed than might be accomplished if the enzymes were taken orally. An enzyme retention enema is a means of instilling some enzymes into the colon to be retained. This allows the enzymes to be absorbed into the body's systems. Physicians use products specially formulated for these uses.

❏ See the list of Resource Groups in the Appendix for an organization that can provide you with more information about lung cancer.

Help for Smokers

Smoking is a major risk factor in the development of lung and other cancers. It also increases free-radical formation and causes killer diseases such as heart disease, atherosclerosis, and emphysema. Three new all-natural dietary products show great promise in easing or eliminating nicotine addiction.

Cig-No is an all-natural smoking deterrent created to help the smoker break the nicotine habit or to control it. Dr. Mary Cody formulated this product with the use of a homeopathic remedy called Plantago major, which reportedly works by creating an immediate aversion to tobacco, making the smoker feel as if he or she smoked too much. The product, which is FDA registered, is available in the form of a spray, capsules, and liquid drops from M.E. Cody Products, Inc./ 41 Bergenline Avenue/ Westwood, New Jersey 07675/ (800) 431-2582.

Metacalm-S is a new drug-free dietary supplement that helps manage anxiety, irritability, weight gain, and cravings associated with nicotine withdrawal. Metacalm-S is a tablet containing a patented, nontoxic blend of vitamins, nutrients, and herbs that reportedly can diminish or eliminate the craving for nicotine. It is available through Metabolic Technologies, Incorporated/ Northwestern University/ Evanston Research Park/ 1840 Oak Avenue/ Evanston, Illinois 60102/ (800) 923-CALM.

A third product Nico-Stop is an aid for those who want to break the smoking habit. This homeopathic product reportedly reduces the craving for nicotine and eases withdrawal symptoms. It is available from Enzymatic Therapy/ 825 Challenger Drive/ Green Bay, Wisconsin 54311/ (414) 469-1313.

❏ *See also* CANCER.

Lung Infection

See EMPYEMAS; PLEURISY; PNEUMONIA.

Lupus

See SYSTEMIC LUPUS ERYTHEMATOSUS.

Lymphedema

Lymphedema is chronic swelling due to an obstruction of the lymphatic vessels, preventing lymph drainage back into the blood. One either is born with lymphedema or acquires it later in life.

Congenital lymphedema occurs when one is born with too few lymphatic vessels to handle all of the lymph in the body. This problem most typically affects the legs and gets worse as time progresses. Rarely, it affects the arms. Congenital lymphedema is usually not evident until later in life, as there is less lymph produced in infants and the vessels can handle that amount. This condition affects women much more often than it does men.

Acquired lymphedema usually occurs after major surgery, particularly after breast cancer surgery, where the lymph nodes and vessels are removed or irradiated. If the lymphatic vessels get infected repeatedly, scarring may result, which may also cause lymphedema. This is very rare, however. Acquired lymphedema is more common than congenital lymphedema.

Symptoms

Lymphedema causes swelling in the extremities. Pain may result from the stretching of the skin caused by the swelling. Acquired lymphedema may affect the skin, causing it to become thick and ridged like elephant skin (elephantiasis).

Enzyme Therapy

According to experienced oncologists in Europe, the incidence of acquired lymphedema can be reduced through anti-inflammatory systemic enzyme therapy. According to Dr. Wolfgang Scheef of the world-famous Janker Clinic in Bonn, Germany, lymphedemas appeared in only 4.5 percent of the breast cancer patients taking enzymes after surgery while some 26 percent of the patients not receiving enzyme therapy developed lymphedema.

Systemic enzyme therapy is used to reduce inflammation, stimulate the immune system, improve circulation, help speed tissue repair, bring nutrients to the damaged area, remove waste products, enhance wellness, strengthen the body as a whole, and build general resistance.

Digestive enzyme therapy is used to improve the digestion of food, reduce stress on the gastrointestinal mucosa, help maintain normal pH levels, detoxify, and promote the growth of healthy intestinal flora, thereby making the body healthier so that it can combat the effects of lymphedema. Digestive enzymes also serve as replacements for the body's pancreatic enzymes, leaving the pancreatic enzymes free to increase circulation and decrease inflammation.

Enzyme Absorption System Enhancers (EASE) maximize enzyme activity. They can also improve the absorption and bioavailability of various nutrients and decrease the drain on the body's own digestive enzymes, thus prolonging their lives.

In addition to the standard multivitamin, multimineral, and multiglandular complexes recommended in the introduction to Part Two, the following supplements are recommended for the treatment of lymphedema. The following recommended dosages are for adults. Children between the ages of two and five should take one-quarter the recommended dosage, those between the ages of six and twelve should take one-half the recommended dosage, and those between the ages of thirteen and seventeen should take three-quarters the recommended adult dosage.

ESSENTIAL ENZYMES

Enzyme	Suggested Dosage	Actions
Formula One—See page 38 of Part One for a list of products. *or*	Take 10 tablets three times per day between meals for six months. Then decrease dosage to 5 tablets three times per day between meals.	Bolsters the immune system; breaks up any circulating immune complexes; fights free-radical formation; fights inflammation.
Papain *or*	Take 500 mg three times per day for six months between meals. Then decrease dosage to 300 mg three times per day between meals.	Stimulates immune function; breaks up circulating immune complexes; fights free radicals; stimulates immune function at the cellular level; fights inflammation.
Bromelain *or*	Take four to six 230–250 mg capsules per day.	Fights inflammation, pain, and swelling; speeds recovery after injury or surgery; stimulates the immune system.
Trypsin with chymotrypsin	Take 250 mg of trypsin and 10 mg of chymotrypsin three times per day between meals for six months. Then decrease dosage to 125 mg of trypsin and	Stimulates immune function; breaks up circulating immune complexes; fights free radicals; stimulates immune function at the cellular level; fights inflammation.

or	5 mg of chymotrypsin three times per day between meals.	
A proteolytic enzyme or combination	Take as indicated on the label.	Stimulates immune function; breaks up circulating immune complexes; fights free radicals; stimulates immune function at the cellular level; fights inflammation.

SUPPORTIVE ENZYMES

Enzyme	Suggested Dosage	Actions
Formula Three—See page 38 of Part One for a list of products.	Take as indicated on the label.	Potent antioxidant enzymes.
A digestive enzyme product containing protease, amylase, and lipase enzymes. See the list of digestive enzyme products in Appendix.	Take as indicated on the label with meals.	Improves the breakdown and absorption of food nutrients; fights toxic build-up.

ENZYME ABSORPTION SYSTEM ENHANCERS (EASE)

Combination	Suggested Dosage	Actions
Bromelain Complex from PhytoPharmica *or*	Take as indicated on the label.	Reduces swelling and inflammation; improves circulation; helps break up any circulating immune complexes.
Cardio-Chelex from Life Plus *or*	Take as indicated on the label.	Supports healthy circulation; improves energy.
Hepazyme from Enzymatic Therapy *or*	Take as indicated on the label.	Stimulates immune function.
Kyolic Formula 102 (with enzymes) *or*	Take 2 capsules or tablets with meals two to three times per day.	Strengthens the immune system; helps alleviate stress; helps fight free radicals; and inflammation.
Metabolic Liver Formula from Prevail *or*	Take as indicated on the label.	Improves liver function, which helps remove toxins from the body.
Protease Concentrate from Tyler Encapsulations	Take as indicated on the label.	Breaks up fibrin; fights inflammation.

ENZYME HELPERS

Nutrient	Suggested Dosage	Actions
Acidophilus	Take as indicated on the label.	Stimulates enzyme activity; improves gastrointestinal function, thus detoxifying the body.
Adrenal glandular	Take 100–400 mg per day.	Helps fight stress and adrenal exhaustion; fights inflammation.
Aged garlic extract	Take 2 capsules three times per day, or 1–3 capsules with meals twice per day.	Stimulates the immune system; fights free radicals.
Bioflavonoids	Take as indicated on the label.	Improve circulation; fight inflammation.
Coenzyme Q$_{10}$	Take 30–60 mg one to two times per day with meals.	Important to immune system function; has antioxidant activity.
Essential fatty acid complex	Take 100–200 mg two to three times per day with meals; or follow the advice of your health-care professional.	Strengthens immunity; helps strengthen cell membranes.
Thymus glandular	Take 100 mg per day, or as indicated on the label.	Assists the thymus gland in immune resistance; improves immune response; fights infections.
Vitamin A	Take up to 50,000 IU for no more than one month. For higher doses, see a physician. If pregnant, do not take more than 10,000 IU per day.	Needed for strong immune system and healthy mucous membranes; potent antioxidant.
Vitamin B complex	Take 100–150 mg three times per day.	B vitamins act as coenzymes.
Vitamin C	Take 10,000–15,000 mg per day; or take 1–2 grams every hour until symptoms improve.	Fights infections; improves wound healing; stimulates the adrenal glands; promotes immune system; powerful antioxidant.
Zinc	Take 50 mg per day. Do not take more than 100 mg per day.	Needed by over 300 enzymes; necessary for white blood cell immune function; improves healing.

Comments

❑ Enzyme-rich fresh fruits and vegetables (particularly green vegetables) should comprise 60 percent of your diet. If your body cannot tolerate raw fruits and vegetables, increase your intake of digestive enzymes, or sauté or steam your produce.

❑ Avoid toxins of all kinds, including alcohol, chlorinated water, cigarettes, caffeine, drugs, food additives and preservatives, refined sugar, and white flour.

❑ For one week, follow the Enzyme Kidney Flush program. The following week, perform the Enzyme Toxin Flush. Use these flushes once per month only. Follow the instructions in Part Three.

❑ Exercises are very important for proper circulation and good health. Exercise five to seven times per week for thirty to forty minutes. The exercises don't have to be elaborate to work—try fast walking (swinging your arms while walking briskly), riding a bicycle, or swimming.

❑ A positive mental attitude and daily meditation (15 minutes every day) can help your body re-energize itself.

Macular Degeneration

The macula is a part of the retina. It has hundreds of groups of nerve endings that serve to sharpen the visual image. Because of all of these nerve endings, the macula is the most light-sensitive part of the retina and the most important part for our vision. If the blood supply to the macula is decreased or interrupted, degeneration results, which greatly affects vision.

There are two types of macular degeneration. *Atrophic* (or dry) macular degeneration occurs when pigment is deposited on the macula without any scarring or fluid leakage. *Exudative* (or wet) macular degeneration occurs when a network of tiny blood vessels develops under the retina and fluid leaks from them. This causes scarring.

Macular degeneration is the leading cause of disturbed vision in the elderly. This condition is more common in white than in black people, but has the same incidence for both males and females. There may be hereditary predisposition for this condition, but the exact cause of it is not known.

Symptoms

Macular degeneration is marked by a sudden or slow loss of central visual acuity, usually in both eyes. It does not affect peripheral or color vision. Visual distortion in one eye may be the first symptom. Though macular degeneration can severely damage vision, it rarely causes total blindness.

Enzyme Therapy

An increase in enzymatic activity with enzyme supplementation should be initiated as soon as possible before further erosion of the blood vessel lining and additional problems develop. If there are deposits in the vessel walls and the condition advances, the vessels will begin to degenerate and enzyme therapy can only partly dissolve the degenerative build-up. Regardless of the patient's health status, an improvement in blood supply will probably cause some reduction in symptoms when oral enzymes are used.

Oral enzyme therapy is of value with macular degeneration because the enzymes and enzyme mixtures inhibit inflammation without suppressing immune function; decrease swelling by breaking up fibrin and by stimulating the body's own fibrinolytic activity; stimulate the break-up of thrombi (blood clots) and inhibit the formation of new thrombi; break up the lipids and fibrin that line the vessel walls, decreasing the blood vessel opening; and improve the blood supply and, therefore, the nutrition of the tissue.

Systemic enzyme therapy is used to slow down the macular degeneration, improve circulation to the eye by breaking up the fibrin and cholesterol deposits, and improve visual acuity. Additionally, enzymes stimulate the immune system, help speed tissue repair, bring nutrients to the damaged area, remove waste products, enhance wellness, and strengthen resistance.

Digestive enzyme therapy is used to improve the digestion and assimilation of food nutrients, reduce stress on the gastrointestinal mucosa, help maintain normal pH levels, detoxify, and promote the growth of healthy intestinal flora, thereby improving the overall health of the body. Digestive enzymes also serve as replacements for the body's pancreatic enzymes, leaving the pancreatic enzymes free to perform other functions in the body, such as improving circulation and boosting immunity.

Enzyme Absorption System Enhancers (EASE) maximize enzyme activity. They can also improve the absorption and bioavailability of various nutrients and decrease the drain on the body's own digestive enzymes, thus prolonging their lives.

The antioxidant enzyme superoxide dismutase and zinc are found in very high concentrations in certain tissues of the eye. Zinc concentration in the eyes is sometimes more than twice that found in other body tissues. When the usually high zinc content is abnormally decreased, it can lead to macular degeneration. It is claimed that macular degeneration patients treated with zinc experience improved vision as the accumulation of the degenerated visual pigment decreases. One reason for this could be that zinc is a cofactor for some 300 different enzymes in the body. Further, the combination of superoxide dismutase and zinc is essential in maintaining visual acuity.

In addition to the standard multivitamin, multimineral, and multiglandular complexes recommended in the introduction to Part Two, the following supplements are recommended for the treatment of macular degeneration.

ESSENTIAL ENZYMES

Enzyme	Suggested Dosage	Actions
Formula Three—See page 38 of Part One for a list of products.	Take 200 mcg three times per day between meals.	Potent antioxidant enzymes.

AND SELECT ONE OF THE FOLLOWING:

Enzyme	Suggested Dosage	Actions
Formula One—See page 38 of Part One for a list of products.	Take 8–10 tablets three times per day between meals for six months or until symptoms improve. Thereafter, take 6 tablets three times per day between meals.	Dissolves blood clots; speeds healing; strengthens capillary walls.
Inflazyme Forte from American Biologics	Take as indicated on the label between meals.	Dissolves blood clots; speeds healing; strengthens capillary walls.
Bromelain	Take four to six 230–250 mg capsules per day between meals.	Speeds healing; stimulates the immune system.
Enzyme complexes	Take as indicated on the	Dissolves blood clots; speeds

rich in proteases, plus amylases and lipases working synergistically — label between meals. — healing; strengthens capillary walls.

SUPPORTIVE ENZYMES

Enzyme	Suggested Dosage	Actions
A digestive enzyme formula containing protease, amylase, and lipase enzymes. See the list of digestive enzyme products in Appendix.	Take as indicated on the label with meals.	Improves the breakdown and absorption of proteins, carbohydrates, and fats; fights toxic build-up.

ENZYME ABSORPTION SYSTEM ENHANCERS (EASE)

Combination	Suggested Dosage	Actions
Antioxidant Complex from Tyler Encapsulations *or*	Take as indicated on the label.	Potent free-radical fighter.
Detox Enzyme Formula from Prevail *or*	Take 1 capsule between meals one or two times per day with a 6- to 8-ounce glass of water or juice.	Helps rid the body of harmful toxins.
Kyolic Formula 102 (with enzymes) *or*	Take 2 capsules or tablets with meals two to three times per day.	Improves circulation; helps fight free radicals; helps fight inflammation.
Oxy-5000 Forte from American Biologics *or*	Take as indicated on the label.	Helps the body excrete toxins; fights biochemical stressors, common infections, and degenerative processes.
Vita-C-1000 from Life Plus	Take as indicated on the label.	Helps maintain good vision.

ENZYME HELPERS

Nutrient	Suggested Dosage	Actions
Aged garlic extract	Take 3 capsules three times per day.	Lowers cholesterol and blood pressure; fights free radicals.
Bioflavonoids, such as rutin and quercetin	Take 500 mg three times per day.	Fights atherosclerosis; potent antioxidant; reduces capillary fragility.
Coenzyme Q_{10}	Take 75 mg one to two times per day with meals.	Potent antioxidant; decreases blood pressure and total serum cholesterol.
Essential fatty acid complex	Take 75-100 mg two to three times per day with meals.	Lowers cholesterol level; strengthens the immune system; helps strengthen cell membranes.
Heart glandular	Take 140 mg, or as indicated on the label.	Improves heart function, thus improving circulation to the eye.
Lecithin	Take 3 capsules with meals, or as indicated on the label.	Emulsifies fat.
Niacin *or*	Take 100–200 mg three times per day.	Improves circulation by dilating blood vessels.

Niacin-bound chromium	Take 20 mg per day, or as indicated on the label.	
Pycnogenol	Take 1,000 mcg per day.	Potent antioxidant.
Vitamin A	Take 15,000 IU per day, or as per your doctor's directions. Dosages over 50,000 IU should be taken only under the direction of a physician. If pregnant, take no more than 10,000 IU per day.	Necessary for skin and cell membrane structure and strong immune system; potent antioxidant.
Vitamin C	Take 5,000–10,000 mg per day.	Essential to vascular function; potent antioxidant; improves the function of many enzymes.
Vitamin E	Take 1,200 IU per day from all sources (including the multivitamin complex in the introduction to Part Two).	Potent antioxidant; stimulates circulation; strengthens capillary walls; prevents blood clots; maintains cell membranes.
Zinc	Take 50 mg per day. Do not take more than 100 mg per day.	Needed for white blood cell immune function; a coenzyme for many enzymes; an antioxidant; a zinc deficiency can lead to macular degeneration.

Comments

❑ Garlic, onions, shallots, chives, leeks, and scallions contain allium compounds, which are potent free-radical fighters that help lower cholesterol and fight circulatory disorders.

❑ Oats, wheat germ, olives, nuts, nut and seed oils, organ meats, and eggs contain alpha-tocopherol (vitamin E), which is known to help prevent cardiovascular disease and prevent blood clots.

❑ Raspberries, cranberries, grapes, red wine, black currants, and red cabbage contain anthocyanidins, which reduce blood vessel plaque formation and maintain blood flow in small vessels.

❑ Oolong and green tea contain epicatechin, which is a potent antioxidant that lowers blood pressure and strengthens capillaries.

❑ The herb bilberry protects and strengthens the blood vessels that feed the eyes.

❑ For additional information on dietary modifications, detoxification methods (such as cleansing, fasting, and juicing), exercises, positive mental attitude, meditation, and general suggestions, see Part Three.

Marie-Strümpell Disease

See ANKYLOSING SPONDYLITIS.

Measles

Measles (rubeola, or nine-day measles) is a highly contagious acute viral infection. German measles (rubella, or three-day measles) is a milder form. Both viruses are spread through airborne droplets from the throat, mouth, or nose of an infected person. The virus has a seven- to fourteen-day incubation period from the time one is infected with the virus to the time symptoms begin to appear. An infected person is contagious from two to four days before the characteristic rash appears until the rash disappears. Measles are most common in children. Once one contracts the virus, he or she is thereafter immune to the virus. Babies inherit a temporary immunity to the disease if their mothers ever had the disease, but immunity only lasts about a year after birth.

Symptoms

Fever (often high), fatigue, and appetitie loss are usually the first symptoms of measles, followed closely by sneezing and a runny nose. Soon after, a hacking cough, sore throat, red eyes, and sensitivity to light develop. Tiny white spots develop in the mouth and throat about two to four days later. A reddish, itchy rash develops about three to five days after the first symptoms began. It begins on the face, head, and neck and spreads to the rest of the body. At this point, the fever subsides and the other symptoms begin to disappear. The rash usually lasts four to seven days.

Enzyme Therapy

Systemic enzyme therapy is used to decrease the viral effects, stimulate the immune system, reduce inflammation, and decrease the potential for a secondary infection. Enzymes improve circulation, help speed tissue repair, bring nutrients to the damaged area, remove waste products, and enhance wellness.

Digestive enzyme therapy improves the digestion of food, reduces stress on the gastrointestinal mucosa, helps maintain normal pH levels, detoxifies the body, and promotes the growth of healthy intestinal flora, thereby making the body healthier so that it can fight an infection of the measles. Digestive enzymes also serve as replacements for the body's pancreatic enzymes, leaving the pancreatic enzymes free to perform other functions in the body such as boosting immunity and decreasing inflammation.

Enzyme Absorption System Enhancers (EASE) maximize enzyme activity. They can also improve the absorption and bioavailability of various nutrients and decrease the drain on the body's own digestive enzymes, thus prolonging their lives.

In addition to the standard multivitamin, multimineral, and multiglandular complexes recommended in the introduction to Part Two, the following supplements are recommended for the treatment of measles. The following recommended dosages are for adults. Children between the ages of two and five should take one-quarter the recommended dosage, those between the ages of six and twelve should take one-half the recommended dosage, and those between the ages of thirteen and seventeen should take three-quarters the recommended adult dosage.

ESSENTIAL ENZYMES

Enzyme	Suggested Dosage	Actions
Formula One (particularly Wobenzym N from Naturally Vitamin Supplements) —See page 38 of Part One for a list of products. *or*	Take 10 tablets three times per day between meals for six months. Then decrease dosage to 5 tablets three times per day between meals.	Bolsters the immune system; breaks up antigen-antibody complexes; fights free-radical formation and inflammation.
Inflazyme Forte from American Biologics *or*	Take 10 tablets three times per day for six months between meals. Then decrease dosage to 5 tablets three times per day between meals.	Bolsters immunity; breaks up circulating immune complexes; fights free-radical formation; fights inflammation.
Bromelain *or*	Take 400 mg three times per day between meals for six months. Then decrease dosage to 250 mg three times per day between meals.	Bolsters the immune system; breaks up antigen-antibody complexes; fights free-radical formation; inflammation.
Papain *or*	Take 500 mg three times per day for six months between meals. Then decrease dosage to 300 mg three times per day between meals.	Stimulates immune function; breaks up circulating immune complexes; fights free radicals and inflammation.
Trypsin with chymotrypsin *or*	Take 250 mg of trypsin and 10 mg of chymotrypsin three times per day between meals for six months. Then decrease dosage to 125 mg of trypsin and 5 mg of chymotrypsin three times per day between meals.	Stimulates immune function; breaks up circulating immune complexes; fights free radicals and inflammation.
Serratiopeptidase *or*	Take 5 mg three times per day between meals.	Fights inflammation; treats respiratory conditions.
A proteolytic enzyme or combination	Take as indicated on the label.	Bolsters the immune system; breaks up antigen-antibody complexes; fights free-radical formation; fights inflammation.

SUPPORTIVE ENZYMES

Enzyme	Suggested Dosage	Actions
Formula Three—See page 38 of Part One for a list of products.	Take as indicated on the label.	Free-radical fighters.

A digestive enzyme product containing protease, amylase, and lipase enzymes. See the list of digestive enzyme products in Appendix.	Take as indicated on the label with meals.	Improves the breakdown and absorption of food nutrients; fights toxic build-up.

ENZYME ABSORPTION SYSTEM ENHANCERS (EASE)

Combination	Suggested Dosage	Actions
Ecology Pak from Life Plus *or*	Take 2 or 3 tablets once or twice per day, preferably in the morning and no later than midday.	Assists with natural detoxification; stimulates immune function.
Hepazyme from Enzymatic Therapy *or*	Take as indicated on the label.	Stimulates immune function.
Inflamzyme from PhytoPharmica *or*	Take as indicated on the label.	Immune system activator; reduces inflammation; fights free radicals.
Kyolic Formula 102 (with enzymes) *or*	Take 2 capsules or tablets with meals two to three times per day.	Strengthens the immune system; helps alleviate stress; helps fight free radicals and inflammation.
Metabolic Liver Formula from Prevail	Take as indicated on the label.	Improves liver function to help detoxify the body.
NESS Formula #7	Take as indicated on the label.	Helps the body utilize essential fatty acids needed to build and maintain cell membranes.

ENZYME HELPERS

Nutrient	Suggested Dosage	Actions
Acidophilus and other probiotics	Take as indicated on the label.	Stimulates enzyme activity; improves gastrointestinal function, thereby improving the overall health of the body.
Adrenal glandular	Take 300 mg per day.	Helps fight stress and adrenal exhaustion; fights inflammation.
Arginine	Take as indicated on the label.	Accelerates healing; enhances thymus activity.
Coenzyme Q_{10}	Take 30–60 mg one to two times per day with meals.	Important to immune system function; has antioxidant activity.
Essential fatty acid complex	Take 100–200 mg two to three times per day with meals; or follow the advice of your healthcare professional.	Strengthens immunity; helps strengthen cell membranes.
Kidney glandular	Take as indicated on the label.	Helps to improve kidney function in eliminating toxins.
Kyolic Aged Garlic Liquid Extract with Vitamins B_1 and B_{12}	Take 10 drops in water or juice three times per day.	Stimulates immune system function.

Selenium	Take 50–100 mcg one or two times per day.	Acts as an enzyme activator; required by glutathione peroxidase.
Thymus glandular extract	Take 90 mg per day, or as indicated on the label.	Assists the thymus gland in immune resistance; improves immune response; fights infections.
Vitamin A or carotenoids	Take 40,000 IU of either per day. Take no more than 50,000 IU from all sources for no more than one month (for higher doses, see a physician). If pregnant, do not take more than 10,000 IU per day.	Potent antioxidants; needed for strong immune system and healthy tissue.
Vitamin C	Take 1–2 grams every hour until symptoms improve.	Fights infections, diseases, and allergies; improves healing; stimulates the adrenal glands; promotes immune system; powerful antioxidant.
Vitamin E	Take 1,200 IU per day from all sources (including the multivitamin complex in the introduction to Part Two).	Acts as an intercellular antioxidant; improves circulation; helps resist diseases at the cellular level.
Zinc (a chewable form may be more effective if you have lesions inside the mouth)	Take 50 mg per day. Do not take more than 100 mg per day from all sources.	Needed by over 300 enzymes; necessary for white blood cell immune function; improves healing.

Comments

❑ Echinacea, ginseng, goldenseal, pau d'arco, as well as maitake, reishi, and shiitake mushrooms fight infections.

❑ Enzyme-rich fresh fruits and vegetables (particularly green vegetables) should comprise 60 percent of your diet. If your body cannot tolerate raw fruits and vegetables, lightly sauté or steam them. Follow dietary recommendations in Part Three.

❑ Since measles are contagious, the patient should be kept isolated until recovered.

❑ Follow the detoxification programs in Part Three to help eliminate poisons and toxins that build up during any disease process.

❑ Pregnant women should avoid anyone with German measles (rubella) during the first three months of pregnancy because of the very high risk of birth defects. (A vaccine against rubella has been available since 1969.)

❑ If you take papain or bromelain, a chewable form can be very helpful. Chewable tablets begin to break down and be absorbed as the enzymes pass into the throat, so enzyme activity in the body begins more quickly.

Menstrual Cramps

See DYSMENORRHEA.

Migraines

See HEADACHES.

Mononucleosis

Mononucleosis is a mildly infectious disease caused by the Epstein-Barr virus (EBV), which infects cells of the immune system. The condition can occur anytime during life; nearly half of all children acquire the virus before they are five years old. It has been dubbed the "kissing disease," since it is often transmitted through kissing. Once acquired, the EBV virus remains in the body for life. This virus has been implicated as a causative factor in chronic fatigue syndrome (see this condition in Part Two).

Symptoms

Symptoms of mononucleosis usually begin to appear thirty to fifty days after infection with the virus. One of the most obvious symptoms of mononucleosis is extreme fatigue. It is usually accompanied by fever, pharyngitis, and swollen lymph glands. Nearly half of all mononucleosis sufferers develop an enlarged spleen or an enlarged liver. Not everyone, of course, experiences all symptoms. It usually takes several weeks to recover from the illness.

Enzyme Therapy

Systemic enzyme therapy is used to improve overall health and, therefore, relieve fatigue, reduce pain and inflammation, stimulate the immune system, and improve circulation. Enzymes transport nutrients throughout the body, remove waste products, and enhance wellness.

Digestive enzyme therapy is used to improve the digestion of food, reduce stress on the gastrointestinal mucosa, help maintain normal pH levels, detoxify the body, and promote the growth of healthy intestinal flora to improve the overall health of the body so that it can fight an infection of mononucleosis. Digestive enzymes also serve as substitutes for the body's pancreatic enzymes, leaving the pancreatic

enzymes free to perform other functions in the body, such as boosting immune function, decreasing inflammation, and improving circulation.

Enzyme Absorption System Enhancers (EASE) maximize enzyme activity. They can also improve the absorption and bioavailability of various nutrients and decrease the drain on the body's own digestive enzymes, thus prolonging their lives.

In addition to the standard multivitamin, multimineral, and multiglandular complexes recommended in the introduction to part Two, the following supplements are recommended for the treatment of mononucleosis. The following recommended dosages are for adults. Children between the ages of two and five should take one-quarter the recommended dosage, those between the ages of six and twelve should take one-half the recommended dosage, and those between the ages of thirteen and seventeen should take three-quarters the recommended adult dosage.

ESSENTIAL ENZYMES

Enzyme	Suggested Dosage	Actions
Formula One—See page 38 of Part One for a list of products. *or*	Take 10 tablets three times per day between meals for six months. Then decrease dosage to 5 tablets three times per day between meals.	Bolsters the immune system; breaks up antigen-antibody complexes; fights free-radical formation and inflammation.
Bromelain *or*	Take 500 mg three times per day between meals for six months. Then decrease dosage to 250 mg three times per day between meals.	Bolsters the immune system; breaks up circulating immune complexes; fights free-radical formation and inflammation.
Pancreatin *or*	Take 500 mg three times per day between meals for six months. Then decrease dosage to 125 mg three times per day between meals.	Stimulates immune function; breaks up antigen-antibody complexes; fights free-radical formation and inflammation.
Papain *or*	Take 600 mg three times per day for six months between meals. Then decrease dosage to 300 mg three times per day between meals.	Stimulates immune function; breaks up circulating immune complexes; fights free-radical formation and inflammation.
Any proteolytic enzyme	Take 10 tablets three times per day between meals for six months. Then decrease dosage to 5 tablets three times per day.	Bolsters the immune system; breaks up antigen-antibody complexes; fights free-radical formation and inflammation.

SUPPORTIVE ENZYMES

Enzyme	Suggested Dosage	Actions
Formula Three—See page 38 of Part One for a list of products.	Take as indicated on the labels.	Potent antioxidant enzymes.

A digestive enzyme product containing protease, amylase, and lipase enzymes. See the list of digestive enzyme products in Appendix.	Take as indicated on the label with meals.	Assists in the breakdown and absorption of food nutrients; fights toxic build-up.

ENZYME ABSORPTION SYSTEM ENHANCERS (EASE)

Combination	Suggested Dosage	Actions
Advanced Nutritional System from Rainbow Light *or*	Take as indicated on the label.	Potent free-radical fighter and immune system stimulator.
AminoLogic from PhysioLogics *or*	Take 1 capsule three times a day between meals; or follow the advice of your health-care professional.	Improves muscle and mental function; provides antioxidant support; improves and energy endurance.
Immuzyme from Tyler Encapsulations *or*	Take as indicated on the label.	Helps stimulate immune function.
NESS Formula #11	Take two capsules immediately after each meal; or take three capsules upon rising and at bedtime.	Helps support the immune system.

ENZYME HELPERS

Nutrient	Suggested Dosage	Actions
Acidophilus or other probiotic	Take as indicated on the label.	Stimulates enzyme activity; improves gastrointestinal function to improve the overall health of the body.
Adrenal glandular	Take 100–400 mg per day.	Helps fight stress and adrenal exhaustion; fights inflammation.
Amino Acid Caps from NOW	Take as indicated on the label.	Amino acids are the building blocks of enzymes.
Calcium	Take 1,000 mg per day.	Plays important roles in healthy cell function and is vital to all body functions.
Coenzyme Q_{10}	Take 30–60 mg one to two times per day with meals.	Important to immune system function; has antioxidant activity.
DHEA	Take 25 mg per day.	Supports immune, cardiovascular, and brain function.
Essential fatty acid complex	Take 100–200 mg two to three times per day with meals; or follow the advice of your health-care professional.	Strengthens immunity; helps strengthen cell membranes.
Kyolic Aged Garlic Liquid Extract with Vitamins B_1 and B_{12}	Take 10 drops in liquid or in food or place directly onto the tongue three times per day.	Stimulates immune system function.

Magnesium	Take 250 mg per day.	Plays important roles in protein building, energy production, and calcium absorption; necessary for the synthesis of fats, proteins, and nucleic acid.
Selenium	Take 50 mcg one or two times per day.	Acts as an enzyme activator; required by glutathione peroxidase; potent antioxidant.
Thymus glandular	Take 100 mg per day, or as indicated on the label.	Assists the thymus gland in immune resistance; improves immune response; fights infection.
Vitamin A	Take up to 50,000 IU for no more than one month. Take doses higher than 50,000 IU per day only under the supervision of a physician. If pregnant, do not take more than 10,000 IU per day.	Needed for strong immune system; potent antioxidant
Vitamin B complex with vitamin B_{12}	Take 200–250 mg three times per day.	B vitamins act as coenzymes; vitamin B_{12} fights chronic fatigue.
Vitamin C	Take 10,000 mg per day, or 1–2 grams every hour until symptoms improve.	Fights infections; improves healing; stimulates the adrenal glands; promotes immune system; powerful antioxidant.
Vitamin E	Take 1,200 IU per day from all sources (including the multivitamin complex in the introduction to Part Two).	Acts as an intercellular antioxidant; improves circulation; helps resist diseases at the cellular level.
Zinc	Take 50 mg per day. Do not take more than 100 mg per day.	Needed by over 300 enzymes; necessary for white blood cell immune function; improves healing.

Comments

❑ If you suffer from mononucleosis, get plenty of rest at least until the fever resolves or you feel more energetic.

❑ Avoid contact sports or heavy lifting for two months after suffering from mononucleosis to avoid the risk of rupturing the spleen.

❑ Eat a well-balanced diet with plenty of fresh fruit and vegetables, including carrots, celery, turnips, radishes, red beets, zucchini, cucumbers, cabbage, onions, green peppers, apples, pears, pineapples, papayas, lemons, oranges, and limes. See dietary recommendations in Part Three.

❑ Herbs and food concentrates that have immune stimulating and antioxidant effects and can help normalize body function include alfalfa, wheat grass, garlic, ginseng, comfrey, goldenseal, aloe vera, burdock, buckthorn, cayenne, and spirulina.

❑ Avoid sugar and other refined carbohydrates; coffee and other caffeine-containing beverages and drugs; tobacco; alcohol; salt; food additives, preservatives, and other chemicals that can weaken the immune system; and physical and emotional stress.

❏ Fasting, juicing, and the Enzyme Toxin Flush and Enzyme Kidney Flush can be used to cleanse and detoxify the body. See Part Three for further information.

❏ A positive mental attitude, relaxation, and meditation are of value. See Part Three for further information.

❏ If symptoms continue, seek the services of a physician well trained in the treatment of mononucleosis.

Multiple Sclerosis

Multiple sclerosis (MS) appears to be an autoimmune condition in which the body's immune system produces antibodies against the myelin sheath, causing inflammation and damage to the sheath. Myelin is a fatty substance that covers and protects our nerve fibers. When scar tissue (sclera) develops on the myelin, it disrupts nerve communication, distorting messages to and from the brain, and may impair speech, sight, or movement. MS is a chronic and sometimes progressive disease, characterized by one or more areas of inflammation and scarring of the myelin sheath.

Between 250,000 and 500,000 Americans have MS, and over 8,000 Americans are diagnosed as having multiple sclerosis each year. The disease more commonly affects women and occurs predominantly among caucasians. MS occurs more often in those born in temperate climates. It occurs much less frequently in those born in tropical climates and rarely, if ever, affects those born near the equator.

Although researchers know that MS occurs when scar tissue forms on the myelin sheath, they don't know why this happens. While it is likely that it is due to an autoimmune reaction, it is not known why this occurs. Heredity seems to play a role, as does environment.

It has been suggested that improper diet might trigger MS. A selenium deficiency may play a role. Selenium, a non-metallic element, is absolutely essential for some metabolic processes and cannot be made in the body. We ingest it in food, breathe it, and absorb it through the skin. We excrete it in urine, stool, perspiration, and exhalations. The intake of too little unsaturated fats may also play a role in the development of multiple sclerosis.

It is probable that viruses are involved in the myelin disturbance. Certain viruses could remain dormant for years in an organism after an initial measles or other viral infection, causing multiple sclerosis months to years after the first infection has disappeared.

The blood serum of MS patients contains higher concentrations of circulating immune complexes than that of healthy individuals (proof of MS's autoimmune nature). It is possible that predisposing factors (i.e., genetic predisposition, selenium deficiency, fatty-acid imbalance, viruses, etc.) all lead to a breakdown of the immune system and an uncontrolled reaction against the body itself. Undissolved immune complexes accumulate, resulting in damage to the nerve tissue.

Symptoms

Symptoms of MS can vary greatly and thus may occur for a number of years before the disease is accurately diagnosed. Symptoms depend upon the area in which the myelin is affected. If the myelin around the motor nerves is affected, then some movement may be affected. If sensory nerves are affected, then there may be problems with sensation. Symptoms usually begin to appear between the ages of 20 and 40. Common early symptoms include vague eye problems, such as blurred or double vision; numbness or tingling in the arms, legs, trunk, or face; weakness and/or loss of dexterity; and emotional or intellectual changes.

Later stage symptoms include increased muscle weakness, difficulty walking and maintaining balance, loss of bladder and bowel control, sexual impotence in men, dizziness or vertigo, tremors, and extreme fatigue. The disease is marked by periods of flare-ups and remissions. Flare-ups may be triggered by infection, or may occur without any trigger. High temperatures, including very warm weather, hot baths or showers, or fever, may intensify symptoms. Unfortunately, it is impossible to predict the severity, symptoms, or progress of MS.

Enzyme Therapy

Traditional MS medications generally fall into two categories. There are those that control or minimize specific symptoms (including fatigue, bladder and bowel problems, and spasticity), while other medications treat the underlying disease itself by slowing its progression with immunosuppressants, lessening the frequency of exacerbations with interferons, or shortening exacerbations with steroids.

Enzyme therapy helps treat both the symptoms and the underlying cause of MS. Corticosteroids, on the other hand, can cause serious long-term side effects. European research has shown enzymes to be more effective and less expensive than conventional medical therapy. A growing number of doctors and biochemists are firmly convinced that the suffering of MS patients can be reduced, their paralysis lessened, and many symptoms alleviated or even eliminated with the use of enzymes.

Systemic enzyme therapy is used to reduce circulating immune complexes, decrease inflammation, and stimulate the immune system. Enzymes improve circulation, help speed tissue repair, bring nutrients to the damaged area, remove waste products, improve health, strengthen the body as a whole, and build general resistance.

Digestive enzyme therapy is used to improve the digestion of food, reduce stress on the gastrointestinal mucosa, help maintain normal pH levels, detoxify the body, and promote the growth of healthy intestinal flora, thereby improving the overall health of the body so that it can deal with the effects of multiple sclerosis. Digestive enzymes also serve as replacements for the body's pancreatic enzymes, leaving the pancreatic enzymes free to perform other functions in the body, such as reducing inflammation and boosting immunity.

Enzyme Absorption System Enhancers (EASE) maximize enzyme activity. They can also improve the absorption and bioavailability of various nutrients and decrease the drain on the body's own digestive enzymes, thus prolonging their lives.

Dr. Christina Neuhofer, an internationally known researcher and medical physician, has used enzyme mixtures to treat multiple sclerosis. An MS patient herself, she first treated herself with the enzyme combinations. She experienced a dramatic improvement in her symptoms. Since that time, Dr. Neuhofer has treated hundreds of MS patients and conducted many studies on the efficacy of enzymes in treating this disease.

She has found that enzymes can improve eye symptoms (including partial blindness, eye pain, and double vision) and sphincter control of the urinary bladder and rectum. She has also found enzymes effective in treating problems of the senses and the cranial nerves. Enzyme therapy appears to have little influence upon spasticity, dizziness, tremors, and muscle coordination problems. Dr. Neuhofer says that after remission, enzyme therapy helps stabilize symptoms and reduces such characteristic symptoms as weariness; burning, tingling, and numbness; sensitivity; and nerve inflammation and degeneration.

During flare-ups, patients being treated with enzyme therapy showed prompt improvement and extensive regression of symptoms. If the disease progresses, it is essential that the enzyme dosage be increased at the first signs of exacerbation.

In her research, Dr. Neuhofer has found that, without exception, those patients whose nerves continued to deteriorate were previously treated with azathioprine, a common treatment for MS, for prolonged periods. According to Dr. Neuhofer, azathioprine seems to interfere with the healing response of enzyme treatment and destroy the body's own immune system when used for an extended period of time.

The enzyme treatment plan works in two phases: the flare-up phase and the maintenance phase. During the flare-up phase, increase the frequency and dosage levels of enzymes. During the maintenance phase, the frequency and dose level can gradually and carefully be decreased.

In addition to the standard multivitamin, multimineral, and multiglandular complexes recommended in the introduction to Part Two, the following supplements are recommended for the treatment of multiple sclerosis. The following recommended dosages are for adults. Children between the ages of two and five should take one-quarter the recommended dosage, those between the ages of six and twelve should take one-half the recommended dosage, and those between the ages of thirteen and seventeen should take three-quarters the recommended adult dosage.

ESSENTIAL ENZYMES

Enzyme	Suggested Dosage	Actions
Mulsal from Mucos	During the flare-up phase, take 10 coated tablets three times per day between meals. During the maintenance phase take 3–5 coated tablets three times per day for 3 to 6 months. Flare-up phase dosages should be taken for one week every three months, whether or not you are having a flare-up of symptoms.	Stimulates immune function; breaks up antigen-antibody complexes; fights free-radical formation; fights inflammation.
or		
Formula One—See page 38 of Part One for a list of products.	During the flare-up phase, take 10 coated tablets three times per day between meals. During the maintenance phase take 5 tablets three times per day between meals.	Stimulates immune function; breaks up antigen-antibody complexes; fights free-radical formation; fights inflammation.
and		
WobeMugos E From Mucos	During the flare-up phase, take 5 tablets twice per day —once in the afternoon and once in the evening— between meals. During the maintenance phase take 4 tablets twice per day.	

SUPPORTIVE ENZYMES

Enzyme	Suggested Dosage	Actions
Formula Three—See page 38 of Part One for a list of products.	Take as indicated on the label.	Free-radical fighters.
A digestive enzyme formula containing protease, amylase, and lipase enzymes. See the list of digestive enzyme products in Appendix.	Take as indicated on the label with meals.	Improves the breakdown and absorption of food nutrients, thus improving the overall health of the body; fights toxic build-up.

ENZYME ABSORPTION SYSTEM ENHANCERS (EASE)

Combination	Suggested Dosage	Actions
AminoLogic from PhysioLogics	Take 1 capsule three times per day between meals; or follow the advice of your health-care professional.	Increases energy and endurance; fights free radicals; improves muscle function.
or		

Hepazyme from Enzymatic Therapy *or*	Take as indicated on the label.	Stimulates immune function.
Inflamzyme from PhytoPharmica *or*	Take as indicated on the label.	Immune system activator; reduces inflammation; fights free radicals.
Oxy-Nectar from Nature's Plus *or*	Take as indicated on the label.	Antioxidant beverage.
Vita-C-1000 from Life Plus	Take as indicated on the label.	Boosts immune function; fights infection.

ENZYME HELPERS

Nutrient	Suggested Dosage	Actions
Acidophilus and other probiotics	Take as indicated on the label.	Stimulates enzyme activity; improves gastrointestinal function, thereby improving the overall health of the body; helps detoxify the body.
Adrenal glandular	Take 100–400 mg per day.	Helps fight stress and adrenal exhaustion; fights inflammation.
Calcium	Take 1,000 mg per day.	Plays important role in nerve transmission.
Coenzyme Q$_{10}$	Take 30–60 mg one to two times per day with meals.	Important to immune system function; has antioxidant activity.
Essential fatty acid complex	Take 100–200 mg two to three times per day with meals; or follow the advice of your health-care professional.	Strengthens immunity; helps strengthen cell membranes.
Kyolic Aged Garlic Liquid Extract with Vitamins B$_1$ and B$_{12}$	Take 10 drops in water three times per day.	Stimulates immune system function.
Lecithin	Take 1,000 mg three times per day.	Emulsifies fat.
Magnesium	Take 500 mg per day.	Plays important role in protein building, muscle contraction, nerve function, energy production, calcium absorption, and synthesis of fats, proteins, and nucleic acid.
Thymus glandular	Take 100 mg per day, or as indicated on the label.	Assists the thymus gland in immune resistance; improves immune response; fights infections.
Vitamin A	Take up to 50,000 IU for no more than one month. (For higher doses, see a physician.) If pregnant, do not take more than 10,000 IU.	Needed for strong immune system and healthy mucous membranes; potent antioxidant. Research indicates that high doses of vitamin A can sometimes improve MS.
Vitamin B complex	Take 100–150 mg three times per day.	B vitamins act as coenzymes and aid in neuromuscular development.
Vitamin C	Take 10,000–15,000 mg per day, or 1–2 grams	Fights infections, diseases, and allergies; improves wound

	every hour until symptoms improve. Reduce dosage if diarrhea occurs.	healing; stimulates the adrenal glands; promotes immune system; powerful antioxidant.
Vitamin E	Take 1,200 IU per day from all sources (including the multivitamin complex in the introduction to Part Two).	Acts as an intercellular antioxidant; improves circulation; helps resist diseases at the cellular level.
Zinc	Take 50 mg per day. Do not take more than 100 mg per day.	Needed by over 300 enzymes; necessary for white blood cell immune function; improves healing.

Comments

❏ Restrict your intake of "hard fats," that is highly saturated fats from meat and butter, as well as man-made fats, including margarine and shortening.

❏ Avoid all red meat. Eat white fish or chicken instead.

❏ Eating fried foods will add even more fat to your diet, so boil, broil, or bake your foods instead.

❏ If high doses of cortisone therapy have been used (for example, 1,000 mg for four days), it is important to begin high doses of enzyme therapy immediately after discontinuing the cortisone.

❏ Regular bowel movements are essential for detoxification. It is best to have two movements a day. In many cases, a juice fast will accomplish this goal (see Part Three for information on the juice fast). If you are constipated, take one teaspoon of an herbal laxative in an eight-ounce glass of water one hour after each meal. Follow this with an additional glass of water one hour later. Colonic irrigations (or enemas) can be used if regularity is not achieved within a week. Irrigations can be continued until movements are regular (but only under the supervision of a well-trained health-care professional). For maintaining regularity, fiber is also important in the diet. Normal bowel movements (once or twice a day) are absolutely essential for this treatment plan and cannot be overstressed.

❏ A coffee retention enema should be administered each day for a week (follow the directions in Part Three).

❏ Avoid exhaust, smoke, foods that have been sprayed with herbicides and fungicides, food additives (especially MSG), and foods with preservatives and artificial colors.

❏ Exercise, diet, adequate rest, and physical therapy can help.

❏ Stress can lead to an exacerbation. Therefore, counseling and stress-reduction techniques may be extremely beneficial in helping an MS patient remain independent and able to lead a full life.

❏ Proper diets help fight the symptoms of MS. This includes a low-fat diet, first discussed by Dr. Roy Swank in *The Multiple Sclerosis Diet Book* (New York: Doubleday, 1987) and

303

the McDougall Plan (*The McDougall Plan*, Clinton, NJ: New Win Publishing, Inc., 1983); the PUFA (polyunsaturated fatty acids) supplement diet (PUFAs include supplements of linolenic acid, linoleic acid, sunflower seed oil, safflower seed oil, and evening primrose oil); gluten-free diets; allergen-free diets; raw foods diets; and high-manganese diets.

❑ Physicians sometimes use enzyme injections and retention enemas in the treatment of multiple sclerosis. These are common practices in Germany and Austria. These methods allow a greater concentration of the enzymes to be absorbed than might be accomplished if the enzymes were taken orally. Physicians use products specially formulated for injections and enemas.

❑ Though MS is neither preventable nor curable, it can be controlled. Treatment can help alleviate some of the symptoms. Many patients choose to treat their disease with holistic therapy, including enzymes and other supplements, specific diets, exercise, and avoiding fatigue and stress. Some patients use applied kinesiology, chiropractic, massage therapy, acupuncture, herbalism, reflexology, and hydrotherapy for symptom relief.

❑ See the list of Resource Groups in the Appendix for a list of organizations that can provide you with more information about multiple sclerosis.

Mumps

Mumps is a contagious viral infection that causes painful swelling of the salivary glands. It sometimes affects other glands, particularly in adults. It most commonly affects children between the ages of 5 and 15, though it may affect anyone at any age. It is spread through infected airborne droplets of moisture or through direct contact with the saliva of an infected person; however, mumps is less contagious than chickenpox or measles. An infection with mumps usually affords one a lifelong immunity to the virus.

Symptoms

Symptoms of mumps usually begin fourteen to twenty-four days after infection with the virus. The first symptoms for most are usually headaches, chills, a loss of appetite, and malaise. About twelve to twenty-four hours later, the salivary glands show signs of infection, which may include swelling, pain upon chewing and swallowing, tenderness, and fever.

If the infection occurs after puberty, other glands may be affected, and complications may ensue. In men, one or both testes may become inflamed, which in rare cases may result in permanent damage. In women, the ovaries may become inflamed causing abdominal pain and, rarely, sterility.

Pancreatitis (inflammation of the pancreas) may also occur early in the infection. This condition is usually benign and disappears in about a week. Meningitis sometimes occurs as a complication of mumps. Most who develop it recover completely. Encephalitis rarely occurs as a complication, but if it does, it is likely to cause some type of permanent nerve or brain damage.

Enzyme Therapy

Systemic enzyme therapy is used to fight the viral infection, reduce swelling and inflammation, and stimulate the immune system. Enzymes improve the circulation, help speed tissue repair, bring nutrients to the damaged area, remove waste products, improve resistance, and enhance wellness.

Digestive enzyme therapy is used to improve the digestion of food, reduce stress on the gastrointestinal mucosa, help maintain normal pH levels, detoxify the body, promote the growth of healthy intestinal flora, and strengthen the body as a whole, thereby improving the overall health of the body so that it can fight a mumps infection. Digestive enzymes also serve as replacements for the body's pancreatic enzymes, leaving the pancreatic enzymes free to perform other functions in the body, such as boosting immunity and decreasing inflammation.

Enzyme Absorption System Enhancers (EASE) maximize enzyme activity. They can also improve the absorption and bioavailability of various nutrients and decrease the drain on the body's own digestive enzymes, thus prolonging their lives.

In addition to the standard multivitamin, multimineral, and multiglandular complexes recommended in the introduction to Part Two, the following supplements are recommended for the treatment of mumps. The following recommended dosages are for adults. Children between the ages of two and five should take one-quarter the recommended dosage, those between the ages of six and twelve should take one-half the recommended dosage, and those between the ages of thirteen and seventeen should take three-quarters the recommended adult dosage.

ESSENTIAL ENZYMES

Enzyme	Suggested Dosage	Actions
Formula One—See page 38 of Part One for a list of products. *or*	Take 10 tablets three times per day between meals for six months. Then decrease dosage to 5 tablets three times per day between meals.	Bolsters the immune system; breaks up antigen-antibody complexes; fights free-radical formation and inflammation.
Papain *or*	Take 500 mg three times per day for six months between meals. Then decrease dosage to 300 mg three times per day between meals.	Stimulates immune function; breaks up circulating immune complexes; fights free-radical formation and inflammation.

Bromelain	Take 400 mg three times per day between meals for six months. Then decrease dosage to 250 mg three times per day between meals.	Bolsters the immune system; breaks up antigen-antibody complexes; fights free-radical formation and inflammation.
or		
Trypsin with chymotrypsin	Take 250 mg of trypsin and 10 mg of chymo-trypsin three times per day between meals for six months. Then decrease dosage to 125 mg of trypsin and 5 mg of chymotrypsin three times per day between meals.	Stimulates immune function; breaks up circulating immune complexes; fights free-radical formation and inflammation.
or		
A proteolytic enzyme or combination	Take as indicated on the label.	Bolsters the immune system; breaks up antigen-antibody complexes; fights free-radical formation and inflammation.

SUPPORTIVE ENZYMES

Enzyme	Suggested Dosage	Actions
Formula Three—See page 38 of Part One for a list of products.	Take as indicated on the label.	Free-radical fighters.
A digestive enzyme product containing protease, amylase, and lipase enzymes. See the list of digestive enzyme products in Appendix.	Take as indicated on the label with meals.	Improves the breakdown and absorption of food nutrients; fights toxic build-up.

ENZYME ABSORPTION SYSTEM ENHANCERS (EASE)

Combination	Suggested Dosage	Actions
Detox Enzyme Formula from Prevail	Take 1 capsule between meals one or two times per day with a 6- to 8-ounce glass of water or juice.	Helps rid the body of harmful toxins.
or		
Ecology Pak from Life Plus	Take 2 or 3 tablets once or twice per day, preferably in the morning and no later than midday.	Assists with natural detoxification; stimulates immune function.
or		
Hepazyme from Enzymatic Therapy	Take as indicated on the label.	Stimulates immune function.
or		
Inflamzyme from PhytoPharmica	Take as indicated on the label.	Immune system activator; reduces inflammation; fights free radicals.

ENZYME HELPERS

Nutrient	Suggested Dosage	Actions
Acidophilus and other probiotics	Take as indicated on the label.	Stimulates enzyme activity; improves gastrointestinal func-

tion, thus improving the overall health of the body.

Adrenal glandular	Take 100–400 mg per day.	Helps fight stress and adrenal exhaustion; fights inflammation.
Aged garlic extract	Take 2 capsules three times per day, or 1–3 capsules twice per day with meals.	Stimulates the immune system; fights free radicals.
Coenzyme Q_{10}	Take 30–60 mg one to two times per day with meals.	Important to immune system function; has antioxidant activity.
Essential fatty acid complex	Take 100–200 mg two to three times per day with meals; or follow the advice of your health-care professional.	Strengthens immunity; helps strengthen cell membranes.
Thymus glandular	Take 100 mg per day, or as indicated on the label.	Assists the thymus gland in immune resistance; improves immune response; fights infections.
Vitamin A	Take up to 50,000 IU for no more than one month. (For higher doses, see a physician.) If pregnant, do not take more than 10,000 IU per day.	Needed for strong immune system and healthy mucous membranes; potent antioxidant.
Vitamin B complex	Take 100–150 mg three times per day.	B vitamins act as coenzymes.
Vitamin C	Take 10,000–15,000 mg per day, or 1–2 grams every hour until symptoms improve.	Fights infections, diseases, and allergies; improves wound healing; stimulates the adrenal glands; promotes immune system; powerful antioxidant.
Vitamin E	Take 1,000–2,000 IU per day.	Acts as an intercellular antioxidant; improves circulation; helps resist diseases at the cellular level; works closely with selenium to increase antioxidant activity.
Zinc	Take 50 mg per day. Do not take more than 100 mg per day.	Needed by over 300 enzymes; necessary for white blood cell immune function; improves healing.

Comments

❏ Eat foods that are immune system stimulators. Elderberry juice is good for swollen salivary glands. It is especially useful for children suffering from mumps, measles, and chickenpox, according to *John Heinerman's New Encyclopedia of Fruits and Vegetables* (Parking Publishing Co., 1995).

❏ To improve the overall health of the body, avoid alcohol, white flour, and refined sugar.

❏ Avoid chlorinated water, use bottled, distilled, or well-water instead.

❏ Perform the Enzyme Kidney Flush and the Enzyme Toxin Flush as outlined in Part Three.

Muscle Cramping

Muscle cramps are tight, painful, involuntary muscle contractions. They can be caused by an electrolyte imbalance (including calcium, magnesium, potassium, and sodium imbalances), nutritional imbalances, lack of water, or insufficient blood flow to the muscles. Muscle cramps can also be caused by very cold temperatures, overexercise, muscle irritation, or dehydration. Poor conditioning habits, such as exerting muscles before sufficiently warming them up, can cause cramping.

Symptoms

Symptoms of a muscle cramp include a sharp, cramping pain in a muscle, usually the legs. The cramps may be severe enough to keep you from walking. If left unattended with continued exercise, muscle cramps can lead to a torn muscle.

Enzyme Therapy

Systemic enzyme therapy is used to relieve the muscle cramps, decrease the pain, swelling, and inflammation, and improve circulation. Enzymes stimulate the immune system, help speed tissue repair, bring nutrients to the damaged area and remove waste products, enhance wellness, strengthen the body as a whole, and build general resistance.

Digestive enzyme therapy is used to improve digestion of food, reduce stress on the gastrointestinal mucosa, help maintain normal pH levels, detoxify, and promote the growth of healthy intestinal flora, thereby improving the overall health of the body to prevent muscle cramping. Digestive enzymes also serve as replacements for the body's pancreatic enzymes, leaving the pancreatic enzymes free to perform other functions in the body, such as improving circulation and decreasing inflammation.

Enzyme Absorption System Enhancers (EASE) maximize enzyme activity. They can also improve the absorption and bioavailability of various nutrients and decrease the drain on the body's own digestive enzymes, thus prolonging their lives.

In addition to the standard multivitamin, multimineral, and multiglandular complexes recommended in the introduction to Part Two, the following supplements are recommended for the prevention and treatment of muscle cramping. The following recommended dosages are for adults. Children between the ages of two and five should take one-quarter the recommended dosage, those between the ages of six and twelve should take one-half the recommended dosage, and those between the ages of thirteen and seventeen should take three-quarters the recommended adult dosage.

ESSENTIAL ENZYMES

Enzyme	Suggested Dosage	Actions
Formula One—See page 38 of Part One for a list of products.	Take a single dose of 10 tablets. An hour later, take 2 to 4 tablets hourly until cramping subsides. As you improve, take 3 to 5 tablets two times per day for as long as cramping persists. Continue taking 2–3 tablets per day between meals as a maintenance dose.	Decreases inflammation and pain; improves circulation; increases rate of healing; improves the oxygenation of tissues.
or		
Inflazyme Forte from American Biologics	Take as indicated on the label between meals.	Decreases inflammation, pain, and swelling; increases rate of healing; stimulates the immune system; improves oxygenation of the tissues.
or		
Enzyme complexes rich in proteases, plus amylases and lipases, working synergistically	Take as indicated on the label between meals.	Decreases inflammation, pain, and swelling; increases rate of healing; stimulates the immune system; improves oxygenation of the tissues.
or		
Bromelain	Take three to four 230–250 mg capsules per day until symptoms subside. If severe, take 6–8 capsules per day.	Fights inflammation, pain, and swelling; speeds recovery after injury; stimulates the immune system.

SUPPORTIVE ENZYMES

Enzyme	Suggested Dosage	Actions
Formula Three—See page 38 of Part One for a list of products.	Take 50 mcg three times per day between meals.	Potent antioxidant enzymes.
A digestive enzyme product containing protease, amylase, and lipase enzymes. See the list of digestive enzyme products in Appendix.	Take as indicated on the label.	Improves digestion by breaking down proteins, carbohydrates, and fats, thus improving the overall health of the body.

ENZYME ABSORPTION SYSTEM ENHANCERS (EASE)

Combination	Suggested Dosage	Actions
AminoLogic from PhysioLogics	Take 1 capsule three times per day between meals; or follow the advice of your health-care professional.	Improves muscle function; lends antioxidant support; improves energy and endurance.
or		
Connect-All from Nature's Plus	Take as indicated on the label.	Supports muscle and other tissue.
or		
Inflamzyme from PhytoPharmica	Take as indicated on the label.	Promotes tissue repair.
or		

| Magnesium Plus from Life Plus | Take one or two tablets one to three times per day. | Important for muscle relaxation, calcium utilization, energy production, and many enzyme reactions. |

ENZYME HELPERS

Nutrient	Suggested Dosage	Actions
Adrenal glandular	Take 100–400 mg per day.	Fights inflammation and adrenal exhaustion.
Aged garlic extract	Take 2 capsules three times per day.	Detoxifies and provides energy.
Alpha-linolenic acid	Take as indicated on the label.	Stimulates immune system; decreases inflammation.
Amino-acid complex that contains L-ornithine, L-glutamine, L-arginine, L-proline, and L-lysine	Take as indicated on the label on an empty stomach.	Builds muscle tissue.
An antioxidant combination, such as Maxi Flav or Phyto Vita Boost from Naturally Vitamin Supplements	Take as indicated on the label.	Fights free radicals.
Bioflavonoids	Take 150–500 mg three times per day.	Strengthen capillaries; potent antioxidants.
Branched-chain amino acids (isoleucine, leucine, and valine)	Take as indicated on the label between meals.	Often found together in a supplement, branched-chain amino acids aid in healing of muscle and act as fuel for muscles.
Calcium	Take 1,000 mg per day.	Plays important roles in muscle contraction and nerve transmission.
Chondroitin sulfate	Take 250 mg two to three times per day with meals; or follow the advice of your health-care professional.	Supports connective tissue; promotes cell hydration.
Coenzyme Q$_{10}$	Take 30–60 mg one to two times per day with meals; or follow the advice of your health-care professional.	Potent antioxidant.
Green Barley Extract from Green Foods Corporation or Kyo-Green from Wakunaga	Take as indicated on the label.	Potent free-radical fighter; rich in vitamins, minerals, and enzyme activity.
Magnesium	Take 500 mg per day.	Plays important role in muscle contraction, nerve function, and calcium absorption.
Vitamin B complex	Take 50 mg three times per day, or as indicated on the label.	B vitamins act as coenzymes and support neuromusculoskeletal function.
Vitamin C	Take 1,000–10,000 mg per day.	Helps to produce collagen and normal connective tissue; promotes immune system; powerful antioxidant.
Vitamin E	Take 1,200 IU per day from all sources (including the multivitamin complex in the introduction to Part Two).	Potent antioxidant; supplies oxygen to cells; improves level of energy.
Zinc	Take 15–50 mg per day. Do not take more than 100 mg per day.	Needed by over 300 enzymes.

Comments

❑ A well-balanced diet with high-protein foods is important. But reduce your intake of red meats and dairy products. Rely on vegetables for your protein. Eat plenty of enzyme-rich fresh fruits and vegetables, including citrus fruits, pineapple, papaya, figs, soybeans, sweet potatoes, whole grains, sprouted seeds, and sunflower seeds. See Part Three for more dietary recommendations

❑ Eliminate toxins, pesticides, food additives, and preservatives from your diet.

❑ Drink at least two 8-ounce glasses of fresh vegetable and fruit juices per day. For further information on detoxification methods (cleansing, fasting, and juicing), see Part Three.

❑ Initially, stretching exercises can help. Then, as the patient progresses, walking, bicycling, and swimming can keep the muscles flexible.

❑ Ice should be used for the first one to three days. Fill two or three small paper cups with water and freeze. When frozen, take one out, tear off the top half inch of paper revealing the ice. Place a paper towel or dish towel over the ice and apply to the painful area. Rub the affected area in a constant circulatory motion with the ice. Do not apply the ice directly to the skin. Apply the ice for five minutes, then keep the ice off the area for five minutes. Repeat this every one to two hours. The purpose is to stimulate blood flow to, and remove waste from, the area.

❑ If muscle cramping continues for longer than three days, alternate ice with heat packs every fifteen minutes until the cramping stops. "Blue ice" or a frozen vegetable packet works better for this application. Wrap it with a dish towel and apply to the painful area. For heat, use a hot water bottle or heating pad. If a hot water bottle is used, wrap in a towel and apply. If a heating pad is used, put a warm, moist towel on the patient, then place a sheet of wax paper over it, then the heating pad, and finally, a large bath towel. Make sure the wax paper is wide and long enough to cover the entire heating pad with approximately one inch more extending out on every side. The heating pad should never touch the moist towel or the patient's skin. Great care should be exercised to assure this does not happen.

❑ *See also* DYSMENORRHEA.

Myofascial Pain Syndrome

Myofascial pain syndrome (MPS) refers to pain in the muscles and fasciae (the tissue surrounding the muscle bundles). The term myofascial pain syndrome is sometimes used interchangeably with fibrositis, fibromyositis, myofasciitis, and fibromyalgia and describes musculoskeletal pain, fatigue, and multiple tender points. These points, called trigger points, are areas of extreme pain and irritability when pressure is applied.

MPS can be caused and perpetuated by mechanical stresses (including poor posture, a short leg, scoliosis, or a pinched nerve), nutritional inadequacies (primarily of the B vitamins, vitamin C, calcium, iron, and potassium), metabolic and endocrine inadequacies, psychological factors (including depression, anxiety, and tension), infections, allergies, and sleep disturbances.

Symptoms

Symptoms of myofascial pain syndrome include muscle spasm and pain in areas or points of hypersensitivity. There is also muscle stiffness and weakness, limited movement, and sometimes dysfunction of the autonomic nerves in the affected area. Although local pain may be present, symptoms are also usually felt in areas away from the trigger points. This musculoskeletal condition most commonly involves the neck, shoulder, upper and lower back, and sometimes the ribs and chest. Quite often, this condition is misdiagnosed as arthritis, bursitis, or even a superficial tissue disorder.

Enzyme Therapy

Systemic enzyme therapy is used to reduce pain, inflammation, tenderness, and muscle stiffness. Enzymes stimulate the immune system, improve circulation, help speed tissue repair, bring nutrients to the damaged area, remove waste products, strengthen the body as a whole, and build general resistance.

Digestive enzyme therapy is used to improve digestion of food, reduce stress on the gastrointestinal mucosa, help maintain normal pH levels, detoxify, and promote the growth of healthy intestinal flora, thereby improving the overall health of the body. Digestive enzymes also serve as replacements for the body's pancreatic enzymes, leaving the pancreatic enzymes free to perform other functions in the body, such as improving circulation and decreasing inflammation.

Enzyme Absorption System Enhancers (EASE) maximize enzyme activity. They can also improve the absorption and bioavailability of various nutrients and decrease the drain on the body's own digestive enzymes, thus prolonging their lives.

In addition to the standard multivitamin, multimineral, and multiglandular complexes recommended in the introduction to Part Two, the following supplements are recommended for the treatment of myofascial pain syndrome. The following recommended dosages are for adults. Children between the ages of two and five should take one-quarter the recommended dosage, those between the ages of six and twelve should take one-half the recommended dosage, and those between the ages of thirteen and seventeen should take three-quarters the recommended adult dosage.

ESSENTIAL ENZYMES

Enzyme	Suggested Dosage	Actions
Formula One (particularly Wobenzym N from Naturally Vitamin Supplements) —See page 38 of Part One for a list of products. *or*	Take a single dose of 10 tablets. An hour later, take 2–4 tablets hourly until the symptoms go away. As you improve, take 3–5 tablets two times per day for as long as symptoms persist. Continue taking a maintenance dose of 2–3 tablets twice per day between meals.	Decreases inflammation, pain, and swelling; increases rate of healing; stimulates the immune system.
Inflazyme Forte from American Biologics *or*	Take as indicated on the label between meals.	Decreases inflammation, pain, and swelling; increases rate of healing; stimulates the immune system.
Bromelain *or*	Take three to four 230–250 mg capsules per day until symptoms subside. If severe, take 6–8 capsules per day.	Fights inflammation, pain, and swelling; speeds recovery after injury or surgery; stimulates the immune system.
A proteolytic enzyme	Take as indicated on the label between meals.	Decreases inflammation, pain, and swelling; increases rate of healing; stimulates the immune system.

SUPPORTIVE ENZYMES

Enzyme	Suggested Dosage	Actions
Formula Three—See page 38 of Part One for a list of products.	Take 50 mcg three times per day between meals, or as indicated on the label.	Fights free radicals.
A digestive enzyme product containing protease, amylase, and lipase enzymes. See the list of digestive enzyme products in Appendix.	Take as indicated on the label with meals.	Improves the breakdown and absorption of food nutrients; fights toxic build-up.

ENZYME ABSORPTION SYSTEM ENHANCERS (EASE)

Combination	Suggested Dosage	Actions
AminoLogic from PhysioLogics *or*	Take 1 capsule three times per day between meals; or follow the advice of your health-care professional.	Improves muscle function; lends antioxidant support; improves energy and endurance.

Connect-All from Nature's Plus *or*	Take as indicated on the label.	Supports tissue repair.
Inflamzyme from PhytoPharmica *or*	Take as indicated on the label.	Promotes tissue repair.
Joint-Ease from Nature's Life *or*	Take as indicated on the label.	Decreases inflammation.
Magnesium Plus from Life Plus *or*	Take one or two tablets one to three times per day.	Important for muscle relaxation, calcium utilization, energy production, and many enzyme reactions.
Mobil-Ease from Prevail	Take one capsule between meals two to three times per day.	Has anti-inflammatory activity.

ENZYME HELPERS

Nutrient	Suggested Dosage	Actions
Adrenal glandular	Take 100–400 mg per day.	Fights inflammation and adrenal exhaustion.
Aged garlic extract	Take 2 capsules three times per day.	Detoxifies and energizes; helps strengthen muscle; decreases inflammation and blood pressure.
Bioflavonoids	Take 150–500 mg three times per day.	Strengthen capillaries; potent antioxidants.
Branched-chain amino acids (isoleucine, leucine, and valine)	Take as indicated on the label between meals.	Often found together in a supplement, branched-chain amino acids aid in healing of muscle and act as fuel for muscles.
Calcium	Take 1,000 mg per day.	Plays important role in muscle contraction and nerve transmission.
Chondroitin sulfate	Take 250 mg two to three times per day with meals; or follow the advice of your health-care professional.	Supports connective tissue; promotes cell hydration.
Magnesium	Take 500 mg per day.	Plays important role in muscle contraction, nerve function, and calcium absorption.
Vitamin A	Take 10,000 IU per day, or as indicated on label.	Potent antioxidant.
Vitamin B complex	Take 50 mg three times per day, or as indicated on the label.	B vitamins act as coenzymes and support neuromusculoskeletal function.
Vitamin C	Take 5,000–10,000 mg per day.	Helps to produce collagen and normal connective tissue; promotes immune system; powerful antioxidant.
Vitamin E	Take 1,200 IU per day from all sources (including the multivitamin complex in the introduction to Part Two).	Potent antioxidant; supplies oxygen to cells; improves level of energy.

Whey to Go from Solgar (a combination of free-form amino acids and protein)	Take as indicated on the label.	Maintains muscles; improves muscle function.
Zinc	Take 15–50 mg per day. Do not take more than 100 mg per day.	Needed by over 300 enzymes; important cofactor for SOD; essential for immune function and improved healing.

Comments

❑ Eat enzyme- and antioxidant-rich fresh foods, such as vegetables (spinach, parsley, broccoli), fruits (like pineapple and citrus fruits), nuts, and whole-wheat foods.

❑ Drink at least two 8-ounce glasses of fresh fruit and vegetable juice per day.

❑ Treating MPS involves interrupting the pain cycle. This can be done through a technique called "stretch and spray," where the skin over the trigger area is sprayed with a liquid coolant (such as ethyl chloride) and then stretched. Pressure, massage, ice massage, and electrical stimulation are also of benefit.

❑ Ice should be used for the first one to three days. Fill two or three small paper cups with water and freeze. When frozen, take one out, tear off the top half inch of paper revealing the ice. Place a paper towel or dish towel over the ice and apply to the painful area. Rub the affected area in a constant circulatory motion with the ice. Do not apply the ice directly to the skin. Apply the ice for five minutes, then keep the ice off the area for five minutes. Repeat this every one to two hours. The purpose is to stimulate blood flow to the area, remove waste from the area, and reduce inflammation.

❑ After three days, alternate ice with heat packs every fifteen minutes until the pain disappears. "Blue ice" or a frozen vegetable packet works better for this application. Wrap it with a dish towel and apply to the painful area. For heat, use a hot water bottle or heating pad. If a hot water bottle is used, wrap in a towel and apply. If a heating pad is used, put a warm, moist towel on the patient, then place a sheet of wax paper over it, then the heating pad, and finally, a large bath towel. Make sure the wax paper is wide and long enough to cover the entire heating pad with approximately one inch more extending out on every side. The heating pad should never touch the moist towel or the patient's skin. Great care should be exercised to assure this does not happen.

❑ Exercises (walking, bicycling, swimming, muscle strengthening, and stretching exercises), positive mental attitude, and meditation can speed recovery. See Part Three for further information.

❑ After conducting a history, physical examination, and X-rays, your physician will develop a treatment program. Your treatment program might include nutritional counseling;

chiropractic adjustments; manipulation; massage; physiotherapy, including pressure point therapy; acupressure/acupuncture; electrical stimulation; ice and/or heat application; ice spray; exercises; relaxation exercises; and postural training (including proper lifting, bending, standing, walking, driving, sitting, and sleeping techniques).

❏ See the list of Resource Groups in the Appendix for a list of organizations that can provide you with more information about myofascial pain syndrome.

Nausea

Nausea is the unpleasant feeling that you are about to vomit. Nausea can have several causes, including viral infections (such as the flu) and bacterial infections (including food poisoning). Nausea can also occur due to alcohol abuse, overeating, gallstones, an excess or a deficiency of stomach acid, and indigestion. It can be a reaction to strong, unpleasant smells or tastes, or it can be brought on by motion sickness, pregnancy, or migraines. Nausea can also be due to such emotional stressors as anxiety. Nausea is a symptom of several disorders, including certain forms of cancer, influenza, chickenpox, Crohn's disease, Meniere's disease, meningitis, mononucleosis, pancreatitis, Reye's syndrome, diabetes, scarlet fever, and stomach ulcers.

Symptoms

Symptoms of nausea include the feeling that you are going to vomit, excessive salivation, and stomach cramping.

Enzyme Therapy

Digestive enzyme therapy is used to improve the digestion of food, reduce stress on the gastrointestinal mucosa, help maintain normal pH levels, detoxify the body, and promote the growth of healthy intestinal flora. Systemic enzyme therapy is used to reduce any inflammation and stimulate the immune system. Enzymes can improve circulation, transport nutrients throughout the body, remove waste products, and improve health.

Enzyme Absorption System Enhancers (EASE) maximize enzyme activity. They can also improve the absorption and bioavailability of various nutrients and decrease the drain on the body's own digestive enzymes, thus prolonging their lives.

In addition to the standard multivitamin, multimineral, and multiglandular complexes recommended in the introduction to Part Two, the following supplements are recommended for the treatment of nausea. The following recommended dosages are for adults. Children between the ages of two and five should take one-quarter the recommended

dosage, those between the ages of six and twelve should take one-half the recommended dosage, and those between the ages of thirteen and seventeen should take three-quarters the recommended adult dosage.

ESSENTIAL ENZYMES

Enzyme	Suggested Dosage	Actions
A digestive enzyme product containing protease, amylase, and lipase enzymes. For the list of digestive enzyme products, see Appendix. *or*	Take as indicated on the label.	Improves digestion by breaking down proteins, carbohydrates, and fats.
Formula One—See page 38 of Part One for a list of products. *or*	Take 5 coated tablets or one heaping tablespoon of a granulated enzyme mixture three times per day between meals.	Fights inflammation and pain; speeds healing; stimulates immune function; fights free radical and immune complex formation; improves digestion.
If you have trouble digesting protein, try a protease, such as bromelain, carboxypeptidase, a microbial protease, papain, pepsin, peptidase, rennin, or trypsin and chymotrypsin. *or*	Take as indicated on the label with meals.	Break down proteins.
If you have trouble digesting carbohydrates, try an amylase, such as alpha-galactosidase, beta-glucosidase, carbohydrase, cellulase, diastase, glucoamylase, hemicellulase, invertase, lactase, maltase, phytase, or sucrase. *or*	Take as indicated on the label with meals.	Break down carbohydrates.
If you have trouble digesting fats, try an enzyme supplement containing lipase. *or*	Take as indicated on the label with meals.	Breaks down fats.
If you have trouble swallowing tablets or pills, try chewable enzyme supplements, such as bromelain or papain (papaya).	Take as indicated on the label.	Improves digestion.

SUPPORTIVE ENZYMES

Enzyme	Suggested Dosage	Actions
Formula Three—See page 38 of Part One for a list of products.	Take as indicated on the labels.	Potent antioxidant enzymes.

ENZYME ABSORPTION SYSTEM ENHANCERS (EASE)

Combination	Suggested Dosage	Actions
Biodias A Granules from Amano Pharmaceutical Co., Ltd.	Adults should take one packet in 8 ounces of water between or after meals three times per day. Children ages 11 to 14 should take two-thirds of a packet; children between the ages of 8 and 10 should take half of the packet; and children between the ages of 5 and 7 should take one-third of a packet.	Helps improve digestion of nutrients; relieves inflammation; promotes regeneration of mucous membranes; decreases stress on the stomach and intestines; reduces gastric pain.

ENZYME HELPERS

Nutrient	Suggested Dosage	Actions
Acidophilus or other probiotics, such as bifidobacterium; or Acidophilasé from Wakunaga, which also contains amyloytic, lipolytic, and proteolytic enzymes.	Take as indicated on the label.	Improves gastrointestinal function; needed for production of biotin, niacin, folic acid, and pyridoxine; helps produce antibacterial substances; stimulates enzyme activity.
Aged garlic extract	Take 2 capsules three times per day.	Potent antioxidant; improves digestion; stimulates healthy intestinal bacteria.
Betaine hydrochloric acid	Take 150 mg three times per day with meals.	Provides hydrochloric acid to improve digestion. If gastric upset occurs, discontinue use.
Fructooligosaccharides	Take as indicated on the label.	Increases peristaltic action of intestines; improves liver function.
Lecithin	Take 1,000 mg three times per day with meals.	Emulsifies fat.
Ox bile extract	Take 120 mg per day.	Aids digestion.
Vitamin A	Take 5,000 IU per day.	Needed by digestive glands to produce enzymes; promotes normal taste and appetite; required for healthy mucous linings and skin and cell membrane structure and a strong immune system; potent antioxidant.
Vitamin B complex	Take 50 mg three times per day.	B vitamins act as coenzymes to enzymes essential for digestion.
Vitamin B₁ (thiamin)	Take 150 mg three times per day.	Essential for carbohydrate metabolism, proper energy production, and appetite; improves production of hydrochloric acid, which improves digestion.
Vitamin C	Take 1,000–2,000 mg per day.	Stimulates the adrenal glands to produce hormones (which also act as coenzymes); aids in

(continued)

		detoxification; a potent antioxidant; improves the function of many enzymes; required for converting the coenzyme folic acid to its active form.
Whey	Take 1 heaping tablespoon three times per day with meals.	Improves digestion.
Zinc	Take 10–50 mg per day. Do not take more than 50 mg per day from all sources.	Needed by over 300 enzymes. Many zinc enzymes work to metabolize carbohydrates, alcohol, and essential fatty acids; synthesize proteins; and dispose of free radicals; important to maintain normal digestion.

Comments

❑ When possible, identify and eliminate any obvious cause of nausea (such as motion sickness) and treat any underlying condition.

❑ Light drinks (including carbonated drinks) and dry foods (such as crackers) can help reduce the stress on the stomach.

❑ Eat cultured foods (such as kefir, yogurt, miso), high-fiber foods (like whole grains), enzyme-rich foods (including pineapple and papaya), and fresh fruits and vegetables.

❑ Eat four to five small meals per day. Chew food longer to put less stress on your gastrointestinal tract.

❑ Avoid spicy and fatty foods, red meats, fried foods, dairy products, sugary foods, and caffeine. Decrease acid-forming food intake.

❑ See Part Three for directions for the Enzyme Kidney Flush and the Enzyme Toxin Flush programs.

Nephrotic Syndrome

See GLOMERULONEPHRITIS.

Neuritis

Neuritis is inflammation of a nerve or nerves. Motor nerves, sensory nerves, and cranial nerves can all be affected. Although the highest incidence of neuritis occurs in men between the ages of thirty and fifty years, neuritis may occur in individuals of any age and in both sexes.

The cause of neuritis can be trauma. Most frequently, there is an injury to a single nerve in a localized area. Pressure from casts, tumors, crutches, hemorrhage, or being in certain positions for an extended period of time (such as a squat position while gardening, or certain prolonged positions of the hand due to occupation) can cause nerve inflammation, as can exposure to radiation, cold, toxic agents (such as sulfonamides, heavy metals, or industrial poisons), or microorganisms (such as *Corynebacterium diphtheriae*, the bacterium that causes *diphtheria*), or an autoimmune reaction (as in Guillain-Barré Syndrome).

In some instances, neuritis can be caused by immune complexes that attach to healthy nerve tissues, causing tissue lesions and chronic inflammation, according to Dr. Franz Klaschka in his book *Oral Enzymes—New Approach to Cancer Treatment* (Munich, Forum Medizin, 1996). The waste products and toxins that result cause the development of additional immune complexes, and a vicious cycle develops.

Nutritional deficiencies and metabolic disorders can also result in neuritis. Alcoholism, pernicious anemia, beriberi, malabsorption syndromes, and pyridoxine deficiency are just a few causes and types of vitamin B-complex deficiencies that may cause neuritis. Neuritis can also be caused by metabolic disorders, such as diabetes mellitus and hypothyroidism.

Symptoms

Pain, swelling, tenderness, heat, and possible loss of function are the symptoms of neuritis. It can result in loss of sensation, muscle weakness and atrophy, and loss of reflexes or decreased deep tendon reflexes. One nerve or many nerves can be affected. Sometimes the skin over the affected area becomes tight and shiny and, in some instances, may not perspire normally.

When neuritis of the optic nerve occurs, there may be inflammation, pain, and sudden or gradual blurring of the eye with some loss of vision (or even blindness in severe cases). The blindness can be temporary if treated immediately. Atrophy of the involved muscles and possible damage to the eye can result.

Enzyme Therapy

Systemic enzyme therapy is used to reduce pain, swelling, and inflammation in the nerves; help speed tissue repair; stimulate the immune system; improve circulation; bring nutrients to the damaged area; remove waste products; strengthen the body as a whole; build general resistance; and enhance wellness.

Digestive enzyme therapy is used to improve digestion of food, reduce stress on the gastrointestinal mucosa, help maintain normal pH levels, detoxify the body, and promote the growth of healthy intestinal flora, thereby improving the overall health of the body. Digestive enzymes also serve as replacements for the body's pancreatic enzymes, leaving the pancreatic enzymes free to perform other functions in the body, such as decreasing inflammation and boosting immunity.

Enzyme Absorption System Enhancers (EASE) maximize enzyme activity. They can also improve the absorption and bioavailability of various nutrients and decrease the drain on the body's own digestive enzymes, thus prolonging their lives.

Successful treatment of this condition requires close cooperation between enzymes and enzyme helpers (especially the B complex vitamins), plus bioflavonoids and essential fatty acids.

In addition to the standard multivitamin, multimineral, and multiglandular complexes recommended in the introduction to Part Two, the following supplements are recommended for the treatment of neuritis. The following recommended dosages are for adults. Children between the ages of two and five should take one-quarter the recommended dosage, those between the ages of six and twelve should take one-half the recommended dosage, and those between the ages of thirteen and seventeen should take three-quarters the recommended adult dosage.

ESSENTIAL ENZYMES

Enzyme	Suggested Dosage	Actions
Superoxide dismutase	Take 150 mcg per day.	Fights inflammation; potent antioxidant.

AND SELECT ONE OF THE FOLLOWING:

Enzyme	Suggested Dosage	Actions
Bromelain	Take three to four 230–250 mg capsules per day until pain and inflammation subside. If pain is severe, take 6–8 capsules per day.	Fights inflammation, pain, and swelling; speeds recovery after injury or surgery; stimulates the immune system.
Serratiopeptidase	Take 10 mg two times per day between meals.	Fights inflammation; reduces pain and swelling.
Trypsin with chymotrypsin	Take 250 mg of trypsin and 10 mg of chymotrypsin three times per day between meals.	Fights inflammation, pain, and swelling; increases rate of healing; stimulates the immune system.
Pancreatin	Take 500 mg six times per day on an empty stomach.	Fights inflammation, pain, and swelling; increases rate of healing; stimulates the immune system; strengthens capillary walls.
Papain or a microbial protease	Take 300 mg two times per day between meals.	Fights inflammation; speeds recovery from injuries and surgery.

SUPPORTIVE ENZYMES

Enzyme	Suggested Dosage	Actions
Catalase and glutathione peroxidase	Take as indicated on the labels.	Potent antioxidant enzymes.
A digestive enzyme product containing protease, amylase, and lipase enzymes. See the	Take as indicated on the label with meals.	Improves the breakdown and absorption of food nutrients; fights toxic build-up.

list of digestive enzyme products in Appendix.

ENZYME ABSORPTION SYSTEM ENHANCERS (EASE)

Combination	Suggested Dosage	Actions
Anti-Flam Caps from Crystal Star Herbal Nutrition *or*	Take as indicated on the label.	Fights inflammation; reduces swelling.
Bromelain Plus from Enzymatic Therapy *or*	Take as indicated on the label.	Decreases swelling; improves circulation; helps break up any circulating immune complexes.
Inflamzyme from PhytoPharmica	Take as indicated on the label.	Reduces inflammation.

ENZYME HELPERS

Nutrient	Suggested Dosage	Actions
Acidophilus and other probiotics, such as Acidophilasé from Wakunaga, which also contains enzymes.	Take as indicated on the label.	Improve gastrointestinal function; needed for production of biotin, niacin, folic acid, and pyridoxine, all coenzymes; stimulate enzyme activity.
Adrenal glandular	Take 100–400 mg per day.	Fights inflammation and adrenal exhaustion.
Bioflavonoids, including rutin and quercetin	Take 500 mg three times per day.	Strengthen capillaries; potent antioxidants.
Calcium	Take 1,000 mg per day.	Plays important roles in muscle and nerve cell function.
Coenzyme Q$_{10}$	Take 30–60 mg one to two times per day with meals; or follow the advice of your health-care professional.	Potent antioxidant.
Essential fatty acids	Take 100–200 mg two to three times per day with meals; or follow the advice of your health-care professional.	Help strengthen cell membranes; support immune function.
Magnesium	Take 500 mg per day.	Plays important role in muscle contraction, nerve function, and calcium absorption.
Niacin *or*	Take 250 mg per day.	Dilates blood vessels, improving circulation, which improves the transport of nutrients to, and wastes from, the inflamed areas.
Niacin-bound chromium	Take 20 mg per day, or as indicated on the label.	
Thymus glandular	Take 100 mg per day.	Assists the thymus gland in immune resistance, improving immune response, which can improve neuritis caused by immune complexes.
Vitamin A	Take 5,000 IU per day.	Needed for healthy tissues; potent antioxidant.
Vitamin B$_1$	Take 100–200 mg per day.	Essential for the metabolism of carbohydrates, which are needed as an energy source for all cells and for nerve function.
Vitamin B$_5$	Take 100–200 mg per day.	Important in the metabolism of protein, fats, and carbohydrates; involved in synthesis of neurotransmitters (such as acetylcholine).
Vitamin B$_6$	Take 100–200 mg per day.	Needed for healthy nerves; assists enzymes that metabolize amino acids and fats; necessary for proper immune function; involved in the synthesis of several neurotransmitters and neurohormones (including epinephrine, serotonin, dopamine, and melatonin); assists in transport of potassium into cell, which assists nerve transmission.
Vitamin B$_{12}$	Take 900 mcg per day.	Needed for activation of folic acid (and vice versa); essential for the synthesis of choline and methionine and for brain and nerve cell function.
Vitamin C	Take 10,000–15,000 mg per day.	Promotes immune system; powerful antioxidant.
Vitamin E	Take 1,200 IU per day from all sources (including the multivitamin complex in the introduction to Part Two).	Potent antioxidant; supplies oxygen to cells; improves circulation and level of energy.

Comments

❏ Anthocyanidins, phytochemicals found in such foods as raspberries, cranberries, red wine, grapes, black currants, and red cabbage, can fight inflammation and maintain blood flow in the capillaries.

❏ Anti-inflammatory herbs include bilberry, calendula, marshmallow root, pine bark, St. John's wort, yarrow, and yucca.

❏ To relax the muscles, use blue vervain, feverfew, hops, kava kava, lobelia, rosemary, skullcap, valerian, white willow bark, wild lettuce, or wood betony.

❏ Fruits and vegetables (particularly green vegetables) should comprise over 60 percent of your diet. Beets, celery, onion, cauliflower, artichokes, turnips, and other vegetables should be eaten raw, blended, or juiced. If your body cannot tolerate raw fruits and vegetables, increase your intake of digestive enzymes, or sauté or steam your produce.

❏ To improve your overall health, avoid alcohol, refined white flour, and sugar. Small amounts of artificial sweeteners and honey are allowed. See dietary recommendations in Part Three.

❏ Gargle with a combination of four ounces of fresh juiced pineapple and four ounces of lemon or grapefruit juice in an 8-ounce glass, or with the Enzymatic Gargle (see Part Three for instructions for its preparation). This mixture is loaded with enzymes and vitamin C to help control inflammation.

❏ Mild daily exercise for thirty to forty minutes can help oxygenate the tissues and relieve trauma to the nerves. Exercises could include walking, bicycling, or swimming.

❏ If you have visual problems, see an eye specialist immediately. Prompt treatment is critical.

Obesity

Obesity is excessive body fat causing a person to weigh 20 percent or more over the normal weight for his or her height, build, age, and sex. It occurs when one consumes more calories than are used by the body. In the United States, the prevalence of obesity has increased dramatically over the last twenty years. In fact, according to one report, the obese now represent one-third of the adult population, and 58 million Americans are now overweight. Americans weigh about eight pounds more on average than they did ten years ago. How can this happen when there are diet centers of one kind or another on nearly every street corner?

There are many factors, in addition to a poor diet, that contribute to obesity. Throughout history, most societies suffered through periods of famine or food scarcity. For this reason, the human body has evolved so that it holds on to excess fat in order to get us through the lean times. Genetics play a large role in whether one becomes obese or not. Recent research seems to conclude that genetic factors affect up to 33 percent of body weight on average.

One's physical activity, of course, also plays a role. Our society as a whole is much more sedentary than it was 100 years ago, which probably accounts for the increase in the prevalence of obesity. We now have machines to do the things that were once done manually. We no longer have to walk to get to our destinations, and we have elevators and escalators to transport us to different stories of buildings.

Socioeconomic factors also play a role in obesity. Women in lower socioeconomic classes are twice as likely to be obese as those in higher socioeconomic groups. The exact reason for this correlation is unknown; however, those in more affluent societies tend to have more time and resources to concentrate on weight-loss measures than do those in less affluent societies, and obesity is more frowned upon in higher socioeconomic groups.

Many psychological factors can also play a role in obesity. Emotional upset and stress can lead to binge eating disorder, which, of course, can lead to obesity. Often, these strong emotions occur in those who are already overweight, as a result of societal prejudices against, and intolerance toward, them. This often leads to a vicious cycle of overeating leading to obesity leading to overeating.

Certain disorders and drugs can also lead to obesity. Brain damage and certain hormonal disorders can cause obesity, as can such drugs as corticosteroids and many used to treat psychiatric disorders.

Obesity also predisposes one to a number of conditions, including cardiovascular diseases, hypoglycemia, diabetes, glandular dysfunction, high blood pressure, and elevated blood cholesterol levels. Certain cancers, such as cancer of the breast, ovaries, uterus, colon, rectum, and prostate, are also more common in the obese, as are menstrual disorders. Statistics prove that those who avoid obesity live longer and healthier lives.

Enzyme Therapy

Recently, I visited Japan for a week. My hosts were incredibly gracious, and even though we ate constantly, I lost ten pounds. How could this happen? Because we were eating live, enzyme-rich foods. My diet consisted mainly of fresh, raw, or very lightly cooked vegetables, fermented foods, and rice. Everyone I encountered in Japan was thin and had gorgeous skin (especially the women). Why? "Our diet and lifestyle," responded my host. "We eat fermented foods, as many fresh vegetables as possible, rice, fresh fish, and only occasionally, meat." As mentioned in Chapter 4 of Part One, fermented foods are rich in enzymes, as are fresh vegetables. The rice was cooked, so was not a significant source of enzymes, but it was low in fat. The fresh fish, uncooked, still had all of its natural "live" enzymes.

Because I have serious reservations about eating raw fish (which can often contain parasites), I ingest my enzymes in the form of fresh fruits and vegetables and in supplements (upon arising in the morning, before every meal, and at bedtime). I also take an enzyme-rich vegetable and grain food extract throughout the day (and don't feel hungry). Enzyme-rich food extracts help my body run more efficiently. I don't have the usual mid-morning and mid-afternoon energy lows. I feel more energetic throughout the day . . . with less desire to eat.

Digestive enzyme therapy is used to improve the digestion and absorption of food nutrients, reduce stress on the gastrointestinal mucosa, help maintain normal pH levels, detoxify the body, and promote the growth of healthy intestinal flora.

Systemic enzyme therapy is used to stimulate the immune system, improve circulation, transport nutrients throughout the body, remove waste products, and improve health. Systemic enzyme therapy also helps lower oxidized low-density lipoproteins. This type of cholesterol is the most dangerous form and can cause atherosclerosis and possibly lead to fat build-up and obesity. Antioxidant enzymes, including superoxide dismutase, catalase, and glutathione peroxidase, can help break up LDL cholesterol. Proteolytic enzymes help break up any circulating antigen-antibody complexes associated with oxidized LDL cholesterol, according to Dr. Karl Ransberger. Studies also indicate that pancreatin supplementation leads to a decreased appetite, which can result in significant weight loss.

Enzyme Absorption System Enhancers (EASE) maximize enzyme activity. They can also improve the absorption and bioavailability of various nutrients and decrease the drain on the body's own digestive enzymes, thus prolonging their lives.

In addition to the standard multivitamin, multimineral, and multiglandular complexes recommended in the introduction to Part Two, the following supplements are recommended for the treatment of obesity. The following recommended dosages are for adults. Children between the ages of two and five should take one-quarter the recommended dosage, those between the ages of six and twelve should take one-half the recommended dosage, and those between the ages of thirteen and seventeen should take three-quarters the recommended adult dosage.

ESSENTIAL ENZYMES

Enzyme	Suggested Dosage	Actions
Pancreatin *or*	Take 300 mg three times per day with meals and at bedtime.	Contains proteases, amylases, and lipases, so digests proteins, carbohydrates, and fats; helps burn body fat.
A digestive enzyme formula containing protease, amylase, and lipase enzymes. For a list of digestive enzyme products, see Appendix.	Take as indicated on the label.	Improves digestion by breaking down proteins, carbohydrates, and fats; helps burn body fat.

SUPPORTIVE ENZYMES

Enzyme	Suggested Dosage	Actions
Formula One—See page 38 of Part One for a list of products. *or*	Take 10 tablets three times per day between meals.	Helps burn body fat; improves fat, protein, and carbohydrate digestion.
Phlogenzym from Mucos *or*	Take 10 tablets three times per day between meals.	Helps burn body fat; improves protein and fat digestion.
Bromelain *or*	Take ten 230–250 mg tablets three times per day between meals.	Helps burn body fat.
Mulsal from Mucos	Take 10 tablets three	Helps burn body fat; improves
	times per day between meals.	fat and protein digestion.
Formula Three—See page 38 of Part One for a list of products.	Take as indicated on the label.	Free-radical fighters.

ENZYME ABSORPTION SYSTEM ENHANCERS (EASE)

Combination	Suggested Dosage	Actions
Biodias A Granules from Amano Pharmaceutical Co., Ltd. *or*	Adults should take one packet in 8 ounces of water between or after meals three times per day. Children ages 11 to 14 should take two-thirds of a packet; children between the ages of 8 and 10 should take half of the packet; and children between the ages of 5 and 7 should take one-third of a packet.	Helps improve digestion of nutrients.
Citrimax Shake from Nature's Plus *or*	Take as indicated on the label.	Combines vegetarian protein, vitamins, minerals, amino acids, Ayurvedic herbs, and enzymes to reduce production and storage of fat.
Garcinia-Max Diet Shake from Rainbow Light *or*	Take as indicated on the label.	Reduces production and storage of fat.
Slender Now from Life Plus *or*	Take as indicated on the label.	Helps in weight loss.
Sugar Check from Enzymatic Therapy *or*	Take as indicated on the label.	Contains nutrients for improved sugar and carbohydrate metabolism and absorption.
Any of the three Thermo Tropic Formulas from Nature's Plus; available in tablets, bars, and shakes *or*	Take as indicated on the label.	Helps stimulate the natural weight loss processes in the body.
Trim-Zyme from Rainbow Light	Take as indicated on the label.	Increases potency of lipase with Citrimax, GTF chromium, and herbs to balance blood sugar levels and improve digestion.

ENZYME HELPERS

Nutrient	Suggested Dosage	Actions
Acidophilus or other probiotics	Take as indicated on the label.	Improves gastrointestinal function; needed for production of biotin, niacin, folic acid, and pyridoxine—coenzymes necessary for fat metabolism; stimulates enzyme activity.
Beta-carotene or vitamin A	Take 15,000–50,000 IU per day. If pregnant, take no more than 10,000 IU per day.	Helps detoxify waste products from fat metabolism; helps burn fat.

Betaine hydrochloric acid	Take 150 mg three times per day with meals.	Hydrochloric acid helps improve digestion of food, thus assisting in weight loss. If gastric upset occurs, discontinue use.
Choline	Take 250–750 mg with each meal.	Dissolves and mobilizes stores of fat in the liver, thus helping you to lose weight.
Chromium	Take 400 mcg per day.	Improves insulin effectiveness; chromium picolinate appears to play a role in body fat reduction.
Coenzyme Q$_{10}$	Take 30–60 mg per day.	Increases cellular energy production.
DHEA	Take 10–30 mg three times per day before meals.	Interferes with fat storage; helps burn fat.
Fiber	Take 3–4 capsules before each meal with an 8-ounce glass of water.	Reduces hunger; decreases fat absorption; burns fat.
Gamma-linolenic acid (GLA)	Take two to four 500 mg capsules two times per day with meals.	Promotes fat burning in the body.
L-arginine	Take as indicated on the label.	Works overnight to help improve weight loss.
L-carnitine	Take 500–2,000 mg in divided doses between meals.	Transports fatty acids across cell membranes; removes cellular toxins; helps fight fat storage.
L-methionine	Take as indicated on the label.	Aids betaine hydrochloride and lipase in fat digestion.
L-ornithine	Take as indicated on the label.	Works overnight to help improve weight loss.
L-tyrosine	Take as indicated on the label.	Suppresses appetite.
Lecithin	Take 1,000 mg three times per day with meals.	Emulsifies fat.

Ox-bile extract	Take 120 mg per day.	Aids digestion.
Spirulina or blue-green algae	Take as indicated on the label; or mix with 8 ounces of water or juice and drink for breakfast or with breakfast.	Reduces hunger.
Vitamin B complex including vitamin B$_6$	Take 250 mg three times per day.	Acts as a coenzyme for many enzymes; necessary for fat metabolism; acts as a diuretic.
Vitamin C	Take 5,000 mg per day.	Helps reduce body fat; helps detoxify waste products from fat metabolism.
Vitamin E	Take 1,200 IU per day from all sources (including the multivitamin complex in the introduction to Part Two).	Helps detoxify the waste products of fat metabolism; accelerates rate of fat burning.
Whey	Take 1 heaping tablespoon three times per day with meals.	Improves digestion.

Comments

❏ See the "Five-Step Jump-Start Enzyme Program" in Part Three.

❏ Increase your intake of fiber. It can help reduce appetite and fat absorption.

❏ Hydroxycitric acid (HCA) is an extract from the Garcinia cambogia fruit that can help suppress appetite, reduce food intake, and inhibit the synthesis of fat and cholesterol.

❏ Detoxification is important while you are losing weight. Follow the Enzyme Kidney Flush and the Enzyme Toxin Flush programs in Part Three.

❏ Fruitein from Nature's Plus is a high-protein, low-calorie energy shake that can be very beneficial in weight loss. See also "The Ten Commandments of Weight Loss" below.

The Ten Commandments of Weight Loss

The ten commandments of weight loss is your guide to taking the weight off and keeping it off. Reviewing these commandments every morning will help keep them foremost in your mind. If you fall off the wagon and cheat, don't let your guilt consume you. As soon as possible, return to your weight-loss program.

1. Eat fewer calories but consume smaller meals more frequently.

2. Eat more whole fiber-rich foods and fewer refined foods.

3. Increase your intake of fresh fruits and vegetables.

4. If you must eat out, order carefully. Be particularly careful of fast food restaurants, where the food tends to be high in fat, carbohydrates, sodium, and calories.

5. Don't eat just before retiring.

6. Drink plenty of pure water (at last eight 8-ounce glasses every day).

7. Exercise regularly to burn calories.

8. Get plenty of rest. Research shows that sleep deprivation may actually increase your appetite and, therefore, your consumption of food.

9. Take enzymes, vitamins, minerals, and herbs to help your body function optimally.

10. Lose weight gradually to ensure the loss of fat, not muscle.

❏ Take herbal lipotropics (products that mobilize and dissolve liver fat stores), including turmeric (curcumin), milk thistle (silymarin), and Oregon grape root.

❏ See the list of Resource Groups in the Appendix for a list of organizations that can provide you with more information about obesity.

Osteoarthritis

Osteoarthritis (degenerative joint disease) is known as "wear and tear" arthritis, though that name may be somewhat of a misnomer, since "wear and tear" is not always the cause. It is a chronic disorder characterized by degeneration of the cartilage surrounding a joint or joints, which causes damage to the bone. It most commonly affects the joints in the fingers, big toes, neck, lower back, hips, and knees. There are two forms of osteoarthritis: primary, in which the cause is unknown, and secondary, in which the cause is another disease, an infection, an injury, or overuse of the joint. Though osteoarthritis is seen most commonly in the aged, it is not a signal of aging or an inevitable part of the aging process. Osteoarthritis can occur in people of all ages.

Symptoms

In its early stages, osteoarthritis often produces no symptoms. Symptoms usually develop gradually, starting with pain, which is usually made worse by exercise. There may be stiffness after periods of rest and upon rising in the morning. As the damage gets worse, the joint may become more difficult to move, and may even freeze in a bent position. Bone spurs may develop on some joints, which may cause cracking or grating noises upon movement of the joint. Obesity may worsen symptoms. Though osteoarthritis is generally a progressively degenerative disease, often causing some degree of disability, sometimes joint degeneration stops altogether and even reverses.

Enzyme Therapy

Enzyme therapy has been used successfully with active osteoarthritis for over twenty years. Enzymes have been used individually (papain, trypsin, and bromelain) and in combinations (such as bromelain, papain, trypsin, chymotrypsin, and pancreatin, with the bioflavonoid rutin). In one five-week study, seven coated tablets of Wobenzym were given daily, and their effects were compared with those of the 50 mg daily drug diclofenac (a common arthritis medication). Both the drug and the enzymes improved mobility of the knees and reduced morning stiffness, pain on movement, and tenderness on pressure. What was sig-

nificant was that the enzyme combination improved the symptoms with no side effects, though there were side effects with treatment with diclofenac. This finding is extremely significant because long-term applications of either the enzyme product or the drug are a necessity in order to control osteoarthritis.

Systemic enzyme therapy is used to reduce inflammation (see the inset "Inflammation" on page 318), stimulate the immune system, improve circulation, help speed tissue repair, bring nutrients to the damaged area and remove waste products. Enzymes strengthen the body as a whole, build general resistance, and enhance wellness.

Digestive enzyme therapy is used to improve the digestion and absorption of food nutrients, reduce stress on the gastrointestinal mucosa, help maintain normal pH levels, detoxify, and promote the growth of healthy intestinal flora, thereby improving the overall health of the body. Digestive enzymes also serve as replacements for the body's pancreatic enzymes, leaving the pancreatic enzymes free to perform other functions in the body, such as decreasing inflammation.

Enzyme Absorption System Enhancers (EASE) maximize enzyme activity. They can also improve the absorption and bioavailability of various nutrients and decrease the drain on the body's own digestive enzymes, thus prolonging their lives.

In addition to the standard multivitamin, multimineral, and multiglandular complexes recommended in the introduction to Part Two, the following supplements are recommended for the treatment of osteoarthritis. The following recommended dosages are for adults. Children between the ages of two and five should take one-quarter the recommended dosage, those between the ages of six and twelve should take one-half the recommended dosage, and those between the ages of thirteen and seventeen should take three-quarters the recommended adult dosage.

ESSENTIAL ENZYMES

Enzyme	Suggested Dosage	Actions
Formula One (particularly Wobenzym N from Naturally Vitamin Supplements) —See page 38 of Part One for a list of products.	Take 7 tablets three times per day between meals. As symptoms subside (which should occur in approximately one week), decrease dosage to 3–5 tablets three times per day between meals. When pain subsides, begin taking a maintenance dose of 2–3 tablets after each meal.	Decreases inflammation, pain, and swelling; increases rate of healing; stimulates the immune system; strengthens capillary walls.
or		
Inflazyme Forte from American Biologics	Take as indicated on the label between meals.	Decreases inflammation, pain, and swelling; increases rate of healing; stimulates the immune system; strengthens capillary walls.
or		

Inflammation

Regardless of where or how an injury occurs, the body's reparative process is the same. At the moment of injury, the body initiates a series of defense measures, called inflammation. Inflammation is the underlying pathologic process in virtually all injuries—whether they be bruises, cuts, sprains, fractures, or burns. On injury, a series of biochemical changes takes place in a predictable fashion. There is an increase in the permeability of the capillaries—which allows excess body fluid to accumulate in the injured area, leading to swelling. The end result is a wall-like deposit of insoluble proteins, particularly fibrin.

This sealing-off process is an important defense mechanism that can prevent or retard the spread of infectious agents. In the absence of infection, however, this process has a negative effect because it delays recovery. Blood flow in the injured area stops; the body's own reparative agents cannot reach the damaged area, and dead tissue debris cannot be removed. Pain and discomfort are probably caused by the pressure of swelling or liberation of lactic acid from destroyed white blood cells. When the swelling goes down, pain will decrease.

The therapeutic use of enzymes is most effective at this stage of inflammation because they help dissolve the thrombotic plugs (fibrin). This fribrin reduction allows more oxygen to reach and revive the tissue cells. Excess fluid is also re-absorbed, reducing inflammation. The result is that dead tissue debris and disintegrated blood cells are absorbed.

With enzymes, the course of inflammation is not stopped or blocked, but supported and quickened. The body's own clearance can regenerate the disturbed structures more quickly. The healing time will be reduced.

In acute injuries, we use a special group of enzymes, called the hydrolases. These include pancreatin, trypsin, chymotrypsin, papain, bromelain, and microbial proteases. Although individual enzymes are extremely effective, for synergistic purposes, combinations of enzymes are used and are far more therapeutically effective than single enzymes.

Enzyme preparations are generally well-tolerated. Undesired side effects have seldom been reported for orally ingested enzymes even in higher doses or over extended periods of time. These side effects are mainly slight disturbances in the gastrointestinal tract, stool softening, flatulence, or fullness. Usually these side effects can easily be controlled by lowering the dosage.

Therefore, research indicates that oral enzyme therapy is successful in the treatment of inflammation in disorders such as osteoarthritis, and even traumatic injuries with minimal side effects. The advantages are substantial. If you so desire, a well-trained alternative-health-care physician can give advice in developing an enzyme program.

Enzyme complexes rich in proteases, plus amylases and lipases, working synergistically *or*	Take as indicated on the label between meals.	Decreases inflammation, pain, and swelling; increases rate of healing; stimulates the immune system; strengthens capillary walls.
Bromelain	Take three to four 230–250 mg capsules per day until symptoms subside. If severe, take 6–8 capsules per day.	Fights inflammation, pain, and swelling; speeds recovery after injury or surgery; stimulates the immune system.

SUPPORTIVE ENZYMES

Enzyme	Suggested Dosage	Actions
Formula Three—See page 38 of Part One for a list products.	Take 50 mcg three times per day between meals.	Free-radical fighters.
A digestive enzyme formula containing protease, amylase, and lipase enzymes. See the list of digestive enzyme products in Appendix.	Take as indicated on the label.	Improves digestion by breaking down proteins, carbohydrates, and fats, thereby improving the overall health of the body; frees pancreatic enzymes to fight osteoarthritis.

ENZYME ABSORPTION SYSTEM ENHANCERS (EASE)

Combination	Suggested Dosage	Actions
Connect-All from Nature's Plus *or*	Take as indicated on the label.	Supports healthy tissue.
CTR Support from PhysioLogics *or*	Take as indicated on the label.	Nourishes the body during tissue recovery and healing from injury; helps maintain healthy connective tissue; diminishes damage caused by swelling and inflammation.
Inflamzyme from PhytoPharmica *or*	Take as indicated on the label.	Promotes tissue repair.
Joint Lube RX 2200 from Phyto-Therapy *or*	Take as indicated on the label.	Helps to support cartilage and keep healthy joints functioning.
Joint-Ease from Nature's Life *or*	Take as indicated on the label.	Decreases inflammation and swelling.
Magnesium Plus from Life Plus *or*	Take one or two tablets one to three times per day.	Important for calcium utilization and many enzyme reactions.

318

| Mobil-Ease from Prevail *or* | Take as indicated on the label. | Reduces joint pain, swelling, and inflammation. |
| Zymain from Anabolic Labs | Take as indicated on the label. | Provides nutrition essential for tissue and joint repair and maintenance. |

ENZYME HELPERS

Nutrient	Suggested Dosage	Actions
Amino-acid complex	Take as indicated on the label on an empty stomach.	Amino acids are the building blocks of enzymes and are essential for the reduction of inflammation, for immune system stimulation, and for fighting free radicals.
Bioflavonoids	Take 150–500 mg three times per day.	Strengthen capillaries; potent antioxidants.
Branched-chain amino acids (isoleucine, leucine, and valine)	Take as indicated on the label between meals.	Often found together in a supplement, branched-chain amino acids aid in healing of muscle, bone, and skin and act as fuel for muscles.
Calcium	Take 1,000 mg per day.	Plays an important role in the strengthening of bone.
Chondroitin sulfate	Take 250 mg two to three times per day with meals; or follow the advice of your health-care professional.	Supports connective tissue; promotes cell hydration.
Coenzyme Q$_{10}$	Take 30–60 mg one to two times per day with meals; or follow the advice of your health-care professional.	Potent antioxidant.
Essential fatty acids	Take 75–100 mg two to three times per day with meals; or follow the advice of your health-care professional.	Help strengthen cell membranes.
Glucosamine hydrochloride *and/or*	Take 500 mg two to four times per day.	Stimulates cartilage repair.
Glucosamine sulfate	Take 500 mg one to three times per day.	
Magnesium	Take 500 mg per day.	Plays important role in the maintenance of bone structure and calcium absorption.
Mucopolysaccharides found in bovine cartilage extract and shark cartilage extract	Take as indicated on the label.	Supports healthy, flexible joints; controls inflammation; helps keep joint membranes fluid; helps the body fight inflammation.
Vitamin B complex	Take 50 mg three times per day, or as indicated on the label.	B vitamins act as coenzymes and support neuromusculoskeletal function.
Vitamin C	Take 5,000–10,000 mg per day.	Helps to produce collagen and normal connective tissue; promotes immune system; powerful antioxidant.

| Vitamin E | Take 1,200 IU per day. | Potent antioxidant; supplies oxygen to cells; improves level of energy. |
| Zinc | Take 50 mg per day. Do not take more than 100 mg per day. | Needed by over 300 enzymes. |

Comments

❑ Enzyme-rich fresh fruits and vegetables (particularly green vegetables) should comprise 60 percent of your diet. If your body cannot tolerate raw fruits and vegetables, increase your intake of digestive enzymes, or sauté or steam your produce.

❑ To improve overall health, avoid alcohol, white flour, and refined sugar.

❑ Avoid chlorinated water. Use bottled, distilled, or well water.

❑ Detoxification through cleansing, fasting, and juicing can help improve symptoms. See Part Three for instructions.

❑ Follow the Enzyme Kidney Flush and the Enzyme Toxin Flush programs as outlined in Part Three.

❑ Exercise thirty to forty minutes per day for increased circulation and enzymatic activity. Brisk walking and bicycling are excellent exercises, as is swimming. Swimming takes pressure off the painful joints. Many communities have heated therapeutic pools and water exercise classes.

❑ Meditate fifteen minutes per day. Void your mind and think of nothing. Allow your body to re-energize itself.

❑ See the list of Resource Groups in the Appendix for a list of organizations that can provide you with more information about osteoarthritis.

Osteoporosis

Osteoporosis is a generalized gradual but progressive loss of bone density that is most common in women. This decrease in density causes weakening of bone strength, which makes bones more likely to fracture. This condition can also result in loss of height, pain in the hips and back, and curvature of the spine.

Calcium, phosphorus, and other minerals are essential components of bones. Generally, one knows that calcium deficiency can lead to osteoporosis. But even if one consumes enough calcium, there are other factors that affect the bone's absorption of calcium and can lead to osteoporosis.

Postmenopausal osteoporosis may occur when estrogen production slows after women reach menopause. Estrogen helps regulate bone absorption of calcium. *Senile osteoporosis* seems to occur as a result of age-related calcium deficiency. After the age of 30, bone density slowly begins to decrease. If the body is unable to maintain the necessary mineral content in the bones to maintain them, osteoporosis results. *Secondary osteoporosis* is a rare type that is caused by some other medical condition, such as kidney failure or hormonal disorders, or the use of such drugs as corticosteroids and barbiturates. *Idiopathic juvenile osteoporosis* is another rare type that occurs in children who seem to have normal hormone and vitamin levels in their blood. The reason for their osteoporosis is unknown.

Symptoms

At first, those with osteoporosis experience no symptoms; in fact, some never experience any. When the decrease in bone density is extensive, one may experience bones that fracture easily, pain in the back and hips, and curvature of the spine, resulting in loss of height.

Enzyme Therapy

Systemic enzyme therapy is used to help increase the absorption of calcium, reduce any inflammation, and stimulate the immune system. Enzymes strengthen the body as a whole, building general resistance, improving circulation and tissue repair, and enhancing wellness. Enzymes improve the transport of nutrients to the osteoporotic area and the removal of waste products.

Digestive enzyme therapy is used to improve the digestion of food, reduce stress on the gastrointestinal mucosa, help maintain normal pH levels, detoxify, and promote the growth of healthy intestinal flora, thereby improving the overall health of the body. Digestive enzymes also serve as replacements for the body's pancreatic enzymes, leaving the pancreatic enzymes free to perform other functions in the body, such as boosting immunity and improving circulation.

Enzyme Absorption System Enhancers (EASE) maximize enzyme activity. They can also improve the absorption and bioavailability of various nutrients and decrease the drain on the body's own digestive enzymes, thus prolonging their lives.

In addition to the standard multivitamin, multimineral, and multiglandular complexes recommended in the introduction to Part Two, the following supplements are recommended for the treatment of osteoporosis. The following recommended dosages are for adults. Children between the ages of two and five should take one-quarter the recommended dosage, those between the ages of six and twelve should take one-half the recommended dosage, and those between the ages of thirteen and seventeen should take three-quarters the recommended adult dosage.

ESSENTIAL ENZYMES

Enzyme	Suggested Dosage	Actions
Formula One—See page 38 of Part One for a list of products.	Take 7–10 coated tablets three times per day between meals for 2–3 months. Then decrease dosage to 3–5 tablets three times per day between meals.	Decreases inflammation, pain, and swelling; increases rate of healing; strengthens capillary walls.
or		
Inflazyme Forte from American Biologics	Take as indicated on the label between meals.	Decreases inflammation, pain, and swelling; increases rate of healing; strengthens capillary walls.
or		
Enzyme complexes rich in proteases, plus amylases and lipases, working synergistically	Take as indicated on the label between meals.	Decreases inflammation, pain, and swelling; increases rate of healing; strengthens capillary walls.

SUPPORTIVE ENZYMES

Enzyme	Suggested Dosage	Actions
Formula Three—See page 38 of Part One for a list of products.	Take 50 mcg three times per day between meals.	Potent antioxidant enzymes.
A digestive enzyme formula containing protease, amylase, and lipase enzymes. For the list of digestive enzyme products, see Appendix.	Take as indicated on the label with meals.	Improves digestion by breaking down proteins, carbohydrates, and fats.

ENZYME ABSORPTION SYSTEM ENHANCERS (EASE)

Combination	Suggested Dosage	Actions
Cal-Mag Enzyme Complex from Tyler Encapsulations	Take 2–4 capsules three times per day.	Provides calcium and magnesium to increase bone density.
or		
Connect-All from Nature's Plus	Take as indicated on the label.	Supports connective tissue.
or		
Magnesium Plus from Life Plus	Take one or two tablets one to three times per day.	Important for muscle relaxation, calcium utilization, energy production, and many enzyme reactions.
or		
Osteo Formula from Prevail	Take as indicated on the label.	Nutritional support for individuals at risk of bone loss.

ENZYME HELPERS

Nutrient	Suggested Dosage	Actions
Aged garlic extract	Take 2 capsules three times per day.	Detoxifies and provides energy.
Bioflavonoids	Take 150–500 mg three times per day.	Strengthen capillaries; potent antioxidants.

Branched-chain amino acids (isoleucine, leucine, and valine)	Take as indicated on the label between meals.	Often found together in a supplement, branched-chain amino acids aid in healing of muscle, bone, and skin.
Calcium	Take 1,000 mg per day.	Plays important role in maintaining healthy bone structure.
Magnesium	Take 500 mg per day.	Plays important role in maintaining healthy bone structure and in calcium absorption.
Mucopolysaccharides, found in bovine cartilage extract and shark cartilage extract	Take as indicated on the label.	Support healthy, flexible joints; control and fight inflammation; help keep joint membranes fluid; help strengthen cartilage and bone.
Vitamin A	Take 5,000 IU per day.	Needed for digestive glands to produce enzymes; required for healthy tissue and bone growth and development; potent antioxidant.
Vitamin B complex	Take 200–500 mg per day.	Needed for activation of many enzymes; essential for the growth and development of healthy bones.
Vitamin C	Take 5,000–10,000 mg per day.	Helps to produce collagen and normal connective tissue and bone; promotes immune system; antioxidant and coenzyme.
Vitamin D	Take 400 IU per day.	Involved in regulating the transport and absorption of calcium and phosphorus.
Vitamin E	Take 1,200 IU per day from all sources (including the multivitamin complex in the introduction to Part Two).	Potent antioxidant; supplies oxygen to cells; improves level of energy.
Zinc	Take 50 mg per day. Do not take more than 100 mg per day.	Needed by over 300 enzymes.

Comments

❏ Eat fresh foods rich in enzymes and antioxidants, including fruits (such as pineapple, papaya, and figs), vegetables (including broccoli and corn), whole wheat, and sprouted seeds. Fruits and vegetables (particularly green vegetables) should comprise 60 percent of your diet. If your body cannot tolerate raw fruits and vegetables, increase your intake of digestive enzymes, or sauté or steam your produce.

❏ Increase your intake of calcium-rich foods, including all dairy products, red meat, chicken, sardines, tuna, herring, oysters, all fruits, soybeans, nuts, broccoli, corn, rice, and whole wheat. Other calcium sources include dark green vegetables (collard greens, kale, parsley, broccoli, and spinach), carrots, yellow squash, sweet potatoes, pumpkins, tomatoes, cantaloupe, apricots, and peaches.

❏ Your magnesium intake should also be increased. Magnesium helps the body to absorb calcium. Magnificent

magnesium sources include blackberries, collard greens, parsley, and other leafy, green vegetables.

❏ To improve the overall health of your body avoid alcohol, white flour, and refined sugar.

❏ Avoid chlorinated water. Distilled, bottled, or spring water is recommended.

❏ Detoxify using the Enzyme Toxin Flush as outlined in Part Three.

❏ Exercises are essential for the good health of those with osteoporosis. It increases circulation, which stimulates normal bone metabolism. Exercise by walking, bicycling, or swimming five to six times per week for thirty to forty minutes. The exercises don't have to be elaborate to work.

❏ Drink at least two glasses of vitamin D-fortified milk per day.

❏ Increase stretching and weight-bearing exercises (including walking and bicycling).

❏ Women might consider estrogen-replacement therapy, as there is some indication that this can stop further bone loss.

❏ See the list of Resource Groups in the Appendix for an organization that can provide you with more information about osteporosis.

Pancreatic Cancer

Cancer of the pancreas is the fourth leading cause of cancer death in men and the fifth leading cause of cancer death in women in the United States. Unfortunately, researchers don't yet know the cause of this disease or how to prevent it. However, there are certain risk factors, including smoking (smokers have more than twice the risk of acquiring pancreatic cancer than do nonsmokers). There may also be a connection between the disease and diabetes, cirrhosis, chronic pancreatitis, and high fat intake. The risk of this disease increases after the age of 50, and most individuals diagnosed with the disease are between the ages of 65 and 79.

About 95 percent of pancreatic cancers are adenocarcinomas, or tumors that begin in the pancreatic duct cell lining. *Adenocarcinoma* of the pancreas usually does not produce any symptoms until the tumor has grown very large and spread, leading to a very poor prognosis. In fact, fewer than 2 percent of such patients survive for five years after the diagnosis. Current research is centering on better diagnostic techniques, since earlier detection would dramatically improve the chance of survival.

Cystadenocarcinoma is a rarer type of pancreatic cancer. These cancers spread more slowly; in fact, most cystadenocarcinomas have not spread by the time they are detected.

Those with this type of cancer have a much better prognosis than do those with adenocarcinomas.

Symptoms

As mentioned above, most pancreatic cancers are adenocarcinomas, which do not cause symptoms until the cancer is well-advanced. When symptoms do develop, the first are usually abdominal pain that may travel to the back and weight loss. Jaundice is another common symptom. Enlargement of the spleen and bleeding varicose veins may also occur.

Enzyme Therapy

Primary therapy for pancreatic cancer usually includes surgery, chemotherapy, and radiation. Systemic enzyme therapy is used as a complementary therapy to stimulate immune function, decrease inflammation and free-radical and fibrin formation, improve circulation, bring nutrients to and remove waste products from the area, and improve health. Studies show that with enzyme therapy, pain can be reduced and "a clear prolongation in life expectancy could be attained," according to Franz Klaschka in *Oral Enzymes—New Approach to Cancer Treatment* (Munich: Forum Medizin, 1996).

Digestive enzyme therapy is used to improve the digestion of food, help maintain normal pH levels, reduce stress on the enzyme-producing pancreas and on the gastrointestinal mucosa, detoxify the body, and promote the growth of healthy intestinal flora. Digestive enzymes also serve as replacements for the body's pancreatic enzymes, leaving the pancreatic enzymes free to perform other functions in the body, such as boosting immunity and improving circulation.

Enzyme Absorption System Enhancers (EASE) maximize enzyme activity. They can also improve the absorption and bioavailability of various nutrients and decrease the drain on the body's own digestive enzymes, thus prolonging their lives.

Proteolytic enzymes are successfully used as complementary therapy with pancreatic cancer before and after surgery; before, during, and after chemotherapy or radiation; and as long-term and palliative therapy.

In addition to the standard multivitamin, multimineral, and multiglandular complexes recommended in the introduction to Part Two, the following supplements are recommended for the treatment of pancreatic cancer. The following recommended dosages are for adults. Children between the ages of two and five should take one-quarter the recommended dosage, those between the ages of six and twelve should take one-half the recommended dosage, and those between the ages of thirteen and seventeen should take three-quarters the recommended adult dosage.

ESSENTIAL ENZYMES

Enzyme	Suggested Dosage	Actions
Before and after surgery:		
WobeMugos E from Mucos	Take 3 tablets three times per day.	Decreases the possibility of suppressed immune function after surgery or anesthesia; breaks down fibrin; decreases inflammation; improves wound healing; decreases risk of blood clot complications.
Before, during, and after chemotherapy or radiotherapy:		
WobeMugos E from Mucos	Take 4 tablets twice per day.	Breaks down fibrin; decreases inflammation; improves effectiveness of chemotherapy or radiation; decreases radiation sickness and nausea; stimulates immune function.
Long-term and palliative therapy:		
WobeMugos E from Mucos	Take 2–5 tablets 3 times per day.	Reduces malnutrition, weakness, and muscle wasting (cachexia), anorexia, depression, and pain; prevents metastasis; eliminates or prevents disturbances of lymphatic drainage.

AND SELECT ONE OF THE FOLLOWING:

Enzyme	Suggested Dosage	Actions
Formula One (particularly Wobenzym N from Naturally Vitamin Supplements) —See page 38 of Part One for a list of products.	Take 8–10 tablets three times per day between meals.	Stimulates immune system function; fights pain, inflammation, and swelling; speeds healing; strengthens capillary walls.
Inflazyme Forte from American Biologics	Take as indicated on the label between meals.	Stimulates immune system function; fights pain, inflammation, and swelling; speeds healing; strengthens capillary walls.
Bromelain	Take four to six 230–250 mg capsules per day.	Stimulates immune system function; fights pain, inflammation, and swelling; speeds healing; strengthens capillary walls.
Enzyme complexes rich in proteases, plus amylases and lipases, working synergistically	Take as indicated on the label between meals.	Stimulates immune system function; fights pain, inflammation, and swelling; speeds healing; strengthens capillary walls.

SUPPORTIVE ENZYMES

Enzyme	Suggested Dosage	Actions
Formula Three—See page 38 of Part One for a list of products.	Take as indicated on the label.	Potent antioxidant enzymes.
A digestive enzyme product containing protease, amylase, and lipase enzymes. See the list of digestive enzyme products in Appendix.	Take as indicated on the label with meals.	Improves the breakdown and absorption of food nutrients; fights toxic build-up.

ENZYME ABSORPTION SYSTEM ENHANCERS (EASE)

Combination	Suggested Dosage	Actions
Antioxidant Complex from Tyler Encapsulations *or*	Take as indicated on the label.	Potent free-radical fighter and immune system stimulator.
Detox Enzyme Formula from Prevail *or*	Take as indicated on the label.	Helps rid the body of harmful toxins.
Hepazyme from Enzymatic Therapy *or*	Take as indicated on the label.	Stimulates immune function.
Kyolic Formula 102 (with enzymes) *or*	Take 2 capsules or tablets with meals two to three times per day.	Strengthens the immune system; helps alleviate stress; helps fight free radicals; helps fight inflammation; improves digestion; helps fight cancer.
Oxy-5000 Forte from American Biologics *or*	Take as indicated on the label.	Helps the body excrete toxins; fights biochemical stressors, common infections, and degenerative processes.
Vita-C-1000 from Life Plus	Take as indicated on the label.	Helps fight cancer; boosts immune function.

ENZYME HELPERS

Nutrient	Suggested Dosage	Actions
Acidophilus and other probiotics	Take as indicated on the label.	Stimulates enzyme activity; improves gastrointestinal function; improves immunity.
Adrenal glandular	Take 100–400 mg per day.	Helps fight stress and adrenal exhaustion; fights inflammation.
Coenzyme Q_{10}	Take 30–60 mg one to two times per day with meals.	Important to immune system function; has antioxidant activity.
Kyolic Aged Garlic Liquid Extract with Vitamins B_1 and B_{12}	Take 10 drops in water or juice three times per day.	Stimulates immune system function.
Thymus glandular	Take 100 mg per day, or as indicated on the label.	Assists the thymus gland in immune resistance; improves immune response; fights infections.
Vitamin A	Take up to 50,000 IU per day. (For higher doses, see a physician. Increased doses of 100,000 IU or higher may be required.) If pregnant, do not take more than 10,000 IU per day.	Needed for strong immune system and healthy mucous membranes; potent antioxidant.
Vitamin B complex	Take 100–150 mg three times per day.	B vitamins act as coenzymes.
Vitamin C	Take 10,000–15,000 mg per day; or 1–2 grams every hour until symptoms improve.	Fights infections and diseases that can lead to cancer; stimulates the adrenal glands; promotes immune system; powerful antioxidant.
Vitamin D	Take 3,000–6,000 IU per day.	Stimulates the immune system; hinders the advance of tumors;
		promotes the break-up of degenerated cells and activates cancer-fighting monocytes.
Vitamin E	Take 3,000–6,000 IU per day.	Acts as an intercellular antioxidant; improves circulation; helps resist diseases at the cellular level; works closely with selenium to fight free radicals.
Zinc	Take 50 mg per day. Do not take more than 100 mg per day.	Needed by over 300 enzymes; necessary for white blood cell immune function; improves healing.

Comments

❑ Decrease your intake of dietary fat.

❑ According to enzyme expert Dr. Franz Klaschka, vitamin A and proteolytic enzymes should be taken at the same time for the treatment of cancer.

❑ Stop smoking. Not only does it greatly increase your risk of pancreatic cancer, but it is also linked to a number of conditions, including lung cancer, heart disease, emphysema, and atherosclerosis (to name just a few).

❑ For further information on dietary modifications, fasting, and juicing, see Part Three.

❑ Increase your intake of food extracts and designer foods rich in enzymes, phytochemicals, and antioxidants.

❑ Follow the Enzyme Kidney Flush and the Enzyme Toxin Flush programs in Part Three.

❑ One week per month, during a week you are not using the Enzyme Kidney Flush or Enzyme Toxin Flush, use the coffee enema as described in Part Three. Coffee enemas are used to stimulate the release of bile from the liver and facilitate the removal of waste metabolites from the body and into the intestines for excretion.

❑ Physicians sometimes use enzyme injections or enzyme retention enemas in the treatment of cancer. Both methods allow a greater concentration of the enzymes to be absorbed than might be accomplished if the enzymes were taken orally. An enzyme retention enema is a means of instilling some enzymes into the colon to be retained. This allows the enzymes to be absorbed into the body's systems. Physicians use products specially formulated for these uses. Caution should be exercised, however, since repeated rectal administration of very high doses of enzymes could lead to irritation of the anus or the skin around the anus. This can be avoided by applying a lubricating ointment, such as vitamin E oil or aloe vera, before administering the enema.

❑ See the list of Resource Groups in the Appendix for a list of organizations that can provide you with more information about pancreatic cancer.

❑ *See also* CANCER.

323

Pancreatic Insufficiency

The pancreas is both an endocrine (without ducts) and exocrine (with ducts) gland. As discussed in Chapter 2 of Part One, it also has two roles. The endocrine part produces insulin and glucagon, which regulate and control blood sugar levels. The exocrine part produces digestive enzymes, including proteolytic, amylolytic, and lipolytic enzymes. Pancreatic insufficiency refers to the exocrine portion of the pancreas. A shortage of any of these enzymes can keep you from properly digesting the food you eat.

Pancreatic insufficiency can occur for many reasons. Pancreatitis, excessive alcohol intake, intake of fast foods, a stressful lifestyle with little exercise and rest, and an enzyme-poor diet can overtax the pancreas' enzyme reserve and its ability to produce enzymes. Diseases, such as cancer, and injury from surgery or trauma can also interfere with the flow of enzymes down the pancreatic duct. Cystic fibrosis patients' exocrine glands produce too much mucus, which often blocks the duct. Certain nutritional deficiencies can lead to an insufficiency of enzymes. Sometimes, for an unknown reason, the pancreas fails to produce enough of a particular enzyme. In addition, as we age, the pancreatic machinery wears out, and the pancreas loses its ability to produce enzymes, and those enzymes produced have a reduced activity level. The end result can be pancreatic insufficiency.

Symptoms

As the protease, amylase, and lipase enzymes decrease, those with pancreatic insufficiency will have more and more problems digesting protein, carbohydrates, and fats, which, of course, will lead to malabsorption of nutrients. The patient may begin to produce fatty, bulky, light-colored, foul-smelling stools and will lose weight. Pain may also be present, especially when one has pancreatitis.

Enzyme Therapy

Pancreatic enzyme extracts have been used for several decades to decrease pancreatic insufficiency, poor digestion, and the abdominal symptoms (including pain) pancreatic insufficiency may create. Digestive enzyme therapy supplies some of the missing enzymes, reduces stress on the pancreas and gastrointestinal mucosa, helps maintain normal pH levels, detoxifies the body, and promotes the growth of healthy intestinal flora, thus improving digestion. A number of enzymes have proven to be effective in treating pancreatic insufficiency, including plant and microbial amylase, bromelain, microbial cellulase, enterokinase, microbial hemicellulase, lipase, pepsin, proteases, and papain.

Systemic enzyme therapy is used to reduce any pancreatic inflammation, stimulate the immune system, improve circulation, help speed pancreatic tissue repair, transport nutrients throughout the body, remove waste products, and improve health.

Enzyme Absorption System Enhancers (EASE) maximize enzyme activity. They can also improve the absorption and bioavailability of various nutrients and decrease the drain on the body's own digestive enzymes, thus prolonging their lives.

In addition to the standard multivitamin, multimineral, and multiglandular complexes recommended in the introduction to Part Two, the following supplements are recommended for the treatment of pancreatic insufficiency. The following recommended dosages are for adults. Children between the ages of two and five should take one-quarter the recommended dosage, those between the ages of six and twelve should take one-half the recommended dosage, and those between the ages of thirteen and seventeen should take three-quarters the recommended adult dosage.

ESSENTIAL ENZYMES

Enzyme	Suggested Dosage	Actions
Pancreatin *or*	Take 600–1,000 mg every time you eat.	Replaces pancreatic enzymes; improves digestion.
Pancrelipase *or*	Take 600–1,000 mg every time you eat.	Replaces pancreatic enzymes; improves digestion.
A digestive enzyme formula containing protease, amylase, and lipase enzymes. For a list of digestive enzyme products, see Appendix. *or*	Take as indicated on the label.	Improves digestion by breaking down proteins, carbohydrates, and fats.
Bromelain	Start with one 230–250 mg capsule per day and increase dosage gradually, if necessary. Take no more than 2 or 3 capsules per day. If diarrhea develops, decrease dosage.	Improves digestion; fights inflammation, pain, and swelling; stimulates the immune system.

SUPPORTIVE ENZYMES

Enzyme	Suggested Dosage	Actions
Formula Three—See page 38 of Part One for a list of products.	Take as indicated on the label.	Potent antioxidant enzymes.
Formula One—See page 38 of Part One for a list of products.	Take 4 tablets three times per day between meals.	Improves immune function; reduces inflammation.

ENZYME ABSORPTION SYSTEM ENHANCERS (EASE)

Combination	Suggested Dosage	Actions
Biodias A Granules from	Adults should take	Helps improve digestion of

Amano Pharmaceutical Co., Ltd.	one packet in 8 ounces of water between or after meals three times per day. Children ages 11 to 14 should take two-thirds of a packet; children between the ages of 8 and 10 should take half of the packet; and children between the ages of 5 and 7 should take one-third of a packet.	nutrients; relieves inflammation; promotes regeneration of mucous membranes; decreases stress on the stomach and intestines; reduces gastric pain.
Pancreas Formula from Life Plus	Take one to two tablets three times per day; or take as indicated on the label.	Supports the pancreas; important in carbohydrate metabolism; improves digestion.

ENZYME HELPERS

Nutrient	Suggested Dosage	Actions
Acidophilus and other probiotics	Take as indicated on the label.	Improve gastrointestinal function; needed for production of biotin, niacin, folic acid, and pyridoxine, all important coenzymes; help produce antibacterial substances; stimulate enzyme activity.
Amino-acid complex	Take as indicated on the label on an empty stomach.	Required for enzyme, peptide, and hormone production.
Coenzyme Q$_{10}$	Take 30–60 mg one to two times per day with meals; or follow the advice of your health-care professional.	Potent antioxidant.
Lecithin	Take as indicated on the label.	Emulsifies fat.
Pancreas glandular	Take 200–300 mg per day.	Fights steatorrhea; aids digestion; decreases food allergies; assists the pancreatic enzymes.
Pantothenic acid (vitamin B$_5$)	Take 50 mg three times per day.	Important in the metabolism of protein, fats, and carbohydrates; involved in synthesis of hormones, hemoglobin, steroids, and neurotransmitters; needed for a well-functioning digestive system.
Vitamin A	Take up to 50,000 IU per day. If pregnant, do not take more than 10,000 IU. Doses of 100,000 IU or higher may be recommended by your physician, but take doses over 50,000 IU only under the supervision of a physician.	Stimulates the immune system; Free-radical scavenger.
Vitamin C	Take 8,000–13,000 mg per day.	Fights infections, diseases, and allergies; improves wound healing; stimulates the adrenal glands; promotes immune system; powerful antioxidant.

Zinc	Take 15–50 mg per day. Do not take more than 100 mg per day.	Needed by over 300 enzymes; necessary for white blood cell immune function; improves wound healing.

Comments

❑ Do not overtax the pancreas. Eat like a grazing animal, consuming four or five smaller meals a day instead of one or two large meals. Fruits and vegetables (particularly green vegetables) should comprise over 60 percent of your diet. If your body cannot tolerate raw fruits and vegetables, increase your intake of digestive enzymes, or sauté or steam your produce.

❑ Eat fermented foods, such as yogurt, buttermilk, and kefir; probiotics, including *Lactobacillus acidophilus* and *Bifidobacteria*; and whey.

❑ Follow the enzyme-rich Cave Man Diet as outlined in Part Three.

❑ Eat very little protein.

❑ Consume no poisons (additives, preservatives, artificial colors or flavors) and avoid alcohol, refined white flour, and sugar. These produce undesirable fermentation in the bowel and stimulate improper metabolism, causing increased stress on the pancreas. In addition, alcohol puts tremendous stress on the pancreas and is the leading cause of pancreatitis.

❑ Avoid chlorinated water. Distilled or bottled spring water, or even well water is recommended.

❑ To detoxify the body, follow the Enzyme Toxin Flush and the Enzyme Kidney Flush programs as outlined in Part Three.

❑ If you think you may have pancreatic insufficiency, be sure to see a doctor to avoid serious complications.

❑ See the list of Resource Groups in the Appendix for an organization that can provide you with more information about pancreatic insufficiency.

Pancreatitis

Pancreatitis is an inflammation of the pancreas. The pancreas is like two organs in one because it has two functions: The exocrine (with ducts) part produces digestive enzymes, and the endocrine (ductless) part produces the hormones glucagon and insulin, which regulate blood sugar levels.

If the duct leading from the pancreas becomes blocked for whatever reason, inactive pancreatic enzymes may become activated in the pancreas and begin to digest the cells of the pancreas. Severe inflammation ensues. According to Dr. Otto Pecher, an enzyme expert in Germany, autoimmune processes may also play a role in pancreatitis.

There are two types of pancreatitis: acute and chronic.

Acute pancreatitis is usually caused by gallstones or alcoholism. Less commonly, it may be caused by viruses, pancreatic cancer, high blood lipid levels, trauma, or certain drugs. Acute pancreatitis produces severe abdominal pain accompanied by nausea and vomiting. Many develop a fever, fast pulse, rapid breathing, low blood pressure accompanied by fainting spells upon standing, and swelling of the abdomen. Some may develop jaundice, and some may even become unconscious if the illness is allowed to progress. Acute pancreatitis requires hospitalization.

In contrast to acute pancreatitis with its sudden and severe appearance, chronic pancreatitis is a continuing inflammatory disease of the pancreas. In the United States, it is most commonly due to alcoholism. It may also result from some type of genetic narrowing of the pancreatic duct, or narrowing of the duct due to pancreatic cancer or an attack of severe acute pancreatitis.

A person suffering from chronic pancreatitis may either have persistent abdominal pain of varying intensity, or have periodic episodes of symptoms resembling mild to moderate acute pancreatitis. Ultimately, the pancreas becomes so damaged that it can no longer produce enzymes. This causes the pain to cease, since the enzymes are no longer inflaming the pancreas. But it also leads to malabsorption problems, since enzymes are not being secreted to adequately digest food. This leads to weight loss and bulky, foul-smelling, light-colored, fatty stools. Ultimately, the endocrine part of the pancreas may also be destroyed, leading to diabetes.

Enzyme Therapy

Data from studies support the effectiveness of certain proteolytic enzyme combinations in successfully treating pancreatitis. Proteolytic enzymes decrease inflammation of the pancreas, decrease the pain and severity of rapidly progressive pancreatitis, assist in avoiding complications, and shorten the duration of hospitalization.

Systemic enzyme therapy is used in chronic pancreatitis to reduce pain, swelling, and inflammation in the pancreas. Enzymes stimulate the immune system, improve circulation, help speed tissue repair, bring nutrients to the damaged area and remove waste products, strengthen the body as a whole, build general resistance, and enhance wellness. Improvement in the pancreas could help prevent cancer of the pancreas and other disorders. Stimulation of the immune system is also important when autoimmune factors are thought to play a role in pancreatitis.

Digestive enzymes are used to ease the stress on the pancreas, serving as replacements for the body's pancreatic enzymes; improve the digestion of food; reduce stress on the gastrointestinal mucosa; help maintain normal pH levels; detoxify the body; and promote the growth of healthy intestinal flora.

Enzyme Absorption System Enhancers (EASE) maximize enzyme activity. They can also improve the absorption and bioavailability of various nutrients and decrease the drain on the body's own digestive enzymes, thus prolonging their lives.

A number of studies have been conducted using enzymes in the treatment of the pain of pancreatitis. In one double-blind study the effects of using proteolytic enzymes were studied with patients suffering from chronic pancreatitis. For one week, nineteen patients took a granulated proteolytic enzyme preparation daily while nineteen pancreatitis patients took a harmless sugar pill (a placebo) each day. Neither those administering the pills nor the patients knew which pill was being dispensed. The group of patients taking the enzymes experienced an average 30-percent reduction in pain. No reduction in abdominal pain was noted in those taking the placebo. In another study, it was noted that the use of systemic enzyme therapy resulted in pain reduction, a decrease in inflammation, and fewer complications, even in those patients who had other concurrent illnesses involving limited renal function, diabetes mellitus, or respiratory disturbances.

The therapy for chronic pancreatitis includes a special diet coupled with high doses of digestive enzyme supplements that should release in the stomach and small intestine. For this reason, non-enterically coated enzymes are preferred for managing pain, while enterically coated enzymes seem to be best for treating the steatorrhea that accompanies pancreatitis. The use of vitamins and minerals is also highly recommended, according to Professors Heinrich Wrba and Otto Pecher, in their book *Enzyme-Wirkstoffe der Zukunft Mit der Enzymtherapie das Immunsystem Stärken* (Vienna, Austria: Verlag Orac im Verlag Kremayr & Scheriau, 1993).

In addition to the standard multivitamin, multimineral, and multiglandular complexes recommended in the introduction to Part Two, the following supplements are recommended for the treatment of pancreatitis. The following recommended dosages are for adults. Children between the ages of two and five should take one-quarter the recommended dosage, those between the ages of six and twelve should take one-half the recommended dosage, and those between the ages of thirteen and seventeen should take three-quarters the recommended adult dosage.

ESSENTIAL ENZYMES

Enzyme	Suggested Dosage	Actions
Formula One—See page 38 of Part One for a list of products. *or*	Take 8–10 tablets three times per day between meals until symptoms subside. Thereafter, take 5 tablets three times per day between meals.	Fights inflammation, pain, and swelling; speeds recovery after injury or surgery; stimulates the immune system; strengthens capillary walls.
Bromelain	Take three to four 230–250 mg capsules per day until symptoms subside.	Fights inflammation, pain, and swelling; speeds recovery after injury or surgery; stimulates

or	If severe, take 6–8 capsules per day.	the immune system.
Serratiopeptidase or	Take 10 mg two times per day between meals.	Fights inflammation; reduces pain and swelling.
Trypsin with chymotrypsin or	Take 250 mg of trypsin and 10 mg of chymotrypsin three times per day between meals.	Fights inflammation, pain, and swelling; increases rate of healing; stimulates the immune system.
Pancreatin or	Take 500 mg six times per day on an empty stomach.	Fights inflammation, pain, and swelling; increases rate of healing; stimulates the immune system; strengthens capillary walls.
Papain or a microbial protease	Take 300 mg two times per day between meals.	Fights inflammation; speeds recovery from injuries and surgery.

SUPPORTIVE ENZYMES

Enzyme	Suggested Dosage	Actions
Superoxide dismutase	Take 150 mcg per day.	Fights inflammation; potent antioxidant.
Catalase and glutathione peroxidase	Take as indicated on the label.	Potent antioxidant enzymes.
A digestive enzyme product containing protease, amylase, and lipase enzymes. See the list of digestive enzyme products in Appendix.	Take as indicated on the label with meals.	Improves the breakdown and absorption of food nutrients; fights toxic build-up.

ENZYME ABSORPTION SYSTEM ENHANCERS (EASE)

Combination	Suggested Dosage	Actions
Biodias A Granules from Amano Pharmaceutical Co., Ltd. or	Adults should take one packet in 8 ounces of water between or after meals three times per day. Children ages 11 to 14 should take two-thirds of a packet; children between the ages of 8 and 10 should take half of the packet; and children between the ages of 5 and 7 should take one-third of a packet.	Helps improve digestion of nutrients; relieves inflammation; promotes regeneration of mucous membranes; decreases stress on the stomach and intestines; reduces gastric pain.
Parazyme from Enzymatic Therapy	Take as indicated on the label.	Reduces inflammation and swelling; strengthens tissue integrity; fights free radicals; improves digestion.

ENZYME HELPERS

Nutrient	Suggested Dosage	Actions
Acidophilus and other probiotics	Take as indicated on the label.	Improve gastrointestinal function; needed for production of biotin, niacin, folic acid, and pyridoxine,
		all important coenzymes; help produce antibacterial substances; stimulate enzyme activity.
Adrenal glandular	Take 100–400 mg per day.	Fights inflammation and adrenal exhaustion.
Aged garlic extract	Take 2 capsules three times per day.	Fights inflammation; stimulates the immune system; potent antioxidant.
Amino-acid complex containing such amino acids as L-ornithine, L-glutamine, L-arginine, L-proline, L-lysine, and L-cysteine	Take as indicated on the label on an empty stomach.	Required for enzyme, peptide, and hormone production; helps the body maintain proper fluid and acid/base balance.
Coenzyme Q_{10}	Take 30–60 mg one to two times per day with meals; or follow the advice of your health-care professional.	Potent antioxidant.
Lecithin	Take as indicated on the label.	Emulsifies fat.
Niacin or	Take 50 mg the first day, then increase dosage daily by 50 mg to 1,000 mg. If adverse effects occur, decrease dosage. Take dosages higher than 1,000 mg only under the supervision of a physician.	Improves circulation to the pancreas; a coenzyme.
Niacin-bound chromium	Take 20 mg per day, or as indicated on the label.	
Pancreas glandular	Take 200–300 mg per day.	Fights steatorrhea; aids digestion; decreases food allergies; assists the pancreatic enzymes.
Pantothenic acid	Take 50 mg three times per day.	Important in the metabolism of protein, fats, and carbohydrates; involved in synthesis of hormones, hemoglobin, steroids, and neurotransmitters; needed for healthy digestive system.
Spirulina	Take as indicated on the label.	Rich in minerals, vitamins, enzymes, proteins, essential amino acids, and essential fatty acids, especially gamma-linolenic acid; good source of beta-carotene; improves immune function.
Thymus glandular	Take 100 mg per day.	Assists the thymus gland in immune resistance; improves immune response; fights infections.
Vitamin A	Take 5,000 IU per day.	Needed for strong immune system and healthy mucous membranes; potent antioxidant.
Vitamin B complex	Take 50 mg three times per day, or as indicated on the label.	B vitamins act as coenzymes for many enzymes that help decrease inflammation; support tissue.

Vitamin C	Take 8,000–13,000 mg per day.	Fights infections, diseases, and allergies; improves wound healing; stimulates the adrenal glands; promotes immune system; powerful antioxidant.
Vitamin E	Take 1,200 IU per day from all sources (including the multivitamin complex in the introduction to Part Two).	Potent antioxidant; supplies oxygen to cells; improves circulation and level of energy.
Zinc	Take 15–50 mg per day. Do not take more than 100 mg per day.	Needed by over 300 enzymes; necessary for white blood cell immune function; improves wound healing.

Comments

❏ According to enzyme expert Dr. Franz Klaschka, vitamin A and proteolytic enzymes should be taken together for the treatment of pancreatitis, as the two work synergistically.

❏ Fruits and vegetables (particularly green vegetables) should comprise over 60 percent of your diet. Beets, celery, onion, cauliflower, artichokes, turnips, and other vegetables should be eaten raw or blended or juiced. If your body cannot tolerate raw fruits and vegetables, increase your intake of digestive enzymes, or sauté or steam your produce.

❏ Eat fermented foods, such as yogurt, buttermilk and kefir; probiotics, including *Lactobacillus acidophilus* and *Bifidobacteria*; and whey.

❏ Do not overtax the pancreas. Eat like a grazing animal, consuming four or five smaller meals a day instead of one or two big meals.

❏ Follow the enzyme-rich Cave Man Diet, as outlined in Part Three.

❏ Eat very little protein.

❏ Consume no poisons (additives, preservatives, artificial colors or flavors).

❏ Avoid alcohol, refined white flour, and sugar. These produce undesirable fermentation in the bowel and stimulate improper metabolism. In addition, alcohol puts tremendous stress on the pancreas and is the leading cause of pancreatitis.

❏ Avoid chlorinated water. Distilled or bottled spring water, or even well water is recommended.

❏ Pancreatitis requires a physician's care. Therefore, any program should be conducted under a doctor's supervision. At times, surgery may be required, such as when there is little or no pancreatic secretion flowing from the pancreas.

❏ Physicians sometimes use enzyme retention enemas in the treatment of pancreatitis. Both methods allow a greater concentration of the enzymes to be absorbed than might be accomplished if the enzymes were taken orally. An enzyme retention enema is a means of instilling some enzymes into the colon to be retained. This allows the enzymes to be absorbed into the body's systems. Physicians use products specially formulated for this use. Caution should be exercised, however, since repeated rectal administration of very high doses of enzymes could lead to irritation of the anus or the skin around the anus. This can be avoided by applying a lubricating ointment, such as vitamin E oil or aloe vera, before administering the enema.

❏ See the list of Resource Groups in the Appendix for an organization that can provide you with more information about pancreatitis.

Peptic Ulcers

A peptic ulcer is a sore that occurs in the lining of the stomach or the duodenum (the first part of the small intestine) due to stomach acid and/or pepsin eating away at the lining. A shallow ulcer is called an erosion. The stomach and duodenum are almost always exposed to acids. The body, however, has certain defense mechanisms to prevent the acids from harming the stomach and duodenum linings. It is not known what causes these defense mechanisms to fail in some people, causing ulcers.

There are different types of peptic ulcers, depending on their location and how they were formed. *Duodenal ulcers* are the most common type. They are often caused by *Helicobacter pylori* bacteria. *Esophageal ulcers* occur as a result of gastroesophageal reflux—the regurgitation of stomach contents (including acid) into the lower part of the esophagus. *Gastric ulcers* usually occur in the upper portion of the stomach. They are often caused by certain drugs, such as aspirin and other nonsteroidal anti-inflammatory drugs. *Marginal ulcers* may occur when part of the stomach is removed. *Stress ulcers* occur as a result of illness, trauma, or other stressors.

Symptoms

The symptoms of peptic ulcers vary according to location. Most tend to heal and recur. Duodenal ulcers tend to produce a gnawing, burning, and/or aching pain that occurs when the stomach is empty. The pain may be relieved by eating or drinking milk, but it returns in a few hours. This may happen several times a day for weeks, and then the pain disappears, only to return in a few years. Only about 50 percent of patients with duodenal ulcers experience these symptoms. Gastric ulcers may cause pain after eating. They may also produce bloating, nausea, and vomiting after eating. The pain of esophageal ulcers is usually felt when swallowing or lying down.

Enzyme Therapy

Digestive enzyme therapy is used to provide any missing enzymes and thereby improve the digestion of food. This will reduce stress on the gastrointestinal mucosa, help maintain normal pH levels, detoxify the body, and promote the growth of healthy intestinal flora. Digestive enzymes also serve as replacements for the body's pancreactic enzymes, leaving the pancreatic enzymes free to perform other functions in the body, such as boosting immunity and decreasing inflammation.

Systemic enzyme therapy is used to reduce any inflammation, stimulate the immune system, improve circulation, help transport nutrients throughout the body, remove waste products, and improve health.

Enzyme Absorption System Enhancers (EASE) maximize enzyme activity. They can also improve the absorption and bioavailability of various nutrients and decrease the drain on the body's own digestive enzymes, thus prolonging their lives.

In addition to the standard multivitamin, multimineral, and multiglandular complexes recommended in the introduction to Part Two, the following supplements are recommended for the treatment of peptic ulcers. The following recommended dosages are for adults. Children between the ages of two and five should take one-quarter the recommended dosage, those between the ages of six and twelve should take one-half the recommended dosage, and those between the ages of thirteen and seventeen should take three-quarters the recommended adult dosage.

ESSENTIAL ENZYMES

Enzyme	Suggested Dosage	Actions
Formula Two—See page 38 of Part One for a list of products. *or*	Take 1–2 capsules three times per day after each meal.	Digests fats, carbohydrates, and fiber.
Formula One—See page 38 of Part One for a list of products.	Take 5 coated tablets or one heaping tablespoon of a granulated enzyme mixture three times per day.	Fights inflammation, swelling, and pain; speeds healing; stimulates immune function; fights free-radical and immune complex formation; improves digestion.

SUPPORTIVE ENZYMES

Enzyme	Suggested Dosage	Actions
Formula Three—See page 38 of Part One for a list of products.	Take as indicated on the label.	Potent antioxidant enzymes.

ENZYME ABSORPTION SYSTEM ENHANCERS (EASE)

Combination	Suggested Dosage	Actions
Biodias A Granules from Amano Pharmaceutical Co., Ltd.	Adults should take one packet in 8 ounces of water between or after meals three times times per day. Children ages 11 to 14 should take two-thirds of a packet; children between the ages of 8 and 10 should take half of the packet; and children between the ages of 5 and 7 should take one-third of a packet.	Helps improve digestion of nutrients; relieves inflammation; coats and promotes regeneration of mucous membranes; decreases stress on the stomach and intestines; reduces gastric pain.

ENZYME HELPERS

Nutrient	Suggested Dosage	Actions
Acidophilus or other probiotics, such as bifidobacterium; or Acidophilasé from Wakunaga, which also contain amyloytic, lipolytic, and proteolytic enzymes	Take as indicated on the label.	Improves gastrointestinal function; needed for production of the coenzymes biotin, niacin, folic acid, and pyridoxine; helps produce antibacterial substances; stimulates enzyme activity.
Aged garlic extract	Take 2 capsules three times per day.	Potent antioxidant; improves digestion; stimulates healthy intestinal bacteria.
Fructooligo-saccharides	Take as indicated on the label.	Increases peristaltic action of intestines; improves liver function.
Lecithin	Take 1,000 mg three times per day with meals.	Emulsifies fat.
Ox bile extract	Take 120 mg per day.	Aids digestion.
Vitamin A	Take 5,000 IU per day.	Needed by digestive glands to produce enzymes; promotes normal taste and appetite; required for healthy mucous linings and skin and cell membrane structure and a strong immune system; potent antioxidant.
Vitamin B complex	Take 50 mg three times per day.	B vitamins act as coenzymes.
Vitamin B_1 (thiamin)	Take 150 mg three times per day.	Essential for carbohydrate metabolism and normal digestion.
A basic (rather than acidic) form of Vitamin C, such as calcium ascorbate or magnesium ascorbate	Take 1,000–2,000 mg per day.	Stimulates the adrenal glands to produce hormones (which also act as coenzymes); aids in detoxification; a potent antioxidant; improves the function of many enzymes; required for converting the coenzyme folic acid into its active form; necessary for proper digestion.

Whey	Take 1 heaping tablespoon three times per day with meals.	Improves digestion.
Zinc	Take 10–50 mg per day. Do not take more than 100 mg per day from all sources.	Needed by over 300 enzymes. Many zinc enzymes work to metabolize carbohydrates, alcohol, and essential fatty acids and synthesize proteins.

Comments

❑ Licorice is as an effective herb in fighting stomach ulcers.

❑ Identify and eliminate all possible factors that can cause the development of peptic ulcers.

❑ Avoid drugs (especially nonsteroidal anti-inflammatory medications and aspirin), cigarette smoking, and stress.

❑ Supplements that heal and prevent further ulcer development include enzymes, vitamins A and E, zinc, and selenium.

❑ Several foods and herbs can help, including cabbage, echinacea, goldenseal, marshmallow, okra, and slippery elm.

Pharyngitis

See SORE THROAT.

Phlebitis

See THROMBOSIS.

Pinkeye

See CONJUNCTIVITIS.

Pleurisy

Pleurisy is an inflammation of the double membrane, called the pleura, that covers each lung and lines the chest cavity. It usually develops when a virus or bacterium irritates the pleura, causing inflammation. Pleurisy is not a disease in and of itself, but rather occurs as a complication of such conditions as pneumonia, cancer, tuberculosis, autoimmune conditions, irritation by such irritants as asbestos, allergic reactions to certain drugs, and infections.

There are two types of pleurisy. When a pleural effusion occurs (an abnormal accumulation of fluid between the two pleural membranes), it is called wet pleurisy. When there is no effusion, it is called dry pleurisy. In dry pleurisy, the two layers of pleura often rub against each other with each breath, making breathing very painful. In wet pleurisy, there is fluid between the two layers of pleura, which reduces the pain but makes breathing more difficult. Sometimes, an empyema (pus in the pleural space) develops. If the effusion is small, treatment of the cause of the pleurisy and effusion may be all that is necessary. If, however, the effusion is large, drainage of the fluid may be necessary.

Symptoms

Symptoms of pleurisy include sudden chest pain that may range from minor discomfort felt only upon deep breathing or coughing to intense stabbing pain. There may be fever. If there is fluid between the pleura, breathing may become very difficult.

Enzyme Therapy

Systemic enzyme therapy is used to decrease the pain and inflammation in the pleura and the lungs and to stimulate the immune system. Enzymes improve circulation, help speed tissue repair, bring nutrients to the damaged area, remove waste products, and enhance wellness. Digestive enzyme therapy is used to improve the digestion of food, reduce stress on the gastrointestinal mucosa, help maintain normal pH levels, detoxify, and promote the growth of healthy intestinal flora.

Enzyme Absorption System Enhancers (EASE) maximize enzyme activity. They can also improve the absorption and bioavailability of various nutrients and decrease the drain on the body's own digestive enzymes, thus prolonging their lives.

In addition to the standard multivitamin, multimineral, and multiglandular complexes recommended in the introduction to Part Two, the following supplements are recommended for the treatment of pleurisy. The following recommended dosages are for adults. Children between the ages of two and five should take one-quarter the recommended dosage, those between the ages of six and twelve should take one-half the recommended dosage, and those between the ages of thirteen and seventeen should take three-quarters the recommended adult dosage.

ESSENTIAL ENZYMES

Enzyme	Suggested Dosage	Actions
Formula One—See page 38 of Part One for a list of products.	Take 10 tablets three times per day between meals for six months. Then decrease dosage to	Stimulates immune function; breaks up circulating immune complexes; fights free-radical formation and inflammation.

	5 tablets three times per day between meals.	
Papain or	Take 500 mg three times per day for six months between meals. Then decrease dosage to 300 mg three times per day between meals.	Bolsters the immune system; breaks up antigen-antibody complexes; fights free-radical formation and inflammation.
Bromelain or	Take 400 mg three times per day between meals for six months. Then decrease dosage to 250 mg three times per day between meals.	Stimulates immune function; breaks up circulating immune complexes; fights free-radical formation and inflammation.
Trypsin with chymotrypsin	Take 200 mg three times per day between meals for six months. Then decrease dosage to 125 mg three times per day between meals.	Bolsters the immune system; breaks up antigen-antibody complexes; fights free-radical formation and inflammation.

SUPPORTIVE ENZYMES

Enzyme	Suggested Dosage	Actions
Formula Three—See page 38 of Part One for a list of products.	Take as indicated on the label.	Potent antioxidant enzymes.
A digestive enzyme product containing protease, amylase, and lipase enzymes. See the list of digestive enzyme products in Appendix.	Take as indicated on the label with meals.	Improves the breakdown and absorption of food nutrients; fights toxic build-up.

ENZYME ABSORPTION SYSTEM ENHANCERS (EASE)

Combination	Suggested Dosage	Actions
Combat from Life Plus or	Take as indicated on the label.	Helps support the respiratory system; stimulates immune system function; helps control infections.
Inflamzyme from PhytoPharmica or	Take as indicated on the label.	Gives support for tissues; immune system activator.
Protease Concentrate from Tyler Encapsulations	Take as indicated on the label.	Breaks up fibrin; fights inflammation.

ENZYME HELPERS

Nutrient	Suggested Dosage	Actions
Acidophilus and other probiotics	Take as indicated on the label.	Stimulates enzyme activity; improves gastrointestinal function.
Adrenal glandular	Take 100–400 mg per day.	Helps fight stress and adrenal exhaustion; fights inflammation.
Coenzyme Q$_{10}$	Take 30–60 mg one to two times per day with meals.	Important to immune system function; has antioxidant activity.

Kyolic Aged Garlic Liquid Extract with Vitamins B$_1$ and B$_{12}$	Take 10 drops in juice or water three times per day.	Stimulates immune system function.
Thymus glandular	Take 100 mg per day, or as indicated on the label.	Assists the thymus gland in immune resistance; improves immune response; fights infections.
Vitamin A	Take up to 50,000 IU for no more than one month (for higher doses, see a physician). If pregnant, do not take more than 10,000 IU per day.	Needed for strong immune system and healthy mucous membranes; potent antioxidant.
Vitamin B complex	Take 100–150 mg three times per day.	B vitamins act as coenzymes.
Vitamin C	Take 10,000–15,000 mg per day; or 1–2 grams every hour until symptoms improve.	Fights infections, diseases, and allergies; improves wound healing; stimulates the adrenal glands; promotes immune system; powerful antioxidant.
Vitamin E	Take 1,200 IU per day from all sources (including the multivitamin complex in the introduction to Part Two).	Acts as an intercellular antioxidant; improves circulation; helps resist diseases at the cellular level; works closely with selenium as a free-radical fighter.
Zinc	Take 15–50 mg per day. Do not take more than 100 mg per day.	Needed by over 300 enzymes; necessary for white blood cell immune function; improves healing.

Comments

❏ A number of herbs can help reduce inflammation, including echinacea, ginger root, ginkgo biloba, licorice, and white willow bark.

❏ Enzyme-rich fresh fruits and vegetables (particularly green vegetables) should comprise 60 percent of your diet. If your body cannot tolerate raw fruits and vegetables, increase your intake of digestive enzymes, or sauté or steam your produce.

❏ Avoid alcohol, refined sugar, and white flour.

❏ Drink plenty of bottled, distilled, or well water. Avoid chlorinated water.

❏ Detoxify periodically following the guidelines in Part Three.

❏ Gargle with the Enzymatic Gargle, the Echinacea/Enzyme Gargle, or the Garlic/Enzyme Gargle, found in Part Three, twice per day.

❏ Chest pain may be relieved by wrapping the chest each day with two or three wide (6-inch) non-adhesive elastic bandages. By using non-adhesive elastic bandages, you can avoid the skin irritation that can develop if using adhesive strapping. While the practice of binding the chest may relieve chest pain, it can, however, increase the risk for pneumonia, since the lungs are unable to fully expand. Therefore, this practice must be used with caution.

331

Pneumonia

Pneumonia is an acute lung infection. Each year, about 2 million people in the United States get pneumonia and 40,000 to 70,000 die (it ranks sixth among all diseases as a cause of death). There are several different types of pneumonia. Some types are caused by bacteria, some are caused by viruses, and some are caused by fungi. Pneumonia can also be caused by inhaling toxic chemicals or even objects into the airways.

Some people are more prone to falling ill with pneumonia than others. The very old and the very young are more susceptible to pneumonia than the average population; as are those with suppressed immune systems and those who are bedridden or unconscious. Alcoholics, those who smoke, and those with diabetes, heart failure, or lung diseases are also more prone to catching pneumonia. If pneumonia is suspected, see your doctor, or go to the hospital.

Symptoms

The symptoms of pneumonia vary depending upon what is causing it and how severe the disease is; however, general symptoms of pneumonia include chills followed by a fever (the temperature may increase rapidly), a sputum-producing cough, chest pain, and shortness of breath. Fatigue, sore throat, muscle aches, and nausea may also be present, depending upon the type and severity of pneumonia.

Enzyme Therapy

Systemic enzyme therapy is used to decrease inflammation and infection and stimulate the immune system. Enzymes improve circulation, help speed tissue repair, bring nutrients to the damaged area, remove waste products, and enhance wellness.

In order to have a healthy body, the digestive system must be functioning properly. Therefore, digestive enzyme therapy is used to improve the digestion of food, reduce stress on the gastrointestinal mucosa, help maintain normal pH levels, detoxify, and promote the growth of healthy intestinal flora.

Enzyme Absorption System Enhancers (EASE) maximize enzyme activity. They can also improve the absorption and bioavailability of various nutrients and decrease the drain on the body's own digestive enzymes, thus prolonging their lives.

In addition to the standard multivitamin, multimineral, and multiglandular complexes recommended in the introduction to Part Two, the following supplements are recommended for the treatment of pneumonia. The following recommended dosages are for adults. Children between the ages of two and five should take one-quarter the recommended dosage, those between the ages of six and twelve should take one-half the recommended dosage, and those between the ages of thirteen and seventeen should take three-quarters the recommended adult dosage.

ESSENTIAL ENZYMES

Enzyme	Suggested Dosage	Actions
Bromelain or	Take 500 mg three times per day between meals.	Fights inflammation; improves respiratory conditions; bolsters the immune system; breaks up mucus.
Serratiopeptidase or	Take 5 mg three times per day between meals.	Fights inflammation; improves respiratory conditions; bolsters the immune system; breaks up mucus.
Formula One—See page 38 of Part One for a list of products. or	Take 8–10 tablets three times per day between meals for six months or until symptoms improve. Thereafter, take 6 tablets three times per day between meals.	Fights inflammation, pain, and swelling; stimulates immune function; fights free radicals.
Inflazyme Forte from American Biologics or	Take as indicated on the label between meals.	Fights inflammation, pain, and swelling; speeds healing; strengthens capillary walls.
Enzyme complexes rich in proteases, plus amylases and lipases, working synergistically or	Take as indicated on label between meals.	Fights inflammation, pain, and swelling; speeds healing, strengthens capillary walls.
Trypsin and chymotrypsin	Take 500 mg of trypsin and 5–10 mg of chymotrypsin three times per day between meals.	Fights inflammation; improves respiratory conditions; bolsters the immune system; breaks up mucus.

SUPPORTIVE ENZYMES

Enzyme	Suggested Dosage	Actions
Formula Three—See page 38 of Part One for a list of products.	Take 200 mcg three times per day between meals.	Potent antioxidant enzymes.
A digestive enzyme formula containing protease, amylase, and lipase enzymes. See the list of digestive enzyme products in Appendix.	Take as indicated on the label with meals.	Improves the breakdown and absorption of proteins, carbohydrates, and fats; fights toxic build-up.

ENZYME ABSORPTION SYSTEM ENHANCERS (EASE)

Combination	Suggested Dosage	Actions
Cold Zzap from Naturally Vitamin Supplements or	Take as indicated on the label.	Fights infection from viruses or bacteria; boosts immunity; has antioxidant activity.
Combat from Life Plus	Take as indicated on the label.	Helps support the respiratory system; stimulates immune

or		system function; helps control infections.
Connect-All from Nature's Plus *or*	Take as indicated on the label.	Supports tissue repair.
Mucous Dissolver Liquezyme from Enzyme Process Laboratories *or*	Take as indicated on the label.	Dissolves mucus.
Protease Concentrate from Tyler Encapsulations *and*	Take as indicated on the label.	Breaks up fibrin; fights inflammation.
Bromelain Complex from PhytoPharmica	Take as indicated on the label.	Reduces swelling and inflammation.
Sinease from Prevail	Take as indicated on the label.	Decongestant.

ENZYME HELPERS

Nutrient	Suggested Dosage	Actions
Adrenal glandular	Take as indicated on the label.	Supports adrenal function.
Aged garlic extract	Take 1–3 capsules with meals twice per day.	Detoxifies and provides energy.
Amino-acid complex	Take as indicated on the label.	Amino acids are the building blocks of protein and enzymes.
Bioflavonoids	Take 100–300 mg twice per day.	Strengthen capillaries; potent antioxidants.
Lung glandular	Take as indicated on the label.	Improves lung function.
Vitamin A *or*	Take 10,000–50,000 IU per day for no more than one month (for higher doses see a physician). If pregnant, take no more than 10,000 IU per day.	Necessary for healthy mucous linings and skin and cell membrane structure and for a strong immune system; potent antioxidant.
Carotenoid complex	Take 25,000 IU per day.	Potent antioxidant; enhances immunity.
Vitamin B complex	Take 100–150 mg three times per day.	B vitamins act as coenzymes and support tissue.
Vitamin C	Take 1–2 grams every hour until symptoms improve; or mix a heaping teaspoon of powdered or crystalline vitamin C with 6 to 8 ounces of water or orange or lemon juice (if lemon juice is used, you may sweeten with honey if desired), and gargle a mouthful and swallow. Repeat every hour.	Fights infections, diseases, allergies, and common cold; potent antioxidant.
Zinc or zinc gluconate lozenges	Take 50 mg per day.	Needed for white blood cell immune function; involved with more than 300 enzymes.

Comments

❑ Eat plenty of enzyme-rich fresh fruits and vegetables (particularly green vegetables). They should comprise 60 percent of your diet. If your body cannot tolerate raw fruits and vegetables, increase your intake of digestive enzymes, or sauté or steam your produce. See Part Three for more dietary recommendations.

❑ Drink plenty of fluids.

❑ Gargle with the Enzymatic Gargle, the Echinacea/Enzyme Gargle, or the Garlic/Enzyme Gargle at least twice a day. See Part Three for directions for their preparation.

❑ Put five to six drops of eucalyptus in a sink or bowl of very hot water. Cover your head and shoulders with a large bath towel over the sink or bowl, and inhale the vapors.

❑ Follow the Enzyme Kidney Flush and the Enzyme Toxin Flush programs to detoxify. See Part Three for instructions.

❑ Hot broth or chicken or vegetable soup is helpful for cleaning out the body and sweating out the toxins. Before making the soup, be sure to remove the skin from the chicken (chicken skins contain many toxins). Sip the broth or soup as frequently as possible.

❑ See the list of Resource Groups in the Appendix for an organization that can provide you with more information about pneumonia.

Polymyositis

Polymyositis is a chronic connective tissue disorder marked by inflammation and degeneration of the muscles. The exact cause of polymyositis is unknown, although it seems that hypersensitivity or autoimmune reactions play a role in the development of the disease. Certain viruses, the use of certain drugs, and cancer also seem to play a role in the development of the disease. Polymyositis usually develops in adults between the ages of 40 and 60, or in children between the ages of 5 and 15. It is twice as likely to affect women as it is to affect men.

Symptoms

The symptoms of polymyositis may begin during or directly following an infection. They include muscle weakness, pain, and tenderness; joint pain; Raynaud's phenomenon (a circulatory disorder that causes the fingers and toes to turn white, especially when exposed to cold); fatigue; weight loss; and fever. This condition can be quite disabling and may confine the patient to a wheelchair. When the disorder affects the muscles of the mouth and esophagus, one may even have difficulties speaking, chewing, and/or swallowing. The onset

of symptoms may be sudden, or the symptoms may develop
slowly.

Enzyme Therapy

Systemic enzyme therapy is used to reduce muscular inflammation, stimulate the immune system, and improve circulation. Enzymes help speed tissue repair, bring nutrients to the
damaged area, remove waste products, and enhance wellness. Digestive enzyme therapy is used to improve the digestion of food, reduce stress on the gastrointestinal mucosa,
help maintain normal pH levels, promote the growth of
healthy intestinal flora, and detoxify the body, thereby
improving the overall health of the body. Digestive enzymes
also serve as replacements for the body's pancreatic
enzymes, leaving the pancreatic enzymes free to perform
other functions in the body, such as boosting immunity and
decreasing inflammation.

Enzyme Absorption System Enhancers (EASE) maximize
enzyme activity. They can also improve the absorption and
bioavailability of various nutrients and decrease the drain on
the body's own digestive enzymes, thus prolonging their lives.

In addition to the standard multivitamin, multimineral,
and multiglandular complexes recommended in the introduction to Part Two, the following supplements are recommended for the treatment of polymyositis. The following
recommended dosages are for adults. Children between the
ages of two and five should take one-quarter the recommended dosage, those between the ages of six and twelve
should take one-half the recommended dosage, and those
between the ages of thirteen and seventeen should take
three-quarters the recommended adult dosage.

ESSENTIAL ENZYMES

Enzyme	Suggested Dosage	Actions
Formula One—See page 38 of Part One for a list of products.	Take 10 tablets three times per day between meals. As symptoms subside (which should occur in approximately one week) decrease dosage to 3–5 tablets three times per day between meals. When pain disappears, begin taking a maintenance dose of 2–3 tablets after each meal.	Decreases inflammation, pain, and swelling; increases rate of healing; stimulates the immune system; strengthens capillary walls.
or		
Inflazyme Forte from American Biologics	Take as indicated on the label between meals.	Decreases inflammation, pain, and swelling; increases rate of healing; stimulates the immune system; strengthens capillary walls.
or		
Enzyme complexes rich in proteases, plus amylases and	Take as indicated on label between meals.	Decreases inflammation, pain, and swelling; increases rate of healing; stimulates the immune

lipases, working
synergistically

or

		system; strengthens capillary walls.
Bromelain	Take three to four 230–250 mg capsules per day until symptoms subside. If severe, take 6 to 8 capsules per day.	Fights inflammation, pain, and swelling; speeds recovery after injury or surgery; stimulates the immune system.

SUPPORTIVE ENZYMES

Enzyme	Suggested Dosage	Actions
Formula Three—See page 38 of Part One for a list of products.	Take 50 mcg three times per day between meals.	Potent antioxidant enzymes.
A digestive enzyme formula containing protease, amylase, and lipase enzymes. For the list of digestive enzyme products, see Appendix.	Take as indicated on the label with meals.	Improves the breakdown and absorption of food nutrients; fights toxic build-up.

ENZYME ABSORPTION SYSTEM ENHANCERS (EASE)

Combination	Suggested Dosage	Actions
AminoLogic from PhysioLogics

or | Take 1 capsule three times per day between meals; or follow the advice of your health-care professional. | Improves muscle function; lends antioxidant support and basic nutrition; improves energy and endurance. |
| Connect-All from Nature's Plus

or | Take as indicated on the label. | Supports tissue repair. |
| Magnesium Plus from Life Plus

or | Take one or two tablets one to three times per day. | Important for muscle relaxation, calcium utilization, energy pro-duction, and many enzyme reactions. |
| Mobil-Ease from Prevail | Take one capsule between meals two to three times per day. | Has anti-inflammatory activity. |

ENZYME HELPERS

Nutrient	Suggested Dosage	Actions
Adrenal glandular	Take 100–400 mg per day.	Fights inflammation and adrenal exhaustion.
Aged garlic extract	Take 2 capsules three times per day.	Detoxifies and provides energy.
Amino-acid complex, including such amino acids as L-ornithine, L-glutamine, L-arginine, L-proline, and L-lysine	Take as indicated on the label on an empty stomach.	Amino acids are the building blocks of protein and enzymes.
Bioflavonoids	Take 150–500 mg three times per day.	Strengthen capillaries; potent antioxidants.
Branched-chain amino acids (isoleu-cine, leucine, and	Take as indicated on the label between meals.	Often found together in a supplement, branched-chain amino acids aid in healing of

valine)		muscle, bone, and skin and act as fuel for muscles.
Calcium	Take 1,000 mg per day.	Plays important roles in muscle contraction and nerve transmission.
Chondroitin sulfate	Take 250 mg two to three times per day with meals; or follow advice of your health-care professional.	Supports connective tissue.
Coenzyme Q$_{10}$	Take 30–60 mg one to two times per day with meals; or follow the advice of your health-care professional.	Potent antioxidant.
Essential fatty acid complex	Take 75–100 mg two to three times per day with meals; or follow the advice of your health-care professional.	Helps strengthen cell membranes.
Magnesium	Take 500 mg per day.	Plays important roles in muscle contraction, nerve function, and calcium absorption.
Vitamin B complex	Take 50 mg three times per day, or as indicated on the label.	B vitamins act as coenzymes and support neuromusculoskeletal function.
Vitamin C	Take 5,000–10,000 mg per day.	Helps to produce collagen and normal connective tissue; promotes immune system; powerful antioxidant.
Zinc	Take 50 mg per day. Do not take more than 100 mg per day.	Needed by over 300 enzymes.

Comments

❑ For those individuals whose lungs are affected, polymyositis seems to be more severe and resistant to treatment. Therefore, consult with a physician.

❑ After a history, physical examination, and X-rays are conducted, your physician will develop a treatment program. Your treatment program might include chiropractic adjustments; manipulation; massage; and physiotherapy, including electrical simulation, ice and heat applications, exercises, and postural training (which includes proper lifting, bending, standing, walking, driving, sitting, and sleeping positions).

❑ Most physical activities should be curtailed until inflammation subsides.

❑ Corticosteroids are frequently used with immunosuppressants (such as methotrexate). However, these drugs may weaken the body's immune system, allowing foreign antigens to attack the body. If enzymes are used, it is possible that the level of immunosuppressive drugs can be reduced (or eventually eliminated).

❑ *See also* DERMATOMYOSITIS.

Post-Childbirth Complications

An *episiotomy* is a small surgical incision through the vaginal wall that may be performed during the last stages of childbirth. An episiotomy enlarges the opening of the vagina, making birth of the baby easier. This procedure prevents tearing of the tissue and is easier to repair than a tear.

If the incision is not repaired properly, it may cause weakening of the muscular and fascial supports of the pelvic floor. Without proper support, the bladder could push downward through the front wall of the vagina to form a hernia (known as a cystocele).

Symptoms

Symptoms following childbirth with an episiotomy include pain, swelling, redness, and heat (inflammation) in the area between the vagina and the rectum. There may be itching, and an infection may develop that can migrate up the vagina or toward the anal opening. Symptoms usually disappear as the incision heals, but a scar can remain.

Enzyme Therapy

Systemic enzyme therapy is used to reduce inflammation, relieve pain, and decrease scar tissue formation. Enzymes enhance wellness by stimulating the immune system, improving circulation, and speeding tissue repair by bringing nutrients to the damaged area and removing waste products.

Digestive enzyme therapy is used to improve the digestion of food, reduce stress on the gastrointestinal mucosa, help maintain normal pH levels, detoxify the body, and promote the growth of healthy intestinal flora, thereby improving the overall health of the body. Digestive enzymes also serve as replacements for the body's pancreatic enzymes, leaving the pancreatic enzymes free to perform other functions in the body, such as boosting immunity, improving circulation, and decreasing inflammation.

Enzyme Absorption System Enhancers (EASE) maximize enzyme activity. They can also improve the absorption and bioavailability of various nutrients and decrease the drain on the body's own digestive enzymes, thus prolonging their lives.

Topical enzyme creams are used to dissolve dead and damaged skin cells and exfoliate the skin, while leaving healthy skin cells intact. A proteolytic enzyme salve can be effectively used to break up fibrin (scar tissue) formation. In many instances, scars will practically vanish if the enzyme salve is used properly and in time.

In addition to the standard multivitamin, multimineral, and multiglandular complexes recommended in the intro-

duction to Part Two, the following supplements are recommended for the treatment of post-childbirth complications.

ESSENTIAL ENZYMES

Enzyme	Suggested Dosage	Actions
Proteolytic enzyme salve	Apply as needed.	Reduces inflammation.
Superoxide dismutase	Take 150 mcg between meals for one to two weeks.	Fights inflammation; potent antioxidant.

AND SELECT ONE OF THE FOLLOWING:

Formula One—See page 38 of Part One for a list of products.	Take 5 tablets three times per day for ten days between meals on an empty stomach.	Decreases inflammation and swelling; reduces scar tissue formation; increases rate of healing; stimulates immune system in fighting infection.
Bromelain	Take 180–200 mg two times per day between meals for 7–14 days.	Fights inflammation, pain, and swelling; speeds recovery after injury or surgery; stimulates the immune system; strengthens capillary walls.
Serratiopeptidase	Take 10 mg two times per day between meals for one to two weeks.	Fights inflammation; reduces pain and swelling.
Trypsin with chymotrypsin	Take 100 mg of trypsin and 4 mg of chymotrypsin three times per day between meals for one to two weeks.	Fights inflammation, pain, and swelling; increases rate of healing; stimulates the immune system.
Pancreatin	Take 400 mg six times per day on an empty stomach for one to two weeks.	Fights inflammation, pain, and swelling; increases rate of healing; stimulates the immune system; strengthens capillary walls.
Papain or a microbial protease	Take 240 mg two times per day between meals for one to two weeks.	Fights inflammation; speeds recovery from injuries and surgery.

SUPPORTIVE ENZYMES

Enzyme	Suggested Dosage	Actions
Catalase and glutathione peroxidase	Take as indicated on the label.	Potent antioxidant enzymes.
A digestive enzyme product containing protease, amylase, and lipase enzymes. See the list of digestive enzyme products in Appendix.	Take as indicated on the label with meals.	Improves the breakdown and absorption of food nutrients; fights toxic build-up.

ENZYME ABSORPTION SYSTEM ENHANCERS (EASE)

Combination	Suggested Dosage	Actions
Inflamzyme from PhytoPharmica or	Take as indicated on the label.	Promotes tissue repair.
Connect-All from Nature's Plus or	Take as indicated on the label.	Supports connective tissue.
CTR Support from PhysioLogics	Take as indicated on the label.	Nourishes the body during tissue recovery and healing from injury; diminishes damage caused by swelling and inflammation.

ENZYME HELPERS

Nutrient	Suggested Dosage	Actions
Adrenal glandular	Take 100–400 mg per day.	Fights inflammation and adrenal exhaustion.
Aloe vera extract	Take as indicated on the label.	Aids in healing.
Amino-acid complex, including L-ornithine, L-glutamine, L-arginine, L-proline, and L-lysine	Take as indicated on the label on an empty stomach.	Required by enzymes that increase immune function and fight infection and inflammation.
Bioflavonoids, such as rutin or quercetin	Take 500 mg three times per day between meals.	Strengthen capillaries; potent antioxidants.
Branched-chain amino acids (isoleucine, leucine, and valine)	Take as indicated on the label between meals.	Often found together in a supplement, branched-chain amino acids aid in healing of muscle, bone, and skin; act as fuel for muscles.
Calcium	Take 1,000 mg per day.	Plays important roles in muscle contraction; involved in the activation of several enzymes.
Coenzyme Q_{10}	Take 30–60 mg one to two times per day with meals; or follow the advice of your health-care professional.	Potent antioxidant.
Magnesium	Take 500 mg per day.	Plays important role in muscle contraction, nerve function, calcium absorption, and blood clotting; catalyst in enzyme activity; protects arterial lining from stressors, such as edema and inflammation.
Pituitary glandular	Take 100 mg per day.	Helps form new blood vessels; fights tumor growth; aids wound healing.
Thymus glandular	Take 100 mg per day.	Assists the thymus gland in immune resistance; improves immune response; fights infections.
Vitamin A	Take 5,000 IU per day.	Needed for strong immune system and healthy mucous membranes; potent antioxidant.
Vitamin B complex	Take 150 mg three times per day; or take as indicated on the label.	B vitamins act as coenzymes and support tissue repair.
Vitamin C	Take 10,000–15,000 mg per day.	Fights infections, diseases, and allergies; improves wound

		healing; stimulates the adrenal glands; promotes immune system; powerful antioxidant.
Vitamin E	Take 1,200 IU per day from all sources (including the multivitamin complex in the introduction to Part Two).	Potent antioxidant; supplies oxygen to cells; improves circulation and level of energy.
Zinc	Take 15–50 mg per day. Do not take more than 100 mg per day.	Needed by over 300 enzymes; necessary for white blood cell immune function; improves wound healing.

Comments

❑ A hot sitz bath several times a day can help relieve the pain of an episiotomy.

❑ Enzyme salves containing individual enzymes (such as papain) or enzyme mixtures (including bromelain, papain, trypsin, chymotrypsin, pancreatin, lipase, and amylase), and sometimes vitamins A and E can be of help in relieving pain and reducing inflammation. These creams can dissolve damaged skin. Once the dead cells are dissolved, the body manufactures new collagen and elastin, thus rebuilding healthy skin with less scarring.

❑ Place ten drops of liquid aged garlic extract on a cotton swab or gauze. Dab gently along the scar area between the vagina and the anus. Repeat at least twice per day.

❑ Break and spread the contents of two to four vitamin A and E soft gel capsules on the episiotomy scar twice per day.

Post-Thrombotic Syndrome

See THROMBOSIS.

Premenstrual Syndrome

Premenstrual syndrome (PMS) is a group of physical, emotional, and psychological symptoms that begins in women anywhere from seven to fourteen days before their menstrual period begins and usually ends once the period begins. It appears to be associated with fluctuations in levels of estrogen and progesterone that occur during the menstrual cycle. At some time between the onset of menstruation and menopause, some 40 percent of all women will experience PMS. Nearly 10 percent of these women will have symptoms so severe that they interfere with their professional and personal lives.

Estrogen affects fluid retention, and temporary increases

in body fluid in some tissues seem to explain some of the symptoms of PMS, such as edema, weight gain, bloating, and breast tenderness. Increased levels of prostaglandin, a hormonelike chemical, in the bloodstream also appears to play a role.

Some other possible causes of PMS might include nutritional deficiencies, stress, hormone imbalances, and low blood sugar (hypoglycemia).

Symptoms

Dozens of physical, emotional, and digestive symptoms have been associated with PMS. Some of the most common emotional symptoms include anxiety, mood swings, panic attacks, tension, nervousness, irritability, depression, and hostility. Mental symptoms include mental confusion, forgetfulness, and difficulty concentrating. Physical symptoms might include backache; breast tenderness and swelling; fluid retention; headache; tissue swelling and joint pain; muscle spasms; weight gain; cravings for alcohol, sweets, chocolates, or carbohydrates; fatigue; dizziness; changes in appetite; fainting; sinus problems; acne; asthma attacks; allergies; cold sores; and seizures. Digestive complaints might include abdominal bloating, diarrhea, and constipation. The symptoms and their severity vary, of course, among different women, and sometimes in the same woman from month to month. Usually, the symptoms occur during the week to fourteen days before menstruation and disappear once menstruation begins.

Enzyme Therapy

Digestive enzyme therapy is used to improve the digestion of food and improve some of the gastrointestinal complaints often associated with PMS. Enzymes can reduce stress on the gastrointestinal mucosa, normalize blood sugar levels, encourage the growth of beneficial microorganisms in the intestine, detoxify the body, and help maintain normal pH levels. Digestive enzymes also serve as replacements for the body's pancreatic enzymes, leaving the pancreatic enzymes free to perform other functions in the body, such as improving circulation and decreasing inflammation.

Systemic enzyme therapy is used to support digestive enzyme therapy to reduce the inflammation that accompanies PMS and to reduce pain, discomfort, and abdominal cramps. Enzymes can stimulate the immune system, improve circulation, transport nutrients throughout the body, remove waste products, and enhance wellness.

Enzyme Absorption System Enhancers (EASE) maximize enzyme activity. They can also improve the absorption and bioavailability of various nutrients and decrease the drain on the body's own digestive enzymes, thus prolonging their lives.

Documentation shows that bromelain, an enzyme extracted from pineapples can relieve menstrual cramping. Since bromelain decreases the spasms of the cervix in PMS, it is believed that bromelain is a smooth muscle relaxant. Evidently, bromelain also helps to balance the synthesis of prostaglandin (which is important in causing premenstrual tension), according to Dr. Michael Murray and Joseph Pizzorno in their book *An Encyclopedia of Natural Medicine* (Rocklin, CA: Prima Publishing, 1991).

In addition to the standard multivitamin, multimineral, and multiglandular complexes recommended in the introduction to Part Two, the following supplements are recommended for the treatment of premenstrual syndrome. The following recommended dosages are for adults. Girls under the age of seventeen should take three-quarters the recommended adult dosage.

ESSENTIAL ENZYMES

Enzyme	Suggested Dosage	Actions
Bromelain

or | Take 750 mg two times per day between meals. | Fights inflammation, pain, and swelling; stimulates the immune system; strengthens capillary walls. |
| Formula One—See page 38 of Part One for a list of products.

or | Take 5 tablets twice per day. | Fights inflammation, pain, and swelling; stimulates the immune system; strengthens capillary walls. |
| Pancreatin

or | Take 500 mg six times per day on an empty stomach. | Fights inflammation, pain, and swelling; increases rate of healing; stimulates the immune system; strengthens capillary walls. |
| Papain

or | Take 300 mg two times per day between meals. | Fights inflammation; speeds recovery from injuries and surgery. |
| Trypsin with chymotrypsin

or | Take 250 mg of trypsin and 10 mg of chymotrypsin three times per day between meals. | Fights inflammation, pain, and swelling; increases rate of healing; stimulates the immune system. |
| Microbial protease | Take 300 mg two times per day between meals. | Fights inflammation. |

SUPPORTIVE ENZYMES

Enzyme	Suggested Dosage	Actions
Formula Three—See page 38 of Part One for a list of products.	Take as indicated on the labels.	Fights inflammation; potent antioxidant enzymes.
A digestive enzyme product containing protease, amylase, and lipase enzymes. See the list of digestive enzyme products in Appendix.	Take as indicated on the label with meals.	Improves the breakdown and absorption of food nutrients; fights toxic build-up.

ENZYME ABSORPTION SYSTEM ENHANCERS (EASE)

Combination	Suggested Dosage	Actions
Hepatic Complex from Tyler Encapsulations

or | Take 2–4 capsules between meals three times per day. | Helps improves the symptoms of PMS. |
| NESS Formula #501

or | Take as indicated on the label. | Reduces inflammation; aids relief of menstrual cramping. |
| Soy SuperComplex from Rainbow Light | Take as indicated on the label. | Contains isoflavonoids, which are weak plant estrogens that can help relieve the symptoms of PMS. |

ENZYME HELPERS

Nutrient	Suggested Dosage	Actions
Acidophilus and other probiotics	Take as indicated on the label.	Stimulates enzyme activity; improves gastrointestinal function, thereby improving the overall health of the body; needed for production of biotin, niacin, folic acid, and pyridoxine, all coenzymes.
Adrenal glandular	Take 300 mg per day.	Fights inflammation and adrenal exhaustion.
Aged garlic extract	Take 2 capsules three times per day.	Fights inflammation; stimulates the immune system; potent antioxidant.
Amino-acid complex	Take as indicated on the label on an empty stomach.	Required for enzyme, peptide, and hormone production; helps the body maintain proper fluid and acid/base balance.
Bioflavonoids	Take 500 mg per day.	Strengthen capillaries; potent antioxidants.
Calcium	Take 1,000 mg per day.	Essential for enzymes; involved in muscle contraction, energy production, and healing; plays important roles in muscle and nerve cell function.
Coenzyme Q_{10}	Take 30–60 mg one to two times per day with meals; or follow the advice of your health-care professional.	Potent antioxidant.
Essential fatty acids (found in flaxseed oil and evening primrose oil)	Take 100 mg of flaxseed oil or evening primrose oil two to three times per day with meals; or follow the advice of your health-care professional.	Help strengthen cell membranes; support immune function.
Magnesium	Take 500 mg per day.	Plays important roles in muscle contraction, energy production, nerve function, and calcium absorption; necessary for glucose metabolism, blood clotting, and synthesis of fats, proteins, and nucleic acid.

Niacin	Take 50 mg the first day, then increase dosage daily by 50 mg to 1,000 mg. If adverse effects occur, discontinue use. Take dosages higher than 1,000 mg only under the supervision of a physician.	Dilates blood vessels, improving circulation; a coenzyme for many enzymes.
or		
Niacin-bound chromium	Take 20 mg per day, or as indicated on the label.	
Thymus glandular	Take 90 mg per day.	Assists the thymus gland in immune resistance; improves immune response.
Vitamin B_6	Take 50 mg three times per day.	Assists enzymes that metabolize amino acids and fats; necessary for proper immune function; involved in the synthesis of several neurotransmitters and neurohormones; needed for healthy nerves, muscle, blood, and skin.
Vitamin C	Take 8,000–13,000 mg per day.	Fights infections, diseases, and allergies; improves wound healing; stimulates the adrenal glands; promotes immune system; powerful antioxidant.

Comments

❏ Herbs, including alfalfa, echinacea, ginger root, ginkgo biloba, licorice, and white willow bark, can help reduce inflammation.

❏ A proper diet and supplementation program can help balance the body and reduce PMS. For further information, see Part Three.

Prostate Cancer

The prostate gland secretes a thin fluid, a large component of semen, that helps the motility and fertility of sperm. Cancer of this organ is the most common cancer in men in the United States and the second leading cause of cancer death (exceeded only by lung cancer).

Prostate cancer incidence increases with age. In fact, over 80 percent of all prostate cancers are diagnosed in men over the age of 65. Although the disease is quite common in North America and Europe, it is rare in other parts of the world (such as Africa and South America). According to the American Cancer Society, Black Americans have the highest incidence of prostate cancer in the world, but the reason is not known. The disease seems to run in families, but researchers aren't sure if this is due to genetic factors or environmental factors. Diet may also play a part, as some studies indicate high intakes of dietary fat increase prostate cancer risk.

Symptoms

Prostate cancer generally grows very slowly and does not produce any symptoms until it is well-advanced. When symptoms do occur, they are often very similar to those of prostatitis, benign prostatic hyperplasia, and other disorders of the prostate, making it difficult to diagnose. Such symptoms include weak or interrupted urine flow, a frequent need to urinate, inability to urinate, or difficulty starting or stopping the urine flow. There may be pain or burning upon urination, blood in the urine, lower back pain, or pain in the upper thighs or hips. A blood test (called a PSA) and a physical exam can help check for prostate cancer. Sometimes, however, pros-tate cancer is not diagnosed until it spreads to the bone, causing painful and weakened bones prone to fracture.

Enzyme Therapy

In the treatment of cancer, systemic enzyme therapy is a complementary therapy and should be integrated with the overall therapy program. The primary therapy for prostate cancer includes surgery, chemotherapy, and radiation. Enzyme therapy is used to stimulate and stabilize the immune system, thus increasing the opportunities to destroy the residual tumor cells and decrease the rate of tumor growth; help differentiate the healthy from the cancerous tissue and better identify the tumor for surgical removal; reduce inflammation; improve wound healing and help speed tissue repair by bringing nutrients to the damaged area and removing waste products; improve the long-term prognosis; improve the patient's quality of life; decrease the adverse side effects of the primary therapy (radiation, chemotherapy, and/or surgery); reduce the cost of therapy by minimizing its use (if enzyme therapy has begun early enough); reduce the risk of complications, such as blood clots; and decrease immune suppression caused by surgery and anesthesia.

Digestive enzymes (from microbial, plant, and animal sources) are essential for the gastrointestinal tract and the body as a whole in helping to prevent the onset of cancer and improve the cancer patient's health because they aid the breakdown, digestion, and absorption of nutrients; decrease the enzyme production load of the pancreas, thereby helping to reduce stress on the pancreas and increasing its productive life; help prevent toxin build-up, especially in the small and large intestine; stimulate friendly microflora production in the intestine; and facilitate elimination of potential cancer-causing waste from the food and cells of the body. Digestive enzymes also serve as replacements for the body's pancreatic enzymes, leaving the pancreatic enzymes free to perform other functions in the body, such as boosting immunity and improving circulation.

Enzyme Absorption System Enhancers (EASE) maximize enzyme activity. They can also improve the absorption and

bioavailability of various nutrients and decrease the drain on the body's own digestive enzymes, thus prolonging their lives.

In addition to the standard multivitamin, multimineral, and multiglandular complexes recommended in the introduction to Part Two, the following supplements are recommended for the treatment of prostate cancer.

ESSENTIAL ENZYMES

Enzyme	Suggested Dosage	Actions
Before and after surgery; before, during, and after chemotherapy or radiotherapy; and as long-term and palliative therapy:		
WobeMugos E from Mucos	Take 2–4 tablets three times per day.	Speeds wound healing; leads to earlier response to chemotherapy and reduced side effects; reduces anorexia, muscle wasting, weakness, and malnutrition (cachexia), depression, and pain; prevents metastasis; eliminates or prevents disturbances of lymphatic drainage.

AND SELECT ONE OF THE FOLLOWING:

Formula One—See page 38 of Part One for a list of products.	Take 8–10 tablets three times per day between meals.	Stimulates immune function; fights cancer, inflammation, pain, and swelling; speeds healing; fights free radicals and immune complexes.
Inflazyme Forte from American Biologics	Take as indicated on the label between meals.	Stimulates immune function; fights cancer, inflammation, pain, and swelling; speeds healing; fights free radicals and immune complexes.
Bromelain	Take four to six 230–250 mg capsules per day.	Stimulates immune function; fights cancer, inflammation, pain, and swelling; speeds healing; fights free radicals and immune complexes.
Enzyme complexes rich in proteases, plus amylases and lipases, working synergistically	Take as indicated on the label between meals.	Stimulates immune function; fights cancer, inflammation, pain, and swelling; speeds healing; fights free radicals and immune complexes.

SUPPORTIVE ENZYMES

Enzyme	Suggested Dosage	Actions
A digestive enzyme formula containing protease, amylase, and lipase, enzymes. For a list of digestive enzyme products, see Appendix.	Take as indicated on the label.	Improves digestion by improving the breakdown of proteins, carbohydrates, and fats.
Formula Three—See page 38 of Part One for a list of products.	Take as indicated on the label.	Potent antioxidant enzymes.

ENZYME ABSORPTION SYSTEM ENHANCERS (EASE)

Combination	Suggested Dosage	Actions
Antioxidant Complex from Tyler Encapsulations *or*	Take as indicated on the label.	Potent free-radical fighter and immune system stimulator.
Detox Enzyme Formula from Prevail *or*	Take as indicated on the label.	Helps rid the body of harmful toxins.
Hepazyme from Enzymatic Therapy *or*	Take as indicated on the label.	Stimulates immune function.
Kyolic Formula 102 (with enzymes) *or*	Take 2 capsules or tablets with meals two to three times per day.	Strengthens the immune system; helps alleviate stress; helps fight free radicals; helps fight inflammation; improves digestion.
Oxy-5000 Forte from American Biologics *or*	Take as indicated on the label.	Helps the body excrete toxins; fights biochemical stressors, common infections, and degenerative processes.
Vita-C-1000 from Life Plus	Take as indicated on the label.	Helps fight cancer; boosts immune function.

ENZYME HELPERS

Nutrient	Suggested Dosage	Actions
Acidophilus or other probiotics	Take as indicated on the label.	Stimulates enzyme activity; improves gastrointestinal function, thereby improving the overall health of the body.
Bioflavonoids, such as rutin or quercetin	Take 150–500 mg three times per day.	Strengthen capillaries; potent antioxidants.
Magnesium	Take 250 mg per day.	Plays important roles in protein building, muscle contraction, nerve function, energy production, and calcium absorption; necessary for synthesis of fats, proteins, and nucleic acid.
Thymus glandular	Take 30 mg per day.	Assists thymus gland (important in immune resistance); helps improve immune response; fights cancer.
Vitamin A	Take 50,000 IU per day. Do not take more than 10,000 IU if you are pregnant. Up to 100,000 IU per day may be taken under the direction of a physician.	Stabilizes the cell membrane; improves response to therapy; needed for strong immune system and healthy tissue; potent antioxidant.
Vitamin C	Take 2–10 grams per day. For amounts over 10 grams, consult a physician.	Prevents capillary weakness; fights infection; improves wound healing; stimulates the adrenal glands; promotes immune system; powerful antioxidant.
Vitamin D	Take 3,000–6,000 IU per day.	Helps stimulate the immune system; hinders the advance of

		tumors; activates cancer-fighting monocytes; helps break up waste products and toxins.
Vitamin E	Take 3,000–6,000 IU per day.	Causes more efficient consumption of oxygen; acts as an intercellular antioxidant; improves circulation; helps resist diseases at the cellular level; works closely with selenium to fight free radicals.
Zinc	Take 50 mg per day. Do not take more than 100 mg from all sources.	Needed for white blood cell immune function, and tissue healing; helps protect the body from toxins that cause free-radical formation; a cofactor for more than 300 enzymes.

Comments

❏ The American Cancer Society recommends that men receive a PSA blood test beginning at age 50. But if you have a family history of prostate cancer, or are an African-American male, have your first test by age 45.

❏ Cut down on your intake of dietary fat. A low-fat diet can slow the progression of the disease.

❏ According to enzyme expert Dr. Franz Klaschka, vitamin A and proteolytic enzymes should be taken at the same time for the treatment of cancer.

❏ See Part Three for instructions on the Enzyme Kidney Flush and the Enzyme Toxin Flush programs, as well as additional information on diet, juicing, fasting, and exercise.

❏ During the weeks you are not using the Enzyme Kidney Flush or Enzyme Toxin Flush, use the coffee enema as described in Part Three. Coffee enemas are used to stimulate the release of bile from the liver and facilitate the removal of waste metabolites from the body and into the intestines for excretion.

❏ Physicians sometimes use enzyme injections or enzyme retention enemas in the treatment of cancer. Both methods allow a greater concentration of the enzymes to be absorbed than might be accomplished if the enzymes were taken orally. An enzyme retention enema is a means of instilling some enzymes into the colon to be retained. This allows the enzymes to be absorbed into the body's systems. Physicians use products specially formulated for these uses. Caution should be exercised, however, since repeated rectal administration of very high doses of enzymes could lead to irritation of the anus or the skin around the anus. This can be avoided by applying a lubricating ointment, such as vitamin E oil or aloe vera, before administering the enema.

❏ See the list of Resource Groups in the Appendix for an organization that can provide you with more information about prostate cancer.

❏ *See also* CANCER; PROSTATE DISORDERS.

Prostate Disorders

The prostate is a chestnut-sized, doughnut-shaped male gland located just under the bladder, and it surrounds the urethra where it joins the bladder. Both urine and sperm pass through the prostate. It gradually enlarges as men age. In fact, enlargement of the prostate gland (*benign prostate hyperplasia*) is commonly seen in some 60 percent of men over 50 years of age. Why the prostate enlarges is not known, but it may be related to hormonal changes. As it enlarges, the prostate narrows the opening of the urethra and interferes with the excretion of urine. When the patient urinates, the bladder may not empty completely. Urine may stagnate in the bladder, leading to infection. If enough urine builds up, kidney damage may ensue. If the bladder overfills, urinary incontinence results.

Prostatitis is inflammation of the prostate gland. Sometimes, this inflammation may be caused by infection—usually bacterial, although sometimes fungal, viral, or protozoal. Bacterial infections often spread from the urinary tract to the prostate, causing inflammation. Usually, however, prostatitis is not caused by infection. In these cases, its cause is unknown.

The inflammation causes pain in the groin, the area between the penis and anus, and the lower back. It may also produce fever, chills, joint and muscle aches, and burning with urination, sometimes with blood in the urine. The man may also experience a frequent urge to urinate, and difficulty starting urination and completely emptying the bladder.

The symptoms of most prostate disorders, including prostate cancer—the second leading cause of cancer death in men, are very similar. Your physician must make a diagnosis with a rectal examination and blood tests to rule out prostate cancer.

Enzyme Therapy

Systemic enzyme therapy is used to reduce prostatic swelling and inflammation, stimulate the immune system, improve circulation, help speed tissue repair, bring nutrients to the damaged area, remove waste products, improve health, strengthen the body as a whole, and build general resistance.

Digestive enzyme therapy is used to improve the digestion of food, reduce stress on the gastrointestinal mucosa, help maintain normal pH levels, detoxify the body, and promote the growth of healthy intestinal flora, thereby improving the health of the overall body. Digestive enzymes also serve as replacements for the body's pancreatic enzymes, leaving the pancreatic enzymes free to perform other functions in the body, such as decreasing inflammation and boosting immunity.

Enzyme Absorption System Enhancers (EASE) maximize enzyme activity. They can also improve the absorption and

bioavailability of various nutrients and decrease the drain on the body's own digestive enzymes, thus prolonging their lives.

In addition to the standard multivitamin, multimineral, and multiglandular complexes recommended in the introduction to Part Two, the following supplements are recommended for the treatment of prostate disorders. Those under the age of seventeen should take three-quarters the recommended adult dosage.

ESSENTIAL ENZYMES

Enzyme	Suggested Dosage	Actions
Superoxide dismutase	Take 150 mcg between meals.	Fights inflammation; potent antioxidant.

AND SELECT ONE OF THE FOLLOWING:

Formula One—See page 38 of Part One for a list of products.	Take 10 tablets three times per day between meals for six months. Then decrease dosage to 5 tablets three times per day between meals.	Fights inflammation, pain, and swelling; stimulates the immune system.
Phlogenzym from Mucos	Take as indicated on the label.	Fights inflammation, pain, and swelling; stimulates the immune system.
Serratiopeptidase	Take 10 mg three times per day between meals.	Fights inflammation; reduces pain and swelling.
Bromelain	Take three to four 230–250 mg capsules per day until symptoms subside. If severe, take 6 to 8 capsules per day.	Fights inflammation, pain, and swelling; speeds recovery after injury or surgery; stimulates the immune system.
A microbial protease (individually or in a combination)	Take 500 mg three times per day between meals, or as indicated on the label.	Breaks down proteins; speeds recovery from injury and inflammation.

SUPPORTIVE ENZYMES

Enzyme	Suggested Dosage	Actions
Catalase and glutathione peroxidase	Take as indicated on the labels.	Potent antioxidant enzymes.
A digestive enzyme product containing protease, amylase, and lipase enzymes. See the list of digestive enzyme products in Appendix.	Take as indicated on the label with meals.	Improves the breakdown and absorption of food nutrients; fights toxic build-up.

ENZYME ABSORPTION SYSTEM ENHANCERS (EASE)

Combination	Suggested Dosage	Actions
Prostate Enzyme Formula from Prevail *or*	Take as indicated on the label.	Stimulates immune function; reduces inflammation.
Prostate Formula from Life Plus	Take as indicated on the label.	Helps fight prostate disorders; stimulates immune function; reduces inflammation.

ENZYME HELPERS

Nutrient	Suggested Dosage	Actions
Acidophilus and other probiotics	Take as indicated on the label.	Improve gastrointestinal function; needed for production of biotin, niacin, folic acid, and pyridoxine, all coenzymes; help produce antibacterial substances; help prevent infection; stimulate enzyme activity.
Adrenal glandular	Take 100–400 mg per day.	Fights inflammation and adrenal exhaustion.
Aged garlic extract	Take 2 capsules three times per day.	Fights inflammation; stimulates the immune system; potent antioxidant.
Amino-acid complex that includes alanine, tyrosine, and glutamic acid	Take as indicated on the label on an empty stomach.	Required for enzyme, peptide, and hormone production; helps the body maintain proper fluid and acid/base balance.
Bioflavonoids, such as rutin and quercetin	Take 500–1,000 mg three times per day between meals.	Strengthens capillaries; potent antioxidants.
Branched-chain amino acids (isoleucine, leucine, and valine)	Take as indicated on the label between meals.	Often found together in a supplement, branched-chain amino acids aid in healing of muscle, bone, and skin and act as fuel for muscles.
Coenzyme Q$_{10}$	Take 30–60 mg one to two times per day with meals; or follow the advice of your health-care professional.	Potent antioxidant.
Folic acid	Take 400–800 mcg per day.	Required for proper cell division, important for red blood cell formation, body growth, protein metabolism, and cell division; coenzyme for many enzymes.
Niacin	Take 50 mg the first day, then increase dosage daily by 50 mg to 1,000 mg. If adverse effects occur, decrease dosage. Take dosages higher than 1,000 mg only under the supervision of a physician.	Dilates blood vessels, improving circulation.
or Niacin-bound chromium	Take 20 mg per day, or as indicated on the label.	
Prostate glandular	Take 130 mg per day.	Improves prostate function.
Thymus glandular	Take 100 mg per day.	Assists the thymus gland in immune resistance; improves immune response; fights infections.

Vitamin A	Take 10,000–15,000 IU per day. If pregnant, take more than 10,000 IU per day.	Needed for strong immune system and healthy mucous membranes; potent antioxidant.
Vitamin B complex	Take 150 mg three times per day, or as indicated on the label.	B vitamins act as coenzymes and support neuromusculoskeletal function.
Vitamin C	Take 10,000–15,000 mg per day.	Fights infections, diseases and allergies; stimulates the adrenal glands; promotes immune system; powerful antioxidant.
Vitamin E	Take 1,200 IU per day.	Potent antioxidant; supplies oxygen to cells; improves circulation and level of energy.
Zinc	Take 15–50 mg per day. Do not take more than 100 mg per day.	Needed by over 300 enzymes; necessary for white blood cell immune function; improves wound healing; potent antioxidant.

Comments

❑ Fruits and vegetables (particularly green vegetables) should comprise over 60 percent of your diet. If your body cannot tolerate raw fruits and vegetables, increase your intake of digestive enzymes, or steam or sauté your produce.

❑ Pumpkin seeds are helpful with prostatic problems because they are rich in zinc.

❑ To improve overall health, avoid alcohol, refined white flour, and sugar.

❑ Saw palmetto is important in maintaining the health of the prostate. It is widely used in Europe (and gaining attention in the United States) as a cure for prostate problems, especially benign prostatic hyperplasia.

❑ Herbs, including echinacea, ginger root, ginkgo biloba, licorice, and white willow bark can help reduce inflammation.

❑ *See also* PROSTATE CANCER.

Prostatitis

See PROSTATE DISORDERS.

Psoriasis

Psoriasis is a chronic skin disease marked by red, scaly patches. It affects over six million people in the United States and an estimated 1–3 percent of the world's population, according to the National Psoriasis Foundation. Every year, 150,000 to 260,000 new cases of psoriasis are diagnosed. The disease primarily affects Caucasians, and females and males are equally affected. Although psoriasis may appear at any time in life, most people first develop the lesions as young adults (the average age of onset is 22.5 years of age).

As our skin cells age, they are constantly sloughed off and replaced by new cells. In psoriasis, cells form too quickly and are replaced by new cells before they can mature. Skin cells are formed in the basal layer (the deepest layer) of the skin. Normally, it takes skin cells about twenty-eight days or so to move from the basal layer to the topmost layer, where they are ultimately sloughed off. In psoriasis, skin cells move to the top in three or four days. It is unknown why this skin cell turnover takes place at such an abnormally rapid rate, but for some reason the skin loses its ability to regulate the production of skin cells. Its cause may be an autoimmune reaction.

Psoriasis is not communicable—you can't catch it from someone who has it. However, it does seem to run in families. But having someone in the family with the disease does not guarantee you will also suffer, since studies show that nearly one-third of those who are genetically predisposed to psoriasis never develop it. Certain factors may trigger psoriasis, such as injury to the skin, certain drugs (such as lithium), stress, cold weather, staphylococcal or streptococcal bacterial infections, surgery, or sunburn. However, a certain amount of solar radiation can also lead to symptom improvement.

Symptoms

Psoriasis is relatively easy to diagnose because of the hereditary/genetic factors and the presence of red patches of various sizes on the skin. They are usually well-defined and dry and may be covered with loose, silvery scales that flake off. Sometimes, the skin may burn, itch, and bleed easily. Although lesions can occur anywhere on the body, they most often are found on the elbows, wrists, forearms, legs, knees, ankles, and scalp. They may also be found on the soles of the feet, abdomen, chest, and back. The fingernails and toenails can develop ridges and pits, losing their luster.

Psoriasis can be mild, moderate, or severe and disabling. It is marked by episodes of remissions and recurrences. Usually, psoriasis requires ongoing treatment to achieve periods of clear skin, maintain health, and restore function. At present, there is no universally effective therapy.

While most who suffer from psoriasis have a mild, limited case of the disease, some suffer from extensive psoriasis or psoriasis with complications. *Psoriatic arthritis* is psoriasis with joint inflammation, producing symptoms similar to those of rheumatoid arthritis. *Pustular psoriasis* is a rare form of psoriasis in which pustules form on the palms of the hands and soles of the feet, and sometimes on other parts of the body. *Exfoliative psoriatic dermatitis* is another rare type in which all of the skin becomes inflamed. This condition can be very serious because it prevents the skin from serving as a protective barrier against infection.

343

Enzyme Therapy

Systemic enzyme therapy is used to help decrease the over-production of skin cells and decrease the accompanying inflammation and pain. Enzymes can stimulate the immune system, improve circulation, help speed tissue repair, bring nutrients to the damaged area, remove waste products, and enhance wellness.

Digestive enzyme therapy is used to improve the digestion of food, reduce stress on the gastrointestinal mucosa, help maintain normal pH levels, detoxify the body, and promote the growth of healthy intestinal flora, thereby improving the overall health of the body. Digestive enzymes also serve as replacements for the body's pancreatic enzymes.

Enzyme Absorption System Enhancers (EASE) maximize enzyme activity. They can also improve the absorption and bioavailability of various nutrients and decrease the drain on the body's own digestive enzymes, thus prolonging their lives.

In addition to the standard multivitamin, multimineral, and multiglandular complexes recommended in the introduction to Part Two, the following supplements are recommended for the treatment of psoriasis. The following recommended dosages are for adults. Children between the ages of two and five should take one-quarter the recommended dosage, those between the ages of six and twelve should take one-half the recommended dosage, and those between the ages of thirteen and seventeen should take three-quarters the recommended adult dosage.

ESSENTIAL ENZYMES

Enzyme	Suggested Dosage	Actions
Formula One—See page 38 of Part One for a list of products. or	Take 10 tablets three times per day between meals for six months. Then decrease dosage to 5 tablets three times per day between meals.	Bolsters the immune system; breaks up antigen-antibody complexes; fights free-radical formation and inflammation.
Inflazyme Forte from American Biologics or	Take as indicated on the label.	Stimulates immune function; breaks up circulating immune complexes; fights free-radical formation and inflammation.
Bromelain or	Take three to four 230–250 mg capsules per day until symptoms subside. If severe, take 6–8 capsules per day.	Fights inflammation, pain, and swelling; speeds recovery after injury or surgery; stimulates the immune system.
Papain or	Take 500 mg three times per day for six months between meals. Then decrease dosage to 300 mg three times per day between meals.	Stimulates immune function; breaks up circulating immune complexes; fights free-radical formation and inflammation.
Trypsin with chymotrypsin	Take 200 mg of trypsin and 5–10 mg of chymotrypsin three times per day between meals for six months.	Stimulates immune function; breaks up circulating immune complexes; fights free-radical formation and inflammation.

Enzyme	Suggested Dosage	Actions
	Then decrease dosage to 125 mg of trypsin and 2–5 mg of chymotrypsin three times per day between meals.	
Any proteolytic enzyme	Take 10 tablets three times per day for six months. Then decrease dosage to 5 tablets three times per day.	Stimulates immune function; breaks up circulating immune complexes; fights free-radical formation and inflammation.

SUPPORTIVE ENZYMES

Enzyme	Suggested Dosage	Actions
Formula Three—See page 38 of Part One for a list of products.	Take as indicated on the label.	Potent antioxidant enzymes.
A digestive enzyme product containing protease, amylase, and lipase enzymes. See the list of digestive enzyme products in Appendix.	Take as indicated on the label with meals.	Improves the breakdown and absorption of food nutrients; fights toxic build-up.

ENZYME ABSORPTION SYSTEM ENHANCERS (EASE)

Combination	Suggested Dosage	Actions
Akne-Zyme from Enzymatic Therapy or	Take 1 or 2 capsules daily as a dietary supplement.	Maintains skin integrity.
Bio-Musculoskeletal Pak from Biotics or	Take as indicated on the label.	Decreases inflammation; fights free-radical formation; stimulates immune function.
Bioprotect from Biotics or	Take as indicated on the label.	Contains antioxidants; improves immune function and tissue repair.
Connect-All from Nature's Plus or	Take as indicated on the label	Supports connective tissue.
CTR Support from PhysioLogics or	Take 2–3 capsules per day preferably on an empty stomach; or follow the advice of your health-care professional.	Nourishes the body during tissue recovery and healing from injury; helps maintain healthy connective tissue.
Detox Enzyme Formula from Prevail or	Take 1 capsule between meals one or two times per day with a 6- to 8-ounce glass of water or juice.	Helps rid the body of harmful toxins.
Kyolic Formula 102 (with enzymes)	Take 2 capsules or tablets with meals two to three times per day.	Strengthens collagen and skin integrity; strengthens the immune system; helps alleviate stress; helps fight free radicals; helps fight inflammation.

ENZYME HELPERS

Nutrient	Suggested Dosage	Actions
Acidophilus or	Take as indicated on	Stimulates enzyme activity;

other probiotics	the label.	improves gastrointestinal function, thus improving the overall health of the body; bolsters immunity.
Adrenal glandular	Take 300 mg per day.	Helps fight stress and adrenal exhaustion; fights inflammation.
Arginine	Take as indicated on the label.	Accelerates healing; enhances thymus activity.
Coenzyme Q$_{10}$	Take 30–60 mg one to two times per day with meals.	Important to immune system function; has antioxidant activity.
DHEA	Take 25 mg per day.	Supports immune function.
Essential fatty acid complex	Take 100 mg two to three times per day with meals; or follow the advice of your health-care professional.	Strengthens immunity; helps strengthen cell membranes.
Kyolic Aged Garlic Liquid Extract with Vitamins B$_1$ and B$_{12}$	Take 10 drops in water or juice three times per day.	Stimulates immune system function.
Lecithin	Take 3 tablespoons per day.	Emulsifies fat.
Thymus glandular	Take 90 mg per day.	Assists the thymus gland in immune resistance; improves immune response; fights infections.
Vitamin A	Take up to 50,000 IU from all sources for no more than one month. (For higher doses, see a physician.) If pregnant, do not take more than 10,000 IU.	Needed for strong immune system and healthy mucous membranes; potent antioxidant.
Vitamin B complex	Take 250 mg per day.	B vitamins are coenzymes for many of the body's enzymes.
Vitamin C	Take 8,000–13,000 mg per day.	Fights infections, diseases, and allergies; improves wound healing; stimulates the adrenal glands; promotes immune system; powerful antioxidant.
Vitamin E	Take 1,200 IU per day from all sources (including the multivitamin complex in the introduction to Part Two).	Acts as an intercellular antioxidant; helps resist diseases at the cellular level; works closely with selenium to fight free radicals.
Zinc	Take 50 mg per day. Do not take more than 100 mg per day from all sources.	Needed by more than 300 enzymes; necessary for white blood cell immune function; improves healing.

Comments

❑ Several herbs can stimulate the immune system, including garlic (try aged garlic extract), astragalus, chlorella, echinacea, ginseng, licorice, royal jelly, spirulina, and maitake, reishi, and shiitake mushrooms.

❑ Apply an enzymatic skin salve three times daily. See Part Three and the appendix for more information.

❑ Place ten drops of liquid aged garlic extract on a cotton swab or gauze. Dab gently along the affected area.

❑ Break open and spread the contents of two to four vitamin A and vitamin E capsules on the affected area.

❑ Ultraviolet radiation can sometimes help psoriasis. Sunburn should be avoided, but sunlight is often helpful.

❑ The immune system may play a key role in the defense against the development of psoriasis. Psoriatic skin is loaded with immune cells trying to fight the condition. Those with psoriasis should take natural anti-inflammatory and immune stimulating supplements to assist the immune system.

❑ See the list of Resource Groups in the Appendix for an organization that can provide you with more information about psoriasis.

Radiation Sickness

One of the major problems of radiation therapy, which is used frequently in cancer treatment, is radiation sickness, or post-radiation "hangover." This can be characterized by nausea, vomiting, fatigue, headache, dry mouth, loss of taste and appetite, diarrhea, malaise, rapid heartbeat, shortness of breath, hair loss, dry cough, inflammation of the skin and possibly the heart, sexual impotence, and low blood cell counts.

Enzyme Therapy

Systemic enzyme therapy is used as a complementary therapy with medical therapy to decrease the side effects of radiation sickness and to decrease inflammation and free-radical formation. Enzymes stimulate the immune system to improve circulation, help speed tissue repair, bring nutrients to the damaged area, remove waste products, and improve health.

Digestive enzyme therapy is used to improve the digestion of food, reduce stress on the gastrointestinal mucosa, help maintain normal pH levels, detoxify, and promote the growth of healthy intestinal flora, thereby improving the overall health of the body. Digestive enzymes also serve as replacements for the body's pancreatic enzymes, leaving the pancreatic enzymes free to perform other functions in the body, such as boosting immunity, improving circulation, and decreasing inflammation.

Topical enzyme creams can be used to dissolve dead and damaged skin cells and exfoliate the skin, while leaving healthy skin cells intact.

Enzyme Absorption System Enhancers (EASE) maximize enzyme activity. They can also improve the absorption and bioavailability of various nutrients and decrease the drain on the body's own digestive enzymes, thus prolonging their lives.

In addition to the standard multivitamin, multimineral,

and multiglandular complexes recommended in the introduction to Part Two, the following supplements are recommended for use before, during, and after chemo- or radiotherapy to treat and prevent radiation sickness. The following recommended dosages are for adults. Children between the ages of two and five should take one-quarter the recommended dosage, those between the ages of six and twelve should take one-half the recommended dosage, and those between the ages of thirteen and seventeen should take three-quarters the recommended adult dosage.

ESSENTIAL ENZYMES

Enzyme	Suggested Dosage	Actions
WobeMugos E from Mucos	Take two tablets twice per day.	Breaks down fibrin; decreases inflammation; improves effectiveness of chemotherapy or radiation therapy; decreases radiation sickness; stimulates immune function.
and		
Formula One—See page 38 of Part One for a list of products.	Take 5 coated tablets or one heaping tablespoon of a granulated enzyme mixture three times a day between meals.	Fights inflammation, swelling, and pain; speeds healing; stimulates immune function; fights free radical formation.

SUPPORTIVE ENZYMES

Enzyme	Suggested Dosage	Actions
Formula Three—See page 38 of Part One for a list of products.	Take as indicated on the label.	Potent antioxidant enzymes.
Bromelain	Take eight to ten 230–250 mg capsules per day. As soon as tumor growth is under control, the dose can be reduced gradually to a maintenance dose of 4 capsules per day.	Fights inflammation, pain, and swelling; stimulates the immune system; improves digestion to improve absorption of nutrients.
A digestive enzyme product containing protease, amylase, and lipase. See the list of digestive enzyme products in Appendix.	Take as indicated on the label with meals.	Improves the breakdown and absorption of food nutrients; fights toxic build-up.

ENZYME HELPERS

Nutrient	Suggested Dosage	Actions
Acidophilus and other probiotics	Take as indicated on the labels.	Stimulate enzyme activity; improve gastrointestinal function; improves immunity.
Adrenal glandular	Take 100–400 mg per day.	Helps fight stress and adrenal exhaustion; fights inflammation.
Coenzyme Q_{10}	Take 30–60 mg one to two times per day with meals.	Important to immune system function; has antioxidant activity.
Kyolic Aged Garlic Liquid Extract with Vitamins B_1 and B_{12}	Take 10 drops in water or juice three times per day.	Stimulates immune system function.
Thymus glandular	Take 100 mg per day, or as indicated on the label.	Assists the thymus gland in immune resistance; improves immune response; fights infections.
Vitamin A	Take up to 50,000 IU per day. Take doses higher than 50,000 IU only under the supervision of a physician. If pregnant, do not take more than 10,000 IU per day.	Needed for a strong immune system and healthy mucous membranes; potent antioxidant.
Vitamin B complex	Take 100–150 mg three times per day.	B vitamins are coenzymes.
Vitamin C	Take 10,000–15,000 mg per day.	Improves healing; stimulates the adrenal glands; promotes the immune system; powerful antioxidant.
Vitamin D	Take 3,000–6,000 IU per day.	Involved in regulating the transport and absorption of calcium and phosphorus; plays a part in tissue formation.
Vitamin E	Take 3,000–6,000 IU per day.	An antioxidant; improves circulation; helps resist diseases at the cellular level.
Zinc	Take 50 mg per day. Do not take more than 100 mg per day.	Needed by over 300 enzymes; necessary for white blood cell immune function; improves healing; antioxidant.

Comments

❏ Physicians sometimes use enzyme injections or enzyme retention enemas in the treatment of radiation sickness. Both methods allow a greater concentration of the enzymes to be absorbed than might be accomplished if the enzymes were taken orally. An enzyme retention enema is a means of instilling some enzymes into the colon to be retained. This allows the enzymes to be absorbed into the body's systems. Physicians use products specially formulated for these uses. Caution should be exercised, however, since repeated rectal administration of very high doses of enzymes could lead to irritation of the anus or the skin around the anus. This can be avoided by applying a lubricating ointment, such as vitamin E oil or aloe vera, before administering the enema.

❏ *See* CANCER and the various individual cancers (including BRAIN TUMORS, BREAST CANCER, COLO-RECTAL CANCER, LUNG CANCER, PANCREATIC CANCER, PROSTATE CANCER, and SKIN CANCER) for specific cancer treatment plans.

Raynaud's Disease and Phenomenon

Raynaud's disease and Raynaud's phenomenon are circulatory disorders in which the blood supply to the hands and feet is interrupted due to spasm of arterioles (the smallest blood vessels). The condition is termed Raynaud's phenomenon when it is an effect of a disorder, the use of certain drugs, or injury. It is termed Raynaud's disease when an underlying cause cannot be found. A large majority of Raynaud's disease and phenomenon sufferers are women—between 60 and 90 percent.

Some causes of Raynaud's phenomenon include rheumatoid arthritis, systemic lupus, erythematosus, sclerodoma, decreased thyroid activity, atherosclerosis and other circulatory disorders, and nerve disorders. Certain drugs, such as ergot preparations, methysergide, alpha- and beta-adrenergic blockers, and calcium-channel blockers, can cause the condition, as can injury. Some with Raynaud's phenomenon also experience migraines, variant angina, and pulmonary hypertension—all conditions that may be caused by arterial spasms—which suggests that their causes may be the same. Smoking greatly increases the incidences of attacks, as it impairs circulation to the extremities.

Symptoms

Arteriolar spasms may be triggered by any stimulation of the nervous system, primarily exposure to cold. Spasms occur quickly and may last anywhere from minutes to hours. The spasms cause the affected areas to turn first white, then blue, then red as circulation returns. Numbness, tingling, a pins-and-needles sensation, burning, and sometimes pain accompany the color changes. Since the spasms are usually caused by cold, rewarming the affected area usually relieves the symptoms. Over time, the skin of the fingers or toes may begin to change, becoming smooth, shiny, and taut. The sufferer may also develop chronic infections around and under fingernails and toenails, and sores on the tips of fingers and toes due to poor blood circulation to the extremities.

Enzyme Therapy

Systemic enzyme therapy is used to improve circulation, reduce inflammation, and relieve discomfort and pain. Enzyme therapy also stimulates the immune system, helps speed tissue repair, brings nutrients to the damaged area, removes waste products, and enhances wellness.

Digestive enzyme therapy is used to improve the digestion of food, reduce stress on the gastrointestinal mucosa, help maintain normal pH levels, detoxify, and promote the growth of healthy intestinal flora. Digestive enzymes also serve as replacements for the body's pancreatic enzymes, leaving the pancreatic enzymes free to perform other functions in the body, such as improving circulation.

Enzyme Absorption System Enhancers (EASE) maximize enzyme activity. They can also improve the absorption and bioavailability of various nutrients and decrease the drain on the body's own digestive enzymes, thus prolonging their lives.

In addition to the standard multivitamin, multimineral, and multiglandular complexes recommended in the introduction to Part Two, the following supplements are recommended for the treatment of Raynaud's disease and phenomenon. The following recommended dosages are for adults. Children between the ages of two and five should take one-quarter the recommended dosage, those between the ages of six and twelve should take one-half the recommended dosage, and those between the ages of thirteen and seventeen should take three-quarters the recommended adult dosage.

ESSENTIAL ENZYMES

Enzyme	Suggested Dosage	Actions
Formula One—See page 38 of Part One for a list of products. *or*	Take 8–10 tablets three times per day between meals for six months or until symptoms improve. Thereafter, take 6 tablets three times per day between meals.	Fights inflammation, pain, and swelling; speeds healing; strengthens capillary walls; improves circulation.
Inflazyme Forte from American Biologics *or*	Take as indicated on the label between meals.	Fights inflammation, pain, and swelling; speeds healing; strengthens capillary walls; improves circulation.
Enzyme complexes rich in proteases, plus amylases and lipases, working synergistically	Take as indicated on the label between meals.	Fight inflammation, pain, and swelling; speed healing; strengthen capillary walls; improves circulation.

SUPPORTIVE ENZYMES

Enzyme	Suggested Dosage	Actions
Formula Three—See page 38 of Part One for a list of products.	Take 200 mcg three times per day between meals.	Potent antioxidant enzymes.
A digestive enzyme formula containing protease, amylase, and lipase enzymes. See the list of digestive enzyme products in Appendix.	Take as indicated on the label with meals.	Improves the breakdown and absorption of proteins, carbohydrates, and fats; fights toxic build-up.

ENZYME ABSORPTION SYSTEM ENHANCERS (EASE)

Combination	Suggested Dosage	Actions
Ecology Pak from Life Plus	Take 2 or 3 tablets once or twice per	Stimulates immune function.

day, preferably in the morning and no later than midday.		
or		
Hepazyme from Enzymatic Therapy *or*	Take as indicated on the label.	Stimulates immune function.
Inflamzyme from PhytoPharmica *or*	Take as indicated on the label.	Immune system activator; reduces inflammation; fights free radicals.
Kyolic Formula 102 (with enzymes)	Take 2 capsules or tablets with meals two to three times per day.	Strengthens the immune system; helps alleviate stress; helps fight free radicals; helps fight inflammation; improves digestion and absorption of nutrients.

ENZYME HELPERS

Nutrient	Suggested Dosage	Actions
Bioflavonoids, such as quercetin, rutin, or pycnogenol	Take 500 mg three times per day.	Strengthen capillaries; potent antioxidants.
Calcium	Take 1,000 mg per day.	Plays important roles in muscle contraction and nerve transmission.
Carnitine	Take 500 mg two times per day.	Increases energy; lowers cholesterol.
Coenzyme Q_{10}	Take 75 mg one to two times per day with meals.	Potent antioxidant; decreases blood pressure and total serum cholesterol.
DHEA	Take as indicated on the label with meals.	Supports cardiovascular function.
Essential fatty acids (found in flaxseed oil, chasteberry, saw palmetto)	Take 100 mg of flaxseed oil, chasteberry, or saw palmetto two to three times per day with meals.	Lowers cholesterol level; strengthens the immune system; helps strengthen cell membranes.
Heart glandular	Take 140 mg, or as indicated on the label.	Improves heart function, improving circulation to the extremities.
Lecithin	Take 3 capsules with meals, or as indicated on the label.	Emulsifies fat.
Magnesium	Take 500 mg per day.	Plays important roles in nerve function and calcium absorption.
Niacin	Take 50 mg the first day, then increase dosage daily by 50 mg to 1,000 mg. If adverse effects occur, decrease dosage. Take doses higher than 1,000 mg only under the supervision of a physician.	Improves circulation by dilating blood vessels.
or		
Niacin-bound chromium	Take up to 20 mg per day, or as indicated on the label.	
Vitamin A	Take 10,000–15,000 IU per day. Take doses higher than 15,000 IU only under	Needed for a strong immune system and healthy tissues; potent antioxidant.

	the supervision of a physician. If pregnant, take no more than 10,000 IU per day under the supervision of a physician.	
Vitamin C	Take 8,000 mg per day.	Essential to vascular function; potent antioxidant; improves the function of many enzymes; helps decrease plaquing on the arterial walls by assisting in converting cholesterol to bile salt.
Vitamin E	Take 1,200 IU per day from all sources (including the multivitamin complex in the introduction to Part Two).	Potent antioxidant; stimulates circulation; strengthens capillary walls; prevents blood clots; maintains cell membranes.

Comments

❑ Several herbs are known to improve circulation or strengthen blood vessels, including astragalus, bilberry, ginkgo biloba, gotu kola, hawthorn, and wheatgrass powder.

❑ Garlic, onions, shallots, chives, leeks, and scallions contain allium compounds, which are potent free-radical fighters that help lower cholesterol and fight circulatory disorders.

❑ Raspberries, cranberries, grapes, red wine, black currants, and red cabbage contain anthocyanidins, which reduce blood vessel plaque formation and maintain blood flow in small vessels.

❑ Oolong and green tea contain epicatechin, which is a potent antioxidant that lowers blood pressure and strengthens capillaries.

❑ Since nicotine is a vasoconstrictor, it is important that those suffering from Raynaud's disease or phenomenon stop smoking.

❑ Relaxation techniques (such as meditation and biofeedback) may help decrease vasospastic episodes.

❑ Protect the extremities and the body from cold.

❑ The effects of drugs formerly used for treatment have been inconsistent and varied, therefore a conservative approach seems most logical.

❑ See the list of Resource Groups in the Appendix for an organization that can provide you with more information about Raynaud's phenomenon.

Reactive Arthritis

See REITER'S SYNDROME.

Reiter's Syndrome

Reiter's syndrome (or reactive arthritis) is an inflammatory disorder that affects the joints, eyes, mucous membranes, skin, and urethra. One with Reiter's syndrome suffers from inflammation of the joints (arthritis) along with inflammation of the conjunctiva of the eye (conjunctivitis); inflammation of the mucous membranes of the urethra (urethritis), mouth, vagina, and penis; and a distinctive skin rash. The condition appears to be a reaction to an infection in some place in the body other than the joints, yet the reaction occurs in the joints. Thus, it is called reactive arthritis.

Reiter's syndrome occurs as a result of two types of infection. It can occur as a reaction to a sexually transmitted infection, such as chlamydia, or as a result of an intestinal bacterial infection. It appears that one must have a genetic predisposition to Reiter's syndrome in order to become afflicted with the disorder after one of these infections, because most who become infected with a sexually transmitted disease or an intestinal infection do not develop Reiter's syndrome. It is most common in males between the ages of 12 and 40. It is very rare in women.

Symptoms

Inflammation of the urethra usually occurs seven to fourteen days after the infection. This causes uncomfortable urination. In men it causes discharge from the penis and an inflamed prostate. Women may experience a vaginal discharge, if they experience any symptoms at all. Soon after, conjunctivitis ensues, causing red, itching, burning, and tearing eyes. Arthritis, which may affect several joints—especially the toes, knees, hips, and back—at once, also develops. It may be mild or severe. Small, painless sores develop in the mouth, on the tongue, and on the tip of the penis. The patient may also develop a low-grade fever and a rash that resembles psoriasis.

Although the illness usually resolves in three to four months, nearly half of all patients suffer from recurrent arthritis attacks and/or other components of the syndrome over a period of many years. Recurrent Reiter's syndrome may lead to joint deformity.

Enzyme Therapy

Enzyme therapy can substantially improve many of the inflammatory symptoms and reduce the need for other medications. The additional administration of vitamins, minerals, and other nutrients is also advantageous. In addition, systemic enzyme therapy can stimulate the immune system, break up circulating immune complexes, improve circula-

tion, help speed tissue repair, bring nutrients to the damaged area, remove waste products, and improve health.

Digestive enzyme therapy is used to improve the digestion of food, reduce stress on the gastrointestinal tract, maintain normal pH levels, detoxify the body, and promote the growth of healthy intestinal flora. Digestive enzymes also serve as replacements for the body's pancreatic enzymes, leaving the pancreatic enzymes free to perform other functions in the body, such as boosting immunity, improving circulation, and decreasing inflammation.

Enzyme Absorption System Enhancers (EASE) maximize enzyme activity. They can also improve the absorption and bioavailability of various nutrients and decrease the drain on the body's own digestive enzymes, thus prolonging their lives.

In addition to the standard multivitamin, multimineral, and multiglandular complexes recommended in the introduction to Part Two, the following supplements are recommended for the treatment of Reiter's syndrome. The following recommended dosages are for adults. Children between the ages of two and five should take one-quarter the recommended dosage, those between the ages of six and twelve should take one-half the recommended dosage, and those between the ages of thirteen and seventeen should take three-quarters the recommended adult dosage.

ESSENTIAL ENZYMES

Enzyme	Suggested Dosage	Actions
Formula One—See page 38 of Part One for a list of products. *or*	Take 10 tablets three times per day between meals for six months. Then decrease dosage to 5 tablets three times per day between meals.	Bolsters the immune system; breaks up antigen-antibody complexes; fights free-radical formation and inflammation.
Papain *or*	Take 500 mg three times per day for six months between meals. Then decrease dosage to 300 mg three times per day between meals.	Stimulates immune function; breaks up circulating immune complexes; fights free radicals and inflammation.
Bromelain *or*	Take three to four 230–250 mg capsules per day until symptoms subside. If severe, take 6 to 8 capsules per day.	Fights inflammation, pain, and swelling; speeds recovery after injury or surgery; stimulates the immune system.
Mulsal from Mucos *or*	Take 8 tablets twice per day between meals.	Stimulates immune function; breaks up circulating immune complexes; fights free radicals and inflammation.
Serratiopeptidase *or*	Take 5 mg three times per day between meals.	Fights inflammation; treats arthritis.
A proteolytic enzyme	Take as indicated on the label.	Bolsters the immune system; breaks up antigen-antibody complexes; fights free-radical formation and inflammation.

SUPPORTIVE ENZYMES

Enzyme	Suggested Dosage	Actions
Formula Three—See page 38 of Part One for a list of products.	Take as indicated on the label.	Potent antioxidant enzymes.
A digestive enzyme product containing protease, amylase, and lipase enzymes. See the list of digestive enzyme products in Appendix.	Take as indicated on the label with meals.	Improves the breakdown and absorption of food nutrients; fights toxic build-up.

ENZYME ABSORPTION SYSTEM ENHANCERS (EASE)

Combination	Suggested Dosage	Actions
Ecology Pak from Life Plus _or_	Take 2 or 3 tablets once or twice per day, preferably in the morning and no later than midday.	Assists with natural detoxification; supports the immune system.
Hepazyme from Enzymatic Therapy _or_	Take as indicated on the label.	Stimulates immune function.
Inflamzyme from PhytoPharmica _or_	Take as indicated on the label.	Immune system activator; reduces inflammation; fights free radicals.
Kyolic Formula 102 (with enzymes) _or_	Take 2 capsules or tablets with meals two to three times per day.	Maintains skin integrity; strengthens the immune system; helps alleviate stress; helps fight free radicals and inflammation; improves digestion and absorption of nutrients.
Metabolic Liver Formula from Prevail	Take as indicated on the label.	Improves liver function.

ENZYME HELPERS

Nutrient	Suggested Dosage	Actions
Acidophilus and other probiotics	Take as indicated on the label.	Stimulate enzyme activity; improve gastrointestinal function.
Adrenal glandular	Take 100–400 mg per day.	Helps fight stress and adrenal exhaustion; fights inflammation.
Bioflavonoids, such as rutin and quercetin	Take 500 mg three times per day.	Strengthen capillaries.
Calcium	Take 1,000 mg per day.	Plays important roles in tissue health.
Chondroitin sulfate	Take 250 mg two to three times per day with meals; or follow the advice of your health-care professional.	Supports connective tissue; promotes cell hydration.
Coenzyme Q_{10}	Take 30–60 mg one to two times per day with meals.	Important to immune system function; has antioxidant activity.
Essential fatty acid complex	Take 100–200 mg two to three times per day with meals; or follow the advice of your health-care professional.	Strengthens immunity; helps strengthen cell membranes.
Glucosamine hydrochloride _and/or_	Take 500 mg two to four times per day.	Stimulate cartilage repair.
Glucosamine sulfate	Take 500 mg one to three times per day.	
Kyolic Aged Garlic Liquid Extract with Vitamins B_1 and B_{12}	Take 10 drops in water or juice three times per day.	Stimulates immune system function.
Magnesium	Take 500 mg per day.	Important for healthy skin and mucous membranes; plays important roles in energy production and calcium absorption; necessary for synthesis of fats, proteins, and nucleic acid.
Thymus glandular	Take 100 mg per day, or as indicated on the label.	Assists the thymus gland in immune resistance; improves immune response; fights infections.
Vitamin A	Take up to 50,000 IU for no more than one month. (For higher doses, see a physician.) If pregnant, do not take more than 10,000 IU per day.	Needed for strong immune system and healthy mucous membranes; potent antioxidant.
Vitamin B complex	Take 100–150 mg three times per day.	B vitamins act as coenzymes.
Vitamin C	Take 10,000–15,000 mg per day; or 1–2 grams every hour until symptoms improve.	Fights infections, diseases, and allergies; improves wound healing; stimulates the adrenal glands; promotes immune system; powerful antioxidant.
Vitamin E	Take 1,200 IU per day from all sources (including the multivitamin complex in the introduction to Part Two).	Acts as an intercellular antioxidant; improves circulation; helps resist diseases at the cellular level; works closely with selenium to fight free radicals.
Zinc	Take 15–50 mg per day. Do not take more than 100 mg per day.	Needed by over 300 enzymes; necessary for white blood cell immune function; improves healing.

Comments

❑ A number of herbs can help bolster the immune system, including astragalus, echinacea, ginger root, and ginkgo biloba.

❑ Follow the detoxification plans in Part Three.

❑ The diet should be low in calories and fat. Decrease or eliminate dairy products and meats.

❑ Eat plenty of enzyme-rich fruits and vegetables, including apples, apricots, beet greens, broccoli, cabbage, carrots, figs, garlic, ginger root, grapes, kale, lettuce, oranges, papayas, parsley, pineapple, spinach, tomatoes, turnips, dark green vegetables, and sprouting seeds.

❏ Increase your intake of fresh cold-water fish, such as salmon, mackerel, sardines, and tuna. These foods contain eicosapentaenoic acid (EPA), which strengthens immunity and helps strengthen cell membranes.

❏ Avoid or eliminate from your diet refined white flour, refined sugar, salt, corn flour, alcohol, coffee, tea, and strong spices.

❏ Exercises, such as walking, bicycling, and swimming; a positive mental attitude; and meditation can be of help.

❏ See the list of Resource Groups in the Appendix for a list of organizations that can provide you with more information about Reiter's syndrome.

Retinopathy

The retina contains lots of nerve endings, making it extremely sensitive to light. Light enters the eye through the cornea and travels through the pupil to the lens. The lens focuses light onto the retina, which turns the light into electrical impulses and sends them to the brain through the optic nerve. The brain then translates these electrical impulses into visual images. The retina also has a supply of many blood vessels carrying blood and oxygen to and from it. When the blood supply to the retina is inadequate, the condition is called retinopathy. Retinopathy can permanently damage the retina and can lead to blindness.

Arteriosclerotic retinopathy occurs when the walls of the small arteries to the eye become thickened, narrowing the vessel opening. This condition in and of itself is usually not very serious and does not threaten vision; however, it is an indication of cardiovascular disease and should be taken care of before the flow of blood becomes blocked.

Hypertensive retinopathy occurs as a result of very high blood pressure in the blood vessels. This may cause some breakage of retinal blood vessels and retinal damage due to lack of blood flow. This condition can impair vision and, if left untreated, can cause permanent vision loss.

Diabetic retinopathy results when high blood sugar levels thicken, yet weaken blood vessels. This can lead to a narrowing of the blood vessels and to breakage. It generally does not develop until a person has diabetes for at least ten years. There are two types of diabetic retinopathy. In *proliferative retinopathy*, new blood vessels to the retina begin to form; however, they grow abnormally and cause scarring on the retina. This type of diabetic retinopathy is the more

severe form and can lead to total blindness. In *nonproliferative*, or *background*, *retinopathy*, small capillaries to the retina break and leak, causing swelling. This may distort vision, but generally does not lead to vision loss.

In *retinopathy of prematurity*, the blood vessels of the retina have not had an opportunity to finish developing in a premature infant. As they continue growing after birth, they develop abnormally, which may cause the vessels to bleed or the retina to detach. This can lead to vision loss or blindness.

Enzyme Therapy

Systemic enzyme therapy is used to reduce pain, swelling, and inflammation in the eye, to stimulate the immune system, and to improve circulation to the eye. Enzymes help speed tissue repair, bring nutrients to the damaged area, remove waste products, and enhance wellness.

Digestive enzyme therapy is used to improve the digestion and assimilation of food nutrients, reduce stress on the gastrointestinal mucosa, help maintain normal pH levels, detoxify the body, and promote the growth of healthy intestinal flora, thus making the body healthier so that it is better able to combat retinopathy. Digestive enzymes also serve as replacements for the body's pancreatic enzymes, leaving the pancreatic enzymes free to perform other functions in the body, such as decreasing inflammation and improving circulation.

Enzyme Absorption System Enhancers (EASE) maximize enzyme activity. They can also improve the absorption and bioavailability of various nutrients and decrease the drain on the body's own digestive enzymes, thus prolonging their lives.

In addition to the standard multivitamin, multimineral, and multiglandular complexes recommended in the introduction to Part Two, the following supplements are recommended for the treatment of retinopathy. The following recommended dosages are for adults. Children between the ages of two and five should take one-quarter the recommended dosage, those between the ages of six and twelve should take one-half the recommended dosage, and those between the ages of thirteen and seventeen should take three-quarters the recommended adult dosage.

ESSENTIAL ENZYMES

Enzyme	Suggested Dosage	Actions
Superoxide dismutase	Take 150 mcg per day.	Fights inflammation; potent antioxidant.

AND SELECT ONE OF THE FOLLOWING:

Formula One—See page 38 of Part One for a list of products.	Take 5 tablets twice per day.	Fights inflammation, pain, and swelling; speeds healing; stimulates the immune system; strengthens capillary walls.
Bromelain	Take three to four 250 mg capsules per day until symptoms subside. If the	Fights inflammation and swelling; speeds healing; stimulates the immune system.

	condition is advanced, take 6–8 capsules per day.	
Serratiopeptidase	Take 10 mg two times per day between meals.	Fights inflammation; reduces swelling.
Pancreatin	Take 500 mg six times per day on an empty stomach.	Fights inflammation and swelling; increases rate of healing; stimulates the immune system; strengthens capillary walls.
Trypsin with chymotrypsin	Take 250 mg of trypsin and 10 mg of chymotrypsin three times per day between meals.	Fights inflammation and swelling; increases rate of healing; stimulates the immune system.
Papain	Take 300 mg two times per day between meals.	Fights inflammation; speeds healing.
Microbial protease	Take 300 mg two times per day between meals.	Fights inflammation; speeds recovery from surgery.

SUPPORTIVE ENZYMES

Enzyme	Suggested Dosage	Actions
Catalase and glutathione peroxidase	Take as indicated on the label.	Potent antioxidant enzymes.
A digestive enzyme product containing protease, amylase, and lipase enzymes. See the list of digestive enzyme products in Appendix.	Take as indicated on the label with meals.	Improves the breakdown and absorption of food nutrients; fights toxic build-up.

ENZYME ABSORPTION SYSTEM ENHANCERS (EASE)

Combination	Suggested Dosage	Actions
Antioxidant Complex from Tyler Encapsulations *or*	Take as indicated on the label.	Potent free-radical fighter and immune system stimulator.
Detox Enzyme Formula from Prevail *or*	Take 1 capsule between meals one or two times per day with a 6- to 8-ounce glass of water or juice.	Helps rid the body of harmful toxins.
Hepazyme from Enzymatic Therapy *or*	Take as indicated on the label.	Stimulates immune function.
Kyolic Formula 102 (with enzymes) *or*	Take 2 capsules or tablets with meals two to three times per day.	Strengthens the immune system; helps alleviate stress; helps reverse circulatory disorders; helps fight free radicals; helps fight inflammation.
Oxy-5000 Forte from American Biologics *or*	Take as indicated on the label.	Helps the body excrete toxins; fights biochemical stressors, infections, and degenerative processes.
Vita-C-1000 from Life Plus	Take as indicated on the label.	Helps maintain vision; boosts immune function.

ENZYME HELPERS

Nutrient	Suggested Dosage	Actions
Acidophilus or other probiotic	Take as indicated on the label.	Improves gastrointestinal function and aids in the digestion and absorption of nutrients, which improves the health of the body; stimulates enzyme activity.
Adrenal glandular	Take 100–400 mg per day.	Fights inflammation and adrenal exhaustion.
Aged garlic extract	Take 2 capsules three times per day.	Fights inflammation; stimulates the immune system; potent antioxidant.
Bioflavonoids, such as rutin or quercetin	Take 500 mg three times per day between meals.	Strengthen capillaries; potent antioxidants.
Coenzyme Q_{10}	Take 30–60 mg one to two times per day with meals; or follow the advice of your health-care professional.	Potent antioxidant.
Essential fatty acids (found in flaxseed oil)	Take 100–200 mg two to three times per day with meals; or follow the advice of your health-care professional.	Help strengthen cell membranes; support immune function.
Niacin *or*	Take 50 mg the first day, then increase dosage daily by 50 mg to 1,000 mg. If adverse effects occur, decrease dosage. Take doses higher than 1,000 mg only under the supervision of a physician.	Dilates blood vessels; improves circulation.
Niacin-bound chromium	Take 20 mg per day or as indicated on the label.	
Pycnogenol	Take as indicated on the label.	Potent antioxidant; improves capillary wall strength.
Vitamin A *or*	Take 10,000–15,000 IU per day. If pregnant, take no more than 10,000 IU per day.	Needed for strong immune system and healthy mucous membranes; potent antioxidant.
Carotenoids	Take 5,000–10,000 IU per day, or as indicated on the label.	Potent antioxidants; enhance immunity.
Vitamin B complex	Take 150 mg three times per day, or as indicated on the label.	B vitamins act as coenzymes.
Vitamin C	Take 10,000–15,000 mg per day.	Promotes immune system; powerful antioxidant.
Vitamin E	Take 1,200 IU per day from all sources (including the multivitamin complex in the introduction to Part Two).	Potent antioxidant; supplies oxygen to cells; improves circulation and level of energy.
Zinc	Take 15–50 mg per day. Do not take more than 100 mg per day from all sources.	Needed by over 300 enzymes; necessary for white blood cell immune function.

Comments

❏ Bilberry protects and strengthens the blood vessels that feed the eyes and supports arteries and capillaries.

❏ Garlic, onions, shallots, chives, leeks, and scallions contain allium compounds that are potent free-radical fighters, help lower cholesterol, and fight circulatory disorders.

❏ Raspberries, cranberries, grapes, red wine, black currants, and red cabbage contain anthocyanidins, which reduce blood vessel plaque formation and maintain blood flow in small vessels.

❏ See the list of Resource Groups in the Appendix for a list of organizations that can provide you with more information about retinopathy.

Rheumatic Fever

Rheumatic fever is a complication of a streptococcus A infection (usually strep throat) that causes joint, heart, and skin inflammation. It is probably an example of the body's antibodies, produced to fight off the bacteria, going haywire and attacking the body's tissues.

Rheumatic fever is becoming less common in the United States, probably due to the use of antibiotics to treat infections. It is most common in children between the ages of 4 and 18. Though the infections that cause it are contagious, rheumatic fever is not.

Symptoms

The symptoms of rheumatic fever appear several weeks after the strep infection. Symptoms can include a fever; a rash on the back, abdomen, or chest; joint inflammation that causes swelling, pain, and warmth, and that may move from joint to joint; heart inflammation, which can cause chest pain, palpitations, permanent damage of the valves that leads to rheumatic heart disease, and possible heart failure; malaise; nodules under the skin; and jerky, involuntary movements (called Sydenham's chorea). Rheumatic fever usually subsides in two to twelve weeks.

Enzyme Therapy

Systemic enzyme therapy is used to stimulate the immune system, reduce inflammation, improve circulation, help speed tissue repair, bring nutrients to the damaged area, and remove waste products. Enzymes strengthen the body as a whole, build general resistance, and improve health.

Digestive enzyme therapy is used to improve the digestion of food, reduce stress on the gastrointestinal mucosa, help maintain normal pH levels, detoxify the body, and promote the growth of healthy intestinal flora, thereby improving the overall health of the body. Digestive enzymes also serve as replacements for the body's pancreatic enzymes, leaving the pancreatic enzymes free to perform other functions in the body, such as boosting immunity and decreasing inflammation.

Enzyme Absorption System Enhancers (EASE) maximize enzyme activity. They can also improve the absorption and bioavailability of various nutrients and decrease the drain on the body's own digestive enzymes, thus prolonging their lives.

In addition to the standard multivitamin, multimineral, and multiglandular complexes recommended in the introduction to Part Two, the following supplements are recommended for the treatment of rheumatic fever. The following recommended dosages are for adults. Children between the ages of two and five should take one-quarter the recommended dosage, those between the ages of six and twelve should take one-half the recommended dosage, and those between the ages of thirteen and seventeen should take three-quarters the recommended adult dosage. Rheumatic fever is a serious condition. Therefore, any treatment should be conducted under the supervision of a well-trained physician.

ESSENTIAL ENZYMES

Enzyme	Suggested Dosage	Actions
Formula One—See page 38 of Part One for a list of products.	Take 10 tablets three times per day between meals. As symptoms subside, decrease dosage to 3–5 tablets three times per day between meals. When pain disappears, begin taking a maintenance dose of 2–3 tablets after each meal.	Decreases inflammation, pain, and swelling; increases rate of healing; stimulates the immune system; strengthens capillary walls.
or		
Inflazyme Forte from American Biologics	Take as indicated on the label between meals.	Decreases inflammation, pain, and swelling; increases rate of healing; stimulates the immune system; strengthens capillary walls.
or		
Enzyme complexes rich in proteases, plus amylases and lipases, working synergistically	Take as indicated on label between meals	Decreases inflammation, pain, and swelling; increases rate of healing; stimulates the immune system; strengthens capillary walls.
or		
Bromelain	Take three to four 230–250 mg capsules per day until symptoms subside. If severe, take 6–8 capsules per day.	Fights inflammation, pain, and swelling; speeds recovery after injury or surgery; stimulates the immune system.

SUPPORTIVE ENZYMES

Enzyme	Suggested Dosage	Actions
Formula Three—See page 38 of Part One for a list of products.	Take 50 mcg of each three times per day between meals.	Potent antioxidant enzymes.

A digestive enzyme formula containing protease, amylase, and lipase enzymes. For a list of digestive enzyme products, see Appendix.	Take as indicated on the label with meals.	Improves the breakdown and absorption of food nutrients; fights toxic build-up.

ENZYME ABSORPTION SYSTEM ENHANCERS (EASE)

Combination	Suggested Dosage	Actions
Antioxidant Complex from Tyler Encapsulations *or*	Take as indicated on the label.	Potent free-radical fighter and immune system stimulator.
Bromelain Complex from PhytoPharmica *or*	Take as indicated on the label.	Reduces swelling and inflammation; improves circulation; helps break up circulating immune complexes.
Detox Enzyme Formula from Prevail *or*	Take as indicated on the label.	Helps rid the body of harmful toxins.
Hepazyme from Enzymatic Therapy *or*	Take as indicated on the label.	Stimulates immune function.
Joint-Ease from Nature's Life *or*	Take as indicated on the label.	Decreases inflammation and swelling.
Mobil-Ease from Prevail *or*	Take one capsule between meals two to three times per day.	Has anti-inflammatory activity and helps decrease joint pain.
Oxy-5000 Forte from American Biologics *or*	Take as indicated on the label.	Helps the body excrete toxins; fights biochemical stressors, common infections, and degenerative processes.
Protease Concentrate from Tyler Encapsulations *or*	Take as indicated on the label.	Breaks up fibrin; fights inflammation.
Vita-C-1000 from Life Plus	Take as indicated on the label.	Boosts immune function; fights infection.

ENZYME HELPERS

Nutrient	Suggested Dosage	Actions
Adrenal glandular	Take 100–400 mg per day.	Fights inflammation and adrenal exhaustion.
Aged garlic extract	Take 2 capsules three times per day.	Detoxifies and provides energy.
Amino-acid complex	Take as indicated on the label on an empty stomach.	Amino acids are the building blocks of proteins and enzymes.
Bioflavonoids	Take 150–500 mg three times per day.	Strengthen capillaries; potent antioxidants.
Chondroitin sulfate	Take 250 mg two to three times per day with meals; or follow the advice of your health-care professional.	Supports connective tissue; promotes cell hydration.

Coenzyme Q10	Take 30–60 mg one to two times per day with meals; or follow the advice of your health-care professional.	Potent antioxidant.
Essential fatty acids (found in flaxseed oil)	Take 75–100 mg of flaxseed oil two to three times per day with meals; or follow the advice of your health-care professional.	Help strengthen cell membranes.
Glucosamine hydrochloride *and/or*	Take 500 mg two to four times per day.	Stimulate cartilage repair.
Glucosamine sulfate	Take 500 mg one to three times per day.	
Vitamin A *or*	Take 5,000–10,000 IU per day, or as indicated on the label. If pregnant, do not take more than 10,000 IU per day.	Potent antioxidants; enhance immunity.
Carotenoids	Take 10,000–20,000 IU per day.	
Vitamin B complex	Take 50 mg three times per day, or as indicated on the label.	B vitamins act as coenzymes and support neuromusculoskeletal function.
Vitamin C	Take 1,000–10,000 mg per day.	Powerful antioxidant; helps to produce collagen and normal connective tissue; promotes immune system.
Vitamin E	Take 1,200 IU per day from all sources (including the multivitamin complex in the introduction to Part Two).	Potent antioxidant; supplies oxygen to cells; improves level of energy.
Zinc	Take 15–50 mg per day. Do not take more than 100 mg per day.	Needed by over 300 enzymes; improves healing.

Comments

❑ A number of herbs can help bolster the immune system, including astragalus, echinacea, ginger root, and ginkgo biloba.

❑ Garlic, shallots, and onions contain allium compounds, which are known to fight free radicals and stimulate immunity.

Rheumatoid Arthritis

Rheumatoid arthritis (RA) is an autoimmune disease; that is, the body, for some unknown reason, sees itself as the enemy. The immune system attacks the tissues surrounding the joints (muscles, tendons, and ligaments). This results in the formation of antigen-antibody complexes (circulating immune complexes) that enter the synovial joints and soft tissue, causing inflammation, swelling, pain, and gradual deterioration of the joint. Like many autoimmune diseases, rheumatoid arthritis is characterized by flare-ups and remissions.

Rheumatoid arthritis affects women three times as often as it does men. It usually first appears somewhere between the ages of 20 and 50, but can appear at any age. For many, rheumatoid arthritis is disabling.

Symptoms

Symptoms of rheumatoid arthritis may begin suddenly but usually have a slow onset. They include inflammation of the joints that is usually symmetric—for example, if it affects one shoulder, it will affect the other. This inflammation causes pain, stiffness, swelling, heat, redness, and muscle and joint weakness. This inflammation usually begins in the small joints of the hands and feet, before affecting the other larger joints of the body. Some joints may lock in one position. The fingers of each hand tend to bend toward the little finger. The inflammation may eventually lead to permanent deformity.

Other possible symptoms of rheumatoid arthritis include low-grade fever; fatigue; the development of nodules under the skin; inflammation of the blood vessels, eyes, and membranes around the heart and lungs; swollen lymph nodes; and Sjögren's syndrome.

Enzyme Therapy

Enzyme combinations, including bromelain, papain, and pancreatic enzymes are effective anti-inflammatory agents. They can help decrease pain, swelling, redness, heat, and loss of function (signs of inflammation). They can stimulate the immune system and help the body eliminate immune complexes that are deposited in the joints (or anywhere in the body). When used systemically as anti-inflammatory agents, enzymes should be taken between meals.

The scientific investigations and publications collected by the Medical Enzyme Research Association indicate that enzyme mixtures are not only virtually free of side effects, but enzymes are as effective as oral gold or other commonly used arthritis therapies. Enzymes not only alleviate the arthritic symptoms, but they also degrade the fibrin built up by the immune complexes, and thus remove the cloak of invisibility from the immune complexes to ensure that the immune complexes are detected, broken down, dissolved, and subsequently eliminated. In this way, the mechanisms leading to inflammation are more rapidly stopped, and the potential for further deterioration is reduced.

One study measured the effectiveness of an enzyme combination containing pancreatin, trypsin, chymotrypsin, bromelain, and papain on over one thousand rheumatic patients. Researchers found that between 76 and 96 percent of the patients improved while being treated with enzymes, while only 10 to 24 percent of the patients showed no response,

and 2 percent deteriorated. Nearly all (99 percent) of the patients reported that they had suffered no side effects.

Enzymes are effective. However, enzyme treatment of rheumatoid arthritis and other rheumatic conditions demands a great deal of patience and perseverance because it may take weeks or even months before positive changes are seen. This means the patient will continue to experience pain and limitations in mobility while continuing to take enzyme tablets. Some patients are resistant to undergoing enzyme therapy because the required dose can be up to thirty tablets per day.

Enzyme therapy is effective only if the patient follows the program faithfully. It is absolutely essential when treating rheumatic disorders that the patient immediately increase the enzyme dose at the first sign of a new active phase, which can be triggered by the onset of a cold, influenza, or other infection; or when feeling run down or otherwise unhealthy. Higher doses might be required under these circumstances. European doctors report excellent results if higher doses of enzyme mixtures are taken at the first signs of a new deterioration phase.

The reason for the greatly increased dose is that in the acute phase, increased immune complexes are found in the joint fluid, in the cartilage of the affected joints, and in the blood. If the body is given a massive dose of enzymes, the enzymes can attack the immune complexes and thus conduct a sort of "spring cleaning." Enzymes ensure the swift conclusion of inflammation. This is one of the decisive advantages of enzyme therapy.

Systemic enzyme therapy is used to reduce pain, swelling, and inflammation, strengthen the body as a whole, build general resistance, and stimulate the immune system. Enzymes improve circulation, help speed tissue repair, bring nutrients to the damaged area, remove waste products, and improve health.

Digestive enzyme therapy is used to improve the digestion of food, reduce stress on the gastrointestinal mucosa, help maintain normal pH levels, detoxify the body, and promote the growth of healthy intestinal flora, thereby improving the overall health of the body. Digestive enzymes also serve as replacements for the body's pancreatic enzymes, leaving the pancreatic enzymes free to perform other functions in the body, such as boosting immunity and decreasing inflammation.

Enzyme Absorption System Enhancers (EASE) maximize enzyme activity. They can also improve the absorption and bioavailability of various nutrients and decrease the drain on the body's own digestive enzymes, thus prolonging their lives.

In addition to the standard multivitamin, multimineral, and multiglandular complexes recommended in the introduction to Part Two, the following supplements are recommended for the treatment of rheumatoid arthritis. The following recommended dosages are for adults. Children between the ages of two and five should take one-quarter the recommended dosage, those between the ages of six and twelve should take one-half the recommended dosage, and those between the ages of thirteen and seventeen should take three-quarters the recommended adult dosage.

ESSENTIAL ENZYMES

Enzyme	Suggested Dosage	Actions
Formula One (particularly Wobenzym N)— See page 38 of Part One for a list of products.	Take 10 tablets three times per day between meals. As symptoms subside (which should occur in approximately one week) decrease dosage to 3–5 tablets three times per day between meals. When symptoms disappear, begin taking a maintenance dose of 2–3 tablets after each meal.	Decreases inflammation, pain, and swelling; increases rate of healing; stimulates the immune system; strengthens capillary walls.
or		
Inflazyme Forte from American Biologics	Take as indicated on the label between meals.	Decreases inflammation, pain, and swelling; increases rate of healing; stimulates the immune system; strengthens capillary walls.
or		
Bromelain	Take three to four 230–250 mg capsules per day until symptoms subside. If severe, take 6 to 8 capsules per day.	Fights inflammation, pain, and swelling; speeds recovery after injury or surgery; stimulates the immune system.
or		
A proteolytic enzyme	Take as indicated on label between meals.	Decreases inflammation, pain, and swelling; increases rate of healing; stimulates the immune system; strengthens capillary walls.

SUPPORTIVE ENZYMES

Enzyme	Suggested Dosage	Actions
Formula Three—See page 38 of Part One for a list of products.	Take 50 mcg three times per day between meals.	Potent antioxidant enzymes.
A digestive enzyme formula containing protease, amylase, and lipase enzymes. For a list of digestive enzyme products, see Appendix.	Take as indicated on the label with meals.	Improves the breakdown and absorption of food nutrients; fights toxic build-up.

ENZYME ABSORPTION SYSTEM ENHANCERS (EASE)

Combination	Suggested Dosage	Actions
Bromelain Complex from PhytoPharmica	Take as indicated on the label.	Reduces swelling and inflammation; improves circulation; helps break up circulating immune complexes.
or		

Joint-Ease from Nature's Life	Take as indicated on the label.	Decreases inflammation and swelling.
or		
Mobil-Ease from Prevail	Take one capsule between meals two to three times per day.	Has anti-inflammatory activity; helps decrease joint pain.
or		
Protease Concentrate from Tyler Encapsulations	Take as indicated on the label.	Breaks up fibrin; fights inflammation.

ENZYME HELPERS

Nutrient	Suggested Dosage	Actions
Aged garlic extract	Take 2 capsules three times per day.	Detoxifies and provides energy.
Alpha-linolenic acid	Take as indicated on the label.	Stimulates immune system; decreases inflammation.
Amino-acid complex	Take as indicated on the label on an empty stomach.	Amino acids are the building blocks of proteins and enzymes.
Bioflavonoids	Take 150–500 mg three times per day.	Strengthen capillaries; potent antioxidants.
Chondroitin sulfate	Take 250 mg two to three times per day with meals; or follow the advice of your health-care professional.	Supports connective tissue; promotes cell hydration.
Calcium	Take 1,000 mg per day.	Promotes healthy tissues.
Coenzyme Q_{10}	Take 30–60 mg one to two times per day with meals; or follow the advice of your health-care professional.	Potent antioxidant.
Glucosamine hydrochloride *and/or*	Take 500 mg two to four times per day.	Stimulate cartilage repair.
Glucosamine sulfate	Take 500 mg one to three times per day.	
Magnesium	Take 500 mg per day.	A cofactor for many enzymes; plays important role in connective tissue function, and calcium absorption.
Niacin *or*	Take 50 mg the first day, then increase dosage 50 mg daily to 1,000 mg. If adverse effects occur, decrease dosage. Take dosages above 1,000 mg only under the supervision of a physician.	Dilates blood vessels; improves circulation.
Niacin-bound chromium	Take 20 mg per day; or take as indicated on the label.	
Vitamin B complex	Take 50 mg three times per day, or as indicated on the label.	B vitamins act as coenzymes and support neuromusculoskeletal function.
Vitamin C	Take 1,000–10,000 mg per day.	Helps to produce collagen and normal connective tissue; pro-

		motes immune system; powerful antioxidant.
Zinc	Take 50 mg per day. Do not take more than 100 mg per day.	Needed by over 300 enzymes.

Comments

❑ Your diet should be low in calories and fat. Decrease or eliminate dairy products and meats.

❑ Eat plenty of enzyme-rich fruits and vegetables, including apricots, apples, figs, grapes, oranges, papayas, pineapple, parsley, broccoli, spinach, carrots, tomatoes, lettuce, cabbage, kale, ginger root, beet greens, turnips, garlic, dark green vegetables, and sprouting seeds.

❑ Increase intake of fresh, cold-water fish, such as salmon, mackerel, sardines, and tuna. These foods contain eicosapentaenoic acid (EPA), effective in the treatment of rheumatoid arthritis.

❑ Avoid or eliminate from your diet refined white flour, refined sugar, salt, corn flour, alcohol, coffee, tea, and strong spices.

❑ Exercises (walking, bicycling, swimming), a positive mental attitude, and meditation can help you deal with the pain of rheumatoid arthritis.

❑ You may want to seek the services of a physician who will develop a treatment program. Your treatment program might include chiropractic adjustments, manipulation, massage, physiotherapy, including electrical simulation, ice and heat applications, exercises, postural training (which includes proper lifting, bending, standing, walking, driving, sitting, and sleeping positions), and a dietary program to balance the body's metabolism.

Ringworm

See FUNGAL SKIN INFECTIONS.

Rosacea

Rosacea is a chronic inflammatory skin disorder, characterized by redness, tiny pimples, and broken blood vessels, usually on the face. W.C. Fields was probably the best-known victim of this chronic skin problem.

Rosacea is referred to as acne rosacea because the inflammation resembles acne; however, it is not a type of acne, as it produces no whiteheads or blackheads. This disorder afflicts approximately one in twenty Americans. It ordinarily begins in middle age or after. It is more common in women, but it affects men more severely. Dark-skinned individuals are much less likely to be afflicted with this disorder than are fair-skinned people.

The cause is not well understood, however, alcohol, spicy foods, smoking, vitamin deficiencies, hot liquids, exposure to extremes in temperature, stress, the use of makeups containing alcohol or other irritants, and infections can all worsen the condition.

Symptoms

In rosacea, the small capillaries near the surface of the skin dilate and rupture, causing the characteristic redness. It most often affects the nose. In mild cases, one experiences recurring flushing of the face that ultimately becomes permanent. In many men the skin, particularly around the nose, thickens. This causes the nose to look swollen and be tender to the touch. This condition is called *rhinophyma*. Sometimes, small raised bumps, and pus-filled blisters form on the affected skin. Occasionally, rosacea affects other areas of the body instead of the face.

Enzyme Therapy

Enzymes work from the outside in and from the inside out. Enzyme supplements can be used topically, as digestive therapy, and as systemic enzyme therapy. Topical enzyme creams are used to dissolve dead and damaged skin cells and exfoliate the skin while leaving healthy skin cells intact.

Digestive enzyme therapy is used to improve the digestion of food, decrease the swelling and inflammation, reduce stress on the gastrointestinal mucosa, help maintain normal pH levels, detoxify the body, promote the growth of healthy intestinal flora, strengthen the body as a whole, and build general resistance, thereby improving the overall health of the body. Digestive enzymes also serve as replacements for the body's pancreatic enzymes, leaving the pancreatic enzymes free to perform other functions in the body, such as boosting immunity and decreasing inflammation.

Systemic enzyme therapy is used as supportive care to reduce inflammation, stimulate the immune system, improve circulation, help speed tissue repair, bring nutrients to the damaged area, remove waste products, and enhance wellness.

Enzyme Absorption System Enhancers (EASE) maximize enzyme activity. They can also improve the absorption and bioavailability of various nutrients and decrease the drain on the body's own digestive enzymes, thus prolonging their lives.

In addition to the standard multivitamin, multimineral, and multiglandular complexes recommended in the introduction to Part Two, the following supplements are recommended for the treatment of rosacea. The following recommended dosages are for adults. Children between the ages of two and five should take one-quarter the recommended dosage, those between the ages of six and twelve should take one-half the recommended dosage, and those between the ages of thirteen and seventeen should take three-quarters the recommended adult dosage.

ESSENTIAL ENZYMES

Enzyme	Suggested Dosage	Actions
Formula One—See page 38 of Part One for a list of products. *or*	Take 10 tablets three times per day between meals for six months. Then decrease dosage to 5 tablets three times per day between meals.	Bolsters the immune system; breaks up antigen-antibody complexes; fights free-radical formation; fights inflammation.
Papain *and*	Take 500 mg three times per day for six months between meals. Then decrease dosage to 300 mg three times per day between meals.	Stimulates immune function; breaks up circulating immune complexes; fights free radicals and inflammation.
Bromelain *and*	Take 400 mg three times per day between meals for six months. Then decrease dosage to 250 mg three times per day between meals.	
Trypsin with chymotrypsin *or*	Take 200 mg three times per day between meals for six months. Then decrease dosage to 125 mg three times per day between meals.	
Bromelain	Take three to four 230–250 mg capsules three times a day between meals. If symptoms are severe, take 6–7 capsules three times a day.	Stimulates immune function; breaks up circulating immune complexes; fights free radicals and inflammation.

SUPPORTIVE ENZYMES

Enzyme	Suggested Dosage	Actions
Formula Three—See page 38 of Part One for a list of products.	Take as indicated on the label.	Free-radical fighters.
A digestive enzyme product containing protease, amylase, and lipase enzymes. See the list of digestive enzyme products in Appendix.	Take as indicated on the label with meals.	Improves the breakdown and absorption of food nutrients; fights toxic build-up.

ENZYME ABSORPTION SYSTEM ENHANCERS (EASE)

Combination	Suggested Dosage	Actions
Akne-Zyme from Enzymatic Therapy *or*	Take 1 or 2 capsules daily as a dietary supplement.	Maintains skin integrity.
Bio-Musculo-Skeletal from Biotics *or*	Take as indicated on the label.	Decreases inflammation; fights free radicals; stimulates immune function.
Connect-All from Nature's Plus *or*	Take as indicated on the label.	Supports connective tissue.
CTR Support from PhysioLogics *or*	Take as indicated on the label.	Nourishes the body during tissue recovery and healing from injury; helps maintain healthy connective tissue; diminishes damage caused by swelling and inflammation.
Detox Enzyme Formula from Prevail *or*	Take 1 capsule between meals one or two times per day with a 6- to 8-ounce glass of water or juice.	Helps rid the body of harmful toxins.
Kyolic Formula 102 (with enzymes)	Take 2 capsules or tablets with meals two to three times per day.	Maintains skin integrity; strengthens the immune system; helps alleviate stress; helps fight free radicals and inflammation; improves digestion and absorption of nutrients.

ENZYME HELPERS

Nutrient	Suggested Dosage	Actions
Acidophilus	Take as indicated on the label.	Stimulates enzyme activity; improves gastrointestinal function, improving the overall health of the body.
Adrenal glandular	Take 100–400 mg per day.	Helps fight stress and adrenal exhaustion; fights inflammation.
Aged garlic extract	Take two 300 mg capsules three times per day, or 1–3 capsules with meals twice per day.	Stimulates the immune system; fights free radicals.
Bioflavonoids, including rutin and quercetin	Take 500 mg three times per day.	Strengthen capillaries.
Coenzyme Q_{10}	Take 30–60 mg one to two times per day with meals.	Important to immune system function; has antioxidant activity.
Essential fatty acid complex	Take 100–200 mg two to three times per day with meals; or follow the advice of your health-care professional.	Strengthens immunity; helps strengthen cell membranes.
Thymus glandular	Take 100 mg per day, or as indicated on the label.	Assists the thymus gland in immune resistance; improves immune response; fights infections.
Vitamin A	Take up to 50,000 IU for no more than one month (for higher doses, see a physician). If pregnant, do not take more than 10,000 IU.	Needed for strong immune system and healthy skin; potent antioxidant.
Vitamin B complex	Take 100–150 mg three times per day.	B vitamins act as coenzymes.

Vitamin C	Take 10,000–15,000 mg per day, or 1–2 grams every hour until symptoms improve.	Fights infections, diseases, and allergies; improves healing; stimulates the adrenal glands; promotes immune system; powerful antioxidant.
Vitamin E	Take 1,200 IU per day from all sources (including the multivitamin complex in the introduction to Part Two).	Acts as an intercellular antioxidant; improves circulation; helps resist diseases at the cellular level.
Zinc	Take 15–50 mg per day. Do not take more than 100 mg per day.	Needed by over 300 enzymes; necessary for white blood cell immune function; improves healing.

Comments

❑ Apply an enzymatic salve twice daily. See Part Three for more information.

❑ Place ten drops of liquid aged garlic extract on a cotton swab or gauze. Dab gently along the affected area.

❑ Break open and spread the contents of two to four vitamin A and vitamin E capsules on the affected area.

❑ Zinc oxide salve can be spread over the affected area once daily.

❑ Follow the Enzyme Toxin Flush and the Enzyme Kidney Flush programs once per month for six months. See Part Three for instructions.

Rubella

See MEASLES.

Ruptured Disc

See SLIPPED DISC.

Scars

A scar is the mark left on tissues, such as the skin, after an injury, wound, burn, or sore has healed. Scars are made up of fibrous material and damaged collagen. The skin is primarily composed of collagen (collagen constitutes 75 percent of the dry weight of skin tissue). Collagen is important to our bodies because its fibers interlace, forming an elastin network (an essential segment of connective tissue), which lends skin its elasticity, smoothness, and strength.

In undamaged skin, collagen molecules apparently slide over each other. With scarring, wounds, and burns (as well as with the wrinkles of aging), however, collagen molecules fuse together (cross link) making the damaged tissue rigid and leathery. The collagen fibers stick together. This results in increased fibrin formation, weakened tissue, and decreased flexibility in the affected tissue. Therefore, tearing of adjacent tissue is quite possible.

In most instances, small scars will fade as time passes. Large scars may not fade but can often be made less noticeable and reduced in size by skin grafting and/or plastic surgery. However, large, dome-shaped scars, known as keloids, may be difficult to treat.

When scars occur inside the body, they can stick to neighboring tissues and organs, sometimes interfering with normal function. In this instance, they are called *adhesions*. However, most of the time, internal scars rarely cause trouble.

In a young, healthy individual, scar tissue formation is decreased because the body's proteolytic/fibrinolytic activity is at its height. But as we get older, or as a result of stress, injury, or illness, the number of proteolytic enzymes produced and the activity of those produced is decreased. This is why proteolytic enzyme supplementation is helpful.

Stretch marks can also be treated with this enzyme therapy program. Stretch marks are shiny lines that appear on the skin, particularly on the breasts, abdomen, thighs, and buttocks, due to stretching and weakening of the skin. It is often caused by pregnancy, obesity, growth spurts in adolescents, and body building.

Symptoms

A scar can be a reddened mark of elevated or depressed tissue. Sometimes the scar can be skin-colored or it may be striated (striped).

Enzyme Therapy

Systemic enzyme therapy is used to break down the fibrin (scar tissue), decrease cross linking of collagen fibers, break up free radicals, and increase flexibility. They can also reduce inflammation, stimulate the immune system, improve circulation, help speed tissue repair, bring nutrients to the damaged area, remove waste products, and improve health.

Digestive enzyme therapy is used to improve the digestion and assimilation of food nutrients, reduce stress on the gastrointestinal mucosa, help maintain normal pH levels, detoxify the body, and promote the growth of healthy intestinal flora, thereby improving the overall health of the body to prevent scarring. Digestive enzymes also serve as replacements for the body's pancreatic enzymes, leaving the pancreatic enzymes free to perform other functions in the body, such as improving circulation.

Enzyme Absorption System Enhancers (EASE) maximize enzyme activity. They can also improve the absorption and bioavailability of various nutrients and decrease the drain on the body's own digestive enzymes, thus prolonging their lives.

In addition to the standard multivitamin, multimineral, and multiglandular complexes recommended in the introduction to Part Two, the following supplements are recommended for the treatment of scars. The following recommended dosages are for adults. Children between the ages of two and five should take one-quarter the recommended dosage, those between the ages of six and twelve should take one-half the recommended dosage, and those between the ages of thirteen and seventeen should take three-quarters the recommended adult dosage.

ESSENTIAL ENZYMES

Enzyme	Suggested Dosage	Actions
Superoxide dismutase	Take 150 mcg per day.	Fights inflammation; potent antioxidant.
WobeMugos E Ointment from Mucos	Apply at least twice per day. Do not apply to open wounds.	Exfoliates the skin and breaks up cross linking.

AND SELECT ONE OF THE FOLLOWING:

Bromelain	Take 750 mg two times per day between meals.	Fights inflammation, pain, and swelling; speeds recovery after injury or surgery; stimulates the immune system; strengthens capillary walls.
Formula One—See page 38 of Part One for a list of products.	Take 5 tablets twice per day between meals.	Fights inflammation, pain, and swelling; speeds recovery after injury or surgery; stimulates the immune system; strengthens capillary walls.
WobeMugos E from Mucos	Take 3–4 tablets three times per day between meals.	Fights inflammation, pain, and swelling; increases rate of healing; stimulates the immune system.
Pancreatin	Take 500 mg six times per day on an empty stomach and between meals.	Fights inflammation, pain, and swelling; increases rate of healing; stimulates the immune system; strengthens capillary walls.
Serratiopeptidase	Take 10 mg two times per day between meals.	Fights inflammation; reduces pain and swelling.
Microbial protease	Take 300 mg two times per day between meals.	Fights inflammation; speeds recovery from injuries and surgery.

SUPPORTIVE ENZYMES

Enzyme	Suggested Dosage	Actions
Catalase and glutathione peroxidase	Take as indicated on the labels.	Potent antioxidant enzymes.
A digestive enzyme product containing protease,	Take as indicated on the label with meals.	Improves the breakdown and absorption of food nutrients;

amylase, and lipase. See the list of digestive enzyme products in Appendix.

fights toxic build-up.

ENZYME ABSORPTION SYSTEM ENHANCERS (EASE)

Combination	Suggested Dosage	Actions
Connect-All from Nature's Plus or	Take as indicated on the label.	Supports connective tissue.
CTR Support from PhysioLogics or	Take as indicated on the label.	Nourishes the body during tissue recovery and healing from injury; diminishes damage caused by swelling and inflammation.
Inflamzyme from PhytoPharmica or	Take as indicated on the label.	Promotes tissue repair.
Magnesium Plus from Life Plus	Take one or two tablets one to three times per day.	Important for calcium utilization and many enzyme reactions.

ENZYME HELPERS

Nutrient	Suggested Dosage	Actions
Acidophilus or other probiotic	Take as indicated on the label.	Improves gastrointestinal function, thereby improving the overall health of the body; needed for production of the coenzymes biotin, niacin, folic acid, and pyridoxine; stimulates enzyme activity.
Amino-acid complex including L-ornithine, L-glutamine, L-arginine, L-proline, and L-lysine	Take as indicated on the label on an empty stomach.	Required for enzyme production; promotes healthy tissue.
Bioflavonoids, such as pycnogenol, rutin, or quercetin	Take 500 mg three times per day between meals.	Strengthen capillaries; potent antioxidants.
Coenzyme Q_{10}	Take 30–60 mg one to two times per day with meals; or follow the advice of your health-care professional.	Potent antioxidant.
Vitamin A	Take 5,000–10,000 IU per day. If pregnant, take no more than 10,000 IU per day.	Needed for strong immune system; promotes healthy skin; potent antioxidant.
Vitamin B complex	Take 50 mg three times per day, or as indicated on the label.	B vitamins act as coenzymes and promote healthy tissues.
Vitamin C	Take 8,000–13,000 mg per day.	Improves wound healing; stimulates the adrenal glands; promotes immune system; powerful antioxidant.
Vitamin E	Take 1,200 IU per day from all sources (including the multivitamin complex	Potent antioxidant; supplies oxygen to cells; improves circulation and level of energy.

	in the introduction to Part Two).	
Zinc	Take 15–50 mg per day. Do not take more than 100 mg per day.	Needed by over 300 enzymes; necessary for white blood cell immune function; improves wound healing; promotes healthy skin.

Comments

❑ A number of herbs can decrease inflammation, including alfalfa concentrate, aloe vera, echinacea, ginger, ginkgo biloba, licorice, and white willow bark.

❑ A number of topical applications can be of benefit (do not apply to open wounds), including the following:

• Apply an enzymatic salve three times per day as an alternative to the salve mentioned in the treatment program. See Part Three for more information.

• Place ten drops of liquid aged garlic extract on a cotton swab or gauze. Dab gently along the affected area.

• Break open and spread the contents of two to four vitamin A and vitamin E capsules on the affected area.

• Zinc oxide salve can be spread over the affected area once daily.

Sciatica

Sciatica is not a disease but a symptom of pressure on the nerve roots in the spinal cord. It is marked by pain radiating down the back of the leg (following the course of the sciatic nerve). The spinal cord is the means of communication between the brain and the rest of the body. It is a group of nerves that extends downward from the brain. It is protected by the bones of the spinal column—the vertebrae. The nerves branch off from the spinal cord at certain points and pass from the protective vertebrae to different parts of the body through openings called intervertebral foramen. The sciatic nerve, the longest nerve in the body, travels down to the legs. Any type of trauma, inflammation, infection, or swelling that places pressure on the sciatic nerve can cause sciatica.

The most common causes of sciatica are an intervertebral disc protrusion pressing on one or more nerve roots, or a subluxation (a misalignment of one or more vertebrae) decreasing the size of the intervertebral foramen and pinching or putting pressure on the nerve roots. The end result is inflammation and a shooting-type pain down the affected leg.

Symptoms

Symptoms of sciatica are pain that radiates along the course of the sciatic nerve to the buttocks, thigh, calf, and foot.

Enzyme Therapy

Systemic enzyme therapy is used to decrease swelling, heat, pain, and inflammation at the nerve root. Enzymes can improve circulation, help speed tissue repair, bring nutrients to the damaged area, remove waste products, stimulate the immune system, and improve health.

Digestive enzyme therapy is used to improve the digestion of food, reduce stress on the gastrointestinal mucosa, help maintain normal pH levels, detoxify the body, and promote the growth of healthy intestinal flora, thereby improving the overall health of the body. Digestive enzymes also serve as replacements for the body's pancreatic enzymes, leaving the pancreatic enzymes free to perform other functions in the body, such as improving circulation and decreasing inflammation.

Enzyme Absorption System Enhancers (EASE) maximize enzyme activity. They can also improve the absorption and bioavailability of various nutrients and decrease the drain on the body's own digestive enzymes, thus prolonging their lives.

In addition to the standard multivitamin, multimineral, and multiglandular complexes recommended in the introduction to Part Two, the following supplements are recommended for the treatment of sciatica. The following recommended dosages are for adults. Children between the ages of two and five should take one-quarter the recommended dosage, those between the ages of six and twelve should take one-half the recommended dosage, and those between the ages of thirteen and seventeen should take three-quarters the recommended adult dosage.

ESSENTIAL ENZYMES

Enzyme	Suggested Dosage	Actions
Superoxide dismutase	Take 150 mcg per day.	Fights inflammation; potent antioxidant.

AND SELECT ONE OF THE FOLLOWING:

For Acute Sciatica:

Bromelain	Take three to four 230–250 mg tablets three times per day between meals for 6 to 10 days, or until symptoms disappear.	Fights inflammation, pain, and swelling; speeds recovery; stimulates the immune system.
Phlogenzym from Mucos	Take 10 tablets three times per day between meals until symptoms disappear.	Fights inflammation, pain, and swelling; speeds recovery; stimulates the immune system.
Serratiopeptidase	Take 10 mg two times per day between meals.	Fights inflammation; reduces pain and swelling.
Formula One—See page 38 of Part One for a list of products.	Take 10 tablets three times per day between meals until symptoms disappear.	Fights inflammation; speeds recovery.

361

A microbial proteolytic combination	Take as indicated on the label.	Fights inflammation, pain, and swelling; speeds recovery.
For Chronic Sciatica:		
Mulsal from Mucos	Take 8 tablets twice per day between meals for one month, then reduce dosage to 6 tablets twice per day between meals for the second month.	Fights inflammation, pain, and swelling; speeds recovery; stimulates the immune system.

SUPPORTIVE ENZYMES

Enzyme	Suggested Dosage	Actions
Catalase and glutathione peroxidase	Take as indicated on the labels.	Potent antioxidant enzymes.
A digestive enzyme product containing protease, amylase, and lipase. See the list of digestive enzyme products in Appendix.	Take as indicated on the label with meals.	Improves the breakdown and absorption of food nutrients; fights toxic build-up.

ENZYME ABSORPTION SYSTEM ENHANCERS (EASE)

Combination	Suggested Dosage	Actions
Bromelain Complex from Enzymatic Therapy or	Take as indicated on the label.	Decreases swelling; improves circulation; helps break up any circulating immune complexes.
Disc-Zyme from Nutri-West or	Take as indicated on the label.	Fights inflammation; decreases swelling; helps break up circulating immune complexes.
Joint-Ease from Nature's Life or	Take as indicated on the label.	Decreases inflammation and swelling.
Mobil-Ease from Prevail or	Take one capsule between meals two to three times per day.	Has anti-inflammatory activity.
Protease Concentrate from Tyler Encapsulations	Take as indicated on the label.	Breaks up fibrin; fights inflammation.

ENZYME HELPERS

Nutrient	Suggested Dosage	Actions
Acidophilus and other probiotics	Take as indicated on the label.	Improves gastrointestinal function; needed for production of the coenzymes biotin, niacin, folic acid, and pyridoxine; stimulates enzyme activity in fighting inflammation.
Adrenal glandular	Take 100–400 mg per day.	Fights inflammation and adrenal exhaustion.
Aged garlic extract	Take 2 capsules three times per day.	Fights inflammation; stimulates the immune system; potent antioxidant.
Bioflavonoids, such as pycnogenol, rutin, or quercetin	Take 500 mg three times per day between meals.	Potent antioxidants; improve capillary wall strength.

Calcium	Take 1,000 mg per day.	Plays important roles in nerve transmission.
Magnesium	Take 500 mg per day.	Plays important role in nerve function and calcium absorption.
Niacin	Take 50 mg the first day, then increase dosage by 50 mg daily to 1,000 mg. If adverse effects occur, decrease dosage. Take dosages higher than 1,000 mg only under the supervision of a physician.	Improves circulation to affected areas.
or Niacin-bound chromium	Take up to 20 mg per day, or as indicated on the label.	
Vitamin B complex	Take 50 mg three times per day, or as indicated on the label.	B vitamins act as coenzymes and support neuromusculoskeletal function.
Vitamin C	Take 8,000–13,000 mg per day.	Improves wound healing; stimulates the adrenal glands; promotes immune system; powerful antioxidant.

Comments

❏ A number of foods, including raspberries, cranberries, red wine, grapes, black currants, and red cabbage contain anthocyanidins. These phytochemicals can reduce inflammation, maintain flow in the capillaries, and fight free radicals.

❏ Herbs, including echinacea, ginger root, ginkgo biloba, licorice, and white willow bark, can fight inflammation, helping to relieve pain.

❏ See a well-trained back specialist. A history, physical examination, and X-rays should be conducted in order to develop an effective treatment program.

❏ If you are overweight, a weight-loss program should be initiated.

❏ See Part Three for further information on diet and detoxification methods, such as cleansing, fasting, and juicing.

Scoliosis

Scoliosis is an abnormal curvature of the spine. When viewed from the back, the spine should run in a straight line from the head and neck to the hips. But with scoliosis, there is a spinal imbalance. The spine curves to one side or the other (it is usually in the shape of the letter "S"), instead of being straight. This can cause the discs (the shock-absorbing pillows between each vertebra of the spine) to bulge, wear out, or rupture. The nerves that carry messages from the brain through the vertebrae can become irritated or pinched. The condition can also cause muscles to become unbalanced on one side. That is, one side of the body may be stronger or weaker than the other side.

Scoliosis most commonly develops in children between the ages of ten and fourteen. It is more common in girls than in boys. In the majority of cases (about 75 percent), the cause of scoliosis is unknown. In those cases where the cause is known, it is usually a congenital defect; a disorder of the central nervous system, such as cerebral palsy, muscular dystrophy, or polio; or uneven leg length. This condition requires the care of a well-trained physician.

Symptoms

Initially, scoliosis produces no symptoms. A parent may notice that one of the child's shoulders or hips is higher than the other. The child may also experience fatigue in the back when sitting or standing. If the scoliosis is allowed to get worse, signs include a forward tilt of the head and neck, a hunching of the mid-back, a protruding shoulder blade, a sunken chest, and a gross forward curvature of the low back. The individual may also suffer from fatigue in the back (often felt after long periods of sitting or standing), generalized fatigue, stiffness and soreness in the back, and localized pain. This condition may also irritate and inflame the joints and ligaments. If left untreated, this condition can result in poor posture and pain in the spine and surrounding areas, as well as other signs of inflammation (including swelling, heat, and loss of function).

Enzyme Therapy

Systemic enzyme therapy is used to decrease pain, swelling, and inflammation. With decreased swelling, the fixated vertebrae may move more easily. Enzymes can help build general resistance, stimulate the immune system, improve circulation, help speed tissue repair, bring nutrients to the damaged area, remove waste products, improve health, and strengthen the body as a whole.

Digestive enzyme therapy is used to improve the digestion of food, reduce stress on the gastrointestinal mucosa, help maintain normal pH levels, detoxify the body, and promote the growth of healthy intestinal flora, thereby improving the overall health of the body. Digestive enzymes also serve as replacements for the body's pancreatic enzymes, leaving the pancreatic enzymes free to perform other functions in the body, such as improving circulation and decreasing inflammation.

Enzyme Absorption System Enhancers (EASE) maximize enzyme activity. They can also improve the absorption and bioavailability of various nutrients and decrease the drain on the body's own digestive enzymes, thus prolonging their lives.

In addition to the standard multivitamin, multimineral, and multiglandular complexes recommended in the introduction to Part Two, the following supplements are recommended for the treatment of scoliosis. The following recommended dosages are for adults. Children between the ages of two and five should take one-quarter the recommended dosage, those between the ages of six and twelve should take one-half the recommended dosage, and those between the ages of thirteen and seventeen should take three-quarters the recommended adult dosage.

ESSENTIAL ENZYMES

Enzyme	Suggested Dosage	Actions
Formula One—See page 38 of Part One for a list of products. *or*	Take 10 tablets three times per day between meals. As symptoms subside, decrease dosage to 3–5 tablets three times per day between meals. When symptom-free, begin taking a maintenance dose of 2–3 tablets after each meal.	Decreases inflammation, pain, and swelling; increases rate of healing; stimulates the immune system; strengthens capillary walls.
Bromelain *or*	Take three to four 230–250 mg capsules per day until symptoms subside. If severe, take 6–8 capsules per day.	Fights inflammation, pain, and swelling; speeds recovery after injury or surgery; stimulates the immune system.
Phlogenzym from Mucos *or*	Take 10 tablets three times per day between meals. As symptoms subside, decrease dosage to 3–5 tablets three times per day between meals. When symptom-free, begin taking a maintenance dose of 2–3 tablets after each meal.	Decreases inflammation, pain, and swelling; increases rate of healing; stimulates the immune system; strengthens capillary walls.
Inflazyme Forte from American Biologics *or*	Take as indicated on the label between meals.	Decreases inflammation, pain, and swelling; increases rate of healing; stimulates the immune system; strengthens capillary walls.
Enzyme complexes rich in proteases, plus amylases and lipases, working synergistically	Take as indicated on the label between meals.	Decreases inflammation, pain, and swelling; increases rate of healing; stimulates the immune system; strengthens capillary walls.

SUPPORTIVE ENZYMES

Enzyme	Suggested Dosage	Actions
Formula Three—See page 38 of Part One for a list of products.	Take 50 mcg three times per day between meals.	Potent antioxidant enzymes.
A digestive enzyme product containing protease, amylase, and lipase enzymes. See the list of digestive enzyme products in Appendix.	Take as indicated on the label with meals.	Improves the breakdown and absorption of food nutrients; fights toxic build-up.

ENZYME ABSORPTION SYSTEM ENHANCERS (EASE)

Combination	Suggested Dosage	Actions
Bromelain Plus from Enzymatic Therapy *or*	Take as indicated on the label.	Decreases swelling; improves circulation; helps break up any circulating immune complexes.
Disc-Zyme from Nutri-West *or*	Take as indicated on the label.	Fights inflammation; decreases swelling; helps break up circulating immune complexes.
Joint-Ease from Nature's Life *or*	Take as indicated on the label.	Decreases inflammation and swelling.
Mobil-Ease from Prevail	Take one capsule between meals two to three times per day.	Has anti-inflammatory activity.
Protease Concentrate from Tyler Encapsulations	Take as indicated on the label.	Breaks up fibrin; fights inflammation.

ENZYME HELPERS

Nutrient	Suggested Dosage	Actions
Bioflavonoids	Take 150–500 mg three times per day.	Strengthen capillaries; potent antioxidants.
Calcium	Take 1,000 mg per day.	Plays important roles in muscle contraction and nerve transmission.
Coenzyme Q_{10}	Take 30–60 mg one to two times per day with meals; or follow the advice of your health-care professional.	Potent antioxidant.
Magnesium	Take 500 mg per day.	Plays important role in muscle contraction, nerve function, and calcium absorption.
Mucopolysaccharides found in bovine cartilage extract and shark cartilage extract	Take as indicated on the label.	Support healthy, flexible joints; control inflammation; help keep joint membranes fluid.
Niacin *or*	Take 50 mg the first day, then increase dosage daily by 50 mg to 1,000 mg. If adverse effects occur, decrease dosage. Take doses higher than 1,000 mg only under the supervision of a physician.	Dilates blood vessels, improving circulation to affected areas.
Niacin-bound chromium	Take 20 mg per day; or take as indicated on the label.	
Vitamin B complex	Take 50 mg three times per day, or as indicated on the label.	B vitamins act as coenzymes and support neuromusculoskeletal function.
Vitamin C	Take 1,000–10,000 mg per day.	Helps to produce collagen and normal connective tissue; promotes immune system; powerful antioxidant.
Whey to Go from Solgar (combination of free-form amino acids and protein)	Take as indicated on the label.	Maintains muscles; improves muscle metabolism.

Comments

❑ Herbs, such as white willow bark, can help relieve inflammation and pain.

❑ See your physician or chiropractor to learn the proper posture to avoid back problems. Posture training should include proper lifting, bending, standing, walking, driving, sitting, and sleeping techniques.

❑ If you are overweight, a weight-loss program should be initiated.

❑ Anyone with scoliosis should be under the care of a well-trained spinal specialist, such as a chiropractor, medical doctor, osteopath, or naturopath. Your physician will develop special back exercises as part of your treatment. Mobilizing exercises plus spinal adjustments or manipulation can help overcome the rigidity of the spine and improve spinal balance. Other therapies could include the application of ice packs, warm moist packs, orthopedic back support, and/or traction.

❑ If scoliosis is discovered early, treatment that will help prevent the condition from worsening can be initiated.

❑ See Part Three for further information on dietary modifications and detoxification methods (through cleansing, fasting, and juicing).

Shingles

See HERPES ZOSTER.

Sinusitis

Sinusitis is an inflammation of the sinuses. There are four paired sinus cavities in the bones of the skull that connect with the nose. The frontal sinuses are located above the eyes, the maxillary sinuses are inside the cheeks on either side of the nose, the ethmoid sinuses are located in the area above the nose on either side, and the sphenoid sinuses are located behind the ethmoid sinuses. The sinuses serve to reduce the weight of the face and add resonance to the voice. Mucus from the sinuses normally drains into the nose; however, if for any reason, the sinuses get blocked, the mucus cannot drain, which opens the gate for infection. Sinusitis may be acute or chronic.

Viral, bacterial, or fungal infections, and allergic reactions are the most common causes of sinusitis. But it can also develop due to a nasal injury, a deviated septum (the septum is the separator between the two nasal passages), nasal polyps, or narrow sinuses. Sinusitis can also be caused by tooth or tonsil infections, or inhaling such irritants as cigarette smoke, dust, or dry air.

Symptoms

The symptoms of sinusitis depend on which sinus group is affected. Ethmoid sinusitis produces a severe headache and pain behind and between the eyes. Frontal sinusitis produces a headache in the center of the forehead. Maxillary sinusitis produces a headache, toothache, and pain in the cheeks and just below the eyes. Sphenoid sinusitis may produce pain in the front or back of the head. Its effects do not always occur in well-defined areas. In addition, sinusitis sufferers may suffer from fatigue, malaise, nasal congestion, and cough. The nasal membrane is red and swollen, and there may be a yellow or green discharge from the nose. Postnasal drip may develop as mucus from the sinuses drips down the back of the throat instead of draining through the nose. If the mucus affects the bronchial tubes, chronic bronchitis may result. If a fever is present, the infection has spread beyond the sinuses.

Enzyme Therapy

Systemic enzyme therapy is used to reduce inflammation, relieve pain, and help break up the mucus. Enzymes stimulate the immune system, improve circulation, help speed tissue repair, bring nutrients to the damaged area, remove waste products, and improve health.

Digestive enzyme therapy is used to improve the digestion of food, reduce stress on the gastrointestinal mucosa, help maintain normal pH levels, detoxify the body, and promote the growth of healthy intestinal flora. Enzymes are also used to strengthen the body as a whole.

Enzyme Absorption System Enhancers (EASE) maximize enzyme activity. They can also improve the absorption and bioavailability of various nutrients and decrease the drain on the body's own digestive enzymes, thus prolonging their lives.

Oral enzymes are extremely effective in treating acute or chronic sinusitis. Research shows that enzyme preparations decrease the thickness of the respiratory mucus, have an anti-inflammatory effect, and may possibly break down complex protein substances creating the congestion. This is well supported by the fact that many enzymes, such as pancreatin, trypsin, chymotrypsin, bromelain, papain, and serratiopeptidase, are absorbed from intestinal mucosa into the bloodstream and are enzymatically active in their absorbed form.

In addition to the standard multivitamin, multimineral, and multiglandular complexes recommended in the introduction to Part Two, the following supplements are recommended for the treatment of sinusitis. The following recommended dosages are for adults. Children between the ages of two and five should take one-quarter the recommended dosage, those between the ages of six and twelve should take one-half the recommended dosage, and those between the ages of thirteen and seventeen should take three-quarters the recommended adult dosage.

ESSENTIAL ENZYMES

Enzyme	Suggested Dosage	Actions
Formula Three—See page 38 of Part One for a list of products.	Take 200 mcg of each three times per day between meals.	Potent antioxidant enzymes.

AND SELECT ONE OF THE FOLLOWING:

Enzyme	Suggested Dosage	Actions
Bromelain	Take three to four 230–250 mg capsules per day until symptoms subside. If severe, take 6–8 capsules per day.	Fights inflammation, pain, and swelling; stimulates the immune system.
Serratiopeptidase	Take 5 mg three times per day between meals.	Fights inflammation; improves respiratory conditions; bolsters the immune system; breaks up mucus.
Sfericase	Take 500 mg three times per day between meals.	Fights inflammation; improves respiratory conditions; bolsters the immune system; breaks up mucus.
Wobenzym N from Naturally Vitamin Supplements	Take 2–4 tablets three times per day between meals for six months or until symptoms improve. Thereafter, take 2 tablets three times per day between meals or as needed.	Fights inflammation, pain, and swelling; speeds healing; strengthens capillary walls.
Mega-Zyme from Enzymatic Therapy	Take as indicated on the label.	Fights inflammation, pain, and swelling; speeds healing; strengthens capillary walls.
Inflazyme Forte from American Biologics	Take as indicated on the label between meals.	Fights inflammation, pain, and swelling; speeds healing; strengthens capillary walls.
Lysozyme	Take as indicated on the label.	Fights bacteria and infection.
Enzyme complexes rich in proteases, plus amylases and lipases, working synergistically	Take as indicated on the label between meals.	Fight inflammation, pain, and swelling; speed healing; strengthen capillary walls.

SUPPORTIVE ENZYMES

Enzyme	Suggested Dosage	Actions
A digestive enzyme formula, containing protease, amylase, and lipase. See the list of digestive enzyme products in Appendix.	Take as indicated on the label with meals.	Improves the breakdown and absorption of proteins, carbohydrates, and fats, improving the overall health of the body; fights toxic build-up.

ENZYME ABSORPTION SYSTEM ENHANCERS (EASE)

Combination	Suggested Dosage	Actions
Cold Zzap from Naturally Vitamin Supplements or	Take as indicated on the label.	Fights infection from viruses or bacteria; boosts immunity; has antioxidant activity.
Combat from Life Plus or	Take as indicated on the label.	Helps support the respiratory system; stimulates immune system function; helps control infections.
Mucous Dissolver Liquezyme from Enzyme Process Laboratories or	Take as indicated on the label.	Dissolves mucus.
Sinease from Prevail or	Take as indicated on the label.	Decongestant.
Sinus Ease from Nature's Life or	Take as indicated on the label.	Stimulates immune function; reduces inflammation.
Zym-Eeze Lysozyme Lozenges from Future Foods	Take as indicated on the label.	Natural, enzymatic antimicrobial agent; fights sore throat, colds, and strep throat.

ENZYME HELPERS

Nutrient	Suggested Dosage	Actions
Bioflavonoids	Take 100–300 mg twice per day.	Strengthen capillaries; potent antioxidants.
Green Barley Extract or Kyo-Green	Take 1 heaping teaspoon in 8 ounces of liquid three times per day, or as often as possible.	Fights free radicals.
Pantothenic acid (vitamin B$_5$)	Take 25–50 mg per day.	Important in protein, fat, and carbohydrate metabolism, improving overall health; needed for healthy digestive system and adrenal glands; a coenzyme for many enzymes.
Vitamin A or	Take 10,000–50,000 IU from all sources for no more than one month. For doses higher than 50,000 IU, see a physician. Pregnant women should not take more than 10,000 IU per day.	Necessary for healthy mucous linings, skin and cell membrane structure, and immune system; potent antioxidant.
Carotenoid complex	Take 15,000 IU per day.	Potent antioxidant; enhances immunity.
Vitamin B complex	Take 50 mg three times per day.	B vitamins act as coenzymes.
Vitamin C	Take 1–2 grams every hour until symptoms improve; or mix a heaping teaspoon of powdered or crystalline vitamin C with water or orange or lemon juice (if lemon juice is used, sweeten with honey, if	Fights infections, diseases, allergies, and common cold; essential for wound healing; potent antioxidant.
	desired). Gargle a mouthful and swallow. Repeat every hour.	
Zinc or zinc gluconate lozenges	Take 50 mg per day. Do not exceed 100 mg per day from all sources.	Needed for white blood cell immune function; involved with more than 300 enzymes.

Comments

❑ Hot broth or chicken or vegetable soup (made with plenty of garlic and onions) is helpful for cleaning out the body and sweating out toxins. Remove the skin from the chicken before cooking, since many toxins are stored in and just under the skin. Sip the broth or soup as frequently as possible.

❑ Herbs, including aged garlic extract, astragalus, chlorella, echinacea, elderberry, ginger, ginkgo biloba, ginseng, goldenseal, licorice, pau d'arco, slippery elm, white willow bark, and mushrooms (including maitake, reishi, and shiitake), can be very helpful in fighting sinusitis and assisting enzymes.

❑ The Enzyme Toxin Flush and the Enzyme Kidney Flush programs are of value to detoxify the body. See instructions in Part Three.

❑ A vaporizer may be of help, or place five to six drops of eucalyptus in a sink or bowl of very hot water. Cover your head and shoulders with a large bath towel over the sink or bowl and inhale the vapors.

❑ For recurrent or chronic sinusitis, also check for the presence of allergies, thyroid dysfunction, and immune system deficiencies.

❑ Because the infection can spread to neighboring bones, the ears (leading to mastoiditis), or the brain (in rare cases), prompt treatment is essential.

❑ Stop smoking. Cigarette and other smoke irritates the sinus passages.

❑ Hot moist packs, a hot water bottle, or a heating pad placed over the sinuses can sometimes help relieve the pain of sinusitis.

❑ Use NSAIDs (nonsteroidal anti-inflammatory drugs), such as aspirin, ibuprofen, or acetaminophen, very carefully. They have side effects.

❑ According to Dr. Linda Rector Page, N.D., Ph.D. in *Healthy Healing*, (Healthy Healing Publications, 1996), "Supportive over-the-counter sinus medications can both trigger an infection by not allowing the draining of infective material, and aggravate it by driving the infection deeper in the sinuses."

❑ In extreme cases, minor surgery is used to improve sinus drainage. This should be considered only as a last resort.

Sjögren's Syndrome

Sjögren's (pronounced show-grins) syndrome is probably an autoimmune condition in which the white blood cells damage the fluid-secreting glands, particularly the tear glands in the eyes and the salivary glands in the mouth. This causes dry eyes and dry mouth, also known as sicca syndrome. It may also affect the mucous membranes of the gastrointestinal tract, respiratory tract, trachea, vagina, and vulva. In severe cases, it may damage vital organs. Sjögren's syndrome often occurs with such autoimmune illnesses as lupus and rheumatoid arthritis.

Sjögren's syndrome affects 2 to 4 million Americans, 90 percent of whom are women. The illness is incurable, although many treatments can help improve the symptoms. The condition was named for Swedish ophthalmologist Hendrick Sjögren, who pioneered study of the condition and developed the Rose Bengal stain for diagnosis.

Symptoms

Many only experience mild symptoms of dry mouth and eyes. Dryness of the mouth may cause difficulty swallowing and talking. It can also dull the senses of taste and smell, cause increased thirst, and cause mouth ulcers and dental cavities to form. Dryness of the eyes can cause such eye discomfort as the feeling of a foreign body in the eye, a gritty feeling, increased sensitivity to light, redness, burning, and itching. It may also cause blurred vision. Dry eyes can lead to permanent eye damage. Dryness of the vagina can cause painful intercourse. Dryness of the upper respiratory tract can lead to hoarseness, a chronic cough, and increased susceptibility to infection, such as pneumonia. The parotid glands often swell, lending a chipmunklike appearance to the sufferer. Joint inflammation is also very common. Other common symptoms include fatigue, muscle pain, hair loss, low-grade fever, and itching. Sometimes vital organs are damaged. Lymphomas are forty-four times more common in those with Sjögren's syndrome than in the those unaffected. Sjögren's syndrome is marked by periods of exacerbations and remissions.

Enzyme Therapy

Systemic enzyme therapy is used to reduce inflammation and stimulate the immune system. Enzymes improve circulation, help speed tissue repair, bring nutrients to the damaged area, remove waste products, and enhance wellness. Digestive enzyme therapy is used to improve the digestion of food, reduce stress on the gastrointestinal mucosa, help maintain normal pH levels, detoxify, and promote the growth of healthy intestinal flora. In Europe, enzymes have been used as adjunctive therapy for Sjögren's syndrome for some time.

Enzyme Absorption System Enhancers (EASE) maximize enzyme activity. They can also improve the absorption and bioavailability of various nutrients and decrease the drain on the body's own digestive enzymes, thus prolonging their lives.

In addition to the standard multivitamin, multimineral, and multiglandular complexes recommended in the introduction to Part Two, the following supplements are recommended for the treatment of Sjögren's syndrome. The following recommended dosages are for adults. Children between the ages of two and five should take one-quarter the recommended dosage, those between the ages of six and twelve should take one-half the recommended dosage, and those between the ages of thirteen and seventeen should take three-quarters the recommended adult dosage.

ESSENTIAL ENZYMES

Enzyme	Suggested Dosage	Actions
Formula One—See page 38 of Part One for a list of products. _or_	Take 10 tablets three times per day between meals for six months. Then decrease dosage to 5 tablets three times per day between meals.	Bolsters the immune system; breaks up antigen-antibody complexes; fights free-radical formation and inflammation.
Phlogenzym from Mucos _or_	Take 10 tablets three times per day between meals; decrease dosage as symptoms improve.	Stimulates immune function; breaks up circulating immune complexes; fights free-radical formation and inflammation.
Bromelain _or_	Take three to four 230–250 mg capsules per day until symptoms subside. If symptoms are severe, take 6–8 capsules per day.	Fights inflammation, pain, and swelling; stimulates the immune system.
A proteolytic enzyme	Take as indicated on the label.	Bolsters the immune system; breaks up antigen-antibody complexes; fights free-radical formation and inflammation.

SUPPORTIVE ENZYMES

Enzyme	Suggested Dosage	Actions
Formula Three—See page 38 of Part One for a list of products.	Take as indicated on the labels.	Potent antioxidant enzymes.
A digestive enzyme product containing protease, amylase, and lipase. See the list of digestive enzyme products in Appendix.	Take as indicated on the label with meals.	Relieves the drain and stress on the body's enzymes; improves breakdown and absorption of food nutrients; fights toxic build-up.

ENZYME ABSORPTION SYSTEM ENHANCERS (EASE)

Combination	Suggested Dosage	Actions
Ecology Pak from	Take 2 or 3 tablets	Assists with natural detoxifica-

Life Plus *or*	once or twice per day, preferably in the morning and no later than midday.	tion; stimulates immune function.
Hepazyme from Enzymatic Therapy *or*	Take as indicated on the label.	Stimulates immune function.
Inflamzyme from PhytoPharmica *or*	Take as indicated on the label.	Immune system activator; reduces inflammation; fights free radicals.
Kyolic Formula 102 (with enzymes)	Take 2 capsules or tablets with meals two to three times per day.	Strengthens the immune system; helps alleviate stress; helps fight free radicals; helps fight inflammation; improves digestion and absorption of nutrients.

ENZYME HELPERS

Nutrient	Suggested Dosage	Actions
Acidophilus and other probiotics	Take as indicated on the label.	Stimulates enzyme activity; improves gastrointestinal function, improving the overall health of the body.
Adrenal glandular	Take 300 mg per day.	Helps fight stress and adrenal exhaustion; fights inflammation.
Arginine	Take as indicated on the label.	Accelerates healing; enhances thymus activity.
Essential fatty acid complex	Take 100–200 mg two to three times per day with meals; or follow the advice of your health-care professional.	Strengthens immunity; helps strengthen cell membranes.
Kyolic Aged Garlic Liquid Extract with Vitamins B_1 and B_{12}	Take 10 drops in water or juice three times per day.	Stimulates immune system function.
Thymus glandular	Take 90 mg per day.	Assists the thymus gland in immune resistance; improves immune response; fights infections.
Vitamin A	Take up to 40,000 IU for no more than one month. Take doses higher than 40,000 IU only under the supervision of a physician. If pregnant, do not take more than 10,000 IU per day.	Needed for strong immune system and healthy mucous membranes; potent antioxidant.
Vitamin C	Take 8,000–13,000 mg per day.	Stimulates the adrenal glands; promotes immune system; powerful antioxidant.
Vitamin E	Take 1,200 IU per day from all sources (including the multivitamin complex in the introduction to Part Two).	Acts as an intercellular antioxidant; improves circulation; helps resist diseases at the cellular level; works closely with selenium to fight free radicals.

Comments

❏ At present, there is no known cure for Sjögren's syndrome. However, to ease the symptoms and discomfort of dryness, moisture replacement therapies are often used.

❏ Sipping fluids throughout the day or chewing gum can help relieve mouth dryness.

❏ A number of herbs can help bolster the immune system, including astragalus, echinacea, ginseng, goldenseal, licorice, maitake mushrooms (as well as reishi and shiitake mushrooms), and pau d'arco.

❏ See the list of Resource Groups in the Appendix for an organization that can provide you with more information about Sjögren's syndrome.

Skin Cancer

Of all the cancer cases diagnosed this year, one in three will be skin cancer. In fact, one in every five Americans will develop skin cancer during his or her lifetime. Although skin cancer is the most common form of cancer in the United States, if the lesions are detected and treated early, the chance of beating the disease is high. The three most common types of skin cancer are basal cell carcinomas, squamous cell carcinomas, and melanomas.

Basal cell carcinoma is the most common kind of skin cancer. Nearly one million Americans each year are affected by this type of cancer. It is also the least dangerous form. It occurs most frequently on the face, ears, neck, and backs of the hands, shoulders, and arms—those areas most often exposed to sunlight. It develops in the basal layer of the skin—the deepest layer of the epidermis. Basal cell carcinomas grow very slowly. They begin as very small shiny raised skin lesions that do not hurt or itch. As they grow and flatten, they resemble scars or sores that may alternately bleed and scab. The border of the lesion has a "pearly" appearance. Basal cell carcinomas rarely spread to other parts of the body, but they will destroy surrounding tissue if not treated.

The second most common skin cancer is *squamous cell carcinoma*, which affects 100,000 Americans each year. This type of cancer also develops most commonly in areas exposed to the sun, although it can occur anywhere on the body, including the mucous membranes. It originates in the middle layer of the epidermis and develops as a small, red, scaly, crusted raised bump on the skin that may resemble a wart. It does not hurt or itch. It eventually becomes an open sore that does not heal. Squamous cell carcinoma may invade deeper tissue, but it rarely spreads to other parts of the body. If it does, it can be fatal.

The most dangerous form of skin cancer is *melanoma*. It develops in the melanocytes, the pigment-producing skin cells. Most melanomas develop in moles or other pre-existing skin lesions, although they may also develop in otherwise normal skin. Melanomas can change the appearance of an already existing mole, or it may appear as a skin lesion that can be black, brown, red, white, blue, or a mixture of colors. Unlike the other two skin cancers, melanomas will easily spread to other parts of the body, which is why it is very important to catch melanomas early in order to completely excise the cancer. Once the melanoma spreads, it is considered incurable and is soon fatal. If detected early, even this type of cancer is treatable.

The risk for all kinds of skin cancer increases with exposure to sunlight and excessive ultraviolet radiation. Those with fair complexions are at increased risk for skin cancers.

Enzyme Therapy

Enzymes are used as complementary therapy with primary forms of cancer treatment. Systemic enzyme therapy can be used as preventative care and to decrease skin cancer growth, inflammation, and free-radical formation. Enzymes stimulate the immune system, improve circulation, help speed tissue repair, bring nutrients to the damaged area, remove waste products, and improve health.

Digestive enzyme therapy is used to improve the digestion of food, reduce stress on the gastrointestinal mucosa, help maintain normal pH levels, detoxify, and promote the growth of healthy intestinal flora, thereby improving the overall health of the body. Digestive enzymes also serve as replacements for the body's pancreatic enzymes, leaving the pancreatic enzymes free to perform other functions in the body, such as boosting immunity and improving circulation.

Topical enzyme creams are used to dissolve dead and damaged skin cells and exfoliate the skin, while leaving healthy skin cells intact.

Enzyme Absorption System Enhancers (EASE) maximize enzyme activity. They can also improve the absorption and bioavailability of various nutrients and decrease the drain on the body's own digestive enzymes, thus prolonging their lives.

Complementary enzyme therapy with proteolytic enzymes is successful before and after surgery; before, during, and after chemotherapy or radiation therapy; and as long-term and palliative therapy.

In addition to the standard multivitamin, multimineral, and multiglandular complexes recommended in the introduction to Part Two, the following supplements are recommended for the treatment of skin cancer. The following recommended dosages are for adults. Children between the ages of two and five should take one-quarter the recommended dosage, those between the ages of six and twelve should take one-half the recommended dosage, and those between the ages of thirteen and seventeen should take three-quarters the recommended adult dosage.

ESSENTIAL ENZYMES

Enzyme	Suggested Dosage	Actions
Before and after surgery:		
Phlogenzym from Mucos	Take 3 tablets three times per day.	Decreases the possibility of suppressed immune function after surgery or anesthesia; breaks down fibrin; decreases inflammation; improves wound healing; decreases risk of blood clot complications.
Before, during, and after chemotherapy or radiotherapy:		
Phlogenzym from Mucos	Take 2 tablets twice per day.	Breaks down fibrin; decreases inflammation; improves effectiveness of chemotherapy or radiation; decreases radiation sickness; stimulates immune function.
Long-term and palliative therapy:		
Phlogenzym from Mucos	Take 2–4 tablets three times per day.	Reduces malnutrition, weakness, and muscle wasting (cachexia), anorexia, depression, and pain; prevents metastasis; eliminates or prevents disturbances of lymphatic drainage.

AND SELECT ONE OF THE FOLLOWING:

Enzyme	Suggested Dosage	Actions
Formula One (particularly Wobenzym N from Naturally Vitamin Supplements)—See page 38 of Part One for a list of products.	Take 8–10 tablets three times per day between meals.	Stimulates immune function to fight cancer; fights inflammation, pain, and swelling; speeds healing; strengthens capillary walls.
Inflazyme Forte from American Biologics	Take as indicated on the label.	Stimulates immune function to fight cancer; fights inflammation, pain, and swelling; speeds healing; strengthens capillary walls.
Bromelain	Take eight to ten 230–250 mg capsules per day.	Stimulates immune function to fight cancer; fights inflammation, pain, and swelling; speeds healing; strengthens capillary walls.
Enzyme complexes rich in proteases, plus amylases and lipases, working synergistically	Take as indicated on the label.	Stimulates immune function to fight cancer; fights inflammation, pain, and swelling; speeds healing; strengthens capillary walls.

SUPPORTIVE ENZYMES

Enzyme	Suggested Dosage	Actions
Formula Three—See page 38 of Part One for a list of products.	Take as indicated on the labels.	Potent antioxidant enzymes.
A digestive enzyme product containing protease, amylase, and lipase enzymes. See the list of digestive enzyme products in Appendix.	Take as indicated on the label with meals.	Improves the breakdown and absorption of food nutrients; fights toxic build-up.

Skin Cancer Warning Signs

According to the Skin Cancer Foundation, the following are warning signs of skin cancer:

❑ A skin growth that increases in size and is pearly, translucent, tan, brown, black, red, pink, or multi-colored in appearance.

❑ A growth that continues to crust, scab, erode, or bleed.

❑ A mole that changes color or texture, becomes irregular in shape, changes in size, or is bigger than a pencil eraser.

❑ An open sore that lasts for more than four weeks or heals and then reopens.

❑ A scaly or crusty bump that is horny, dry, and rough and may produce a prickling or tender sensation.

ENZYME ABSORPTION SYSTEM ENHANCERS (EASE)

Combination	Suggested Dosage	Actions
Akne-Zyme from Enzymatic Therapy or	Take 1 or 2 capsules daily as a dietary supplement.	Maintains skin integrity.
Antioxidant Complex from Tyler Encapsulations or	Take as indicated on the label.	Potent free-radical fighter and immune system stimulator.
Bio-Musculoskeletal Pak from Biotics or	Take as indicated on the label.	Decreases inflammation; fights free-radical formation; stimulates immune function.
Bioprotect from Biotics or	Take as indicated on the label.	Potent antioxidant combination; improves immune function and tissue repair.
Connect-All from Nature's Plus or	Take as indicated on the label.	Supports connective tissue.
CTR Support from PhysioLogics or	Take as indicated on the label.	Nourishes the body during tissue recovery and healing; helps maintain healthy connective tissue; diminishes damage caused by swelling and inflammation.
Detox Enzyme Formula from Prevail or	Take as indicated on the label.	Helps rid the body of harmful toxins.
Kyolic Formula 102 (with enzymes) or	Take 2 capsules or tablets with meals two to three times per day.	Maintains skin integrity; strengthens the immune system; helps alleviate stress; helps fight free radicals; improves digestion and absorption of nutrients.
Oxy-5000 Forte from American Biologics or	Take as indicated on the label.	Helps the body excrete toxins; fights biochemical stressors, common infections, and degenerative processes.
Vita-C-1000 from Life Plus	Take as indicated on the label.	Helps fight cancer; boosts immune function.

ENZYME HELPERS

Nutrient	Suggested Dosage	Actions
Acidophilus and other probiotics	Take as indicated on the label.	Stimulates enzyme activity; improves gastrointestinal function, thereby improving the overall health of the body; improves immunity.
Adrenal glandular	Take 100–400 mg per day.	Helps fight stress and adrenal exhaustion; fights inflammation.
Coenzyme Q_{10}	Take 30–60 mg one to two times per day with meals.	Important to immune system function; has antioxidant activity.
Kyolic Aged Garlic Liquid Extract with Vitamins B_1 and B_{12}	Take 10 drops in water or juice three times per day.	Stimulates immune system function.
Thymus glandular	Take 100 mg per day, or as indicated on the label.	Assists the thymus gland in immune resistance; improves immune response.
Vitamin A	Take up to 50,000 IU per day. Take doses higher than 50,000 IU only under the direction and supervision of a physician. If pregnant, do not take more than 10,000 IU per day.	Needed for strong immune system and healthy skin; potent antioxidant.
Vitamin B complex	Take 100–150 mg three times per day.	B vitamins act as coenzymes.
Vitamin C	Take 10,000–15,000 mg per day, or 1–2 grams every hour until cancer symptoms improve.	Improves wound healing; stimulates the adrenal glands; promotes immune system; powerful antioxidant.
Vitamin D	Take 3,000–6,000 IU per day.	Vitamin D combined with cancer may have anticancer properties.
Vitamin E	Take 1,200 IU per day from all sources (including the multivitamin complex in the introduction to Part Two).	Acts as an intercellular antioxidant; improves circulation; helps resist diseases at the cellular level; works closely with selenium to fight free radicals.
Zinc	Take 50 mg per day. Do not take more than 100 mg per day.	Needed by over 300 enzymes; necessary for white blood cell immune function; improves healing.

Comments

❏ Stay out of the sun or cover up and wear sunscreen. According to the American Cancer Society, almost all of the cases of squamous and basal cell skin cancers diagnosed each year in this country are sun-related.

❏ Protect yourself and your children from the sun's rays. Most of us receive the majority (up to 80 percent) of our lifetime sun exposure before the age of 18.

❏ For information on fasting, juicing, diets, and flushes, see Part Three.

❏ If skin cancer is suspected, see a cancer specialist immediately. (See "Skin Cancer Warning Signs," on page 370.)

❏ Coffee enemas are often used with cancer patients to stimulate the release of bile from the liver and facilitate the removal of waste metabolites from the body and into the intestines for excretion. See Part Three for instructions.

❏ Physicians sometimes use enzyme injections or enzyme retention enemas in the treatment of cancer. Both methods allow a greater concentration of the enzymes to be absorbed than might be accomplished if the enzymes were taken orally. An enzyme retention enema is a means of instilling some enzymes into the colon to be retained. This allows the enzymes to be absorbed into the body's systems. Physicians use products specially formulated for these uses.

❏ See the list of Resource Groups in the Appendix for a list of organizations that can provide you with more information about Skin cancer.

❏ *See also* Cancer.

Skin Problems

See ABSCESS; ACNE; BOILS; BRUISES AND HEMATOMAS; BURNS; CANKER SORES; CHICKENPOX, DERMATITIS; FUNGAL SKIN INFECTIONS; HIVES; INSECT BITES AND STINGS; MEASLES; PSORIASIS; ROSACEA; SCARS, SKIN CANCER; SKIN RASH; SKIN ULCERS; STAPHYLOCOCCAL INFECTIONS; VARICOSE VEINS; WARTS.

Skin Rash

A skin rash is a temporary eruption on the skin. A rash is usually a symptom of an underlying condition and can indicate an allergic reaction (to foods, cosmetics, or other aller-

gens) or disease (measles, rubella, chickenpox, infectious mononucleosis, scarlet fever, Lyme disease, psoriasis, and shingles). Insect bites, dry skin, seborrhea, dermatitis, athlete's foot, and even nutritional deficiencies, such as pellagra, can also cause a skin rash. If your rash is the result of an underlying condition, find that condition in Part Two and follow its enzyme therapy program.

The skin is a reflection of health (or disease) and is the first area in the body exposed to environmental toxins, radiation, and trauma (from a bruise or a cut to a car accident). The skin is composed of three layers: the epidermis (the outermost, superficial layer), the dermis (the middle layer), and an underlying layer of fat called the subcutaneous layer. Although any of the layers may be involved, most rashes occur on the superficial layer.

Symptoms

A skin rash usually looks like small red or pink bumps. It may or may not itch. There may be scaly, round, or oval patches on the skin.

Enzyme Therapy

Systemic enzyme therapy is used to reduce inflammation, stimulate the immune system, improve circulation, help speed tissue repair, bring nutrients to the damaged area, remove waste products, improve health, strengthen the body as a whole, and build general resistance.

Digestive enzyme therapy is used to improve the digestion of food, reduce stress on the gastrointestinal mucosa, help maintain normal pH levels, promote the growth of healthy intestinal flora, and flush the body of toxins.

Topical enzyme creams are used to dissolve dead and damaged skin cells, while leaving healthy skin cells intact.

Enzyme Absorption System Enhancers (EASE) maximize enzyme activity. They can also improve the absorption and bioavailability of various nutrients and decrease the drain on the body's own digestive enzymes, thus prolonging their lives.

The red starburst, rashlike appearance and sallow complexion of many alcoholics is an indication that liver enzymes are overtaxed. Liver spots on the skin relate to reduced enzyme production (a sign of aging and decreased enzyme production in the body). More than any other nutrient, enzymes are essential to improve skin and body functions.

Even stress-related rashes, such as diaper rash, "raspberries" (from sliding on your hip in baseball) and other friction-type rashes, sunburn (and other radiation burns), or wind burn can be helped and heal faster with enzymes. Enzymes can even help slow the free-radical damage (and resulting cross linking) that the sun and other sources of radiation cause on the skin.

In addition to the standard multivitamin, multimineral, and multiglandular complexes recommended in the intro-

duction to Part Two, the following supplements are recommended for the treatment of skin rashes. The following recommended dosages are for adults. Children between the ages of two and five should take one-quarter the recommended dosage, those between the ages of six and twelve should take one-half the recommended dosage, and those between the ages of thirteen and seventeen should take three-quarters the recommended adult dosage.

ESSENTIAL ENZYMES

Enzyme	Suggested Dosage	Actions
Superoxide dismutase	Take 150 mcg per day.	Fights inflammation; potent antioxidant.
WobeMugos E Ointment from Mucos	Apply at least twice per day. Do not apply to open wounds.	Exfoliates the skin and breaks up cross linking.

AND SELECT ONE OF THE FOLLOWING:

Formula One—See page 38 of Part One for a list of products.	Take 5 tablets twice per day between meals.	Fights inflammation, pain, and swelling; speeds recovery after injury or surgery; stimulates the immune system; strengthens capillary walls.
Bromelain	Take 750 mg two times per day between meals.	Fights inflammation, pain, and swelling; speeds recovery after injury; stimulates the immune system; strengthens capillary walls.
Serratiopeptidase	Take 10 mg two times per day between meals.	Fights inflammation; reduces pain and swelling.
WobeMugos E from Mucos	Take 3–4 tablets three times per day between meals.	Fights inflammation, pain, and swelling; increases rate of healing; stimulates the immune system.
Pancreatin	Take 500 mg six times per day on an empty stomach and between meals.	Fights inflammation, pain, and swelling; increases rate of healing; stimulates the immune system; strengthens capillary walls.
Microbial protease	Take 300 mg two times per day between meals.	Fights inflammation; speeds recovery from injuries.

SUPPORTIVE ENZYMES

Enzyme	Suggested Dosage	Actions
Catalase and glutathione peroxidase	Take as indicated on the labels.	Potent antioxidant enzymes.
A digestive enzyme product containing protease, amylase, and lipase. See the list of digestive enzyme products in Appendix.	Take as indicated on the label with meals.	Improves the breakdown and absorption of food nutrients; fights toxic build-up.

ENZYME ABSORPTION SYSTEM ENHANCERS (EASE)

Combination	Suggested Dosage	Actions
Akne-Zyme from Enzymatic Therapy or	Take 1 or 2 capsules daily as a dietary supplement.	Maintains skin integrity.
Bioprotect from Biotics or	Take one to three capsules per day.	Potent antioxidant combination; improves immune function and tissue repair.
CTR Support from PhysioLogics or	Take 2–3 capsules per day, preferably on an empty stomach; or follow the advice of your health-care professional.	Nourishes the body during tissue recovery and healing; helps maintain healthy connective tissue.
Detox Enzyme Formula from Prevail or	Take one capsule between meals one or two times per day with a 6- to 8-ounce glass of water or juice.	Helps rid the body of harmful toxins.
Inflamzyme from PhytoPharmica	Take as indicated on the label.	Promotes tissue repair.

ENZYME HELPERS

Nutrient	Suggested Dosage	Actions
Acidophilus or other probiotic	Take as indicated on the label.	Improves gastrointestinal function to improve overall health; needed for the production of the coenzymes biotin, niacin, folic acid, and pyridoxine; helps produce antibacterial substances; stimulates enzyme activity.
Amino-acid complex	Take as indicated on the label on an empty stomach.	Required for enzyme production; helps the body maintain proper fluid and acid/base balance to improve the overall health of the body.
Bioflavonoids, such as pycnogenol, rutin, or quercetin	Take 500 mg three times per day between meals.	Strengthen capillaries; potent antioxidants.
Coenzyme Q$_{10}$	Take 30–60 mg one to two times per day with meals; or follow the advice of your health-care professional.	Potent antioxidant.
Vitamin A	Take 5,000–10,000 IU per day. If pregnant, take no more than 10,000 IU per day.	Needed for strong immune system and healthy skin; potent antioxidant.
Vitamin B complex	Take 50 mg three times per day, or as indicated on the label.	B vitamins act as coenzymes and support healthy skin.
Vitamin C	Take 8,000–13,000 mg per day.	Fights infections, diseases, and allergies; improves healing; stimulates the adrenal glands; promotes immune system; powerful antioxidant.
Vitamin E	Take 1,200 IU per day from all sources (including	Potent antioxidant; supplies oxygen to cells; improves cir-

	the multivitamin complex in the introduction to Part Two).	culation and level of energy.
Zinc	Take 15–50 mg per day. Do not take more than 100 mg per day.	Needed by over 300 enzymes; necessary for white blood cell immune function; improves wound healing; promotes healthy skin.

Comments

❑ Eat plenty of fresh enzyme-rich fruits and vegetables, such as pineapple, papaya, and figs. Follow the dietary guidelines in Part Three.

❑ Several herbs can improve digestion, including alfalfa concentrate, chlorella, chlorophyll, ginger root extract, and green foods (such as Kyo-Green from Wakunaga).

❑ Detoxify the body with juicing and fasting and other detoxification techniques described in Part Three, including the Enzyme Kidney Flush and the Enzyme Toxin Flush programs.

❑ Apply an enzymatic salve three times a day. See Part Three for more information.

❑ Place ten drops of liquid aged garlic extract on a cotton swab or gauze. Dab gently along the affected area.

❑ Break open and spread the contents of two to four vitamin A and vitamin E capsules on the affected area.

❑ Zinc oxide salve can be spread over the affected area once daily.

❑ A positive mental attitude can help stimulate antioxidants, which fight free-radical formation.

❑ Eat foods high in vitamin B$_6$, including wheat germ and brown rice.

❑ Avoid any foods that seem to worsen your skin condition. In many people, this may include seafood, dairy products (especially milk), eggs, and nuts. Keep a food diary to help you identify any food allergies.

❑ Avoid any clothing that may irritate the skin or cause an allergic reaction, such as wool.

❑ The skin will turn an orange-like hue, flake, and/or peel if vitamin A intake is excessive and liver enzymes are overtaxed. This jaundice color is the sign of liver problems, toxin-build up and/or elimination problems. If this occurs, stop taking vitamin A until the orange-like hue disappears and flaking stops.

❑ Use no salves or ointments over open or infected rashes.

❑ *See also* ALLERGIES; DERMATITIS; HIVES.

Skin Ulcers

A skin ulcer is the loss of part or all of the skin that penetrates all the way down to the dermis (the deeper layer of the skin). Skin ulcers usually develop due to damage of the skin caused by infection, prolonged pressure, irritation, or extreme temperatures.

Ulcers can form in the anus, as a result of large or hard bowel movements (*anal fissures*). They may also occur as a result of a deep vein thrombosis; diabetes mellitus, which can impair circulation; infections, such as tularemia; and sexually transmitted infections, such as syphilis, chancroid, and lymphogranuloma venereum.

A *decubitus ulcer* (also called a bed sore) occurs as a result of lack of blood flow to, or irritation of skin over, a bony, weight-bearing area of the body. It is caused by prolonged pressure applied to the area by a bed, wheelchair, cast, or other hard object. They occur most often in those who cannot move, such as people who are paralyzed and people in comas, and those who cannot feel the discomfort or pain that would normally signal one to change position, such as those who suffer from some type of nerve damage.

Symptoms

Skin ulcers are marked by swelling, redness, pain, heat, and inflammation of an open skin sore. The sore may be infected and full of pus. Bed sores can be classified in four stages based on their symptoms. In stage 1, the skin appears red. In stage 2, the skin is soft and swollen, and there may be blistering. The upper layers of the skin begin to die. In stage 3, the sore extends to the deepest layers of the skin. In stage 4, the necrosis (tissue death) extends through the skin and fat to the muscle. Stage 5 is marked by progressive necrosis of fat and muscle. Bone destruction occurs in stage 6, and there is an increased risk of blood poisoning, septic arthritis, and fracture.

Enzyme Therapy

Systemic enzyme therapy is used to decrease pain, swelling, and inflammation; improve circulation; and help speed repair of the skin ulcer. Enzymes stimulate the immune system, bring nutrients to the damaged area, remove waste products, improve health, and fight infection.

Digestive enzyme therapy is used to improve the digestion of food, reduce stress on the gastrointestinal mucosa, help maintain normal pH levels, detoxify the body, and promote the growth of healthy intestinal flora, thereby improving the overall health of the body. Digestive enzymes also serve as replacements for the body's pancreatic enzymes, leaving the pancreatic enzymes free to perform other func-

373

tions in the body, such as improving circulation, boosting immunity, and decreasing inflammation.

Enzyme Absorption System Enhancers (EASE) maximize enzyme activity. They can also improve the absorption and bioavailability of various nutrients and decrease the drain on the body's own digestive enzymes, thus prolonging their lives.

In addition to the standard multivitamin, multimineral, and multiglandular complexes recommended in the introduction to Part Two, the following supplements are recommended for the treatment of skin ulcers. The following recommended dosages are for adults. Children between the ages of two and five should take one-quarter the recommended dosage, those between the ages of six and twelve should take one-half the recommended dosage, and those between the ages of thirteen and seventeen should take three-quarters the recommended adult dosage.

ESSENTIAL ENZYMES

Enzyme	Suggested Dosage	Actions
Formula Three—See page 38 of Part One for a list of products.	Take 200 mcg of each three times per day between meals.	Potent antioxidant enzymes.

AND SELECT ONE OF THE FOLLOWING:

Enzyme	Suggested Dosage	Actions
Formula One—See page 38 of Part One for a list of products.	Take 8–10 tablets three times per day between meals until symptoms improve.	Dissolves blood clots; fights inflammation, pain, and swelling; speeds healing.
Phlogenzym from Mucos	Take 10 tablets three times per day between meals until symptoms improve.	Dissolves blood clots; fights inflammation, pain, and swelling; speeds healing.
Bromelain	Take 1,800 mg three times per day between meals until symptoms improve.	Dissolves blood clots; fights inflammation, pain, and swelling; speeds healing.
A proteolytic enzyme (from microbial, animal, or plant sources)	Take as indicated on the label between meals until symptoms improve.	Dissolves blood clots; fights inflammation, pain, and swelling; speeds healing; strengthens capillary walls.

SUPPORTIVE ENZYMES

Enzyme	Suggested Dosage	Actions
A digestive enzyme formula containing protease, amylase, and lipase enzymes. See the list of digestive enzyme products in Appendix.	Take as indicated on the label with meals.	Improves the breakdown and absorption of proteins, carbohydrates, and fats; fights toxic build-up.

ENZYME ABSORPTION SYSTEM ENHANCERS (EASE)

Combination	Suggested Dosage	Actions
Bioprotect from Biotics *or*	Take one to three capsules each day.	Potent antioxidant combination; improves immune function and skin and tissue repair.
Connect-All from Nature's Plus *or*	Take as indicated on the label.	Supports connective tissue, including the skin.
CTR Support from PhysioLogics *or*	Take 2–3 capsules per day, preferably on an empty stomach; or follow the advice of your health-care practitioner.	Nourishes the body during tissue recovery and healing; helps maintain healthy connective tissue and skin.
Detox Enzyme Formula from Prevail *or*	Take 1 capsule between meals one or two times per day with a 6- to 8-ounce glass of water or juice.	Helps rid the body of harmful toxins.
Inflamzyme from PhytoPharmica	Take as indicated on the label.	Promotes skin and tissue repair.

ENZYME HELPERS

Nutrient	Suggested Dosage	Actions
Acidophilus or other probiotics	Take as indicated on the label.	Improves gastrointestinal function to improve the overall health of the body; needed for the production of the coenzymes biotin, niacin, folic acid, and pyridoxine; helps produce antibacterial substances; stimulates enzyme activity.
Bioflavonoids, such as rutin and quercetin	Take 500 mg three times per day.	Fights atherosclerosis; potent antioxidant; reduces capillary fragility.
Essential fatty acid complex	Take 75–100 mg two to three times per day with meals.	Lowers cholesterol level; strengthens the immune system; helps strengthen cell membranes.
Green Barley Extract	Take 1–2 teaspoons in water or juice three times per day.	Improves circulation; fights free radicals.
Niacin *or*	Take 50 mg the first day, then increase dosage by 50 mg daily to 1,000 mg. If adverse effects occur, discontinue use. Take doses higher than 1,000 mg only under the supervision of a physician.	Improves circulation by dilating blood vessels.
Niacin-bound chromium	Take 20 mg per day, or as indicated on the label.	
Vitamin A	Take 15,000 IU per day, or as per your doctor's directions. If pregnant, take no more than 10,000 IU per day.	Necessary for skin and cell membrane structure and strong immune system; potent antioxidant.
Zinc	Take 50 mg per day. Do not take more than 100 mg per day.	Needed for white blood cell immune function; interacts with platelets in blood clotting.

Comments

❏ Dead tissue and pus can delay the repair process. Debriding the wound with enzymes is particularly effective. Several enzymatic debriders contain trypsin, others contain streptokinase/streptodornase (such as Varnase), fibrinolysin/deoxyribonuclease (Fibrolan, Elase), collagenase (Iruxol), or combinations of trypsin, chymotrypsin, bromelain, papain, and rutin (Wobenzym N). A potent debriding enzyme was recently discovered in the digestive system of the Antarctic krill (*Euphausia superba*—a small shrimp). Studies show that krill enzymes have definite debriding potential over non-enzymatic treatment. Further, there are no apparent side effects from the use of these enzymes, and they are effective in debriding many conditions, such as venous leg ulcers.

❏ Drink plenty of water. This will keep the skin well hydrated.

❏ To improve circulation, massage the skin in the affected area frequently.

Slipped Disc

"Slipped disc" is not a medical term, but a lay term to describe a disc protrusion or rupture. The vertebrae of the spine are separated by disc pads made of cartilage. These pads absorb the shock of movement and the weight of the body. Each disc is like a jelly-filled doughnut with an outer ring (annulus fibrosis) and an inner jelly-like segment (nucleus pulposus). When the disc receives undue pressure or stress, or with increased thinning and hardening and decreased space, the inner portion of the jelly doughnut pushes out through the outer fibrous portion, pressing on nearby nerves and arteries. The result is pain, swelling, heat, and loss of function.

A protruded disc can occur anywhere along the spine. This condition usually occurs in the low back (called the lumbar or lumbosacral region). However, this condition can also occur in the neck (cervical) area. Usually, discs protrude or rupture in areas of greatest back movement and are compounded by increased weight exerted on the area. The midback area seldom suffers from a protruded disc, since the ribs help to protect the vertebrae and restrict excessive movement.

Symptoms

The most obvious sign of a slipped disc is a sharp, stabbing pain. If it occurs in the lower back, pain may radiate down the back of usually one leg. There may be "foot drop" (the patient cannot walk on his or her toes on the affected side). Low back pain is increased when coughing, sneezing, or straining during bowel movements. When in an upright position, the patient may bend to one side (usually away from the affected side) and possibly forward (antalgic posture).

When a disc in the cervical spine is affected, symptoms affect the arm (usually only one arm is affected). It usually causes a shooting pain in the shoulder blade and arm that travels down to the fingers. The arm muscles may become weak, and there may be numbness and tingling in the hands and fingers.

X-rays and a CT scan or MRI can confirm a "slipped disc" diagnosis. This is not a condition to treat lightly. If the disc is protruded (like a portion of rubber pushing out in a weak spot of a balloon), it can rupture. The consequences could be serious. Those muscles supplied by the "pinched" nerve root would eventually become flaccid and weak and waste away. A compression of the cervical or lumbar spinal cord may cause spastic partial paralysis of the limbs. Incontinence may develop because of loss of sphincter function. Further, the acute condition can become chronic and may lead to bacterial or viral infection. If you have symptoms of a protruded disc, see a well-trained back specialist immediately for examination, X-rays (or CT scans or an MRI), and treatment. You then may be referred to an orthopedist or neurologist for consultation.

Enzyme Therapy

Systemic enzyme therapy is used to reduce pain, swelling, and inflammation; strengthen the body as a whole; build general resistance; and stimulate the immune system. Enzymes improve circulation, help speed tissue repair, bring nutrients to the damaged area, remove waste products, and improve health.

Digestive enzyme therapy is used to improve the digestion of food, reduce stress on the gastrointestinal mucosa, help maintain normal pH levels, detoxify the body, and promote the growth of healthy intestinal flora, thereby improving the overall health of the body. Digestive enzymes also serve as replacements for the body's pancreatic enzymes, leaving the pancreatic enzymes free to perform other functions in the body, such as improving circulation and decreasing inflammation. Both systemic and digestive enzyme therapy are essential in the care of a disc condition.

Enzyme Absorption System Enhancers (EASE) maximize enzyme activity. They can also improve the absorption and bioavailability of various nutrients and decrease the drain on the body's own digestive enzymes, thus prolonging their lives.

Enzymes' benefits in fighting back pain and intervertebral disc protrusions can be illustrated by the experience of a university soccer coach. The coach was playing goalie on Monday, while scrimmaging. A shot was taken on goal, which he blocked with the inside of his right thigh. The force of the impact extended his thigh to the side. It also caused severe low back pain.

In the hospital emergency room, he received pain medication and muscle relaxers. The diagnosis was a separation of the vertebrae, complicated by a lumbar disc protrusion with nerve root involvement in the right leg. He was told to stay in bed for two weeks. By Thursday, he experienced no reduction in symptoms and literally had to crawl to the bathroom. The back and leg pain worsened. On Friday, in desperation, he sought another physician who told him to begin taking proteolytic enzymes (including pancreatin, bromelain, papain, trypsin, and chymotrypsin, with the bioflavonoid rutin). The following Tuesday morning, he noted that almost all of the pain had disappeared, and he could stand erect. According to the coach, the reduction of pain was phenomenal.

Some nine weeks after the injury, an orthopedic surgeon evaluated the injured coach and reported that there was little doubt this man had suffered a lumbar disc injury (slipped disc) with right leg nerve pain. At the time of the examination, the surgeon was amazed at the patient's "remarkable" recovery and that he was essentially symptom-free. The orthopedist concluded that the patient needed no further medical treatment.

As a researcher and a scientist (with a doctorate in biomechanics), the coach realizes that studies would need to be conducted to form a definitive link between the cause-and-effect relationship of proteolytic enzymes and the reduction in low back pain, but he is convinced that the enzymes relieved his symptoms.

It should be noted that this soccer coach's underlying condition (separation of the vertebra) is not going to be cured with enzymes. However, it seems evident that the enzymes decreased the pain by reducing inflammation, swelling, and pain.

In addition to the standard multivitamin, multimineral, and multiglandular complexes recommended in the introduction to Part Two, the following supplements are recommended for the treatment of slipped discs. The following recommended dosages are for adults. Children between the ages of two and five should take one-quarter the recommended dosage, those between the ages of six and twelve should take one-half the recommended dosage, and those between the ages of thirteen and seventeen should take three-quarters the recommended adult dosage.

ESSENTIAL ENZYMES

Enzyme	Suggested Dosage	Actions
Formula One—See page 38 of Part One for a list of products.	Take 10 tablets three times per day between meals. As symptoms subside (which should occur in approximately one week) decrease dosage to 3–5 tablets three times per day between meals. When symptom-free, begin taking a maintenance dose of 2–3 tablets after each meal.	Decreases inflammation, pain, and swelling; increases rate of healing; stimulates the immune system.
or		
Phlogenzym from Mucos	Take 10 tablets three times per day between meals.	Decreases inflammation, pain, and swelling; increases rate of healing; stimulates the immune system.
or		
Bromelain	Take three to four 230–250 mg capsules per day until symptoms subside. If severe, take 6 to 8 capsules per day.	Fights inflammation, pain, and swelling; speeds recovery after injury or surgery; stimulates the immune system.
or		
A proteolytic enzyme (from microbial, animal, or plant sources)	Take as indicated on the label between meals.	Decreases inflammation, pain, and swelling; increases rate of healing; stimulates the immune system.

SUPPORTIVE ENZYMES

Enzyme	Suggested Dosage	Actions
Formula Three—See page 38 of Part One for a list of products.	Take 50 mcg three times per day between meals.	Potent antioxidant enzymes.
A digestive enzyme formula containing protease, amylase, and lipase enzymes. For a list of digestive enzyme products, see Appendix.	Take as indicated on the label with meals.	Improves the breakdown and absorption of food nutrients; fights toxic build-up.

ENZYME ABSORPTION SYSTEM ENHANCERS (EASE)

Combination	Suggested Dosage	Actions
Bromelain Complex from PhytoPharmica	Take as indicated on the label.	Reduces swelling and inflammation; improves circulation.
or		
Joint-Ease from Nature's Life	Take as indicated on the label.	Decreases inflammation and swelling.
or		
Joint Lube Rx "2200" from Phyto-Therapy	Take as indicated on the label.	Helps to support fibrous tissue and keep healthy joints functioning.
or		
Mobil-Ease from Prevail	Take one capsule between meals two to three times per day.	Has anti-inflammatory activity.
or		
Protease Concentrate from Tyler Encapsulations	Take as indicated on the label.	Breaks up fibrin; fights inflammation.
or		
Zymain from Anabolic Labs	Take as indicated on the label.	Provides nutrition essential for tissue and joint repair and maintenance.

ENZYME HELPERS

Nutrient	Suggested Dosage	Actions
Amino-acid complex	Take as indicated on the	Required for enzyme production;

	label on an empty stomach.	helps the body maintain proper fluid and acid/base balance, essential for overall health; helps maintain collagen and tissue important for healthy discs.
Bioflavonoids	Take 150–500 mg three times per day.	Strengthen capillaries; potent antioxidants.
Calcium	Take 1,000 mg per day.	Plays important roles in collagen and tissue health.
Coenzyme Q$_{10}$	Take 30–60 mg one to two times per day with meals; or follow the advice of your health-care professional.	Potent antioxidant.
Glucosamine sulfate	Take as indicated on the label.	Supports healthy, flexible joint tissue; controls inflammation; helps keep joint membranes fluid.
Green Barley Extract or Kyo-Green	Take as indicated on the label.	Potent free-radical fighter; rich in vitamins, minerals, and enzyme activity.
Magnesium	Take 250–500 mg per day.	Plays important role in tissue strength and calcium absorption.
Mucopoly-saccharides (found in bovine cartilage extract or shark cartilage extract)	Take as indicated on the label.	Support healthy tissues; control inflammation; help keep joint membranes fluid.
Niacin or Niacin-bound chromium	Take 50 mg the first day, then increase dosage daily by 50 mg to 1,000 mg. If adverse effects occur, decrease dosage. Take doses higher than 1,000 mg only under the supervision of a physician. / Take 20 mg per day, or as indicated on the label.	Dilates blood vessels to improve circulation to damaged area; coenzyme for many enzymes.
Vitamin A	Take 10,000–50,000 IU per day. If pregnant, do not take more than 10,000 IU per day.	Stimulates the immune system; fights free radicals.
Vitamin B complex	Take 50 mg three times per day, or as indicated on the label.	B vitamins act as coenzymes to enzymes that support neuro-musculoskeletal function.
Vitamin C (use in combination with enzymes and bio-flavonoids, such as rutin)	Take 8,000 mg per day.	Helps to produce collagen and normal connective tissue; promotes immune system; powerful antioxidant.
Vitamin E	Take 1,200 IU per day from all sources (including the multivitamin complex in the introduction to Part Two).	Potent antioxidant; supplies oxygen to cells; improves level of energy.
Zinc	Take 50 mg per day.	A cofactor for more than 300 enzymes; potent antioxidant.

Comments

❑ If your physician diagnoses a ruptured disc, you will probably be referred to an orthopedist or neurologist for immediate consultation. In many cases, surgery or an injection of the enzymes chymopapain or collagenase is performed.

❑ If surgery is indicated, oral proteolytic enzymes can be used pre-operatively and post-operatively to decrease inflammation, help stimulate the immune system, reduce the possibility of infection, and improve recovery time. Be sure to let your surgeon know that you are taking enzymes or would like to take enzymes.

❑ Initial treatment of a slipped disc is usually conservative. That is, stay off your feet and remain in bed, lying in a fetal position (on your side with your knees drawn up). You may find it more comfortable to lie on your back with the knees bent and your head and lower legs propped up with pillows.

❑ Ice should be used for the first one to three days. Fill two or three small paper cups with water and freeze. When frozen, take one out, tear off the top half inch of paper revealing the ice. Place a paper towel or dish towel over the ice and apply to the painful area. Rub the affected area in a constant circulatory motion with the ice. Do not apply the ice directly to the skin. Apply the ice for five minutes, then keep the ice off the area for five minutes. Repeat this every one to two hours. The purpose is to stimulate blood flow to the area, remove waste from the area, and reduce inflammation.

❑ After three days, alternate ice with heat packs every fifteen minutes until symptoms disappear. "Blue ice" or a frozen vegetable packet works better for this application. Wrap it with a dish towel and apply to the painful area. For heat, use a hot water bottle or heating pad. If a hot water bottle is used, wrap in a towel and apply. If a heating pad is used, put a warm, moist towel on the patient, then place a sheet of wax paper over it, then the heating pad, and finally, a large bath towel. Make sure the wax paper is wide and long enough to cover the entire heating pad with approximately one inch more extending out on every side. The heating pad should never touch the moist towel or the patient's skin. Great care should be exercised to assure this does not happen.

❑ For further information on back care, read *Introduction to Chiropractic Health* (Cichoke. New Canaan, CT: Keats Publishing, Inc., 1996).

❑ *See also* BACKACHE.

Sore Throat

Pharyngitis (sore throat) is inflammation of the throat, usually caused by viruses, such as the common cold, flu, or mononucleosis, but also caused by bacterial infections, such as streptococcal infections. Typically, a sore throat is a minor problem that resolves in a short time. Environmental toxins and irritants (such as dust, tobacco smoke, and polluted air) and screaming or very loud talking can also irritate and traumatize the throat. Even extremely hot or spicy foods and drinks can irritate the membranes of the throat.

A sore throat is usually a symptom of an underlying condition, such as flu, a cold, bronchitis, laryngitis, tonsillitis, mononucleosis, herpes simplex, an abscess, some forms of cancer (such as laryngeal), chronic fatigue syndrome, epiglottitis, diphtheria, gingivitis, or childhood diseases, including chickenpox and measles.

Symptoms

Symptoms of pharyngitis may include pain on swallowing and a hoarse, harsh voice. The throat may also be inflamed and red.

Enzyme Therapy

Systemic enzyme therapy is used to reduce or eliminate the inflammation and speed tissue repair. Enzymes bring nutrients to the damaged area, remove waste products, stimulate the immune system, improve health, strengthen the body as a whole, and build general resistance.

Digestive enzyme therapy is used to improve the digestion of food, reduce stress on the gastrointestinal mucosa, help maintain normal body pH levels, detoxify the body, and promote the growth of healthy intestinal flora, thereby improving the overall health of the body so that it can combat pharyngitis. Digestive enzymes also serve as replacements for the body's pancreatic enzymes, leaving the pancreatic enzymes free to perform other functions in the body, such as boosting immunity and decreasing inflammation.

Enzyme Absorption System Enhancers (EASE) maximize enzyme activity. They can also improve the absorption and bioavailability of various nutrients and decrease the drain on the body's own digestive enzymes, thus prolonging their lives.

In addition to the standard multivitamin, multimineral, and multiglandular complexes recommended in the introduction to Part Two, the following supplements are recommended for the treatment of pharyngitis. The following recommended dosages are for adults. Children between the ages of two and five should take one-quarter the recommended dosage, those between the ages of six and twelve should take one-half the recommended dosage, and those between the ages of thirteen and seventeen should take three-quarters the recommended adult dosage.

ESSENTIAL ENZYMES

Enzyme	Suggested Dosage	Actions
Formula Three—See page 38 of Part One for a list of products.	Take 200 mcg three times per day between meals.	Potent antioxidant enzymes.
Chewable papain	Take as indicated on the label.	Fights inflammation, infection, and free radicals.

AND SELECT ONE OF THE FOLLOWING:

Enzyme	Suggested Dosage	Actions
Bromelain	Take three to four 230–250 mg capsules per day until symptoms subside. If severe, take 6–8 capsules per day.	Fights inflammation, pain, and swelling; speeds recovery; stimulates the immune system.
Serratiopeptidase	Take 5 mg three times per day between meals.	Fights inflammation; improves respiratory conditions; bolsters the immune system; breaks up mucus.
Formula One—See page 38 of Part One for a list of products.	Take 3–4 tablets three times per day between meals until symptoms improve.	Fights inflammation, pain, and swelling; speeds healing.
Mulsal from Mucos	Take 3 tablets three times per day between meals.	Fights inflammation; improves respiratory conditions; bolsters the immune system; breaks up mucus.
A proteolytic enzyme	Take as indicated on the label between meals.	Fights inflammation, pain, and swelling; speeds healing; strengthens capillary walls.

SUPPORTIVE ENZYMES

Enzyme	Suggested Dosage	Actions
A digestive enzyme formula containing protease, amylase, and lipase enzymes. See the list of digestive enzyme products in Appendix.	Take as indicated on the label with meals.	Improves the breakdown and absorption of proteins, carbohydrates, and fats, improving the overall health of the body; fights toxic build-up; frees the body's enzymes to fight inflammation.

ENZYME ABSORPTION SYSTEM ENHANCERS (EASE)

Combination	Suggested Dosage	Actions
Combat from Life Plus	Take as indicated on the label.	Helps support the respiratory system; stimulates immune system function; helps control infections.
or		
Defense Formula from Prevail	Take as indicated on the label.	Enhances the body's natural resistance to viral infections.
or		
Kyolic Formula 102 (with enzymes)	Take 2 capsules or tablets with meals two to three times per day.	Strengthens the immune system; helps alleviate stress; helps fight free radicals and inflammation; improves digestion and absorption of nutrients.
or		

Mucous Dissolver Liquezyme from Enzyme Process Laboratories *or*	Take as indicated on the label.	Dissolves mucus.
Sinus Ease from Nature's Life *or*	Take as indicated on the label.	Stimulates immune function; reduces inflammation.
Zym-Eeze Lysozyme Lozenges from Future Foods	Take as indicated on the label.	Natural, enzymatic antimicrobial agent; fights sore throat, colds, and strep throat.

ENZYME HELPERS

Nutrient	Suggested Dosage	Actions
Adrenal glandular	Take as indicated on the label.	Helps fight stress and adrenal exhaustion; fights inflammation.
Bioflavonoids, including quercetin and rutin	Take 100–300 mg twice per day.	Fights inflammation; reduces capillary fragility; has antihistamine activity.
Kyolic Aged Garlic Liquid Extract with Vitamins B_1 and B_{12}	Take as indicated on the label.	Stimulates antioxidant activity and immune system function.
Lung glandular	Take as indicated on the label.	Improves lung function.
Thymus glandular	Take as indicated on the label.	Stimulates thymus gland activity; stimulates the immune system.
Vitamin A	Take 10,000–40,000 IU per day for no more than one month. Take doses higher than 40,000 IU only under the supervision of a physician. If pregnant, take no more than 10,000 IU per day.	Necessary for mucous linings, skin and cell membrane structure, and strong immune system; potent antioxidant.
Vitamin C *or*	Take 1–2 grams every hour until symptoms improve. Or mix a heaping teaspoon of powdered or crystalline vitamin C with water or orange or lemon juice (if lemon juice is used, sweeten with honey if desired); gargle a mouthful and swallow; repeat every hour.	Fights infections, diseases, allergies, and common cold; potent antioxidant.
Winterized C from Naturally Vitamin Supplements	Take as indicated on the label.	Fights colds and flu.
Vitamin E	Take 1,200 IU per day from all sources (including the multivitamin complex in the introduction to Part Two).	Free-radical scavenger.
Zinc or zinc gluconate lozenges	Take 50 mg per day. (Take no more than 100 mg per day from all sources.)	Needed for white blood cell immune function; involved with more than 300 enzymes.

Comments

❑ Slippery elm lozenges can soothe sore throats. Echinacea, ginger root, ginkgo biloba, and white willow bark can help alleviate the inflammation.

❑ Hot broth or chicken or vegetable soup is helpful for cleaning out the body and sweating out toxins. Before making the soup, be sure to remove the skin from the chicken (chicken skins contain many toxins). Sip the broth or soup as frequently as possible.

❑ Fruits and vegetables (particularly green vegetables) should comprise over 60 percent of your diet. If your body cannot tolerate raw fruits and vegetables, increase your intake of digestive enzymes, or sauté or steam your produce.

❑ Milk and milk products are mucus-formers and should be strictly limited. These foods cause increased congestion.

❑ Gargle three times every day with four ounces of fresh pineapple juice blended with four ounces of lemon or grapefruit juice, or use the Enzymatic Gargle as outlined in Part Three.

❑ If you have a flu or cold, have your doctor check for the presence of allergies, thyroid dysfunction, and immune system deficiencies. If any such disorder is present, see that disorder in Part Two of this book for its enzyme therapy program.

❑ NESS makes a helpful enzyme combination containing protease, amylase, lipase, disaccharidase, and cellulase enzymes along with bioflavonoids, wheat germ, acerola cherries, and rose hips. Follow directions on the label. Another NESS product contains vitamin C, bioflavonoids, acerola cherries, and rose hips, high in vitamin C and antioxidant action to fight sore throats, and amylase enzymes to improve the absorption.

❑ Follow the Enzyme Kidney Flush and the Enzyme Toxin Flush programs as outlined in Part Three.

❑ *See also* COLDS; EAR INFECTIONS; LARYNGITIS; TONSILLITIS.

Sports Injuries

Physical exercise helps improve cardiorespiratory fitness, tones muscles, burns calories, and can help us lose weight. But with exercise, sooner or later comes a pulled muscle, torn ligament, or other injury. According to the American Academy of Orthopaedic Surgeons, over one-half million Americans were injured last year playing basketball, and more than 10 million sports injuries occur in the United States each year. (See the inset "Sports Injury Incidence" on page 380.) Most sports injuries are due to poor training techniques, such as working through pain, inadequate stretching and warming up before exercise, and not allowing enough recovery time between workouts; structural abnormalities, such as unequal leg lengths; and connective tissue weakness. The most frequent sports injuries are bone fractures or soft tissue damage of the muscles, ligaments, or tendons. These injuries can involve inflammation, tears, ruptures, strains, sprains, edema, bruises, and hematomas.

For athletes, trainers, coaches, and physicians treating a sports injury presents a challenge because not only does the athlete want complete recovery, but he or she "demands" that it be accomplished faster than is normally possible. This is understandable due to today's pressures. Sports have limited seasons, and athletes have limited careers.

Doctors and trainers usually prescribe ice or heat, rest, and any number of anti-inflammatory products and pain relievers to treat strains, sprains, and other sports injuries. But enzymes can bring faster recovery (sometimes twice as fast) without the serious side effects of drugs and may actually help prevent injuries from occurring. (See the inset "Therapeutic Programs for Different Phases of Recovery" on page 383.) Enzymes can also facilitate energy increase (see "Enzymes Increase Energy" on page 381), weight loss (*see* OBESITY in Part Two), and muscle mass build up (see "Muscle Build-up" on Page 382).

Enzyme Therapy

Numerous studies have repeatedly shown enzyme mixtures to be effective in trauma and injuries from karate, ice hockey, soccer, football, boxing, judo, running, and other sports. Systemic enzyme therapy is used in sports injuries to decrease pain, swelling, and inflammation. Enzymes improve circulation, help speed tissue repair, bring nutrients to the damaged area, remove waste products, stimulate the immune system, improve health, strengthen the body as a whole, and build general resistance. Digestive enzyme therapy is used to improve the digestion of food, reduce stress on the gastrointestinal mucosa, help maintain normal pH levels, detoxify the body, promote the growth of healthy intestinal flora, and make a healthier athlete.

Enzyme Absorption System Enhancers (EASE) maximize enzyme activity. They can also improve the absorption and bioavailability of various nutrients and decrease the drain on the body's own digestive enzymes, thus prolonging their lives.

Topical applications of enzymes through a treatment technique called iontophoresis is used in the treatment of traumatic injuries, including sprains and strains, bursitis, tendonitis, etc. Iontophoresis uses an electric current to introduce therapeutic particles to the skin (a form of electrotherapy).

In addition to the proteolytic enzymes' own anti-inflammatory abilities, they stimulate the body's own natural enzymatic processes without causing the immune system to be suppressed. They increase tissue permeability, as well as the rate of degradation of inflammatory and toxic products, and increase the rate of microthrombi breakdown, therefore reducing swelling. Consequently, they improve the supply of nutrients and oxygen to the tissues and the removal of the end products of normal metabolism. Therefore, the duration of the inflammatory process is reduced, pain stops, and the healing process progresses more quickly with reduced scar tissue formation.

A study I conducted at Portland State University (with Leo Marty, the director of sports medicine there) documented the improved rate of recovery from football injuries

---

Sports Injury Incidence

When playing a sport, there is often the risk of injury. In some sports, there is more of a risk for injury than in others. Following are estimates of the numbers of injuries that occurred for individual recreational sports in the United States in 1996. These estimates were compiled by the U.S. Consumer Product Safety Commission, based upon injuries treated in hospital emergency rooms that year.

SPORT	INJURIES	SPORT	INJURIES
Baseball, softball	351,708	Horseback riding	60,071
Basketball	652,350	Martial Arts	26,757
Bicycling	565,169	Skating	214,192
Boxing	6,763	Soccer	156,312
Diving	10,790	Swimming	25,900
Football	363,217	Tennis	25,523
Golf	36,239	Trampolining	83,312
Gymnastics	31,563	Volleyball	74,522
Hockey	77,537	Weightlifting	53,862
		Wrestling	43,814

Enzymes Increase Energy

Many sports supplements now contain enzymes, vitamins, minerals, and/or herbs. For increased and sustained energy, some sports supplements are based not upon starch, but on the glucose polymers collectively termed maltodextrins and oligosaccharides. These soluble polymers are more desirable than the simple starches because they yield their energy more slowly, resulting in a smoother rise in blood glucose.

There are enzyme combinations that convert insoluble dietary starches into soluble, energy-rich, branched-chain maltodextrins and oligosaccharides in the stomach.

Studies at the University of Rhode Island show that taking these enzymes before eating insoluble dietary starches mediates the release of glucose from starches. This sustained release action may be examined in future URI studies with insulin response and blood glucose elevation curves.

This type of enzyme combination increases the amount of time it takes for athletes to "hit the wall," thus helping the athletes to work harder and longer. Using this type of enzyme combination, athletes could consistently maintain higher energy levels throughout competition.

---

through the use of an oral enzyme mixture including bromelain, papain, pancreatin, trypsin, chymotrypsin, amylase, lipase, and the bioflavonoid rutin. This double-blind study was conducted on sixty-four PSU football players. A double-blind study is one in which neither the participants nor the researchers know who is taking which product. Half the players took enzymes, while the remainder took a placebo. Results indicated that when injured, those football players using enzymes before the injury recovered twice as fast. Injuries that responded faster with enzyme therapy not only included soft tissue injuries, such as sprains, strains, and hematomas, but also fractures.

Studies in both Germany and the United States reveal that hematomas and swelling disappeared in less than seven days when the athletes were treated prophylactically with enzymes. On the contrary, those athletes without enzymes took almost sixteen days to recover. Restrictions of movement as a result of pain and injury disappeared after five days in the enzyme group, but lasted more than twelve days in the control group.

Proteolytic enzymes are used individually or in enzyme mixtures. For example, Carica papaya (papain) obtained marked results in 87 percent of college and professional athletes and moderate results in 10 percent of the cases, particularly in the reduction of edema, associated pain, and time ordinarily required for full mobility of the injured area.

In another study conducted over two consecutive seasons researchers found a 31-percent reduction in time lost from playing of a sport in those athletes taking oral proteolytic enzymes (trypsin and chymotrypsin). Bromelain is used effectively to decrease pain and swelling with boxers, soccer players, and other athletes.

In a double-blind randomized study, Dr. M. Baumuller (a famous German sports medicine physician) examined the effectiveness and tolerance of an enzyme mixture containing papain, bromelain, pancreatin, trypsin, chymotrypsin, amylase, lipase, and the bioflavonoid rutin in various types of soft-

tissue sports injuries. Results of the study show that pain and blood clots due to soft tissue injury are decreased significantly and more rapidly with oral enzymes. The advantage of enzyme therapy is reflected in the reduced need for painkillers and the earlier return to activity in the athletes taking the enzymes. Finally, the resulting time loss from injury was substantially shortened.

For treatment plans to help heal specific sports injuries, turn to the necessary conditions in this book, including BACKACHE; BONE FRACTURES; BRUISES AND HEMATOMAS; BURNS; BURSITIS; MUSCLE CRAMPING; SCIATICA; SPRAINS AND STRAINS; SUBLUXATIONS; SURGERY; SYNOVITIS; TENDONITIS; and WHIPLASH.

In addition to the standard multivitamin, multimineral, and multiglandular complexes recommended in the introduction to Part Two, the following supplements are recommended for the prevention of sports injuries. The following recommended dosages are for adults. Children between the ages of two and five should take one-quarter the recommended dosage, those between the ages of six and twelve should take one-half the recommended dosage, and those between the ages of thirteen and seventeen should take three-quarters the recommended adult dosage.

ESSENTIAL ENZYMES

Enzyme	Suggested Dosage	Actions
Throughout the season as preventative maintenance use:		
Formula One—See page 38 of Part One for a list of products.	Take 5–6 tablets twice per day between meals.	Speeds recovery time from any injury or inflammation.
or		
Phlogenzym from Mucos	Take 5 tablets twice per day between meals.	
or		
Bromelain	Take five 180–200 mg tablets twice per day between meals.	

During the off-season, use Formula One—See page 38 of Part One for a list of products. | Take 2–3 tablets three times per day between meals. | Maintains health.

SUPPORTIVE ENZYMES

Enzyme	Suggested Dosage	Actions
A digestive enzyme formula containing protease, amylase, and lipase enzymes. For a list of digestive enzyme products, see Appendix.	Take as indicated on the label.	Digests proteins, carbohydrates, and fats.

ENZYME ABSORPTION SYSTEM ENHANCERS (EASE)

Combination	Suggested Dosage	Actions
Connect-All from Nature's Plus or	Take as indicated on the label.	Supports connective tissue.
CTR Support from PhysioLogics or	Take as indicated on the label.	Nourishes the body during tissue recovery and healing from injury; helps maintain healthy connective tissue; diminishes damage caused by swelling and inflammation.

Inflamzyme from PhytoPharmica or	Take as indicated on the label.	Promotes tissue repair.
Joint-Ease from Nature's Life or	Take as indicated on the label.	Decreases joint inflammation and swelling.
Magnesium Plus from Life Plus or	Take one or two tablets one to three times per day.	Important for calcium utilization and many enzyme reactions.
Mobil-Ease from Prevail	Take one capsule between meals two to three times per day.	Has anti-inflammatory activity.

ENZYME HELPERS

Nutrient	Suggested Dosage	Actions
Adrenal glandular	Take as indicated on the label.	Helps fight stress and adrenal exhaustion; fights inflammation.
Amino acid complex	Take as indicated on the label on an empty stomach.	Required for enzyme, protein, and hormone production; helps the body maintain proper fluid and acid/base balance, necessary for health; maintains health of tissues.
Bioflavonoids	Take 100–500 mg twice per day.	Fights inflammation; reduces capillary fragility.
Calcium	Take 1,000 mg per day.	Plays important roles in collagen and tissue health.

Muscle Build-up

The body uses protein's component amino acids to manufacture enzymes, hormones, and other compounds and to build and repair its tissues, including muscles. In fact, a significant amount of muscle tissue is composed of the branched-chain amino acids (BCAAs) leucine, isoleucine, and valine. Unfortunately, these and other amino acids are broken down during exercise. It is for this reason that athletes supplement their diets with protein powders or amino acids.

Certain enzyme combinations might increase protein/amino acid absorption, thus allowing improved amino acid utilization and protein availability and facilitating muscle build-up. Combinations of very specialized proteases can liberate free-form amino acids from protein-containing food. Research studies performed on a simulated human gastrointestinal model at the University of Rhode Island have demonstrated that a combination of proteases has the power to generate almost 250 times its weight in free amino acids.

These specialized proteases are a series of substrate-specific bio-active peptides (SSBAPs). These specialized peptides have the ability to search the length of a protein molecule until they locate the branched-chain amino acids valine, leucine, and isoleucine. They then latch onto the protein chain, breaking free these essential amino acids from their protein bond, making them more available to the muscles.

This protease combination is highly resistant to the destructive effects of the stomach's enzymes and acid. Unaffected by bile, it is even more effective at the harsh alkaline pH levels typical of the intestine. In the human body, protein conversion is both energy-intensive and tedious. Certain enzyme mixtures can possibly assist in accelerating amino acid production and in so doing, spare valuable body energy for such uses as muscle synthesis and repair.

Although many common digestive enzyme-based supplements can help to reduce some long-chain proteins into smaller parts (proteoses and peptones), certain enzyme mixtures combine for fast release of free-form amino acids from any food sources at most pH levels encountered. This gives an interesting alternative to free amino acids for athletes interested in "kick-starting" muscle building stimulation.

Coenzyme Q$_{10}$	Take 30–60 mg one to two times per day with meals; or follow the advice of your health-care professional.	Potent antioxidant.
Glucosamine sulfate	Take as indicated on the label.	Supports healthy, flexible joints; controls inflammation.
Green Barley Extract or Kyo-Green	Take as indicated on the label.	Free-radical fighter; rich in vitamins, minerals, and enzyme activity.
Kyolic Aged Garlic Liquid Extract with Vitamins B$_1$ and B$_{12}$	Take as indicated on the label.	Stimulates antioxidant activity and immune system function.
Magnesium	Take 250–500 mg per day.	Plays important role in calcium absorption and promoting tissue strength.
Mucopolysaccharides (found in bovine cartilage extract and shark cartilage extract)	Take as indicated on the label.	Support healthy joints and tissues; control inflammation.
Niacin _or_	Take 50 mg the first day, then increase dosage by 50 mg daily to 1,000 mg. Decrease dosage if adverse effects occur. Take doses higher than 1,000 mg only under the supervision of a physician.	Dilates blood vessels to improve circulation to the damaged area; coenzyme for many enzymes.
Niacin-bound chromium	Take 20 mg per day, or as indicated on the label.	
Thymus glandular	Take as indicated on the label.	Stimulates the thymus gland and the immune system.

Vitamin A	Take 10,000–50,000 IU per day. Take doses higher than 50,000 IU only under the supervision of a physician. If pregnant, do not take more than 10,000 IU per day.	Stimulates the immune system; fights free radicals.
Vitamin B complex	Take 50 mg three times per day, or as indicated on the label.	B vitamins act as coenzymes to enzymes that support neuro-musculoskeletal function.
Vitamin C (use in combination with enzymes and bio-flavonoids, such as rutin)	Take 8,000 mg per day.	Helps to produce collagen and normal connective tissue; promotes immune system; powerful antioxidant.
Vitamin D	Take 3,000–6,000 IU per day.	Stimulates the immune system; helps fight infection.
Vitamin E	Take 1,200 IU per day from all sources (including the multivitamin complex in the introduction to Part Two).	Potent antioxidant; supplies oxygen to cells; improves level of energy.
Zinc	Take 50 mg per day.	A cofactor for more than 300 enzymes; potent antioxidant.

Comments

❑ Never stop taking enzyme supplements. Regardless of your age, if you play sports or are physically active, there is tissue breakdown ongoing in your body. Enzymes will help maintain repair.

Therapeutic Programs for Different Phases of Recovery

There are several phases in recovery from sports injuries. Each phase has different therapy requirements. Understanding these phases and applying the appropriate therapy can speed recovery from sports injury.

Phase I (Acute: 0 to 30 hours after injury)

• Oral enzyme therapy.
• Additional supplementation with enzyme helpers (vitamins, minerals, phytochemicals, and herbs).
• Therapy with ice or cold packs.
• Compression of the area (wrapping for support).

Phase II (Post-acute: 20 to 60 hours after injury)

• Oral enzyme therapy.
• Additional supplementation with enzyme helpers (vitamins, minerals, phytochemicals, and herbs).
• Therapy with ice or cold packs, alternating with warm moist heat.

• Compression of the area (wrapping for support).
• Lymphatic drainage (pressure point therapy to help drain lymphatic toxin build-up).

Phase III (Rehabilitation: 40+ hours after injury)

• Oral enzyme therapy.
• Additional supplementation with enzyme helpers (vitamins, minerals, phytochemicals, and herbs).
• Physiotherapy and physical therapy (such as electrical stimulation, massage, and hydrotherapy).
• Dressing (if necessary).

Phase IV (Protective phase)

• Oral enzyme therapy.
• Additional supplementation with enzyme helpers (vitamins, minerals, phytochemicals, and herbs).
• Rehabilitative exercises.
• Muscular stabilization and protective dressing (wraps and supports, if needed).

❑ In conjunction with enzymes, you should be on a program of vitamins, minerals, designer foods, plant nutrients, and herbs.

❑ Eat regularly (three to five times per day).

❑ Follow a detoxification plan. See Part Three for directions for the Enzyme Kidney Flush and the Enzyme Toxin Flush programs.

❑ Depending on your level of recovery after injury, gradually increase stretching and weight-bearing exercises to increase muscle strength and flexibility, circulation, and detoxification.

❑ Ice should be used as soon as possible after the injury and for the next one to three days. Fill two or three small paper cups with water and freeze. When frozen, take one out, tear off the top half inch of paper revealing the ice. Place a paper towel or dish towel over the ice and apply to the painful area. Rub the affected area in a constant circulatory motion with the ice. Do not apply the ice directly to the skin. Apply the ice for five minutes, then keep the ice off the area for five minutes. Repeat this every one to two hours. The purpose is to stimulate blood flow to the area, remove waste from the area, and reduce inflammation.

❑ After three days, alternate ice with heat packs every fifteen minutes until pain and other symptoms subside. "Blue ice" or a frozen vegetable packet works better for this application. Wrap it with a dish towel and apply to the painful area. For heat, use a hot water bottle or heating pad. If a hot water bottle is used, wrap in a towel and apply. If a heating pad is used, put a warm, moist towel on the patient, then place a sheet of wax paper over it, then the heating pad, and finally, a large bath towel. Make sure the wax paper is wide and long enough to cover the entire heating pad with approximately one inch more extending out on every side. The heating pad should never touch the moist towel or the patient's skin. Great care should be exercised to assure this does not happen.

❑ You may want to seek the services of a health-care professional who will develop a treatment program. Your treatment program might include chiropractic adjustments, manipulation, massage, physiotherapy, including electrical simulation, ice and heat applications, exercises, and postural training (including proper lifting, bending, standing, walking, driving, sitting, and sleeping techniques).

❑ Cortisone and other steroids are frequently administered to traumatically injured individuals. The effectiveness of this therapy should be weighed against the potential for a number of unwanted reactions. Cortisone can affect protein and carbohydrate metabolism; cause circulatory, central nervous system, gastrointestinal, and skeletal changes; and suppress gonadal activity. When very large doses of cortisone are used in therapy, the resistance of the patient to certain infections may be reduced. The usual signs of the infection (fever, malaise, etc.) may be suppressed by these drugs so that the presence and progress of the disease is unrecognized. In contrast to cortisone and other prescribed anti-inflammatory medicines, there are only a few contraindications for enzyme therapy.

Sprains and Strains

Muscle strains and ligamentous sprains are the most frequent soft tissue injuries, whether they occur on the job, at home, or during sports or recreational activities. A *sprain* is a stretched or torn ligament (the tissue that connects bones to one another) that results from a severe wrenching of a joint or from placing too much weight on a joint. This causes overstretching, twisting, or tearing of the supporting ligaments. Most sprains are caused by trauma, such as a fall or accident. They most commonly occur in the ankle, but may also occur in the knees, hips, back, shoulder, neck, fingers, or wrist.

A *strain* is a stretched or torn muscle or tendon (the tissue that connects muscle to bone). A strain can be relatively minor (a pulled muscle) or more serious (with tearing or rupturing of the muscle). It is often difficult to tell the difference between a simple pulled muscle or a more serious torn muscle. When muscles are injured, they release a number of substances, including enzymes, into the bloodstream. By measuring any increased levels of these enzymes, your physician can tell if you have a pulled or torn muscle.

Symptoms

Symptoms of sprains and strains include pain, swelling, and bruising (red or purple discoloration), with possible heat. Movement in the affected area is often limited because of the pain and/or swelling. Mild sprains and strains may heal without complications. But more severe injuries without proper nutritional supplementation and care can become chronic, develop scar tissue, limit motion, and ultimately cause problems in surrounding tissues, nerves, vessels, and organs.

Enzyme Therapy

Systemic enzyme therapy is used to relieve pain and reduce swelling and inflammation. Enzyme therapy stimulates the immune system, improves circulation, and helps speed tissue repair by bringing nutrients to the damaged area and removing waste products.

Digestive enzyme therapy is used to improve the breakdown and absorption of food nutrients, reduce stress on the gastrointestinal mucosa, help maintain normal body pH levels, detoxify the body, and promote the growth of healthy intestinal flora, thereby improving the overall health of the body. Research shows that sprains and strains can heal twice as fast with enzyme supplementation.

Enzyme Absorption System Enhancers (EASE) maximize enzyme activity. They can also improve the absorption and bioavailability of various nutrients and decrease the drain on the body's own digestive enzymes, thus prolonging their lives.

An example of the effectiveness of enzymes in treating sprains and strains can be seen in the case of a high school football player who injured both knees during a game. After suffering a violent sprain to both knees, the young man had to be carried off the field. Examination and X-rays revealed severe knee sprains (but luckily, no tears or fractures). However, according to the team physician, this was a season-ending injury.

To reduce the inflammation, pain, and swelling, the patient was given over thirty enzyme tablets containing pancreatin, trypsin, chymotrypsin, bromelain, and papain, plus the bioflavonoid rutin per day between meals on the day of the accident and for the following two days. On the fourth day, the football player returned to the team physician for a status examination. An amazed physician found the player was nearly symptom-free and able to move his knees with no pain. He was, therefore, allowed to return to play. The young man took ten tablets each day between meals for the remaining part of the season and had no further pain or swelling and helped lead his team to the high school state playoffs. Subsequently, he played both university and professional football.

Enzymes have a long history of fast and effective treatment of acute soft tissue injuries with high school, university, professional, and Olympic athletes and coaches in this country and throughout the world. Many sprains and strains heal within a few days, where a week or two would be normal for other treatment programs. Even fractures heal faster than normal.

In addition to the standard multivitamin, multimineral, and multiglandular complexes recommended in the introduction to Part Two, the following supplements are recommended for the treatment of sprains and strains. The following recommended dosages are for adults. Children between the ages of two and five should take one-quarter the recommended dosage, those between the ages of six and twelve should take one-half the recommended dosage, and those between the ages of thirteen and seventeen should take three-quarters the recommended adult dosage.

ESSENTIAL ENZYMES

Enzyme	Suggested Dosage	Actions
Formula One—See page 38 of Part One for a list of products.	Take a single dose of 10 tablets. An hour later, take 2–4 tablets hourly for a period of one to two days or as long as necessary. As you improve, take 3–5 tablets two times	Decreases inflammation, pain, and swelling; increases rate of healing; stimulates the immune system.
	per day as long as symptoms persist. When symptoms disappear, begin taking a maintenance dose of 2–3 tablets between meals.	
Phlogenzym from Mucos *or*	Take 10 tablets three times per day between meals.	Decreases inflammation, pain, and swelling; increases rate of healing; stimulates the immune system.
Bromelain *or*	Take three to four 230–250 mg capsules per day until symptoms subside. If severe, take 6–8 capsules per day.	Fights inflammation, pain, and swelling; speeds recovery after injury or surgery; stimulates the immune system.
A proteolytic enzyme	Take as indicated on the label between meals.	Decreases inflammation, pain, and swelling; increases rate of healing; stimulates the immune system.

SUPPORTIVE ENZYMES

Enzyme	Suggested Dosage	Actions
Formula Three—See page 38 of Part One for a list of products.	Take as indicated on the labels between meals.	Potent antioxidant enzymes.
A digestive enzyme formula containing protease, amylase, and lipase enzymes. See the list of digestive enzyme products in Appendix.	Take as indicated on the label with meals.	Improves the breakdown and absorption of food nutrients, improving the overall health of the body; fights toxic build-up.

ENZYME ABSORPTION SYSTEM ENHANCERS (EASE)

Combination	Suggested Dosage	Actions
Connect-All from Nature's Plus *or*	Take as indicated on the label.	Supports connective tissue.
CTR Support from PhysioLogics *or*	Take as indicated on the label.	Nourishes the body during tissue recovery and healing from injury; helps maintain healthy connective tissue; diminishes damage caused by swelling and inflammation.
Inflamzyme from PhytoPharmica *or*	Take as indicated on the label.	Promotes tissue repair.
Joint-Ease from Nature's Life *or*	Take as indicated on the label.	Decreases inflammation and swelling.
Magnesium Plus from Life Plus *or*	Take one or two tablets one to three times per day.	Important for calcium utilization and many enzyme reactions.
Mobil-Ease from Prevail	Take one capsule between meals two to three times per day.	Has anti-inflammatory activity.

ENZYME HELPERS

Nutrient	Suggested Dosage	Actions
Adrenal glandular	Take 300 mg per day.	Fights inflammation and adrenal exhaustion.
Bioflavonoids (such as rutin and quercetin)	Take 500 mg three times per day between meals.	Strengthen capillaries; potent antioxidants.
Calcium	Take 1,000 mg per day.	Plays important role in muscle contraction.
Coenzyme Q_{10}	Take 30–60 mg one to two times per day with meals; or follow the advice of your health-care professional.	Potent antioxidant.
Magnesium	Take 500 mg per day.	Plays important role in muscle contraction and calcium absorption.
Vitamin C	Take 8,000 mg per day.	Helps to produce collagen and normal connective tissue.
Vitamin E	Take 1,200 IU per day from all sources (including the multivitamin complex in the introduction to Part Two).	Supplies oxygen to the cells and is a potent antioxidant.

Comments

❏ Several herbs can help fight inflammation, including chlorella, gotu kola, hawthorn, garlic, ginkgo biloba, white willow bark, and green barley extract.

❏ Ice should be used for the first one to three days. Fill two or three small paper cups with water and freeze. When frozen, take one out, tear off the top half inch of paper revealing the ice. Place a paper towel or dish towel over the ice and apply to the painful area. Rub the affected area in a constant circulatory motion with the ice. Do not apply the ice directly to the skin. Apply the ice for five minutes, then keep the ice off the area for five minutes. Repeat this every one to two hours. The purpose is to stimulate blood flow to the area, remove waste from the area, and reduce inflammation.

❏ After three days, alternate ice with heat packs every fifteen minutes until symptoms disappear. "Blue ice" or a frozen vegetable packet works better for this application. Wrap it with a dish towel and apply to the painful area. For heat, use a hot water bottle or heating pad. If a hot water bottle is used, wrap in a towel and apply. If a heating pad is used, put a warm, moist towel on the patient, then place a sheet of wax paper over it, then the heating pad, and finally, a large bath towel. Make sure the wax paper is wide and long enough to cover the entire heating pad with approximately one inch more extending out on every side. The heating pad should never touch the moist towel or the patient's skin. Great care should be exercised to assure this does not happen.

❏ Enzymes combined with bioflavonoids (such as rutin or quercetin), vitamins (especially vitamin C), minerals (such as zinc, manganese, and magnesium), and other supplements are incredibly effective in the treatment of sprains and strains, and research studies prove it.

❏ To further boost your body's healing capacity, increase your consumption of enzyme-rich foods (including fruits, vegetables, and fermented foods). See Chapter 4 of Part One for guidelines for maximizing your enzyme intake.

❏ Bilberry extract, raspberries, cranberries, red wine, grapes, hawthorn, black currants, and red cabbage contain anthocyanidins. These phytochemicals fight free radicals and inflammation and maintain blood flow in the capillaries.

❏ Injuries cause stress to the body. As a result, there is an increase in free-radical formation and a breakdown of metabolism at both the organ and cellular levels. Toxins form and can interfere with the healing process. Eliminate these toxins by following one or more of the detoxification plans in Part Three of this book.

❏ The injured area may need to be bandaged for support. As the injury heals, mild stretching exercises (for a week or two) might help lessen stiffness and improve recovery. If you have a severe sprain or strain, you may need to see a musculoskeletal specialist (chiropractor, medical doctor, naturopath, or osteopath).

❏ To reduce the risk of suffering a sprain or strain, don't overwork your muscles, and be sure to warm up adequately before exercising. Cold muscles, tendons, and ligaments are much more prone to injury.

❏ *See also* BACKACHE; BRUISES AND HEMATOMAS; SPORTS INJURIES; TENDONITIS.

Sprue, Nontropical

See CELIAC DISEASE.

Staphylococcal Infections

Staphylococcal bacteria are normally carried on the skin of about 20 percent and just inside the nostrils of about 30 percent of healthy adults. They are normally harmless; however, injury, such as a break in the skin, may allow the bacteria to enter the body where they can cause infection.

Some people are especially susceptible to staphylococcal infections, including newborn babies; nursing mothers; and individuals suffering from flu, cancer, burns, chronic skin disorders, diabetes, or chronic broncho pulmonary disorders (such as emphysema and cystic fibrosis). Also at risk are those receiving radiation treatments, chemotherapy, and immunosuppressive or adrenal steroids, because these treatments reduce the body's immune reaction.

Staph infections are often spread in hospitals. A staph infection can be especially troublesome when it develops after surgery. In fact, postoperative infections are commonly caused by staphylococci. These infections may not appear until several weeks after an operation or may appear within just a few days. If the patient receives antibiotics at the time of surgery, these infections are likely to have a delayed onset. Unfortunately, staph bacteria have become smarter over the years and have developed a resistance to nearly all antibiotics. In fact, most strains of staphylococci are resistant to ampicillin, penicillin G, streptomycin, the tetracyclines, and carbenicillin. Recent news from Japan details the emergence of a strain of staph bacteria that is resistant to all forms of antibiotics. But enzymes can offer hope.

Symptoms

Staphylococcal infections can be mild or severe. The symptoms of a staphylococcus infection vary depending on the location of the infection, but commonly include furuncles, abscesses, and carbuncles. It can also cause bacteremia, pneumonia, osteomyelitis, endocarditis, and gastroenteritis. Staphylococcal bacteremia is an infection of the blood. A chronic high fever and possibly shock may result. If the bacteria infect the blood, the infection can travel to other parts of the body. Staphylococcal pneumonia causes a high fever and severe symptoms of pneumonia, such as shortness of breath. Osteomyelitis (bone infection) can cause fever, chills, and pain in the bones. It is most common in children and the elderly. Endocarditis is infection of the heart's lining and its valves. This can cause heart failure and possibly death. Infection of the gastrointestinal tract may cause fever, bloating, diarrhea, and other gastrointestinal complications. Laboratory tests, including culture specimens can help identify the particular staph strain.

Enzyme Therapy

Systemic enzyme therapy is typically used to facilitate the action of antibiotics. Papain was studied alone as well as in combination with a number of antibiotics (such as streptomycin, benzylpenicillin, tetracycline, chloramphenicol, novobiocin, and erythromycin) in the treatment of mice with septicemia (bacterial blood poisoning) caused by antibiotic-resistant staphylococci bacteria. Researchers found that enzyme therapy increased the effects of the antibiotics an average of 50 percent. When papain was used alone, it prevented death in 20 percent of the animals. In Russia, antibi-

otic combinations with papain are recommended to treat infections caused by antibiotic-resistant staphylococci.

Enzymes are also used as supportive care to reduce inflammation, stimulate the immune system, improve circulation, help speed tissue repair, bring nutrients to the damaged area, remove waste products, and improve health.

Digestive enzyme therapy is used to improve the digestion of food, reduce stress on the gastrointestinal mucosa, help maintain normal pH levels, detoxify, and promote the growth of healthy intestinal flora. Digestive enzymes also serve as replacements for the body's pancreatic enzymes, leaving the pancreatic enzymes free to perform other functions in the body, such as boosting immunity.

Enzyme Absorption System Enhancers (EASE) maximize enzyme activity. They can also improve the absorption and bioavailability of various nutrients and decrease the drain on the body's own digestive enzymes, thus prolonging their lives.

In addition to the standard multivitamin, multimineral, and multiglandular complexes recommended in the introduction to Part Two, the following supplements are recommended for the treatment of staphylococcal infections. The following recommended dosages are for adults. Children between the ages of two and five should take one-quarter the recommended dosage, those between the ages of six and twelve should take one-half the recommended dosage, and those between the ages of thirteen and seventeen should take three-quarters the recommended adult dosage.

ESSENTIAL ENZYMES

Enzyme	Suggested Dosage	Actions
Papain or	Take 500 mg three times per day for six months between meals. Then decrease dosage to 300 mg three times per day between meals.	Effective with antibiotics; bolsters immunity; breaks up circulating immune complexes; fights free radicals and inflammation.
WobeMugos E from Mucos or	Take 5 tablets three times per day for six months between meals. Then take as needed between meals.	Stimulates immune function; breaks up circulating immune complexes; fights free radicals and inflammation.
Formula One—See page 38 of Part One for a list of products. or	Take 10 tablets three times per day between meals for six months. Then decrease dosage to 5 tablets three times per day between meals.	Bolsters the immune system; breaks up antigen-antibody complexes; fights free-radical formation; stimulates immune function at the cellular level; fights inflammation.
Mulsal from Mucos or	Take 8 tablets three times per day between meals.	Stimulates immune function; breaks up circulating immune complexes; fights free radicals and inflammation.
Bromelain or	Take 230–250 mg six times per day between meals for 6–10 days or until symptoms improve.	Stimulates immune function; breaks up antigen-antibody complexes; fights free-radical formation and inflammation.
A proteolytic enzyme	Take 10 tablets three times per day for six	Stimulates immune function; breaks up circulating immune

months. Then decrease dosage to 5 tablets three times per day.

complexes; fights free radicals and inflammation.

SUPPORTIVE ENZYMES

Enzyme	Suggested Dosage	Actions
Formula Three—See page 38 of Part One for a list of products.	Take as indicated on the labels.	Potent antioxidant enzymes.
A digestive enzyme product containing protease, amylase, and lipase. See the list of digestive enzyme products in Appendix.	Take as indicated on the label with meals.	Improves the breakdown and absorption of food nutrients to improve the overall health of the body; fights toxic build-up.

ENZYME ABSORPTION SYSTEM ENHANCERS (EASE)

Combination	Suggested Dosage	Actions
Ecology Pak from Life Plus	Take 2 or 3 tablets once or twice per day, preferably in the morning and no later than midday.	Assists with natural detoxification; stimulates immune function.
or		
Hepazyme from Enzymatic Therapy	Take as indicated on the label.	Stimulates immune function.
or		
Inflamzyme from PhytoPharmica	Take as indicated on the label.	Immune system activator; reduces inflammation; fights free radicals.
or		
Kyolic Formula 102 (with enzymes)	Take 2 capsules or tablets with meals two to three times per day.	Strengthens the immune system; helps alleviate stress; helps fight free radicals; helps fight inflammation; improves digestion and absorption of nutrients.
or		
Vita-C-1000 from Life Plus	Take as indicated on the label.	Boosts immune function; fights infection.

ENZYME HELPERS

Nutrient	Suggested Dosage	Actions
Acidophilus and other probiotics	Take as indicated on the label.	Stimulates enzyme activity; improves gastrointestinal function.
Adrenal glandular	Take 100–400 mg per day.	Helps fight stress and adrenal exhaustion; fights inflammation.
Coenzyme Q_{10}	Take 30–60 mg one to two times per day with meals.	Important to immune system function; has antioxidant activity.
Kidney glandular	Take as indicated on the label.	Improves kidney function for detoxification.
Kyolic Aged Garlic Liquid Extract with Vitamins B_1 and B_{12}	Take 10 drops in juice or water three times per day.	Stimulates immune system function.
Thymus glandular	Take 100 mg per day, or as indicated on	Assists the thymus gland in immune resistance; improves

the label.

immune response; fights infections.

Vitamin A	Take up to 50,000 IU for no more than one month. Take doses higher than 50,000 IU only under the supervision of a physician. If pregnant, do not take more than 10,000 IU.	Needed for strong immune system and healthy mucous membranes; potent antioxidant.
Vitamin B complex	Take 100–150 mg three times per day.	B vitamins act as coenzymes.
Vitamin C	Take 10,000–15,000 mg per day, or 1–2 grams every hour until symptoms improve.	Fights infections, diseases, and allergies; improves healing; stimulates the adrenal glands; promotes immune system; powerful antioxidant.
Zinc	Take 15–50 mg per day. Do not take more than 100 mg per day from all sources.	Needed by over 300 enzymes; necessary for white blood cell immune function; improves healing.

Comments

❑ Topical applications of tea tree oil (*Melaleuca alternifolia*) have been found to be effective against antibiotic-resistant *Staphylococcus aureus*.

❑ Topical applications of garlic may help, since garlic contains a sulfur compound called ajoene that has been shown to inhibit the growth of *S. aureus*.

❑ If you suffer from a staph infection, it is especially important to frequently and thoroughly wash your hands. This will limit the spread of the infection.

❑ Infected individuals and their bedding should be isolated from other individuals.

Steatorrhea

Steatorrhea is the existence of excessive fat in the feces. It is absolute evidence of a malabsorption problem—that your digestive system is not adequately breaking down the fat in your diet. Steatorrhea can occur if you are deficient in certain enzymes (particularly the lipases), and it is a symptom of several conditions, including celiac disease, cystic fibrosis, pancreatic disease, and tropical sprue.

Symptoms

Steatorrhea is more a symptom than a disease. This condition is marked by soft, pale, malodorous, bulky, and fatty stools. They may be difficult to flush away or may stick to the side of the toilet bowl. This is because they are high in undigested fat.

Enzyme Therapy

Digestive enzyme therapy is used to replace any missing enzymes and improve the digestion and absorption of food nutrients. Enzymes help reduce stress on the gastrointestinal mucosa, encourage the growth of beneficial microorganisms in the intestine, detoxify the body, and help maintain normal pH levels.

Systemic enzyme therapy is used to reduce any inflammation, stimulate the immune system, improve circulation, transport nutrients throughout the body, remove waste products, improve health, and strengthen the body as a whole.

Enzyme Absorption System Enhancers (EASE) maximize enzyme activity. They can also improve the absorption and bioavailability of various nutrients and decrease the drain on the body's own digestive enzymes, thus prolonging their lives.

Because each person is different and because of the varying causes of steatorrhea, enzyme treatment may vary widely between individuals, you may need more or less of the enzymes mentioned in this treatment program. If symptoms do not improve within one day, increase enzyme intake gradually until a change is noted. Successfully treating steatorrhea may require large amounts of enzymes. This can sometimes interfere with therapy because most people don't like swallowing large numbers of pills. However, enzyme therapy is necessary to decrease the fatty stools and normalize the body.

In addition to the standard multivitamin, multimineral, and multiglandular complexes recommended in the introduction to Part Two, the following supplements are recommended for the treatment of steatorrhea. The following recommended dosages are for adults. Children between the ages of two and five should take one-quarter the recommended dosage, those between the ages of six and twelve should take one-half the recommended dosage, and those between the ages of thirteen and seventeen should take three-quarters the recommended adult dosage.

ESSENTIAL ENZYMES

Enzyme	Suggested Dosage	Actions
Pancrelipase or pancreatin or	Take 1–3 capsules or tablets before each snack or meal, or as directed by your physician.	Contains proteases, amylases, and lipases, so is able to digest proteins, carbohydrates, and fat.
A digestive enzyme formula containing protease, amylase, and lipase enzymes. See the list of digestive enzyme products in Appendix. or	Take as indicated on the label.	Improves digestion by breaking down proteins, carbohydrates, and fats.
Formula One—See page 38 of Part One for a list of products.	Take 10 tablets three times per day with meals.	Contains proteases, amylases, and lipases, so is able to digest proteins, carbohydrates, and fat.

SUPPORTIVE ENZYMES

Enzyme	Suggested Dosage	Actions
Formula Three—See page 38 of Part One for a list of products.	Take as indicated on the label.	Potent antioxidant enzymes.

ENZYME ABSORPTION SYSTEM ENHANCERS (EASE)

Combination	Suggested Dosage	Actions
Biodias A Granules from Amano Pharmaceutical Co., Ltd. or	Adults should take one packet in 8 ounces of water between or after meals three times per day. Children ages 11 to 14 should take two-thirds of a packet; children between the ages of 8 and 10 should take half of the packet; and children between the ages of 5 and 7 should take one-third of a packet.	Helps improve digestion of nutrients; promotes regeneration of mucous membranes; decreases stress on the stomach and intestines; reduces gastric pain.
Kyolic Formula 102 (with enzymes) or	Take 2 capsules or tablets with meals two to three times per day.	Improves fat, protein, and carbohydrate digestion and absorption; strengthens the immune system; helps alleviate stress; helps fight free radicals; helps fight inflammation.
Lipozyme from Enzymatic Therapy or	Take as indicated on the label.	Helps digest fats.
Supergest from Enzyme Process Laboratories	Take as indicated on the label.	Improves digestion; helps break down fat.

ENZYME HELPERS

Nutrient	Suggested Dosage	Actions
Acidophilus or other probiotics, such as bifidobacterium; or Acidophilasé from Wakunaga, which contains amylolytic, lipolytic, and proteolytic enzymes.	Take as indicated on the label.	Improves gastrointestinal function; needed for production of the coenzymes biotin, niacin, folic acid, and pyridoxine; helps produce antibacterial substances; stimulates enzyme activity.
Adrenal glandular	Take 100–400 mg per day.	Fights adrenal exhaustion; restores normal carbohydrate balance.
Amino-acid complex such as AminoLogic from PhysioLogics	Take as indicated on the label.	Amino acids serve as building blocks for the body's enzymes.
Fiber	Take as indicated on the label.	Provides bulk; needed for proper elimination.
Fructooligo-saccharides	Take as indicated on the label.	Increases peristaltic action of intestines; improves liver function.

Green food product, such as Green Kamut, Kyo-Green, or Green Barley Extract	Take as indicated on the label.	Green foods are rich in enzymes, vitamins, minerals, chlorophyll, and amino acids.
Lecithin	Take 1,000 mg three times per day with meals.	Emulsifies fat.
Ox bile extract	Take 120 mg per day.	Aids digestion.
Pancreas glandular	Take 300 mg per day.	Aids digestion; fights steatorrhea; assists the pancreatic enzymes trypsin and chymotrypsin.
Vitamin A	Take 5,000–10,000 IU day. If pregnant, take no more than 10,000 IU per day.	Needed by digestive glands to produce enzymes; required for healthy mucous linings and cell membrane structure.
Vitamin B complex	Take 50 mg three times per day.	B vitamins act as coenzymes.
Vitamin C	Take 1,000–2,000 mg per day.	Stimulates the adrenal glands to produce hormones (which also act as coenzymes); aids in detoxification; is a potent antioxidant; improves the function of many enzymes; assists in converting cholesterol to bile salts.
Whey	Take 1 heaping table-spoon three times per day with meals.	Improves digestion.
Zinc	Take 10–50 mg per day.	Needed by over 300 enzymes (including pancreatic enzymes and superoxide dismutase).

Comments

❑ Several herbs can help improve digestion, including alfalfa concentrate (also rich in enzymes), chlorella, chlorophyll, ginger root extract, and green foods (such as Green Barley Extract and Kyo-Green).

❑ As mentioned, steatorrhea is a symptom. Identifying and treating the underlying cause of the steatorrhea should be the primary goal.

❑ See also CYSTIC FIBROSIS.

Sties

See BOILS.

Stress

Stress is any physically, mentally, or emotionally upsetting condition caused by external factors, usually characterized by muscular tension, irritability, increased heart rate and blood pressure, and possibly depression. While most usually think of emotional stress when they hear the word stress, stress can also be biochemical (pollution, toxins, pesticides, or poor nutrition) or physical (a car accident, sports injury, surgery, or other trauma). What creates emotional or psychological stress for one individual may not for another—it may even be enjoyable. Stress causes a physiological reaction in the body. The body produces increased amounts of adrenaline in response to stress, which increases the heart rate and blood pressure and tenses the muscles.

Symptoms

As mentioned above, stress causes increased heart rate and blood pressure, muscle tension, irritability, depression, loss of appetite, and gastrointestinal complications. The muscle tension can lead to such symptoms as head- and backache and pain in the neck and other areas of the body. Stress can also produce illness, although the exact mechanism by which this happens is unclear. It can cause such physical symptoms as reduced immunity and trigger or worsen such illnesses as diabetes mellitus, leukemia, multiple sclerosis, and systemic lupus erythematosus. Stress can also promote free-radical activity in the body.

Stress, regardless of the cause, can result in vitamin and mineral losses, due to metabolic changes and loss of appetite, and enzyme depletion. It is for this reason that good nutrition is particularly important. Research shows that healthy, well-nourished individuals handle stress better than those who are poorly nourished.

Enzyme Therapy

Digestive enzyme therapy is used to improve the digestion and absorption of food nutrients. Without sufficient digestive enzymes, your body will not fully utilize the nutrients in your diet. Enzymes can reduce stress on the gastrointestinal mucosa, help maintain normal pH levels, detoxify the body, and promote the growth of healthy intestinal flora. Antioxidant enzymes are important for fighting free-radical activity within the body.

Systemic enzyme therapy is used to stimulate the immune system, break down circulating immune complexes, improve circulation, transport nutrients throughout the body, remove waste products, improve health, and balance the body as a whole.

Enzyme Absorption System Enhancers (EASE) maximize enzyme activity. They can also improve the absorption and bioavailability of various nutrients and decrease the drain on the body's own digestive enzymes, thus prolonging their lives.

In addition to the standard multivitamin, multimineral, and multiglandular complexes recommended in the intro-

Personal Stress Scale (PSS)

Certain levels of stress can be stimulating to the body. However, too much stress can overwork and ultimately fatigue the body and the body's enzymes. The following fifty questions can help you determine some of the stressors in your life. Answer "Yes" or "No" to the following questions. For each "yes" answer, give yourself one point.

QUESTION	YES	NO
Do you smoke?	☐	☐
Do you drink an average of one or more soft drinks per day?	☐	☐
Do you frequently use aspirin or other over-the-counter pain medications?	☐	☐
Do you have a chronic health condition, such as asthma, colitis, diabetes, arthritis, etc.?	☐	☐
Do you feel weak or shaky between meals?	☐	☐
Do you drink more than three cups of coffee per day?	☐	☐
Do you skip breakfast?	☐	☐
Do you frequently suffer from diarrhea and/or constipation?	☐	☐
Do you drink alcohol?	☐	☐
Do you take drugs (prescription, over-the-counter, or recreational)?	☐	☐
Do you ever crave sweets, liquor, or chocolate?	☐	☐
Are you irritable before breakfast or your first cup of coffee?	☐	☐
Do you have poor eating habits?	☐	☐
Do you eat a high-fat diet?	☐	☐
Do you have difficulty sleeping?	☐	☐
Do you ever feel lightheaded?	☐	☐
Do you find it hard to concentrate at times, especially in the afternoon?	☐	☐
Do you frequently have headaches upon arising?	☐	☐
Do you have high blood pressure?	☐	☐
Do you have weight problems?	☐	☐
Are you a "couch potato"?	☐	☐
Do you sometimes have muscle cramps?	☐	☐
Did you experience pain today, such as backache, chest pain, or headache?	☐	☐
Do you feel you need to be more assertive?	☐	☐
Have you had a recent death in your family?	☐	☐
Do you have conflicts with your relatives, children, or spouse (or significant other)?	☐	☐
Do you have a short temper, always feeling on edge?	☐	☐
Do you feel you need more recreation?	☐	☐
Does noise bother you?	☐	☐
Is your work environment noisy?	☐	☐

Do you awaken tired?	☐	☐
Do you need more time for yourself?	☐	☐
Are work deadlines and time pressures getting to you?	☐	☐
Do you feel depressed and down?	☐	☐
Do you feel you just can't say no and find yourself doing things you do not want to do?	☐	☐
Are you having problems with your employer?	☐	☐
Were you nervous or on edge today?	☐	☐
Do you have sexual difficulties?	☐	☐
Do you have trouble making decisions?	☐	☐
Do you have conflicts with your neighbors?	☐	☐
Do financial difficulties bother you?	☐	☐
Do you have feelings of guilt?	☐	☐
Are you bored and feel your day was pointless?	☐	☐
Do you feel dissatisfied with your job?	☐	☐
Do you feel that you have no job security?	☐	☐
Did you have a tension-filled day where things got out of hand at times?	☐	☐
Did you worry today and are you concerned about it?	☐	☐
Are you concerned today about a pending or ongoing divorce or separation?	☐	☐
Are you generally unhappy and feel that most areas of your life are out of balance?	☐	☐
Do you feel you need more self-discipline?	☐	☐
TOTAL	____	____

Identify the biochemical, physical, and emotional factors causing you stress and work on those areas for which you have answered yes. Try to eliminate them from your life. Review this questionnaire every month and compare your previous answers. This will help indicate the areas in which you have improved, and those areas that need more work. Your goal should be to develop personal techniques to lower your stress level and thus lower your score over time.

duction to Part Two, the following supplements are recommended for the treatment of stress symptoms. The following recommended dosages are for adults. Children between the ages of two and five should take one-quarter the recommended dosage, those between the ages of six and twelve should take one-half the recommended dosage, and those between the ages of thirteen and seventeen should take three-quarters the recommended adult dosage.

ESSENTIAL ENZYMES

Enzyme	Suggested Dosage	Actions
A digestive enzyme formula containing protease, amylase, and lipase enzymes.	Take as indicated on the label.	Improves digestion by breaking down proteins, carbohydrates, and fats.

For the list of digestive enzyme products, see Appendix.

Formula Three—See page 38 of Part One for a list of products.	Take as indicated on the labels.	Potent antioxidant enzymes.

SUPPORTIVE ENZYMES

Enzyme	Suggested Dosage	Actions
Bromelain	Take three to four 230–250 mg capsules per day until symptoms subside. If under extreme stress, take 6–8 capsules per day.	Fights inflammation, pain, and swelling; speeds recovery after injury or surgery; stimulates the immune system.

Papain	Take 300 mg three times per day between meals.	Bolsters the immune system; breaks up antigen-antibody complexes; fights free-radical formation and inflammation.
WobeMugos E from Mucos	Take 3–4 tablets three times per day between meals.	Bolsters the immune system; breaks up antigen-antibody complexes; fights free-radical formation; fights inflammation.

ENZYME ABSORPTION SYSTEM ENHANCERS (EASE)

Combination	Suggested Dosage	Actions
Tranquilon from Life Plus	Take one tablet two or three times per day.	Stimulates immune function; naturally relaxes; helps reduce anxiety and nervousness.

ENZYME HELPERS

Nutrient	Suggested Dosage	Actions
Acidophilus or other probiotics, such as bifidobacterium; or Acidophilasé from Wakunaga, which contains amylolytic, lipolytic, and proteolytic enzymes.	Take as indicated on the label.	Improves gastrointestinal function; needed for production of the coenzymes biotin, niacin, folic acid, and pyridoxine; helps produce antibacterial substances; stimulates enzyme activity.
Adrenal glandular	Take 100–400 mg per day.	Fights adrenal exhaustion; restores normal carbohydrate balance.
Amino-acid complex such as AminoLogic from PhysioLogics	Take as indicated on the label.	Amino acids serve as building blocks for the body's enzymes.
Betaine hydrocholoric acid	Take 150 mg three times per day with meals.	Provides hydrochloric acid to improve digestion. If gastric upset occurs, discontinue use.
Calcium	Take 1,000 mg per day.	Essential for the activity of certain enzymes involved in energy production and several enzymes involved in the healing process.
Fiber	Take as indicated on the label.	Provides bulk; needed for proper elimination.
Fructooligo-saccharides	Take as indicated on the label.	Increases peristaltic action of intestines; improves liver function for detoxification.
Lecithin	Take 1,000 mg three times per day with meals.	Emulsifies fat.
Magnesium	Take 500 mg per day.	Plays important roles in protein building, nerve function, energy production, and calcium absorption; necessary for synthesis of fats, proteins, and nucleic acid.
Ox bile extract	Take 120 mg per day.	Aids digestion.
Vitamin B complex	Take 50 mg three times per day.	B vitamins act as coenzymes.
Whey	Take 1 heaping tablespoon three times per day with meals.	Improves digestion.

Zinc	Take 50 mg per day. Do not exceed 100 mg per day from all sources.	Stress depletes your body of zinc, which is needed by over 300 enzymes. Many zinc enzymes work to metabolize carbohydrates, alcohol, and essential fatty acids; synthesize proteins; and dispose of free radicals.

Comments

❏ See Part Three for dietary recommendations.

❏ It is also important to treat specific manifestations of stress, such as anger and high blood pressure.

❏ Eliminate (as much as possible) any physical, emotional, and nutritional stressors. (See "Personal Stress Scale" on page 391 to help you identify the stressors in your life.) Along with enzyme therapy, stress management techniques can help bring balance to your life.

❏ Exercise daily for at least thirty to forty minutes per day. Brisk walking, bicycling, and swimming are all great. You should try to work up a sweat. Physical activity can relax the body and stimulate positive enzymatic activity. Do simple things like walking your dog, parking the car a block away from the store, or climbing the stairs of your office building. See Part Three for further information on exercise.

❏ *See also* HIGH BLOOD PRESSURE; STROKE.

Stroke

A stroke is death of brain tissue due to lack of blood flow and oxygen to the brain. It can be either hemorrhagic or ischemic. In a *hemorrhagic stroke*, the blood flow is cut off because a blood vessel to the brain bursts. This may be caused by high blood pressure or the rupture of an aneurysm in an artery to the brain. In an *ischemic stroke*, the lack of blood flow is due to a blockage of a blood vessel to the brain. The blockage can occur as a result of an atheroma (fatty deposits on the walls of blood vessels); blood clots that may form elsewhere in the body and travel to the brain; narrowing of the vessels due to inflammation, infection, or the use of certain drugs; or a severe and prolonged drop in blood pressure.

In this country, a stroke occurs every sixty seconds. Strokes killed 149,740 people in the United States in 1993, which accounted for about one of every fifteen deaths. It is the third leading cause of death (ranking behind diseases of the heart and cancer), according to 1996 figures from the National Center for Health Statistics. Strokes and other cerebrovascular diseases are the most common causes of neurologic disability in Western countries.

Symptoms

Strokes usually occur suddenly. However, they may also develop over time, progressing over hours or days. Symptoms vary, depending upon the area of the brain that is affected. Symptoms may include blindness in one eye; double vision; dizziness; weakness, loss of sensation, or paralysis in an arm or leg on one side of the body; slurred speech; partial loss of hearing; loss of bladder control; difficulty thinking; unusual movements; stupor; brain swelling; and possibly coma. A stroke can also cause depression or an inability to control emotions.

Enzyme Therapy

Systemic enzyme therapy is used to help decrease the cause of the stroke by breaking up cholesterol and fibrin in the blood vessels, improving circulation and reducing inflammation. Enzymes can stimulate the immune system, help speed tissue repair, bring nutrients to any damaged area, remove waste products, and strengthen the body as a whole.

Digestive enzyme therapy is used to improve the digestion of food, reduce stress on the gastrointestinal mucosa, help maintain normal pH levels, detoxify the body, and promote the growth of healthy intestinal flora. This takes stress off the body's own enzymes.

Enzyme Absorption System Enhancers (EASE) maximize enzyme activity. They can also improve the absorption and bioavailability of various nutrients and decrease the drain on the body's own digestive enzymes, thus prolonging their lives.

In addition to the standard multivitamin, multimineral, and multiglandular complexes recommended in the introduction to Part Two, the following supplements are recommended for the treatment of strokes. The following recommended dosages are for adults. Children between the ages of two and five should take one-quarter the recommended dosage, those between the ages of six and twelve should take one-half the recommended dosage, and those between the ages of thirteen and seventeen should take three-quarters the recommended adult dosage.

ESSENTIAL ENZYMES

Enzyme	Suggested Dosage	Actions
Formula Three—See page 38 of Part One for a list of products.	Take 200 mcg three times per day between meals.	Potent antioxidant enzymes.

AND SELECT ONE OF THE FOLLOWING:

Enzyme	Suggested Dosage	Actions
Phlogenzym from Mucos	Take 10 tablets three times per day between meals.	Dissolves blood clots; fights inflammation, pain, and swelling; speeds healing.
Formula One—See page 38 of Part	Take 8–10 tablets three times per day between	Dissolves blood clots; fights inflammation, and

One for a list of products.	meals for six months or until symptoms improve. Thereafter, take 6 tablets three times per day between meals.	swelling; speeds healing.
Bromelain	Take four to six 230–250 mg capsules per day.	Fights inflammation, pain, and swelling; speeds recovery after injury or surgery; stimulates the immune system.
A proteolytic enzyme	Take as indicated on the label between meals.	Dissolves blood clots; fights inflammation, pain, and swelling; speeds healing.

SUPPORTIVE ENZYMES

Enzyme	Suggested Dosage	Actions
A digestive enzyme formula containing protease, amylase, and lipase enzymes. See the list of digestive enzyme products in Appendix.	Take as indicated on the label with meals.	Improves the breakdown and absorption of proteins, carbohydrates, and fats; fights toxic build-up.

ENZYME ABSORPTION SYSTEM ENHANCERS (EASE)

Combination	Suggested Dosage	Actions
Bromelain Complex from PhytoPharmica or	Take as indicated on the label.	Reduces swelling; improves circulation; helps fight circulating immune complexes.
Cardio-Chelex from Life Plus or	Take as indicated on the label.	Supports healthy circulation; improves energy.
Kyolic Formula 102 (with enzymes) or	Take 2 capsules or tablets with meals two to three times per day.	Helps improve circulation; strengthens the immune system; helps alleviate stress; helps fight free radicals and inflammation; improves digestion and the absorption of nutrients.
Protease Concentrate from Tyler Encapsulations	Take as indicated on the label.	Breaks up fibrin; fights inflammation.

ENZYME HELPERS

Nutrient	Suggested Dosage	Actions
Adrenal glandular	Take as indicated on the label.	Helps fight stress and adrenal exhaustion.
Bioflavonoids, including pycnogenol, rutin, or quercetin	Take 500 mg three times per day between meals.	Strengthen capillaries and blood vessel walls; potent antioxidants.
Brain glandular	Take as indicated on the label.	Helps improve brain function.
Calcium	Take 1,000 mg per day.	Lowers high blood pressure.
Coenzyme Q_{10}	Take 75 mg one to two times per day with meals.	Potent antioxidant; decreases blood pressure and total serum cholesterol.

Heart glandular	Take 140 mg or as indicated on the label.	Improves heart function, improving circulation.
Magnesium	Take 500 mg per day.	Lowers blood pressure and fights cardiovascular disease; plays important roles in calcium absorption.
Niacin *or*	Take 50 mg the first day, then increase dosage daily by 50 mg to 1,000 mg. If adverse effects occur, decrease dosage. Take doses higher than 1,000 mg only under the supervision of a physician.	Improves circulation by dilating blood vessels.
Niacin-bound chromium	Take up to 20 mg per day, or as indicated on the label.	
Vitamin B complex	Take 50 mg three times per day, or as indicated on the label.	B vitamins act as coenzymes and support neuromuscu-loskeletal function.
Vitamin C	Take 8,000–10,000 mg per day.	Essential to vascular function; potent antioxidant; improves the function of many enzymes; assists in converting cholesterol to bile salt.
Vitamin E	Take 1,200 IU per day from all sources (including the multivitamin complex in the introduction to Part Two).	Potent antioxidant; helps prevent cardiovascular disease; stimulates circulation; strengthens capillary walls; prevents blood clots; maintains cell membranes.
Zinc	Take 50 mg per day. Do not take more than 100 mg per day from all sources.	Needed for white blood cell immune function; works with enzymes to reduce inflammation and from build-up to normalize blood flow.

Comments

❏ Increase your intake of garlic, onions, shallots, chives, leeks, and scallions. These foods contain allium compounds, which are potent free-radical fighters that help lower cholesterol and fight circulatory disorders.

❏ Oats, wheat germ, olives, nuts, nut and seed oils, organ meats, and eggs contain alpha-tocopherol (vitamin E), known to help prevent blood clots.

❏ Raspberries, cranberries, grapes, red wine, black currants, and red cabbage contain anthocyanidins, which reduce blood vessel plaque formation and maintain blood flow in small vessels.

❏ Decrease your intake of cholesterol, as well as foods high in saturated fats.

❏ Restrict sodium consumption.

❏ If you are overweight, follow a weight-reduction program.

❏ For further information on dietary modifications, detoxification, exercises, positive mental attitude, meditation, and general suggestions, please turn to Part Three.

❏ See the list of Resouce Groups in the Appendix for an organization that can provide you with more information about strokes.

❏ *See also* ARTERIOSCLEROSIS; CARDIOVASCULAR DISORDERS.

Subluxation

A subluxation occurs when a spinal vertebra is rotated beyond its normal range of movement and cannot return without help. That is, the vertebra is out of alignment. A subluxation causes a decrease in the intervertebral opening (called the foramen). This results in swelling, which presses or pinches the nerve and blood vessels that pass through the foramen at the level of the subluxation, and pain.

As we progress from crawling on all fours to an upright, two-legged posture, our spine adjusts accordingly. At birth, the spine is shaped like a big backward "C." However, as we first begin to raise our head, the neck starts to curve forward (this position is called convex forward). Then, as we shimmy up the side of the crib and take a two-legged stance, the vertebrae of the lower back also begin to curve forward. By the time we're completely upright, the spine's movable vertebrae have developed three natural curves. The neck is curved forward; the mid-back (thoracic spine) is curved toward the back; and the lower back (lumbar spine) is curved forward. Because of gravity's pull, our struggle to sustain an upright posture causes many biomechanical difficulties.

Possible causes and/or complications of a subluxation could be poor posture, being overweight, lack of exercise, trauma, tension and emotional problems, stress, mechanical weaknesses (such as a scoliosis or a missing portion of a vertebra), osteoarthritis, back sprains and strains, protruded or ruptured discs, degenerative changes, or spondylolysthesis (a separation in a portion of a vertebra, where one vertebra shifts forward over the vertebra below it). The result can be nerve and muscle atrophy or organ impairment with possible accompanying disease.

Symptoms

The symptoms of a subluxation include back pain and decreased (or lack of) motion of one or more vertebrae. There may be swelling, heat, and inflammation of the tissues surrounding the vertebrae. The pain can be a sharp, stabbing pain felt when you move, or a dull, aching-type pain without movement. It may be localized in a specific area of the spine and/or referred to neighboring tissues. There may also be muscle spasms and stiffness.

Enzyme Therapy

Systemic enzyme therapy is used to reduce inflammation, swelling, and pain. This will allow the neighboring muscles to become more relaxed and the vertebrae to move more easily. Therefore, an adjustment, manipulation, or mild pressure can help realign the vertebrae and relieve irritation of the nerve. The result is decreased pain and swelling, plus a free flow of nerve energy to the affected organs and tissues. Enzymes stimulate the immune system, improve circulation, help speed tissue repair, bring nutrients to the damaged area, remove waste products, and improve health.

Digestive enzyme therapy is used to improve the digestion of food, reduce stress on the gastrointestinal mucosa, help maintain normal pH levels, detoxify the body, and promote the growth of healthy intestinal flora. As a result of pain and swelling from some subluxations, you might be constipated. Therefore, detoxification and return to normal bowel habits are essential and can be achieved through the use of digestive enzymes.

Enzyme Absorption System Enhancers (EASE) maximize enzyme activity. They can also improve the absorption and bioavailability of various nutrients and decrease the drain on the body's own digestive enzymes, thus prolonging their lives.

In addition to the standard multivitamin, multimineral, and multiglandular complexes recommended in the introduction to Part Two, the following supplements are recommended for the treatment of subluxations. The following recommended dosages are for adults. Children between the ages of two and five should take one-quarter the recommended dosage, those between the ages of six and twelve should take one-half the recommended dosage, and those between the ages of thirteen and seventeen should take three-quarters the recommended adult dosage.

ESSENTIAL ENZYMES

Enzyme	Suggested Dosage	Actions
Formula One—See page 38 of Part One for a list of products. or	Take a single dose of 10 tablets. An hour later, take 2–4 tablets hourly for one to two days. As symptoms improve, take 3–5 tablets two times per day for as long as symptoms persist. When symptoms disappear, begin taking a maintenance dose of 2–3 tablets between meals.	Decreases inflammation, pain, and swelling; increases rate of healing; stimulates the immune system.
Bromelain or	Take three to four 230–250 mg capsules per day until symptoms subside. If severe, take 6–8 capsules per day.	Fights inflammation, pain, and swelling; speeds recovery after injury or surgery; stimulates the immune system.

Enzyme	Suggested Dosage	Actions
Phlogenzym from Mucos or	Take 10 tablets three times per day between meals.	Decreases inflammation, pain, and swelling; increases rate of healing; stimulates the immune system.
Serratiopeptidase or	Take 10 mg three times per day between meals.	Decreases inflammation, pain, and swelling; increases rate of healing; stimulates the immune system.
A proteolytic enzyme	Take as indicated on the label between meals.	Decreases inflammation, pain, and swelling; increases rate of healing; stimulates the immune system.

SUPPORTIVE ENZYMES

Enzyme	Suggested Dosage	Actions
Formula Three—See page 38 of Part One for list of products.	Take 50 mcg of each three times per day between meals.	Potent antioxidant enzymes.
A digestive enzyme product containing protease, amylase, and lipase enzymes. For the list of digestive enzyme products, see Appendix.	Take as indicated on the label.	Improves the breakdown and absorption of food nutrients.

ENZYME ABSORPTION SYSTEM ENHANCERS (EASE)

Combination	Suggested Dosage	Actions
Connect-All from Nature's Plus or	Take as indicated on the label.	Supports connective tissue.
CTR Support from PhysioLogics or	Take as indicated on the label.	Nourishes the body during tissue recovery and healing from injury; helps maintain healthy connective tissue; diminishes damage caused by swelling and inflammation.
Inflamzyme from PhytoPharmica or	Take as indicated on the label.	Promotes tissue repair.
Joint Lube Rx "2200" from Phyto-Therapy or	Take as indicated on the label.	Helps to support cartilage and keep healthy joints functioning.
Joint-Ease from Nature's Life or	Take as indicated on the label.	Decreases inflammation and swelling.
Magnesium Plus from Life Plus or	Take one or two tablets one to three times per day.	Important for calcium utilization and many enzyme reactions.
Mobil-Ease from Prevail or	Take as indicated on the label.	Reduces pain, swelling, and inflammation.
Zymain from Anabolic Labs	Take as indicated on the label.	Provides nutrition essential for tissue and joint repair and maintenance.

ENZYME HELPERS

Nutrient	Suggested Dosage	Actions
Acidophilus and other probiotics	Take as indicated on the label.	Helps the body stay in balance; improves digestive tract function to improve the overall health of the body.
Adrenal glandular	Take 100–400 mg per day.	Fights inflammation and adrenal exhaustion.
Aged garlic extract	Take two 300 mg capsules three times per day.	Detoxifies and provides energy.
Amino-acid complex	Take as indicated on the label on an empty stomach.	Amino acids are the building blocks of proteins and enzymes and build muscle and connective tissue.
Bioflavonoids	Take 150–500 mg three times per day.	Strengthen capillaries; potent antioxidants.
Branched-chain amino acids (isoleucine, leucine, and valine)	Take as indicated on the label between meals.	Often found together in a supplement, branched-chain amino acids aid in healing of muscle, bone, and skin and act as fuel for muscles.
Calcium	Take 1,000 mg per day.	Plays important roles in muscle contraction and nerve transmission.
Chondroitin sulfate	Take 250 mg two to three times per day with meals; or follow advice of your health-care professional.	Supports connective tissue; promotes cell hydration.
Coenzyme Q_{10}	Take 30–60 mg one to two times per day with meals; or follow the advice of your health-care professional.	Potent antioxidant.
Essential fatty acids (found in flaxseed oil)	Take 75–100 mg of flaxseed oil two to three times per day with meals; or follow the advice of your health-care professional.	Help strengthen cell membranes.
Glucosamine hydrochloride *and/or* Glucosamine sulfate	Take 500 mg two to four times per day. Take 500 mg one to three times per day.	Stimulate cartilage repair.
Magnesium	Take 500 mg per day.	Plays important role in muscle contraction, nerve function, and calcium absorption.
Vitamin B complex	Take 50 mg three times per day, or as indicated on the label.	B vitamins act as coenzymes and support neuromusculoskeletal function.
Vitamin C	Take 5,000–10,000 mg per day.	Helps to produce collagen and normal connective tissue; promotes immune system; powerful antioxidant.
Vitamin E	Take 1,200 IU per day from all sources (including the multivitamin complex in the introduction to Part Two).	Potent antioxidant; supplies oxygen to cells; improves level of energy.
Zinc	Take 15–50 mg per day. Do not take more than 100 mg per day.	Needed by over 300 enzymes.

Comments

❑ Several herbs can help fight inflammation, including echinacea, ginger root, ginkgo biloba, licorice, and white willow bark.

❑ Exercise thirty to forty minutes every day. Depending on the location of your subluxation, your exercise can include walking, swimming, or riding a bicycle. Exercise increases antioxidant activity, stimulates the immune system and enzyme activity, and increases circulation.

❑ A positive mental attitude and meditation are excellent tools in freeing the mind, reducing free-radical activity, and coping with pain.

❑ Your condition may require the care of a well-trained back specialist (such as a chiropractor, medical doctor, naturopath, or osteopath). After conducting a history, physical examination, and an X-ray screening, your doctor will develop a treatment program.

❑ *See also* BACKACHE; SLIPPED DISC.

Surgical-Related Problems

Every day, tens of thousands of people undergo surgery in hospitals or in doctors' offices. Unfortunately, surgery often results in swelling, pain, hematomas, and inflammation, which makes recovery more difficult and can cause serious problems. Postoperative swelling interferes with the essential supply of nutrients to the tissues. This increases the possibility of wound infection and delays tissue regeneration. Further, there is an increased risk of blood clots because of impaired circulation. Some areas of the body have a particularly rich blood supply, including the head and mouth. Surgery in these areas can sometimes result in particularly massive postoperative swelling and delayed recovery.

Enzyme Therapy

Systemic enzyme therapy is used to reduce pain, swelling, and inflammation and return the patient to activity as quickly as possible. Enzymes improve circulation, decrease the possibility of blood clots, help speed tissue repair, stimulate the immune system, bring nutrients to the damaged area, remove waste products, and strengthen the body as a whole.

Digestive enzyme therapy is used to improve the digestion of food, reduce stress on the gastrointestinal mucosa, help maintain normal pH levels, detoxify the body, and promote the growth of healthy intestinal flora. Digestive enzymes also

Transplant Surgery

Transplant surgery is a technique that is increasingly used to replace damaged, diseased, or defective aortas, livers, hearts, kidneys, and other body organs with healthy organs or artificial ones. However, this technique is dangerous and organ rejection is common. Further, atherosclerosis and fibrosis frequently result.

Professor A. Heidland (head of the department of nephrology, University of Wurzberg, Germany) has recently demonstrated, using animal experiments, that partial kidney removal, plus heart, aorta, and liver transplants, can be successfully performed without marked sclerosis, fibrosis, atherosclerosis, or coronary heart disease when enzymes are used. In preliminary studies, Professor Heidland surgically removed 80 percent of the kidneys in rats. In the rats not receiving enzymes, severe atherosclerosis developed, plus kidney insufficiency, which led to kidney failure and eventual death. Those given oral enzymes had no atherosclerosis

formation in the remaining tissue and did not develop kidney insufficiency. The enzyme treated animals remained healthy and survived.

In another experiment, transplanted aortas in the non-enzyme group sclerosed (scarred) and blood vessel linings were six to eight times thicker with very severe atherosclerosis as opposed to the oral enzyme group. This is also true for heart and liver transplants.

Because of these results, clinical human studies of heart, liver, kidney, and aortic transplants are presently underway in Germany, Austria, Hungary, and Poland, according to Dr. Karl Ransberger of the Medical Enzyme Research Society. Preliminary results show that enzymes, including bromelain, papain, trypsin, chymotrypsin, and pancreatin, decrease atherosclerosis and result in a marked reduction of fibrosis, sclerosis, and coronary heart disease.

---

serve as replacements for the body's pancreatic enzymes, leaving the pancreatic enzymes free to perform other functions in the body, such as improving circulation and decreasing inflammation.

Enzyme Absorption System Enhancers (EASE) maximize enzyme activity. They can also improve the absorption and bioavailability of various nutrients and decrease the drain on the body's own digestive enzymes, thus prolonging their lives.

Proteolytic enzymes improve microcirculation and decrease swelling, the stickiness of the blood, hematomas, and fibrin formation. They improve the healing process without undesirable side effects. The success of oral enzymes in trauma and surgery has been documented extensively in controlled clinical studies. (See "Surgical Applications of Enzyme Therapy" on page 399.)

In addition to the standard multivitamin, multimineral, and multiglandular complexes recommended in the introduction to Part Two, the following supplements are recommended for pre- and postoperative surgery treatment. The following recommended dosages are for adults. Children between the ages of two and five should take one-quarter the recommended dosage, those between the ages of six and twelve should take one-half the recommended dosage, and those between the ages of thirteen and seventeen should take three-quarters the recommended adult dosage.

ESSENTIAL ENZYMES

Enzyme	Suggested Dosage	Actions
Superoxide dismutase	Take 150 mcg per day.	Fights inflammation; potent antioxidant.

AND SELECT ONE OF THE FOLLOWING:

Formula One—See page 38 of Part One for a list of products.	Take 10 tablets three times per day between meals for 6–10 days following surgery or until symptoms disappear.	Fights inflammation, pain, and swelling; speeds recovery after injury or surgery; stimulates the immune system.
Phlogenzym from Mucos	Take 10 tablets three times per day between meals.	Fights inflammation, pain, and swelling; speeds recovery after injury or surgery; stimulates the immune system.
Bromelain	Take three to four 230–250 mg capsules per day until symptoms subside. If severe, take 6–8 capsules per day.	Fights inflammation, pain, and swelling; speeds recovery after injury or surgery; stimulates the immune system.
Serratiopeptidase	Take 10 mg two times per day between meals.	Fights inflammation, pain, and swelling; speeds recovery after injury or surgery; stimulates the immune system.
Papain	Take 800 mg twice per day between meals.	Fights inflammation, pain, and swelling; speeds recovery after injury or surgery; stimulates the immune system.
A proteolytic enzyme	Take as indicated on the label.	Fights inflammation, pain, and swelling; speeds recovery after injury or surgery; stimulates the immune system.

SUPPORTIVE ENZYMES

Enzyme	Suggested Dosage	Actions
Catalase and glutathione peroxidase	Take as indicated on the labels.	Potent antioxidant enzymes.

398

Surgical Applications of Enzyme Therapy

Research has repeatedly demonstrated that systemic enzyme therapy is effective as a preoperative and postoperative treatment for a wide variety of surgeries. The following is a sampling of the more frequent surgical applications using enzyme therapy.

TYPE OF SURGERY	BENEFITS OF ENZYME THERAPY
Burn surgery	Enzymes reduce scar formation and are used orally and topically. They can reduce inflammation and risk of inflammation and pain.
Bypass surgery of the leg	When enzyme therapy was begun three days prior to surgery and continued for up to 14 days postoperatively, there was approximately a 50-percent reduction in postoperative swelling and inflammation.
Cancer surgery	Decreases inflammation and swelling; reduces metastatic potential; stimulates immune function; decreases adverse effects of chemotherapy and radiation; improves long-term prognosis.
Dental surgery	Studies show that oral supplementation with enzymes decreases the postoperative swelling and inflammation more quickly.
Hysterectomy	Reduces postoperative inflammation, swelling, and pain.
Meniscectomy and other knee surgery	Swelling, pain, and inflammation all improve significantly, and mobility to the knee joint improves nearly twice as fast.
Proctological surgery	After surgery the anal region is very sensitive to pain, bowel movements can be difficult, and infections are common. The topical use of an enzyme ointment and oral enzyme supplementation reduce swelling, pain, secretion, hemorrhage, and scar formation. Studies show the rate of healing was more than 50-percent faster and clinical complaints disappeared completely after 40 to 50 days for individuals using enzymes, compared with up to 100 days for those not taking enzymes.
Surgery to repair broken bones	Results show that pain relief, decrease in swelling, and speed of repair are substantially improved with enzyme therapy.

A digestive enzyme product containing protease, amylase, and lipase enzymes. See the llist of digestive enzyme products in Appendix.	Take as indicated on the label with meals.	Improves the breakdown and absorption of food nutrients; fights toxic build-up.

ENZYME ABSORPTION SYSTEM ENHANCERS (EASE)

Combination	Suggested Dosage	Actions
Connect-All from Nature's Plus	Take as indicated on the label.	Supports connective tissue.
or		
CTR Support from PhysioLogics	Take as indicated on the label.	Nourishes the body during tissue recovery and healing from injury; helps maintain healthy connective tissue; diminishes damage caused by swelling and inflammation.
or		

Inflamzyme from PhytoPharmica	Take as indicated on the label.	Promotes tissue repair.
or		
Magnesium Plus from Life Plus	Take one or two tablets one to three times per day.	Important for calcium utilization and many enzyme reactions.
or		
Mobil-Ease from Prevail	Take one capsule between meals two to three times per day.	Reduces pain, swelling, and inflammation.

ENZYME HELPERS

Nutrient	Suggested Dosage	Actions
Acidophilus and other probiotics	Take as indicated on the label.	Improves gastrointestinal function; needed for production of the coenzymes biotin, niacin, folic acid, and pyridoxine; helps produce antibacterial substances; stimulates enzyme activity.

Adrenal glandular	Take 100–400 mg per day.	Fights inflammation and adrenal exhaustion.
Aged garlic extract	Take 2 capsules three times per day.	Fights inflammation; stimulates the immune system; potent antioxidant.
Amino-acid complex	Take as indicated on the label on an empty stomach.	Required for enzyme, peptide, and hormone production; helps the body maintain proper fluid and acid/base balance; builds muscle and connective tissue.
Bioflavonoids, such as rutin or quercetin	Take 500 mg three times per day between meals.	Strengthen capillaries; potent antioxidants.
Calcium	Take 500 mg two times per day.	Plays important roles in muscle contraction and nerve transmission.
Magnesium	Take 500 mg per day.	Plays important role in muscle contraction, nerve function, and calcium absorption.
Niacin	Take 50 mg the first day, then increase dosage daily by 50 mg to 1,000 mg. If adverse effects occur, decrease dosage. Take doses higher than 1,000 mg only under the supervision of a physician.	Dilates blood vessels; improves circulation; a coenzyme for many enzymes.
or		
Niacin-bound chromium	Take 20 mg per day, or as indicated on the label.	
Pituitary glandular	Take 100 mg per day.	Helps form new blood vessels; aids wound healing.
Thymus glandular	Take 100 mg per day.	Assists the thymus gland in immune resistance; improves immune response; fights infections.
Vitamin A	Take 10,000–50,000 IU per day. If pregnant, do not take more than 10,000 IU per day.	Needed for strong immune system and healthy mucous membranes; potent antioxidant.
Vitamin B complex	Take 150 mg three times per day, or as indicated on the label.	B vitamins act as coenzymes and support neuromusculoskeletal function.
Vitamin C	Take 10,000–15,000 mg per day.	Fights infections, diseases, and allergies; improves wound healing; stimulates the adrenal glands; promotes immune system; powerful antioxidant.
Vitamin E	Take 1,200 IU per day from all sources (including the multivitamin complex in the introduction to Part Two).	Potent antioxidant; supplies oxygen to cells; improves circulation and level of energy.
Zinc	Take 15–50 mg per day. Do not take more than 100 mg per day.	Needed by over 300 enzymes; necessary for white blood cell immune function; improves wound healing.

Comments

❑ Bilberry extract, raspberries, cranberries, red wine, grapes, hawthorn, black currants, and red cabbage contain anthocyanidins. These phytochemicals fight inflammation and

free radicals and improve circulation to the capillaries. Less inflammation will mean less pain and an earlier return to normal activities.

❑ See Part Three for information on the Enzyme Kidney Flush and the Enzyme Toxin Flush programs.

Synovitis

See BURSITIS AND SYNOVITIS.

Systemic Lupus Erythematosus

Systemic lupus erythematosus (SLE) is an autoimmune disorder (see "Immune Disorders" on page 402) that causes inflammation in the connective tissues, blood vessels, and joints. In this condition, the immune system becomes confused and develops antibodies that attack its own tissues. Probably more than 2 million Americans are affected with SLE. It affects young women more than men, in fact, 90 percent of the cases are women, and more than 80 percent of these are Hispanic or Black women. It can also occur in older people and children.

Lupus is a chronic condition, marked by periods of remissions and flare-ups. The long-term prognosis for lupus is good if the flare-ups are controlled. In most Western countries, the ten-year survival rate is over 95 percent.

Symptoms

Though most symptoms of lupus depend upon which antibodies are being produced against which systems, there are a few general symptoms of lupus. Such symptoms include a prolonged low-grade fever, joint inflammation, weakness, fatigue, ulcers in the mouth and nose, a rash, usually on the cheeks, swelling of the lymph glands, sun-sensitivity, malaise, achiness, chills, weight loss, and anemia. If the skin is affected, one may develop lesions, rashes, and red patches on the skin. One may experience hair loss due to scarring of the scalp. If the lupus affects the chest cavity, one may experience chest pain due to pericarditis, pleurisy, or inflammation of the rib cage or abdominal muscles. One may also experience shortness of breath and a rapid heartbeat. Symptoms of digestive system involvement include abdominal pain, nausea, vomiting, diarrhea, and constipation. If the circulatory system is affected, one may experience swelling of the extremities and Raynaud's phenomenon. If any part of the nervous system is affected, one may experience head-aches, seizures, temporary paralysis, psychotic behavior, or stroke. One of the most serious, yet fairly common, problems for one to have with lupus is kidney problems. Kidney problems affect about 75 percent of those with lupus. If allowed to progress, kidney complications can lead to kidney failure—a life-threatening situation.

Enzyme Therapy

Natural therapies can help rebuild the immune system and decrease inflammation and pain. Systemic enzyme therapy is used to reduce inflammation, stimulate the immune system, break down circulating antigen-antibody complexes, improve circulation, help speed tissue repair, bring nutrients to the damaged area, remove waste products, and improve health.

Digestive enzyme therapy is used to improve the breakdown and assimilation of food nutrients, reduce stress on the gastrointestinal mucosa, help maintain normal pH levels, detoxify the body, and promote the growth of healthy intestinal flora. Digestive enzymes also serve as replacements for the body's pancreatic enzymes, leaving the pancreatic enzymes free to perform other functions, such as boosting immunity and decreasing inflammation.

Enzyme Absorption System Enhancers (EASE) maximize enzyme activity. They can also improve the absorption and bioavailability of various nutrients and decrease the drain on the body's own digestive enzymes, thus prolonging their lives.

In addition to the standard multivitamin, multimineral, and multiglandular complexes recommended in the introduction to Part Two, the following supplements are recommended for the treatment of systemic lupus erythematosus. The following recommended dosages are for adults. Children between the ages of two and five should take one-quarter the recommended dosage, those between the ages of six and twelve should take one-half the recommended dosage, and those between the ages of thirteen and seventeen should take three-quarters the recommended adult dosage.

ESSENTIAL ENZYMES

Enzyme	Suggested Dosage	Actions
Formula One—See page 38 of Part One for a list of products.	Take 10 tablets three times per day between meals. As symptoms subside (which should occur in approximately one week) decrease dosage to 3–5 tablets three times per day between meals. When symptoms disappear, begin taking a maintenance dose of 2–3 tablets after each meal.	Decreases inflammation, pain, and swelling; increases rate of healing; stimulates the immune system.
or		
Bromelain	Take three to four 230–250 mg capsules per day until symptoms subside.	Fights inflammation, pain, and swelling; speeds recovery after injury or surgery; stimulates

or	If symptoms are severe, take 6–8 capsules per day.	the immune system.
A proteolytic enzyme	Take as indicated on the label between meals.	Decreases inflammation, pain, and swelling; increases rate of healing; stimulates the immune system.

SUPPORTIVE ENZYMES

Enzyme	Suggested Dosage	Actions
Formula Three—See page 38 of Part One for a list of products.	Take 50 mcg of each three times per day between meals.	Potent antioxidant enzymes.
A digestive enzyme formula containing protease, amylase, and lipase enzymes. For a list of digestive enzyme products, see Appendix.	Take as indicated on the label with meals.	Improves the breakdown and absorption of food nutrients; fights toxic build-up.

ENZYME ABSORPTION SYSTEM ENHANCERS (EASE)

Combination	Suggested Dosage	Actions
Ecology Pak from Life Plus	Take 2 or 3 tablets once or twice per day, preferably in the morning and no later than midday.	Stimulates immune function.
or		
Hepazyme from Enzymatic Therapy	Take as indicated on the label.	Stimulates immune function.
or		
Inflamzyme from PhytoPharmica	Take as indicated on the label.	Immune system activator; reduces inflammation; fights free radicals.
or		
Kyolic Formula 102 (with enzymes)	Take 2 capsules or tablets with meals two to three times per day.	Strengthens the immune system; helps alleviate stress; helps fight free radicals and inflammation; improves digestion and absorption of nutrients.

ENZYME HELPERS

Nutrient	Suggested Dosage	Actions
Acidophilus and other probiotics	Take as indicated on the label.	Needed for the production of the coenzymes biotin, folic acid, niacin, and pyridoxine to stimulate enzyme activity.
Adrenal glandular	Take 100–400 mg per day.	Fights inflammation and adrenal exhaustion.
Aged garlic extract	Take 2 capsules three times per day.	Detoxifies and provides energy.
Amino-acid complex	Take as indicated on the label on an empty stomach.	Amino acids are the building blocks for protein, enzymes, and connective tissue.
Bioflavonoids	Take 150–500 mg three times per day.	Strengthen capillaries; potent antioxidants.
Calcium	Take 1,000 mg per day.	Plays important roles in muscle contraction and nerve transmission.

Immune Disorders

Our bodies' immune systems are essential to our very survival and are among the main areas for which enzyme therapy is effective. The immune system is a complex network of specialized cells and organs that defends our bodies against attacks by foreign invaders. When functioning properly, it seeks out, finds, and destroys pathogens and infectious agents, such as bacteria, viruses, fungi, and parasites. When it malfunctions, however, a number of diseases can develop, from allergies to arthritis to cancer to AIDS; or the immune system could see its own cells as foreign and attack them.

When functioning normally, the immune system can distinguish between the body and foreign invaders and attack only the nonself foreign invaders. It can remember previous attacks and prevent you from succumbing to certain diseases (such as chickenpox) again. It can recognize many millions of distinctive nonself molecules and produce molecules and cells to attack millions of different foreign invaders.

Virtually every body cell carries distinctive molecules that identify it as self. The body's immune defenses do not normally attack tissues that carry a self marker. Rather, immune cells and other body cells coexist peacefully in a state known as self-tolerance. But when immune defenders encounter cells or organisms carrying molecules that are foreign, the immune troops move quickly to eliminate the intruders.

Any substance capable of triggering an immune response is called an *antigen*. An antigen can be a virus, a bacterium, a fungus, a parasite, or a portion or product of one of these organisms. Tissues or cells from another individual (except an identical twin whose cells carry identical self-markers), also act as antigens. Because the immune system recognizes transplanted tissues as foreign, it rejects them.

When an enemy pathogen invades the body, the immune system sends out soldiers called *macrophages* to destroy them. These macrophages call for two other types of immune cells, called *B cells* and *T cells*, that directly fight the pathogen invaders. T cells signal other immune cells that an invader is present, destroy the invaders, and signal when to stop after the job is accomplished. B cells produce *antibodies* to fight the invader directly.

Immune complexes are clusters of interlocking antigens and Y-shaped antibodies. Under normal conditions, immune complexes are rapidly removed from the bloodstream by macrophages in the spleen and in the liver. In some circumstances, however, immune complexes remain undetected and continue to circulate. Eventually they become trapped in such tissues as the

kidneys, lungs, skin, joints, or blood vessels. There, they set off reactions that lead to inflammation and tissue damage.

Immune complexes can cause many diseases. Sometimes, as is the case with malaria and viral hepatitis, they lead to persistent low-grade infections. Sometimes they arise in response to environmental antigens, such as the moldy hay that causes the disease known as farmer's lung. Frequently, immune complexes develop in autoimmune diseases where the continuous production of autoantibodies overloads the immune complex removal system.

Depending on the tissue to which the immune complex adheres, specific inflammatory disorders result with tissue damage. For example, if the immune complexes are directed toward the nucleic acids of the kidneys, then glomerulonephritis with kidney damage can result. If the immune complexes are directed toward joint cartilage or collagen, then the resulting conditions may be arthritis. Should the immune complexes be directed toward the myelin of nerve cells, then the result is demyelination resulting in multiple sclerosis. Immune complex diseases do not occur as long as macrophages are active and retain their ability to detoxify.

Sometimes the immune system's recognition apparatus breaks down, and the body begins to manufacture antibodies and T cells directed against itself, against the body's own cells, cell components, or specific organs. Such antibodies are known as *autoantibodies*, and the diseases they produce are called autoimmune diseases.

Autoimmune reactions contribute to many puzzling diseases. For instance, autoantibodies to red blood cells can cause anemia, autoantibodies to pancreas cells contribute to juvenile diabetes, and autoantibodies to nerve and muscle cells are found in patients with the chronic muscle weakness known as myasthenia gravis. An autoantibody known as rheumatoid factor is common in persons with rheumatoid arthritis. Many of these conditions involve circulating immune complexes and can be helped with systemic enzyme therapy.

Persons with systemic lupus erythematosus (SLE), which affects many systems, have antibodies to many types of cells and cellular components. These include antibodies directed against substances found in the cell's nucleus—DNA, RNA, or proteins known as antinuclear antibodies, or ANAs. These antibodies can cause serious damage when they link up with self antigens to form circulating immune complexes, which become lodged in body tissues and set off inflammatory reactions.

Several factors are likely to be involved in autoimmune diseases. These may include viruses and environmental factors such as exposure to sunlight, certain chemicals, and some drugs, all of which may damage or alter body cells so that they are no longer recognizable as "self." For instance, many metabolic toxins and addictive substances (such as cocaine and morphine) cause a powerful inhibitory effect on macrophage activity, thus allowing formation of high concentrations of immune complexes and occurrence of autoimmune diseases. Sex hormones may be important too, since most autoimmune diseases are far more common in women than in men. Heredity also appears to play a role.

Many types of therapies are being used to combat autoimmune diseases. These include corticosteroids, immunosuppressive drugs developed as anticancer agents, radiation of the lymph nodes, plasmapheresis (a sort of "blood washing" that removes diseased cells and harmful molecules from the circulation), and enzyme therapy.

To date, organized medicine has not been very successful in the treatment of autoimmune diseases. Since the relationship between immune complexes and certain diseases has become more apparent, attempts have been made to prevent immune complex formation and to suppress inflammation. Many drugs currently prescribed actually weaken the action of the immune system, while reducing the symptoms of inflammation. Therefore, enzyme therapy should be a welcomed alternative.

If a person is taking drugs that weaken the body's defenses (such as cortisone), he or she might initially feel better. This is because some medications (such as cortisone) decrease the numbers of antibodies. However, by taking drugs that lower our immune defenses, we are weakening the body to a point where it is defenseless against the next attack from enemy antigens. It is easy for bacteria, bacilli, viruses, or toxic substances to attack the body. Further, the risk of cancer is much higher when such drugs are taken over a long period of time. In addition, these drugs can interfere with the mechanisms of inflammation and treat the symptoms and not their causes.

There is an ever-increasing number of studies on systemic enzyme therapy and autoimmune diseases. This therapy is able to break down and remove pathogenic immune complexes, stimulate the body's endogenous defenses, and accelerate the mechanisms of inflammation, thus resolving the condition more quickly.

The enzyme mixtures are able to break up the immune complexes (deposited in the tissues) to smaller sizes and thus bring them back into the bloodstream for elimination. The increased presence of immune complexes in the blood stream can temporarily cause an increase in the severity of disease symptoms. However, the enzymes (if introduced into the body in sufficient quantities) can soon break down the immune complexes in the bloodstream (sometimes within a few hours) and the disease symptoms should subside.

Chondroitin sulfate	Take 250 mg two to three times per day with meals; or follow advice of your health-care professional.	Supports connective tissue; promotes cell hydration.
Coenzyme Q$_{10}$	Take 30–60 mg one to two times per day with meals; or follow the advice of your health-care professional.	Potent antioxidant.
Essential fatty acids (found in flaxseed oil)	Take 75–100 mg of flaxseed two to three times per day with meals; or follow the advice of your health-care professional.	Helps strengthen cell membranes.
Glucosamine hydrochloride *and/or*	Take 500 mg two to four times per day.	Stimulates cartilage repair.
Glucosamine sulfate	Take 500 mg one to three times per day.	
Magnesium	Take 500 mg per day.	Plays important role in muscle contraction, nerve function, and calcium absorption.
Vitamin A	Take 10,000 IU per day. If pregnant, take no more than 10,000 IU per day	Needed for strong immune system and healthy mucous membranes; potent antioxidant.
Vitamin B complex	Take 50 mg three times per day, or as indicated on the label.	B vitamins act as coenzymes and support neuromusculoskeletal function.
Vitamin C	Take 5,000–10,000 mg per day.	Helps to produce collagen and normal connective tissue; promotes immune system; powerful antioxidant.
Zinc	Take 50 mg per day. Do not take more than 100 mg per day.	Needed by over 300 enzymes.

Comments

❏ The key factor in the treatment of SLE is to increase your intake of enzymes and enzyme-rich foods and supplements, plus herbs that effectively aid the body's cleansing systems, thus helping to eliminate toxins. Such herbs may include licorice root, red clover, and echinacea.

❏ Fruits and vegetables (particularly green vegetables) should comprise over 60 percent of your diet. Try citrus fruits, broccoli, soybeans, turnip greens, sweet potatoes, and spinach. If your body cannot tolerate raw fruits and vegetables, increase your intake of digestive enzymes, or sauté or steam your produce.

❏ To improve the overall health of the body, avoid alcohol, refined white flour, and sugar.

❏ Avoid chlorinated water. Distilled, bottled spring water, or well water is recommended.

❏ Use sea salt instead of ordinary salt because it contains important minerals.

❏ Drink four ounces of fresh pineapple juice combined with four ounces of lemon or grapefruit juice three times daily.

❏ Exercising can be very helpful because it stimulates circulation, immune function, and enzyme activity and aids in detoxification. Exercise thirty to forty minutes per day, five to seven days per week. See Part Three for further information.

❏ Seek the services of a rheumatologist well trained in the treatment of lupus, who will develop a treatment program for you.

❏ See the list of Resource Groups in the Appendix for an organization that can provide you with more information about lupus.

Temporomandibular Joint Dysfunction

The temporomandibular joints are formed by the temporal bone (a bone on the side of the skull) and the mandible (the lower jawbone). Temporomandibular joint dysfunction or disorder (TMJ) is a collective term for problems occurring in these joints and the muscles that surround them. Temporomandibular joint dysfunction can affect anyone, but is most common in women between the ages of 20 and 50. It is most commonly caused by muscle tightness around the jaw, often due to overuse or bruxism (clenching or grinding of the teeth).

Another common cause of TMJ disorders is internal derangement, in which the disk in the temporomandibular joint that prevents that lower jawbone from rubbing against the skull slips out of place. In internal derangement without reduction, the disk does not return to its normal position, limiting jaw movement. More commonly, in internal derangement with reduction, the disk slips back into place when the jaw is opened, causing a clicking noise at it does.

Arthritis is another cause of TMJ. Osteoarthritis is the most common cause as rheumatoid arthritis does not often affect the temporomandibular joint. Injury to, or inflammation of, the temporomandibular joint is another cause of TMJ. Often an injury to the head or neck can cause a misalignment in the temporomandibular joint, resulting in pain.

Hypermobility and hypomobility of the temporomandibular joint are less common causes of dysfunction. Hypomobility may result from ankylosis, a condition in which the bones of the joints fuse together. Hypermobility may result from stretched ligaments of the temporomandibular joint, which causes the joint to dislocate when opening the mouth too wide.

Symptoms

Symptoms of TMJ include pain in one or both of the temporomandibular joints that may radiate to the neck, shoulders, and back; headache; toothache; tenderness in the muscles used for chewing; clicking or popping sounds upon opening the mouth; and inability to open the jaw completely.

Enzyme Therapy

Systemic enzyme therapy is used to decrease pain, swelling, and inflammation. Enzymes stimulate the immune system and help speed tissue repair because they improve circulation, bringing nutrients to the damaged area and removing waste products. They strengthen the body as a whole, build general resistance, and improve health.

Digestive enzyme therapy is used to improve the digestion of food, reduce stress on the gastrointestinal mucosa, help maintain normal pH levels, detoxify the body, and promote the growth of healthy intestinal flora, thereby improving the overall health of the body, so that the body can fight the effects of TMJ. Digestive enzymes also serve as replacements for the body's pancreatic enzymes, leaving the pancreatic enzymes free to perform other functions in the body, such as improving circulation and decreasing inflammation.

Enzyme Absorption System Enhancers (EASE) maximize enzyme activity. They can also improve the absorption and bioavailability of various nutrients and decrease the drain on the body's own digestive enzymes, thus prolonging their lives.

In addition to the standard multivitamin, multimineral, and multiglandular complexes recommended in the introduction to Part Two, the following supplements are recommended for the treatment of temporomandibular joint disorders. The following recommended dosages are for adults. Children between the ages of two and five should take one-quarter the recommended dosage, those between the ages of six and twelve should take one-half the recommended dosage, and those between the ages of thirteen and seventeen should take three-quarters the recommended adult dosage.

ESSENTIAL ENZYMES

Enzyme	Suggested Dosage	Actions
Formula One—See page 38 of Part One for a list of products.	Take 10 tablets three times per day between meals. As symptoms subside (which should occur in approximately one week), decrease dosage to 3–5 tablets three times per	Decreases inflammation, pain, and swelling; increases rate of healing; stimulates the immune system.

TMJ Self-Test

This test may help you determine whether or not you have a temporomandibular joint disorder such as internal derangement. Place a finger in each ear, and then open and close your mouth. If you can feel or hear the temporomandibular joint clicking, popping, or grinding, you probably have a temporomandibular disorder. Seek the services of a TMJ specialist (an orthodontist, dentist, chiropractor, medical doctor, naturopath, or osteopath), who can develop a treatment program for you. Your treatment program might include chiropractic adjustments; manipulation; massage therapy; applied kinesiology; physiotherapy (including electrical stimulation and ice and heat applications); exercises for the face, jaw, and neck; and training in chewing and sleeping techniques. Exercises can help decrease the severity of pain and build up muscle groups.

or	day between meals. When symptoms disappear, begin taking a maintenance dose of 2–3 tablets after each meal.	
Bromelain *or*	Take three to four 230 mg capsules per day until symptoms subside. If severe, take 6–8 capsules per day.	Fights inflammation, pain, and swelling; speeds recovery after injury or surgery; stimulates the immune system.
Mulsal from Mucos *or*	Take 8 tablets twice per day between meals until symptoms disappear.	Decreases inflammation, pain, and swelling; increases rate of healing; stimulates the immune system.
Phlogenzym from Mucos *or*	Take 10 tablets three times per day between meals until symptoms disappear.	Decreases inflammation, pain, and swelling; increases rate of healing; stimulates the immune system.
A proteolytic enzyme	Take as indicated on the label between meals.	Decreases inflammation, pain, and swelling; increases rate of healing; stimulates the immune system.

SUPPORTIVE ENZYMES

Enzyme	Suggested Dosage	Actions
Formula Three—See page 38 of Part One for a list of products.	Take 50 mcg of each three times per day between meals.	Potent antioxidant enzymes.
A digestive enzyme formula containing protease, amylase, and lipase enzymes. See the list of digestive enzyme products in Appendix.	Take as indicated on the label with meals.	Improves the breakdown and absorption of food nutrients to improve the overall health of the body; fights toxic build-up.

ENZYME ABSORPTION SYSTEM ENHANCERS (EASE)

Combination	Suggested Dosage	Actions
Bromelain Complex from PhytoPharmica *or*	Take as indicated on the label.	Reduces swelling and inflammation.
Connect-All from Nature's Plus *or*	Take as indicated on the label.	Supports connective tissue.

CTR Support from PhysioLogics *or*	Take as indicated on the label.	Nourishes the body during tissue recovery and healing from injury; helps maintain healthy connective tissue; diminishes damage caused by swelling and inflammation.
Inflamzyme from PhytoPharmica *or*	Take as indicated on the label.	Promotes tissue repair.
Joint Renewal Rx "1500" from Phyto-Therapy *or*	Take as indicated on the label.	Helps to support healthy cartilage and joint function.
Joint-Ease from Nature's Life *or*	Take as indicated on the label.	Decreases inflammation and swelling.
Mobil-Ease from Prevail *or*	Take one capsule between meals two to three times per day.	Has anti-inflammatory activity.
Protease Concentrate from Tyler Encapsulations *or*	Take as indicated on the label.	Breaks up fibrin; fights inflammation.
Zymain from Anabolic Labs	Take as indicated on the label.	Provides nutrition essential for tissue and joint repair and maintenance.

ENZYME HELPERS

Nutrient	Suggested Dosage	Actions
Adrenal glandular	Take 100–400 mg per day.	Fights inflammation and adrenal exhaustion.
Bioflavonoids	Take 150–500 mg three times per day.	Potent antioxidants; have anti-inflammatory action.
Calcium	Take 250–500 mg two times per day.	Plays important roles in muscle contraction and the maintenance of strong, healthy bones.
Chondroitin sulfate	Take 250 mg two to three times per day with meals; or follow the advice of your health-care professional.	Supports connective tissue; promotes cell hydration.
Coenzyme Q_{10}	Take 30–60 mg one to two times per day with meals; or follow the advice of your health-care professional.	Potent antioxidant.

Glucosamine hydrochloride *and/or*	Take 500 mg two to four times per day.	Stimulates cartilage repair.
Glucosamine sulfate	Take 500 mg one to three times per day.	
Magnesium	Take 250–500 mg three times per day.	Plays important role in muscle contraction, nerve function, and calcium absorption.
Mucopolysac-charides found in bovine cartilage extract and shark cartilage extract	Take as indicated on the label.	Supports healthy, flexible joints; controls inflammation.
Vitamin B complex	Take 50 mg three times per day, or as indicated on the label.	B vitamins act as coenzymes and support neuromusculoskeletal function.
Vitamin C	Take 3,000–10,000 mg per day.	Helps to produce collagen and normal connective tissue; promotes immune system; powerful antioxidant.

Comments

❑ Eliminate foods that are difficult to chew.

❑ A number of herbs can fight the inflammation that occurs in TMJ, including echinacea, ginger root, ginkgo biloba, licorice, and white willow bark.

❑ Drink eight ounces of fresh fruit or vegetable juices three times per day. Try pineapples, papayas, dark green leafy vegetables (such as parsley, broccoli, spinach, kale, and beet greens), carrots, tomatoes, lettuce, cabbage, ginger root, apples, turnips, oranges, grapes, and garlic.

❑ Using a fitted mouthpiece (available from an orthodontist or dentist) while sleeping may reduce pressure on the temporomandibular joint and the incidence of grinding.

Tendonitis

Tendonitis (or tendinitis) is an inflammation of the tendon. Tendons are tough, strong strands of fibers that are bound together and connect bones to muscle. The muscle moves the bone by means of the tendon. Tendonitis most commonly occurs in the Achilles' tendon, elbow, hamstring, knee, and shoulder. It most commonly affects those in middle or older age, those who exercise vigorously, and those who perform repetitive tasks.

Symptoms

Tendonitis usually causes pain, swelling, and weakness in the affected tendon, making movement of the area difficult. The tendon sheaths may be noticeably swollen due to inflammation and fluid accumulation, or the sheaths may remain dry and cause friction that may be heard with a stethoscope or felt when the tendon moves in its sheath.

Enzyme Therapy

Systemic enzyme therapy is used to reduce pain, swelling, heat, and inflammation. Enzymes stimulate the immune system, improve circulation, help speed tissue repair, bring nutrients to the damaged area, remove waste products, improve health, build general resistance, and strengthen the body as a whole. Individual enzymes and enzyme combinations are of value in fighting inflammation. For example, bromelain and other enzymes have well demonstrated efficacy in treating virtually all inflammatory conditions. Many top European sports medicine specialists, such as Dr. Hans W. Müller-Wohlfahrt, the world-famous doctor for the Bavarian (Germany) Professional Football Club, use enzyme combinations including trypsin, chymotrypsin, pancreatin, bromelain, and papain to effectively treat soft tissue injuries, such as tendonitis.

Digestive enzyme therapy is used to improve the digestion and absorption of food nutrients and to reduce stress on the gastrointestinal mucosa. This helps maintain normal pH levels, detoxifies the body, and promotes the growth of healthy intestinal flora, thereby improving the overall health of the body to prevent tendonitis. Digestive enzymes also serve as replacements for the body's pancreatic enzymes, leaving the pancreatic enzymes free to perform other functions in the body, such as improving circulation and decreasing inflammation.

Enzyme Absorption System Enhancers (EASE) maximize enzyme activity. They can also improve the absorption and bioavailability of various nutrients and decrease the drain on the body's own digestive enzymes, thus prolonging their lives.

In addition to the standard multivitamin, multimineral, and multiglandular complexes recommended in the introduction to Part Two, the following supplements are recommended for the treatment of tendonitis. The following recommended dosages are for adults. Children between the ages of two and five should take one-quarter the recommended dosage, those between the ages of six and twelve should take one-half the recommended dosage, and those between the ages of thirteen and seventeen should take three-quarters the recommended adult dosage.

ESSENTIAL ENZYMES

Enzyme	Suggested Dosage	Actions
Formula One—See page 38 of Part One for a list of products.	Take a single dose of 10 tablets. An hour later, take 2–4 tablets hourly, for one to two days, or as long as necessary. With	Decreases inflammation, pain, and swelling; increases rate of healing; stimulates the immune system.

improvement, take 3–5 tablets two times per day for as long as symptoms persist. Maintenance dose is 2–3 tablets between meals (one to 2 hours after each meal).

or

Phlogenzym from Mucos	Take 10 tablets three times per day between meals.	Decreases inflammation, pain, and swelling; increases rate of healing; stimulates the immune system.

or

Bromelain	Take three to four 230–250 mg capsules per day until symptoms subside. If severe, take 6–8 capsules per day.	Fights inflammation, pain, and swelling; speeds recovery after injury or surgery; stimulates the immune system.

or

Mulsal from Mucos	Take 8 tablets twice per day between meals.	Decreases inflammation, pain, and swelling; increases rate of healing; stimulates the immune system.

or

A proteolytic enzyme	Take as indicated on the label between meals.	Decreases inflammation, pain, and swelling; increases rate of healing; stimulates the immune system.

SUPPORTIVE ENZYMES

Enzyme	Suggested Dosage	Actions
Formula Three—See page 38 of Part One for a list of products.	Take 50 mcg of each three times per day between meals.	Potent antioxidant enzymes.
A digestive enzyme product containing protease, amylase, and lipase enzymes. See the list of digestive enzyme products in Appendix.	Take as indicated on the label with meals.	Improves the breakdown and absorption of food nutrients; fights toxic build-up.

ENZYME ABSORPTION SYSTEM ENHANCERS (EASE)

Combination	Suggested Dosage	Actions
Bromelain Complex from PhytoPharmica	Take as indicated on the label.	Reduces swelling and inflammation; improves circulation.
or		
Connect-All from Nature's Plus	Take as indicated on the label.	Supports connective tissue.
or		
CTR Support from PhysioLogics	Take as indicated on the label.	Nourishes the body during tissue recovery and healing from injury; helps maintain healthy connective tissue.
or		
Inflamzyme from PhytoPharmica	Take as indicated on the label.	Promotes tissue repair.
or		
Joint-Ease from Nature's Life	Take as indicated on the label.	Decreases inflammation and swelling.
or		
Mobil-Ease from	Take one capsule	Has anti-inflammatory activity.

Prevail	between meals two to three times per day.	
or		
Protease Concentrate from Tyler Encapsulations	Take as indicated on the label.	Breaks up fibrin; fights inflammation.

ENZYME HELPERS

Nutrient	Suggested Dosage	Actions
Adrenal glandular	Take 100–400 mg per day.	Fights inflammation and adrenal exhaustion.
Amino-acid complex	Take as indicated on the label on an empty stomach.	Builds and maintains connective tissue.
Bioflavonoids	Take 150–500 mg three times per day.	Potent antioxidants; have anti-inflammatory actions.
Branched-chain amino acids (isoleucine, leucine, and valine)	Take as indicated on the label between meals.	Often found together in a supplement, branched-chain amino acids aid in healing of tendons and act as fuel for muscles.
Calcium	Take 1,000 mg per day.	Plays important roles in muscle contraction and nerve transmission.
Chondroitin sulfate	Take 250 mg two to three times per day with meals; or follow the advice of your health-care professional.	Supports connective tissue; promotes cell hydration.
Glucosamine hydrochloride *and/or*	Take 500 mg two to four times per day.	Stimulates cartilage repair.
Glucosamine sulfate	Take 500 mg one to three times per day.	
Magnesium	Take 500 mg per day.	Plays important role in muscle contraction, nerve function, and calcium absorption.
Mucopolysaccharides, found in bovine cartilage extract and shark cartilage extract	Take as indicated on the label.	Support healthy, flexible joints; control inflammation; help keep joint membranes fluid.
Vitamin B complex	Take 50 mg three times per day, or as indicated on the label.	B vitamins act as coenzymes and support neuromusculoskeletal function.
Vitamin B$_{12}$	Take 500–1,000 mcg for seven to ten days, or as indicated on the label.	Part of the coenzymes methylcobalamin and deoxyadenosylcobalamin, which play parts in the synthesis of new cells, assist in the breakdown of some amino acids and fatty acids, and help maintain nerve cell function.
Vitamin C	Take 2,000–10,000 mg per day, or as recommended by your doctor.	Helps to produce collagen and normal connective tissue; promotes immune system; powerful antioxidant.
Zinc	Take 50 mg per day. Do not take more than 100 mg per day.	Needed by over 300 enzymes; a cofactor for superoxide dismutase; necessary for immune function; facilitates healing; potent antioxidant.

Comments

❏ A number of herbs can fight inflammation, including alfalfa, echinacea, ginger root, ginkgo biloba, licorice root, and white willow bark.

❏ For information on dietary modifications and detoxification methods (through cleansing, fasting, and juicing), see Part Three.

❏ Ice should be used for the first one to three days. Fill two or three small paper cups with water and freeze. When frozen, take one out, tear off the top half inch of paper revealing the ice. Place a paper towel or dish towel over the ice and apply to the painful area. Rub the affected area in a constant circulatory motion with the ice. Do not apply the ice directly to the skin. Apply the ice for five minutes, then keep the ice off the area for five minutes. Repeat this every one to two hours. The purpose is to stimulate blood flow to the area, remove waste from the area, and reduce inflammation.

❏ After three days, alternate ice with heat packs every fifteen minutes until symptoms disappear. "Blue ice" or a frozen vegetable packet works better for this application. Wrap it with a dish towel and apply to the painful area. For heat, use a hot water bottle or heating pad. If a hot water bottle is used, wrap in a towel and apply. If a heating pad is used, put a warm, moist towel on the patient, then place a sheet of wax paper over it, then the heating pad, and finally, a large bath towel. Make sure the wax paper is wide and long enough to cover the entire heating pad with approximately one inch more extending out on every side. The heating pad should never touch the moist towel or the patient's skin. Great care should be exercised to assure this does not happen.

❏ *See also* Bursitis and Synovitis; Sports Injuries; Sprains and Strains.

Testicular Inflammation

See Epididymitis.

Thrombophlebitis

See Thrombosis.

Thrombosis

Thrombosis is the formation or presence of a blood clot. Blood clots can cause plaquing in the blood vessel walls, decreasing the vessel opening. In some instances, the clots can break away from the blood vessel wall and move to another location, usually a smaller blood vessel (a detached thrombus is called an embolus). This can completely block the blood vessel and cause obstruction of blood flow to a vital organ.

Thrombophlebitis is inflammation of a vein, resulting from a blood clot. Thrombophlebitis affects the superficial veins—those near the surface of the skin. It can be caused by infection, trauma, obesity, lack of exercise, birth control pills, drugs, smoking, pregnancy, certain allergies, or low blood levels of proteolytic enzymes. Though thrombophlebitis can affect any superficial vein, it is most common in the legs. This is not a dangerous condition and occurs more often in women than men, but can occur in anyone. It usually occurs in people with varicose veins.

Thrombophlebitis may cause the veins to feel hard, like a cord, and there may be pain, swelling, redness, and tenderness over the affected area.

When a thrombus is in a deep vein, it is called *deep-vein thrombosis* (DVT). It affects the intermuscular veins deep below the surface of the skin. DVT can be serious because the affected veins are located deep inside the leg muscles, are much larger than their superficial counterparts, and transport 90 percent of the blood back to the heart.

Risk factors that contribute to deep vein thrombosis include an increased tendency of blood to clot, which can result from injury to the lining of a vein, surgery, recent childbirth, the use of oral contraceptives, obesity, and diabetes; and slowing of blood flow, which can result from prolonged bed rest and sitting for long periods of time because the calf muscles are not contracting and forcing blood flow. The risk of deep vein thrombosis increases measurably after the age of 40, and triples each additional twenty years.

A deep vein thrombosis can become life-threatening if a blood clot breaks off from the lining of the vein and passes through the venous system to the heart, brain, or lungs, where it may plug a blood vessel, thus cutting off circulation to a vital organ.

Often, despite its killer potential, a deep vein thrombosis can have no symptoms. However, when it causes a large degree of inflammation, it may cause swelling, pain, and heat. Chills and fever can also occur. Over time, the skin over the affected area may become discolored. Pain is increased on walking or standing and improves when lying down with legs elevated (a pillow or two can be beneficial). When in an upright position, the superficial veins become more pronounced. The skin over the affected area may also be vulnerable to ulcers.

Post-thrombotic syndrome (PTS) is considered to be one of the most serious circulatory diseases. It often develops within two to twelve months as a complication of a deep-vein thrombosis or increased pressure in the blood vessels. As pressure increases within the vein, the vessel is stretched and expands outward, and large protein molecules and plasma are forced into the surrounding tissues. This is the beginning of a chronic inflammatory reaction and ultimately results in scar tissue formation. This results in swelling of the leg.

While this condition may begin as a simple cosmetic concern, without the proper treatment, it can lead to a serious health problem. The protein that escapes from the blood vessel into the surrounding tissue is now seen by the body as foreign material (the enemy). This is the beginning of an autoimmune condition. Anyone with PTS should be under the care and supervision of a health care professional well trained in the treatment of this condition.

Thrombosis can occur in seemingly otherwise healthy people of any age, but is most common in adults. Proteolytic enzymes naturally produced by the body help to maintain the balance between clot formation and clot break up (called fibrinolysis). But as we age, our bodies do not produce as many enzymes and those produced have decreased activity levels. For this reason, enzyme supplements become increasingly essential to fight excessive clot formation. (See a more extensive discussion of this topic under CARDIOVASCULAR DISORDERS in Part Two.)

Enzyme Therapy

Systemic enzyme therapy is used to dissolve the clots and reduce or prevent future clot formation in the blood vessels. Enzymes reduce inflammation, stimulate the immune system, improve circulation, help speed tissue repair, bring nutrients to the damaged area, and remove waste products. They strengthen the body as a whole, build general resistance, and improve health.

Enzymes taken orally are an effective therapy for thrombophlebitis. Many studies using oral enzyme combinations, including pancreatin, trypsin, chymotrypsin, bromelain, and papain, plus the bioflavonoid rutin, for the treatment of phlebitis, deep-vein thrombosis, and post-thrombotic syndrome produced very favorable results. In one study, 216 patients were given eight enterically coated enzyme tablets per day for an average of twenty-seven days. Results indicated that 31 percent were cured, 62 percent improved, and 4 percent remained unchanged after enzyme therapy. No side effects were noted.

In another study of thrombophlebitis, the period of healing was significantly shortened from several months to several weeks in those taking oral enzymes. Of those patients in the study, 86 percent either achieved total reduction or a minimum of 50-percent improvement in venous swelling. Post-phlebitis ulcers all healed. In deep-venous thrombosis, 92 percent had measurable improvement.

Enzyme therapy is used to restore equilibrium in blood circulation by increasing the fibrinolytic activity of the blood, which dissolves the deposits of fibrin and cholesterol and decrease swelling; and by improving blood fluidity and, therefore, blood flow.

Digestive enzyme therapy is used to improve the digestion of food, reduce stress on the gastrointestinal mucosa, help maintain normal pH levels, detoxify the body, and promote the growth of healthy intestinal flora, thereby improving the overall health of the body to prevent the formation of thrombosis. Digestive enzymes also serve as replacements for the body's pancreatic enzymes, leaving the pancreatic enzymes free to perform other functions in the body, such as improving circulation and decreasing inflammation.

Enzyme Absorption System Enhancers (EASE) maximize enzyme activity. They can also improve the absorption and bioavailability of various nutrients and decrease the drain on the body's own digestive enzymes, thus prolonging their lives.

If you believe you have a thrombus (blood clot), immediately see a physician for examination and consultation. The following treatment plan can be shared with your physician and can help prevent clot formation. Remember, prevention is the best "cure" for any condition.

In addition to the standard multivitamin, multimineral, and multiglandular complexes recommended in the introduction to Part Two, the following supplements are recommended for the treatment of thrombosis. The following recommended dosages are for adults. Children between the ages of two and five should take one-quarter the recommended dosage, those between the ages of six and twelve should take one-half the recommended dosage, and those between the ages of thirteen and seventeen should take three-quarters the recommended adult dosage.

ESSENTIAL ENZYMES

Enzyme	Suggested Dosage	Actions
Formula Three—See page 38 of Part One for a list of products.	Take 200 mcg of each three times per day between meals.	Potent antioxidant enzymes.

AND SELECT ONE OF THE FOLLOWING:

Enzyme	Suggested Dosage	Actions
Formula One (particularly Wobenzym N from Naturally Vitamin Supplements) —See page 38 of Part One for a list of products.	Take 10 tablets three times per day between meals for six months or until symptoms improve. Thereafter, take 6 tablets three times per day between meals.	Dissolves blood clots; fights inflammation, pain, and swelling; speeds healing.
Phlogenzym from Mucos	Take 10 tablets three times per day between meals.	Dissolves blood clots; fights inflammation, pain, and swelling; speeds healing.
Inflazyme Forte from American Biologics	Take as indicated on the label between meals.	Dissolves blood clots; fights inflammation, pain, and swelling; speeds healing; strengthens capillary walls.

| Bromelain | Take six to eight 230–250 mg capsules per day for as long as pain persists, then reduce dosage to a maintenance dose of 3–4 capsules day. | Fights inflammation, pain, and swelling; helps dissolve blood clots; stimulates the immune system. |
| A proteolytic enzyme | Take as indicated on the label between meals. | Dissolves blood clots; fights inflammation, pain, and swelling; speeds healing; strengthens capillary walls. |

SUPPORTIVE ENZYMES

Enzyme	Suggested Dosage	Actions
A digestive enzyme formula containing protease, amylase, and lipase enzymes. See the list of digestive enzyme products in Appendix.	Take as indicated on the label with meals.	Improves the breakdown and absorption of proteins, carbohydrates, and fats; fights toxic build-up.

ENZYME ABSORPTION SYSTEM ENHANCERS (EASE)

Combination	Suggested Dosage	Actions
Bromelain Complex from PhytoPharmica *or*	Take as indicated on the label.	Reduces swelling and inflammation; improves circulation.
Cardio-Chelex from Life Plus *or*	Take as indicated on the label.	Supports healthy circulation; improves energy.
Kyolic Formula 102 (with enzymes) *or*	Take 2 capsules or tablets with meals two to three times per day.	Improves circulatory disorders; helps alleviate stress, lower cholesterol, and fight free radicals and inflammation; improves digestion and absorption of nutrients.
Protease Concentrate from Tyler Encapsulations	Take as indicated on the label.	Breaks up fibrin; fights inflammation.

ENZYME HELPERS

Nutrient	Suggested Dosage	Actions
Adrenal glandular	Take as indicated on the label.	Helps fight stress and adrenal exhaustion.
Aged garlic extract	Take 2 capsules three times per day between meals.	Fights inflammation; potent antioxidant; improves circulation.
Aorta glandular	Take 50 mg per day.	Helps improve circulation; protects against atherosclerosis.
Bioflavonoids, such as rutin and quercetin	Take 500 mg three times per day.	Fights atherosclerosis; potent antioxidant; reduces capillary fragility.
Calcium	Take 1,000 mg per day.	Plays important roles in muscle contraction.

Carnitine	Take 500 mg two times per day.	Increases energy; lowers cholesterol; reduces risk of heart disease.
Coenzyme Q$_{10}$	Take 75 mg one to two times per day with meals.	Potent antioxidant; decreases blood pressure and total serum cholesterol.
Heart glandular	Take 140 mg, or as indicated on the label.	Improves heart function, thereby improving circulation.
Lecithin	Take 3 capsules with meals, or as indicated on the label.	Emulsifies fat.
Magnesium	Take 500 mg per day.	Plays important roles in protein building, muscle contraction, energy production, calcium absorption, and blood clotting.
Niacin *or*	Take 50 mg the first day, then increase dosage daily by 50 mg to 1,000 mg. If adverse effects occur, decrease dosage. Take doses higher than 1,000 mg only under the supervision of a physician.	Improves circulation by dilating blood vessels.
Niacin-bound chromium	Take 20 mg per day, or as indicated on the label.	
Pycnogenol or grape seed and skin extract	Take 1,000 mcg per day.	Potent antioxidant; fights heart disease.
Vitamin A	Take 5,000 IU per day, or as per your doctor's directions.	Necessary for skin and cell membrane structure and strong immune system; potent antioxidant.
Vitamin C	Take 5,000–10,000 mg per day.	Essential to vascular function; potent antioxidant; improves the function of many enzymes; assists in converting cholesterol to bile salt.
Vitamin E	Take 1,200 IU per day from all sources (including the multivitamin complex in the introduction to Part Two).	Potent antioxidant; helps prevent cardiovascular disease; stimulates circulation; strengthens capillary walls; prevents blood clots; maintains cell membranes.
Zinc	Take 50 mg per day. Do not take more than 100 mg per day.	Needed for white blood cell immune function; helps prevent blood clotting.

Comments

❏ Thrombolytic therapy using enzymes such as urokinase or streptokinase in conjunction with anticoagulants is a significant improvement in the treatment of acute DVT and is used in many hospitals. Usually, within 24 to 48 hours, partial or complete break up of thrombi will occur. Successful treatment prevents damage to the valves and lessens the complication of chronic venous insufficiency by restoring normal venous structure.

❏ Garlic, onions, shallots, chives, leeks, and scallions contain allium compounds, which are potent free-radical fighters that help lower cholesterol and fight circulatory disorders.

❏ Oats, wheat germ, olives, nuts, nut and seed oils, organ meats, and eggs contain alpha-tocopherol (vitamin E), which is known to help prevent cardiovascular disease and prevent blood clots.

❏ Raspberries, cranberries, grapes, red wine, black currants, and red cabbage contain anthocyanidins, which reduce blood vessel plaque formation and maintain blood flow in small vessels.

❏ Oolong and green tea contain epicatechin, which is a potent antioxidant that lowers blood pressure and strengthens capillaries.

❏ Eat plenty of enzyme-rich fresh fruits and vegetables, especially pineapples, papayas, apples, oranges, grapes, green leafy vegetables (including parsley, broccoli, spinach, kale, and beet greens), carrots, tomatoes, turnips, and garlic.

❏ Your diet should be low in calories and fat. Decrease (or better, eliminate) dairy products and meats.

❏ Increase your intake of fresh cold-water fish, such as salmon, mackerel, sardines, and tuna.

❏ Avoid or eliminate refined flour and sugar, salt, corn flour, alcohol, coffee, tea, and strong spices.

❏ Warm, moist compresses can be helpful when applied over the affected veins.

❏ Aspirin or other analgesics that interfere with normal platelet function should be used with caution. Proteolytic enzyme therapy can improve the patient's condition without causing the serious side effects of many medications.

❏ Wear firm stockings (or compression dressings) to give leg support.

❏ Exercise is important to maintain the flow of blood and to stimulate enzyme and antioxidant activity. Walk, swim, or bicycle for thirty to forty minutes per day.

❏ For further information on dietary modifications, detoxification methods (through cleansing, fasting, and juicing), exercises, positive mental attitude, meditation, and general suggestions, see Part Three.

❏ *See also* EMBOLISM.

Thrush

See FUNGAL SKIN INFECTIONS.

Tinnitus

See EAR INFECTIONS.

Tonsillitis

The tonsils are composed of lymphatic tissue and are located at the back of the mouth on either side of the throat. They help fight infections. They are largest during childhood and gradually shrink with age. Tonsillitis is an inflammation of the tonsils and occurs when an infection, usually streptococcal but sometimes viral, attacks the tonsils. The condition is most frequently found in children, but it is possible to have tonsillitis at any age.

Since tonsils are part of the immune system, tonsillitis in adults can be a sign of a weakened immune system. An improper diet (low in proteins and other nutrients, yet high in refined carbohydrates) can decrease the body's resistance to infection. The more bouts one has with inflamed tonsils, the more one's body becomes weakened against bacterial, viral, and toxic assault.

Symptoms

Symptoms of tonsillitis can include a sore throat; pain, usually felt in the ears particularly when swallowing; swelling; and redness in the back of the throat. Additional symptoms may include coughing, hoarseness, earache, headache, chills, fever, malaise, vomiting, and nausea. Other lymph nodes in the body may also be enlarged.

Enzyme Therapy

Systemic enzyme therapy is used to relieve pain and swelling, reduce inflammation, and stimulate the lymphatic and immune systems. Enzymes improve circulation, help speed tissue repair, bring nutrients to the damaged area, remove waste products, improve health, strengthen the body as a whole, and build general resistance.

Digestive enzyme therapy is used to improve the digestion of food, reduce stress on the gastrointestinal mucosa, help maintain normal pH levels, detoxify the body, and promote the growth of healthy intestinal flora, thereby improving the overall health of the body so that the body can fight tonsillitis. Digestive enzymes also serve as replacements for the body's pancreatic enzymes, leaving the pancreatic enzymes free to perform other functions in the body, such as boosting immunity and decreasing circulation.

Enzyme Absorption System Enhancers (EASE) maximize enzyme activity. They can also improve the absorption and bioavailability of various nutrients and decrease the drain on the body's own digestive enzymes, thus prolonging their lives.

In addition to the standard multivitamin, multimineral, and multiglandular complexes recommended in the introduction to Part Two, the following supplements are recommended for the treatment of tonsillitis. The following recommended dosages are for adults. Children between the ages of two and five should take one-quarter the recommended dosage, those between the ages of six and twelve should take one-half the recommended dosage, and those between the ages of thirteen and seventeen should take three-quarters the recommended adult dosage.

ESSENTIAL ENZYMES

Enzyme	Suggested Dosage	Actions
Formula Three—See page 38 of Part One for a list of products.	Take 200 mcg of each three times per day between meals.	Potent antioxidant enzymes.

AND SELECT ONE OF THE FOLLOWING:

Bromelain	Take three to four 230–250 mg capsules per day until symptoms subside. If severe, take 6–8 capsules per day.	Fights inflammation, pain, and swelling; speeds recovery after injury or surgery; stimulates the immune system.
Formula One—See page 38 of Part One for a list of products.	Take 5–10 coated tablets or 1–2 heaping tablespoons of a granulated enzyme mixture three times per day between meals.	Fights inflammation, swelling, and pain; speeds healing; stimulates immune function; fights free-radical and immune complex formation.
Serratiopeptidase	Take 5 mg three times per day between meals.	Fights inflammation; improves respiratory conditions; bolsters the immune system; breaks up mucus.
Phlogenzym from Mucos	Take 4–6 tablets three times per day between meals.	Fights inflammation; improves respiratory conditions; bolsters the immune system; breaks up mucus.
Enzyme complexes rich in proteases, plus amylases and lipases, working synergistically	Take as indicated on the label between meals.	Fights inflammation, pain, and swelling; speeds healing; strengthens capillary walls.

SUPPORTIVE ENZYMES

Enzyme	Suggested Dosage	Actions
A digestive enzyme formula containing protease, amylase, and lipase enzymes. See the list of digestive enzyme products in Appendix.	Take as indicated on the label with meals.	Improves the breakdown and absorption of proteins, carbohydrates, and fats; fights toxic build-up.

ENZYME ABSORPTION SYSTEM ENHANCERS (EASE)

Combination	Suggested Dosage	Actions
Cold Zzap from Naturally Vitamin Supplements *or*	Take as indicated on the label.	Fights infection from viruses or bacteria; boosts immunity; has antioxidant activity.
Combat from Life Plus *or*	Take as indicated on the label.	Helps support the respiratory system; stimulates immune system function; helps control infections.
Connect-All from Nature's Plus *or*	Take as indicated on the label.	Supports tissue repair.
Protease Concentrate from Tyler Encapsulations *and*	Take as indicated on the label.	Breaks up fibrin; fights inflammation.
Bromelain Complex from PhytoPharmica	Take as indicated on the label.	Reduces swelling and inflammation; improves circulation; helps break up any circulating immune complexes.

ENZYME HELPERS

Nutrient	Suggested Dosage	Actions
Acidophilus or other probiotic *or*	Take as indicated on the label.	Improves gastrointestinal function, thereby improving the overall health of the body; stimulates enzyme activity.
Adrenal glandular *or*	Take 100–400 mg per day.	Fights inflammation and adrenal exhaustion.
Bioflavonoids *or*	Take 100–300 mg twice per day.	Potent antioxidant.
Kyolic Aged Garlic Liquid Extract with Vitamins B_1 and B_{12} *or*	Take as indicated on the label.	Stimulates antioxidant activity and immune system function.
Pantothenic acid (vitamin B_5) *or*	Take 125 mg per day.	Important in protein, fat, and carbohydrate metabolism and a healthy digestive system to improve the overall health of the body.
Pituitary glandular *or*	Take 100 mg per day.	Aids healing.
Thymus glandular *or*	Take 100 mg per day.	Assists the thymus gland in immune resistance; improves immune response; fights infections.
Vitamin A *or*	Take 10,000–50,000 IU for no more than one month. (For higher doses, see a physician.) If pregnant, take no more than 10,000 IU per day.	Necessary for healthy mucous linings and skin and cell membrane structure and a strong immune system; potent antioxidant.
Vitamin B complex *or*	Take 100–150 mg three times per day.	B vitamins act as coenzymes and support neuromusculoskeletal function.

Vitamin C	Take 1–2 grams every hour until symptoms improve. Or mix a heaping teaspoon of powdered or crystalline vitamin C with 8 ounces of water or orange or lemon juice (if lemon juice is used, you may sweeten with honey if desired); gargle a mouthful and swallow; repeat every hour.	Fights infections, diseases, allergies, and common cold; essential for healing; potent antioxidant.
Vitamin E	Take 1,200 IU per day from all sources (including the multivitamin complex in the introduction to Part Two).	Potent antioxidant; resists diseases at the cellular level; effective against allergies and sinusitis.
Zinc or zinc gluconate lozenges	Take 50 mg per day. (Take no more than 100 mg per day from all sources.)	Needed for white blood cell immune function; involved with more than 300 enzymes.

Comments

❑ The best way to prevent tonsillitis is to eat a well-balanced diet, adequate in protein, vitamins, minerals, and enzymes. Take high doses of enzymes and bioflavonoids. A strong mind and body means a healthy, enzyme-active body.

❑ Fruits and vegetables (particularly green vegetables) should comprise over 60 percent of your diet. If your body cannot tolerate raw fruits and vegetables, increase your intake of digestive enzymes, or sauté or steam your produce.

❑ To improve the overall health of the body, avoid alcohol, refined white flour, and sugar.

❑ Milk and milk products are mucus-formers and should be strictly limited. These foods cause increased congestion, which can irritate the mouth and throat and aggravate tonsillitis.

❑ Avoid chlorinated water. Distilled, bottled spring water, or well water is recommended.

❑ In place of ordinary salt, use sea salt because it contains minerals important to nutrition.

❑ Twice daily, gargle with the Enzymatic Gargle. For instructions for its preparation, turn to Part Three.

❑ Take ten drops, or one tablespoon, of echinacea three times per day, along with one to four chewable papain tablets.

❑ Liver and kidney flushes are of value in detoxification (see Part Three for instructions for the Enzyme Kidney Flush and the Enzyme Toxin Flush programs).

❑ Hot broth or chicken or vegetable soup is helpful for cleaning out the body and sweating out toxins. Before making soup, be sure to remove the skin from the chicken (chicken skins contain many toxins). Sip the broth or soup as frequently as possible.

❑ Surgical removal of the tonsils may be necessary in cases of severe infection. However, the tonsils are part of the immune system and act to guard against outside invasion of enemy bacteria or viruses. When the tonsils are removed, these lymphatic glands can no longer assist the immune system in guarding our mouth, throat, lungs, and digestive passage. Consider conservative health care before surgery.

Tooth Decay

Tooth decay can only occur under the ideal circumstances for its occurrence. The tooth must have sufficient grooves and pits to hold plaque, and it must have little fluoride; the "bad" bacteria that promote decay must be present in the mouth; and there must be food available for these bacteria to sustain themselves. When all of these conditions are present, tooth decay (also called caries or cavities) can occur. Decay-causing bacteria feed on food remnants in the mouth, usually sugar, and produce an acid that dissolves tooth enamel. This combination of sugars, acid-producing bacteria, and chemicals in the saliva forms a hard substance called plaque that deposits in grooves in the teeth. The plaque continues producing acid, eroding the teeth. Plaque can only be removed with instruments used by your dentist. Left untreated, tooth decay can lead to tooth loss and inflammation and infection of the gums and teeth.

Symptoms

Decay in the enamel, the hard outer covering of the tooth, usually causes no pain. Pain begins when the decay reaches the dentin, the much softer second layer of the tooth. The pain may be felt only when eating or drinking very cold or sugary foods. If the decay is allowed to progress and reach the pulp, it causes irreversible damage. Pain becomes persistent, and an abscess may form. The tooth will have to be removed.

Enzyme Therapy

When applied topically, enzymes can help inhibit plaque formation and subsequent tooth decay. Many toothpastes contain enzymes. Proteases break down the proteins in saliva that help form the first layer of plaque (called the pellicle). Dextranases reduce the formation of caries and the development of periodontal disease because they break down certain sugars. Dextranase is so effective that it has been added to a number of dental rinses, chewing gums, and toothpastes. Even antioxidant enzymes are added to toothpastes to supplement or activate the body's own salivary peroxidase system.

Digestive enzyme therapy is used to improve the digestion of food, reduce stress on the gastrointestinal mucosa,

413

help maintain normal pH levels, detoxify the body, and promote the growth of healthy intestinal flora, thereby improving the overall health of the body to help the body prevent tooth decay. Digestive enzymes also serve as replacements for the body's pancreatic enzymes, leaving the pancreatic enzymes free to perform other functions in the body, such as decreasing inflammation.

Systemic enzyme therapy is used to reduce inflammation, fight plaque formation, stimulate the immune system, improve circulation, help speed tissue repair, bring nutrients to the damaged area, remove waste products, and improve health.

Enzyme Absorption System Enhancers (EASE) maximize enzyme activity. They can also improve the absorption and bioavailability of various nutrients and decrease the drain on the body's own digestive enzymes, thus prolonging their lives.

In addition to the standard multivitamin, multimineral, and multiglandular complexes recommended in the introduction to Part Two, the following supplements are recommended for the treatment of tooth decay. The following recommended dosages are for adults. Children between the ages of two and five should take one-quarter the recommended dosage, those between the ages of six and twelve should take one-half the recommended dosage, and those between the ages of thirteen and seventeen should take three-quarters the recommended adult dosage.

ESSENTIAL ENZYMES

Enzyme	Suggested Dosage	Actions
Superoxide dismutase	Take 150 mcg per day.	Fights inflammation; potent antioxidant.

AND SELECT ONE OF THE FOLLOWING:

Bromelain	Take 720–750 mg two times per day between meals.	Fights inflammation, pain, and swelling; speeds recovery; stimulates the immune system.
Serratiopeptidase	Take 10 mg two times per day between meals.	Fights inflammation; reduces pain and swelling.
Trypsin with chymotrypsin	Take 250 mg of trypsin and 10 mg of chymotrypsin three times per day between meals.	Fights inflammation, pain, and swelling; increases rate of healing; stimulates the immune system.
Pancreatin	Take 500 mg six times per day on an empty stomach.	Fights inflammation, pain, and swelling; increases rate of healing; stimulates the immune system.
Papain	Take 300 mg two times per day between meals.	Fights inflammation; speeds recovery.
Microbial protease	Take 300 mg two times per day between meals.	Fights inflammation; speeds recovery.
Formula One—See page 38 of Part One for a list of products.	Take 5 tablets twice per day between meals.	Fights inflammation, pain, and swelling; speeds recovery.

SUPPORTIVE ENZYMES

Enzyme	Suggested Dosage	Actions
Catalase and glutathione peroxidase	Take as indicated on the labels.	Potent antioxidant enzymes.
A digestive enzyme product containing protease, amylase, and lipase. See the list of digestive enzyme products in Appendix.	Take as indicated on the label with meals.	Improves the breakdown and absorption of food nutrients to improve the overall health of the body; fights toxic build-up.
Papain chewable tablets	Take 50 mg 30 minutes before eating and approximately one hour after each meal.	Breaks up protein; fights inflammation.

ENZYME ABSORPTION SYSTEM ENHANCERS (EASE)

Combination	Suggested Dosage	Actions
Detox-Zyme from Rainbow Light *or*	Take as indicated on the label.	Helps cleanse the body and eliminate toxins.
Catimune from Life Plus *or*	Take 1 tablet twice daily between meals.	Helps detoxify the kidney and liver; flushes out toxins; supports blood and circulatory system.
Hepatic Complex from Tyler Encapulations *or*	Take as indicated on the label.	Helps detoxify the liver.
Bioprotect from Biotics *or*	Take as indicated on the label.	Antioxidant combination with free-radical fighting capabilities.
Metabolic Liver Formula from Prevail *or*	Take as inciated on the label.	Improves liver function.
Bio-Immunozyme Forte from Biotics *or*	Take as indicated on the label.	Stimulates immune function.
Advanced Nutritional System from Rainbow Light *or*	Take as indicated on the label.	Potent free-radical fighter and immune system stimulator.
NESS Formula #416 *or* Protease Concentrate from Tyler *and*	Take as indicated on the label.	Breaks down fibrin; fights inflammation.
Bromelain Complex from PhytoPharmica *and*	Take as indicated on the label.	Decreases swelling; improves circulation; helps break up circulating immune complexes.
Cardio–Chelex from Life Plus	Take five tablets once a day, first thing in the morning.	Supports healthy circulation; improves removal of toxins.

ENZYME HELPERS

Nutrient	Suggested Dosage	Actions
Acidophilus and	Take as indicated	Improve gastrointestinal func-

other probiotics	on the label.	tion to improve the overall health of the body; needed for production of the coenzymes biotin, niacin, folic acid, and pyridoxine; help produce antibacterial substances; stimulate enzyme activity.
Calcium	Take 500 mg two times per day.	Plays important role in maintaining strong, healthy teeth.
Magnesium	Take 250–500 mg per day.	Plays important role in nerve function and calcium absorption.
Niacin or	Take 50 mg the first day, then increase dosage daily by 50 mg to 1,000 mg. If adverse effects occur, decrease dosage. Take doses higher than 1,000 mg only under the supervision of a physician.	Dilates blood vessels to improve circulation to the affected area.
Niacin-bound chromium	Take 20 mg per day, or as indicated on the label.	
Vitamin A	Take 10,000–15,000 IU per day. If pregnant, take no more than 10,000 IU per day.	Needed for strong immune system; potent antioxidant.
Vitamin B complex	Take 150 mg three times per day, or as indicated on the label.	B vitamins act as coenzymes and support neuromusculoskeletal function.
Vitamin C	Take 10,000–15,000 mg per day.	Fights infections; improves healing; stimulates the adrenal glands; promotes immune system; powerful antioxidant.
Vitamin D	Take 400 IU per day.	Involved in regulating the transport and absorption of calcium and phosphorus.
Vitamin E	Take 1,200 IU per day from all sources (including the multivitamin complex in the introduction to Part Two).	Potent antioxidant; supplies oxygen to cells; improves circulation and level of energy.
Zinc	Take 50 mg per day. Take no more than 100 mg per day from all sources.	A cofactor for more than 300 enzymes; potent antioxidant.

Comments

❏ To avoid tooth decay, eat primarily fresh fruits and vegetables, rich in enzymatic activity.

❏ Cut down or eliminate your consumption of refined sugar and white flour.

❏ Floss and brush after every meal and visit your dentist regularly.

❏ Use the Enzyme Toxin Flush and the Enzyme Kidney Flush programs as detailed in Part Three.

❏ Buy a toothpaste or dental rinse containing enzymes. Many large companies, including Proctor & Gamble and Colgate-Palmolive, own patents for dental products contain-

ing enzymes. Many formulas for dental creams, tooth powder, oral gels, toothpastes, mouthwashes, dental floss, and chewing gums contain dextranases, peroxidases, and proteases.

❏ Check your health-food store, pharmacy, or dentist for products, such as chewing gums, dental rinses, and toothpastes, containing dextranase, proteases, and antioxidant enzymes.

Torticollis

Torticollis is a neurologic disorder marked by constant or intermittent spasms of the neck muscles. This spasm can cause a rotation of the head and neck (if the spasm is on one side) or a forward or backward tilting of the head (if the spasm is on both sides).

Often, its cause is unknown. It may be caused by such conditions as hyperthyroidism, neck tumors, nervous system disorders, or tardive dyskinesia. The disorder often seems to be caused by overactivity in the brain. It may also be caused by injury to the neck or spine or inflammation of the neck muscles.

Congenital torticollis is rare. In this condition, the infant's neck and head twist and tilt to one side. This is usually due to trauma to the neck muscles during the birthing process. Without treatment, this condition becomes permanent.

Symptoms

The symptoms of torticollis include continuous or intermittent painful neck spasms that may begin suddenly. The neck and head twist and tilt. The direction of the twist depends on the side of the neck that is affected. Usually, only one side of the neck is affected.

Torticollis may be anywhere from mild to severe. Depending on its cause, it may be a lifelong affliction, causing pain and deformity, or it may be cured completely. In about 10 to 20 percent of the cases, spontaneous recovery occurs within some five years of onset. In about one-third of the cases, spasms occur elsewhere in the body, such as the face, jaw, eyelids, or extremities, including the hands. Women have a higher incidence of torticollis than men. (It occurs one-and-a-half times more frequently in women.)

Enzyme Therapy

Enzyme therapy is most helpful with torticollis caused by injury, although it may also help ease the symptoms of a more chronic case of torticollis, depending upon the severity of the condition.

Systemic enzyme therapy is used to decrease muscle stiffness, pain, swelling, and inflammation. By decreasing the inflammation, the tight muscles relax and the head returns to a more upright position. Enzymes stimulate the immune system, improve circulation, help speed tissue repair, bring nutrients to the damaged area, remove waste products, improve health, strengthen the body as a whole, and build general resistance.

Digestive enzyme therapy is used to improve the digestion of food, reduce stress on the gastrointestinal mucosa, help maintain normal pH levels, detoxify the body, and promote the growth of healthy intestinal flora, thereby improving the overall health of the body. Digestive enzymes also serve as replacements for the body's pancreatic enzymes, leaving the pancreatic enzymes free to perform other functions in the body, such as improving circulation and decreasing inflammation.

Enzyme Absorption System Enhancers (EASE) maximize enzyme activity. They can also improve the absorption and bioavailability of various nutrients and decrease the drain on the body's own digestive enzymes, thus prolonging their lives.

In addition to the standard multivitamin, multimineral, and multiglandular complexes recommended in the introduction to Part Two, the following supplements are recommended for the treatment of torticollis. The following recommended dosages are for adults. Children between the ages of two and five should take one-quarter the recommended dosage, those between the ages of six and twelve should take one-half the recommended dosage, and those between the ages of thirteen and seventeen should take three-quarters the recommended adult dosage.

ESSENTIAL ENZYMES

Enzyme	Suggested Dosage	Actions
Formula One—See page 38 of Part One for a list of products.	Take 10 tablets three times per day between meals. As symptoms subside (which should occur in approximately one week) decrease dosage to 3–5 tablets three times per day between meals. When symptom-free, begin taking a maintenance dose of 2–3 tablets after each meal.	Decreases inflammation, pain, and swelling; increases rate of healing; stimulates the immune system.
or		
Phlogenzym from Mucos	Take 10 tablets three times per day between meals until symptoms disappear.	Decreases inflammation, pain, and swelling; increases rate of healing; stimulates the immune system.
or		
Bromelain	Take three to four 230–250 mg capsules per day until symptoms subside. If symptoms are severe, take 6–8 capsules per day.	Fights inflammation, pain, and swelling; speeds recovery after injury or surgery; stimulates the immune system.
or		
Mulsal from Mucos	Take 8 tablets twice per day between meals until symptoms disappear.	Decreases inflammation, pain, and swelling; increases rate of healing; stimulates the immune

		system; strengthens capillary walls.
or		
A proteolytic enzyme	Take as indicated on the label between meals.	Decreases inflammation, pain, and swelling; increases rate of healing; stimulates the immune system.

SUPPORTIVE ENZYMES

Enzyme	Suggested Dosage	Actions
Formula Three—See page 38 of Part One for a list of products.	Take 50 mcg of each three times per day between meals.	Potent antioxidant enzymes.
A digestive enzyme product containing protease, amylase, and lipase enzymes. For the list of digestive enzyme products, see Appendix.	Take as indicated on the label with meals.	Improves digestion by breaking down protein, carbohydrates, and fats.

ENZYME ABSORPTION SYSTEM ENHANCERS (EASE)

Combination	Suggested Dosage	Actions
Bromelain Complex from PhytoPharmica	Take as indicated on the label.	Reduces swelling and inflammation; improves circulation.
or		
Joint-Ease from Nature's Life	Take as indicated on the label.	Decreases inflammation and swelling.
or		
Mobil-Ease from Prevail	Take one capsule between meals two to three times per day.	Has anti-inflammatory activity.
or		
Protease Concentrate from Tyler Encapsulations	Take as indicated on the label.	Breaks up fibrin; fights inflammation.

ENZYME HELPERS

Nutrient	Suggested Dosage	Actions
Adrenal glandular	Take 100–400 mg per day.	Fights inflammation and adrenal exhaustion.
Aged garlic extract	Take two 300 mg capsules three times per day.	Detoxifies and provides energy.
Alpha-linolenic acid	Take as indicated on the label.	Stimulates immune system; decreases inflammation.
Amino-acid complex	Take as indicated on the label on an empty stomach.	Amino acids are the building blocks of protein and enzymes.
Bioflavonoids	Take 150–500 mg three times per day.	Strengthen capillaries; potent antioxidant.
Branched-chain amino acids (isoleucine, leucine, and valine)	Take as indicated on the label between meals.	Often found together in a supplement, branched-chain amino acids aid in healing of muscle, bone, and skin and act as fuel for muscles.
Calcium	Take 1,000 mg per day.	Plays important roles in muscle contraction and nerve transmission.

Chondroitin sulfate	Take 250 mg two to three times per day with meals; or follow the advice of your health-care professional.	Supports connective tissue; promotes cell hydration.
Coenzyme Q$_{10}$	Take 30–60 mg one to two times per day with meals; or follow the advice of your health-care professional.	Potent antioxidant.
Glucosamine hydrochloride *and/or*	Take 500 mg two to four times per day.	Stimulates cartilage repair.
Glucosamine sulfate	Take 500 mg one to three times per day.	
Magnesium	Take 500 mg per day.	Plays important role in muscle contraction, nerve function, and calcium absorption.
Vitamin B complex	Take 50 mg three times per day, or as indicated on the label.	B vitamins act as coenzymes and support neuromusculoskeletal function.
Vitamin C	Take 1,000–10,000 mg per day.	Helps to produce collagen and normal connective tissue; promotes immune system; powerful antioxidant.
Zinc	Take 50 mg three times per day.	Important for soft tissue integrity and repair; a cofactor for more than 300 enzymes; potent antioxidant.

Comments

❏ You may want to seek the services of a well-trained specialist, such as a chiropractor, naturopath, or osteopath. Your physician will develop a treatment program that might include chiropractic adjustments, manipulation, massage, physiotherapy, including electrical simulation, ice and heat applications, exercises, and postural training (including proper lifting, bending, standing, walking, driving, sitting, and sleeping positions). At times, a cervical collar and traction are used.

❏ If your torticollis is the result of neck or spine trauma, your enzyme/enzyme helper program should help to decrease inflammation regardless of which other treatment is used. If left untreated, the stiff neck can become chronic and possibly cause degenerative changes in the bones and soft tissue of the neck.

❏ Congenital torticollis might leave the individual with shortened neck muscles on one side, which may need to be surgically repaired.

❏ See the list of Resource Groups in the Appendix for an organization that can provide you with more information about torticollis.

Truck Driver's Syndrome

See BELL'S PALSY.

Ulcers

See PEPTIC ULCERS; SKIN ULCERS.

Underweight

Some individuals have a genetic predisposition to being thin. These people might find it difficult to gain weight. Fortunately for them, research shows that thin people live longer. However, when a formerly healthy person of normal weight suddenly becomes underweight, it may be a symptom of aging, stress, or illness.

Many people lose weight as they age because taste buds change with age, and food does not taste as good. In addition, many older individuals live alone. Research shows that those living alone have poorer diets and eat less. Drugs (recreational or prescription), smoking, and alcohol can also cause weight loss. Sudden weight loss can also be a result of a very serious disease, such as AIDS, cancer, diabetes, or an emotional disorder, such as anorexia nervosa or bulimia. Understanding the underlying cause of the weight loss is critical to ensure successful weight gain. If sudden, severe, or unexplained weight loss occurs, see a physician.

Enzyme Therapy

Research shows that many underweight individuals can be helped with enzyme therapy. Many patients lose weight as a result of such conditions as cancer or AIDS or of chemotherapy or radiation therapy. Enzyme therapy helps restore a feeling of well-being, reduces pain, and helps to normalize the body. Further, enzymes can decrease the side effects of radiation "hangover" (*see* RADIATION SICKNESS in Part Two). Enzymes help cancer and AIDS patients to regain appetite, put on weight, and feel more energetic—both mentally and physically.

Digestive enzyme therapy is used to improve the digestion and assimilation of food nutrients, reduce stress on the gastrointestinal mucosa, help maintain normal pH levels, detoxify the body, and promote the growth of healthy intestinal flora. Systemic enzyme therapy is used as supportive care to stimulate the immune system, improve circulation, transport nutrients throughout the body, remove waste products, and improve health.

Enzyme Absorption System Enhancers (EASE) maximize enzyme activity. They can also improve the absorption and bioavailability of various nutrients and decrease the drain on the body's own digestive enzymes, thus prolonging their lives.

In addition to the standard multivitamin, multimineral, and multiglandular complexes recommended in the introduction to Part Two, the following supplements are recommended for those who are underweight. The following recommended dosages are for adults. Children between the ages of two and five should take one-quarter the recommended dosage, those between the ages of six and twelve should take one-half the recommended dosage, and those between the ages of thirteen and seventeen should take three-quarters the recommended adult dosage.

ESSENTIAL ENZYMES

Enzyme	Suggested Dosage	Actions
A digestive enzyme formula containing protease, amylase, and lipase enzymes. For a list of digestive enzyme products, see Appendix. *or*	Take as indicated on the label.	Improves digestion by breaking down proteins, carbohydrates, and fats; helps balance the body.
Pancreatin	Take 300 mg three times per day with meals.	Contains proteases, amylases, and lipases, so digests proteins, carbohydrates, and fats.

SUPPORTIVE ENZYMES

Enzyme	Suggested Dosage	Actions
Formula Three—See page 38 of Part One for a list of products.	Take as indicated on the labels.	Antioxidant enzymes.

AND SELECT ONE OF THE FOLLOWING:

Enzyme	Suggested Dosage	Actions
Formula Three—See page 38 of Part One for a list of products.	Take 10 tablets three times per day between meals.	Strengthens immunity; fights inflammation and free radicals.
Phlogenzym from Mucos	Take 10 tablets three times per day between meals.	Strengthens immunity; fights inflammation and free radicals.
Bromelain	Take ten 230–250 mg tablets three times per day between meals.	Strengthens immunity; fights inflammation and free radicals.

ENZYME ABSORPTION SYSTEM ENHANCERS (EASE)

Combination	Suggested Dosage	Actions
Bean & Vegi Enzyme	Take as indicated on	Facilitates carbohydrate
Formula from Prevail *or*	the label.	digestion and absorption.
Biodias A Granules from Amano Pharmaceutical Co., Ltd. *or*	Adults should take one packet in 8 ounces of water between or after meals three times per day. Children ages 11 to 14 should take two-thirds of a packet; children between the ages of 8 and 10 should take half of the packet; and children between the ages of 5 and 7 should take one-third of a packet.	Helps improve digestion of nutrients; promotes regeneration of mucous membranes; decreases stress on the stomach and intestines.
Kyolic Formula 102 (with enzymes) *or*	Take 2 capsules or tablets with meals two to three times per day.	Improves fat, carbohydrate, and protein digestion and absorption; helps alleviate stress; fights free radicals; helps fight inflammation.
NESS Formula #10 *or*	Take 2 capsules immediately after each meal.	Helps the body to absorb carbohydrates and sugars.
Pro-Gest-Ade from Enzymatic Therapy	Take as indicated on the label.	Contains nutrients essential for proper digestive function, plus enzymes to enhance digestion and absorption of proteins, fats, and carbohydrates.

ENZYME HELPERS

Nutrient	Suggested Dosage	Actions
Acidophilus or other probiotics	Take as indicated on the label.	Improves gastrointestinal function; needed for production of the coenzyems biotin, niacin, folic acid, and pyridoxine; helps produce antibacterial substances; stimulates enzyme activity.
Amino-acid complex	Take as indicated on the label.	Amino acids serve as building blocks for the body's enzymes.
Betaine hydrochloric acid	Take 150 mg three times per day with meals.	Provides hydrochloric acid to improve digestion. If gastric upset occurs, discontinue use.
Lecithin	Take 1,000 mg three times per day with meals.	Emulsifies fat.
Ox-bile extract	Take 120 mg per day.	Aids digestion.
Whey	Take 1 heaping tablespoon three times per day with meals.	Improves digestion.

Comments

❑ Enzymes help normalize the body, but it is important to treat the underlying cause of the sudden weight loss (such as cancer, AIDS, etc.).

❑ See Part Three for dietary recommendations and Chapter 4 of Part One for ways to maximize enzymes in your diet.

Varicella

See CHICKENPOX.

Varicose Veins

Varicose veins are enlarged veins that appear close to the skin's surface. They occur mainly in the superficial veins of the legs and their branches. They may also appear in the anus (hemorrhoids) or in the lips of the vagina during pregnancy.

In the human body, blood moves from the heart to the extremities by way of the arteries. The veins return the blood to the heart. The blood pressure in the veins is much lower than that in the arteries and is not adequate to transport the blood from the legs up to the heart against the force of gravity. This is partly because the walls of the veins are much thinner and have less muscle tissue than arterial walls.

One-way valves (or cusps) appear periodically in the veins. The purpose of these valves is to keep the blood flow from stopping or flowing downward. The valves ensure that the blood flows toward the heart (the venous return of blood is also supported by the contraction of surrounding muscles). If the valves leak, the blood is no longer being forced upward and accumulates in the veins of the legs, engorging them. This causes the veins to stretch. As they widen, the valves separate, making it no longer possible for them to prevent any backflow of the blood. Ultimately, the blood drains back into a connecting vein, causing the same occurrences there.

Varicose veins seem to occur due to weakness of the walls of the superficial veins. This weakness of the walls seems to be what causes the leakage in the first place. There are a number of other factors that can contribute to the development of varicose veins, including some type of venous obstruction; heredity; and pressure put on the affected area by long periods of standing, obesity, lack of exercise, pregnancy, prolonged constipation, prolonged sitting with legs crossed, bed rest following surgery, and repeated heavy lifting.

Symptoms

Varicose veins are generally bulging, bluish veins, most apparent when standing. The affected legs often feel fatigued and ache, especially after standing; although many feel no pain. The lower leg and ankle may itch. Symptoms are usually worse during the period that the varicose veins are developing. Rarely, bleeding externally or under the skin, dermatitis, or phlebitis may develop in the affected leg.

Enzyme Therapy

Systemic enzyme therapy is used to reduce inflammation, stimulate the immune system, improve circulation, help speed tissue repair, bring nutrients to the damaged area, remove waste products, and enhance wellness.

Because fibrin is deposited in the area near the varicose veins, the tissue becomes lumpy and hard. Further, there is an increased risk of blood clot formation because of decreased fibrinolytic activity. This may result in myocardial infarction, stroke, pulmonary embolism, or thrombophlebitis. A decreased ability to break down fibrin is often found in individuals with varicose veins. But proteolytic enzymes can increase the blood's fibrinolytic activity, improve blood fluidity, and help normalize blood flow. Bromelain, papain, and enzymes from animal sources (pancreatin, trypsin, and chymotrypsin) have been effective in treating in varicose veins.

Digestive enzyme therapy is used to improve the digestion of food, reduce stress on the gastrointestinal mucosa, help maintain normal pH levels, detoxify the body, and promote the growth of healthy intestinal flora, thereby improving the overall health of the body to prevent the formation of varicose veins. Digestive enzymes also serve as replacements for the body's pancreatic enzymes, leaving the pancreatic enzymes free to perform other functions in the body, such as improving circulation and decreasing inflammation.

Enzyme Absorption System Enhancers (EASE) maximize enzyme activity. They can also improve the absorption and bioavailability of various nutrients and decrease the drain on the body's own digestive enzymes, thus prolonging their lives.

In addition to the standard multivitamin, multimineral, and multiglandular complexes recommended in the introduction to Part Two, the following supplements are recommended for the treatment of varicose veins. The following recommended dosages are for adults. Children between the ages of two and five should take one-quarter the recommended dosage, those between the ages of six and twelve should take one-half the recommended dosage, and those between the ages of thirteen and seventeen should take three-quarters the recommended adult dosage.

ESSENTIAL ENZYMES

Enzyme	Suggested Dosage	Actions
Formula Three—See page 38 of Part One for a list of products.	Take 200 mcg three times per day between meals.	Potent antioxidant enzymes.

AND SELECT ONE OF THE FOLLOWING:

Formula One—See page 38 of Part One for a list of products.	Take 10 tablets three times per day between meals for six months or until symptoms improve. Thereafter, take 6 tablets	Dissolves blood clots; fights inflammation, pain, and swelling; speeds healing.

	three times per day between meals.	
Phlogenzym from Mucos	Take 10 tablets three times per day between meals.	Dissolves blood clots; fights inflammation, pain, and swelling; speeds healing.
Bromelain	Take four to six 230–250 mg capsules per day.	Fights inflammation, pain, and swelling; speeds healing; stimulates the immune system.
Mulsal from Mucos	Take 8 tablets twice per day between meals for one month. The next month, reduce dosage to 6 tablets twice per day.	Dissolves blood clots; fights inflammation, pain, and swelling; speeds healing.
A proteolytic enzyme	Take as indicated on label between meals.	Dissolves blood clots; fights inflammation, pain, and swelling; speeds healing.

SUPPORTIVE ENZYMES

Enzyme	Suggested Dosage	Actions
A digestive enzyme formula containing protease, amylase, and lipase enzymes. See the list of digestive enzyme products in Appendix.	Take as indicated on the label with meals.	Improves the breakdown and absorption of proteins, carbohydrates, and fats to improve the overall health of the body; fights toxic build-up.

ENZYME ABSORPTION SYSTEM ENHANCERS (EASE)

Combination	Suggested Dosage	Actions
Bromelain Complex from PhytoPharmica *or*	Take as indicated on the label.	Reduces swelling and inflammation; improves circulation; helps break up any present immune complexes.
Cardio-Chelex from Life Plus *or*	Take 5 tablets twice per day—first thing in the morning and at midday.	Supports healthy circulation; improves energy.
Kyolic Formula 102 (with enzymes) *or*	Take 2 capsules or tablets with meals two to three times per day.	Helps reverse circulatory disorders; helps alleviate stress; helps fight free radicals and lower serum cholesterol levels; helps fight inflammation; improves digestion and absorption of nutrients.
Protease Concentrate from Tyler Encapsulations	Take as indicated on the label.	Breaks up fibrin; fights inflammation.

ENZYME HELPERS

Nutrient	Suggested Dosage	Actions
Aged garlic extract	Take three 300 mg capsules three times per day.	Lowers cholesterol and blood pressure; fights free radicals.
Bioflavonoids, such as rutin and quercetin	Take 500 mg three times per day.	Fights atherosclerosis; potent antioxidant; reduces capillary fragility.
Calcium	Take 1,000 mg per day.	Plays important roles in muscle contraction and nerve transmission.

Coenzyme Q$_{10}$	Take 75 mg one to two times per day with meals.	Potent antioxidant; decreases blood pressure and total serum cholesterol.
Heart glandular	Take 140 mg, or as indicated on the label.	Improves heart function to improve circulation.
Lecithin	Take 3 capsules with meals, or as indicated on the label.	Emulsifies fat.
Magnesium	Take 500 mg per day.	Plays important roles in protein building, muscle contraction, nerve function, energy production, calcium absorption and maintaining blood vessel health.
Niacin *or* Niacin-bound chromium	Take 50 mg the first day, then increase dosage daily by 50 mg to 1,000 mg. If adverse effects occur, decrease dosage. Take doses higher than 1,000 mg only under the supervision of a physician. Take 20 mg per day, or as indicated on the label.	Improves circulation by dilating blood vessels.
Selenium	Take 200 mcg per day.	Potent antioxidant; works with vitamin E to promote blood circulation.
Vitamin A	Take 15,000 IU per day, or as per your doctor's directions. If pregnant, take no more than 10,000 IU per day.	Necessary for skin and cell membrane structure and a strong immune system; potent antioxidant.
Vitamin B complex	Take 50 mg three times per day, or as indicated on the label.	B vitamins act as coenzymes.
Vitamin C	Take 5,000–10,000 mg per day.	Essential to vascular function; potent antioxidant; improves the function of many enzymes; assists in converting cholesterol to bile salt.
Vitamin E	Take 1,200 IU per day from all sources (including the multivitamin complex in the introduction to Part Two).	Potent antioxidant; helps prevent cardiovascular disease; stimulates circulation; strengthens capillary walls; prevents blood clots; maintains cell membranes.
Wheat grass powder	Take 1–2 teaspoons three times per day.	Stimulates blood circulation; guards against degenerative disease.
Zinc	Take 50 mg per day. Do not take more than 100 mg per day.	Needed for white blood cell immune function; helps prevent blood clots.

Comments

❑ Garlic, onions, shallots, chives, leeks, and scallions contain allium compounds, which are potent free-radical fighters that help lower cholesterol and fight circulatory disorders.

❑ Oats, wheat germ, olives, nuts, nut and seed oils, organ meats, and eggs contain alpha-tocopherol (vitamin E), which is known to help prevent cardiovascular disease and prevent blood clots.

❑ Green barley extract and wheat germ oil are very effective in assisting normal circulation.

❑ Raspberries, cranberries, grapes, red wine, black currants, and red cabbage contain anthocyanidins, which reduce blood vessel plaque formation and maintain blood flow in small vessels.

❑ Oolong and green tea contain epicatechin, which is a potent antioxidant that lowers blood pressure and strengthens capillaries.

❑ Varicose veins of the rectum or anus are called hemorrhoids. (*See* HEMORRHOIDS in Part Two.)

❑ For further information on dietary modifications, detoxification methods (through cleansing, fasting, and juicing), exercises, positive mental attitude, meditation, and general suggestions, see Part Three.

Verrucae

See WARTS.

Vertigo

Vertigo is the false unpleasant sensation that you and/or objects in your surroundings are moving or spinning. This is actually the true definition of dizziness—not the feeling of lightheadedness or faintness generally associated with dizziness.

The inner ear controls the body's sense of balance and equilibrium. Vertigo usually results from abnormalities in the inner ear, such as infection, nerve inflammation, or tumors. High or low blood pressure, head injuries, lack of oxygen to the brain, nutritional deficiencies, the use of certain drugs, fever, neurological disorders, such as multiple sclerosis, deficiency of blood circulating to the brain, changes in atmospheric pressure, and even excessive amounts of wax in the ear canal can all cause vertigo. Problems with vision may also cause vertigo, as the eyes also play a role in the body's maintenance of balance. Motion sickness is another cause of vertigo.

Symptoms

Vertigo causes the false sensation that one is spinning, falling, or moving in some other way; or that one's surroundings are moving. It is often accompanied by nausea, loss of balance, and sometimes hearing loss. The feeling may last only a few moments, or it may endure for several hours or even days.

The sensation may be relieved by lying down, but may persist even then.

Enzyme Therapy

Systemic enzyme therapy is used to help improve vertigo when caused by circulatory problems or cancer. Enzymes decrease atherosclerosis and plaquing in the blood vessels, reduce inflammation, stimulate the immune system, improve circulation, help speed tissue repair, bring nutrients to the damaged area, remove waste products, and improve health.

Digestive enzyme therapy is used to improve the digestion of food, reduce stress on the gastrointestinal mucosa, help maintain normal pH levels, detoxify the body, and promote the growth of healthy intestinal flora, thereby improving the overall health of the body to prevent the symptoms of vertigo. Digestive enzymes also serve as replacements for the body's pancreatic enzymes, leaving the pancreatic enzymes free to perform other functions in the body, such as improving circulation.

Enzyme Absorption System Enhancers (EASE) maximize enzyme activity. They can also improve the absorption and bioavailability of various nutrients and decrease the drain on the body's own digestive enzymes, thus prolonging their lives.

In addition to the standard multivitamin, multimineral, and multiglandular complexes recommended in the introduction to Part Two, the following supplements are recommended for the treatment of vertigo. The following recommended dosages are for adults. Children between the ages of two and five should take one-quarter the recommended dosage, those between the ages of six and twelve should take one-half the recommended dosage, and those between the ages of thirteen and seventeen should take three-quarters the recommended adult dosage.

ESSENTIAL ENZYMES

Enzyme	Suggested Dosage	Actions
Formula Three—See page 38 of Part One for a list of products.	Take 200 mcg of each three times per day between meals.	Potent antioxidant enzymes.

AND SELECT ONE OF THE FOLLOWING:

Formula One—See page 38 of Part One for a list of products.	Take 8–10 tablets three times per day between meals for one to two months or until symptoms begin to improve. Then take 6 tablets three times per day between meals until symptoms are gone. Thereafter take 3–4 tablets three times per day.	Improves circulation; speeds healing.
Phlogenzym from Mucos	Take 10 tablets three times per day between meals.	Improves circulation; speeds healing.

Bromelain	Take four to six 230–250 mg capsules per day.	Improves circulation; speeds healing; stimulates the immune system.
Mulsal from Mucos	Take 8 tablets twice per day between meals.	Improves circulation; speeds healing.
A proteolytic enzyme	Take as indicated on the label between meals.	Improves circulation; speeds healing.

SUPPORTIVE ENZYMES

Enzyme	Suggested Dosage	Actions
A digestive enzyme formula containing protease, amylase, and lipase enzymes. See the list of digestive enzyme products in Appendix.	Take as indicated on the label with meals.	Improves the breakdown and absorption of proteins, carbohydrates, and fats to improve the overall health of the body; fights toxic build-up.

ENZYME ABSORPTION SYSTEM ENHANCERS (EASE)

Combination	Suggested Dosage	Actions
Bromelain Complex from PhytoPharmica *or*	Take as indicated on the label.	Reduces swelling and inflammation; improves circulation.
Cardio-Chelex from Life Plus *or*	Take 5 tablets twice per day—first thing in the morning and at midday.	Supports healthy circulation; improves energy.
Kyolic Formula 102 (with enzymes) *or*	Take 2 capsules or tablets with meals two to three times per day.	Improves circulation; helps alleviate stress; helps fight free radicals and inflammation; improves digestion and absorption of nutrients.
Protease Concentrate from Tyler Encapsulations	Take as indicated on the label.	Breaks up fibrin; fights inflammation.

ENZYME HELPERS

Nutrient	Suggested Dosage	Actions
Adrenal glandular	Take 100 mg per day.	Helps fight stress and adrenal exhaustion.
Aged garlic extract	Take 3 capsules three times per day.	Lowers cholesterol and blood pressure; fights free radicals.
Bioflavonoids, such as rutin or quercetin	Take 500 mg three times per day.	Improve circulation; potent antioxidants; reduce capillary fragility.
Calcium	Take 1,000 mg two times per day.	Plays important roles in nerve transmission.
Carnitine	Take 500 mg two times per day.	Increases energy; lowers cholesterol; improves circulation.
Coenzyme Q_{10}	Take 100 mg one to three times per day with meals.	Potent antioxidant; decreases blood pressure and total serum cholesterol.
Heart glandular	Take 140 mg, or as indicated on the label.	Improves heart function to improve circulation.

Magnesium	Take 500 mg twice per day.	Plays important roles in nerve function, energy production, and calcium absorption.
Niacin *or* Niacin-bound chromium	Take 300 mg the first day, then increase dosage daily by 50 mg to 1,000 mg. If adverse effects occur, decrease dosage. Take doses higher than 1,000 mg only under the supervision of a physician. Take 20 mg per day, or as indicated on the label.	Improves circulation by dilating blood vessels.
Vitamin B complex	Take 50 mg three times per day, or as indicated on the label.	B vitamins act as coenzymes and support neuromusculoskeletal function.
Vitamin C	Take 5,000–10,000 mg per day.	Essential to vascular function; potent antioxidant; improves the function of many enzymes.
Vitamin E	Take 1,200 IU per day from all sources (including the multivitamin complex in the introduction to Part Two).	Potent antioxidant; stimulates circulation; strengthens capillary walls; maintains cell membranes.

Comments

❏ Fruits and vegetables (particularly green vegetables) should comprise over 60 percent of your diet. If your body cannot tolerate raw fruits and vegetables, increase your intake of digestive enzymes, or sauté or steam your produce.

❏ To improve your overall health, avoid alcohol, refined white flour, and sugar.

❏ Natural spices are fine, but chemical preservatives, particularly nitrites, should be avoided.

❏ Avoid chlorinated water. Distilled, bottled spring water, or well water is recommended.

❏ Cleanse the body through fasting, juicing, and detoxification. See Part Three for instructions for perfoming the Enzyme Toxin Flush and the Enzyme Kidney Flush.

❏ Exercise (such as walking, bicycling, or swimming) is important to stimulate enzyme activity and fight free-radical formation. Exercise thirty to forty minutes per day. A positive mental attitude and meditation are important to balance the body and stimulate antioxidant activity.

Viral Infections

A virus is an extremely small organism, capable of growth

and reproduction only when inside a living cell. Viruses may be small, but they can cause a vast array of deadly diseases. Once inside the cell, viruses can disturb cell function, replicating during normal cell division, sometimes causing the cell to become cancerous. Some viruses kill the host cell. Viruses can also incorporate their hereditary materials into the normal cell, basically taking over the cell's metabolic resources. These viruses may remain dormant for years before being reactivated by such stressors as psychological stress or illness.

If the virus is able to get past such physical barriers against infection as the skin and the mucous membranes, the immune system is capable of recognizing and destroying viruses before they infect the cell. Sometimes, however, the immune system is unable to fend off the infection. Respiratory infections, such as the common cold and influenza, are the most common type of viral infection.

Some viruses are spread by coughing, sneezing, or other respiratory excretions. Other viruses are sexually transmitted (including HIV/AIDS), while still others are spread by direct contact or contamination with human excrement.

In general, the orthodox medical approach to the control of viral diseases is dependent on the development of vaccines. In this way, a few diseases (such as rubella and polio) have been brought under control. Antibiotics have no effect on the viral diseases.

Symptoms

Symptoms of a viral disease can vary, depending upon the condition. Viruses are responsible for a host of diseases and conditions, including the common cold, shingles (herpes zoster), influenza, mumps, measles, cold sores, rabies, chickenpox, smallpox, Epstein-Barr virus, infectious mononucleosis, and HIV/AIDS. Further, there is suggestive evidence that certain types of cancer (such as leukemia), may be caused by viruses. Researchers have known for a long time that certain viruses can cause cancer in animals.

Enzyme Therapy

The defense against viruses in our body is executed by macrophages or by the natural killer (NK) cells, which destroy the virus-infected cells. Oral enzyme combinations (such as papain, trypsin, and chymotrypsin) synergistically increase the antiviral effects of the immune system and the breakdown of circulating immune complexes.

Systemic enzyme therapy is used to fight the virus and strengthen the body as a whole. Enzymes also reduce inflammation, stimulate the immune system, improve circulation, help speed tissue repair, bring nutrients to the damaged area, remove waste products, and enhance wellness.

Digestive enzyme therapy is used to improve digestion of food, reduce stress on the gastrointestinal mucosa, help maintain normal pH levels, detoxify the body, and promote the growth of healthy intestinal flora, thereby improving the overall health of the body so that the body can fight off viral infections. Digestive enzymes also serve as replacements for the body's pancreatic enzymes, leaving the pancreatic enzymes free to perform other functions in the body, such as boosting immunity, improving circulation, and decreasing inflammation.

Enzyme Absorption System Enhancers (EASE) maximize enzyme activity. They can also improve the absorption and bioavailability of various nutrients and decrease the drain on the body's own digestive enzymes, thus prolonging their lives.

In addition to the standard multivitamin, multimineral, and multiglandular complexes recommended in the introduction to Part Two, the following supplements are recommended for the treatment of viral infections. The following recommended dosages are for adults. Children between the ages of two and five should take one-quarter the recommended dosage, those between the ages of six and twelve should take one-half the recommended dosage, and those between the ages of thirteen and seventeen should take three-quarters the recommended adult dosage.

ESSENTIAL ENZYMES

Enzyme	Suggested Dosage	Actions
WobeMugos E from Mucos *or*	Take 15 tablets per day.	Bolsters the immune system; breaks up antigen-antibody complexes; fights free radicals and inflammation.
Formula One—See page 38 of Part One for a list of products. *or*	Take 10 tablets three times per day.	Stimulates immune function; breaks up circulating immune complexes; fights inflammation.
Bromelain	Take 10 tablets three times per day.	Stimulates immune function; breaks up circulating immune complexes; fights inflammation.

SUPPORTIVE ENZYMES

Enzyme	Suggested Dosage	Actions
Formula Three—See page 38 of Part One a list of products.	Take as indicated on the label.	Potent antioxidant enzymes.
A digestive enzyme product containing protease, amylase, and lipase. See the list of digestive enzyme products in Appendix.	Take as indicated on the label with meals.	Improves the breakdown and absorption of food nutrients; fights toxic build-up.

ENZYME ABSORPTION SYSTEM ENHANCERS (EASE)

Combination	Suggested Dosage	Actions
Ecology Pak from Life Plus	Take 2 or 3 tablets once or twice per day, preferably in	Assists with natural detoxification; stimulates immune function.

or	the morning and no later than midday.	
Hepazyme from Enzymatic Therapy *or*	Take as indicated on the label.	Stimulates immune function.
Inflamzyme from PhytoPharmica *or*	Take as indicated on the label.	Immune system activator; reduces inflammation; fights free radicals.
Kyolic Formula 102 (with enzymes) *or*	Take 2 capsules or tablets with meals two to three times per day.	Strengthens the immune system; helps alleviate stress; helps fight free radicals and inflammation; improves digestion and absorption of nutrients.
Metabolic Liver Formula from Prevail *or*	Take as indicated on the label.	Improves liver function, thus assisting detoxification.
Vita-C-1000 from Life Plus	Take as indicated on the label.	Boosts immune function; fights infection.

ENZYME HELPERS

Nutrient	Suggested Dosage	Actions
Acidophilus and other probiotics	Take as indicated on the label.	Stimulates enzyme activity; improves gastrointestinal function to improve the overall health of the body.
Adrenal glandular	Take 100–400 mg per day.	Helps fight stress and adrenal exhaustion; fights inflammation.
Bioflavonoids, such as rutin or quercetin	Take 500 mg three times per day.	Potent antioxidants; stimulate immune function.
Coenzyme Q_{10}	Take 30–60 mg one to two times per day with meals.	Important for immune system function; has antioxidant activity.
Kyolic Aged Garlic Liquid Extract with Vitamins B_1 and B_{12}	Take 10 drops in water or juice three times per day.	Stimulates immune system function.
Niacin	Take 50 mg the first day, then increase dosage daily by 50 mg to 1,000 mg. If adverse effects occur, decrease dosage. Take doses higher than 1,000 mg only under the supervision of a physician.	Dilates blood vessels, improving circulation; a coenzyme for many enzymes.
or Niacin-bound chromium	Take 20 mg per day, or as indicated on the label.	
Thymus glandular	Take 100 mg per day; or take as indicated on the label.	Stimulates thymus gland activity and immune system function.
Vitamin A	Take up to 50,000 IU for no more than one month. Take doses higher than 50,000 IU only under the supervision of a physician. If pregnant, do not take more than 10,000 IU.	Needed for a strong immune system and healthy mucous membranes; potent antioxidant.
Vitamin B complex	Take 100–150 mg three times per day.	B vitamins act as coenzymes.
Vitamin C	Take 10,000–15,000 mg per day; or take 1–2 grams every hour until symptoms improve.	Fights infections; improves healing; stimulates the adrenal glands; promotes immune system function; powerful antioxidant.
Vitamin E	Take 1,200 IU per day from all sources (including the multivitamin complex in the introduction to Part Two).	Acts as an intercellular antioxidant; improves circulation; helps resist diseases at the cellular level.
Zinc	Take 50 mg per day. Do not take more than 100 mg per day from all sources.	Needed by over 300 enzymes; necessary for white blood cell immune function; improves healing.

Comments

❏ Enzyme-rich fresh fruits and vegetables (particularly green vegetables) should comprise 60 percent of your diet. If your body cannot tolerate raw fruits and vegetables, increase your intake of digestive enzymes, or sauté or steam your produce. For additional dietary recommendations, see Part Three.

❏ To improve your overall health, avoid alcohol, white flour, and refined sugar. Also avoid drugs (except for those prescribed by your physician), cigarettes, food additives and preservatives, caffeine, and chlorinated water (use bottled, distilled, or well water instead)

❏ According to Joan Priestley, M.D. (a world-renowned HIV/AIDS specialist), maitake mushrooms can be helpful in the treatment of certain viral disorders, such as HIV.

❏ Use the Enzyme Kidney Flush for one week. The following week, use the Enzyme Toxin Flush. Use them only once per month. See Part Three for instructions. It is essential to restore the intestinal flora and re-establish intestinal peristalsis in the treatment of viral infections. Cleansing methods, including juicing, fasting, detoxification diets, and exercise are helpful. See Part Three for more information.

❏ Physicians sometimes use enzyme injections or enzyme retention enemas in the treatment of viral infections. Both methods allow a greater concentration of the enzymes to be absorbed than might be accomplished if the enzymes were taken orally. An enzyme retention enema is a means of instilling some enzymes into the colon to be retained. This allows the enzymes to be absorbed into the body's systems. Physicians use products specially formulated for these uses. Caution should be exercised, however, since repeated rectal administration of very high doses of enzymes could lead to irritation of the anus or the skin around the anus. This can be avoided by applying a lubricating ointment, such as vitamin E oil or aloe vera, before administering the enema.

❏ With more serious viral infections, such as HIV/AIDS, seek the services of a health-care practitioner.

❏ *See also* AIDS; CHICKENPOX; COLDS; HEPATITIS AND SYNOVITIS; HERPES SIMPLEX VIRUS; HERPES ZOSTER; INFLUENZA; MEASLES; WARTS

Warts

Warts (or verrucae) are small benign growths on the skin caused by human papillomaviruses. Warts are most common in children and young adults, but can occur in anyone. Generally, warts are not very contagious from person to person (except genital warts), although they can be spread from one area of the body to another. They are usually painless, and often disappear within about two years. Some warts, however, remain for years, and some return after they disappear.

There are five types of warts. *Common warts* generally appear on the fingers, face, scalp, and knees. They are usually small (less than a half-inch in diameter); roundish; and grayish, yellow, or brown. *Filiform warts* are small thin and long warts that usually appear on the eyelids, face, lips, or neck. *Flat warts* usually occur on the face and in groups. They appear as smooth, yellow-brown spots. *Plantar warts* are flat tender growths on the soles of feet. They are usually surrounded by callused skin. *Genital warts* are sexually transmissible and appear in and around the vagina, cervix, penis, and rectum. They appear as tiny soft, moist, pink growths with rough surfaces that may resemble cauliflower. They grow very quickly, often in groups. When they appear on the cervix, they can lead to cervical cancer. Genital warts must be treated by a physician.

Enzyme Therapy

Systemic enzyme therapy is used to fight viruses, decrease inflammation, stimulate the immune system, and break up antigen-antibody complexes. Enzymes improve circulation, help speed tissue repair, bring nutrients to the damaged area, remove waste products, improve health, strengthen the body as a whole, and build general resistance.

Digestive enzyme therapy is used to improve the digestion of food, reduce stress on the gastrointestinal mucosa, help maintain normal pH levels, detoxify the body, and promote the growth of healthy intestinal flora, thereby improving the overall health of the body. Digestive enzymes also serve as replacements for the body's pancreatic enzymes, leaving the pancreatic enzymes free to perform other functions in the body, such as boosting immunity. Enzymes can also be used topically to treat warts.

Enzyme Absorption System Enhancers (EASE) maximize enzyme activity. They can also improve the absorption and bioavailability of various nutrients and decrease the drain on the body's own digestive enzymes, thus prolonging their lives.

In addition to the standard multivitamin, multimineral, and multiglandular complexes recommended in the introduction to Part Two, the following supplements are recommended for the treatment of warts. The following recommended dosages are for adults. Children between the ages of two and five should take one-quarter the recommended dosage, those between the ages of six and twelve should take one-half the recommended dosage, and those between the ages of thirteen and seventeen should take three-quarters the recommended adult dosage.

ESSENTIAL ENZYMES

Enzyme	Suggested Dosage	Actions
WobeMugos E Ointment from Mucos	Apply to the affected area three times per day.	Helps fight warts.

AND SELECT ONE OF THE FOLLOWING:

Enzyme	Suggested Dosage	Actions
Formula One—See page 38 of Part One for a list of products.	Take 10 tablets three times per day between meals for six months. Then decrease dosage to 5 tablets three times per day between meals.	Bolsters the immune system; breaks up antigen-antibody complexes; fights free-radical formation and inflammation.
Mulsal From Mucos	Take 10 tablets three times per day for six months between meals. Then decrease dosage to 2 tablets between each meal.	Stimulates the immune system; breaks up antigen-antibody complexes; fights free-radical formation and inflammation.
Phlogenzym from Mucos	Take 10 tablets three times per day between meals.	Stimulates the immune system; breaks up antigen-antibody complexes; fights free-radical formation and inflammation.
Bromelain	Take seven to eight 230–250 mg tablets three times per day between meals until symptoms disappear.	Stimulates the immune system; breaks up antigen-antibody complexes; fights free-radical formation and inflammation.
Proteolytic enzymes	Take as indicated on the label.	Bolsters the immune system; breaks up antigen-antibody complexes; fights free-radical formation and inflammation.

SUPPORTIVE ENZYMES

Enzyme	Suggested Dosage	Actions
Formula Three—See page 38 of Part One for a list of products.	Take as indicated on the labels.	Potent antioxidant enzymes.
A digestive enzyme product containing protease, amylase, and lipase enzymes. See the list of digestive enzyme products in Appendix.	Take as indicated on the label with meals.	Improves the breakdown and absorption of food nutrients to improve the overall health of the body; fights toxic build-up.

ENZYME ABSORPTION SYSTEM ENHANCERS (EASE)

Combination	Suggested Dosage	Actions
Ecology Pak from Life Plus *or*	Take 2 or 3 tablets once or twice per day, preferably in the morning and no later than midday.	Assists with natural detoxification; stimulates immune function.
Hepazyme from Enzymatic Therapy *or*	Take as indicated on the label.	Stimulates immune function.
Inflamzyme from PhytoPharmica *or*	Take as indicated on the label.	Immune system activator; reduces inflammation; fights free radicals.
Kyolic Formula 102 (with enzymes) *or*	Take 2 capsules or tablets with meals two to three times per day.	Strengthens the immune system; helps alleviate stress; helps fight free radicals; helps fight inflammation; improves digestion and absorption of nutrients.
Metabolic Liver Formula from Prevail *or*	Take as indicated on the label.	Improves liver function, thus improving detoxification.
Vita-C-1000 from Life Plus	Take as indicated on the label.	Improves liver function, thus boosts immune function; fights infection.

ENZYME HELPERS

Nutrient	Suggested Dosage	Actions
Acidophilus and other probiotics	Take as indicated on the label.	Stimulates enzyme activity; improves gastrointestinal function to improve overall health; improves immunity.
Adrenal glandular	Take 100–400 mg per day.	Helps fight stress and adrenal exhaustion; fights inflammation.
Kidney glandular	Take as indicated on the label.	Supports immune defense.
Kyolic Aged Garlic Liquid Extract with Vitamins B₁ and B₁₂	Take 10 drops in juice or water three times per day.	Stimulates immune system function.
Liver glandular	Take 40 mg one to three times per day.	Improves immune defense.
Selenium	Take 50 mcg one or two times per day.	Acts as an enzyme activator; required by glutathione peroxidase.
Thymus glandular	Take 90 mg per day.	Stimulates activity of the thyroid gland.
Vitamin A	Take up to 50,000 IU for no more than one month. Take doses higher than 50,000 IU only under the supervision of a physician. If pregnant, do not take more than 10,000 IU per day.	Needed for strong immune system and healthy mucous membranes; potent antioxidant.
Vitamin B complex	Take 100–150 mg three times per day.	B vitamins act as coenzymes.

Vitamin C	Take 10,000–15,000 mg per day, or 1–2 grams every hour until symptoms improve.	Fights infections; improves healing; stimulates the adrenal glands; promotes immune system; powerful antioxidant.
Vitamin E	Take 1,200 IU per day from all sources (including the multivitamin complex in the introduction to Part Two).	Acts as an intercellular antioxidant; improves circulation; helps resist diseases at the cellular level.
Zinc	Take 50 mg per day. Do not take more than 100 mg per day from all sources.	Needed by over 300 enzymes; necessary for white blood cell immune function; improves healing.

Comments

❑ To improve overall health, avoid alcohol, refined white flour, and sugar.

❑ A number of herbs can bolster the immune system, including astragalus; echinacea; green barley extract; ginseng; licorice; maitake, reishi, and shiitake mushrooms; pau d'arco; and royal jelly.

❑ Follow the Enzyme Kidney Flush program for one week. The following week, perform the Enzyme Toxin Flush program. See Part Three for instructions on these flushes. Repeat the Enzyme Toxin Flush only once per month.

❑ Combine four ounces of pineapple juice with four ounces of lemon or grapefruit juice. Gargle and swallow. Do this three times per day. Follow directions in Part Three for juicing.

Whiplash

A whiplash results from a sudden injury or accident that hurls your head violently backward (called hyperextension) and then forward (called hyperflexion). These injuries often occur after your car is rear-ended by another car. Muscles and ligaments tying the front portion of the neck are stretched and torn as the head moves backward. Then, as the head moves forward, the muscles in the back of the neck are also stretched and torn, plus the vertebrae and discs are jammed together. There may be rupturing of the small blood vessels (called capillaries), swelling, heat, and resulting pain. Nerve damage is quite likely. The force of the accident can cause a reversal of your neck's normal curve. In extreme cases, there may be ruptured intervertebral discs, fractured vertebrae, and/or a concussion. Spinal cord or other nerve damage is possible.

Symptoms

The symptoms of whiplash include mild to severe neck pain,

numbness, and tingling, which may radiate up the back of the head and/or down the neck, into the shoulders and sometimes down the arms. Headache may be present, and there may be dizziness, ringing in the ears, blurred or double vision, vomiting, and nausea. Movement of the head and neck may be limited (see "Neck Exercises to Strengthen and Return Normal Range of Motion" on page 428).

The symptoms of a whiplash may not become evident for some twelve to twenty-four hours after the accident. It takes this long for fluid from the capillaries to seep into the surrounding tissue and cause the inflammatory process to occur. This is why immediate health care is critical. Every moment you waste can add weeks, months, or even years to your recovery time.

Frequently, the pain subsides, but the ligaments, discs, and/or blood vessels may not return to their pre-injury state. Further, the torn ligaments and muscles may cause cervical instability. A ligament holding two vertebrae in alignment is a lot like a rubber band. When the band is overstretched, it will not return to its prestretched, pre-injury length and strength.

A whiplash injury or frequent microtrauma (small traumas) can cause the vertebrae to deteriorate faster than normal. It can also cause a decrease in the spaces between the discs and lead to spinal derangements. If your injury resulted from a car accident, the speed the vehicles were going may have little or no relationship to the severity of your neck injury. In some instances, a mild collision may cause greater damage than one at a higher speed.

With improper or delayed treatment, the symptoms of a whiplash injury can become chronic. Residual problems can be double or blurred vision, ringing in the ears, impaired taste, headaches, neck, shoulder and back pain, tingling down the arms and possibly into the hands, depression, inability to sleep, and emotional problems.

Enzyme Therapy

Systemic enzyme therapy is effective at reducing inflammation, swelling, and pain in the injured area and rehabilitating the injured patient. Enzymes can decrease scar tissue build-up and improve circulation, speeding tissue repair by bringing nutrients to, and removing waste products from, the damaged area. They can also stimulate the immune system and enhance wellness.

Digestive enzyme therapy will improve the breakdown and absorption of food nutrients, reduce stress on the gastrointestinal mucosa, promote the growth of healthy intestinal flora, help maintain normal body pH levels, and detoxify the body, thereby improving the overall health of the body. Digestive enzymes also serve as replacements for the body's pancreatic enzymes, leaving the pancreatic enzymes free to perform other functions in the body, such as decreasing inflammation and improving circulation.

Enzyme Absorption System Enhancers (EASE), that is, enzymes combined with vitamins, minerals, phytochemicals, and other nutrients maximize enzyme activity. They can also improve the absorption and bioavailability of various nutrients, and decrease the drain on the body's own digestive enzymes, thus prolonging their lives.

In addition to the standard multivitamin, multimineral, and multiglandular complexes recommended in the introduction to Part Two, the following supplements are recommended for the treatment of whiplash. The following recommended dosages are for adults. Children between the ages of two and five should take one-quarter the recommended dosage, those between the ages of six and twelve should take one-half the recommended dosage, and those between the ages of thirteen and seventeen should take three-quarters the recommended adult dosage.

ESSENTIAL ENZYMES

Enzyme	Suggested Dosage	Actions
Formula One—See page 38 of Part One for a list of products.	Take 10 tablets three times per day between meals. As symptoms subside (which should occur in approximately one week) decrease dosage to 3–5 tablets three times per day between meals. When symptoms disappear, begin taking a maintenance dose of 2–3 tablets after each meal.	Decreases inflammation, pain, and swelling; increases rate of healing; stimulates the immune system; strengthens capillary walls.
or		
Bromelain	Take three to four 230–250 mg capsules per day until symptoms subside. If severe, take 6–8 capsules per day.	Fights inflammation, pain, and swelling; speeds recovery after injury or surgery; stimulates the immune system.
or		
Trypsin with chymotrypsin	Take 250 mg of trypsin and 10 mg of chymotrypsin three times per day between meals. As symptoms subside (which should occur in approximately one week), decrease dosage to 125 mg of trypsin and 5 mg of chymotrypsin tablets three times per day between meals.	Decreases inflammation, pain, and swelling; increases rate of healing; stimulates the immune system; strengthens capillary walls.
or		
Papain	Take 600 mg three times per day between meals. As symptoms subside (which should occur in approximately one week), decrease dosage to 200 mg three times per day between meals.	Decreases inflammation, pain, and swelling; increases rate of healing; stimulates the immune system; strengthens capillary walls.
or		
Pancreatin	Take 1,000 mg three times per day between meals. As symptoms subside (which should occur in approximately	Decreases inflammation, pain, and swelling; increases rate of healing; stimulates the immune system; strengthens capillary walls.

Neck Exercises to Strengthen and Return Normal Range of Motion

These exercises strengthen the ligaments and muscles that support the head and spine and help to maintain normal range of motion. The exercises should be performed only with your doctor's approval and should not be overdone to the point of exhaustion or pain.

To warm up the neck muscles, start slowly. Gradually increase the number of exercises, but stay within your comfort range. Do not go beyond five to ten repetitions. If the neck begins to hurt, stop! Do not overdo!

1. Gentle flexion and extension: First, tuck your chin into your chest. Then, bend your head forward gently. Repeat as your doctor prescribes (initially, five times).

2. Lateral bending: Bend the head to the left as far as possible, and then to the right, attempting to touch your ear to your shoulder on each side. Do five times.

3. Turn head laterally: Turn your head as far as possible to the left and then to the right five times each.

4. Lateral flexion against resistance: Place the palm of your hand against the side of your head just above the ear. Bend the head laterally, while applying resistive pressure. Hold for five counts. Repeat this exercise five times on each side.

5. Flexion against resistance: Place the palms of both hands against your forehead. Apply mild resistance as you slightly lower your head for a count of five. Repeat five times.

6. Extension against resistance: Lock your hands behind your head. Begin with the head and neck in their normal position. Bend your neck backwards, while applying mild resistance to the back of your head. Do five times.

or	one week), decrease dosage to 100 mg three times per day between meals.	
Enzyme complexes rich in proteases plus amylases and lipases, working synergistically.	Take as indicated on the label between meals.	Decreases inflammation, pain, and swelling; increases rate of healing; stimulates the immune system; strengthens capillary walls.

SUPPORTIVE ENZYMES

Enzyme	Suggested Dosage	Actions
Formula Three—See page 38 of Part One for a list of products.	Take 50 mcg three times per day between meals.	Potent antioxidant enzymes.
A digestive enzyme product containing protease, amylase, and lipase. See the list of digestive enzyme products in Appendix.	Take as indicated on the label with meals.	Improves breakdown and absorption of food nutrients to improve the overall health of the body; fights toxic build-up.

ENZYME ABSORPTION SYSTEM ENHANCERS (EASE)

Combination	Suggested Dosage	Actions
Bromelain Complex from PhytoPharmica *or*	Take as indicated on the label.	Reduces swelling and inflammation; improves circulation.
Connect-All from Nature's Plus *or*	Take as indicated on the label.	Supports connective tissue.
CTR Support from	Take 2–3 capsules per	Nourishes the body during tissue

PhysioLogics *or*	day, preferably on an empty stomach; or follow the advice of your health-care professional.	recovery and healing from injury; helps maintain healthy connective tissue.
Joint Renewal Rx "1500" from Phyto-Therapy *or*	Take as indicated on the label.	Helps to support healthy cartilage and joint function.
Joint-Ease from Nature's Life *or*	Take as indicated on the label.	Decreases inflammation and swelling.
Mobil-Ease from Prevail *or*	Take one capsule between meals two to three times per day.	Has anti-inflammatory activity.
Protease Concentrate from Tyler Encapsulations *or*	Take as indicated on the label.	Breaks up fibrin; fights inflammation.
Zymain from Anabolic Labs	Take as indicated on the label.	Provides nutrition essential for tissue and joint repair and maintenance.

ENZYME HELPERS

Nutrient	Suggested Dosage	Actions
Amino-acid complex including branched-chain amino acids (isoleucine, leucine, and valine)	Take as indicated on the label on an empty stomach.	Required for enzyme production and function; helps build muscles, tendons, and ligaments.
Calcium	Take 1,000 mg per day.	Plays important roles in nerve transmission and muscle contraction.

Magnesium	Take 500 mg per day.	Needed for proper calcium absorption.
Selenium	Take 300 mcg per day.	Acts as an enzyme activator and is required by the antioxidant enzyme glutathione peroxidase.
Vitamin C	Take 5,000–10,000 mg per day.	Helps to produce collagen and normal connective tissue.
Vitamin E	Take 600 IU per day.	Supplies oxygen to the cells.
Zinc	Take 50 mg per day. Do not take more than 100 mg per day from all sources.	A cofactor for more than 300 enzymes; potent antioxidant.

Comments

❑ Ice should be used for the first one to three days. Fill two or three small paper cups with water and freeze. When frozen, take one out, tear off the top half inch of paper revealing the ice. Place a paper towel or dish towel over the ice and apply to the painful area. Rub the affected area in a constant circulatory motion with the ice. Do not apply the ice directly to the skin. Apply the ice for five minutes, then keep the ice off the area for five minutes. Repeat this every one to two hours. The purpose is to stimulate blood flow to the area, remove waste from the area, and reduce inflammation.

❑ After three days, alternate ice with heat packs every fifteen minutes until symptoms go away. "Blue ice" or a frozen vegetable packet works better for this application. Wrap it with a dish towel and apply to the painful area. For heat, use a hot water bottle or heating pad. If a hot water bottle is used, wrap in a towel and apply. If a heating pad is used, put a warm, moist towel on the patient, then place a sheet of wax paper over it, then the heating pad, and finally, a large bath towel. Make sure the wax paper is wide and long enough to cover the entire heating pad with approximately one inch more extending out on every side. The heating pad should never touch the moist towel or the patient's skin. Great care should be exercised to assure this does not happen.

❑ Your diet should contain fruits and vegetables high in enzymatic activity, such as pineapples, papayas, figs, broccoli, parsley, and sprouting vegetables. Turn to Part Three for dietary recommendations.

❑ Increase your intake of protein. An amino-acid complex, such as AminoLogic by PhysioLogics, can supply the amino acids (the building blocks of protein and enzymes) your body needs to build muscle and connective tissue.

❑ Bioflavonoids, such as rutin, quercetin, or hesperidin, can strengthen capillaries and fight free radicals.

❑ A number of herbs, including ginkgo biloba and white willow bark, can reduce inflammation and enhance healing.

❑ Glucosamine (either as a hydrochloride or sulfate) and chondroitin sulfate can stimulate cartilage repair.

❑ Mucopolysaccharides, found in bovine cartilage extract and shark cartilage extract, support healthy, flexible joints; control inflammation; and help fight inflammation.

❑ Detoxification can help eliminate some of the toxins and free radicals that result from an injury. Follow the Enzyme Toxin Flush programs as outlined in Part Three of this book. Do no more than once per month.

❑ Seek the services of an alternative health care professional, such as a chiropractor, osteopath, or naturopath. These doctors are trained to restore normal function to the spine and to encourage a healthy lifestyle.

❑ A cervical collar is sometimes used to support the neck. A cervical pillow can be used at night to help return the neck to its normal curve.

Wounds

See SCARS; SKIN ULCERS.

Wryneck

See TORTICOLLIS.

X-Ray Hangover

See RADIATION SICKNESS.

Yeast Infections

See FUNGAL SKIN INFECTIONS.

PART THREE

COMPLEMENTARY THERAPIES

COMPLEMENTARY THERAPIES

Part Two discussed enzyme treatment programs for specific conditions. In Part Three, I will explain a number of therapies that can augment the enzyme treatment programs. These therapies, including dietary modifications, detoxification methods, exercise, and various stress reduction techniques, will assist enzymes and enzyme helpers. After reviewing your options, choose the programs that can be of most benefit to you.

Baths

Warm baths are beneficial because they aid circulation by dilating blood vessels. The blood cleanses, transports, and disposes of debris and other wastes. Improving circulation will help improve detoxification.

Warm baths are also relaxing and help to create a general feeling of well-being. Hot showers (to tolerance) with a pulsating showerhead may be even more beneficial, as the pressure exerted on the body by the streams of water creates a massage effect that stimulates circulation. If you have access to a sauna or whirlpool, it would be wise to take advantage of this.

Detoxification Methods

As mentioned in Chapter 3 of Part One, enzyme depleters in our air, water, soil, and food are constantly assaulting the very enzymes that keep us alive. Because enzymes regulate your body's chemical reactions, anything that interferes with their ability to function will diminish their strength and activity, leading to a build-up of toxins, resulting in fatigue and illness. Enzyme-dead foods overtax your body's own enzyme supply, forcing the pancreas and other organs to produce more enzymes in an attempt to digest the food you eat. Any energy your body expends digesting food is energy that cannot be utilized to maintain your immune system, your circulation, or any other process—including your body's own detoxification process.

The human body possesses many methods for detoxification and regeneration. A great deal of effort goes into maintaining the circulation of blood and lymph. The blood, supplied by the arteries, not only provides oxygen and nutrients responsible for metabolism in all cells, it also cleanses and transports and disposes of debris and other wastes. Part of the disposal process takes place through excretion via the intestines, kidneys, lungs, and skin, and part is accomplished by degradation, detoxification, and catabolism in organs, including the liver, spleen, and lymph.

In addition, phagocytes (cells of the immune system that ingest bacteria and other foreign particles) constitute one of the main systems for internal detoxification, as they have the ability to invade microorganisms and chemical toxins. The phagocytes also help eliminate wastes produced by the body itself (including dead cells). The best known phagocytic cells are the macrophages, whose action is critical in maintaining health and avoiding chronic degenerative diseases. Unfortunately, many modern medicines inhibit macrophages. This causes a gradual accumulation of metabolic wastes and toxins in the fluids and tissues of the body, leading to chronic disorders.

The primary goals of detoxification are to break up and eliminate toxins and waste products in the body, to improve excretion, to stimulate the body's detoxification mechanism, and to give relief when the immune system is suppressed.

Detoxifying the body is basically a twofold process. The first step is to avoid toxins in the environment as much as possible. See the inset on page 435, and Chapter 3 of Part One to refresh your memory on where and how you are exposed to these toxins.

The second step is to aid the body in voiding itself of toxins. Often these toxins are the reason for a disease process. You can help the body by helping along intestinal peristalsis (the movement of food through the alimentary canal), thus improving elimination. The large intestine is probably the most common place for toxins to linger. Improving elimination can benefit a number of conditions, including chronic kidney disorders, headaches, rheumatic disorders, depressions, and allergies.

There are many ways to improve elimination. Increase fiber intake by eating more fresh fruits and vegetables or by taking a fiber supplement. Drink plenty of water. If constipated, an enema may be helpful. It is wise to take an enema for one to two days to keep the bowel and rectum free of debris. By using just warm water, you will facilitate removal of much potentially dangerous material. Cold water may cause spasms and discomfort. It is not wise to use soap or any other additive (other than coffee—*see* Coffee Retention Enema below) in the water. This could cause irritation to the bowel wall. In addition, some of these additives can be absorbed into the bloodstream and form toxins.

It is also beneficial to improve and stimulate kidney function. This can be accomplished by increasing liquid intake and acidifying the urine. It may also be necessary to stimulate bile secretion through the use of the coffee enema (see below). After elimination is improved, it is important to restore intestinal flora with probiotic supplements.

Although important, improving elimination and restoring intestinal flora may not be sufficient to restore health. Over the years, toxins and waste accumulate on the walls of the colon and throughout the body's blood vessels. Flushing the body periodically with coffee enemas, the Enzyme Kidney Flush, the Enzyme Toxin Flush, juice detoxifications, and baths can help eliminate this build up of waste.

COFFEE RETENTION ENEMA

When there are toxins in the body, enzyme activity is decreased. Therefore, periodic detoxification enemas are essential for an optimally functioning body and enzymatic activity. Coffee retention enemas are quite beneficial in stimulating the liver into detoxifying itself. Caffeine stimulates secretion of bile, which helps to restore or maintain the necessary alkaline pH in the small intestine.

Note: The caffeine in the coffee enema could cause some nausea. If it is too great, reduce the amount of coffee used.

Use a coffee retention enema each day for seven days. Use twice per day if your body is extremely toxic. To prepare, add one cup of brewed coffee (do not use instant or decaffeinated coffee) to one quart of warm tap or distilled water; or add three tablespoons of ground coffee to one quart of water and boil for three minutes, then simmer for twenty minutes. Strain and cool to body temperature. Put the cooled mixture in an enema bag, which can be purchased at most pharmacies. Lubricate the anus and the tip of the enema bag or syringe with vitamin E oil or a surgical lubricant. Do not use petroleum jelly. Lie on your right side with both legs drawn close to the abdomen or kneel with your head down and your buttocks up. Breathe deeply. Administer about one cup into the rectum at a time, massaging the area between the rib cage and the pelvis manually until the urge to expel the liquid subsides. Begin massaging the lower right side of the abdomen and slowly move in a clockwise direction. Allow the coffee to remain in the colon for ten to fifteen minutes (depending on tolerance), then expel into the toilet.

If you experience spasms or tension in the bowel, use a warmer solution (99°–102°F). If the muscle tone of the bowel is weak, make the solution a bit colder—75°–80°F—to help strengthen it.

ENZYME RETENTION ENEMA

An enzyme retention enema is a means of instilling some enzymes into the colon to be retained. This allows the enzymes to be absorbed into the body's systems. They are often used in the treatment of cancers. Enzyme retention enemas can be obtained only through a physician, who will write a prescription for its use. It is important to use an enzyme product that has been formulated specifically for this use, such as Wobe-Mugos Th. Follow the instructions on the label. Caution should be exercised, however, since repeated rectal administration of very high doses of enzymes could lead to irritation of the anus or the skin around the anus. This can be avoided by applying a lubricating ointment, such as vitamin E oil or aloe vera, before administering the enema.

ENZYME KIDNEY FLUSH

The kidney filters some 4,000 quarts of blood per day. In order to maintain the alkaline/acid balance, the elimination of waste is critical for kidney health. Sufficient amounts of liquid (such as spring or distilled water) should be consumed daily. However, fruit and vegetable juices (in addition to raw, whole, enzyme-rich foods, enzyme supplements, and their helpers) should be used if you are nutritionally deficient.

To flush the kidneys, eat at least three cups of watermelon or drink at least three 8-ounce glasses of its juice every day for one week. During this time, your diet should consist almost entirely of fresh fruits and vegetables.

If you have difficulties finding watermelon at a reasonable price outside of the summer months, you can substitute one of the following methods for the above one. Every day for one week, drink six to eight 8-ounce glasses of unsweetened cranberry juice (available at health-food stores) or make your own from freshly crushed cranberries. If desired, you may add fresh apple juice to cranberry juice to cut the bitter taste. Just be sure to begin with eight ounces of cranberry juice. In addition to the juice, take 500 mg of vitamin B_6 three times per day. Also take 10,000 mg of vitamin C and 1,000 IU of vitamin E per day. Take these doses in addition to any nutritional supplement plan you are currently following. Your diet during this time should consist almost entirely of fresh fruits and vegetables.

Your other option is to mix the juice of one lemon in a 6–8 ounce glass of distilled, filtered, or bottled water. If necessary, add a half teaspoon of honey to enhance the flavor. Drink two to three glasses per day for one week. Every day for one week, drink three 8-ounce glasses of unsweetened cranberry juice. In addition to the juice, take 500 mg of vitamin B_6 three times per day. Also take 10,000 mg of vitamin C and 1,000 IU of vitamin E per day. This should be done in addition to any nutritional supplement plan you are currently following.

ENZYME TOXIN FLUSH

This flush is an essential detoxifying tool that can help reestablish gallbladder and liver function, especially in those individuals with chronic degenerative diseases. This flush should be used by adults only. If you have a chronic disorder, check with a well-trained physician before using the flush.

For the first six days, while taking your normal supplements and eating regular meals (three to five per day), drink a minimum of six to eight 8-ounce glasses of apple cider or apple juice per day. To ensure freshness, make your cider or juice from organically grown apples, or buy it from a health-food store. About three hours after lunch on day seven, dissolve two teaspoons of disodium phosphate (DSP—which can be found at health-food stores and pharmacies) in about one ounce of hot water. If the drink is unpleasant, follow it with freshly squeezed citrus juice. If you are on a sodium-free diet, a sodium-free cathartic, such as castor oil or Epsom salts can be substituted for the disodium phosphate. Repeat in two hours. The evening meal should consist only of grapefruit and grapefruit juice or other citrus fruits or juices.

At bedtime, take either one-half cup of lemon juice blended with one-half cup of unrefined olive oil at room temperature (unrefined olive oil can be purchased at any health-food store) or one-half cup of unrefined olive oil followed by four to six ounces of grapefruit juice. Fresh juice is best, although bottled or canned citrus juices are permissible. When you go to bed, lie on your right side with your right knee pulled up close to the chest for the first thirty minutes.

The next day, one hour before breakfast, again take two teaspoons of disodium phosphate dissolved in two ounces of hot water. You may then return to your normal diet and whatever nutritional program you were following.

Note: Mild to moderate nausea may occur when consuming the olive oil and citrus juice. By the time you go to bed, the nausea should dissipate. If the olive oil causes vomiting, do not repeat the procedure.

The day after the flush, those who suffer from gallstones, nausea, and backaches may find small gallstone-like objects in the stool. There might be a large number of these irregularly-shaped gelatinous objects of varying size (from the size of a grape seed to the size of a cherry seed). If you find these objects in the stool, repeat the flush in two weeks.

THE VITAMIN C/ENZYME FLUSH

The Vitamin C/Enzyme Flush is another useful detoxifying tool. The pineapple juice is also effective in soothing sore throats and treating respiratory conditions, due to its high enzyme, fruit acid, and vitamin C content. The vitamin C and the enzymes also stimulate immune function.

Place one peeled and washed orange and two 2 1/2 inch wedges of peeled and washed pineapple in a blender and blend at high speed. Pour into a glass and mix in 5,000 to 10,000 milligrams of crystalline vitamin C. Drink once per day, preferably upon rising or after breakfast.

THE TWO-DAY ENZYME JUICE FAST

Juice fasting is a safe and effective way to detoxify and cleanse the body. Periodically, our bodies need a "house cleaning." Traditionally, various religions have used fasting and prayer as a means of cleansing the body. It is my opinion that some nutrients are needed during a fast. Therefore, I do not recommend a strict water fast. If you are diabetic, before juice fasting, be sure to consult with your physician and thoroughly discuss the program.

The juicing diet can be a big step in aiding the body to detoxify itself. The natural enzymes, vitamins, minerals, and phytochemicals in fresh fruits and vegetables will also aid tremendously. In addition, several fruits, especially cherries, lemons, apples, grapes, bananas, apricots, prunes, peaches, and plums contain malic acid, which cleanses the liver, kidneys, stomach, and intestines. They are also enzyme-rich. Therefore, choosing these fruits when on a juice fast can be extremely beneficial. Also include fruits known to be particularly rich in enzymes, including kiwis, papayas, and pineapples.

Vegetables are loaded with enzymes, vitamins, minerals, and phytochemicals. Make plenty of juice from carrots, celery, parsley, cabbage, and beets.

As mentioned in Chapter 4 of Part One, juices, like their whole fruit and vegetable counterparts, have all-around protective action. Historically, juices have been used to fight a number of degenerative diseases, including cancer, kidney problems, cardiovascular diseases, high blood pressure, obesity, and high cholesterol.

The Two-Day Enzyme Juice Fast decreases toxins in the body. Throughout the fast, maintain your current nutritional supplement plan. On day one, eat no solid foods. Instead, drink plenty of fresh, pure water and juice made from equal parts of carrot and apple juice. On day two, add the juice from several vegetables (including celery, beets, potatoes, parsley, broccoli, spinach, and other green leafy vegetables) and fruits (including pineapple, papaya, citrus, and grapes). On day three, resume eating as normal. Use this fast no more than once per month.

When juicing, do not remove the peels. It is important to juice the entire vegetable or fruit. This is why it is important to use organically grown foods only. If you use organically grown foods, you don't have to worry about the presence of pesticides or other toxins in your juice. Avoid canned juices, since all of their enzymes have been destroyed by the heat used in canning. To assure the greatest enzyme activity and to prevent oxidation, prepare juices daily.

If fresh produce is unavailable, use concentrated vegetable

Ways to Avoid Toxin Development in the Body

Toxins can cause weakness, fatigue, and illness. They can also impair digestion and the absorption and activity of nutrients in the body, which can also adversely affect health. Therefore, knowing how to avoid toxic development in the body is critical for good health. To prevent toxin build-up in the body, avoid the following:

• Air pollution, which generates free-radical activity. Ozone, lead, sulfur dioxide, nitrogen oxide, cadmium, and mercury all inhibit enzyme activity.

• Indoor air pollution from cleaning products, paints, aerosols (which can propel their contents into the lungs, causing serious damage), and tobacco smoke. The tar in tobacco smoke collects in the lungs and interferes with the exchange of oxygen and carbon dioxide, causing irritation and inflammation. Also, tars reduce the proteolytic activity of the blood by as much as 10 percent for a period of time.

• Synthetic cosmetics and skin-care products. Substances applied directly to the body can be absorbed through the skin. Products such as insect repellents, first aid sprays, and most deodorants should also be avoided (unless made from natural ingredients). *See* SKIN CARE in Part Three for more information.

• Excessive sunshine. Not only can it promote free-radical activity, destroy antioxidant enzymes, and age your skin prematurely, it is the leading cause of skin cancer.

• Drugs (unless prescribed by a physician and absolutely essential for health). The proper use of enzyme, vitamin, and mineral supplements can help reduce the need for a number of drugs, including antibiotics. Other drugs, including antihistamines, harsh laxatives, tranquilizers, sleeping pills, stimulants, and the like, should not be used. In almost all cases, these items are synthetic chemicals that are foreign to the body and create a large demand on the liver.

and fruit powders or tablets. To give your juice an even greater enzyme boost, mix in a heaping teaspoon of such vegetable and fruit powders as Kyo-Green from Wakunaga, Green Magma from Green Foods Corporation, Emerald Harvest from Nikken, or similar products (available at your health-food store).

The type of juicer you use can be of importance. There are probably as many juicers on the market as there are juices. Purchase your juicer from a reputable dealer. They may cost anywhere from fifty dollars to over six hundred dollars. Those that fall in the middle price bracket are usually quite good. For more information on juicing, see Chapter 4 of Part One.

Diet

In numerous animal studies, degenerative disorders and the speed of aging have been decreased and life expectancy increased by some 50 percent just by reducing food intake. This may be due to the effect on free-radical activity. Research studies indicate that antioxidant enzyme activity (of superoxide dismutase, catalase, and glutathione peroxidase) increases in laboratory mice and rats fed fewer calories.

During and after World War I and World War II, the German food supply decreased. However, there was a positive side. Atherosclerosis, coronary heart disease, stroke, and diabetes became almost unknown. People were hungrier, but much healthier. When the food supply increased, degen-

ertive diseases returned, especially atherosclerosis, diabetes, coronary heart disease, stroke, and vascular diseases. German scientists found that eating fewer calories and less animal fat, refined sugar, chocolates, and cake helped prolong life.

But reducing your food intake is not enough. It is also important to make every calorie count. A nutritious diet can keep you healthy by improving your immune system's ability to fight disease and improving circulation and digestion, and can help you live longer and fight diseases when they do occur. (See Chapter 4 of Part One and "Increasing Your Enzyme Intake" on page 436 for ways to maximize enzymes in your diet.) Because fresh, raw fruits and vegetables are loaded with enzymes (as well as vitamins, minerals, and phytochemicals), I advocate the following diet:

• Eat enzyme-rich fresh fruits (including pineapples and papayas) and vegetables (especially green, leafy vegetables) in season. These foods should constitute 60 percent or more of your dietary intake. If your body cannot tolerate raw fruits and vegetables, then sauté or steam them. But remember, heat kills enzymes, so cook them very lightly.

• Juicing fruits and vegetables is an easy way to increase your fruit and vegetable intake. Drink large amounts of freshly made juices, especially vegetable juices made from carrots, celery, parsley, broccoli, and beets.

• Eat a diet low in fats, cholesterol, salt, soft drinks, and refined carbohydrates.

• Increase your intake of enzyme enhancers, including sprouts, raw honey, and wheat germ.

Increasing Your Enzyme Intake

There are many factors in our diets and environments that deplete the body of its enzymes or inhibit their activity. In addition, illness, injury, stress, and aging cause our bodies to need increased amounts of enzymes. Supplementation is one way to increase enzyme activity in our bodies, but it is also important to increase your enzyme intake in your food. Following are foods to avoid and foods to add to your diet to increase your enzyme intake.

TYPE OF FOOD	INCLUDE	AVOID
Beverages	Water, herb teas, fresh fruit and vegetable juices.	Coffee, cola and other soft drinks, alcohol, cocoa, artificial fruit drinks, canned juices.
Dairy products	Nonfat, unsweetened yogurt; buttermilk; kefir; kumiss; butter; nonfat cottage cheese; low-fat milk; cheese.	Whole milk, sweetened yogurt, ice cream.
Eggs	Poached, boiled.	Fried.
Fats	Cold-pressed oils, olive oil, flaxseed oil.	Shortening, hydrogenated oils, margarine, saturated oils.
Fruits	Fresh, raw fruits, organically grown and in season, when possible.	Canned, dried, frozen.
Grains	Whole or cracked wheat, rye, soy, barley, buckwheat, cornmeal, oats, millet, brown rice, wheat germ, bran, whole seeds (sunflower, sesame, flax, pumpkin).	White or bleached flour and pasta, white rice.
Juices	Fresh, unsweetened from organically grown fruits and vegetables.	Sweetened, canned, artificially flavored or colored juices.
Meats, fish, and poultry	Fish and poultry. When preparing, first remove the toxin-laden skin.	Red meat; smoked, fried, processed meats (such as hot dogs, lunch meat, bacon, sausage).
Seasonings	Herbs, garlic, onion, sea salt.	Pepper, table salt, hot spices.
Soups	Homemade soups made from scratch (such as chicken, vegetable, millet, barley, brown rice).	Creamed and canned soups, fatty stocks, commercial bouillon.
Sprouts	All, especially alfalfa, mung, wheat, lentil, and pea.	None.
Sweets	Raw honey, carob, unsulfured molasses, pure maple syrup.	Refined sugars (white, turbinado, or brown), candy, chocolate, syrups, jams, jellies, marmalades.
Vegetables	Fresh and raw vegetables, or lightly steamed or sautéed.	Canned, dried, frozen, and fried.
Miscellaneous	Kelp, pickles (refrigerated only), sauerkraut (refrigerated only), green olives, kimchi (fresh only), low-salt versions of soy, miso, tempeh, and natto.	Potato, corn, or other chips; peanut and other butters made with added oil and sugar.

• Eat more fermented foods, including yogurt, kefir, sauerkraut, soy sauce, miso, tempeh, natto, kimchi, and cheese. These foods contain enzymes and beneficial bacteria.

• The remainder of your diet should be composed of whole grains (including whole-wheat bread, whole-grain cereals, brown rice, etc.) and a good quality source of protein, including poultry or fish.

• Include a lot of garlic and onions in your diet. They contain a number of health-giving phytochemicals.

• Eat like a grazing animal, consuming four to five small meals a day, rather than one or two large meals. An individual can eat the same amount of food at one meal and gain weight, or spread the same total calories out over five meals and lose weight. Eating smaller meals will also allow your body's enzymes to function more efficiently and effectively and will not overtax them.

A number of foods and food additives can inhibit or decrease enzymatic activity (see Chapter 3 of Part One for more information on enzyme inhibitors in the diet). For improved health, avoid the following in your diet:

• Avoid diets composed primarily of cooked foods. As mentioned in Chapter 3 of Part One, heat kills enzymes.

• Avoid refined foods, such as those made with white flour and white sugar. These foods are nutrient-poor and enzyme-dead.

• Avoid fast foods, which are typically high in fat, cholesterol, and calories, but low in enzymes, fiber, vitamins, and minerals.

• Limit your intake of foods containing enzyme inhibitors, such as raw soybean products, peanuts, and lentils. However, when heated, soybeans lose their capacity to inhibit enzymes and serve as a low-fat, phytochemically rich protein source.

• Don't eat any foods containing artificial additives, including colors, flavors, and preservatives.

• Avoid all foods that have been irradiated. Irradiation kills the enzymes in food. Also, we just don't know enough about the long-term effects of food irradiation.

• Avoid any foods grown with pesticides, fungicides, herbicides, and so on. Many of these products work by inhibiting enzymes. You should also avoid foods grown close to a highway (the cadmium and other toxins from car tires floats through the air and deposits in the soil).

• Reduce or eliminate your intake of alcohol and coffee and other caffeine-containing drinks.

• Don't cook in aluminum pans or with aluminum cooking utensils. Aluminum is a heavy metal that can cause toxic build-up.

• Avoid chlorinated or fluoridated water. Excessive amounts of chlorine can kill the "good" bacteria living in the intestinal tract, thereby depriving the body of their beneficial effects. Although fluoride can prevent tooth decay, it may also cause an increase in free-radical formation in the blood, leading to tissue damage and accelerated aging. Fluoride inhibits enzymes and depresses immunity. Use distilled, bottled, or well water instead.

• Avoid ordinary table salt. Use sea salt instead because it contains minerals important to nutrition. Salt from any source should be used in moderation, however, because overuse of salt is connected to certain diseases. For seasoning, use garlic, kelp, and potassium. Avoid pungent spices and condiments, which can cause irritation in the gastrointestinal tract.

• Avoid very cold or iced beverages. They can shock the stomach and decrease the amount of hydrochloric acid it produces. This is particularly true right before eating meals. It is wise not to drink a great deal of liquids right before or during a meal, as liquids may dilute the digestive juices. Hydrochloric acid stimulates the production of enzymes. The enzymes produced by the stomach and pancreas are responsible for digestion, as is the hydrochloric acid produced in the stomach.

• Eat one food group in your meal at a time, and then proceed to another. Eat the primarily carbohydrate-containing foods first, and finish with protein-rich foods.

When changing from a typical American fast-food diet to a healthy enzyme-rich diet, containing more fresh fruits, vegetables, and fiber, you may notice a slight increase in gas or other intestinal disturbances. To counteract the unpleasant gas, take an increased amount of digestive enzymes during or up to thirty minutes before each meal. Initially, it may be wise to double or triple the dose noted on the bottle. Let your gastrointestinal distress be your guide. If you are over thirty years of age, digestive enzyme supplements should be your constant companions before each meal. For further information, *see* GAS, ABDOMINAL and HEARTBURN in Part Two.

Exercise

In order to maximize health, include physical activity in your daily life. This may be difficult for an individual in a weakened condition. However, if at all possible, begin exercising in some way, no matter how minimal. Begin with stretching movements, and progress (as you are physically able) to such activities as yoga, walking, swimming, using a stationary bicycle, or dancing. Exercises such as swimming or those where one foot is always on the ground cause less stress to the back.

Many diseases can be conquered more rapidly with exercise. Exercise stimulates enzymatic activity and antioxidant production in the body. It also stimulates circulation to all parts of the body. Improved circulation can aid in delivering oxygen, nutrients, and therapeutic agents to each cell. At the same time, improved circulation facilitates more efficient elimination of waste products and toxins.

Avoid overexertion. Too much exercise can put excessive demands on the body. Increase endurance gradually. To prevent restricting circulation, wear loose fitting clothing. Breathe deeply and freely while exercising. This will facilitate oxygen intake.

Calisthenics and yoga are helpful to improve flexibility. Weight training is an excellent strength builder. For endurance (muscular and cardiovascular), running, jogging, brisk walking, swimming, and bicycling are the best.

A minimum of thirty to forty minutes of exercising per day is very productive. If you live in an area with frequent inclement weather, you can exercise indoors by using a stationary bicycle, StairMaster, or other exercise equipment, such as Total Gym, Health Rider, or NordicTrac, all of which exercise the whole body.

The Five-Step Jump-Start Enzyme Program

Unlike our caveman ancestors, most of us don't die today as a result of a whack to the head with a club. Instead, the majority of deaths in this country are due to degenerative diseases. With these diseases, we deteriorate little by little. Fortunately, because the decline is gradual, we can interrupt the process at any time and slow further breakdown. Certain principles can be used to combat diseases and disorders. My "Five-Step Jump-Start Enzyme Program," based on ancient principles and modern research, is the key.

A wellness program should involve doing all you can nutritionally, physically, and mentally to improve your health. Years ago, I developed the "Five-Step Jump-Start Enzyme Program" to use as a guide in maintaining health and fighting illness. The plan includes five simple but important steps, which should be an integral part of every health plan:

1. Eat a well-balanced "jump-start" diet, high in enzymatic activity.

2. Detoxify regularly to eliminate toxins, which are enzyme inhibitors.

3. Use enzyme, vitamin, and mineral supplements to "jump-start" your day. See Chapters 5 and 6 of Part One, as well as the introduction to Part Two for further information regarding supplementation.

4. Exercise daily.

5. Have a positive mental attitude.

The five steps in the program are part of a multiple approach, caring for the body as a whole (as opposed to its isolated parts), treating the causes of any condition (not just the symptoms), emphasizing prevention, and using natural, nontoxic nutrients, with enzymes as the cornerstone. This program recognizes the detrimental effects of impaired circulation, a weakened immune system, and body chemistry, and how they increase susceptibility to disease. Although viruses and other pathogens may be involved in the development of many degenerative diseases, no illness would be possible with a strong immune system and healthy body chemistry. Diseases are characterized by a breakdown of the body's natural functions.

You may already be following one or more of the five steps in the program. True success comes by integrating all of the steps into your daily life.

Gargles

Gargles can be used to rinse debris from your mouth and throat and to soothe pain in the mouth and throat. However, they also allow the nutrients in the gargle to be absorbed into the tissue and to begin to work before reaching the stomach and small intestine.

ECHINACEA/ENZYME GARGLE

Grind one to two uncoated chewable papain tablets to a powder with the back of teaspoon. Mix with ten drops of echinacea in an eight-ounce glass. Fill the remainder of the glass with distilled or bottled water or juice and stir. Gargle a mouthful of the solution for approximately thirty seconds, and swallow. Repeat the process until the glass is empty. Use the Echinacea/Enzyme Gargle at least three times per day.

ENZYMATIC GARGLE

Place one to two uncoated chewable papain tablets in a cup. Using the back of a teaspoon, grind the tablets to a powder. Combine the powder with one-half teaspoon of calcium ascorbate or magnesium ascorbate crystals or powder and add four ounces of water or juice. Stir thoroughly. Gargle a mouthful of the mixture for thirty seconds, then swallow. Repeat the process until the glass is empty. Use this gargle two to three times per day. Make a new batch each time you use the gargle to ensure that enzyme activity is at its optimum potency.

GARLIC/ENZYME GARGLE

Stir eight to ten drops of garlic extract into an 8-ounce glass of distilled or bottled water or juice (citrus and tomato are best). Crush one to two tablets of chewable papain and add to the mixture. Gargle a mouthful of the mixture for thirty seconds and swallow. Repeat until the glass is empty. Repeat the Garlic/Enzyme Gargle at least three times per day.

Light Therapy

Try to be outside several hours per day. The sun's rays give us vitamin D. However, exercise caution, as too much sun can cause skin cancer. Avoid the midday sun, when the rays are directly overhead and the most intense. Morning or evening sun (before 10 a.m. and after 3 p.m.) is safer, as the rays are not as direct.

Meditation

Meditation helps to free the health-giving energy from within your body and to release and void those deadly toxins that drain your vitality. It decreases stress and allows enzymes to work at a more optimal level. According to Alan Watts, who is credited with introducing meditation to the Western world, "Meditation is the art of suspending verbal and symbolic thinking for a time, somewhat as a courteous audience will stop talking when a concert is about to begin." Meditation can help you achieve total relaxation and discover your inner being.

Meditation is a learned behavior. Each time you meditate you will be more successful. Set aside fifteen to twenty minutes per day. Choose an area where sounds are minimal and that you find peaceful. If you are at work, you might want to try a time when no one is around, such as before or after work hours or lunch time. Play a tape of water trickling down a gentle stream. Place yourself in a relaxed position (either sitting comfortably or lying on a mat with a pillow or blanket rolled under your neck and under your knees). If lying down, allow your arms and hands to lie relaxed on either side of your body. If sitting, cross your hands in a relaxed position, resting them in your lap. With your eyes closed, take a deep breath and let it out through your partially open, relaxed mouth very slowly. As you do this, drag out the word "peace."

Now, visualize yourself lying relaxed on a sandy beach. You can feel the warmth from the sun's rays engulfing you and warming the sand on which you lie. Feel the earth's energy rising up into your body, and feel its warmth giving you life, strength, protection, and peace. The ocean's warm, calming waters move in and gently engulf you. No harm can come to you. Magically, you are floating on the water. You feel new energies in your body, new vitality and forces of life gently moving around you, caressing your body and your soul. You can feel the energies start in the fingers and toes, and then gradually moving up your arms and legs, through your thighs and upper arms—moving slowly and gently like a feather. Feel the energy moving into your shoulders and torso, then into your inner body, your soul. Feel the gentle touch. Feel the energies from the sky and the healing water all around you. Just let yourself float and enjoy. No one can hurt you.

Then, softly, gently, a warm caressing wind begins to blow around you, and you begin to move slowly back to shore. As you move back to shore, you sense a newfound energy and inner peace. You are moving slowly, floating freely. The water subsides, but you still feel the energy, vitality, and peace; and your special beach will always be there for you. You can return every day to be refreshed. And each day, the beach will become more pleasant, and the sun will become more energizing and the water more cleansing.

This is just one of several meditation techniques. There are many fine books and videotapes that describe various techniques. Check with your health-food or book store.

Positive Mental Attitude (PMA)

Think of your mind as a container. When it is full of positive thoughts, there is no room for negative ideas. Negative thoughts, such as anxiety, cynicism, depression, despair, grief, guilt, hopelessness, irritability, and resentment are diseased states of the mind. These mental states can cause every bit as much damage as the physical toxins that continually assault our bodies, causing disease.

A positive mental attitude (PMA) can be acquired. Just as exercise is important to keep muscles strong and healthy, constantly practicing a positive mental attitude can train your mind to see the positive aspects of every situation. This will improve your health and help you to better cope with the stresses of everyday life. Just as with the body's muscles, "If you don't use it, you lose it." Therefore, practice is essential. It is also important to know what you can control and what you cannot control, and to learn the difference. If something is out of your hands, accept it and move on.

Remember, you don't have to climb a mountain, save a

439

child, or save the world! All you have to do is jump-start your own positive attitude and maintain it. The following points may help you improve your mental attitude. Tap the inner power of your belief system. Make a conscious effort to try to find something positive about any situation, rather than seeing the worst in the situation. Find something good to say about a friend, relative, or coworker today. Your subconscious will repeatedly catalyze you to achieve goals heretofore thought unattainable. What the mind conceives, the body achieves. Form a mental picture of yourself achieving a specific goal. Repeat the mental picture over and over.

Relaxation

Stress interferes with normal enzymatic function. Relaxation allows body energy to flow. Relaxation techniques can help you focus and achieve an inner calm. You can free yourself from needless mental (and sometimes physical) suffering through the mental focus and strength achieved through relaxation techniques. Relaxation helps us to cope with life's stresses and can make us better human beings.

First, choose a quiet, preferably dark, location (turn off the telephone). Lie on your back on the bed or the floor. Place pillows under your head and neck and under both knees. Close your eyes and listen to a pre-recorded tape or CD of nature sounds, which can be relaxing. Breathing deeply, relax for at least thirty minutes.

Skin Care

Everyone wants to look good and feel good. Increased emphasis is placed on appearance, and skin care is the key. But many of us ignore our skin until it's too late. The skin is the body's barrier to the outside world, but it's also a reflection of our overall health. Years of exposure to the sun, environmental toxins, poor diet, tobacco smoke, drugs, and alcohol can hasten the appearance of wrinkles and sagging skin.

The skin is a complex organ that serves a number of vital metabolic and protective functions. The skin protects the internal organs and tissues from trauma, abrasion, and other injury, and serves as a barrier against invasion from damaging bacteria, viruses, and toxins. Skin also helps the body maintain normal body temperature and retain valuable moisture.

The skin consists of two main cell layers: a thicker inner layer (dermis) and a thin outer layer (epidermis). The dermis is formed by fibrous, elastic connective tissue, which contains lymph channels, nerve endings, blood vessels, sweat glands, fat cells, hair follicles, and sebaceous glands. The epi-

dermis is composed of flattened epithelial cells and contains a pigment (melanin) that gives the skin color and protects us from ultraviolet radiation. The receptors (or nerve endings) perform an important sensory function by responding to numerous stimuli, such as pressure, pain, heat, and cold.

The skin is the body's largest organ (about 10 percent of the body's weight). It has been said that a piece of skin the size of a quarter has four yards of nerves, 15 sebaceous glands, 100 sweat glands, more than 3 million cells, one yard of blood vessels, and 10 hair follicles. Collagen, the main constituent of skin (75 percent of the dry weight of skin tissue), tendons, and cartilage (as well as the protein component of teeth and bones), is uniquely important to our bodies. Collagen fibers interlace, forming an elastin network (an essential segment of connective tissue), which lends skin its elasticity, smoothness, and strength.

In undamaged young skin, collagen molecules apparently slide over each other. With scarring, wounds, and burns, as well as with the wrinkles of aging, collagen molecules fuse together (cross link), making the damaged tissue rigid and leathery.

New skin cells are formed at the basal layer, the deepest layer of the epidermis, where they start off plump and round. As skin cells move upward through the next three levels, they get flatter and older. Finally, they reach the visible outer layer where they are scraped off or shed (called exfoliation or desquamation). The amount of time it takes the new cells to travel to the outer layer is called transit time.

Skin is loaded with natural enzymes that help cells develop in the lower skin layers. Each enzyme group has a specific job, and they all work in harmony. Some serve as antioxidants, defending the skin against free radicals, which are formed by pollutants or sun exposure and speed the aging process. Other enzymes promote elastin and collagen destruction. Also, the natural exfoliation of dead cells from the skin's surface can be accomplished with enzymatic assistance.

The skin is a very active organ. But with aging, the transit time slows down. The transit time of a 20-year-old is merely two to three weeks, while the transit time in an 80-year-old woman may be four to six weeks. The flatter and older the skin cells, the longer it takes them to reach the outer layer. This is why old skin doesn't glow or shine. Skin cell turnover begins to slow down even as you enter your thirties. This makes your skin seem to lack luster. Laugh lines and fine furrows will intensify with increasing years of exposure to the sun.

Everyone's skin is different. Some people have dry skin, some oily. The health of the skin can be affected by aging, infections, allergies, bites and stings, bruises, burns, sunburns, and injuries. Many of these conditions are covered in this book.

Dry skin is marked by a rough, dry feeling to the skin and loss of elasticity. The skin may crack and peel. Dryness of the skin is basically due to insufficient movement of moisture upward (from the lower layers of the skin) and/or to loss of moisture from the skin's horny outer later. As we age, oil gland production slows down. This vital skin oil (called sebum) is actually a mixture of water and oil and is the body's natural moisturizer.

Sebum is secreted by the sebaceous glands. It contains soaps, fats, cholesterol, inorganic salts, and remnants of epithelial cells. It serves to protect the hairs from becoming brittle and too dry, as well as from becoming too easily saturated with moisture. On the surface of the skin it forms a thin protective layer, which prevents undue evaporation or undue absorption of water from the skin. This secretion keeps the skin pliable and soft.

Some people have drier skin than others. This is because genetics determine our skin texture. But the way you treat your skin can improve or worsen your basic underlying skin tendency.

Oily skin produces too much oil. Oily skin is characterized by enlarged pores and an oily, shiny appearance. If your skin has a greasy feeling when you wake up in the morning, you probably have oily skin. If unattended, oily skin can lead to trapped pores and acne.

Sebaceous glands are found everywhere over the skin surface (with the exception of the soles of the feet and palms of the hands). They are abundant in the face and scalp and around the openings of the mouth, nose, anus, and ears. Each gland is filled with larger cells containing fat and a number of epithelial cells. These cells are constantly being cast off and new cells continuously formed. Some of the largest sebaceous glands are found on the face, including the nose, where they may become enlarged with accumulated secretion (which can become discolored and referred to as blackheads). It also provides an environment for the growth of pus-producing organisms and, therefore, is a common cause of pimples and boils.

There are several conditions that can affect the appearance of skin. The following are discussed in Part Two: ABSCESS, ACNE, BOILS, BRUISES AND HEMATOMAS, BURNS, CANKER SORES, CHICKENPOX, FUNGAL SKIN INFECTIONS, HIVES, INSECT BITES AND STINGS, MEASLES, PSORIASIS, ROSACEA, SCARS, SKIN CANCER, SKIN RASHES; SKIN ULCERS, STAPHYLOCOCCAL INFECTIONS, VARICOSE VEINS, and WARTS.

How are enzymes used to improve skin health and vitality? Enzymes can be taken orally (to improve digestion or fight inflammation) and applied topically directly onto the skin.

ENZYME SKIN EXFOLIANTS

Skin exfoliants and masks are excellent for cleansing and exfoliating dead skin from the face and body. They can be helpful in removing or reducing scars, blemishes, and wrinkles and returning a more vital appearance to the skin. They soften and smooth skin by digesting dead cells and waste material. They can also be ideal as a skin resurfacing product after a regimen with alpha-hydroxy acids, including glycolic acid (from sugar cane), citric acid (from citrus fruits), lactic acid (from buttermilk), and tartaric acid (from red wine). Zia Cosmetics, Richard Arve, and HDS Labs all make exfoliants and masks containing enzymes.

Enzyme salves and creams are excellent skin exfoliants and rejuvenate the skin when used topically. Speeding up your skin cells' transit time is one way to look ten years younger. In this way, younger-looking, newer cells will appear on the surface. This can be accomplished with the help of exfoliants.

An exfoliant made from green papaya (with the enzyme papain) is now available that digests the dead protein cells on your skin, resulting in softer, smoother skin. In fact, many enzyme peels contain papain and work in the same manner as Retin-A. According to Richard Arve, a facial enzyme expert, the most potent papain is found in young, green papaya. However, it loses potency as the fruit matures. He explains green papaya has the ability to dissolve and digest old, debilitated, or dead cells from the outer layer of the skin, without harming the younger living cells.

Unlike Retin-A, papain enzyme peels are available without prescription and do not have any negative side effects. It is their gentleness that makes them so useful for all skin types. Heat destroys the enzymes, therefore, it is important that any papaya skin product be processed at low temperatures and made from green papayas.

According to Zia Wesley-Hosford, esthetician/cosmetologist, skin-care instructor, and author of *Face Value: Skin Care for Women Over 35* (San Francisco: Zia Cosmetics, Inc., 1990), "Enzymes gently yet thoroughly dissolve old cells without harming new ones or irritating the skin." Further, Zia believes that green papaya enzymes assist the healing of uneven pigmentation, fine lines, and brown spots by fighting free-radical damage and boosting cell production and are considered the "natural alternative to Retin-A."

Marilyn Territo, renowned esthetician, cosmetologist, and skin-care consultant, adds that green papaya offers many benefits in helping to prevent and/or control cross-linking, not only because of its exfoliating properties, but also for its abundance of antioxidants (vitamins A, C, and E).

Several enzymes in addition to papain play an essential role in anti-aging and skin care. Bromelain (from pineapple) is also used as a skin exfoliant. Trypsin, pancreatin, and keratinase are used to break down and dissolve dead skin cells, and superoxide dismutase (SOD) can help fight free radicals.

ENZYMATIC SKIN SALVES

Enzymatic skin salves are excellent for reducing inflammation and fighting infection and free-radical formation. They are effective in treating such skin conditions as abscesses, acne, boils, bruises, burns, eczema, herpes, insect bites and stings, scars, shingles, and wrinkles, among others. Salves can also be used to treat conditions that are more than skin deep, including bursitis, tendonitis, and other inflammatory conditions. Wobe-Mugos E by Mucos Pharma GmbH & Co. is an excellent enzyme skin salve.

GENERAL SKIN CARE RECOMMENDATIONS

There are several general guidelines you can follow to keep you skin healthy, in addition to treating it enzymatically:

❏ For healthy skin, drink plenty of pure water (eight to ten glasses every day). Water can keep the skin hydrated and minimize wrinkling.

❏ Eat a healthy diet with plenty of enzyme-rich fresh fruits and vegetables. Decrease your intake of fatty foods, dairy products, and meats.

❏ Get plenty of sleep. Your body produces skin growth factors and human growth hormone when you sleep. These hormones accelerate collagen production (the protein responsible for skin support and elasticity) and give the skin a faster exfoliation rate.

❏ Eliminate toxins, cigarette smoking, alcohol abuse, radical weight swings, and nutrient-poor diets. Routinely flush out toxins through detoxification methods and exercise. See these sections in Part Three for instructions.

❏ Avoid excessive exposure to sunshine (and sunburns), which damages collagen, leads to an increase in free radicals, destroys antioxidant enzymes, is a major cause of skin cancer, and can age you before your time. Remember, the severity of a sunburn depends on a number of factors, such as the time of day you selected for sunning, the number of hours of exposure, and your location (near or far from the equator). For example, at midday, when the sun is directly overhead, the rays of the sun travel a shorter distance and there is a minimal amount of filtering by the earth's atmosphere. On the other hand, later in the afternoon, the sun's rays have to travel farther, so are not as strong.

❏ If you must be in the sun, cover up with protective clothing or a hat and use sunscreen. In the United States, sunscreens are now rated by their sun protection factors, known as SPF. The higher the SPF, the greater the protection from the sun.

❏ Look for skin-care products containing vitamins (including A, B, C, D, and E), minerals (such as zinc), and herbs (such as aloe vera and chamomile) that can lend additional nutritional support to the skin. For example, vitamin A is very important because it improves cell structure and helps destroy free radicals.

❏ Labels list product ingredients in decreasing order of concentration. But regardless of the ingredients, even the best skin-care products can't overcome a nutrient-poor, unhealthy diet, or destructive lifestyle. It is important to improve health from the inside (with vitamins, minerals, enzymes and other supplements) and from the outside with natural, nutritional, and safe skin-care products.

❏ Exfoliate regularly. Arve's Papaya Enzyme Exfoliating Cream dissolves damaged or dead skin cells, while leaving healthy cells intact. Once the dead cells are dissolved, the body manufactures new collagen and elastin, thus rebuilding firmer, more youthful skin. Also try Arve's Poi & Papaya Enzyme Exfoliating Gel (for oily, acne, or sensitive skin), Arve's Papaya Facial Cleanser, and Arve's Poi Facial Cleanser.

❏ Avoid any products containing mineral oil, which is a petroleum derivative. Mineral oil acts as a moisture barrier, but can also trap toxins (including perspiration), smothering the skin. Use products containing essential oils (oils taken from the essences of plants), instead.

❏ Whatever you apply to the skin is absorbed by the body. For this reason, use only natural skin care products that contain no artificial colors, synthetic ingredients, carcinogens, or phosphates, and that are biodegradable. Instead, choose skin-care products with natural additives (such as citric acid), natural colors (from beta-carotene, chlorophyll, or beetroot juice powder), and natural preservatives (Alexandra Avery uses grapefruit seed extract in many of her products). Also, be sure to choose products that support the skin's functions.

❏ Exercise thirty to forty minutes per day to increase blood flow and enzymatic and antioxidant activity. For further information, see EXERCISE in Part Three.

Water

Health books are full of admonitions to drink six to eight glasses of fresh water every day. This is because water is not only essential for many activities in the body, it is also critical for most enzymes, as well. For example, proteases, amylases, and lipases are hydrolytic enzymes and require water to do their jobs. Water activates the enzymes. Without water, these and other enzymes could not function.

Water is actually the most abundant chemical substance in the human body and accounts for approximately 60 percent of our body weight. Gastric juice is mostly water (99.5 percent), while our teeth are only 10-percent water (see the inset "Water Content of Various Body Parts" on page 443).

Water provides the environment necessary for chemical reactions, carries nutrients to each cell, and transports waste from the cells. Water helps us maintain proper body temperature, whether we're suffering from winter's frozen grip or

Water Content of Various Body Parts

Water is essential for the function of every plant and animal, and it is present throughout our bodies. Below is a listing of the water content of some of the components of our bodies.

BODY COMPONENT	WATER CONTENT PERCENTAGE	BODY COMPONENT	WATER CONTENT PERCENTAGE
Bile	86 percent	Lungs	80 percent
Blood plasma	90 percent	Lymph	94 percent
Blood serum	90.7 percent	Muscular tissues	75 percent
Bones	13 percent	Red blood corpuscles	68.7 percent
Brain	80.5 percent	Saliva	95.5 percent
Cartilage	55 percent	Spleen	75.5 percent
Gastric juice	99.5 percent	Teeth	10 percent
Liver	71.5 percent		

summer's intense heat. Water also provides a protective cushion around body organs, such as the brain. We get most of our water (about 60 percent) from fluids. Of the remaining amount, about 30 percent is obtained from our foods and 10 percent from cellular metabolism.

We lose water through perspiration and elimination (in urine and feces), and some vaporizes out of the lungs. Water loss and lack of replenishment can result in dehydration, characterized by decreased urine output; flushed, dry skin; and a "cottony" feeling in the mouth. If dehydration is allowed to continue, mental confusion, fever, weight loss, and death can occur. Water consumption must equal water output if the body is to remain properly hydrated.

Note: The water from most municipal water systems is either contaminated or loaded with chlorine, fluoride, and other chemicals. Therefore, use bottled, distilled, spring, or well water, or purchase a good quality water purification system.

APPENDIX

GLOSSARY

absorption. The uptake of substances into tissue. Absorption can take place in many locations, including the kidney tubules, the intestines, the mouth, and the skin.

active site. The site of the connection between an enzyme and a substrate.

adsorption. The process by which one substance attaches to the surface of another.

amino acid. Organic compounds, composed of nitrogen, carbon, oxygen, and hydrogen, that are the building blocks of proteins and enzymes.

amylases. Carbohydrate-decomposing enzymes.

anabolism. The metabolic process by which complex substances are made from simple substances in living tissue.

antibodies. Large proteins produced by the immune system that combine with and destroy foreign invaders (antigens).

antigen. Any substance that the body sees as foreign and produces antibodies to destroy it.

antioxidant. A substance that prevents or delays deterioration caused by oxidation.

carbohydrate. A group of organic compounds composed of carbon and hydrogen and oxygen in a two-to-one ratio. Carbohydrates are the main source of energy for the body.

catabolism. The metabolic process by which complex substances are broken down into simpler substances, causing energy to be released.

catalyst. A substance that influences the rate of a chemical reaction but remains unchanged by the reaction.

chelation. A chemical reaction in which a metal ion bonds with two or more nonmetal ions.

coenzyme. A nonprotein organic compound that, when combined with a specific protein, forms an active enzyme.

cofactor. A nonprotein substance that must be present for an enzyme to function.

enzyme inhibitor. Any substance that inhibits the activity of an enzyme.

enzymes. Proteins that catalyze biochemical reactions in living organisms. No organism can live without enzymes.

essential nutrients. Those nutrients that cannot be produced by the body but are needed for proper development and function, and so must be obtained from diet.

fat. An organic compound composed of one or more fatty acids. Fat is the main form in which energy is stored by the body.

fiber. The non-digestable portion of food. It provides no nutrients or energy to the body, but dietary fiber helps promote more efficient intestinal function.

fibrin. The basic component of a blood clot, formed by the action of thrombin (an enzyme) on fibrinogen (a protein).

free radical. An atom or molecule that has at least one unpaired electron, making it highly unstable, which causes it to react with another atom or molecule. This can create a lot of damage in the body. If their destruction is not stopped, free radicals can weaken the whole body, causing illness and premature aging.

hydrolases. One of the six main groups of enzymes. Hydrolases break chemical bonds with the addition of water. They are the most frequently used enzymes in enzyme therapy.

immune complex. The result of an antibody linking to an antigen. Immune complexes are normally removed from the body once formed. If they are not removed, illness may result.

induced fit theory. The idea that the shape of an enzyme actually changes to allow the substrate to bind—similar to a hand fitting into a glove. The shape of the hand makes the glove's shape alter a little.

isomerases. One of the six main groups of enzymes. Isomerases catalyze the conversion of one isomer (any of two or more compounds that are composed of the same elements and in the same proportion, but have different properties due to the arrangement of the atoms) into another.

ligases. One of the six main groups of enzymes. Ligases catalyze the formation of a bond between two substrate molecules through the use of an energy source (usually adenosine triphosphate—ATP).

lipases. Fat-decomposing enzymes.

lock and key theory. The idea that a substrate is like a key that must fit into a specific shape on an enzyme in order to activate that enzyme (much the same way a key fits into a lock to unlock a door).

lyases. One of the six main groups of enzymes. Lyases catalyze the formation of double bonds between substances by removing or adding a chemical group.

metabolism. The set of chemical processes, which take place in living organisms that are essential to life. Anabolism and catabolism together are referred to as metabolism.

mineral. A naturally occurring inorganic substance with a characteristic composition and crystalline structure.

oxidation. A reaction in which one reactant transfers electrons to another. Often, this occurs when a molecule reacts with oxygen, hence the name.

oxidoreductases. One of the six main groups of enzymes. They catalyze oxidation-reduction reactions.

peptide. A molecule chain of two or more amino acids.

pH. Potential hydrogen. The pH scale is a logarithmic scale used to define the degree of acidity or alkalinity, which is determined by the number of hydrogen ions in a solution. A pH of 0 is considered acidic, 7.0 is considered neutral, and 14 is alkaline.

pinocytosis. The process by which fluid is taken into a cell.

protease. A protein-decomposing enzyme.

proteins. Any of a group of complex organic nitrogen compounds, composed of amino acids. They are the major structural components of all body tissue and are needed for growth and repair.

retention enema. A means of instilling some enzymes into the colon to be retained, allowing the enzymes to be absorbed into the body's systems. The retention enema is much more effective than a suppository because the active ingredients are already in solution and are more easily absorbed.

substrate. A specific substance upon which an enzyme acts.

transferases. One of the six main groups of enzymes. They catalyze reactions by transferring chemical groups (other than hydrogen) from one substrate to another.

vitamin. An organic nutrient required by the body in minute amounts, but cannot be produced by the body. *Vite* is Latin for life, and vitamins were originally thought to be *amines*.

DIGESTIVE ENZYME PRODUCTS

The following are several digestive enzyme products. Some of these products contain individual enzymes or enzyme combinations. Others are concentrated food extracts (which contain enzymes in addition to other nutrients). These have two benefits: The mixture provides a wide range of nutrients, and the enzymes can improve nutrient absorption. This listing is just a sample of the many enzyme products there are to choose from.

Product Name	Manufacturer/ Distributor	Ingredients
Acidophilasé	Wakunaga	*L. acidophilus, B. bifidum,* and a food enzyme complex (protease, lipase, and amylase).
Acti-Zyme	Nature's Plus	Aminogen (proteolytic enzyme complex), Legumase (oligosaccharidase enzyme complex), amylase, lactase, lipase, cellulase, protease, oxidase, bromelain, diastase, maltase, fructooligosaccharides, bioperin (piper nigrum fruit).
Adult Formula Enzyme	Murdock Madaus Schwabe Professional Products, Inc.	Plant enzymes: amylase, cellulase, lactase, lipase, protease, sucrase.
Advanced Enzyme System	Rainbow Light Nutritional Systems	Amylase, apple pectin, bromelain, cellulase, fennel, ginger, glucoamylase, green papaya, invertase, lipase, maltase, papain, peppermint, protease, sea vegetable complex, turmeric.
After Meal Enz Extract	Crystal Star Herbal Nutrition	Glycerin, hibiscus flower, licorice root, mint mix (peppermint leaf, spearmint leaf, orange peel, lemon peel), papaya seed.
Alkazyme-3	Enzyme Process International	Amylase, protease, lipase, cellulase, green foods mixture, pancreatin.
All-Zyme + Lactase	Rainbow Light Nutritional Systems	Lactase, protease, amylase, lipase, cellulase, fructooligosaccharide (FOS).
All-Zyme Original (also available as Double Strength)	Rainbow Light Nutritional Systems	Protease, amylase, lipase, cellulase.
Alpha-Zyme	Bio-Energy Systems, Inc.	Amylase, cellulase, Cultured Enzymes™ gamma-oryzanol, invertase, lactase, lipase, and maltase, with goldenseal, papaya leaf, slippery elm, and O2Factor™ in a base of pure plant fiber.
Amino Zyme	Crystal Star Herbal Nutrition	Bromelain, carnitine, chickweed herb, DL-phenylalanine, ginger root, gotu kola herb, GTF-chromium, Gymnema sylvestre leaf, L-glycine, L-lysine, L-ornithine, L-tyrosine, lecithin.
Amino-Zyme	Nature's Plus	Aminogen (proteolytic complex), chromium, coenzyme Q_{10}, vanadium, vitamin B_6.
Anavit-F3	CC International, Inc.	Ascorbic acid, bromelain.
Anti-Flam Extract	Crystal Star Herbal Nutrition	Bromelain, butcher's broom herb, curcumin, St. John's wort leaf, white willow bark.
Bean & Vegi Enzyme Formula	Prevail Corporation	Alpha-galactosidase, amylase, cellulase (I and II), lipase, and protease (I, II, and III).
Beano in drops and tablets	AkPharma, Inc.	Alpha-galactosidase.

Product Name	Manufacturer/ Distributor	Ingredients
Beans . . . and more Beans RX	Quad Laboratories, Inc.	Legumase (food enzymes from *Saccharomyces* and *Aspergillus*), with ajowan oil, ginger, peppermint oil, fennel oil.
BeSure	Wakunaga	Food enzymes (Legumase) derived from *Saccharomyces* and *Aspergillus*.
Betaine HCL Caps (with pepsin)	Twin Labs	Betaine hydrochloride, pepsin NF.
Beta Pepsin Tablets	Solgar Vitamin and Herb Co., Inc.	Betaine hydrochloric acid, pepsin.
Betaine	Highland Labs	Betaine hydrochloric acid, pepsin.
Bio-6 Plus	Biotics Research Corporation	Amylase, lipase, raw pancreas concentrate, protease.
Bio-Zyme	PhytoPharmica	Amylase, bromelain, lipase, lysozyme, pancreatic enzymes (including amylase, lipase, protease), papain, trypsin.
BotaniTrim	NuBotanic International, Inc.	Garcinia cambogia, chromium picolinate, ascorbic acid, buchu leaves, potassium, cayenne, enzyme complex omega (with protease I, protease II, protease III, protease IV, fungal lipase) pyridoxine, L-tyrosine, L-carnitine, pantothenic acid, selenium.
Bromelain	Futurebiotics	Bromelain.
Bromelain	HealthSmart Vitamins	Bromelain.
Bromelain	KAL, Inc.	Bromelain with cellulose, magnesium stearate, silica, vegetable stearin.
Bromelain	Solgar Vitamin and Herb Co., Inc.	Bromelain.
Bromelain (in 100, 250, and 500 mg formulations)	Nature's Plus	Bromelain.
Bromelain & Papain Vegetarian Enzymes	Nature's Life	Bromelain and papain.
Bromelain Complex	PhytoPharmica	Bromelain, pantothenic acid, vitamin C.
Bromelain Papain	Professional Health Products, Inc.	Ascorbic acid, betaine (HCl), bromelain, calcium, glutamic acid, pancreatin, papain.
Bromelain Forte Tablets	General Research Laboratories	Bromelain, inositol.
Bromelain Plus	Enzymatic Therapy	Bromelain, pantothenic acid, vitamin C.
Bromelain Plus CLA	Biotics Research Corporation	Amylase, bromelain, cellulase, inositol, L-cysteine, lipase, papain.
Bromelain Tablets	General Research Laboratories	Bromelain, papain.
Bromelain 1200	Highland Labs	Bromelain.
Caretaker	Bioenergy Nutrients	Alfalfa, apple pectin, barley, buckthorn bark, cascara sagrada, fructooligosaccharides, goldenseal, oat fiber, papain, psyllium seed and husk, rice bran, and triphala.
Chewable Bromelain	Nature's Plus	Bromelain.

Product Name	Manufacturer/ Distributor	Ingredients
Children's Digestion Formula	Prevail Corporation	Amylase, cellulase, lactase, lipase, maltase, protease (I, II, III, and IV), and sucrase.
Chymozyme	Ethical Nutrients	Chymotrypsin, trypsin.
Colo-Clen	Enzyme Process International	Psyllium powder, celery powder, aloe vera powder, prune concentrate, bentonite powder, mint, lactobacillus, acidophilus, papain, flax powder, anise, vitamin C, bromelain, magnesium, bile salts.
CTS Infla-EZ	Vitagenics	Bromelain, papain.
D.A. #34 Food Enzymes	J.R. Carlson Laboratories	Ox bile, pancreatin, pepsin.
Daily Enzyme Complex	Futurebiotics	Anise, betaine, bromelain, chamomile, duodenum, liver, niacinamide, ox bile, papaya enzyme, peppermint, protease, vitamin B_1, vitamin B_2.
Dairy Enzyme Formula	Prevail Corporation	Amylase, lactase (I and II), lipase, and protease (I, II, III, and IV).
DairyEase	Bayer Corporation	Lactase.
Detox-Zyme	Rainbow Light Nutritional Systems	Protease, lipase, garlic, milk thistle, red clover, yellow dock, parsley, cleavers, burdock, Oregon grape, bupleurum, beet.
Di-Gest	Wild Rose Herbal Formulas	Glutamic acid HCl, Betaine HCl, calcium ascorbate, pancreatin NF, bromelain, and papain.
Digest Plus	Natrol	Betaine HCl, bromelain, calcium ascorbate, glutamic acid HCl, pancreatin (supplying protease, amylase, lipase activity), papain, protease.
Digest-Able	Highland Labs	Amylase, bromelain, cellulase, lipase, and papain.
Digestal	Vitagenics	Papain, pancreatin, bromelain, amylase, pancrelipase.
DigestChews	Earthrise	Amylase, cellulase, lactase, lipase, and protease in a chewable wafer.
Digest-eze RX	Phyto-Therapy, Inc.	Ajowan oil, bromelain, cellulase, fennel oil, ginger extract, *L. sporogenes*, lactase, legumase, lipase, pancreatin, papain, peppermint oil, proteolytic enzyme complex (Aminogen), spearmint oil, vegetable enzyme complex (Digezyme).
Digestin #987	Progressive Laboratories	Betaine HCl, cellulase, ox bile extract, pancrelipase, papain, pepsin, stomach substance.
Digestive Aid tablets	Solgar Vitamin and Herb Co., Inc.	Pancreatin, ox bile extract, betaine HCL, diastase, papain.
Digestive Enzyme Liquiescence	Professional Health Products, Inc.	*Coriandrum sativum* 2x, *Syzgium jambolanum* 3x, papain 3x, herbal bitters 3x, chamomile 3x, bromelain 3x, *Illicum anisatum* 3x, *Cinnamonum* 3x, parotid, stomach, pancreas, gallbladder 3x, *pancreatinum* 6x.
Digestive Enzymes	Nature's Plus	Malt diastase, ox bile extract, pancreatin, papain, and pepsin.
Digeszyme-V	Enzyme Process International	Acid-stable amylase, protease, lipase, cellulase.
DigeZyme	Sabinsa Corporation	Amylases, protease, lipase, lactase, cellulase, phosphatases, ribonucleases.
Dismuzyme #748	Progressive Laboratories	Superoxide dismutase, catalase.

Product Name	Manufacturer/ Distributor	Ingredients
Dismuzyme Plus 5000	Biotics Research Corporation	Catalase, copper, glutathione peroxidase and other peroxidases, manganese, selenium, superoxide dismutase, zinc.
Enzygen	Rexall Showcase International	Lipase, cellulase, amylase, lactase, protease.
Enzymall	Schiff Products	Pancreatin, amylase, cellulase, and papain.
Enzyme Digestive Supplement	American Dietary Labs	Pancreatin, papaya leaf.
Enzyme Formula	Natural Brands (GNC)	Alpha-amylase, betaine hydrochloric acid, ox bile, pancreatin, papain, pepsin.
Enzyme Forte	Nutri-West	Raw pancreatin (including protease, lipase, and amylase— enterically coated).
Enzyme Pforesis Gel	Nutri-West	Pancreatin, papaya, rutin, bromelain, thymus, trypsin, chymotrypsin, pancrelipase, amylase, zinc gluconate.
Fat Enzyme Formula	Prevail Corporation	Amylase, cellulase I, lipase (I and II), and protease (I, II, III, and IV).
Fiber Enzyme Formula	Prevail Corporation	Amylase, cellulase (I and II), and phytase.
Foodzymes	Earthrise	Amylase, cellulase, lactase, lipase, phosphatase, protease, and ribonuclease.
Gastrazyme	Biotics Research Corp.	Superoxide dismutase, catalase, vitamin A, natural mixed carotenoids, gamma-oryzanol, DL-methionine, S-methyl sulfonium chloride, chlorophyllins.
Gastrogest	Professional Health Products, Inc.	Betaine hydrochloric acid, L-glutamic acid, pancreatin, papain, pepsin.
Green-Zymes	Nikken U.S.A., Inc.	Amylase, lipase, protease.
Hydro-Zyme	Biotics Research Corp.	Pepsin, betaine HCl, glutamic acid HCl, ammonium chloride, vitamin B_6, pancreatin.
Infla-EZ	Vitagenics	Pancreatin, papain, trypsin, amylase, lipase, lysozyme, cellulase.
Inflamzyme	PhytoPharmica	Bromelain, L-cysteine, magnesium, mixed bioflavonoid complexes, quercetin, vitamin C.
Inflazyme Forte	American Biologics	Amylase, bromelain, catalase, chymotrypsin, L-cysteine, lipase, pancreatin, papain, rutin, superoxide dismutase, trypsin, zinc.
Intenzyme Forte	Biotics Research Corporation	Alpha-chymotrypsin amylase, bromelain, lipase, pancreatin, papain, trypsin.
Jarro-Zymes Plus	Jarrow Formulas	Pancreatin with extra strength lipase plus alpha-galactosidase.
Kal-N Zyme	KAL, Inc.	Microbial amylase, cellulase, lipase, and protease enzymes; natural mint extracts, sucrose, and chlorophyll in a vegetable-derived gum base.
Kyolic Garlic Extract Formula 102	Kyolic	Aged garlic extract, amylase, cellulase, lipase, and protease.
Lactaid	McNeil Consumer Products Co.	Lactase.

Product Name	Manufacturer/ Distributor	Ingredients
Lactase Enzyme #493	Progressive Laboratories	Lactase.
Lactase RX	Quad Laboratories, Inc.	Lactase.
Lactase 3500	Solgar Vitamin and Herb Co., Inc.	Lactase.
Lacta-Zyme	Douglas Laboratories	Lactase.
Lipozyme	Enzymatic Therapy	Beef bile salts, bromelain, duodenum extract, L-cysteine, pancrelipase.
Lyso-Lyph-Forte	Nutri-West	Pancreatin, papain, bromelain, trypsin, chymotrypsin, lipase, rutin, raw calf thymus, zinc gluconate, raw parotid concentrate.
Lyso-Lozenge	Nutri-West	Trypsin, chymotrypsin, pancreatin, bromelain, papain, thymus, mannitol, propolis, spearmint.
Manda Hi Kohso	Magnus Enterprises	Countless enzymes and byproducts from 58 different natural fruits, vegetables, grains, beans, seaweeds, and brown rice.
Maxigest	Ethical Nutrients	Pepsin.
Medi-Zyme N	Gero-Vita	Amylase, bromelain, chymotrypsin, lipase, pancreatin, papain, rutin, trypsin.
Mega Bromelain Caps	Twin Labs	Bromelain.
Mega-Zyme	Enzymatic Therapy	Amylase, bromelain, chymotrypsin, lipase, lysozyme, pancreatic enzymes (with protease, amylase, and lipase activity), papain, trypsin.
Milk Digest-Aid	Schiff Products	Lactase.
Milk Enzymes	Highland Labs	*Acidophilus lactobacillus*, bromelain, lactase, pancreatin, papain, and rennin.
Mulsal	Mucos Pharma GmbH & Co.	Bromelain, papain, trypsin.
Multi-Enzyme	Nutrigest (GNC)	Fungal protease, acid protease, lactase, cellulase, amyloglycosidase, fungal amylase, lipase.
Multi-Enzyme Formula	Preventative Nutrition (GNC)	Amylase, betaine hydrochloric acid, bromelain, cellulase, diastase, hemi-cellulase *L. acidophilus*, lactase, lipase, papain, protease.
Multi-Zyme Complex Maximum Potency	Nature's Life	Betaine hydrochloric acid, bromelain, cellulase, lipase, ox bile extract, pancreatin, papain, and pepsin.
Nutri-Essence Broad Spectrum Enzymes in VegiCaps or chewables	Nutri-Essence	Protease, amylase, lactase, sucrase, maltase, glucoamylase, lipase, cellulase, hemicellulase, pectinase, phytase.
Nutri-Zyme	Nature's Plus	Amylase (brown rice fermentation), apple pectin, bromelain, lipase (brown rice fermentation), papain. (This product is chewable.)
Opti-Guard	Optimal Nutrients	Wheat sprouts (supplying superoxide dismutase, catalase, glutathione peroxidase, and methionine reductase).
Oxy-5000 Forte	American Biologics	Catalase, folic acid, glutathione, L-cysteine, N-acetylcysteine, selenium, superoxide dismutase, thioproline, vitamin B_{12} (riboflavin), vitamin E.

Product Name	Manufacturer/ Distributor	Ingredients
Pan 10X #428	Progressive Laboratories	Amylase, lipase, protease.
Pancrea-Lipase Forte Tablets	General Research Laboratories	Pancrealipase (three times the protease activity of Pancreas Compound, plus inositol).
Pancreas Compound with Duodenum (uncoated tablet)	General Research Laboratories	Pancreas substance, duodenum substance.
Pancreas 523	Enzyme Process International	Trypsin, chymotrypsin, amylase, protease, lipase enzymes equivalent to 205 mg of pancreatin.
Pancreatin	KAL, Inc.	Pancreatin plus dicalcium phosphate, magnesium stearate, silica, stearic acid, and vegetable protein coating.
Pancreatin	Nature's Life	Pancreatin (with protease, amylase, lipase).
Pancreatin (Quadruple Strength)	Twin Labs	Pancreatin (supplying amylase, protease, and lipase activity).
Pancreatin Capsules	Solgar Vitamin and Herb Co., Inc.	Pancreatin.
Pancreatin 8x Plus	Professional Health Products, Inc.	Pancreatin, trypsin, chymotrypsin.
Pancreatin 1000 mg	Nature's Plus	Pancreatin (supplying amylase, protease, and lipase).
Papaya Enzyme	American Dietary Labs	Amylase, bromelain, papain, papaya leaves.
Papaya Enzyme	Schiff Products	Papain.
Papaya Enzyme (chewable)	Nature's Plus	Ripe papaya melon and papaya leaves, supplying papain, amylase, and protease.
Papaya Enzyme Plus	Highland Labs	Bromelain, mycozyme (*Aspergillus oryzae*), and papain.
Papaya-Zyme Chewable	KAL, Inc.	Papain, protease, papaya fruit. Additional ingredients: fructose, magnesium stearate, natural papaya flavor, stearic acid, sucrose.
Parazyme	Enzymatic Therapy	Bitter melon extract, bromelain, cayenne pepper extract, cranberry extract, goldenseal root, grapefruit seed extract, haborandi pepper extract, pancreatic enzymes.
Phlogenzym	Mucos Pharma GmbH & Co.	Bromelain, rutin, trypsin.
Poly-Zym 021 Tablets	General Research Laboratories	Pancreatin, papain, bromelain, trypsin, chymotrypsin, rutin, lipase, amylase.
Poly-Zym 022 Tablets	General Research Laboratories	Papain, pancreas, thymus.
Poly-Zym 023 Tablets	General Research Laboratories	Papain, pancreatin, thymus concentrate, bromelain, lipase.
PowerPlus	Herbal Products and Development	Amylase, bromelain, cellulase, invertase, lactase, lipase, maltase, papaya, pectinase/phytase, protease.
Pregest	Bio-Energy Systems, Inc.	Amylase, cellulase, invertase, lactase, lipase, maltase, protease, O2 Factor, papaya leaf.
Pre-Meal Enz Extract	Crystal Star Herbal Nutrition	Catnip herb, cramp bark, fennel seed, fresh ginger root, papaya leaf, peppermint leaf, spearmint leaf, turmeric root.

Product Name	Manufacturer/ Distributor	Ingredients
Pro Vegezyme Complex	Professional Health Products, Inc.	Comfrey leaf, gamma oryzanol, papaya leaf, slippery elm bark.
Pro-E-Sentinal	Bio-Energy Systems, Inc.	Enzyme Enhancement System™, barley concentrate, bioflavonoids, broccoli concentrate, garlic concentrate, grape seed extract, lecithin, rose hip, tomato concentrate.
Pro-Gest-Ade	Enzymatic Therapy	Betaine HCl, bromelain, glutamic acid HCl, mycozyme (fungal amylase), mylase (lipase), niacinamide, ox bile extract, pancreas substance, papain.
Protagest	PhytoPharmica	Ammonium chloride, betaine HCl, glutamic acid HCl, niacinamide, ox bile, pancreatic enzymes, papain, pepsin, potassium chloride, vitamin B_6.
Protazyme	Enzymatic Therapy	Ammonium chloride, betaine HCl, glutamic acid HCl, niacinamide, ox bile extract, pancreatic enzymes, papain, pepsin, potassium chloride, vitamin B_6.
Proteozyme Forte	Douglas Laboratories	Aspartic acid, bioflavonoids, bromelain, calcium, cartilage/ chondroitin sulfate, chymotrypsin, folic acid, hesperidin complex, magnesium, manganese, papain, pepsin, potassium, raw muscle concentrate, raw veal bone, rutin, trypsin, vitamins A, B_1, B_2, B_3, B_5, B_6, B_{12}, C, and zinc.
Retenzyme Forte	Biotics Research Corporation	*Lens esculenta*, neonatal calf thymus, pancreatin, papain, *Pisum sativum*.
Say Yes to Beans	Nature's Plus	Legumase™ (enzyme complex), licorice root, ginger root, parsley seed.
Say Yes to Dairy	Nature's Plus	Lactase.
SOD Complex	Gero-Vita	Chromium picolinate, copper, glutathione, selenium, superoxide dismutase, vitamin A, vitamin E, vitamin C, zinc.
SOD (Superoxide Dismutase)	Douglas Laboratories	Superoxide dismutase.
S.O.D. Lozenge	Nutri-West	Superoxide dismutase, catalase, glutathione peroxidase.
Source of Life Vibra-Gest	Nature's Plus	Amylase, lactase, lipase, cellulase, protease, oxidase, diastase, maltase, phosphatase, bromelain, and papain, plus *lactobacillus acidophilus*, *bifidobacterium longum*, and *lactobacillus bulgaricus*.
Spirazymes	Earthrise	Amylase, cellulase, lactase, lipase, phosphatases, protease, and ribonucleases.
Super Enzymall	Schiff Products	Pancreatin, bromelain, papain, protease, amylase, cellulase, and ox bile extract.
Super Enzyme Caps	Twin Labs	Pancreatin (supplying amylase, protease, and lipase activity), betaine HCl, pepsin, ox bile, bromelain, papain.
Super Enzymes	Highland Labs	Betaine HCl, diastase, fenugreek seed powder, ox bile extract, pancreatin, pancrelipase, papain, and pepsin.
Super-gest +	Enzyme Process International	Pepsin, papain, bromelain, and microbial protease, amylase, and lipase.
Superzymes	AIM International, Inc.	Antioxidant enzymes.
Thurston Compound Tablets	General Research Laboratories	Betaine hydrochloride, pepsin, papain, mycozyme, ox bile.

Product Name	Manufacturer/ Distributor	Ingredients
Trigest	Ethical Nutrients	Amylase, lipase, protease.
Trim-Zyme	Rainbow Light Nutritional Systems	Protease, amylase, cellulase, lipase, Citramax (garcinia cambogia), nettles, *Gymnema sylvestre,* kelp, saw palmetto, ginger, turmeric, GTF chromium.
2 Phase-Zyme	Douglas Laboratories	Cellulase, dehydrochloric acid, enzyme concentrate (with lipolytic, amylolytic, and proteolytic activity), lysine hydrochloride, oleoresin ginger.
Ultra Hair Thick Shake	Nature's Plus	Choline, inositol, PABA, L-methionine, L-cystine, L-taurine, L-glutathione, safflower oil, MCT (medium chain triglyceride), bromelain, papain, *L. acidophilus,* amylase, lipase, protease.
Ultra-Zyme	Nature's Plus	Pancreatin (supplying amylase, protease, and lipase), acidophilus, ox bile, bromelain, pepsin, papain, malt diastase, cellulase, hemicellulase, lactase.
Vegetarian digestive aid tablets	Solgar Vitamin and Herb Co., Inc.	Amylase, amyloglucosidase, lipase, prozyme, protease, cellulase.
Vegetarian Enzyme	Douglas Laboratories	Vegetarian enzyme concentrate (supplying amylase, protease, lipase, cellulase).
Vegetarian Enzyme Complex	Futurebiotics	Alfalfa powder, amylase, betaine HCl, bromelain, cellulase, lactase, lipase, papain, protease.
Vegetarian Enzyme Complex 50Plus	Futurebiotics	Amylase, bromelain, cellulase, fennel seed, gentian root, invertase, lactase, lipase, maltase, papain, protease.
Vegie-zymes	Professional Botanicals	Amylase, cellulase, lipase, protease.
Vitase Digestion Formula	Prevail Corporation	Amylase, cellulase, lactase, lipase, maltase, protease (I, II, III, and IV), and sucrase.
WobeMugos E (ointment)	Mucos Pharma GmbH & Co.	Chymotrypsin, papain, trypsin.
WobeMugos E (oral)	Mucos Pharma GmbH & Co.	Chymotrypsin, papain, trypsin.
WobeMugos Th (rectal tablet)	Mucos Pharma GmbH & Co.	Papain, trypsin, thymus peptides.
Wobenzym	Mucos Pharma GmbH & Co.	Bromelain, chymotrypsin, pancreatin, papain, trypsin, rutin.
Wobenzym N	Naturally Vitamin Supplements Co.	Bromelain, chymotrypsin, pancreatin, papain, rutin, trypsin.
Zygest	PhysioLogics	Papain, *Asp. oryzae,* amylase, protease, acid protease, bromelain, lactase, cellulase, hemicellulase.
Zymain	Anabolic Laboratories, Inc.	Bioflavonoids, bromelain, cartilage, chymotrypsin, manganese, papain, shave grass, trypsin, vitamin C, zinc.
Zyme Dophilus	PhytoPharmica	Amylase, bifidobacterium bifidum live bacteria, colstrex 36, concentrated fructooligosaccharide from Jerusalem artichoke, *Lactobacillus acidophilus,* lipase, protease.

ENZYME COMPANIES

Many companies have quality enzyme products taken from a number of sources. The following list includes manufacturers, compounders, marketers, and distributors of many of the enzyme products in the previous list. Each company's address and telephone number is included so you can contact the company to learn more about its products. This list is not in any way comprehensive, as there are hundreds of companies around the world involved in the manufacture and distribution of enzymes. Be aware that telephone numbers and addresses are subject to change.

ABCO, Incorporated
2377 Stanwell Drive
Concord, California 94520
Tel: (510) 685-1212
Fax: (510) 682-7241

Abkit, Inc.
207 East 94th Street
2nd Floor
New York, New York 10128
Tel: (800) 226-6227
or: (212) 860-8358

Action Labs, Inc.
2851 Via Martens
Anaheim, California 92806
Tel: (714) 630-5941
Fax: (714) 630-8221

ADH Health Products, Inc.
215 N. Route 303
Congers, New York 10920
Tel: (914) 268-0027
Fax: (914) 268-2988

ADM Laboratories, Inc.
American Desert Manufacturing
5536 W. Roosevelt Street, Suite #1
Phoenix, Arizona 85043
Tel: (602) 272-3777
Fax: (602) 272-1500

Advanced Labs
8759 Airport Road, Suite C
Redding, California 96002
Tel: (800) 955-5553
Fax: (916) 223-0699

Advanced Nutrient Science
P.O. Box 668
1157 North Hotsprings Drive
Parker, Colorado 80134
Tel: (303) 840-0555
Fax: (303) 840-0550

AIM (American Image Marketing)
3904 East Flamingo Avenue
Nampa, Idaho 83687
Tel: (208) 465-5116
Fax: (208) 463-2187

AkPharma, Inc.
P.O. Box 111
Pleasantville, New Jersey 08232
Tel: (609) 645-5100
Fax: (609) 645-0767

Alltech, Inc.
Biotechnology Center
3031 Catnip Hill Pike
Nicholasville, Kentucky 40356
Tel: (606) 885-9613
Fax: (606) 885-6736

Alpine Health Industries
1525 W. Business Park Drive
Orem, Utah 84651
Tel: (801) 225-5525
Fax: (801) 225-5899

Amano Enzyme U.S.A. Company, Ltd.
1157 North Main Street
Lombard, Illinois 60148
Tel: (800) 446-7652
Fax: (630) 953-1895

American Biologics
1180 Walnut Avenue
Chula Vista, California 91911
Tel: (619) 429-8200
Fax: (619) 429-8004

American Dietary Labs
14631 Best Avenue
Norwalk, California 90650
Tel: (800) 423-8837

American Health
4320 Veterans Memorial Highway

Holbrook, New York 11741
Tel: (800) 445-7135
Fax: (516) 244-1777

American Health & Herbs
P.O. Box 940
Philomath, Oregon 97370
Tel: (800) 345-4152
Fax: (541) 929-2911

American Laboratories, Inc.
4410 S. 102nd Street
Omaha, Nebraska 68127
Tel: (402) 339-2494
Fax: (402) 339-0801

Anabol Naturals
1550 Mansfield Street
Santa Cruz, California 95062
Tel: (800) 426-2265
Fax: (408) 479-1406

Anabolic Laboratories, Inc.
17801 Gillette Avenue
P.O. Box C19508
Irvine, California 92713
Tel: (714) 863-0340
Fax: (714) 261-2928

Apothecary Products, Inc.
11531 Rupp Drive
Burnsville, Minnesota 55337
Tel: (612) 890-1940
Fax: (612) 890-0418

Archon Vitamin Corporation
209 40th Street
Irvington, New Jersey 07111
Tel: (201) 371-1700
Fax: (201) 371-1277

Arve's Zyming Beauty Products
P.O. Box 1869
Flagler Beach, Florida 32136

Tel: (904) 439-3305
Fax: (904) 439-7303

Ashland Nutritional Products
17751 Mitchell Street
Irvine, California 92714
Tel: (714) 833-9500
Fax: (714) 833-9595

Aveda
4000 Pheasant Ridge Drive
Minneapolis, Minnesota 55449
Tel: (800) 283-3224
Fax: (612) 783-4110

Barth's Nutra Products Corp.
3890 Park Central Blvd., North
Pompano Beach, Florida 33064
Tel: (800) 645-2208
Fax: (954) 978-7093

Bayer Corporation
P.O. Box 3100
Elkhart, Indiana 46515
Tel: (800) 248-2637

Bio-Cat, Inc.
Route 2, Box 1475
Troy, Virginia 22944
Tel: (804) 589-4777
Fax: (804) 589-3301

BioConnections
P.O. Box 20985
Portland, Oregon 97294
Tel: (800) 743-0171
Fax: (503) 254-5773

Bioenergy Nutrients, Inc.
6395 Gunpark Drive, Suite A
Boulder, Colorado 80301
Tel: (800) 553-0227

Bio-Energy Systems, Inc.
157 N. Meridian, Suite 105
Kalispell, Montana 59901
Tel: (800) 929-8328
Fax: (406) 257-9111

Biomed Brokerage Network
32081 Via Flores
San Juan Capistrano, California 92675

Tel: (714) 661-1245
Fax: (714) 661-1254

BioMed Scientific Inc.
P.O. Box 911
124 Hebron Avenue
Glastonbury, Connecticut 06033
Tel: (860) 657-2258
Fax: (860) 657-2800

Bio-Nutritional Products
41 Bergenline Avenue
Westwood, New Jersey 07676
Tel: (800) 431-2582
Fax: (201) 666-2929

BioSan Laboratories, Inc.
P.O. Box 325
8 Bowers Road
Derry, New Hampshire 03038
Tel: (603) 432-5022
Fax: (603) 434-4736

BioStar Nutraceuticals Intl.
475 Gate Five Road, Suite 210A
San Mateo, California 94403
Tel: (415) 331-9699
Fax: (415) 331-9698

Biotec Foods
1 Capitol District
250 S. Hotel Street, Suite 200
Honolulu, Hawaii 96813
Tel: (800) 331-5888
Fax: (808) 529-93423

Biotics Research Corporation
P.O. Box 36888
Houston, Texas 77236
Tel: (713) 240-8010

Bluebonnet Nutrition Corp.
12503 Exchange Drive, Suite 530
Stafford, Texas 77477
Tel: (800) 580-8866
Fax: (713) 240-3535

Body Mechanics
624 Estuary Drive
Bradenton, Florida 34209
Tel: (800) 264-1114

Botanicals International
2550 El Presidio Street
Long Beach, California 90810-1193
Tel: (310) 637-9566
Fax: (310) 637-3644
E-mail: botan@botanicals.com

Canadian Natural Products Ltd.
B15.60020 2nd Street, S.E.
Calgary, Alberta, Canada T2H 2L8
Tel: (403) 252-0177
Fax: (403) 252-0176

Carlson Laboratories, Inc.
15 College Drive
Arlington Heights, Illinois 60004
Tel: (800) 323-4141
or: (708) 255-1600

CC International, Inc.
P.O. Box 2452
Rancho Santa Fe, California 92067
Tel: (800) 775-3575
Fax: (619) 756-1334

C.C. Pollen Company
3627 East Indian School Road
Suite 209
Phoenix, Arizona 85018
Tel: (800) 875-0096
or: (602) 957-0096

C.E. Jamieson & Co., Ltd.
2 St. Clair Avenue West, Suite 1502
Toronto, Ontario, Canada M4V 1L5
Tel: (416) 960-0052
Fax: (416) 960-4803

Cell Tech
1300 Main Street
Klamath Falls, Oregon 97601-5914
Tel: (503) 882-5406
Fax: (503) 884-1869

Connection Source, Inc.
5515 Taylor Road
Alpharetta, Georgia 30202
Tel: (403) 667-1051
Fax: (404) 667-1283

Country Life
101 Corporate Drive
Hauppauge, New York 11788

Tel: (516) 231-1031
Fax: (516) 231-2331

Crystal Star Herbal Nutrition
4609 Wedgeway Court
Earth City, Missouri 63045
Tel: (800) 736-6015

Dr. Goodpet Labs
322-1/2 AE. Beach Avenue
P.O. Box 4728
Inglewood, California 90309
Tel: (800) 222-9932
Fax: (310) 672-4287

Dr. Grandel, Inc.
626 W. Sunset Road
San Antonio, Texas 78216
Tel: (210) 829-1763

Doctor's Pride, Inc.
75 Bi-County Blvd.
Farmingdale, New York 11735
Tel: (800) 645-9909
Fax: (800) 397-4252

Douglas Laboratories, Inc.
600 Boyce Road
Pittsburgh, Pennsylvania 15205
Tel: (888) DOUGLAB
Fax: (412) 494-0155

DowMor Labs
Suite B-111, 1641 E. Sunset Road
Las Vegas, Nevada 89119
Tel: (800) 926-5211
Fax: (702) 896-1120

DynaPro International
P.O. Box 3002
Ogden, Utah 84409
Tel: (800) 877-1413
Fax: (801) 621-8258

Earthrise Company
424 Payran Street
Petaluma, California 94952
Tel: (707) 778-9078
Fax: (707) 778-9028
E-mail: Info@Earthrise.com

Eckhart Corp.
1620 Grant Avenue, Suite #2
Novato, California 94945

Tel: (415) 898-9528
Fax: (415) 898-1917

Eclectic Institute
14385 S.E. Lusted Road
Sandy, Oregon 97055
Tel: (800) 332-HERB

Ecological Formulas
1061-B Shary Circle
Concord, California 94518
Tel: (800) 888-4585
Fax: (510) 676-9231

Energen Products, Inc.
14631 Best Avenue
Norwalk, California 90650
Tel: (800) 423-8837
Fax: (510) 921-0039

Energy Factors, Inc.
6950 Bryan Dairy Road
Largo, Florida 34647
Tel: (813) 544-8866

Enzymatic Therapy, Inc.
825 Challenger Drive
Green Bay, Wisconsin 54311
Tel: (800) 558-7372
Fax: (920) 469-4400
Internet: http://www.enz.com

Enzyme Development Corporation
2 Penn Plaza, Suite 2439
New York, New York 10121-0034
Tel: (212) 736-1580
Fax: (212) 279-0056

Enzyme Process International
2035 East Cedar Street
Tempe, Arizona 85281
Tel: (602) 731-9290
or: (800) 655-9092

Ethical Nutrients
971 Calle Negocio
San Clemente, California 92673
Tel: (800) 668-8743
Fax: (714) 366-2859

Flora, Inc.
805 E. Badger Road
Lynden, Washington 98264

Tel: (800) 498-3610
Fax: (360) 354-5355

Foodscience Laboratories
20 New England Drive, #C-1504
Essex Junction, Vermont 05453-1504
Tel: (800) 874-9444
Fax: (802) 878-0549

Freeda Vitamins, Inc.
36 E. 41st Street
New York, New York 10017
Tel: (800) 777-3737
Fax: (212) 685-7297

Fruit of the Land
14631 Best Avenue
Norwalk, California 90650
Tel: (800) 423-8837

Futurebiotics, Inc.
145 Ricefield Lane
Hauppauge, New York 11788
Tel: (516)273-6300
Fax: (516) 273-1165

Garden State Nutritionals
100 Lehigh Drive
Fairfield, New Jersey 07004
Tel: (800) 526-9095
Fax: (201) 575-6782

Gelda Scientific
6320 Northwest Drive
Mississauga, Ontario, Canada L4V 1J78
Tel: (905) 673-9320
Fax: (905) 673-8114

General Nutrition, Inc.
921 Penn Avenue
Pittsburgh, Pennsylvania 15222
Tel: (412) 288-4713

General Research Laboratories
8900 Winnetka Avenue
Northridge, California 91324
Tel: (800) 421-1856
Fax: (818) 407-8500

Gero Vita
6021 Yonge Street
Toronto, Ontario, Canada M2M 3W2
Tel: (800) 694-8366

Gero Vita International
520 Washington Street
#391
Marina Del Ray, California 90292
Tel: (800) 825-8482

Global Nutritional Research Corp.
4022-1 S. 20th Street
Phoenix, Arizona 85040
Tel: (602) 243-5189
Fax: (602) 243-6551

Global Trading (USA)
1119 Springfield Road
Union, New Jersey 07083
Tel: (908) 964-0900
Fax: (908) 964-0948

Good'N Natural Vitamins
90 Orville Drive
Bohemia, New York 11716
Tel: (516) 244-2041
Fax: (516) 224-2013

Great American Natural Products
4121 16th Street, North
St. Petersburg, Florida 33703
Tel: (813) 521-4372

Greater Continents International
140 Arrowood Lane
San Mateo, California 94403
Tel: (415) 697-4700
Fax: (415) 697-6300

Green Foods Corporation
320 Graves Avenue
Oxnard, California 90505
Tel: (800) 777-4430
or: (805) 983-7470
Fax: (805) 983-8840

Green Leaf Herbs
1080 E. Sandy Lake Road
Coppel, Texas 75019
Tel: (214) 437-9959
Fax: (214) 437-9954

Health Enhancers, Inc.
8139 Corunne Road
Flint, Michigan 48532
Tel: (800) 792-9199
Fax: (810) 659-4949

Health Products Corp.
1060 Nepperhan Avenue
Yonkers, New York 10703
Tel: (914) 423-2900
Fax: (914) 963-6001

HealthSmart Vitamins
1921 Miller Drive
Longmont, Colorado 80501
Tel: (800) 492-3003

Henkel Corporation
5325 South Ninth Avenue
LaGrange, Illinois 60525
Tel: (708) 579-6150

Herbal Products and Development
P.O. Box 1084
Aptos, California 95001
Tel: (408) 688-8706
Fax: (408) 688-8711

Highland Laboratories
110 South Garfield
Mt. Angel, Oregon 97362
Tel: (800) 547-0273

Hillestad International
178 Highway 51 North
P.O. Box 1700
Woodruff, Wisconsin 54568
Tel: (800) 535-7742
Fax: (715) 358-7812

Holistic Animal Care
7334 E. Broadway
Tucson, Arizona 85710
Tel: (800) 497-5665
Fax: (520) 886-1727

Infinity2 Distribution, Inc.
63 E. Main Street #700
Mesa, Arizona 85201
Tel: (602) 668-1856

Interior Design Nutritionals
75 West Center Street
Provo, Utah 84601
Tel: (801) 345-2000
Fax (801) 345-1999

International Enzyme Foundation
P.O. Box 249, Highway 160

Forsyth, Missouri 65653
Tel: (800) 433-8589
Fax: (417) 546-6433

Ion Labs, Inc.
6545 44th Street North
Pinellas Park, Florida 34665
Tel: (800) 275-2653
Fax: (813) 527-6758

Jarrow Formulas
1824 So. Robertson
Los Angeles, California 90035
Tel: (800) 726-0886
Fax: (213) 204-2520

J.R. Carlson Laboratories
15 College
Arlington Heights, Illinois 60004
Tel: (800) 323-4141
Fax: (708) 255-1605

Juice Plus+
NSA
4260 E. Raines Road
Memphis, Tennessee 38118-6977
Tel: (901) 366-9288

KAL, Inc.
P.O. Box 4023
Woodland Hills, California 91365
Tel: (818) 340-3035

Klamath Blue Green, Inc.
P.O. Box 1626
Mt. Shasta, California 96067
Tel: (800) 327-1956
Fax: (916) 926-6685

K.W. Pfannenschmit GmbH
P.B. 610151
22421 Hamburg, Germany
Habichthorst 36
22459 Hamburg, Germany
Tel: (040) 555-8660
Fax (040) 555-3898

Kyolic, Ltd.
23501 Madero
Mission Viejo, California 92691
Tel: (800) 421-2998

Lactaid, Inc.
7050 Camp Hill Road
Fort Washington, Pennsylvania 19034
Tel: (800) LACTAID

Life Plus
P.O. Box 3749
Batesville, Arkansas 72503
Tel: (800) 572-8446

Longevity Network, Ltd.
15 Cactus Drive
Henderson, Nevada 89014
Tel: (702) 454-7000
Fax: (702) 435-4786

Lotus Light Enterprises, Inc.
P.O. Box 1008
Silver Lake, Wisconsin 53170
Tel: (414) 889-8501

Maat Trade Development
1888 Century Park East, Suite 1900
Los Angeles, California 90067
Tel: (310) 284-6808
Fax: (310) 284-6818

Magnus Enterprises
1406 West 178th Street
Gardena, California 90248
Tel: (310) 532-8440
Fax: (310) 515-5263

MAK Wood Inc.
Box 184
Thiensville, Wisconsin 53092
Tel: (414) 242-2323
Fax: (414) 242-9448

Malabar Formulas
28537 Nuevo Valley Drive
Nuevo, California 92567
Tel: (800) 462-6617

Marcor Development Corp.
108 John Street
Hackensack, New Jersey 07601
Tel: (201) 489-5700
Fax: (201) 489-7357

Marlyn Nutraceuticals
14851 North Scottsdale Road

Scottsdale, Arizona 85254
Tel: (800) 462-7596

McNeil Consumer Products Co.
7050 Camp Hill Road
Ft. Washington, Pennsylvania 19034
Tel: (800) 522-8243

Medi-Plex International
520 Washington Street, Suite 391
Marina del Rey, California 90292
Tel: (800) 292-6006

Michael's Naturopathic Programs
6820 Alamo Downs Parkway
San Antonio, Texas 78238
Tel: (210) 647-4700

Miller Pharmacal Group, Inc.
4562 Prime Parkway
McHenry, Illinois 60050
Tel: (800) 323-2915

MiniStar International, Inc.
21118 Commerce Pointe Drive
City of Industry, California 91789
Tel: (909) 598-3963
Fax: (909) 598-5733

Mucos Pharma GmbH & Co.
Malvenweg 2
D-82538 Geretsried
Germany
Tel: 011-49-0-8171-5180
Fax: 011-49-0-8171-52008

Murdock Madaus Schwabe
10 Mountain Springs Parkway
Springville, Utah 84663
Tel: (801) 489-1413
Fax: (801) 489-1700

National Brokerage Network
32081 Via Flores, Suite 300
San Juan Capistrano, California 92675
Tel: (714) 661-1254
Fax: (714) 661-1254

National Enzyme Company
P.O. Box 128
Forsyth, Missouri 65653
Tel: (800) 433-8589
Fax: (417) 546-6433

National Vitamin Company, Inc.
2075 West Scranton Avenue
Porterville, California 93257
Tel: (209) 781-8871
Fax: (209) 781-8878

Natren
3105 Willow Lake
Westlake Village, California 91361
Tel: (800) 992-3323

Natrol, Inc.
20731 Marilla Street
Chatsworth, California 91311
Tel: (800) 326-1520
Fax: (818) 701-0623

Natural Alternatives International, Inc.
1185 Linda Vista Drive
San Marcos, California 92069
Tel: (760) 744-7340
Fax: (760) 744-8402

Natural Factors Nutritional Products
1420 80th Street, S.W., Suite B
Everett, Washington 98203
Tel: (206) 513-8800

Natural Food Systems
8301 Torresdale Avenue
Philadelphia, Pennsylvania 19115
Tel: (215) 624-3559
Fax: (215) 331-8728

Naturally Vitamin Supplements Co.
14851 N. Scottsdale Road
Scottsdale, Arizona 85254
Tel: (800) 899-4499
Fax: (602) 991-0551

Nature Most Labs., Inc.
P.O. Box 721
Middletown, Connecticut 06457
Tel: (800) 234-2112
Fax: (860) 247-2800

Nature's Bounty
90 Orville Drive
Bohemia, New York 11716
Tel: (516) 567-9500
Fax: (516) 563-1623

Nature's Concept
5242 Bolsa Avenue, Suite 3
Huntington Beach, California 92649
Tel: (714) 893-0017
Fax: (714) 897-5677

Nature's Life
7180 Lampson Avenue
Garden Grove, California 92841
Tel: (714) 379-6500
Fax: (714) 379-6501

Nature's Plus
548 Broad Hollow Road
Melville, New York 11747
Tel: (516) 293-0030
Fax: (516) 249-2022

Nature's Products, Inc.
2525 Davie Road
Davie, Florida 33317
Tel: (800) 752-7873
Fax: (305) 474-0989

Nature's Sunshine Products, Inc.
P.O. Box 19005
Provo, Utah 84605-9005
Tel: (801) 342-4300

Nature's Way
10 Mountain Springs Parkway
Springville, Utah 84663
Tel: (800) 926-8883
Fax: (801) 489-1700

Nebraska Cultures, Inc.
6610 Van Dorn
Lincoln, Nebraska 68506
Tel: (602) 230-2758

Nevada Nutritional
4900 Mill Street, #B-32
Reno, Nevada 89502
Tel: (702) 857-2700
Fax: (702) 857-1412

NF Formulas, Inc.
805 S.E. Sherman
Portland, Oregon 97214-4666
Tel: (800) 547-4891

Nikken U.S.A., Inc.
15363 Barranca Parkway

Irvine, California 92718
Tel: (714) 789-2000
Fax: (714) 789-2080

Novo Nordisk A/S
Novo Allé
2880 Bagsvaerd
Denmark
Tel: 011-45-4444-8888
Fax: 011-45-4449-0555

Now Foods
550 Mitchell Road
Glendale Heights, Illinois 60139
Tel: (800) 999-8069
or: (630) 545-9098

NuBotanic International, Inc.
15512 S. Figueroa Street
Gardena, California 90248
Tel: (310) 327-4500

Nutra-Caps Intl.
6106 Avenida Encinas, Suite E
Carlsbad, California 92008
Tel: (619) 930-4220
Fax: (619) 930-8024

NutriCology, Inc.
400 Preda Street
San Leandro, California 94577
Tel: (800) 545-9960
or: (510) 639-4572
Fax: (510) 635-6730

Nutri-Essence
100 N.W. Business Park Lane
Riverside, Missouri 64150
Tel: (800) 647-6377
Fax: (800) 844-1957

Nutrilabs, Inc.
5000 W. Oakey, #D-12
Las Vegas, Nevada 89102
Tel: (702) 878-7376
Fax: (702) 878-4863

Nutrina Company, Inc.
1727 Cosmic Way
Glendale, California 91201
Tel: (818) 790-1776
Fax: (818) 790-9420

Nutriscience, Ltd.
51 Beverly Hills Drive
Toronto, Ontario, Canada M3L 1A2
Tel: (416) 249-1234
Fax: (416) 249-0341

**Nutritional Enzyme Support System
(NESS)**
2903 N.W. Platte Road
Riverside, Missouri 64150
Tel: (800) 637 7893

Nutritional Research Associates
407 E. Broad Street
South Whitley, Indiana 46787
Tel: (219) 723-4931

Nutritional Specialties, Inc.
1452 E. Kateila Avenue
Anaheim, California 92805-6635
Tel: (800) 333-6168
Fax: (714) 634-9347

Nutritionals, Etc.
20105 Nordhoff Street
Chatsworth, California 91311
Tel: (818) 727-9373
Fax: (818) 727-9385

Nutri-West
P.O. Box 950
Douglas, Wyoming 82633
Tel: (307) 358-5066

Oekpharma GmbH
Moosham 29
A-5580 Unterberg, Austria
Tel: (43) 6476-805-0
Fax: (43) 6476-805-40

Optimal Nutrients
1163 Chess Drive, Suite F
Foster City, California 94404
Tel: (800) 966-8874
Fax: (415) 349-1686

PacMarc Packaging
1501 Adrian Road
Burlingame, California 94010
Tel: (415) 697-4700
Fax: (415) 697-6300

Parametric Associates, Inc.
10934 Lin-Valle Drive
St. Louis, Missouri 63123
Tel: (800) 747-1601
Fax: (314) 892-0988

Pharmadass, Ltd.
16 Aintree Road
Ub6 7 La
Greenford, Middlesex, England
Tel: (081) 991-0035
Fax: (081) 997-3490

Pharmavite
15451 San Fernando Boulevard
Mission Hills, California 91345
Tel: (818) 837-3633
Fax: (818) 837-6182

Phillips Nutritionals
27071 Cabot Road, Suite 122
Laguna Hills, California 92653
Tel: (800) 514 5115
Fax: (714) 582-8461

Phoenix Laboratories
140 Lauman Lane
Hicksville, New York 11801
Tel: (516) 822-1230
Fax: (516) 939-0234

PhysioLogics
6565 Odell Place
Boulder, Colorado 80301-3330
Tel: (800) 765-6775

PhytoPharmica
825 Challenger Drive
Green Bay, Wisconsin 54311
Tel: (800) 553-2370
Fax: (414) 469-4418

Phyto-Therapy, Inc.
P.O. Box 555
Franklin Lakes, New Jersey 07417
Tel: (201) 891-1104
Fax: (201) 848-1867

The Pierson Company
14631 Best Avenue
Norwalk, California 90650
Tel: (800) 423-8837

Premier Labs
27475 Ynez Road, Suite 305
Temecula, California 92591
Tel: (800) 887-5227
Fax: (909) 699-8801

Prevail Corporation
2204-8 N.W. Birdsdale
Gresham, Oregon 97030
Tel: (800) 248-0885
Fax: (503) 667-4790

Professional Botanicals
P.O. Box 9822
Ogden, Utah 84409
Tel: (800) 824-8181

Professional Health Products
P.O. Box 80085
Portland, Oregon 97280-1085
Tel: (800) 952-2219
Fax: (503) 452-1239

Progressive Laboratories, Inc.
1701 W. Walnut Hill Lane
Irving, Texas 75038
Tel: (214) 518-9660
Fax: (214) 518-9665

Progressive Research Labs, Inc.
9396 Richmond
Suite 514
Houston, Texas 77063
Tel: (800) 877-0966

Pro-Pharma, Inc.
21 N. Skokie Highway
Lake Bluff, Illinois 60044
Tel: (847) 234-5200
Fax: (847) 234-5298

Quad Laboratories
P.O. Box 555
Franklin Lakes, New Jersey 07417
Tel: (201) 891-1104
Fax: (201) 848-1867

Rainbow Light Nutritional Systems
P.O. Box 600
Santa Cruz, California 95061
Tel: (800) 635-1233

Really Raw Honey
1301 S. Baylis Street
Baltimore, Maryland 21224
Tel: (410) 675-7233
Fax: (410) 675-7411

Reliance Vitamin Co., Inc.
185-B Industrial Parkway
Branchburg, New Jersey 08876
Tel: (201) 218-1221

Rexall Showcase International
851 Broken Sound Parkway, N.W.
Boca Raton, Florida 33487
Tel: (407) 994-2090
Fax: (407) 241-5319

Rexall Sundown, Inc.
851 Broken Sound Parkway, N.W.
Boca Raton, Florida 33487
Tel: (407) 241-9400
Fax: (407) 995-6880

Robinson Pharma
2638 S. Croddy Way
Santa Ana, California 92704
Tel: (714) 241-0235
Fax: (714) 751-6066

The Rockland Corporation (TRC)
12320 E. Skelly Drive
Tulsa, Oklahoma 74128
Tel: (918) 437-7310
or: (918) 437-7311

Sabinsa Corporation
121 Ethel Road West, Unit #6
Piscataway, New Jersey 08854
Tel: (908) 777-1111
Fax: (908) 777-1443

Savin Your Health Products
A division of Enzyme Process
N.Y., Inc.
P.O. Box 30027
Staten Island, New York 10303
Tel: (800) 762-6841
or: (718) 494-8446
Fax: (718) 370-2942

Schiff Products, Inc.
1960 South 4250 West
Salt Lake City, Utah 84104

Tel: (801) 975-5000
Fax: (801) 972-6532

Schweizerhall, Inc.
Chemical Dynamics Corp.
10 Corporate Place South
Piscataway, New Jersey 08854
Tel: (908) 981-8200
Fax: (908) 981-8282

Scientific Consulting Service
466 Whitney Street
San Leandro, California 94577
Tel: (510) 632-2370

Set-N-Me-Free Aloe Vera Co.
19220 S.E. Stark
Portland, Oregon 97233
Tel: (800) 221-9727
Fax: (503) 669-9057

Sigma Chemical Co.
P.O. Box 14508
St. Louis, Missouri 63178
Tel: (800) 325-3010
Fax: (314) 771-5757

Solaray, Inc.
1104 Country Hill Drive, Suite 412
Ogden, Utah 84403
Tel: (801) 626-4956
Fax: (801) 393-8215

Solgar Vitamin and Herb Co.
500 Willow Tree Road
Leonia, New Jersey 07605
Tel: (800) 645-2246
Fax: (201) 944-7351

Source Naturals
P.O. Box 2118
Santa Cruz, California 95063
Tel: (408) 438-1144
Fax: (408) 438-7410

**Specialty Enzymes and
Biochemicals Co.**
5390 La Crescenta
Yorba Linda, California 92687
Tel: (714) 692-3350
Fax: (714) 692-3051

Standard Process
12521 131st Court, N.E.
Kirkland, Washington 98083-2484
Tel: (800) 292-6699

Stow Mills
P.O. Box 301
Chesterfield, New Hampshire 03443
Tel: (800) 451-2525
Fax: (603) 256-6959

Superior Nutraceuticals, Inc.
P.O. Box 979
Elkhorn, Wisconsin 53121-0886
Tel: (800) 842-0924
Fax: (414) 723-5462

Synergy Plus
500 Halls Mill Road
Freehold, New Jersey 07728
Tel: (800) 666-8482
Fax: (908) 866-0850

Thompson Nutritional Products
4031 N.E. 12th Terrace
Ft. Lauderdale, Florida 33334
Tel: (800) 421-1192

Three-Vee Food Co.
110 Bridge Street
Brooklyn, New York 11201
Tel: (718) 858-7333
Fax: (718) 858-7371

Tishcon Corporation
30 New York Avenue
Westbury, New York 11950
Tel: (800) 848-8442
Fax: (516) 997-3660

TKD Cosmos, Inc.
P.O. Box 7659
Torrance, California 90504-9059
Tel: (310) 538-5977
Fax: (310) 538-5978

Total Wellness, Inc.
P.O. Box 553
Parker, Colorado 80134
Tel: (800) 554-0276
Fax: (303) 841-0448

Tri Medica
8321 East Evans Road
Scottsdale, Arizona 85260
Tel: (602) 998-1041

Triarco, Inc.
6 Morris Street
Paterson, New Jersey 07501
Tel: (201) 278-7300
Fax: (201) 278-0317

Twin Laboratories, Inc.
2120 Smithtown Avenue
Ronkonkoma, New York 11779
Tel: (516) 467-3140
Fax: (516) 471-2375

Tyler Encapsulations
2204-8 N.W. Birdsdale
Gresham, Oregon 97030
Tel: (800) 869-9705

Ultimate Nutrition Products
P.O. Box 643
Farmington, Connecticut 06034
Tel: (860) 409-7373
Fax: (860) 409-7377

USA Sports Labs
1438 Highway 96
Burns, Tennessee 37029
Tel: (800) 489-4872
Fax: (615) 446-3788

Valley Research, Inc.
P.O. Box 750
South Bend, Indiana 46624-0750
Tel: (219) 232-5000
Fax: (219) 232-2468

Vinco, Inc.
1519 Mars-Evans City Road
Evans City, Pennsylvania 16033
Tel: (800) 245-1939

Viobin Corporation
700 E. Main Street
P.O. Box 158
Waunakee, Wisconsin 53597
Tel: (608) 849-5944

Vitagenics
240 South Broad Street
P.O. Box 886
Elkhorn, Wisconsin 53121
Tel: (414) 723-4942
Fax: (414) 723-5462

Vitaline Formulas
385 Williamson Way
Ashland, Oregon 96520
Tel: (800) 648-4755
Fax: (503) 482-9112

Vitamin Research Products, Inc.
3579 Highway 50 East
Carson City, Nevada 89701
Tel: (800) 877-2447
Fax: (702) 844-1331

Vita-Pure, Inc.
410 W. 1st Avenue
Roselle, New Jersey 07203
Tel: (908) 245-1212
Fax: (908) 245-1999

VitaScience Health Products, Inc.
P.O. Box 90756
Long Beach, California 90809-0756
Tel: (800) 600-VITA
Fax: (800) 700-VITA

Wakunaga of America Co., Ltd.
23501 Madero
Mission Viejo, California 92691
Tel: (714) 855-2776
Fax: (714) 458-2764

Weider Food Companies
1911 South 3850 West
Salt Lake City, Utah 84104
Tel: (801) 972-0330

Westar Nutrition, Inc.
1239 Victoria Street
Costa Mesa, California 92627
Tel: (714) 645-6100
Fax: (714) 645-9131

Whole Life/Calray Inc.
13234 Weidner Street
Pacoima, California 91331
Tel: (800) 748-5841

Wild Rose Herbal Formulas
#203, 8173-128th Street
Surrey, B.C. V3W 4G1
Canada
Tel: (604) 591-8881
Fax: (604) 597-1784

Wildflower Pharmacal Corp.
174 Herricks Road
Mineola, New York 11501
Tel: (516) 741-3304
Fax: (516) 741-3356

YH Products Corporation
400 North Lombard Street
Oxnard, California 93030
Tel: (805) 983-1130
Fax: (805) 983-3648

Zia Cosmetics
410 Townsend Street, 2nd Floor
San Francisco, California 94122
Tel: (415) 543-7546

TREATMENT CENTERS

The following are clinics, hospitals, physicians, and health practitioners that use enzyme therapy in the treatment of such conditions as allergies, arthritis, chronic fatigue syndrome, multiple sclerosis, lupus, viral infections, cancer, AIDS, cardiovascular disorders, and other immune, inflammatory, and chronic degenerative disorders. Be aware that telephone numbers and addresses are subject to change.

Advanced Medical Group
United States Office:
5862 Cromo Drive
Suite 147
El Paso, Texas 79912
Tel: (915) 581-2273
or (800) 863-7686
Fax: (915) 585-2274

Clinic:
Ave. Lopez Mateos #1281
Cd. Juarez, Mexico 32350
Tel: 011-52-16-13-84-58

American Biologics-Mexico S.A. Medical Center
Tijuana, Mexico
Admissions Office:
1180 Walnut Avenue
Chula Vista, California 91911
Tel: (800) 227-4473
or: (619) 429-8200
Fax: (619) 429-8004

American Metabolic Institute
555 Saturn Boulevard
Building B M/S 432
San Diego, California 92154
Tel: (800) 388-1083
or: (619) 267-1107
Fax: (619) 267-1109
E-mail: ami@connectnet.com
Internet: http://www.ami.health.com

Atkins Center for Complementary
Medicine
152 East 55th Street
New York, New York 10022
Tel: (212) 758-2110
Fax: (212) 754-4284

Daniel Beilin, O.M.D, L.Ac.
9057 Soquel Drive AB
Aptos, California 95003

Tel: (408) 685-1125
Fax: (408) 685-1128

Cancer Control Society
2043 North Berendo
Los Angeles, California 90027
Tel: (213) 663-7801
Fax: (213) 663-2730

Cancer Treatment Centers of America
3455 Salt Creek Lane
Arlington Heights, Illinois 60005
Tel: (800) 615-3055
or: (800) 955-2822
Fax: (847) 342-7477
E-mail: info@cancercenter.com
Internet: http://www.
cancercenter.com

Center for General Medicine and Acupuncture (Centro de Medicina General y Acupuntura)
Avenue Ensenada #110
Tijuana, B.C.
Mexico
Tel: 011-526-634-1412

Mailing Address:
P.O. Box 2757
Chula Vista California 91912
Tel: (800) 390-5610

Europa Institute of Integrated Medicine
Allen W. Lloyd Building
Suite 21
406 Avenue Paseo de Tijuana

International Border Zone
Tijuana, B.C.
Mexico
Tel: 011-526-682-4902
Fax: 011-526-682-4920

United States Office:
P.O. Box 950
Twin Peaks, California 92391
Tel: (909) 336-3671

Gerson Healing Centers of America
Gerson Center at Sedona
Sedona, Arizona
Call or write the Gerson Institute (below) for information and admissions.

Gerson Institute (Hospital Meridien de Playas de Tijuana)
Calle de Lava 2971
Secc. Costa Hermosa
Playas de Tijuana, B.C.
Mexico

For information and admissions:
P.O. Box 430 Bonita California 91902
Tel: (619) 585-7600
Fax: (619) 585-7610

Health Center of Lisbon
Rua de Mesericordia, #137-1
1200 Lisbon
Portugal
Tel: 011-351-1-347-1117
Fax: 011-351-1-347-1111

Hospital Santa Monica
Corporate Offices:
880 Canarios Court
Suite 210
Chula Vista, California 91910
Tel: (800) 359-6547
or: (619) 428-1146

Hosptial:
Avenida Mazathan O/N
Rosarito Beach, B.C.
Mexico

Hufeland Klinik for Holistic Immunotherapy
Loeffelstelzer Str. 1-3
D-97980 Bad Mergentheim
Germany
Tel: 011-49-7931-5360
Fax: 011-49-7931-8185

International Medical Center
16 de Septiembre #2215
3203-Cd. Juarez
Mexico
011-52-16-16-26-01

United States Office:
1501 Arizona Street 1-E
El Paso, Texas 79902
Tel: (800) 621-8924
or: (915) 543-5621

Magaziner Medical Center
1907 Greentree Road
Cherry Hill, New Jersey 08003
Tel: (609) 424-8222
Fax: (609) 424-2599

Mantell Medical Clinic
General and Family Practice
6505 Mars Road
Cranberry Township, Pennsylvania
16066
Tel: (412) 776-5610

Hans A. Nieper, M.D.
Outpatient Office:
Sedan Strasse 21
30161 Hannover
Germany
Tel: 011-0511-3-48-08-08

or: 011-0511-3-31111
Fax: 011-0511-318417

Inpatient Clinic:
Paracelsus Klinik at Silbersee
Oertzeweg 24
30851 Langenhagen
Germany

Oasis Hospital
Paseo Playas No. 19
Playas de Tijuana, B.C., Mexico 22700

Mailing Address:
P.O. Box 439045
San Ysidro, California 92143
Tel: (800) 700-1850
Fax: (619) 428-0994

Optimum Health Centre
54-58 Jardine's Bazaar
Prosperous Commercial Building
2nd Floor
Causeway Bay
Hong Kong
Tel: 011-(852)-2577-3798
Fax: 011-(852)-2890-8469

Panama City Clinic (Formerly Akbar Clinic)
340 West 23rd Street
Suite E
Panama City, Florida 32405
Tel: (904) 763-7689
Fax: (904) 763-5396

Program for Studies of Alternative Medicine
Centro Universitario de Los Altos

Universidad de Guadalajara
Carretera a Yakualica KM. 7
Tepatitlan, Jal.
Mexico
Tel: 011-378-13532
or: 011-378-15133
or: 011-378-15134
Fax: 011-36-370030
or: 011-36-193722

Michael B. Schachter, M.D., P.C., and Associates
Two Executive Boulevard
Suite 202
Suffern, New York 10901
Tel: (914) 368-4700
Fax: (914) 368-4727

Ahmad Shamim, M.D.
200 Fort Meade Road
Laurel, Maryland 20707
Tel: (301) 776-3700 (D.C. area)
or: (410) 792-0333 (Baltimore area)

Stella Maris Clinic
P.O. Box 435123
San Ysidro, California 92143
Tel: (800) 662-1319
or: (800) 973-7909
Fax: 011-52-66-346850

Vital-Life Institute
P.O. Box 294
Encinitas, California 92024
Tel: (619) 943-8485
Fax: (619) 436-9642

RESOURCE GROUPS

The following organizations can provide you with information about specific disorders and situations. They are categorized according to disorder. Be aware that addresses and telephone numbers are subject to change.

Aging

National Institutes of Aging
Public Information Office
Building 31, Room 5C27
31 Center Drive MS C2292
Bethesda, Maryland 20892-2292
Tel: (800) 222-2225

The National Council on the Aging, Inc.
409 Third Street, S.W.
Washington, D.C. 20024
Tel: (202) 479-1200
Fax: (202) 479-0735

National Aging Information Center
7830 Old Georgetown Road
Bethesda, Maryland 20814-2434
Tel: (301) 907-6743
Fax: (301) 907-6987
Internet: naic@ageinfo.org

U.S. Administration on Aging
330 Independence Avenue, S.W.
Washington, DC 20547-0008
Tel: (202) 619-0724

AARP
601 E St., N.W.
Washington, DC 20049-0001
Tel: (800) 424-3410

AIDS

AIDS Action Hotline
Tel: (800) 235-2331

Gay Men's Health Crisis, Inc.
129 West 20th Street
New York, New York 10011-3629
Hotline: (212) 807-6655
TDD (Telecommunication Device for the Deaf):
(212) 645-7470

National AIDS Network
1012 14th Street, N.W., Suite 601
Washington, D.C. 20005
Tel: (202) 347-0390

Project Inform National HIV/AIDS Treatment Hotline
Tel: (800) 822-7422

Alcoholism

Al-Anon Family Groups, Inc.
1600 Corporate Landing Parkway
Virginia Beach, Virginia 23454-5617
Tel: (804) 563-1600

Alcoholics Anonymous
475 Riverside Drive, 11th Floor
New York, New York 10115
Tel: (212) 870-3400
Fax: (202) 870-3003

Mothers Against Drunk Driving
511 E. John Carpenter Freeway, Suite 700
Irving, Texas 75062-8187
Tel: (214) 744-MADD
Fax: (972) 869-2206/2207

**National Clearinghouse for Alcohol
  and Drug Information**
P.O. Box 2345
Rockville, Maryland 20852
Tel: (800) 729-6686

**National Council on Alcoholism
and Drug Dependence**
12 West 21st Street, 8th Floor
New York, New York 10010
Tel: (212) 206-6770
Fax: (212) 645-1690
Internet: http://www.ncadd.org

Allergies

Allergy Research Group
400 Preda Street
P.O. Box 480
San Leandro, California 94577
Tel: (800) 545-9960
Fax: (510) 635-6730
E-mail: info@nutricology.com

American Academy of Allergy and Immunology
611 Wells Street
Milwaukee, Wisconsin 53202
Tel: (414) 272-6071

American Allergy Association
P.O. Box 7273
Menlo Park, California 94026
Tel: (415) 322-1663

Feingold Association of the United States
127 E. Main Street, Suite 106
Riverhead, New York 11901
Tel: (800) 321-3287

Food Allergy Network (FAN)
10400 Eaton Place, Suite 107
Fairfax, Virginia 22030-2208
Tel: (703) 691-3179
Fax: (703) 691-2713
E-mail: fan@worldweb.net
Internet: http://www.foodallergy.org

National Institute of Allergies and Infectious Diseases
National Institutes of Health
Building 31, Room 7A50
9000 Rockville Pike
Bethesda, Maryland 20892
Tel: (301) 496-5717

Alzheimer's Disease

Alzheimer's Association
919 N. Michigan Avenue, Suite 1000
Chicago, Illinois 60611
Tel: (312) 335-8700
or: (800) 272-3900
Fax: (312) 335-1110

Alzheimer's Disease Education and Referral Center
P.O. Box 8250
Silver Springs, Maryland 20907-8250
Tel: (800) 438-4380

Alzheimer's Disease and Related Disorders Association
919 North Michigan Avenue, Suite 1000
Chicago, Illinois 60611-1676
Tel: (800) 621-0379

Medic Alert Foundation
2323 Colorado Avenue
Turlock, California 95381-1009
Tel: (209) 669-2449
or: (800) 432-5378
Fax: (209) 669-2495

Angina Pectoris

American Heart Association
7272 Greenville Avenue
Dallas, Texas 75231-4596
Tel: (800) 242 8721
Internet: http://www.amhrt.org

Asthma

Asthma and Allergy Foundation of America
1717 Massachusetts Avenue, Suite 305
Washington, D.C. 20036
Tel: (202) 265-0265

National Heart, Lung, and Blood Institute (NHLBI)
National Institutes of Health
Public Health Service
Information Center
P.O. Box 30105

Bethesda, Maryland 20824-0105
Tel: (301) 251-1222
Fax: (310) 251-1223
Internet: http://www.nhlbi.nih.gov/nhlbi/nblbi.htm

Brain Tumors

American Brain Tumor Association
2720 River Road
Des Plaines, Illinois 60018
Tel: (847) 827-9910
Fax: (847) 827-9918
E-mail: info@abta.org
Internet: http://www.abta.org

Breast Cancer

American Cancer Society
1599 Clifton Road, N.E.
Atlanta, Georgia 30329-4251
Tel: (800) ACS-2345

National Cancer Institute
Cancer Information Service
Building 31, Room 10A24
9000 Rockville Pike
Bethesda, Maryland 20892
Tel: (800) 4CANCER

Y-Me National Breast Cancer Organization
212 West Van Buren Street
Chicago, Illinois 60607-3908
Hotline: (800) 221-2141
Spanish hotline: (800) 986-9505
Tel: (312) 986-8338
Fax: (312) 294-8597
Internet: http://www.y-me.org

Burns

Burns United Support Group (BUSG)
P.O. Box 36416
Grosse Pointe Farms, Michigan 48236-0416
Tel: (313) 881-5577
Fax: (313) 417-8702
E-mail: 1561@concentric.net

Cancer

American Brain Tumor Association
2720 River Road
Des Plaines, Illinois 60018
Tel: (847) 827-9910
Fax: (847) 827-9918
Patient line: (800) 886-2282
E-mail: ABTA@aol.com
Internet: http://pubweb.acns.nwu.edu/
~lberko/abta_html/abta1.htm

American Cancer Society
1599 Clifton Road, N.E.
Atlanta, Georgia 30329-4251
Tel: (800) ACS-2345

National Cancer Institute
Cancer Information Service
Building 31, Room 10A24
9000 Rockville Pike
Bethesda, Maryland 20892
Tel: (800) 4-CANCER

The National Hospice Organization
1901 North Moore Street, Suite 901
Arlington, Virginia 22209
Hotline: (800) 658-8898
Tel: (703) 243-5900
Fax: (703) 525-5762
Internet: http://www.nho.org

Skin Cancer Foundation
245 Fifth Avenue, Suite 2402
New York, New York 10016
Tel: (800) 754-6490
or: (212) 725-5176

Y-Me National Breast Cancer Organization
212 W. Van Buren Street
Chicago, Illinois 60607-3908
Hotline: (800) 221-2141
Spanish hotline: (800) 986-9505
Tel: (312) 986-8338
Fax: (312) 294-8597
Internet: http://www.y-me.org

Cardiovascular Disorders

American Heart Association
7272 Greenville Avenue
Dallas, Texas 75231-4596
Tel: (800) 242-8721
or: (213) 373-6300
Internet: http://www.americanheart.org

Heart Disease Research Foundation
50 Court Street, Room 306A
Brooklyn, New York 11201
Tel: (718) 649-6210

Cataracts

Eye Care (EC)
13425 Hidden Meadow Court
Herndon, Virginia 22071
Tel: (202) 628-3816
Fax: (703) 904-3965
E-mail: eyecare@mnsinc.com

International Eye Foundation
7801 Norfolk Avenue, Suite 200
Bethesda, Maryland 20814
Tel: (301) 986-1830
E-mail: info@ief.permanet.org

Celiac Disease

American Celiac Society
58 Musano Court
West orange, New Jersey 07052
Tel: (201) 325-8837

Celiac Disease Foundation
13251 Ventura Boulevard, #1
Studio City, California 91604-1838
Tel: (818) 990-2354
Fax: (818) 990-2379

Celiac Sprue Association/USA, Inc.
P.O. Box 31700
Omaha, Nebraska 68131-0700
Tel: (402) 558-0600

Gluten Intolerance Group of North America
P.O. Box 23053
Seattle, Washington 98102-0353
Tel: (206) 325-6980

**National Digestive Diseases Information
 Clearinghouse**
**National Institute of Diabetes and Digestive and
 Kidney Disease**
National Institutes of Health
2 Information Way
Bethesda, Maryland 20892-3570
Tel: (301) 654-3810

Chronic Fatigue Syndrome

The CFIDS Association of America, Inc.
P.O. Box 220398
Charlotte, NC 28222-0398
Tel: (800) 442-3437
Fax: (704) 365-9755
Internet: http://cfids.org/cfids
E-mail: cfids@vnet.net

The National CFS Association
919 Scott Avenue
Kansas City, Kansas 66105
Tel: (913) 321-2278

Cirrhosis

American Association for the Study of Liver Diseases
c/o SLACK, Inc.
6900 Grove Road
Thorofare, New Jersey 08086
Tel: (609) 848-1000

American Liver Foundation
1425 Pompton Avenue
Cedar Grove, New Jersey 07009
Tel: (800) 223-0179

**National Digestive Diseases Information
 Clearinghouse**
**National Institute of Diabetes and Digestive and
 Kidney Disease**
National Institutes of Health
2 Information Way

Bethesda, Maryland 20892-3570
Tel: (301) 654-3810

Colorectal Cancer

National Cancer Institute
Cancer Information Service
Building 31, Room 10A24
9000 Rockville Pike
Bethesda, Maryland 20892
Tel: (800) 4-CANCER

Constipation

National Digestive Diseases Information Clearinghouse
National Institute of Diabetes and Digestive
  and Kidney Disease
National Institutes of Health
2 Information Way
Bethesda, Maryland 20892-3570
Tel: (301) 654-3810

Cystic Fibrosis

Cystic Fibrosis Foundation
6931 Arlington Road
Bethesda, Maryland 20814
Tel: (800) 344-4823

Diabetes

American Diabetes Association
ADA National Service Center
1660 Duke Street
Alexandria, Virginia 22314
(800) 232-3472

Diverticular Disease

National Digestive Diseases
  Information Clearinghouse
National Institute of Diabetes and
  Digestive and Kidney Disease
National Institutes of Health
2 Information Way
Bethesda, Maryland 20892-3570
Tel: (301) 654-3810

Gas, Abdominal

National Digestive Diseases
  Information Clearinghouse
2 Information Way
Bethesda, Maryland 20892-3570
Box NDDIC
9000 Rockville Pike
Bethesda, Maryland 20892
Tel: (301) 654-3810

Glaucoma

Eye Care (EC)
13425 Hidden Meadow Court
Herndon, Virginia 22071
Tel: (202) 628-3816
Fax: (703) 904-3965
E-mail: eyecare@mnsinc.com

Glaucoma Research Foundation
490 Post Street, Suite 830
San Francisco, California 94102
Tel: (800) 826-6693
or: (415) 986-3162
Fax: (415) 986-3763
Internet: http://www.glaucoma.org

International Eye Foundation
7801 Norfolk Avenue, Suite 200
Bethesda, Maryland 20814
Tel: (301) 986-1830
E-mail: info@ief.permanet.org

Glomerulonephritis

American Kidney Fund
6110 Executive Boulevard, Suite 1010
Rockville, Maryland 20852
Tel: (800) 638-8299
or: (301) 881-3052
Fax: (301) 881-0898
Internet: http://www.arbon.com/kidney

The American Society of Nephrology
1101 Connecticut Avenue, N.W., Suite 700
Washington, D.C. 20036
Tel: (202) 857-1190
Fax: (202) 223-4579

Headache

American Association for the Study of Headache
875 Kings Highway, Suite 200
Woodbury, New Jersey 08096-3172
Tel: (609) 845-0322
Fax: (609) 384-5811

National Headache Foundation
428 West St. James Place, 2nd Floor
Chicago, Illinois 60614-2750
Internet: http://www.headaches.org

Hemorrhoids

**National Digestive Diseases Information
  Clearinghouse**
**National Institute of Diabetes and Digestive
  and Kidney Disease**
National Institutes of Health
2 Information Way
Bethesda, Maryland 20892-3570
Tel: (301) 654-3810

Hepatitis

American Liver Foundation
1425 Pompton Avenue
Cedar Grove, New Jersey 07009
Tel: (800) 223-0179

Hepatitis B Coalition
1573 Selby Avenue, Suite 229
Saint Paul, Minnesota 55104-6328
Tel: (612) 647-9009

Hepatitis Foundation International
30 Sunrise Terrace
Cedar Grove, New Jersey 07009-1423
Tel: (201) 239-1035
or: (800) 891-0707

Herpes Simplex Virus

**Herpes Resource Center
  American Social Health Association (HRC)**
P.O. Box 13827
Research Triangle Park, North Carolina 27709
Tel: (800) 230-6039

Inflammatory
Bowel Disease

Crohn's and Colitis Foundation of America, Inc.
386 Park Avenue South, 17th Floor
New York, NY 10016-8804
Tel: (212) 685-3440
or: (800) 343-3637
Fax: (212) 779-4098

Pediatric Crohn's and Colitis Association, Inc.
P.O. Box 188
Newton, Massachusetts 02168
Tel: (617) 290-0902

Irritable Bowel Syndrome

National Digestive Diseases Information
  Clearinghouse
National Institute of Diabetes and Digestive
  and Kidney Disease
National Institutes of Health
2 Information Way
Bethesda, Maryland 20892-3570
Tel: (301) 654-3810

Kidney Disorders

American Foundation for Urologic Disease
300 West Pratt Street, Suite 401
Baltimore, Maryland 21201
Tel: (800) 242-2383
  or: (410) 727 2908

American Kidney Fund
6110 Executive Boulevard, Suite 1010
Rockville, Maryland 20852
Tel: (800) 638-8299
If calling from Maryland: (800) 492-8361
Fax: (301) 811-0898
Internet: http://www.arbon.com/kidney

National Digestive Diseases Information Clearinghouse
National Institute of Diabetes and Digestive
  and Kidney Disease
National Institutes of Health
2 Information Way
Bethesda, Maryland 20892-3570
Tel: (301) 654-3810

National Kidney Foundation
30 East 33rd Street, #1100
New York, New York 10016
Tel: (212) 889-2210
or: (800) 622-9010

Lactase Deficiency

National Digestive Diseases Information
  Clearinghouse
National Institute of Diabetes and Digestive

and Kidney Disease
National Institutes of Health
2 Information Way
Bethesda, Maryland 20892-3570
Tel: (301) 654-3810

Lung Cancer

The American Cancer Society
1599 Clifton Road, N.E.
Atlanta, Georgia 30329-4251
Tel: (800) ACS-2345

Multiple Sclerosis

Multiple Sclerosis Association of America
706 Haddonfield Road
Cherry Hill, New Jersey 08002
Tel: (800) LEARNMS
or: (800) 822-4672

Multiple Sclerosis Foundation, Inc.
6350 N. Andrews Avenue
Ft. Lauderdale, Florida 33309
Tel: (954) 776-6805
or: (800) 441-7055

National Multiple Sclerosis Society
733 Third Avenue
New York, New York 10017-3288
Tel: (212) 986-3240
or: (800) FIGHTMS

Myofascial Pain Syndrome

Fibromyalgia Alliance of America
P.O. Box 21990
Columbus, Ohio 43221-0990
Tel: (614) 457-4222

National Fibromyalgia Research Association
P.O. Box 500
Salem, Oregon 97302
E-mail: nfra@teleport.com

**National Institute of Arthritis and Musculoskeletal
  and Skin Diseases (NIAMS)**
National Institutes of Health
Building 31, Room 4C32
31 Center Drive, MSC 2350
Bethesda, Maryland 20892-2350
Tel: (301) 496-4353 (Stephen I. Katz, M.D., Ph.D., Director)
Fax: (301) 480-6069
Internet: http://www.nih.gov/niams/

Obesity

Overeaters Anonymous
World Service Office
6075 Zenith Court, N.E.
Rio Rancho, New Mexico 87124
Tel: (505) 891-2664
Fax: (505) 891-4320

Weight Watchers International
175 Crossways Park
West Woodbury, New York 11797
Tel: (800) 651-6000

Osteoarthritis

Arthritis Foundation
1330 West Peachtree Street
Atlanta, Georgia 30309
Tel: (800) 283-7800
 or: (404) 872-7100
Fax: (404) 872-8694
Internet: http://www.arthritis.org

**National Arthritis and Musculoskeletal and Skin
  Diseases Information Clearinghouse (NAMSIC)**
NIH 1 Ams Circle
Bethesda, Maryland 20892-2903
Tel: (301) 495-4484
TDD (Telecommunication Device for the Deaf):
(301) 565-2966
Fax: (301) 587-4352

Osteoporosis

National Osteoporosis Foundation
1150 17th Street, Suite 500 NW
Washington, D.C. 20036-4603

Pancreatic Cancer

American Cancer Society
1599 Clifton Road, N.E.
Atlanta, Georgia 30329-4251
Tel: (800) ACS-2345

National Cancer Institute
Cancer Information Service
Building 31, Room 10A24
9000 Rockville Pike
Bethesda, Maryland 20892
Tel: (800) 4-CANCER

Pancreatic Insufficiency

**National Digestive Diseases
  Information Clearinghouse**
**National Institute of Diabetes
  and Digestive and Kidney Disease**
National Institutes of Health
2 Information Way
Bethesda, Maryland 20892-3570
Tel: (301) 654-3810

Pancreatitis

**National Digestive Diseases
  Information Clearinghouse**
**National Institute of Diabetes and
  Digestive and Kidney Disease**
National Institutes of Health
2 Information Way
Bethesda, Maryland 20892-3570
Tel: (301) 654-3810

Pneumonia

American Lung Association
1740 Broadway
New York, New York 10019
Tel: (212) 245-8000

Prostate Cancer

American Prostate Society
1340 Charwood Road
Hanover, Maryland 21076
Tel: (410) 859-3735
Fax: (510) 850-0818
Internet: http://www.ameripros.org

Psoriasis

National Psoriasis Foundation
6600 S.W. 92nd Avenue, Suite 300
Portland, Oregon 97223-7195
Tel: (503) 244-7404
Fax: (503) 245-0626
E-mail: 76135.2746@compuserve.com
Internet: www.psoriasis.org

Raynaud's Disease and Phenomenon

Arthritis Foundation
1314 Spring Street, N.W.
Atlanta, Georgia 30309
Tel: (800) 283-7800
or: (404) 827-7100

Reiter's Syndrome

Arthritis Foundation
1330 West Peachtree Street
Atlanta, Georgia 30309

Tel: (800) 283-7800
or: (404) 872-7100
Fax: (404) 872-8694
Internet: http://www.arthritis.org

The Arthritis Fund (aka The Rheumatoid Disease Foundation)
5106 Old Harding Road
Franklin, Tennessee 37064-9400
Tel/fax: (615) 656-1030

National Arthritis and Musculoskeletal and Skin Diseases Information Clearinghouse (NAMSIC)
NIH 1 Ams Circle
Bethesda, Maryland 20892-2903
Tel: (301) 495-4484
TDD (Telecommunication Device for the Deaf):
(301) 565-2966
Fax: (301) 587-4352

Retinopathy

Eye Care (EC)
13425 Hidden Meadow Court
Herndon, Virginia 22071
Tel: (202) 628-3816
Fax: (703) 904-3965
E-mail: eyecare@mnsinc.com

International Eye Foundation
7801 Norfolk Avenue, Suite 200
Bethesda, Maryland 20814
Tel: (301) 986-1830
E-mail: info@ief.permanet.org

Sjögren's Syndrome

Sjögren's Syndrome Foundation, Inc.
333 North Broadway, Suite 2000
Jericho, New York 11753
Tel: (516) 933-6365
Fax: (516) 933-6368
Internet: http://www.w2.com/ss.html

Skin Cancer

American Cancer Society, Inc.
1599 Clifton Road, N.E.
Atlanta, Georgia 30329-4251
Tel: (800) ACS-2345

Skin Cancer Foundation
245 Fifth Avenue, Suite 2402
New York City, New York 10016
Tel: (800) 754-6490
or: (212) 725-5176

Stroke

National Stroke Association
8480 East Orchard Road, Suite 1000
Englewood, Colorado 80111-5105
Tel: 1-800-STROKES
or: (303) 649-9299
Fax: (303) 649-1328
E-mail: nsa@stroke.org

Systemic Lupus Erythematosus

Lupus Foundation of America, Inc.
1300 Piccard Drive, Suite 200
Rockville, Maryland 20850-4303
Tel: (800) 558-0121
or: (301) 670-9292
Internet: http://www.lupus.org/lupus

Torticollis

National Spasmodic Torticollis Association
P.O. Box 424
Mukwonago, Wisconsin 53149-0424
Tel: (800) 487-8385
Fax: (414) 662-9887

BIBLIOGRAPHY

Over 5,000 books, articles, and manuscripts were reviewed and studied in the writing of this book. Due to space constraints, only a representative sampling of these resources is listed here.

Abel, Ulrich. *Chemotherapy of Advanced Cancer*. Stuttgart, Germany: Hippokrates, 1990.

Aihara, Cornellia and Herman Aihara. *Natural Healing from Head to Toe*. Garden City Park, NY: Avery Publishing Group, Inc., 1994.

Balch, James F. and Phyllis A. Balch. *Prescription for Nutritional Healing*. Garden City Park, NY: Avery Publishing Group, Inc., 1997.

Beger, H.G., M. Büchler, and K. Gyr, ed. *The Role of Enzyme Treatment in Pancreatic Disease*. Basel, Switzerland: Karger, 1993

Berkow, Robert, ed. *The Merck Manual of Diagnosis and Therapy*. 16th ed. Rahway, NJ: Merck Research Laboratories, 1992.

Best, Charles Herbert and Norman Burke Taylor. *The Physiological Basis of Medical Practice*. Baltimore: The Williams and Wilkins Company, 1989.

Bland, Jeffrey. *Medical Applications of Clinical Nutrition*. New Canaan, CT: Keats Publishing, 1983.

Blauer, Stephen. *The Juicing Book*. Garden City Park, NY: Avery Publishing Group, Inc., 1989.

Borgström, Bengt and Howard L. Brockman. *Lipases*. Amsterdam: Elsevier, 1984.

Brooks, G. and T. Fahey. *Exercise Physiology: Human Bioenergetics and Its Applications*. New York: John Wiley and Sons, 1984.

Burton Goldberg Group, The, eds. *Alternative Medicine: The Definitive Guide*. Puyallup, WA: Future Medicine Publishing, Inc., 1993.

Calbom, Cherie and Maureen Keane. *Juicing for Life*. Garden City Park: NY: Avery Publishing Group, Inc., 1992.

Cernaj, Ingeborg and Josef Cernaj. *Gesund und Schön Durch Enzyme*. Munich: Südwest, 1995.

Cichoke, Anthony J. *A New Look at Chronic Disorders and Systemic Enzyme Therapy*. Portland: Seven Seas Publishing, 1993.

_____. *A New Look at Enzyme Therapy*. Portland: Seven Seas Publishing, 1993.

_____. *Acute Trauma and Systemic Enzyme Therapy*. Portland: Seven Seas Publishing Co.), 1993.

_____. *AIDS and Metabolic Therapy*. Portland: Seven Seas Publishing, 1994.

_____. *Enzymes and Enzyme Therapy: How to Jump Start Your Way to Lifelong Good Health*. New Canaan, CT: Keats Publishing, Inc., 1994.

_____. *Enzymes: Nature's Energizers*. New Canaan, CT: Keats Publishing, Inc., 1997.

_____. *Introduction to Chiropractic Health*. New Canaan, CT: Keats Publishing, Inc., 1996.

_____. *Neurologic Considerations in Toxic, Metabolic, and Nutritional Disorders*. Portland: Seven Seas Publishing, 1996.

_____. *New Hope for AIDS*. Portland: Seven Seas Publishing, 1995.

_____. *Nutrition to Give Your Athlete the Winning Edge*. Portland: Seven Seas Publishing Co., 1990.

Clouatre, Dallas. *Anti-Fat Nutrients*. San Francisco: Pax Publishing, 1993.

Cody, V., E. Middleton, Jr., and J.B. Harborne, eds. *Plant Flavonoids in Biology and Medicine: Biochemical, Pharmacological, and Structure-Activity Relationships*. New York: Liss, 1986.

Committee on Food Chemicals Codex. *Food Chemicals Codex*. Washington, D.C.: National Academy Press, 1996.

Dalling, Michael. *Plant Proteolytic Enzymes*. Boca Raton, FL: CRC Press, Inc., 1986.

de Haas, Cherie. *Natural Skin Care*. Garden City Park, NY: Avery Publishing Group, 1989.

Dressler, David and Huntington Potter. *Discovering Enzymes*. New York: Scientific American Library, 1991

Duke, James. CRC Handbook of *Medical Herbs*. Boca Raton, FL: CRC Press, Inc., 1985

_____. *Handbook of Biologically Active Phytochemicals and Their Activities*. Boca Raton, FL: CRC Press, Inc., 1992

_____. *Handbook of Edible Weeds*. Boca Raton, Florida: CRC Press, Inc., 1992

Eisenthal, R. and M.J. Danson. *Enzyme Assays: A Practical Approach*. Oxford, NY: IRL Press, 1992.

Ewing, W.N. and D.J.A. Cole. *The Living Gut*. Trowbridge, Wiltshire, England: Redwood Books, 1994.

Fersht, Alan. *Enzyme Structure and Mechanism*. San Francisco: W.H. Freeman and Company, 1977.

Fink, John. *Third Opinion: An International Directory to Alternative Therapy Centers for the Treatment and*

Prevention of Cancer. Garden City Park, NY: Avery Publishing Group, 1997

Gardner, M.L.G. and K. J. Steffens, ed. *Absorption of Orally Administered Enzymes.* Berlin: Springer-Verlag, 1995.

Glenk, Wilhelm and Sven Neu. *Enzyme.* Munich: Wilhelm Heyne Verlag, 1990.

Godfrey, Tony and Jon Reichelt. *Industrial Enzymology: The Application of Enzymes in Industry.* New York: The Nature Press, 1996.

Goldberg, Israel, ed. *Functional Foods, Designer Foods, Pharmafoods, Nutraceuticals.* New York: Chapman & Hall, 1994.

Greenberg, David M. and Harold A. Harper. *Enzymes in Health and Disease.* Springfield, Illinois: Charles C. Thomas Publisher, 1960.

Hager, E.D. *Komplementäre Onkologie.* Gräfelfing, Germany: Forum Verlagsgesellschaft, 1996.

Heinerman, John. Heinerman's *Encyclopedia of Fruits, Vegetables and Herbs.* West Nyack, NY: Parker Publishing Company, 1988.

_____. *Heinerman's Encyclopedia of Healing Juices.* West Nyack, NY: Parker Publishing Company, 1994.

_____. *Heinerman's Encyclopedia of Juices, Teas, and Tonics.* Englewood Cliffs, NJ: Prentice Hall, 1996.

Holder, Ian A., ed. *Bacterial Enzymes and Virulence.* Boca Raton, FL: CRC Press, Inc., 1985

Homburger, Henry A., ed. *Clinical and Analytical Concepts in Enzymology.* Skokie, IL: College of American Pathologists, 1983

Howell, Edward. *Enzyme Nutrition: The Food Enzyme Concept.* Wayne, NJ: Avery Publishing Group, Inc., 1985.

Kenton, L. and S. Kenton. *Raw Energy.* London: Century Publishing, 1984.

Khalsa, Dharma Singh. *Brain Longevity.* New York: Warner Books, 1997.

Klaschka, F. *Oral Enzymes—New*

Approach to Cancer Treatment. Munich: Forum Medizin, 1996.

Klatz, Ronald and Robert Goldman. *7 Anti-Aging Secrets.* Chicago: Elite Sports Medicine Publications, 1996.

_____. *Stopping the Clock.* New Canaan, CT: Keats Publishing, Inc., 1996.

Kordich, Jay. *The Juiceman's Power of Juicing.* New York: William Morrow and Co., Inc., 1992.

Krämer, J. *Intervertebral Disk Diseases.* Chicago: Year Book Medical Publishers, Inc., 1981.

Kruger, James E., David Lineback, and Clyde E. Stauffer. *Enzymes and Their Role In Cereal Technology.* St. Paul, MN: American Association of Cereal Chemists, Inc., 1987.

Kugler, Hans J. *Tripping the Clock.* Reno: Health Quest Publications, 1993.

Kullman, Willi. *Enzymatic Peptide Synthesis.* Boca Raton, FL: CRC Press, Inc., 1987.

Kunz, R., et al. *Humoral Immunmodulatory Capacity of Proteases in Immuncomplex Decomposition and Formation.* Washington, D.C.: First International Symposium on Combination Therapies, 1991.

Lankisch, Paul G., ed. *Pancreatic Enzymes in Health and Disease.* Berlin: Springer-Verlag, 1991.

Lee, William H. R. *The Book of Raw Fruit and Vegetable Juices and Drinks.* New Canaan, CT: Keats Publishing, Inc., 1982.

Levine, Stephen A. and Parris M. Kidd. *Antioxidant Adaptation: Its Role in Free Radical Pathology.* San Leandro, CA: Allergy Research Group, 1994.

Lopez, D.A., R.M. Williams, and K. Miehlke. *Enzymes: The Fountain of Life.* Charleston, SC: The Neville Press, Inc., 1994.

Lorand, Laszlo, ed. *Methods in Enzymology Volume 80 Proteolytic*

Enzymes. New York: Academic Press, 1981.

Lyons, T.P. *Biotechnology in the Feed Industry: Proceedings of Alltech's Fifth Annual Symposium.* Nicholasville, KY: Alltech Technical Publications, 1989.

Lyons, T.P. and K.A. Jacques. *Biotechnology in the Feed Industry: Proceedings of Alltech's Thirteenth Annual Symposium.* Nottingham, England: Nottingham University Press, 1997.

Maggio, Edward T. *Enzyme Immunoassay.* Boca Raton, FL: CRC Press, Inc., 1980.

Manner, Harold W., Steven J. Disanti, and Thomas L. Michalsen. *The Death of Cancer.* Chicago: Advanced Century Publishing Corp., 1978.

Matthews, D.M. *Protein Absorption.* New York: Wiley-Liss, 1992.

McColl, Ian and G.E.G. Sladen. *Intestinal Absorption in Man.* London: Academic Press, 1975.

McKellar, Robin C. *Enzymes of Psychrotrophs in Raw Food.* Boca Raton, FL: CRC Press, Inc., 1989.

Mowrey, Daniel B. *Herbal Tonic Therapies.* New Canaan, CT: Keats Publishing, Inc., 1993.

Müller-Wohlfahrt, H.W., H.J. Montag, and W. Diesbschlag. *Süsse Pille Sport Verletzt, Was Nun?* München, Germany: Verlag Medical Concept Jochen Knips, 1984.

Murray, Michael T. *The Complete Book of Juicing: Your Delicious Guide to Healthful Living.* Rocklin, CA: Prima Publishing, 1992.

_____. *Encyclopedia of Nutritional Supplements.* Rocklin, CA: Prima Publishing, 1996.

Murray, Michael T. and Joseph E. Pizzorno. *An Encyclopedia of Natural Medicine.* Rocklin, CA: Prima Publishing, 1991.

Oldham, R.K. *Principles of Cancer Biotherapy.* New York: Raven Press, 1991.

Page, M.I. and A. Williams, ed. *Enzyme Mechanisms*. Boca Raton, FL: CRC Press, Inc., 1987.

Palmer, Trevor. *Understanding Enzymes*. 4th ed. New York: Prentice Hall, 1995.

Pastorino, Ugo and Waun Ki Hong. *Chemoimmuno Prevention of Cancer*. Stuttgart, Germany: Georg Thieme Verlag Thieme Medical Publishers, Inc., 1991.

Pottenger, Francis M., Jr. *Pottenger's Cats: A Study in Nutrition*. La Mesa, CA: Price-Pottenger Nutrition Foundation, 1983.

Price, Nicholas C. and Lewis Stevens. *Fundamentals of Enzymology*. 2d ed. Oxford, NY: Oxford University Press, 1989.

Pryor, W.A., ed. *Free Radicals in Biology*. Orlando: Academic Press, 1984.

Ransberger, Karl. *Max Wolf—Ein Leben für die Enzymtherapie*. Gräfelfing, Germany: Forum Medizin, 1994.

Rommel, K. and H. Goebell, eds. *Lipid Absorption: Biochemical and Chemical Aspects*. Lancaster, England: MTP Press, 1976.

Schomburg, D. and M. Salzmann, eds. *Enzyme Handbook: Class 3: Hydrolases*. Berlin, Springer-Verlag, 1991.

Schwimmer, Sigmund. *Source Book of Food Enzymology*. Westport, CT: Te AVI Publishing Company, Inc., 1981.

Selye, Hans. *Stress Without Distress*. Philadelphia: J.B. Lippincott Co., 1974.

Simontacchi, Carol. *Your Fat Is Not Your Fault*. New York: Penguin Putnam, Inc., 1997.

Sinatra, Stephen T. *Heartbreak and Heart Disease*. New Canaan, CT: Keats Publishing, Inc., 1996.

Stay, Flora Parsa. *The Complete Book of Dental Remedies*. Garden City Park, NY: Avery Publishing Group, Inc., 1996.

Takamine, Jokichi. *Documents From the Dawn of Industrial Biotechnology*. Elkhart, IN: Miles, Inc., 1988.

Tietz, Norbert W., Albert Weinstock, and Denis O. Rodgerson, eds. *Proceeedings of the Second International Symposium on Clinical Enzymology*. Chicago: American Association for Clinical Chemists, 1976.

United States Pharmacopeia, The. Rockville, MD: United States Pharmacopeial Convention, Inc.,1995.

Vollmer, Helga. *Enzyme Für die Frau*. Berlin: Verlag Gesundheit, 1995.

Wade, Carlson. *The Pocket Handbook of Juice Power*. New Canaan, CT: Keats Publishing, Inc., 1992.

Wang, Daniel I. *Fermentation and Enzyme Technology*. New York: John Wiley and Sons, 1979.

Webb, Edwin C. *Enzyme Nomenclature*. San Diego: Academic Press, Inc., 1992.

Wenk, C. and M. Boessinger, eds. *Enzymes in Animal Nutrition*— Proceedings of the first symposium held October 13–16, 1993 in Kartause Ittingen, Switzerland

Werbach, Melvyn R. *Nutritional Influences on Illness*. New Canaan, CT: Keats Publishing, Inc., 1990.

White, John Stephen and Dorothy Chong White. *Source Book of Enzymes*. Boca Raton, FL: CRC Press, 1997.

Wigmore, Ann. *The Wheatgrass Book*. Garden City Park, NY: Avery Publishing Group, 1985.

Wiseman, Alan, ed. *Handbook of Enzyme Biotechnology*. Chichester, England: Halsted Press, 1986.

Wolf, Max and Karl Ransberger. *Enzyme Therapy*. Los Angeles: Regent House, 1972.

Wolnak, Bernard. *Industrial Use of Enzymes*. Decatur, MI: Johnson Graphics, 1990.

Wrba, H., et al. *Systemische Enzymtherapie*. Munich: Medizin Verlag, 1996.

Wrba, Heinrich and Otto Pecher. *Enzyme—Wirkstoffe der Zukunft Mit der Enzym-Therapie das Immunsystem*. Zurich: Orac, 1993.

Zand, Janet, Rachel Walton, and Bob Rountree. *Smart Medicine for a Healthier Child*. Garden City Park, NY: Avery Publishing Group, Inc., 1994.

INDEX

Abrasions. *See* Scars; Skin ulcers.
Abscess, 89–90
Absorption, 14
Achlorhydria, 16
Acid stomach. *See* Heartburn.
Acid-forming foods, 110
Acidosis, 91–93
    metabolic, 91
    respiratory, 91
    See also Alkalosis.
Acne, 93–96
Acne rosacea. *See* Rosacea.
Acquired immune deficiency syndrome. *See* AIDS.
Acquired lymphedema, 293
Additives. *See* Food additives.
Adenocarcinoma, 321. *See also* Cancer; Pancreatic cancer.
Adenoiditis, 96–98
Adnexitis, 98–100
Adrenal extract, 70
Age spots, 100–101
Aging, 102–104
AIDS, 104–106
Alanine, 72
Alcohol, 22
Alcoholism, 106–109
Alfalfa, 71
Algae, 71, 73
Alkali-forming foods, 93
Alkaloids, 66
Alkalosis, 109–110
    metabolic, 109
    respiratory, 109
    See also Acidosis.
Allergens, 111
Allergic asthma, 125
Allergies, 110–113
    breast-feeding for prevention of, 112
    food allergies, 111
    See also Asthma; Hay fever.
Allicin, 28
Alliin, 28
Allinase, 28

Allium compounds, 74
Allium sativum. See Garlic.
Aloe vera, 73
Alpha-galactosidase, 45
Alpha-glucosidase. *See* Maltase.
Alport's syndrome, 282
Alveolar cell carcinoma, 290
Alzheimer, Alois, 113
Alzheimer's disease (AD), 113–116. *See also* Aging.
Amino acids, 3, 70–71, 72. *See also* Alanine; Arginine; Asparagine; Aspartic acid; Cysteine; Glutamic acid; Glutamine; Glycine; Histidine; Isoleucine; Leucine; Lysine; Methionine, Phenylalanine; Proline; Serine; Threonine; Tryptophan; Tyrosine; Valine.
Amylase, 2, 11, 27, 37, 38–39, 45
Anabolism, 2
Anal fissures, 373
Anemia, 116–118
    aplastic, 116
    folic acid, 116
    hemolytic, 116
    iron-deficiency, 116
    pernicious, 116
    of pregnancy, 116
    sickle-cell, 116
    vitamin B_{12}, 116
Angina pectoris, 118–120
    unstable, 118
    variant, 118
Ankylosing spondylitis (AS), 120–122
Anthocyanidins, 65, 74
Antigen, 402
Antioxidants, 71, 74
Aorta extract, 70
Aplastic anemia, 116
Appert, Nicolas, 19
Arginine, 72
Arteriosclerosis, 122–124
    atherosclerosis, 122
    Mönckeberg's, 122
    obliterans, 122
    See also Cardiovascular disorders.

Arteriosclerosis obliterans, 122
Arteriosclerotic retinopathy, 351
Arthritis. *See* Ankylosing spondylitis; Gout; Osteoarthritis;
    Reiter's syndrome; Rheumatoid arthritis.
Arthritis, reactive. *See* Reiter's syndrome.
Ascorbic acid. *See* Vitamin C.
Asparagine, 72
Aspartame, 5
Aspartic acid, 5, 72
Aspergillus niger, 27, 38
Aspergillus oryzae, 27, 33
Asthma, 125–127
    allergic, 125
    intrinsic, 125
Astragalus, 73
Atheroembolic kidney disease, 282
Atherosclerosis, 122
Athlete's foot, 225
Atopic dermatitis, 197
Atrophic macular degeneration, 295
Attitude, positive, 439–440

B cells, 402
Bacillus licheniformis, 38
Bacillus natto, 33
Bacillus subtilis, 38
Back posture, proper, 129
Backache, 127–131
    conditions resulting in, 128
    posture to prevent, 129
    See also Scoliosis; Slipped disc; Subluxation.
Background retinopathy, 351
Bacterial infections, 131–133
Bamboo spine. *See* Ankylosing spondylitis.
Bartsch, W., 259
Bartter's syndrome, 282
Basal cell carcinoma, 368
Baths, as therapy, 432
Baumuller, M., 381
Beard, John, 6–7
Bed sores. *See* Skin ulcers.
Bee pollen, 73
Bee stings. *See* Insect bites and stings.
Bell's palsy, 134–135
Benign prostate hyperplasia, 341
Benitez, Helen, 7
Benzene, 24
Benzo[a]pyrene, 29
Bernstein, Richard K., 202
Beta-carotene, 56, 74
Beta-fructofuranosidase. *See* Invertase.
Beta-galactosidase. *See* Lactase.
Beta-glucosidase, 46

Bilberry, 73
Bioflavonoids. *See* Flavonoids.
Biotechnology in the Feed Industry (Lyons), 63
Biotin, 58
Bites, insect. *See* Insect bites and stings.
Bladder infection, 135–137
Blood clots. *See* Cardiovascular disorders; Embolism;
    Thrombosis.
Boils, 137–140
Bone fractures, 140–142
Book of Macrobiotics, The, (Kushi), 33
Brain extract, 70
Brain tumors, 142–147. *See also* Cancer.
Breast cancer, 144–146. *See also* Cancer.
Breast-feeding and allergy prevention, 112
Brewer's yeast, 73
Brinase, 46
British Pharmacopoeia (BP), 43
Bromelain, 46
Bronchitis, 146–148
    acute, 146
    chronic, 146
Bronchogenic carcinoma, 290
Bruises and hematomas, 148–150. *See also* Cardiovascular
    disorders.
Buchner, Eduard, 2
Bueger's disease, 278
Burns, 150–153
Bursitis and synovitis, 153–156. *See also* Tendonitis.

Caffeine, 23
Calcium, 60
Cancer, 29–30, 156–161
    reducing risk of, 160
    warning signs of, 157
    See also Brain tumors; Breast cancer; Lung cancer;
        Pancreatic cancer; Prostate cancer; Skin cancer.
Cancer Research, 32
Candida, 226
Candidiasis, 225, 226
Candidosis. *See* Candidiasis.
Canker sores, 161–163
Canthaxanthines, 66
Carboanhydrase, 4
Carbohydrase. *See* Amylase.
Carbohydrate intolerance, 163–165. *See also* Celiac disease;
    Lactase deficiency.
Carbohydrates, 11
Carboxypeptidase, 13, 46–47
Carcinoma, 156
    adenocarcinoma, 321
    alveolar cell, 290
    basal cell, 368

bronchogenic, 290
cystadenocarcinoma, 321
ductal, 144
squamous cell, 368
See also Cancer.
Cardiovascular disorders, 165–169. *See also* Angina;
Arteriosclerosis; Embolism; High blood pressure;
Raynaud's disease; Stroke; Thrombosis.
Carotenoids, 66, 74
Carpal tunnel syndrome (CTS), 169–172
Catabolism, 2
Catalase, 47, 74
Cataracts, 172–173
Catechins, 65
Celiac disease, 174–176
Cellulase, 47
Chaitlow, Leon, 31
Cheese, 35
Chickenpox, 176–178
Childbirth complications. *See* Post-childbirth
complications.
Chitinase, 27
Chloasma. *See* Age spots.
Chlorella, 73
Chlorophyll, 66, 74
Cholesterol, elevated, 179–181
Choline, 58
Chromium, 60
Chronic fatigue syndrome (CFS), 181–183
Chymopapain, 47
Chymosin. *See* Rennin.
Chymotrypsin, 47
Chymotrypsinogen, 13
Cirrhosis, 183–185
Closed-angle glaucoma, 235
Clostridium botulinum, 19
Clostridium perfringens, 17
Cluster headache, 246
Cobalt, 60–61
Coenzyme Q$_{10}$, 64, 74
Coenzymes, 4. *See also* Vitamins.
Cofactors, 4. *See also* Minerals.
Coffee, 23
Coffee retention enema, 432
Cold sores. *See* Herpes simplex virus.
Colds, 186–188
Colitis, ulcerative, 271
Collagen, 440
Collagenase, 47
Colorectal cancer, 188–190
warning signs of, 189
See also Cancer.
Common warts, 425

Congenital lymphedema, 292
Conjunctivitis, 190–192
allergic, 190
bacterial, 190
viral, 190
Constipation, 192–195
Contact dermatitis, 197
Copper, 61, 74
Cortical necrosis, 282
Coumarins, 66
Cramps, menstrual. *See* Dysmenorrhea.
Cramps, muscle. *See* Muscle cramping.
Crohn's disease, 271
Curcumin, 66, 74
Cuts. *See* Scars; Skin ulcers.
Cyanocobalamin. *See* Vitamin B$_{12}$.
Cystadenocarcinoma, 321. *See also* Cancer.
Cysteine, 72, 74
Cystic fibrosis (CF), 195–197. *See also* Steatorrhea.
Cystic mastitis. *See* Fibrocystic breast disease.
Cystinuria, 282
Cystitis. *See* Bladder infection.
Cytochrome P450-dependent monooxygenases, 29

Dan, Jacob, 236
Decubitus ulcer, 373
Deep-vein thrombosis (DVT), 408
Degenerative joint disease. *See* Osteoarthritis.
Dental cavities. *See* Tooth decay.
Dermatitis, 197–199
atopic, 197
contact, 197
seborrheic, 197
stasis, 197
Dermatomyositis, 200–201. *See also* Polymyositis.
Dermis, 440
Detoxification methods, 432–435
coffee retention enema, 433
enzyme kidney flush, 433
vitamin C/enzyme flush, 434
two-day enzyme juice fast, 434–435
Devlin, Thomas, 25
Diabetes, 201–206
complications of, 203
insulin-dependent (type I), 202
non-insulin dependent (type II), 202
Diabetic retinopathy, 351
Diarrhea, 206–208
exudative, 206
osmotic, 206
secretory, 206
Diastase, 48
Diet, enzyme-rich

fermented foods, role in, 30–36
fruits and vegetables, role in, 26–27, 28–30
honey, raw, role in, 35–36
sprouts, role in, 35
and vegans, 28–29
wheat germ, role in, 36
Diet, recommended, 435–437
Digestion
enzymes' role in, 11–13
in the large intestine, 14–15
in the mouth, 11–12
poor, 10–11, 15
process of, 11–15
in the small intestine, 12–14
in the stomach, 12
Digestive disorders. *See* Carbohydrate intolerance; Celiac disease; Constipation; Diarrhea; Diverticular disease; Gas, abdominal; Gastritis; Heartburn; Hypochlorhydria; Indigestion; Inflammatory bowel disease; Irritable bowel syndrome; Lactase deficiency; Leaky gut syndrome; Pancreatic insufficiency; Pancreatitis; Peptic ulcers; Steatorrhea.
Disaccharides, 11
Diterpenes, 66, 74
Diverticular disease, 208–210
Diverticulitis. *See* Diverticular disease.
Diverticulosis. *See* Diverticular disease.
Dr. Bernstein's Diabetes Solution (Bernstein), 202
Dorrer, R., 258
Drugs, 25
Ductal carcinoma, 144
Dulles, John Foster, 7
Duodenal ulcers, 328
Duodenum, 12
DVT. *See* Deep-vein thrombosis.
Dysmenorrhea, 210–213

E. coli. *See* Escherichia coli.
Ear infection, 213–215
EASE. *See* Enzyme Absorption System Enhancers.
Echinacea, 73, 75
Echinacea/Enzyme Gargle, 438
Eczema. *See* Dermatitis.
EFAs. *See* Essential fatty acids.
Elastase, 48
Eleutherococcus senticosus. See Ginseng.
Eleutherosides, 66
Elimination, improving, 431–432
Ellagic acid, 66, 74
Embolism, 216–218. *See also* Thrombosis.
Empyemas, 218–220
Emulsin. *See* Beta-glucosidase.
Encyclopedia of Natural Medicine, An (Murray and Pizzorno),

175, 338
Encyclopedia of Nutritional Supplements (Murray), 69
Endogluconase. *See* Cellulase.
Endometriosis, 212. *See also* Dysmenorrhea.
Enema
coffee retention, 432
enzyme retention, 432
Enterokinase, 48
Enzymatic Gargle, 437
Enzymatic skin salves, 440
Enzyme (Glenk and Sven), 179
Enzyme (Pecher and Wrba), 257
Enzyme Absorption System Enhancers, 6, 42
Enzyme deficiency, percentage of celiac patients with, 174
Enzyme Nutrition (Howell), 7
Enzyme Program, Five-Step Jump Start, 438
Enzyme retention enema, 432
Enzyme skin exfoliants, 441
Enzyme therapy
cancers helped by, 158
cardiovascular disorders helped by, 166
digestive, 38–39
indigestion and, 270
inflammation and, 318
injury recovery and, 383
muscle build-up and, 382
surgical applications of, 399
systemic, 6, 39
Enzyme Therapy (Ransberger and Wolf), 44
Enzyme Treatment of Cancer (Beard), 7
Enzymes
aging's effects on, 15
alcohol's effects on, 22
canning's effects on, 19
classes of, 3
coffee's effects on, 23
cooking's effects on, 17–18
deficiency of, 5–6, 15–16
description of, 1
digestion and, 10–16
drugs' effects on, 25
enhancers of, 6, 35–36, 42
fat's effects on, 29
flavorings' effects on, 21
food colorings' effects on, 21
food drying techniques' effects on, 19
foods that deplete, 436, 437
free radicals and, 23
freezing's effects on, 19
function of, 4
historical use of, 2, 6–7
increasing dietary intake of, 436
inhibitors of, 5

irradiation's effects on, 19–20, 21
metabolism, their role in, 2–3
milling's effects on, 18
naming of, 3
other uses, 8
Phase II, 29
preservatives' effects on, 20
refining's effects on, 18
smoking's effects on, 24
sources of, 6
speed of, 4
tanning's effects on, 24–25
See also Enzymes, supplemental.
Enzymes, supplemental, 37–54
absorption of, 39–40
from animal sources, 37
choosing, 41–42
combinations of, 38
as digestive aids, 38
dosage considerations, 42
labels of, reading, 43
from microbial sources, 37–38
pH levels and, 39
from plant sources, 37
side effects of, 44
sources of, 37–38
systemic use of, 39
taking, 42–43
temperature's effect upon, 40
uses of, 38–39
See also Enzymes.
Enzymes: The Foundation of Life (Lopez, Miehlke, and Williams), 41, 179
Enzymtherapie (Ransberger and Wolf), 7
Enzyme-Wirkstoffe der Zukunft Mit der Enzymtherapie das Immunsystem Stärken (Pecher and Wrba), 326
Epicatechin, 65, 74
Epidermis, 440
Epididymitis, 220–221
Episiotomy, 335
Epoxyhydrolases, 29
Epstein-Barr virus (EBV), 299
Escherichia coli (E. coli), 17
Esophageal ulcers, 328
Essential amino acids. *See* Amino acids.
Essential fatty acids (EFA), 66, 74
Esterase, 48
Euphorbiaceae, 6
Exercise, 437–438
Exfoliants, enzyme skin, 441
Exfoliative psoriatic dermatitis, 343
Exogluconase. *See* Cellulase.
External otitis, 213. *See also* Ear infection.

Exudative macular degeneration, 295
Eye problems. *See* Cataracts; Conjunctivitis; Macular degeneration; Retinopathy.

Face Value (Wesley-Hosford), 41, 441
Fanconi's syndrome, 282
Fast, Two-Day Enzyme Juice, 434–435
Fatty acids. *See* Essential fatty acids.
Federation Internationale du Pharmaceutiques (FIP), 43
Fermentation, 2
Fermented foods. *See* Foods, fermented.
Fever blisters. *See* Herpes simplex virus.
Fiber, 29
importance of, 193
Fibrinolysin. *See* Plasmin.
Fibrocystic breast disease, 221–223
Fibroids, 223–225
Fibromyalgia. *See* Myofascial pain syndrome.
Ficain. *See* Ficin.
Ficin, 6, 48
Filiform warts, 425
Five-Step Jump-Start Enzyme Program, 438
FIP. *See Federation Internationale du Pharmaceutiques.*
Flat warts, 425
Flatulence. *See* Gas, abdominal.
Flavonglycosides, 66, 74
Flavonoids, 63–64, 65, 74
Flu. *See* Influenza.
Flushes
Enzyme Kidney Flush, 433
Enzyme Toxin Flush, 434
Vitamin C/Enzyme Flush, 434
Folic acid, 57–58
Folic acid anemia, 116
Food, Drug and Cosmetic Act, The, 21
Food additives, 20–21
Food Chemicals Codex (FCC), 43
Foods
acid-forming, 110
alkali-forming, 93
fermented, 30–35
functional, 65–70
pH levels in, 39
preservation methods, 18–20
Foods, fermented
cheese, 35
kefir, 31–32, 33
kimchi (kimchee), 34–35
miso, 33
natto, 33–34
sauerkraut, 32
soy sauce, 32–33
tempeh, 33

yogurt, 30–31, 32
FOS. *See* Fructooligosaccharides.
Fractures, bone, 140–142
Fredericks, Carlton, 266
Free radicals, 23–24
Freund, Ernst, 7
Fructooligosaccharides (FOS), 66
Fruits and vegetables
    cancer and, 29–30
    in the diet, 26–27, 28–30
    enzyme activity in, 27–28, 37
    juicing, 30
Functional foods
    animal, 69–70
    plant, 65–68, 69
Fungal skin infections, 225–228
Furuncles. *See* Boils.

Gallic acid, 66, 74
Gamma-glutamyl allylic cysteines, 66
Ganoderma lucidum. See Reishi mushrooms.
Gargles, 438–439
    Echinacea/Enzyme, 438
    Enzymatic, 438
    Garlic/Enzyme, 439
Garlic, 75
Garlic/Enzyme Gargle, 439
Gas, abdominal, 228–230. *See also* Celiac disease; Heartburn;
    Lactase deficiency; Pancreatic deficiency.
Gastric ulcers, 328. *See also* Peptic ulcers.
Gastritis, 230–232
    acute stress, 230
    atrophic, 230
    bacterial, 230
    chronic erosive, 230
    eosinophilic, 230
    fungal, 230
    plasma cell, 230
    viral, 230
Gastroesophageal reflux disease. *See* Heartburn.
Gastroparesis diabeticorum, 202
Gaucher's disease, 5
Generally recognized as safe (GRAS) list, 20
Genistein, 33–34, 65, 74
Genistin, 34
Genital warts, 425
German measles, 297
Ginger, 75
Gingerols, 66, 74
Gingivitis, 232–234
Ginkgo biloba, 75
Ginkolic acid, 66–67
Ginseng, 75

Gland concentrates, 69–70
Glandular therapy, 69
Glaucoma, 234–237
    closed-angle, 235
    narrow-angle, 235
    open-angle, 235
    pigmentary-dispersion, 235
    secondary, 235
Glenk, Wilhelm, 179
Glomerulonephritis, 237–239. *See also* Kidney disorders.
Glucoamylase, 48
Glucocerebrosidase, 5
Glucosamine sulfate, 76
Glucosinolates, 67
Glutamic acid, 72
Glutamine, 72
Glutathione, 74
Glutathione peroxidase, 28, 48, 74
Glutathione S-transferases (GST), 29
Gluten enteropathy/intolerance. *See* Celiac disease.
Glycine, 72
Glycogenase. *See* Amylase.
Glycyrrhizins, 65, 74
Godfrey, Tony, 37
Goldenseal, 76
Gotu kola, 76
Gout, 239–240
Granulomatous colitis, 271
Granulomatous ileitis, 271
Green barley, 76
Grippe. *See* Influenza.
Growth factor, 12
GST. *See* Glutathione S-transferases.
Guillain-Barré Syndrome, 241–242. *See also* Neuritis.
Gum disease. *See* Gingivitis.

Hagiwara, Yoshihide, 76
Hangover, 242–244
Hartnup disease, 282
Hawthorn, 76
Hay fever, 244–246. *See also* Allergies.
HDLs. *See* High-density lipoproteins.
Headache, 246–249
    cluster, 246
    migraine, 246, 247
    sinus, 246
    tension, 246
Healthy Healing (Rector Page), 366
Heart disease. *See* Cardiovascular disease.
Heart extract, 70
Heartburn, 249–251. *See also* Gas, abdominal; Indigestion.
Heidland, A., 179, 260, 398
Helianthus tuberosus, 27

Hematomas. *See* Bruises and hematomas.

Hemi-cellulase, 48–49

Hemolytic anemia, 116

Hemorrhagic stroke, 393

Hemorrhoids, 251–252

Hepatitis, 252–255

    acute, 253

    chronic, 253

    viruses, 254

Herniated disc. *See* Slipped disc.

Herpes simplex virus, 255–257. *See also* Herpes zoster.

Herpes zoster, 257–259

Hesperidin, 65, 74

High blood pressure, 259–261. *See also* Cardiovascular disorders; Stress; Stroke.

High-density lipoproteins (HDLs), 179

Histidine, 72

HIV. *See* AIDS.

Hives, 261–264. *See also* Allergies.

Hobbs, Christopher, 77

Honey, raw, 35–36

Howell, Edward, 7, 9, 23, 102

Howes, E.L., 151

Human immunodeficiency virus (HIV). *See* AIDS.

Hyaluronidase, 49

Hydrolases, 3–4

Hypercholesterolemia. *See* Cholestelrol, elevated.

Hyperextension, 426

Hyperflexion, 426

Hypericin, 67

Hypericine. *See* Hypericin.

Hypericum. *See* St. John's wort.

Hypertension. *See* High blood pressure.

Hypertensive retinopathy, 351

Hypochlorhydria, 16, 264–265

Hypoglycemia, 266–268

    fasting, 266

    reactive, 266

IBD. *See* Inflammatory bowel disease.

IBS. *See* Irritable bowel syndrome.

Idiopathic juvenile osteoporosis, 320

Ileocolitis. *See* Crohn's disease.

Ileum, 12

Immune disorders, 402–403

Indigestion, 10, 268–271

    frequent causes of, 270

Indoles, 67

Induced fit theory, 4

Industrial Enzymology (Godfrey and West), 37

Infection. *See* Bacterial infections; Fungal skin infections; Viral infections.

Inflammation, as result of injury, 318

Inflammatory bowel disease (IBD), 271–273

Influenza, 273–275

Injuries, sports. *See* Sports injuries.

Inositol, 58

Insect bites and stings, 276–278

Insulin-dependent diabetes, 202

Intermittent claudication, 278–280

Intestinal toxemia, 16

Intrinsic asthma, 125

Introduction to Chiropractic Health (Cichoke), 377

Inulase, 27

Invertase, 49

Iodine, 61

Iron, 61

Iron deficiency anemia, 116

Irritable bowel syndrome (IBS), 280–282

Ischemic stroke, 393

Isoflavonoids, 65

Isoleucine, 72

Isomerases, 3–4

Isothiocyanates, 67

Japanese Pharmacopoeia (JP), 43

Jejunum, 12

Jock itch, 225

John Heinerman's New Encyclopedia of Fruits and Vegetables, 305

Journal of American Medical Association, 205

Journal of Neurochemistry, The, 114

Juice Fast, Two-Day Enzyme, 434–435

Juicing, 30

Kallikrein, 12, 49

Kefir, 31–32, 33

Kelp, 76

Kidney disorders, 282–284. *See also* Glomerulonephritis.

Kidney extract, 70

Kidney failure, 282

Kidney Flush, Enzyme, 433

Kidney infarction, 282

Kimchee. *See* Kimchi.

Kimchi, 34–35

Kininogenin. *See* Kallikrein.

Klaschka, Franz, 312, 322

Koji, 33

Kugler, Hans, 6, 69

Kushi, Michio, 33

LAB. *See* Lactid acid bacteria.

Lactase, 49

Lactase deficiency, 15–16, 284–286

    percentages of populations with, 285

Lactic acid bacteria (LAB), 30

Lactobacillus acidophilus, 77
Lactobacillus bulgaricus, 31
Lactoferrin, 12
Lactones, 67
Lactose intolerance, 5
Laryngitis, 286–288
LDLs. *See* Low-density lipoproteins.
Leaky gut syndrome, 288–290
Leiomyomas. *See* Fibroids.
Lentinula edodes. See Shiitake mushrooms.
Leucine, 72
Leukemia, 156. *See also* Cancer.
Licorice, 76
Liddle's syndrome, 282
Ligament sprains. *See* Sprains and strains.
Ligases, 3–4
Light therapy, 439
Lignans, 67
Limonene, 74
Limonoids, 67
Lin, Robert I-San, 28
Lipases, 11, 27, 37, 38–39
Lipids, 15
Lipoic acid, 67, 74
Live-cell therapy. *See* Glandular therapy.
Liver extract, 70
Liver spots. *See* Age spots.
Lobular carcinoma, 144
Lock and key theory, 4
Lopez, D.A., 41
Low-density lipoproteins (LDLs), 179
Lung cancer, 290–293. *See also* Cancer.
Lung extract, 70
Lung infection. *See* Empyemas; Pleurisy; Pneumonia.
Lupus. *See* Systemic lupus erythematosus.
Lyases, 3–4
Lycopene, 67, 74
Lymphedema, 293–294
  acquired, 293
  congenital, 293
Lymphoma, 156. *See also* Cancer.
Lyons, T. Pearse, 63
Lysine, 72
Lysozyme, 4, 12, 50

Macrophages, 402, 432
Macular degeneration, 295–296
  atrophic, 295
  exudative, 295
Magnesium, 61
Maitake mushrooms, 76
Malabsorption, 16
Malignant nephrosclerosis, 282

Malt diastase. *See* Diastase.
Maltase, 50
Manganese, 61, 74
Marginal ulcers, 328. *See also* Peptic ulcers.
Marie-Strümpell disease. *See* Ankylosing spondylitis.
Marty, Leo, 44
Mastoiditis, acute, 213. *See also* Ear infection.
Mayer-Davis, Elizabeth, 205
McDougall Plan, The, 304
Measles, 297–298
Medicinal Mushrooms (Hobbs), 77
Meditation, 438
Medullary cystic disease, 282
Melanoma, 368–369
Melibiase. *See* Alpha-galactosidase.
Menstrual cramps. *See* Dysmenorrhea.
Mental attitude, positive, 439–440
Metabolic acidosis, 91
Metabolic alkalosis, 109
Metabolism, 2
Methionine, 72, 74
Miehe, H., 27
Miehlke, M., 41
Migraine, 246, 247
Milk thistle, 76–77
Mineral yeasts, 63
Minerals, 59–63. *See also* Calcium; Chromium; Cobalt; Cofactors; Copper; Iodine; Iron; Magnesium; Manganese; Molybdenum; Phosphorus; Potassium; Selenium; Silica; Sodium; Zinc.
Miso, 33
Mistletoe, 77
Molybdenum, 61
Mönckeberg's arteriosclerosis, 122
Moniliasis. *See* Candidiasis.
Mononucleosis, 299–301
Monosaccharides, 11
Monosodium glutamate (MSG), 21
Monoterpenes, 67, 74
MSG. *See* Monosodium glutamate.
Multiple sclerosis (MS), 301–304
Multiple Sclerosis Diet Book, The (Swank), 303
Mumps, 304–305
Muramidase. *See* Lysozyme.
Murray, Michael, 69, 175, 338
Muscle cramping, 306–307. *See also* Dysmenorrhea.
Muscle strains. *See* Sprains and strains.
Muscles, build up of, 382
Myeloma, 156. *See also* Cancer.
Myofascial pain syndrome (MPS), 308–310

Nail ringworm, 225
Naringin, 65, 74

Narrow-angle glaucoma, 235
National Digestive Diseases Information Clearinghouse, 10
Natto, 33–34
Nausea, 310–311
Nephritis, 282. *See also* Glomerulonephritis.
Nephrotic syndrome, 237, 282. *See also* Glomerulonephritis.
Neu, Sven, 179
Neuhofer, Christina, 302
Neuralgia, post-zoster, 258
Neuritis, 311–314
Niacin. *See* Vitamin B$_3$.
Niacinamide. *See* Vitamin B$_3$.
Non-insulin-dependent diabetes, 202
Nonproliferative retinopathy, 351
Nuclease, 50

Obesity, 314–317
Occlusive arterial disease of the limbs. *See* Intermittent claudication.
Oilomycosis. *See* Candidiasis.
Onion, 77
Open-angle glaucoma, 235
Oral Enzymes—New Approach to Cancer Treatment (Klaschka), 312, 322
Organ concentrates, 69–70
Organosulfur compounds, 67, 74
Organotherapy. *See* Glandular therapy.
Ornithine decarboxylase (ODC), 25
Osteoarthritis, 317–318
Osteoporosis, 319–321
    idiopathic juvenile, 320
    postmenopausal, 319
    secondary, 320
    senile, 319–320
Otitis media, 213. *See also* Ear infection.
Ovary extract, 70
Oxidoreductases, 3–4

PABA. *See* Para-aminobenzoic acid.
Paget's disease of the nipple, 144
Panax ginseng. See Ginseng.
Panax quinquefolium. See Ginseng.
Pancreas extract, 70
Pancreatic cancer, 321–323. *See also* Cancer.
Pancreatic Enzymes in Health and Disease (Lankisch), 43
Pancreatic insufficiency, 13–14, 16, 324–325
Pancreatin, 50
Pancreatitis, 16, 325–328
Pancrelipase, 50
Pantothenic acid. *See* Vitamin B$_5$.
Papain, 38, 50–51
Para-aminobenzoic acid (PABA), 58
Pasteur, Louis, 19

Pau d'arco, 77
Pecher, Otto, 257, 326
Pectin, 67, 74
Pectin depolymerase. *See* Pectinase.
Pectinase, 51
Penicillium, 38
Penicillium glaucum, 27
Pepsin, 51
Peptic ulcers, 328–330
Peptidase. *See* Protease.
Pernicious anemia, 116
Peroxidase, 12, 51
Persorption, 40
Peschke, G.J., 43
Pesticides, 21–22
PH levels, 39
    in food, 39
    in household products, 39
Pharyngitis. *See* Sore throat.
Phenolic acids, 67, 74
Phenylalanine, 5, 72
Phenylalanine hydroxylase, 5
Phenylketonurics, 5
Phlebitis. *See* Thrombosis.
Phosphorus, 62
Phthalides, 67
Phytase, 51
Phytic acid, 67, 74
Phytochemicals, 65, 66–68
Phytosterols, 67
Pigmentary dispersion glaucoma, 235
Piles. *See* Hemorrhoids.
Pinkeye. *See* Conjunctivitis.
Pinocytosis, 14, 39
Pituitary extract, 70
Pizzorno, Joseph, 175, 338
Plantar warts, 425
Plants
    enzyme activity in, 27, 37
    enzyme extract from, 64–65
    phytochemicals in, 65
Plasmin, 51–52
Pleurisy, 330–331
PMA. *See* Positive mental attitude.
PMS. *See* Premenstrual syndrome.
Pneumonia, 332–333
Polyacetylenes, 67
Polycystic kidney disease, 282
Polygalacturonase. *See* Pectinase.
Polygalacturonic acid. *See* Pectin.
Polymyositis, 333–335. *See also* Dermatomyositis.
Polyphenols, 74
Polysaccharides, 11, 74

Positive mental attitude (PMA), 439–440
Post-childbirth complications, 335–337
Postmenopausal osteoporosis, 319
Post-thrombotic syndrome (PTS), 408–409
Post-zoster neuralgia, 258
Potassium, 62
Potential hydrogen levels. *See* PH levels.
Premenstrual syndrome (PMS), 337–339
Preservation methods of food, 18–20
Proanthocyanidins, 65, 74
Probiotics, 77
Proliferative retinopathy, 351
Proline, 72
Prostate cancer, 339–341. *See also* Cancer; Prostate disorders.
Prostate disorders, 341–343. *See also* Prostate cancer.
Prostatitis, 341
Protease inhibitors, 23
Proteases, 11, 27, 37, 38–39, 52. *See also* Bromelain; Chymopapain; Chymotrypsin; Pancreatin; Papain; Trypsin.
Proteins, 15
Protomorphogens, 37
Psoriasis, 343–345
Psoriatic arthritis, 343
PTS. *See* Post-thrombotic syndrome.
Ptyalin, 11
Pustular psoriasis, 343
Pycnogenol, 65
Pyelonephritis, 282
Pyridoxine. *See* Vitamin B$_6$.

Quercetin, 65, 74

Radiation sickness, 345–346. *See also* Cancer.
Ransberger, Karl, 7, 44, 157, 179, 202, 260, 315
Raynaud's disease and phenomenon, 347–348
Reactive arthritis. *See* Reiter's syndrome.
Rector Page, Linda, 366
Regional ileitis. *See* Crohn's disease.
Reishi mushrooms, 77
Reiter's syndrome, 349–351
Rejuvelac, 36
Relaxation, 440
Renal glycosuria, 282
Renal vein thrombosis, 282
Rennase. *See* Rennin.
Rennet, 35
Rennin, 2, 52
Resnick, Corey, 288
Respiratory acidosis, 91
Respiratory alkalosis, 109
Retin-A, 441
Retinopathy, 351–353

arteriosclerotic, 351
background, 351
diabetic, 351
hypertensive, 351
nonproliferative, 351
of prematurity, 351
proliferative, 351
Retinopathy of prematurity, 351
Rheumatic fever, 353–354
Rheumatoid arthritis (RA), 354–357
Rhinophyma, 357
Rhizopus niveus, 38
Riboflavin. *See* Vitamin B$_2$.
Ribonuclease, 52
Ringworm, 225
Rosacea, 357–359
Rosemarinic acid, 67, 74
Royal jelly, 77–78
Rubella. *See* German measles.
Rubeola. *See* Measles.
Ruptured disc. *See* Slipped disc.
Rutin, 65, 74

Saccharase. *See* Invertase.
Saccharomyces, 38
St. John's wort, 78
Salicin. *See* Salin.
Salin, 67
Saliva, 11–12
S-allyl cysteine, 74
Salmonella, 17
Salves, enzymatic skin, 441–442
Saponins, 67, 74
Sarcoma, 156. *See also* Cancer.
Sauerkraut, 32
Saw palmetto, 78
Scalp ringworm, 225
Scars, 359–361
Scheef, Wolfgang, 293
Sciatica, 361–362
Scoliosis, 362–363
Sebaceous glands, 440
Seborrheic dermatitis, 197
Sebum, 441
Secondary glaucoma, 235
Secondary osteoporosis, 320
Seifert, J., 40
Selenium, 62, 74
Senile osteoporosis, 319–320
Serine, 72
Serratia protease. *See* Serratiopeptidase.
Serratiopeptidase, 52
Sfericase, 52

Shibamoto, Takayuki, 76

Shiitake mushrooms, 78

Shingles. *See* Herpes zoster.

Sickle-cell anemia, 116

Silica, 62

Silybum marianum, 76–77

Silymarin, 67, 74, 76–77

Sinatra, Stephen T., 64

Sinus headache, 246

Sinusitis, 364–366

Sjögren's syndrome, 367–368

Skin

    care for, 440–442

    exfoliants, 441

    salves, 441–442

Skin cancer, 368–371. *See also* Cancer.

Skin problems. *See* Abscess; Acne; Boils; Bruises and
    hematomas; Burns; Canker sores; Chickenpox;
    Dermatitis; Fungal skin infections; Hives; Insect bites
    and stings; Measles; Psoriasis; Rosacea; Scars; Skin can-
    cer; Skin rash; Skin ulcers; Staphylococcal infections;
    Varicose veins; Warts.

Skin rash, 371–372. *See also* Allergies; Dermatitis; Hives.

Skin ulcers, 373–375.

SLE. *See* Systemic lupus erythematosus.

Slipped disc, 375–377. *See also* Backache.

Smokers, help for, 292

Sodium, 63

Sodium nitrate, 20

Sore throat, 377–379. *See also* Colds; Ear infections;
    Laryngitis; Tonsillitis.

Soy sauce, 32–33

Spirulina, 78

Spleen extract, 70

Sports injuries, 379–384

    estimated number of, 380

    recovery program, 383

    supplements for energy, to prevent 381

    supplements for muscle build-up, to prevent, 382

Sprains and strains, 384–386. *See also* Backache; Bursitis and
    synovitis; Sports injuries; Tendonitis.

Sprouts, 35

Squamous cell carcinoma, 368

Staphylococcal infections, 386–388

Staphylococcus aureus, 17

Stasis dermatitis, 197

Status asthmaticus, 125

Steatorrhea, 388–390. *See also* Cystic fibrosis.

Sties. *See* Boils.

Stings, insect. *See* Insect bites and stings.

Strains. *See* Sprains and strains.

Streichhan, Peter, 38

Streptococcus A infection, 353

Streptococcus thermophilus, 31

Streptodornase, 53

Streptokinase, 53

Streptomyces griseus, 37

Stress, 390–393. *See also* High blood pressure; Stroke.

Stress scale, 391–392

Stress ulcers, 328. *See also* Peptic ulcers.

Stroke, 393

    hemorrhagic, 393

    ischemic, 393

    See also Arteriosclerosis; Cardiovascular disorders; High
      blood pressure; Stress.

Subluxation, 395–397. *See also* Backache; Slipped disc.

Sucrase, 53

Sucrase isomaltase. *See* Sucrase.

Sucrose alpha-glucosidase. *See* Sucrase.

Sulforaphane, 29, 68, 74

Sun's rays, 24–25

Superoxide dismutase (SOD), 28, 53, 74

Surgery, transplant, 398. *See also* Surgical-related problems.

Surgical-related problems, 397–400

Svec, S.V., 184

Swank, Roy, 303

Symptoms and possible causes, troubleshooting for, 83–88

Synovitis. *See* Bursitis and synovitis.

Systemic enzyme therapy, 6, 39

Systemic lupus erythematosus (SLE), 400–404

2-O-GIV. *See* 2″-O-Glycosylisovitexin.

2″-O-Glycosylisovitexin, 68, 76

T cells, 402

Taka-Diastase. *See* Takadiastase.

Takadiastase, 7, 27

Takamine, Jokiche, 7

Tannins, 65, 74

Taussig, Stephen J., 157

Tempeh, 33

Temporomandibular joint dysfunction (TMJ), 404–406

    self-test for, 405

Tendinitis. *See* Tendonitis.

Tendonitis, 406–408. *See also* Bursitis and synovitis; Sports
    injuries; Sprains and strains.

Tension headache, 246

Terpene lactones, 66

Terpenes, 68, 74

Territo, Marilyn, 441

Testicular inflammation. *See* Epididymitis.

Textbook of Biochemistry (Devlin), 25

Thiamin. *See* Vitamin B_1.

Thioallyl compounds, 68

Thiocyanates, 68

Threonine, 72

Thromboangiitis obliterans, 278. *See also* Intermittent

claudication.

Thrombophlebitis. *See* Thrombosis.

Thrombosis, 408–411. *See also* Embolism.
  renal vein, 282

Thrush, 226

Thymus extract, 70

Tinea capitis. See Scalp ringworm.

Tinea corporis. See Ringworm.

Tinea cruris. See Jock itch.

Tinea pedis. See Athlete's foot.

Tinea unguium. See Nail ringworm.

Tinea versicolor, 226

Tinnitus, 214

TMJ. *See* Temporomandibular joint dysfunction.

TMJ self-test, 405

Tobacco smoke, 24

Tocopherol. *See* Vitamin E.

Tonsillitis, 411–413

Tooth decay, 413–415

Torticollis, 415–417

Toxin development, avoiding, 435

Toxin Flush, Enzyme, 434

Transcellular absorption, 14

Transferases, 3–4

Transplant surgery, 395

Trenev, Natasha, 31

Triterpenes, 74

Triterpenoids, 68

Troubleshooting for symptoms and possible causes, 83–88

Truck driver's syndrome. *See* Bell's palsy.

Trypsin, 37, 53–54

Trypsin inhibitor, 13

Trypsinogen, 13

Tryptophan, 72

Two-Day Enzyme Juice Fast, 433–434

Type I diabetes. *See* Insulin-dependent diabetes.

Type II diabetes. *See* Non-insulin-dependent diabetes.

Tyrosine, 72

Ulcerative colitis, 271

Ulcers. *See* Peptic ulcers; Skin ulcers.

Underweight, 417–418

United States Department of Agriculture, 1, 20

Unstable angina, 118

Urokinase, 54

U.S. Pharmacopoeia (USP), 43

UVB radiation, 24–25

Valine, 72

Variant angina, 118

Varicella. *See* Chickenpox.

Varicose veins, 419–421

Vasilenko, A.M., 184

Vegetables. *See* Fruits and vegetables.

Veins, varicose, 419–421

Verrucae. *See* Warts.

Vertigo, 421–422

Viral infections, 422–425. *See also* AIDS; Chickenpox; Colds; Hepatitis; Herpes simplex virus; Herpes zoster; Influenza; Measles; Warts.

Vitamin A, 56, 74

Vitamin B_1, 56

Vitamin B_2, 56

Vitamin B_3, 56–57

Vitamin B_5, 57

Vitamin B_6, 57

Vitamin B_{12}, 57

Vitamin B_{12} anemia, 116

Vitamin C, 58–59, 74, 433

Vitamin D, 59

Vitamin E, 59, 74

Vitamin K, 59

Vitamins. 55–59. *See also* Beta-carotene; Biotin; Choline; Coenzymes; Folic acid; Inositol; Para-aminobenzoic acid; Vitamin A; Vitamin B_1; Vitamin B_2; Vitamin B_3; Vitamin B_5; Vitamin B_6; Vitamin B_{12}; Vitamin C; Vitamin D; Vitamin E; Vitamin K.

Vitexin-2"-rhamnoside, 76

Warts, 425–426
  common, 425
  filiform, 425
  flat, 425
  genital, 425
  plantar, 425

Water, 442–443
  contents of, in various body parts, 443

Watts, Alan, 439

Weight-loss program, 316

Wellness Encyclopedia of Food Nutrition, The, 18

Wesley-Hosford, Zia, 41, 100, 441

West, Stuart, 37

Wheat germ, 36

Wheat grass, 78

Wheatgrass Book, The (Wigmore), 36

Whiplash, 426–429
  exercises for, 428

Whitehead, Ted, 35

Wigmore, Ann, 36

Williams, R.M., 41

Wolf, Max, 7, 44, 102, 157

Wounds. *See* Scars; Skin ulcers.

Wrba, Heinrich, 257, 326

Wryneck. *See* Torticollis.

X-ray hangover. *See* Radiation sickness.

Yaron, Arieh, 236
Yeast infections. *See* Fungal skin infections.
Yogurt, 30–31, 32

Z-9-hexenal, 28
Zinc, 63, 74